Neuropeptide Function in the Gastrointestinal Tract

Editor

Edwin E. Daniel, Ph.D.

Professor
Department of Biomedical Sciences
McMaster University
Hamilton, Ontario, Canada

CRC Press
Taylor & Francis Group
Boca Raton London New York

CRC Press is an imprint of the
Taylor & Francis Group, an **informa** business

CRC Press
Taylor & Francis Group
6000 Broken Sound Parkway NW, Suite 300
Boca Raton, FL 33487-2742

Reissued 2019 by CRC Press

© 1991 by Taylor & Francis Group, LLC
CRC Press is an imprint of Taylor & Francis Group, an Informa business

No claim to original U.S. Government works

A Library of Congress record exists under LC control number:

Publisher's Note
The publisher has gone to great lengths to ensure the quality of this reprint but points out that some imperfections in the original copies may be apparent.

Disclaimer
The publisher has made every effort to trace copyright holders and welcomes correspondence from those they have been unable to contact.

ISBN 13: 978-0-367-25056-0 (hbk)
ISBN 13: 978-0-367-25058-4 (pbk)
ISBN 13: 978-0-429-28576-9 (ebk)

**Visit the Taylor & Francis Web site at http://www.taylorandfrancis.com and the
CRC Press Web site at http://www.crcpress.com**

PREFACE

Research on neuro- and endocrine peptides in the gastrointestinal tract has expanded markedly in the last 20 years. The information available has become so extensive that most workers in the field tend to focus only on a limited portion of it, and even then it is common to observe that significant and relevant information seems to be ignored. Also, despite the proliferation of work in this field, not many physiological or pathophysiological roles of neuropeptides or even new hormonal roles of endocrine peptides have been unequivocally established. New techniques in biotechnology and new immunological and pharmacological tools are likely to become available soon and be applied to gut peptides and their receptors, but it is not clear to the Editor, at least, that these will automatically produce clearer insights into peptide function in the gastrointestinal tract. Thus, the objective of this book is to bring together authors who have contributed extensively to this field and to obtain their critical insights into the history, the current status of understanding of the distribution and actions of these peptides and receptors, the evidence about their physiological and pathophysiological role, and the nature of the unsolved problems and how they might be approached. It is my hope that this information, most of it never assembled and evaluated before in this fashion, will provide a basis for new insights into the roles of gut peptides and a better basis for selection and application of new tools to deepen and expand our understanding of their control functions. It is also conceivable that information, concepts, and hypotheses about the role of peptide signals in the enteric nervous system may be useful to those who work on the central nervous system. The enteric nervous system may prove a useful model for some purposes.

I wish to thank the authors of this volume for their excellent contributions; also, Ms. Denise Scott, whose help in organization of this work was very valuable; and most important, my wife, Virginia Posey-Daniel, who was responsible for resolving many of the innumerable and complex problems that always occur in preparing such a book. The Medical Research Council of Canada supported the work of our group in the study of gut neuropeptides, and I acknowledge their support.

Edwin E. Daniel, Ph.D.
Hamilton, Ontario, Canada
April 1990

THE EDITOR

Edwin E. Daniel, Ph.D., is Professor of Biomedical Sciences, Acting Head of the Division of Physiology and Pharmacology, and Director of the Honours Program in Biology-Pharmacology at McMaster University Health Sciences Centre, Hamilton, Ontario, Canada.

Dr. Daniel received his B.A. *cum laude* in 1947 and his M.A. in 1949 in Biology from Johns Hopkins University, Baltimore, Maryland. Dr. Daniel received his Ph.D. in Pharmacology from the University of Utah in 1952. He was appointed Assistant Professor in the Department of Pharmacology at the University of British Columbia in 1952 and Associate Professor in 1960. He became Professor and Chairman of the Department of Pharmacology at the University of Alberta in 1961. In 1971, Dr. Daniel retained the post of Professor at University of Alberta to center on research activities and in 1975 became Professor in the Department of Neurosciences at McMaster University, where in 1988 he assumed his present position.

Dr. Daniel is a member of the Biophysical Society of Canada, the Canadian Association of Gastroenterology, the Canadian Foundation for Ileitis and Colitis, the Canadian Hypertension Society, the Canadian Lung Association, the Canadian Physiological Society, the Medical Reform Group, the Pharmacological Society of Canada, the American Association for the Advancement of Science, the American Gastroenterological Association, the American Physiology Society, the American Society of Pharmacology and Experimental Therapeutics, the Biophysical Society, the International Society for Heart Research (American Section), and the New York Academy of Science. He has received several Medical Research Council Visiting Scientist Awards. He has been a Riker Scholar and is a past president of the Pharmacological Society of Canada. He was a 1988 Canadian Gastroenterology Association Research Lecturer and Upjohn Lecturer of the Pharmacology Society of Canada. In 1989 Dr. Daniel received the Canadian Journal of Physiology and Pharmacology Best Paper Award and the ASPET Otto Krayer Award. He has been the recipient of many research grants from the Medical Research Council of Canada, the Canadian Heart Foundation, the Ontario Heart Foundation, the Canadian Foundation for Ileitis and Colitis, and private industry and is also currently Program Director of a program grant from the MRC studying the structure and function of smooth muscle.

Dr. Daniel is the author of more than 350 peer-reviewed papers and 75 books or book chapters. His current research interest is in the control of smooth muscle function.

CONTRIBUTORS

Sultan Ahmad, Ph.D.
Biotechnology Research Institute
National Research Council of Canada
Montreal, Quebec
Canada

Hans-Dieter Allescher, M.D.
II Department of Medicine
Klinikum rechts der Isar
Technical University
Munich, West Germany

Frédéric Checler
Centre de Biochimie du CNRS
Faculté des Sciences
Université de Nice
Nice, France

Edwin E. Daniel, Ph.D.
Department of Biomedical Sciences
Division of Physiology/Pharmacology
McMaster University
Hamilton, Ontario
Canada

S. Dion
Department of Pharmacology
University of Sherbrooke
Sherbrooke, Quebec
Canada

G. Drapeau
Department of Pharmacology
University of Sherbrooke
Sherbrooke, Quebec
Canada

E. Ekblad, Ph.D.
Department of Medical Cell Research
University of Lund
Lund, Sweden

Jo-Ann E. T. Fox-Threlkeld, Ph.D.
School of Nursing
Department of Biomedical Sciences
McMaster University
Hamilton, Ontario
Canada

R. Håkanson, Ph.D.
Department of Pharmacology
University of Lund
Lund, Sweden

David J. Hill
Lawson Research Institute
St. Joseph's Health Centre
London, Ontario
Canada

Jan D. Huizinga, Ph.D.
Division of Physiology and Pharmacology
Department of Biomedical Sciences
Intestinal Disease Research Unit
McMaster University
Hamilton, Ontario
Canada

Patrick Kitabgi
Centre de Biochimie du CNRS
Université de Nice
Nice, France

Peter Kostka, Ph.D.
Division of Physiology and Pharmacology
Department of Biomedical Sciences
McMaster University
Hamilton, Ontario
Canada

Chiu-Yin Kwan, Ph.D.
Division of Physiology and Pharmacology
Department of Biomedical Sciences
McMaster University
Hamilton, Ontario
Canada

T. J. McDonald, M.D., F.R.C.P.(C).
Departments of Medicine and
  Biochemistry
University of Western Ontario
London, Ontario
Canada

Julio Pintin-Quezada, M.D.
Division of Physiology and Pharmacology
Department of Biomedical Sciences
Intestinal Disease Research Unit
McMaster University
Hamilton, Ontario
Canada

P. K. Rangachari, Ph.D.
Department of Medicine
McMaster University
Hamilton, Ontario
Canada

Domenico Regoli, M.D.
Department of Pharmacology
University of Sherbrooke
Sherbrooke, Quebec
Canada

Åke Rökaeus, Ph.D.
Department of Biochemistry I
Medical Nobel Institute
Karolinska Institute
Stockholm, Sweden

John S. Smeda, Ph.D.
Department of Anaesthesia
McMaster University
Hamilton, Ontario
Canada

F. Sundler, M.D. , Ph.D.
Department of Medical Cell Research
University of Lund
Lund, Sweden

Jean-Pierre Vincent
Centre de Biochimie du CNRS
Faculté des Sciences
Université de Nice
Nice, France

TABLE OF CONTENTS

Chapter 1
A Historical Overview of the Gastroenteropancreatic Regulatory Peptides 1
T. J. McDonald

Chapter 2
Gastroenteropancreatic Regulatory Peptide Structures: An Overview 19
T. J. McDonald

Chapter 3
Regulation of Gastrointestinal Neuropeptide Gene Expression and Processing........... 87
Åke Rökaeus

Chapter 4
Microanatomy and Chemical Coding of Peptide-Containing Neurons in the Digestive
Tract 131
E. Ekblad, R. Håkanson, and F. Sundler

Chapter 5
Neuropeptide Motor Actions Vary between *In Vivo* and *In Vitro* Experimental
Conditions... 181
Jo-Ann E. T. Fox-Threlkeld

Chapter 6
Receptors for Neuropeptides (Neurokinins): Functional Studies 193
D. Regoli, S. Dion, and G. Drapeau

Chapter 7
Receptors for Neuropeptides: Ligand Binding Studies.................................. 209
Sultan Ahmad, H.-D. Allescher, and C.-Y. Kwan

Chapter 8
Receptors for Neuropeptides: Receptor Isolation Studies and Molecular Biology 231
Jean-Pierre Vincent and Patrick Kitabgi

Chapter 9
Gastrointestinal Neuropeptides and Second Messenger Systems 249
Peter Kostka

Chapter 10
Peptidases and Neuropeptide-Inactivating Mechanisms in the Circulation and in the
Gastrointestinal Tract... 273
Frédéric Checler

Chapter 11
Postulated Physiological and Pathophysiological Roles on Motility 309
Hans-Dieter Allescher and Sultan Ahmad

Chapter 12
Physiological and Pathophysiological Roles of VIP, Somatostatin, Opioids, Galanin, GRP,
and Secretin 401
Jan D. Huizinga and Julio Pintin-Quezada

Chapter 13
Effects of Neuropeptides on Intestinal Ion Transport 429
P. K. Rangachari

Chapter 14
The Role of Neuropeptides in the Normal and Pathophysiological Control of
Blood Flow.. 447
John S. Smeda

Chapter 15
Neuropeptides and Cell Proliferation 479
David J. Hill

Chapter 16
Perspectives for the Study of Gut Neuropeptides 491
E. E. Daniel

Index 499

Chapter 1

A HISTORICAL OVERVIEW OF THE GASTROENTEROPANCREATIC REGULATORY PEPTIDES

T. J. McDonald

TABLE OF CONTENTS

I.   Introduction 2

II.   The Discovery of the GI Hormones 3
        A.   Secretin 3
        B.   Gastrin 4
        C.   Cholecystokinin-Pancreozymin (CCK-PZ) 5
        D.   Pancreatic Glucagon 6
        E.   The Gastric Inhibitory Polypeptide (GIP) 6
        F.   Pancreatic Polypeptide (PP) and Peptide YY (PYY) 7
        G.   Miscellaneous 7

III.   The Discovery of the Neuropeptides 7

IV.   Routes to Discovery of Regulatory Peptides 11

V.   Summary and Conclusions 14

References 16

I. INTRODUCTION

By the turn of the century, Pavlov and his colleagues had presented extensive evidence to suggest that the nervous system was a major factor in the regulation of gastrointestinal (GI) function. Bayliss and Starling's dramatic demonstration in 1902 that secretin was transmitted via the bloodstream from the small intestine to the pancreas, where it stimulated the secretion of water and bicarbonate, altered the focus of investigations on GI regulatory mechanisms from neural to blood-borne mechanisms. Although not the first reported evidence suggesting the existence of humoral mechanisms, the results of Bayliss and Starling's classical experiments were conclusive and achieved rapid and universal acceptance. They immediately recognized that their discovery that secretin was transported via the bloodstream was of general importance, and in the Croonian lectures of 1905, Starling referred to these blood-borne messengers as hormones; the word hormone, coined by W. B. Hardy, was taken from the Greek word usually translated as "I arouse to activity".[1] In the 1915 edition of his textbook *Principles of General Physiology*,[2] Bayliss characterized hormones as substances that were blood-borne messengers and argued against the word being applied in a general sense to any substance with excitatory properties. Investigators since have generally accepted this argument, and many still regard the word hormone as implying a blood-borne mode of delivery for chemical messengers.

In at least one sense, the field of endocrinology was initiated by Bayliss and Starling's discovery; since then, and questionably by definition, endocrinologists have been preoccupied by blood-borne transport mechanisms. This preoccupation was particularly striking for endocrinologists studying the so-called "classical" hormones which are located in and secreted from discrete endocrine glands (i.e., the adrenal gland, the pituitary gland, etc.). GI endocrinologists encountered a number of technical problems which forced them into somewhat different investigative pathways. For those studying "classical endocrinology", the biologically active agents are contained within discrete glands. In contrast, the GI hormones are spread diffusely throughout a complex organ which has multiple and diverse functions. This fundamental anatomical difference had important implications. In contrast to the relatively high tissue concentrations of the classical hormones, GI hormones were present in relatively low concentrations in an organ well suited for proteolytic degradation. Hence, it took approximately 60 years from Bayliss and Starling's experiment in 1902 until the first GI hormones, secretin and gastrin, were isolated and chemically characterized. The experiments employing glandular extirpation followed by exogenous hormonal replacement therapy, which were eminently successful in establishing the physiology of hormones present in the discrete endocrine glands, were, for obvious reasons, unavailable for the study of the GI hormones. Ironically some of these technical difficulties led GI endocrinologists to explore different approaches which led to the recognition of certain unexpected complexities:

1. Some of the GI hormones had unsuspected structural homology with each other which, in certain cases, resulted in overlapping bioactivities.
2. These so-called GI hormones were found to be present not only in the GI tract, but also in a number of tissues throughout the body; furthermore, they were present not only in endocrine-type cells, but also in neural structures.
3. GI hormones probably exert regulatory functions by means of various transport mechanisms; in certain cases, they are delivered to target organs via the bloodstream (i.e., a humoral or hormonal mechanism), but in other instances their delivery probably occurs via strictly local release mechanisms (i.e., paracrine, neurotransmission/neuromodulatory mechanisms).
4. Biological interactions occur between peptide hormones; this is undoubtedly of functional importance since a physiological stimulus to the GI tract (i.e., a mixed meal) obviously changes multiple factors simultaneously.

These points, recognized via a series of unexpected findings in the 1960s and 1970s which occurred either by chance or by the application of new technology, have radically changed our views of the functional nature of both the endocrine and nervous systems.

Because it is now generally recognized that these biologically active peptides, present in the GI tract and elsewhere, may exert regulatory functions by nonhumoral as well as humoral mechanisms, the question arises as to what these peptides should be named. Certain workers refer to these peptides as "peptide hormones", using the term hormone in the general sense of the Greek word from which it was derived.[1] Biologically active peptides present in neural structures have been commonly referred to as neuropeptides. Both terms may be regarded by certain investigators as too exclusive as any given peptide, or at least the biologically essential portion of the molecule, may, in different regions of the body, function as both a hormone and as a neurotransmitter/neuromodulatory agent. For example, cholecystokinin (CCK), which is present in intestinal endocrine cells, is generally accepted to act via secretion into the bloodstream to cause gallbladder contraction and pancreatic enzyme secretion; CCK present in neural structures in the brain and the extrinsic enteric plexuses undoubtedly has completely different functions, and its mechanism of action probably involves delivery to target organs via local release mechanisms. A term in common usage is "regulatory peptide", which does not imply any transport mechanism, thereby avoiding conflict with those who prefer to follow Bayliss' suggestion and reserve the term hormone for those agents whose mechanism of action includes blood-borne transportation; nor does it imply an anatomical location. It does, however, imply the possession of a regulatory function, and for some of these agents a physiological role is yet to be proved. While no term is as yet free from semantic problems, in this chapter the term regulatory peptide will be used to refer to these biologically active peptides, but the choice is arbitrary.

The regulatory peptide field is in a state of rapid change and is probably best considered to be in a primarily descriptive phase. In a short period of time, the study of regulatory peptides has left the restricted areas of the gastrointestinal tract and pancreas and now involves a number of different organ systems, including the central nervous system. A number of previously unrecognized peptides have been identified recently, and more "new" peptides will probably be "discovered" in the future. Because readers of this volume will probably have varied backgrounds, this chapter is designed to provide a brief outline of the manner in which the field developed; wherever possible, reference will be given to review articles rather than to the original literature. This overview is brief and restricted to those regulatory peptides which are common to the gastroenteropancreatic (GEP) tissues and to the peripheral and central nervous systems; it will not address the background of certain related biologically active agents, such as the hypothalamic-releasing factors, plasma-derived factors (i.e., angiotensin), thymic peptides, cardiac-derived peptides (i.e., atrial naturetic peptide), or the burgeoning field of peptide growth factors (i.e., insulin-like growth factors 1 and 2 [IGF-1 and IGF-2], etc.).

II. THE DISCOVERY OF THE GI HORMONES

A. SECRETIN

Bayliss and Starling's demonstration that secretin acts via humoral mechanisms was a seminal event, and certain investigators mark this discovery as the beginning of the field of endocrinology. However, the detection of secretin occurred earlier. In 1825 in France, Leuret and Lassaigne[3] observed that the application of acetic acid (vinegar) to the duodenal mucosa resulted in the secretion of pancreatic juice and the flow of bile. They reached the conclusion that chyme, being acidic, should have the same effect when emptied from the stomach into the intestine. This effect was studied in Pavlov's laboratory, where Dolinsky[3a] wrote a thesis on the subject in 1894 and Popielski[3b] found that duodenal acidification resulted in the stimulation of pancreatic secretion despite prior extensive removal of extrinsic nerves and plexuses. Similar studies were performed by Wertheimer and Lepage[3c] in France, who further demonstrated that pancreatic stimulation occurred if acid was applied to the jejunum and upper ileum, but not if introduced into the distal ileum. Both the Russian and French groups reached the conclusion that this effect was mediated by a peripheral nervous reflex passing directly from the intestinal mucosa to the pancreas. Bayliss and Starling[4] had a different idea, and their landmark experiment was performed on January 16, 1902. On an anesthetized dog, the vagi were cut and the prevertebral ganglia around the superior mesenteric and celiac arteries were extirpated. A loop of jejunum was tied at both ends, and the mesenteric nerves supplying the loop were cut. Introduction of hydrochloric acid into this isolated intestinal loop produced a stimulation of pancreatic secretion similar to that seen after intraduodenal acidification. Since all nervous connections were considered to have been severed, Bayliss and Starling concluded that an active agent was carried by the bloodstream from the jejunum to the pancreas. Next, they scraped off the jejunal mucosa from the isolated loop and extracted the mucosa with sand and acid. After filtration, intravenous injection of this extract was found to produce pancreatic secretion while a similar extract of distal ileal mucosa did not, thereby mimicking the effect of introducing acid into the proximal and distal intestine. Bayliss and Starling's findings were presented to the Royal Society, and their results were rapidly confirmed; of particular importance were cross-circulation experiments which provided a definitive demonstration that secretin was indeed transported via the bloodstream.

In the succeeding decades, several groups attempted to purify secretin from intestinal extracts, and a number of preparations achieved sufficient purity to allow biological studies, including infusions in man, but none approached homogeneity (for review see Reference 5). In the early 1950s, Mutt and Jorpes at the Karolinska Institute, Stockholm, Sweden began their attempts to isolate and characterize secretin. They approached the problem by establishing procedures to process large-scale extractions of porcine small intestinal tissue and applied classic purification procedures to these extracts together with the then newly developed ion-exchange, countercurrent, and gel-filtration chromatographic techniques. A homogeneous preparation of secretin was achieved in 1961,[6] the first isolation of a "GI hormone"; the full amino acid sequence and the synthesis of secretin were announced at a meeting in 1966,[7] and the details of the work were published in 1970.[8] This was a major achievement, but possibly of greater importance was Jorpes and Mutt's establishment of the methodology and facilities to process large-scale intestinal tissue extracts; side fractions derived from the purification of these upper small intestinal extracts have been used subsequently by Mutt and colleagues to isolate a number of previously unrecognized biologically active peptides.

B. GASTRIN

In 1905, Edkins[9] reported that i.v. bolus injections of extracts taken from the porcine gastric antrum, but not from the gastric corpus or from other tissues, stimulated gastric acid secretion in anesthetized cats. He postulated the presence in the gastric antrum of a hormone, designated "gastrin", which stimulated gastric acid secretion. In striking contrast to the rapid acceptance of Bayliss and Starling's results, for technical reasons Edkins' work was not readily confirmed, and the existence of gastrin was long held in doubt. A major problem was the ubiquitous presence of a vasoactive substance which stimulated gastric acid secretion in extracts taken from tissues throughout the body. This substance was subsequently identified by Dale and Laidlaw as histamine. When given subcutaneously, histamine present in tissue extracts stimulates gastric acid secretion. Because this was a popular route of administration, the histamine present in the extracts taken from many tissues produced gastric acid secretion; hence, the specificity of Edkins' findings were not readily confirmed. Since histamine is not effective as a gastric acid secretagogue when given as a bolus i.v. injection, Edkins had unknowingly avoided this problem. For this and other reasons, a great deal of time was to elapse before Edkins' gastrin hypothesis was finally proved correct. In retrospect, a number of subsequent observations had provided strong evidence in favor of Edkins' hypothesis: in 1922 Lim repeated Edkins' experiments exactly and confirmed the original observation; in 1938 Komarov prepared a histamine-free preparation and found that this contained a gastric acid secretagogue; and, importantly, in 1948 Grossman and colleagues[10] provided proof of a hormonal mechanism regulating gastric acid secretion by observing that such secretion occurred following distention of a transplanted antral pouch.

The final proof of Edkins' gastrin hypothesis came with the isolation of gastrin by Gregory and Tracy at the University of Liverpool. Their classic studies culminated in the isolation of two peptides from porcine antral mucosa which differed only in the presence or absence of a sulfate group on a single tyrosine amino acid residue present in both peptides.[11] The presence or absence of the sulfate group did not substantially affect the gastric acid-stimulating activity. Both gastrins were composed of 17 amino acid residues and contained an N-terminal pyroglutamyl residue (a cyclized glutamyl residue) and a C-terminal phenylalanine amide residue. Gregory and Tracy's work produced the first chemical characterization of what was thought to be strictly a GI hormone and first demonstrated the presence of the tyrosine-O-sulfate structure in a peptide hormone. Since this time, a number of N-terminally and C-terminally extended forms of gastrin have been isolated.

Because Edkins' gastrin hypothesis was not as readily accepted as the discovery of secretin by Bayliss and Starling, the work of Pavlov and colleagues suggesting the importance of nerves in controlling gastric acid secretion was not entirely forgotten and, hence, continued to be a focus of investigation. That the regulation of gastric acid secretion was mediated by a close interaction between the nervous and endocrine systems was probably first perceived by the Swedish pharmacologist Uvnas. Uvnas[11a] demonstrated that stimulation of the vagus nerve resulted in the release of gastrin but that, in the absence of gastrin, vagal stimulation was less effective in producing acid secretion. It is well recognized that the physiological release of certain hormones may, in part, be regulated by neural mechanisms (i.e., the so-called "cephalic-phase" responses). Conversely, it is entirely possible that circulating peptides may act on neural receptors to alter neuronal function in the peripheral nervous system or possibly, in certain areas, the central nervous system. Uvnas' perception and demonstrations of the intimate relationship between the humoral entity gastrin and vagal nerve activity was clearly prescient.

C. CHOLECYSTOKININ-PANCREOZYMIN (CCK-PZ)

In a series of experiments in 1927 and 1928, Ivy and Oldberg and their co-workers[12,13] demonstrated that crude extracts of secretin increased bile flow and resulted in gallbladder contraction. Cross-circulation experiments conclusively demonstrated that the gallbladder contraction seen after intraduodenal acidification was mediated by an agent which, like secretin, was transported by the bloodstream. This agent was named cholecystokinin (CCK). Shortly thereafter, a secretin-free intestinal extract containing CCK was prepared. In 1941, Harper and Vass[14] confirmed and extended earlier observations made in Pavlov's laboratory, that food in the intestine resulted in the stimulation of pancreatic enzyme secretion despite prior extensive denervation. Harper and Raper[15] then demonstrated that i.v. administration of a secretin-free intestinal extract mimicked the pancreatic enzyme secretory effect of food in the intestine. These experiments led them to suggest that this pancreozymic effect was mediated by a distinct intestinal hormone which they named pancreozymin (PZ).

In Stockholm, during attempts to isolate CCK and PZ, Jorpes and Mutt noted that both activities were always present in identical chromatographic fractions during their purification procedures and reached the conclusion that both activities were mediated by a single entity. On purification to homogeneity, this entity was found to be a peptide composed of 33 amino acid residues and contained the C-terminal phenylalanine amide structure; it was also found that the C-terminal octapeptide (CCK-8) contained the amino acid residues essential for stimulation of pancreatic enzyme secretion and contraction of the gallbladder.[16] Subsequently, Jorpes and Mutt isolated an N-terminally extended form of cholecystokinin, CCK-39,[17] and further N-terminally extended forms have been characterized recently.[18]

At about the same time, Erspamer and co-workers prepared methanolic extracts of amphibian skin and, after purification of the extracts on alumina columns, screened the effluents for biological activity. By this process, Anastasi and co-workers[19] isolated a peptide composed of ten amino acid residues, which they named caerulein after the species from which the active peptide was isolated. Caerulein had biological activities similar to that of CCK, but surprisingly, sequence analysis of caerulein revealed that it had a C-terminal pentapeptide amide identical to that of gastrin, and it also contained a sulfated tyrosine residue. Subsequently, Mutt and Jorpes sequenced CCK and found that, similar to caerulein, CCK had a C-terminal pentapeptide amide identical to that of gastrin and also contained a sulfated tyrosine residue. These, of course, were disturbing findings at the time, since gastrin was a potent gastric acid secretagogue while CCK and caerulein were potent stimulators of pancreatic enzyme secretion and gallbladder contraction. Subsequent studies revealed that CCK does indeed have gastric acid secretagogue activity, but is much less potent than gastrin, and similarly gastrin has weak CCK-like and PZ-like activities.

D. PANCREATIC GLUCAGON

Immediately after Banting's discovery of insulin in 1922, a number of groups prepared pancreatic extracts to be used in the treatment of diabetes mellitus. Certain extracts prepared in North America were noted to produce an initial rise in blood glucose levels which occurred before the expected decline in blood glucose. This initial hyperglycemic effect did not prove sufficiently reproducible between preparations; consequently, little progress was made in the identification of the active agent over the succeeding two decades. Interest in this hyperglycemic factor was rekindled in 1948 by Sutherland and de Duve, who demonstrated the presence of a potent hyperglycemic glycogenolytic factor in extracts taken from canine gastric corpus tissue extracts. Subsequent studies demonstrated that this factor, glucagon, is indeed present in the mammalian GI tract, but, with the exception of the canine gastric corpus, generally occurs in relatively small amounts. Because a far more abundant source was present in the pancreas, in 1957 Bromer and colleagues[20] isolated a hyperglycemic glycogenolytic factor from pancreatic extracts and determined the structure to be a linear polypeptide composed of 29 amino acid residues, including an N-terminal histidine. This peptide became known as pancreatic glucagon. Doi et al.,[21,22] at a much later date, reported the final demonstration of the identity of the canine gastric hyperglycemia-producing peptide with that of pancreatic glucagon.

E. THE GASTRIC INHIBITORY POLYPEPTIDE (GIP)

During a screen for biological activities in Mutt's chromatographic side fractions obtained during the purification of porcine CCK, Brown and colleagues[23] discovered a potent inhibitor of canine gastric acid secretion using the denervated canine gastric pouch as the bioassay. This acid-inhibiting property was used to purify the active agent, which they named the gastric inhibitory polypeptide (GIP). After chemical characterization, porcine GIP was demonstrated to be a linear polypeptide containing 42 amino acid residues.[24] The acid inhibitory activity of GIP is probably mediated via its stimulation of the release of somatostatin,[25] and studies on humans and dogs have demonstrated GIP to be a weak inhibitor of gastric acid secretion;[26,27] this has led to the suggestion that GIP is unlikely to be an important physiological inhibitor of gastric acid secretion.

In an analogous fashion, Rabinovitch and Dupré noted that impure CCK preparations contained a potent insulin secretagogue, but that homogeneous preparations of CCK had a much weaker effect. Reasoning that during the purification of CCK a potent insulin secretagogue had been eliminated and that this factor might be GIP, Dupré, in collaboration with Brown,[29] demonstrated that GIP was indeed a potent insulin secretagogue in humans. Since this time numerous studies have confirmed this finding, and it is now generally accepted that GIP is probably a physiologically important "incretin". Brown has suggested that GIP be renamed the glucose-dependent insulinotropic polypeptide.

F. PANCREATIC POLYPEPTIDE (PP) AND PEPTIDE YY (PYY)

Kimmel and colleagues[30] noted the presence of a dominant chromatographic peak which occurred during the purification of insulin from avian pancreatic extracts; they subsequently chemically characterized a peptide present in this peak, which was named pancreatic polypeptide. Similarly, Chance and colleagues[31] isolated mammalian peptides from pancreatic extracts which had sequence homology with the avian peptide. Although the precise physiological role of PP is still uncertain, abundant evidence has been obtained which demonstrates that PP is released into the systemic circulation following ingestion of a meal. This release occurs in a biphasic manner; the cephalic-phase release of PP is predominantly under vagal control, and the secondary or "GI phase" of PP release is mediated via local neural reflexes and by humoral mechanisms.[32]

Tatemoto and Mutt,[33] using a chemical assay that detects the presence of peptides having the distinctive C-terminal α-amide structure, noted that porcine small intestinal extracts contained a peptide with the C-terminal tyrosine amide structure. On characterization, the peptide, like PP, was found to be composed of 36 amino acid residues and had an N-terminal tyrosine residue, a C-terminal tyrosine amide structure, and striking sequence homology with PP.[34] The peptide was named after its N- and C-terminal residues, both of which are tyrosine residues (the one-letter identifier for tyrosine being Y). Similar to PP, PYY is released into the systemic circulation following meals, but unlike PP does not appear to be under predominant vagal control. The precise physiological function of PYY remains to be determined. Interestingly, using the same chemical assay technique, Tatemoto[35] isolated from brain extracts another peptide with a C-terminal tyrosine amide structure which was designated NPY. NPY has striking sequence homology with both PP and PYY. All members of this family of structurally related peptides, the pancreatic polypeptide family, were detected strictly by chemical means; subsequent to their isolation, the individual members were shown to possess potent biological activities and to have characteristic tissue distributions.

G. MISCELLANEOUS

Motilin, neurotensin, and somatostatin have also been demonstrated by radioimmunoassay (RIA) techniques to be released into the systemic circulation. Neurotensin was detected serendipitously as the result of its vasoactive properties during the isolation of substance P,[36] and somatostatin was detected in hypothalamic extracts as an "interfering" factor during attempts to isolate the growth hormone-releasing factor.[37] Motilin was isolated by Brown and colleagues[38] from Mutt's porcine intestinal extracts on the basis of its motility-altering properties. These peptide hormones produce distinct spectrums of biological activity. The precise physiological roles of neurotensin and motilin as circulating agents remain subjects of investigation.

III. THE DISCOVERY OF THE NEUROPEPTIDES

Extensive investigations conducted in the last century established that the enteric nervous system was a unique entity. The presence of distinct histological features, the existence of complete reflex pathways within the enteric nervous system, and the considerable autonomy of function exhibited by the GI tract led Langley to classify the autonomic nervous system into three divisions: the sympathetic, the parasympathetic, and the enteric nervous systems (for review see Reference 39). Viewed in this context, Pavlov's doctrine of "nervism" (i.e., that gastrointestinal function was predominantly under nervous control) made "sense" and was widely accepted at the turn of the century. As discussed previously, this doctrine came close to being eclipsed for decades by Bayliss and Starling's demonstration of the importance of humoral mechanisms. We have now come full circle. The recognition that neural mechanisms do indeed play important roles in regulating GI function as part of an integrated endocrine-neural regulatory system has been reestablished. One of the reasons for our current acceptance of the importance of the nervous system in the control of GI tract function and of an integration between the nervous and endocrine systems in regulatory physiology, is the recent recognition of the existence of "neuropeptides".

In 1931, von Euler and Gaddum[40] described the presence, in tissue extracts made from both brain and intestine, of an agent which had an identical spectrum of biological activity. This agent was designated substance P (the P standing for powder), and in retrospect this was the first description of a "brain-gut" peptide. The significance of this finding and that of Pernow in 1953 that substance P-like bioactivity was present not in the mucosal layers of the intestine which contains endocrine cells but in the muscle layers, was not fully appreciated at the time. In 1970, Chang and Leeman[41] isolated a sialogogic peptide from hypothalamic tissue which they characterized as being an undecapeptide (a peptide containing 11 amino acid residues). They quickly recognized that this peptide possessed the biological properties described for substance P by von Euler and Gaddum; soon thereafter, substance P was also characterized from the intestine,[42] confirming von Euler and Gaddum's original observation that substance P was a "brain-gut" peptide. Similarly, but somewhat later, two other peptides, neurotensin and somatostatin, originally isolated and characterized from brain extracts, were demonstrated by immunochemical techniques to be present in the GI tract and the pancreas, respectively.[43,44] For GI endocrinologists, a definite turning point came in 1975 with the report of Vanderhaeghen and co-workers[45] that a peptide reacting with gastrin antisera was present in the central nervous system of vertebrates. This peptide was subsequently reported to be the active C-terminal octapeptide of cholecystokinin (CCK-8) which probably cross-reacted with the gastrin antibody employed in the original study,[46] but nonetheless, these reports clearly demonstrated that a peptide considered to be a classic GI hormone was located outside the GI tract.[45,46] At about the same time, a number of groups reported the presence of vasoactive intestinal polypeptide (VIP)-like immunoreactivity in neurons of the peripheral and central nervous systems. These and other studies established the presence of peptides in neurons of both the peripheral and central nervous systems, and since then it has been demonstrated repeatedly that this is a general rather than an exceptional phenomenon.

The concept that biologically active agents might be released from a cell other than a nerve and delivered to target organs by a strictly local delivery mechanism, probably began with the report in 1938 by Feyrter,[47] who described a diffuse system of "clear cells" (i.e., cells that stain poorly using conventional histological techniques) in the GI tract and in other tissues throughout the body. It was suggested that these clear cells directly influenced adjacent cells by a direct cell-to-cell transmission of an active agent; this delivery mechanism was termed a paracrine mechanism, and Feyrter suggested that these cells formed a "diffuse endocrine" or "paracrine" system. Similarly, Lepage[47a] published his thesis suggesting much the same mechanism. Independently, Pearse and colleagues at the Post-Graduate Medical School in London made a series of important observations. They noted a number of cell types throughout the body that demonstrated certain chemical and ultrastructural characteristics and either contained monoamines or were able to sympathetize these amines from precursors — hence the acronym APUD for *a*mine *p*recursor *u*ptake and *d*ecarboxylation.[48] Pearse correctly predicted that these cells would contain polypeptides and described a number of diverse cells in widespread locations throughout the body (i.e., GI tract and pancreas and C cells of the thyroid, hypothalamus, and pituitary) as having the necessary characteristics to be regarded as part of the APUD system. He originally proposed a common embryological origin of the APUD cells from the neural crest, but this concept has not been supported by experimental evidence, and the significance of the ability to synthesize mono-amines as a link between these cells has been questioned.[49] However, the production of regulatory peptides by these cells has been seen as a more acceptable basis for their being grouped together, and tumors arising from these cells which secrete regulatory peptides that cause recognized clinical syndromes are now commonly referred to as APUDomas. Neuron-specific enolase (NSE), an isoenzyme of the glycolytic enzyme enolase thought to be specific for neural structures, is also present in cells of the APUD series. Thus, the regulatory peptides shared by neurons and the APUD endocrine cells and the common presence of NSE suggest a close relationship between neurons and the cells of the endocrine system; the excitatory agents used as chemical messengers appear to be shared by both systems, but the modes of delivery may differ depending upon location, with each system having the potential to employ humoral or local release mechanisms.

The discovery that biologically active peptides are present in neural structures throughout the body was made possible by the development of immunochemical techniques. Specifically, RIA and immunohistochemistry (IHC) have played seminal roles in changing our concepts regarding both the anatomical locations and the mechanisms of actions of regulatory peptides. Prior to the development of these techniques, estimation of tissue concentrations and the descriptions of tissue distributions of such peptides were based on bioassays performed on crude tissue extracts. RIA greatly increased the sensitivity and specificity of detection and enabled the rapid assay of large numbers of samples, and, most importantly, IHC allowed the elucidation of the anatomical locations of these peptides within tissues at the microscopic level. These immunochemical techniques combined with more efficient methods of peptide isolation and characterization have revolutionized our concepts regarding the nature and function of regulatory peptides. Studies performed in the 1970s utilizing these techniques established not only the existence of regulatory peptides in neural structures, but also the fact that some of these neuropeptides were identical to GI hormones which were present in intestinal endocrine cells. Numerous studies have now demonstrated that more than one such neuropeptide may be present in a single neural structure and that these neuropeptides coexist not only with each other, but also with chemical messengers considered to be classical neurotransmitters (i.e., noradrenaline, acetylcholine).[39,50] In a few years, the volume of evidence generated by these powerful immunochemical techniques rapidly overwhelmed any initial reluctance to accept the concept of neuropeptides functioning in regulatory mecha-nisms. An occasional doubt is still raised, but in general it is accepted that these neuropeptides probably act in concert with classical neurotransmitters to regulate physiological functions throughout the body.

In retrospect, it is curious that there was initial reluctance to accept the evidence sug-gesting the importance of neuropeptides. In a brilliant series of experiments in the 1950s, the antidiuretic hormone (ADH) was isolated, its structure was determined, and it became the first biologically active peptide to be subjected to chemical synthesis.[51,52] It was clearly recognized at the time that ADH was released from neurons in the hypothalamus and from nerve endings in the posterior pituitary into the systemic circulation where it was transmitted to a distant target organ, the distal nephron. Clearly, the concept of a biologically active peptide being present in neural structures and being released from these structures into the systemic circulation to exert a distinct physiological effect had been established at an early date. The realization that this might be a more general phenomenon awaited the technical developments that occurred in the 1970s.

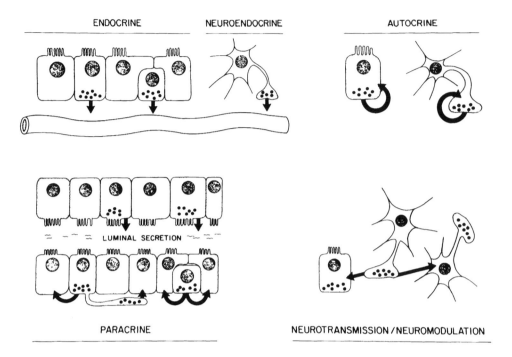

FIGURE 1.   Possible modes of delivery of regulatory peptides to target cells.

Figure 1 illustrates certain possible mechanisms of transport which may be employed by these regulatory peptides. The hormones insulin, glucagon, gastrin, and secretin function in a classical endocrine mode with secretion of the active peptide into the peripheral circulation to be carried to distant target organs. Two types of endocrine cells that occur in the GI tract are shown: (1) an "open-type" cell which contacts the gastrointestinal lumen and contains microvilli at the luminal surface (presumably, this type of cell possesses the capability of functionally responding to luminal contents) and (2) a "closed-type" of endocrine cell present especially in the gastric corpus mucosa which does not contact the gastric lumen. A neuroendocrine mode of transport involves the secretion of regulatory peptides from neurons into the circulation to be carried to distant target organs. Examples of peptides utilizing this mode of transport are ADH and the hypothalamic-releasing factors (i.e., corticotrophin releasing hormone (CRH), thyrotrophin releasing hormone (TRH), etc.), which are secreted from neurons in the hypothalamus into a local (portal) circulation to be carried to the pituitary, where they regulate the release of anterior pituitary hormones.

The term paracrine is generally used to indicate the direct transmission of an agent from a non-neural cell of origin to an adjacent target cell. Somatostatin-containing endocrine cells in the gastric antrum possess long processes that may contact gastrin-containing cells,[53] leading to the suggestion that somatostatin may be released locally to exert a continuous restraint on gastrin secretion via a paracrine mechanism. Similarly, somatostatin-containing cells have been proposed to exert inhibitory influences on pancreatic islet cells via such paracrine mechanisms. Regulatory peptides present in neurons in the peripheral and central nervous systems probably function as neurotransmitters/neuromodulators after being released from nerve endings or varicosities directly onto, or in the vicinity of, target cells; the target cells being, for example, another neuron, an endocrine cell, or a smooth muscle cell. A number of regulatory peptides, such as substance P, VIP, and the bombesin-related gastrin-releasing polypeptide (GRP), are postulated to function in this manner. An autocrine mechanism of action involves secretion of an active agent from either an endocrine cell, a nerve,

TABLE 1
Peripheral Modes of Action — GEP Regulatory Peptides

Circulating peptides	Locally acting peptides
Insulin	VIP
Secretin	GRP
CCK	Substance P
Gastrin	Enkephalins
Pancreatic glucagon	Dynorphin
Pancreatic polypeptide	NPY
GIP	PHI
PYY	Somatostatin
Motilin	Galanin
Enteroglucagon (GLI)	Neuromedin B
Neurotensin	Neurokinin A
Somatostatin	Neurokinin B
	CCK-8

or other type of cell to react with receptors present on their cells of origin, thereby modulating further function. The coexistence of classical neurotransmitters and regulatory peptides in neurons has been confirmed experimentally many times. These coexisting agents within a single cell or nerve potentially may regulate the release of each other, or alternatively, modify the actions of each other at postsynaptic receptors. Evidence for autocrine mechanisms of action exists for a number of growth factor peptides (i.e., IGF-1 and IGF-2). It has been demonstrated that regulatory peptides are present in gastrointestinal luminal secretions and that luminal instillation of regulatory peptides may result in biological activity. Hence, it has been postulated that regulatory peptides may be released from endocrine cells into the gastrointestinal lumen where they are transported with exocrine secretions to reach receptors on distal target cells.

Table 1 lists a number of regulatory peptides categorized by their documented or postulated transport mechanisms. Those peptides listed as circulating agents have been demonstrated to function in a humoral fashion (i.e., they are released into the systemic circulation). Other regulatory peptides have not been definitely demonstrated to be secreted into the peripheral circulation and, hence, are postulated to exert their activities via strictly local release mechanisms. It is not a simple either/or situation because a number of these peptides are present in both endocrine cells and in nerves; therefore, they may function as a humoral agent in one location, but as a paracrine or neurotransmitter agent in another location.

IV. ROUTES TO DISCOVERY OF REGULATORY PEPTIDES

Table 2 lists the recognized mammalian GEP regulatory peptides classified as to the decade in which they were chemically characterized. It should be noted that this list does not include pituitary, parathyroid, cardiac (i.e., atrial naturetic peptide, etc.), thymic, or plasma-derived (i.e., angiotensin) peptides. There has been a tremendous acceleration in the pace at which new peptides are being discovered. This obvious acceleration in the discovery of new regulatory peptides is due to technologic advances in peptide purification and sequence analysis, but also to major changes having occurred in the manner in which previously unrecognized regulatory peptides are discovered. The major determining factor in the discovery of a new peptide is no longer the ability to isolate or characterize the peptide, but resides in our ability to detect the presence of and monitor the purification of a new peptide. An understanding of the ways in which these peptides were first identified is essential in appreciating the field as it has developed and, more importantly, in recognizing potential future investigative possibilities.

TABLE 2

Mammalian Regulatory Peptides — Dates of Characterization

Pre-1960

Insulin	Vasopressin
Pancreatic glucagon	Oxytocin

1960—1969

Gastrin	CCK-PZ
Secretin	β-Lipotropin

1970—1979

VIP	Thyroliberin (TRH)
GIP	Gonadotropin-releasing hormone (GnRH)
Substance P	Enkephalins, endorphins
Neurotensin	Gastrin-releasing polypeptide (GRP)
Motilin	Somatostatin

1980—Present

Glicentin, Oxyntomodulin	Growth hormone-releasing hormone (GRH)
(Glucagon precursors)	Corticotropin-releasing hormone (CRH)
Glucagon-like peptide 1 (GLP-1)	Neurokinin A
Glucagon-like peptide 2 (GLP-2)	Neurokinin-B
PHI/PHM	Neuromedin B
PYY	Neuromedin N
NPY	Neuromedin U
Galanin	Valosin
Pancreastatin	Head activator peptide
Katacalcin	Endothelin
Diazepam-binding inhibitor	Vasoactive intestinal contractor
Calcitonin gene-related peptide	

The first recognized regulatory peptides were the GI hormones which were recognized via a so-called "physiological clue". That is, a biological event following a physiological stimulus was found to be mimicked by i.v. administration of an extract taken from the site of stimulation. The biological event was then generally used as the bioassay to monitor the purification of the active agent. The discovery of secretin, gastrin, and CCK-PZ are classic examples of this method of identification. Discovery via this route has an obvious appeal, as it contains a "built-in" implication of physiological significance. However, only a very few of the regulatory peptides were discovered in this manner. Many were detected as an unexpected biological activity (i.e., serendipitously) present in chromatographic side fractions generated during the purification of another peptide or via a general screening procedure for previously unrecognized biological activities in tissue extracts.

Table 3 lists the different methods by which regulatory peptides have been detected. The early GI hormones (secretin, gastrin, CCK-PZ) were, as discussed previously, discovered via a physiological clue. Probably the first brain-gut peptide discovered was substance P, which was found as an active agent present in both brain and gut extracts. Subsequently, substance P was unexpectedly isolated and characterized from hypothalamic extracts, having been detected by its sialogogic activity.[41] Similarly, during the purification of substance P, the same group identified neurotensin in a chromatographic side fraction by noting the unexpected presence of a vasoactive agent.[36] Somatostatin was discovered as an interfering factor during bioassays designed to detect the presence of growth hormone-releasing factor

TABLE 3
Routes to "Discovery" of Mammalian GEP Regulatory Peptides

Detected via a "physiological clue"
  i.e., Secretin, Gastrin, CCK-PZ
Detected by chance or via a general screening process
  i.e., Gastrin-releasing polypeptide (GRP), Neurotensin, Somatostatin
Detected via chemical techniques
  During chromatography
    i.e., β-Lipotropin, Pancreatic polypeptide (PP)
Detected via immunologic techniques
  i.e., Glicentin, oxyntomodulin
Detected via the C-terminal α-amide structure
  i.e., PHI, PYY, NPY, Galanin
Detected via recombinant DNA techniques
  i.e., Calcitonin gene-related peptide, Glucagon-like peptide 1, Glucagon-like peptide 2

in hypothalamic extracts.[37] The bombesin-related GRP was identified and isolated from porcine gastric corpus extracts as a gastric acid secretagogue during attempts to identify inhibitors of acid secretion in this tissue.[54,55] GIP was identified as an inhibitor of gastric acid secretion present in chromatographic side fractions obtained during the purification of CCK,[23] but subsequently was demonstrated to be an important physiological insulin secretagogue. In the 1960s, Erspamer and colleagues began a series of experiments in which methanolic extracts of amphibian skin were partially purified on alumina columns and the effluents from the alumina columns screened for *in vitro* smooth muscle bioactivity. A number of active agents were identified which, in subsequent studies, were shown to have potent bioactivities in mammals. This series of important experiments led to predictions that analogous mammalian peptides existed. Hence, the identification of caerulein, bombesin, physalaemin, and related tachykinins from amphibian skin predated and, more importantly, predicted the eventual chemical characterization of their probable mammalian counterparts, respectively, CCK, GRP, substance P, and other mammalian tachykinins.[56] Similarly, independent groups in Japan have screened porcine spinal cord tissue extracts for *in vitro* smooth muscle activity, resulting in the characterization of neuromedin B,[57] the neurokinins A and B,[58-60] neuromedin N,[61] etc.

Novel approaches to the recognition of regulatory peptides have been developed recently. Previously unrecognized peptides are now being identified by strictly chemical methods, with their bioactivity being demonstrated only subsequent to their chemical characterization. This approach apparently began with Li's isolation and chemical characterization of β-lipotropin, which occurred in a prominent chromatographic peak obtained during the purification of adrenocorticotropic hormones (ACTH).[62] It is now recognized that β-lipotropin is part of the pro-opiomelanocorticotropin (POMC) molecule and contains within its structure β-endorphin and Met-enkephalin, both of which were subsequently and independently characterized. Similarly, Kimmel and co-workers[30] noted the presence of a dominant chromatographic peak during the purification of avian insulin; this chromatographic peak contained a 36-amino-acid residue peptide they named the pancreatic polypeptide (PP). Similarly, Chance and co-workers[31] isolated the mammalian forms of PP from pancreatic extracts. PP was subsequently shown to possess biological activities and to be contained within endocrine cells of the pancreatic islets. Immunological techniques have played important roles in detecting different molecular forms of known regulatory peptides. For example, the use of RIA has played a central role in determining pathways by which gastrin is processed from its precursors and by which it is metabolized in tissues and plasma. Similarly, RIA identified the probable precursors of glucagon, glicentin and oxyntomodulin.[63,64] Although a precedent has not yet been set, it is conceivable that a previously unrecognized member of a peptide family will be discovered by its cross-reaction with an antisera raised to a recognized regulatory peptide.

Certain regulatory peptides do not contain a free carboxyl group at the C-terminus (i.e., –COOH), but instead contain an amidated carboxyl group (i.e., –CONH$_2$). During sequence studies on CCK, Mutt noted that the C-terminal phenylalanine amide residue could be isolated chromatographically. This observation led Tatemoto and Mutt[33] to devise an assay system to detect, in crude tissue extracts, the presence of peptides containing such C-terminal α-amide structures. While monitoring the purification of CCK and secretin from porcine intestinal extracts with this assay system, they noted the presence of peptides containing C-terminal α-amide structures which could not be attributed to known regulatory peptides. They quickly recognized the significance of this finding and proceeded to isolate a series of previously unrecognized peptides, each of which has been shown subsequently to possess potent biological activities and to have a characteristic tissue distribution and microscopic anatomy. In this manner, the following peptides were isolated:

1.   PHI, a neuropeptide with an N-terminal histidine, the C-terminal isoleucine amide structure, and structural homology with the secretin-glucagon family of peptides[65]
2.   PYY, a hormonal peptide with 36 amino acid residues, an N-terminal tyrosyl residue, the C-terminal tyrosine amide structure, and a striking sequence homology with PP and NPY.[34] PYY is secreted into the systemic circulation from intestinal endocrine cells and inhibits exocrine pancreatic secretion
3.   NPY, a neuropeptide with the C-terminal tyrosine amide structure was isolated from brain extracts and has intense sequence homology with PP and PYY.[35] NPY possesses potent vasoconstrictor properties
4.   Galanin, a neuropeptide with a unique amino acid sequence containing the C-terminal alanine amide structure,[66] potently alters motility and inhibits insulin secretion[67,68]

Each of the above regulatory peptides was isolated and identified using a strictly chemical assay, and only subsequent to its chemical characterizations were its potent bioactivities and characteristic tissue distribution described.

Based on known amino acid sequences of regulatory peptides elucidated by classical peptide chemistry, recombinant DNA technology is now widely used to identify genes coding for regulatory peptides; from the gene, the structures of precursor molecules can be determined. Using this approach, the gene coding for calcitonin was identified and found to encode also for a previously unrecognized peptide structure which was flanked by pairs of basic amino acid residues (such pairs of basic amino acid residues often indicate cleavage sites for active peptides within precursor molecules). A synthetic replicate of this peptide, the so-called calcitonin gene-related peptide (CGRP), was found to have potent central and peripheral biological activities and specific anatomical locations.[69] Similarly, the structures of two peptides with structural homology to pancreatic glucagon, the glucagon-like peptides 1 and 2 (GLP-1 and GLP-2), were described following elucidation of the structure of the gene coding for pancreatic glucagon.[70] Neurokinin A, a member of the tachykinin family of peptides, was identified almost simultaneously by both isolation of the peptide from extracts of porcine spinal cord and by the identification of its presence in a gene coding for substance P.[58,59] There is every possibility that this approach will be the means by which other "new" regulatory peptides will be recognized.

V. SUMMARY AND CONCLUSIONS

The last two decades have produced remarkable changes in our concepts of the manner in which regulatory peptides exert their activities. Biologically active peptides which were thought to be strictly GI hormones are now recognized as being present in different cell types throughout the body, specifically being present both in endocrine-type cells and in neural structures. Furthermore, these peptides, which were once thought to be exclusively humoral agents, also may be delivered very specifically by release from their cell of origin directly onto target cells. Hence, we now recognize that these peptides are no longer strictly GI in origin and are not necessarily hormones if the word hormone is taken to imply a blood-borne mechanism of transport. There has been a recent acceleration in the rate at which previously unrecognized peptides are being identified and characterized, which has made the task of determining physiological roles for regulatory peptides more difficult. There is little reason to suppose that further agents will not be identified in the future. Because a knowledge of the variables participating in biological systems is an essential prerequisite to the design of experiments to determine the physiological function of such systems, the importance of identifying these "new" peptides is self-evident.

Greater efforts will have to be directed at determining the physiology of these regulatory peptides. It has proved somewhat difficult to establish physiological roles for GI peptides which are secreted into the systemic circulation, but for peptides which appear to exert their activity via strictly local release mechanisms an even greater degree of complexity is encountered in attempts to establish their physiological roles. For regulatory peptides that are delivered to target organs by the systemic circulation, a measure of the amount of endogenous peptide released during a physiological event can be assessed by quantitating alterations in plasma levels; such alterations can then be related to the biological responses of the presumed target organs. Exogenous peptides can then be administered in amounts sufficient to match the alterations in plasma levels which were attained during the physiological stimulus and the biological responses of the target organ compared. For locally acting peptides (paracrine, neural release), techniques are not available to assess accurately the magnitude of the physiological release of such peptides for correlation with their putative biological effects. A further fundamental problem is our inability to duplicate experimentally the local release of regulatory peptides; i.v. administration of such peptides elicits multiple events which may not be seen during a strictly local physiological release of the peptide; hence, unwanted modifications of target organ responses may occur. Consequently, a definitive assessment of the physiological importance of these locally acting peptides will probably prove impossible until potent and specific competitive antagonists are developed. The importance of some of these peptides in the production of clinical syndromes during pathological states is clear, and measurement of the plasma levels of these peptides is crucial to the diagnosis of the presence of tumors (APUDomas) producing and secreting regulatory peptides.

Studies on the potential physiology of regulatory peptides generally focus on the use of a regulatory peptide in isolation. This is somewhat unrealistic because during a physiological stimulus, such as the ingestion of a mixed meal, multiple variables are changed simultaneously. We have only begun to explore the potential interactions (potentiative, additive, or inhibitory) between these peptides in physiological situations. The potentiation which occurs between secretin and CCK in stimulating pancreatic secretion is a classic example. That is, the effects seen by the simultaneous administration of CCK and secretin are greater than the sum of the effects of administering the individual peptides. It is clearly recognized that in certain physiological circumstances the release of certain endocrine regulatory peptides (i.e., gastrin, insulin) may be regulated, in part, by neural mechanisms (the so-called cephalic-phase responses). It is conceivable that the reverse mechanism may be important. That is, circulating regulatory peptides may act on neural receptors to alter neuronal function in the peripheral nervous system and possibly, in certain areas, the central nervous system. Future studies of such potential interactions should prove fruitful.

Much attention has been given to the chemical characterization of regulatory peptides, but only recently has technology developed to allow chemical characterization of the specific receptors with which they interact. Following the evolution of the structure and function of both regulatory peptides and their specific receptors should provide useful insights into physiological mechanisms. We have only just begun to investigate the subcellular mechanisms which mediate specific biological events following receptor-regulatory peptide interaction. Second messenger systems have been identified, and the ability of regulatory peptides to elicit responses in these systems is currently being investigated. However, little is known of the mechanisms by which the second messenger systems function to elicit specific biological responses within the cell.

REFERENCES

1. **Mutt, V.,** New approaches to the identification and isolation of hormonal polypeptides, *Trends Neurosci.,* 6, 357, 1983.
2. **Bayliss, W. M.,** *Principles of General Physiology,* Longmans & Green, London, 1915, 706.
3. **Leuret, F. and Lassaigne, J. L.,** *Recherches Physiologiques et Chimiques pour Servir a l'Histoire de la Digestione,* Huzard, Paris, 1825, 141.
3a. **Dolinsky, M. J.,** Études sur l'excitabilité sécrétoire spécifique de la muqueuse du canal digestif, *Arch. Sci. Biol.,* St. Petersburg, IL, 399, 1894.
3b. **Popielski, L.,** Über das periphertsche reflectorische Nervencentrum des Pankreas, *Pflügers Arch. Physiol.,* 86, 215, 1901.
3c. **Wertheimer, E. and Lepage, L.,** Sur le fonctions réflexes des ganglions abdomineaux du sympathique dans l'innervation sécrétoire due pancréas, *J. Physiol. Pathol. Gen.,* 3, 335, 1901.
4. **Bayliss, W. M. and Starling, E. H.,** On the causation of the so-called "peripheral reflex secretion" of the pancreas, *Proc. R. Soc. London,* 69, 352, 1902.
5. **Mutt, V.,** Secretin: isolation, structure and functions, in *Gastrointestinal Hormones,* Glass, G. B. J., Ed., Raven Press, New York, 1980, 85.
6. **Jorpes, J. E. and Mutt, V.,** On the biological activity and amino acid composition of secretin, *Acta Chem. Scand.,* 15, 1790, 1961.
7. **Mutt, V. and Jorpes, J. E.,** Secretin: isolation and determination of structure, in *Proc. I.U.P.A.C. 4th Int. Congr. Natural Products,* Stockholm, Sweden, 1966, Section 2C-3.
8. **Mutt, V. and Magnusson, S.,** Structure of porcine secretin, the amino acid sequence, *Eur. J. Biochem.,* 15, 513, 1970.
9. **Edkins, J. S.,** On the chemical mechanism of gastric secretion, *Proc. R. Soc. London Ser. B,* 76, 376, 1905.
10. **Grossman, M. I., Robertson, C. R., and Ivy, A. C.,** Proof of a hormonal mechanism for gastric secretion. The humoral transmission of the distention stimulus, *Am. J. Physiol.,* 153, 1, 1948.
11. **Gregory, R. A. and Tracy, H. J.,** The constitution and properties of two gastrins extracted from hog antral mucosa, *Gut,* 5, 103, 1964.
11a. **Uvnäs, B.,** The part played by the pyloric region in the cephalic phase of gastric secretion, *Acta Physiol. Scand.,* 4, Suppl. 13, 1, 1942.
12. **Ivy, A. C. and Oldberg, E.,** Contraction and evacuation of the gall bladder by a purified "secretin" preparation, *J. Am. Med. Assoc.,* 90, 445, 1928.
13. **Ivy, A. C. and Oldberg, E.,** Hormone mechanism for gallbladder contraction and evacuation, *Am. J. Physiol.,* 99, 415, 1941.
14. **Harper, A. A. and Vass, C. C. N.,** The control of the external secretion of the pancreas in cats, *J. Physiol. (London),* 86, 599, 1928.
15. **Harper, A. A. and Raper, H. S.,** Pancreozymin, a stimulant of the secretion of pancreatic enzymes in extracts of the small intestine, *J. Physiol. (London),* 102, 115, 1943.
16. **Mutt, V. and Jorpes, J. E.,** Structure of porcine cholecystokinin-pancreozymin. I. Cleavage with thrombin and with trypsin, *Eur. J. Biochem.,* 6, 156, 1968.
17. **Mutt, V.,** Further investigations on intestinal hormonal peptides, *Clin. Endocrinol.,* 5 (Suppl.), 1755, 1976.
18. **Eysselein, V. E., Reeve, J. R., Jr., Shively, J. E., Hawke, D., and Walsh, J. H.,** Partial structure of a large canine cholecystokinin (CCK_{58}): amino acid sequence, *Peptides,* 3, 687, 1982.
19. **Anastasi, A., Erspamer, V., and Endean, R.,** Isolation and amino acid sequence of caerulein, the active decapeptide of the skin of *Hyla caerulea, Arch. Biochem. Biophys.,* 125, 57, 1968.

20. **Bromer, W. W., Sinn, L. G., and Behrens, O. K.**, The amino acid sequence of glucagon. V. Location of the amide groups, acid degradation studies and summary of sequential evidence, *J. Am. Chem. Soc.*, 79, 2807, 1957.
21. **Doi, K., Prentki, M., Yip, C., Muller, W. A., Jeanrenaud, B., and Vranic, M.**, Identical biological effects of pancreatic glucagon and a purified moiety of canine gastric immunoreactive glucagon, *J. Clin. Invest.*, 63, 525, 1979.
22. **Hatton, T. W., Yip, C. C., and Vranic, M.**, Biosynthesis of glucagon (IRG³⁵⁰⁰) in canine gastric mucosa, *Diabetes*, 34, 38, 1985.
23. **Brown, J. C., Mutt, V., and Pederson, R. A.**, Further purification of a polypeptide demonstrating enterogastrone activity, *J. Physiol. (London)*, 209, 57, 1970.
24. **Jörnvall, H., Carlquist, M., Kwauk, S., Otte, S., McIntosh, C., Brown, J., and Mutt, V.**, Amino acid sequence and heterogeneity of gastric inhibitory polypeptide (GIP), *FEBS Lett.*, 123, 205, 1981.
25. **McIntosh, C. H. S., Pederson, R. A., Koop, H., and Brown, J. C.**, Gastric inhibitory polypeptide stimulated secretion of somatostatin like immuno-reactivity from the stomach: inhibition by acetylcholine or vagal stimulation, *Can. J. Physiol. Pharmacol.*, 59, 468, 1980.
26. **Maxwell, V., Shulkes, A., Brown, J. C., Solomon, R., Walsh, J., and Grossman, M.**, Effect of gastric inhibitory polypeptide on pentagastrin-stimulated acid secretion in man, *Dig. Dis. Sci.*, 25, 113, 1980.
27. **Yamagishi, T. and Debas, H. T.**, Gastric inhibitory polypeptide (GIP) is not the primary mediator of the enterogastrone action of fat in the dog, *Gastroenterology*, 78, 931, 1980.
28. **Rabinovitch, A. and Dupré, J.**, Insulinotropic and glucagonotropic activities in crude preparations of cholecystokinin-pancreozymin, *Clin. Res.*, 20, 945, 1972.
29. **Dupré, J., Ross, S. A., Watson, D., and Brown, J. C.**, Stimulation of insulin secretion by gastric inhibitory polypeptide in man, *J. Clin. Endocrinol. Metab.*, 37, 826, 1973.
30. **Kimmel, J. R., Hayden, L. J., and Pollock, H. G.**, Isolation and characterization of a new pancreatic polypeptide hormone, *J. Biol. Chem.*, 250, 9369, 1975.
31. **Chance, R. E. and Jones, W. E.**, Polypeptide from Bovine, Ovine, Human and Porcine Pancreas, U.S. Patent 3,842,063, 1974.
32. **Schwartz, T. W.**, Pancreatic polypeptide: a hormone under vagal control, *Gastroenterology*, 85, 1411, 1983.
33. **Tatemoto, K. and Mutt, V.**, Chemical determination of polypeptide hormones, *Proc. Natl. Acad. Sci. U.S.A.*, 75, 4115, 1978.
34. **Tatemoto, K.**, Isolation and characterization of peptide YY (PYY), a candidate gut hormone that inhibits pancreatic exocrine secretion, *Proc. Natl. Acad. Sci. U.S.A.*, 79, 2514, 1982.
35. **Tatemoto, K.**, Neuropeptide Y: complete amino acid sequence of the brain peptide, *Proc. Natl. Acad. Sci. U.S.A.*, 79, 5485, 1982.
36. **Carraway, R. and Leeman, S. E.**, The amino acid sequence of a hypothalamic peptide neurotensin, *J. Biol. Chem.*, 250, 1907, 1975.
37. **Brazeau, P., Vale, W., Burgus, R., Ling, N., Butcher, M., Rivier, J., and Guillemin, R.**, Hypothalamic polypeptide that inhibits the secretion of immunoreactive pituitary growth hormone, *Science*, 179, 77, 1973.
38. **Schubert, H. and Brown, J. C.**, Correction to the amino acid sequence of porcine motilin, *Can. J. Biochem.*, 52, 7, 1974.
39. **Furness, J. B. and Costa, M.**, *The Enteric Nervous System*, Churchill Livingstone, Edinburgh, 1987.
40. **von Euler, U. S. and Gaddum, J. H.**, An unidentified depressor substance in certain tissue extracts, *J. Physiol. (London)*, 72, 74, 1931.
41. **Chang, M. M. and Leeman, S. E.**, Isolation of a sialogogic peptide from bovine hypothalamic tissue and its characterization as substance P, *J. Biol. Chem.*, 245, 4784, 1970.
42. **Studer, R. O., Trzeciak, A., and Lergier, W.**, Isolierung und Aminosauresequenz von Substanz P aus Pferdearm, *Helv. Chim. Acta*, 56, 860, 1973.
43. **Carraway, R., Kitabgi, P., and Leeman, S. E.**, The amino acid sequence of radioimmunoassayable neurotensin from bovine intestine, *J. Biol. Chem.*, 253, 7996, 1978.
44. **Luft, R., Efendic, S., Hökfelt, T., Johansson, O., and Arimura, A.**, Immunohistochemical evidence for the localization of somatostatin-like immunoreactivity in a cell population of the pancreatic islets, *Med. Biol.*, 52, 428, 1974.
45. **Vanderhaeghen, J. J., Signeau, J. C., and Gepts, W.**, New peptide in the vertebrate CNS reacting with antigastrin antibodies, *Nature (London)*, 257, 604, 1975.
46. **Dockray, G. J., Gregory, R. A., Hutchison, J. B., Harris, J. I., and Runswick, M. J.**, Isolation, structure and biological activity of two cholecystokinin octapeptides from sheep brain, *Nature (London)*, 274, 711, 1978.
47. **Feyrter, F.**, *Über Diffuse Endokrine Epitheliale Organe*, J. A. Barth, Leipzig, 1938, 6.
47a. **Lepage**, Ph.D. thesis, Montpellier, France, 1955.
48. **Pearce, A. G. E.**, The diffuse neuroendocrine system and the APUD concept, in *Gut Hormones*, Bloom, S. R., Ed., Churchill Livingstone, Edinburgh, 1978, 33.

49. **Andrews, A.**, The APUD concept: where has it led us?, *Br. Med. Bull.*, 38, 221, 1982.
50. **Schultzberg, M., Hökfelt, T., and Lundberg, J. M.**, Co-existence of classical transmitters and peptides in the central and peripheral nervous systems, *Br. Med. Bull.*, 38, 309, 1982.
51. **Acher, R. and Chauvet, J.**, La structure de al vasopressine de boeuf, *Biochim. Biophys. Acta*, 12, 487, 1953.
52. **du Vigneaud, V., Lawler, H. C., and Popenoe, E. A.**, Enzymatic cleavage of glycinamide from vasopressin and a proposed structure for this pressor-antidiuretic hormone of the posterior pituitary, *J. Am. Chem. Soc.*, 75, 4880, 1953.
53. **Larsson, L.-I., Goltermann, N., De Magistris, L., Rehfeld, J. F., and Schwartz, T. W.**, Somatostatin cell processes as pathways for paracrine secretion, *Science*, 205, 1393, 1979.
54. **McDonald, T. J., Nilsson, G., Bloom, S. R., Ghatei, M. A., Vagne, M., and Mutt, V.**, A gastrin releasing peptide from the porcine non-antral gastric tissue, *Gut*, 19, 767, 1978.
55. **McDonald, T. J., Jörnvall, H., Nilsson, G., Vagne, M., Ghatei, M., Bloom, S. R., and Mutt, V.**, Characterization of a gastrin-releasing peptide from porcine non-antral gastric tissue, *Biochem. Biophys. Res. Commun.*, 90, 227, 1979.
56. **Erspamer, V. and Melchiorri, P.**, Active polypeptides: from amphibian skin to gastrointestinal tract and brain of mammals, *Trends Pharmacol. Sci.*, 1, 391, 1980.
57. **Minamino, N., Sudoh, T., Kngawa, K., and Matsuo, H.**, Neuromedin B-32 and B-30: two "big" neuromedin B identified in porcine brain and spinal cord, *Biochem. Biophys. Res. Commun.*, 124, 925, 1984.
58. **Kangawa, K., Minamino, N., and Fukuda, A., and Matsuo, H.**, Neuromedin K: a novel mammalian tachykinin identified in porcine spinal cord, *Biochem. Biophys. Res. Commun.*, 114, 533, 1983.
59. **Nawa, H., Hirose, T., Takashima, H., Inayama, S., and Nakanishi, S.**, Nucleotide sequences of cloned cDNAs for two types of bovine brain substance P precursor, *Nature (London)*, 306, 32, 1983.
60. **Kimura, S., Okada, M., Sugita, Y., Kanazawa, I., and Munekata, E.**, Novel neuropeptides, neurokinin, alpha and beta, isolated from porcine spinal cord, *Proc. Jpn. Acad.*, 59(B), 101, 1983.
61. **Minamino, N., Kangawa, K., and Matsuo, H.**, Neuromedin N: a novel neurotensin-like peptide identified in porcine spinal cord, *Biochem. Biophys. Res. Commun.*, 122, 542, 1984.
62. **Li, C. H.**, Beta-lipotropin, a new pituitary hormone, *Arch. Biol. Med. Exp.*, 5, 55, 1968.
63. **Thim, L. and Moody, A. J.**, The primary structure of porcine glicentin (proglucagon), *Regul. Peptides*, 2, 139, 1981.
64. **Bataille, D., Coudray, A. M., Carlquist, M., Rosselin, G., and Mutt, V.**, Isolation of glucagon-37 (bioactive enteroglucagon/oxyntomodulin) from porcine jejunal-ileum: isolation of the peptide, *FEBS Lett.*, 146, 73, 1982.
65. **Tatemoto, K. and Mutt, V.**, Isolation and characterization of the intestinal peptide porcine PHI (PHI-27), a new member of the glucagon-secretin family, *Proc. Natl. Acad. Sci. U.S.A.*, 78, 6603, 1981.
66. **Tatemoto, K., Rökaeus, Å., Jörnvall, H., McDonald, T. J., and Mutt, V.**, Galanin — a novel biologically active peptide from porcine intestine, *FEBS Lett.*, 164, 124, 1983.
67. **Fox, J. E. T., McDonald, T. J., Kostolanska, F., and Tatemoto, K.**, Galanin: an inhibitory neural peptide of the canine small intestine, *Life Sci.*, 39, 103, 1986.
68. **McDonald, T. J., Dupré, J., Tatemoto, K., Greenberg, G. R., Radziuk, J., and Mutt, V.**, Galanin inhibits insulin secretion and induces hyperglycemia in dogs, *Diabetes*, 34, 192, 1985.
69. **Rosenfeld, M. G., Mermod, J.-J., Amara, S. G., Swanson, L. W., Sawchenko, P. E., Rivier, J., Vale, W. W., and Evans, R. M.**, Production of a novel neuropeptide encoded by the calcitonin gene via tissue-specific RNA processing, *Nature (London)*, 304, 129, 1983.
70. **Bell, G. I., Sanchez-Pescador, R., Laybourn, P. J., and Najarian, R. C.**, Exon duplication and divergence in the human preproglucagon gene, *Nature (London)*, 304, 368, 1983.

Chapter 2

GASTROENTEROPANCREATIC REGULATORY PEPTIDE STRUCTURES: AN OVERVIEW

T. J. McDonald

TABLE OF CONTENTS

I.   Introduction 20

II.   The Secretin-Glucagon Family of Peptides 25
        A.   Secretin, Glucagon, and the Glucagon-like Peptides 1 and 2 26
        B.   The Vasoactive Intestinal Polypeptide and the Gastric Inhibitory
                Polypeptide 29
        C.   PHI, Growth Hormone Releasing Factor (GHRF), and
                Nonmammalian Members of the Secretin-Glucagon Family 31

III.   The Gastrin-Cholecystokinin Family 33
        A.   Gastrin 33
        B.   Cholecystokinin 36
        C.   Nonmammalian Members of the Gastrin-CCK Family 40

IV.   The Pancreatic Polypeptide Family 41
        A.   Mammalian and Nonmammalian Pancreatic Polypeptides 43
        B.   PYY and NPY 44
        C.   Nonmammalian Members of the Pancreatic Polypeptide Family 46

V.   The Neurotensin Family 46

VI.   The Tachykinin Family 49

VII.   The Bombesin Family 54

VIII.   The Endothelin Family 59

IX.   GEP Regulatory Peptides Not Yet Demonstrated To Be Part of a
        Family 63
        A.   Galanin 63
        B.   Motilin 65
        C.   Calcitonin Gene-Related Peptides (CGRP) 66

X.   Concluding Remarks 67

References 71

I. INTRODUCTION

This chapter provides a brief overview of the structural features of certain recognized biologically active peptides common to both gastroenteropancreatic (GEP) tissues and to neural structures of the peripheral and central nervous systems. These peptides have been variously referred to as "GI hormones", "neuropeptides", and "regulatory peptides"; for reasons discussed in Chapter 1 of this volume, these peptides will be referred to as regulatory peptides. The objectives of this chapter are to outline (1) the primary structures of certain regulatory peptides and, where known, the differences in amino acid sequences which occur between mammalian and certain nonmammalian species; (2) the structural similarities between certain regulatory peptides which result in their being classified as members of peptide families; and (3) where known, important data on structure-function relationships. Space does not permit the review of the structures of such important biologically active agents as the hypothalamic releasing factors, the plasma-derived factors (i.e., angiotensin), the cardiac-derived and related but brain-derived peptides (i.e., the atrial naturetic peptides), the burgeoning and important field of peptide growth factors (i.e., IGF-1, IGF-2, etc.), the peptides derived from immunologically competent lymphocytes or from the thymus or the insulin super-family of peptides, and somatostatin and the opiate-like peptides which have been the subjects of recent and/or ongoing reviews. Description of the structures of mammalian GEP regulatory peptides will be emphasized, but mention will be made of certain important nonmammalian peptide structures which have predicted the eventual discovery of an analogous mammalian peptide or where immunologic evidence suggests that an analogous mammalian peptide will be found. The purpose of this chapter is to describe the primary structures of these peptides, and reference is made to the original literature where possible. The function of these peptides is mentioned only briefly and, in the interest of brevity, where possible, reference is given to review articles rather than to the original literature.

In contrast to the relatively high concentrations of peptide hormones present in discrete endocrine glands, regulatory peptides of the gastrointestinal tract are present in relatively low concentrations, being spread diffusely throughout an organ well suited to proteolytic degradation. Hence, close to 60 years after Bayliss and Starling's (1902) and J. S. Edkins' (1905) classical experiments were performed,[1,2] preparations of secretin and gastrin were achieved in sufficient purity to establish their primary structures. Two laboratories played key roles leading to the initial isolation and characterization of mammalian regulatory peptides; those of Mutt and Jorpes in the Karolinska Institute, Stockholm, Sweden and of Tracy and Gregory in Liverpool, U.K. Both groups applied certain lessons learned from Bayliss and Starling's experiment and the early attempts to isolate secretin (i.e., that enzymes could be heat inactivated without affecting the activity of secretin, that secretin was stable to acid extraction, etc.) and also developed techniques to process relatively large-scale tissue extractions; this was particularly true of the Karolinska laboratory. In the late 1950s and early 1960s, these groups applied the newly developed ion-exchange, countercurrent distribution, and gel-filtration chromatographic techniques to the purification of peptides from tissue extracts. The Liverpool group found that their principal hormone of interest, gastrin, entered the water phase on boiling tissue at neutral pH; gastrin was therefore extracted from the water phase by adsorption to cellulose-based ion exchangers. In contrast, the Karolinska Institute group found that the principal hormones of their interest, secretin, CCK, and pancreozymin, were predominantly retained in the tissue during the boiling process; hence, the active peptides were liberated from the heat-inactivated tissue by extraction with acid in the cold. The acid extraction of peptides from tissue proved of immense value because a number of previously unrecognized regulatory peptides were subsequently identified and characterized from Mutt's upper small intestinal extracts. Apart from Gregory and Tracy's isolation of gastrin from the water phase obtained during the boiling of gastric antral tissue, little work has been directed toward peptides present in this water phase. Secretin, isolated and characterized by Mutt and Jorpes,[3,4] and gastrin, isolated and characterized by Gregory and Tracy,[5] were the first of the so-called GI hormones to be isolated from gastrointestinal tissue. The isolations of secretin and gastrin were seminal events that (1) established the presence of the C-terminal α-amide structure and the esterification of tyrosyl residues with sulfuric acid as being important posttranslational modifications of amino acid residues in such peptides and (2) established the then necessary techniques for the isolation and structural characterization of other regulatory peptides. At about the same time, Erspamer and colleagues in Rome began their pioneering work on isolating peptides from amphibian skin and demonstrating that the peptides possessed potent bioactivities on administration into mammals. Often, the isolation of these amphibian peptides predicted that a mammalian counterpart would be identified.

During the early attempts to isolate regulatory peptides, the limiting factors were the available chemical techniques for peptide purification and, once a homogeneous preparation was obtained, the techniques for structural characterization. Recent advances in peptide separation technology, particularly the development of high performance liquid chromatography (HPLC) and the development of microsequencing techniques, have virtually eliminated these factors as posing insurmountable problems. The major limiting factor in the isolation and characterization of regulatory peptides at present is probably the methodology available to detect initially a previously unrecognized peptide and to follow its purification. Developments in immunologic methodology greatly enhanced our ability to identify known regulatory peptides. Compared to the tedious and often technically difficult bioassay techniques, the use of radioimmunoassay (RIA) provides considerably greater sensitivity and, at times, greater specificity, and, importantly, allows the rapid assay of large numbers of small amounts of samples during chromatographic procedures. RIA identification of known regulatory peptides has proved invaluable in the elucidation of the processing procedures which generate "mature" regulatory peptides from precursor molecules, in determining the metabolic fate of regulatory peptides, in the isolation and characterization of their variant forms, and in determining the structural differences of regulatory peptides between species. The use of RIA as a detection device allows the isolation and characterization of regulatory peptides from extraordinarily small amounts of tissue. However, for a truly novel peptide, of which nothing is known about its structure, it is self-evident that recourse must be made to the use of alternate methodology (often bioassays) for detection and purification. The process then becomes technically more difficult, necessitating the use of greater amounts of starting material.

In certain cases, the detection device can be a chemical assay which detects specific structures within the peptide molecule. The best-developed example of this technique is the assay developed by Tatemoto and Mutt which detects the presence of peptides containing the C-terminal α-amide structure in crude tissue extracts and identifies the specific amidated residue present.[6] In a few cases, the structures of regulatory peptides have been detected serendipitously during chromatographic procedures; for example, the pancreatic polypeptide and β-lipotropin were characterized from prominent chromatographic peaks noted during the isolation of insulin and adrenocorticotropic hormone (ACTH), respectively.[7-9] Peptides isolated using a strictly chemical approach to detection often have their biological activities elucidated subsequent to their structural characterization. Recently, a number of novel peptides have been unexpectedly discovered due to their presence in the structures of precursor molecules which have been deduced from cDNA nucleotide sequences coding for known regulatory peptides. The technologies for the purification and structural analysis of peptides are no longer the limiting factors in the process of discovering previously unrecognized regulatory peptides; the limiting factor now appears to be the "detection device" employed to recognize the regulatory peptide.

Steiner and colleagues,[10] using the biosynthesis of insulin as a model, found that the general pathways by which regulatory peptides are synthesized are similar to those described for other secretory proteins. Specific mRNAs direct the ribosomal synthesis of regulatory peptides in the forms of precursor molecules generally known as "prepro-" molecules. The N-terminal region of the precursor molecules contains an amino acid residue sequence with recognizable characteristics (the so-called "signal peptide" or "pre-" sequence) which directs the nascent peptide to the membrane of the endoplasmic reticulum. The first proteolytic cleavage of the precursor molecule occurs with removal of this "signal sequence" as the molecule passes through the membrane into the cisterns of the endoplasmic reticulum. The "pro-" form of the precursor molecule is then directed to the Golgi apparatus. Enroute to and in the Golgi, further proteolytic cleavages occur in the precursor molecule (removing the "pro-" sequences). This process of sequential proteolytic cleavages of precursor proteins to yield smaller mature mammalian peptide hormones or neuropeptides was first described for peptide hormones in the 1960s,[10-12] and subsequent investigations have repeatedly confirmed the general nature of this process which parallels that described for other secretory proteins. Cleavage of the "pro-" sequence often occurs at pairs of basic amino acid residues (i.e., Lys–Lys, Arg–Arg, Arg–Lys, Lys–Arg), and it is now generally recognized that this is an important signal for posttranslational proteolytic cleavage. However, it is also clear that cleavage may occur at sites other than pairs of basic amino acid residues to produce mature molecules and that certain pairs of basic residues within the interior of a mature molecule are not cleaved. Consequently, although pairs of basic residues are an important signal for posttranslational proteolytic cleavage to occur, this is by no means the only such signal. The rules governing the cleavage of mature peptides from precursor molecules, although presently not well understood, are currently under investigation.[13,14] The mature peptide hormones or neuropeptides are finally packaged into secretory granules.

Following or concurrent with proteolytic cleavage of the regulatory peptides from their precursor proteins, certain residues within the mature peptide may undergo additional posttranslational processing. One modification is the cyclization of an N-terminal glutamine residue into the N-terminal pyroglutamyl (Pyr) structure. This cyclization has been noted to occur spontaneously in *in vitro* studies and is present in such regulatory peptides as gastrin-17 and gastrin-34, neurotensin, and bombesin. A second and apparently very common posttranslational process which occurs in GEP regulatory peptides is the formation of the C-terminal α-amide structure. This structure is formed by cleavage of the precursor peptide chain such that the amide group is formed via the incorporation of the amino group of a succeeding glycine residue.[15,16] The signal for C-terminal α-amide formation appears to be the presence of a glycine residue followed by a pair of basic amino acids (Arg, Lys), which occurs in the precursor molecule immediately C-terminal to the final residue of the mature peptide.[15,16] Evidence has been presented that the basic amino acid residues must be removed (via sequential cleavage by enzymes with trypsin-like activities followed by carboxypeptidase-like activities) before the amidating enzyme catalyzes the amidation of the C-terminal residue.[17,18] This is a common structure found in GEP regulatory peptides, although arguably this may be partly due to the fact that Tatemoto and Mutt have isolated a number of such peptides based on the detection of the presence of peptides with this structure. A relatively less common modification is the esterification of tyrosine residues with sulfuric acid. This reaction occurs in the trans-Golgi region and is catalyzed by an integral membrane enzyme, tyrosylprotein sulfotransferase, which recognizes tyrosine residues that occur in close proximity to acidic amino acid residues.[19] The tyrosyl-*O*-sulfate structure occurs in 50% of gastrin molecules and in virtually all cholecystokinin molecules formed *in vivo*; the presence of this structure is crucial for potent biological activity of CCK, but does not appear to be necessary for the gastric acid secretagogue activity of gastrin.

As the GEP regulatory peptides were chemically characterized, it became unexpectedly

TABLE 1
Notations for Amino Acid Sequences

Three-Letter Symbol	Amino Acid	One-Letter symbol
Ala	Alanine	A
Cys	Cysteine	C
Asp	Aspartic acid	D
Glu	Glutamic acid	E
Phe	Phenylalanine	F
Gly	Glycine	G
His	Histidine	H
Ile	Isoleucine	I
Lys	Lysine	K
Leu	Leucine	L
Met	Methionine	M
Asn	Asparagine	N
Pro	Proline	P
Gln	Glutamine	Q
Arg	Arginine	R
Ser	Serine	S
Thr	Threonine	T
Val	Valine	V
Trp	Tryptophan	W
Tyr	Tyrosine	Y

Note: Additional symbols used in this chapter are: (1) $-NH_2$ in the three letter notation and ∗ in the one-letter notation indicates amidation of a C-terminal residue; (2) $-SO_3H$ in the three-letter notation and ° in the one-letter notation indicate esterification of a tyrosyl residue with sulfuric acid; (3) Pyr in the three-letter notation and □ in the one-letter notation indicate an N-terminal pyroglutamyl residue; (4) –S–S– indicates the positions of a disulfide bridge.

After IUIPAC-IUB Commission on Biochemical Nomenclature (CBN), A one-letter notation for amino acid sequences. Tentative rules, *Eur. J. Biochem*, 5, 151, 1968.

clear that they often, but not invariably, had structural homology with each other and consequently appeared to form a series of peptide families. In certain cases, these sequence similarities result in overlapping biological activities, while in other cases, the structural similarities occur outside the region of the molecule important for biological activity. In certain families, the sequence similarities occur throughout the molecules; members of the secretin-glucagon family, the pancreatic polypeptide family, and the insulin super-family of peptides are in this category. In other cases, the sequence similarities are confined to specific regions of the molecule; the gastrin-CCK family, the bombesin family, the tachykinin family, the opioid peptide family, the neurotensin family, and the endothelian family of peptides are in this category. Regulatory peptides such as somatostatin, galanin, motilin, etc. appear at present to have unique structures.

In this chapter the structures of these regulatory peptides will be portrayed in the one- or three-letter notation for amino acids. Since certain readers will not be familiar with the one-letter notations for amino acids, these are listed in Table 1. Structural similarities and dissimilarities are more readily portrayed using the one-letter notation; hence, it is used for

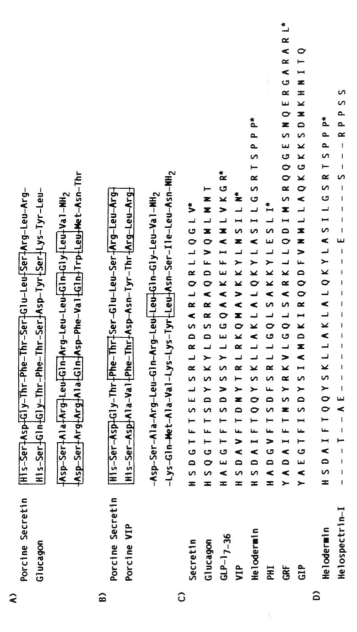

FIGURE 1.   The structures of the glucagon-secretin family of peptides are portrayed for comparison by aligning all peptides at their N-termini. (A) The structures of porcine secretin and glucagon are given using the three-letter notation for amino acid residues. Identical residues occurring at the same position in the two peptides are enclosed within boxes. (B) The structures of porcine secretin and VIP are portrayed using the three-letter notation for amino acid residues. Amino acid residue identities between the two peptides are enclosed within boxes. (C) The amino acid sequences of accepted and putative members of the secretin-glucagon family are portrayed in their entirety using the one-letter notation for amino acid residues. (D) The structures of two peptides isolated from Gila monster venom are portrayed using the one-letter notation for amino acid residues. The amino acid sequence of helodermin is given in its entirety. The amino acid residue differences occurring between helospectrin 1 and helodermin are shown. A dash indicates that an identical amino acid residue occurs in helospectrin 1 and helodermin in that position. See Table 1 for definition of symbols denoting specific chemical modifications of amino acid residues.

this purpose. Where possible, the three-letter notation for amino acids is used because this is probably more familiar to most readers. Symbols used to denote specific modification of amino acid residues are also portrayed in Table 1.

II. THE SECRETIN-GLUCAGON FAMILY OF PEPTIDES

This peptide family is, at present, generally recognized as being composed of at least six, probably seven, mammalian peptides and two nonmammalian peptides, all of which have considerable sequence homology with each other; these structural homologies are spread throughout the molecules and not restricted to a particular region (see Figure 1A,B,C). The sequences of five mammalian peptides (secretin,[3] glucagon,[20] the vasoactive intestinal polypeptide or VIP,[21] the gastric inhibitory polypeptide or GIP,[22] and peptide histidine isoleucine amide or PHI[23]) were first isolated from the gastrointestinal tract, but the hypothalamic-releasing peptide, the growth hormone-releasing factor or somatocrinin, was first isolated from pancreatic tumors[24,25] and subsequently from hypothalamic tissue.[26-28] Peptides isolated from the venom of Gila monsters, helospectrin I and II[29] and helodermin,[30] have striking sequence homology with the mammalian members of this family. The sequences of helospectrin I and II and helodermin are portrayed in Figure 1D, and the sequence of helodermin is compared with the mammalian peptides (Figure 1C) because immunologic evidence suggests the presence of a helodermin-like peptide in mammals.[31]

A. SECRETIN, GLUCAGON, AND THE GLUCAGON-LIKE PEPTIDES 1 AND 2

Pancreatic glucagon was first detected as a contaminant in early insulin preparations, but its existence was held in doubt until the classical studies of Sutherland and DeDuve[32] demonstrated the presence of a potent hyperglycemic glycogenolytic factor in canine gastric corpus tissue. It is now recognized that pancreatic glucagon occurs in the mammalian gastrointestinal tract, but, with the exception of the canine gastric corpus, is generally present in relatively small amounts. In both mammalian and nonmammalian species, glucagon is present in far greater concentrations in pancreatic tissue; hence, glucagon was isolated and characterized by Bromer and colleagues[20] from pig pancreas. Pancreatic glucagon was found to be a linear polypeptide composed of 29 amino acids and to have an N-terminal residue of histidine and a C-terminal threonine residue (see Figure 1A).

A few years later, Mutt and Jorpes in Stockholm succeeded in isolating the first described GI hormone, secretin.[3] Sequence analysis revealed that secretin was a linear polypeptide composed of 27 amino acid residues with an N-terminal histidine residue, but the C-terminus contained an amidated valine residue. Of the 27 amino acid residues in secretin, 14 occurred in positions identical to those of pancreatic glucagon; even more surprising, 7 of the 8 N-terminal residues were identical in the two molecules (see Figure 1A). This was a totally unexpected finding because the major biological activities attributed to the two peptides are exceedingly different; the main activity of glucagon is considered to be the elevation of blood glucose levels; and that of secretin is considered to be the stimulation of pancreatic secretion of water and bicarbonate. Furthermore, it had been recognized that the N-terminal histidine residue, present in both peptides, was of critical importance for the biological activity of glucagon. This finding raised difficulties in terms of structure-function relationships and was the first demonstration that two GEP peptides could have considerable sequence homology, even in an area important for the expression of bioactivity, yet have different putative physiological roles. This surprise was the first of many which occurred in the 1960s and early 1970s demonstrating that the GEP regulatory peptides, based on structural homology, formed families of peptides.

Immunohistochemical studies have established the presence of glucagon in the pancreatic islet A cells and also in endocrine-like cells of the canine gastric fundal mucosa which are

A)

Glucagon

	1	10	20
Mammalian	H S Q G T F T S D Y S K Y L D S R R A Q D F V Q W L M N T
Guinea pig	- - - - - - - - - - - - - - - - - Q - L K - - L - V
Frog	- S
Opossum, chicken, turkey	- S -
Duck	- - - - - - - - - - - - - - - - T - - - - - - - - - S -
Alligator (Gar)	- - - - - - - N - - - - - - - T - - - - - - - - - S -

B)

Porcine Glucagon and Glucagon Precursors

Glicentin   R S L Q N T E E K S R S F P A P Q T D P L D D P D Q M T E D K R–

H S Q G T F T S D Y S K Y L D S R R A Q D F V Q W L M N T K R N K N N I A

Oxyntomodulin   H S Q G T F T S D Y S K Y L D S R R A Q D F V Q W L M N T K R N K N N I A

Glucagon   H S Q G T F T S D Y S K Y L D S R R A Q D F V Q W L M N T

C)

Human Glucagon and the Glucagon-like (GLP) Peptides

Glucagon   H S Q G T F T S D Y S K Y L D S R R A Q D F V Q W L M N T

GLP-1   H D E F E R H A E G T F T S D V S S Y L E G Q A A K E F I A W L V K G R*

GLP-2   H A D G S F S D E M N T I L D N L A A R D F I N W L I Q T K I T D R

FIGURE 2. (A) The complete structure of the pancreatic glucagon molecule common to pig, cow, rabbit, rat, human, dog, camel, and hamster is given in its entirety using the one-letter notation for amino acid residues. Differences in structure between the common mammalian structure and that of the guinea pig molecule and certain nonmammalian glucagon molecules are portrayed; dashes indicate amino acid residue identities with the common mammalian molecule. (B) The structural relationships between pancreatic glucagon, oxyntomodulin, and their precursor molecule, glicentin, are portrayed using the one-letter notation for amino acid residues. (C) The structural homologies between human glucagon and the human glucagon-like peptides 1 and 2 are portrayed using the one-letter notation for amino acid residues. For comparison, the histidine residue in position 7 of GLP-1 is aligned with the N-termini of glucagon and GLP-2. Identities are enclosed within boxes. See Table 1 for definition of special symbols indicating specific chemical modifications of amino acid residues.

morphologically indistinguishable from pancreatic islet A cells; the identity of the gastric and pancreatic forms of glucagon has been established.[33-35] An accepted physiological role for glucagon is the elevation of plasma glucose levels, predominantly via the activation of glycogenolysis and gluconeogenesis. Glucagon is released, predominantly from the A cells of the pancreatic islets, into the systemic circulation to exert its effects on the liver and other distant organs. A number of other effects of glucagon have been reported, but are probably pharmacological; the physiological and pharmacological roles of glucagon have been the subject of recent reviews.[36,37] In species other than dog, there is little pancreatic glucagon present in gastrointestinal tissue as determined by C-terminally directed pancreatic glucagon antisera (the so-called pancreatic "specific" antisera). N-terminally or mid-molecule-directed "nonspecific" glucagon antisera detect glucagon-like immunoreactivity (GLI) which occurs as a predominantly larger molecular weight form and is present in endocrine-like cells (designated L cells) concentrated in the distal small intestine and the colon. The relationship of pancreatic glucagon to its presumed precursor molecules is shown in Figure 2B.

In 1973 Tager and Steiner[38] isolated from pancreatic extracts a molecular form of glucagon which contained an eight amino-acid residue C-terminal extension. Recently, Bataille and co-workers[39] isolated a similar peptide from porcine intestinal extracts which they designated oxyntomodulin due to its specific but previously unrecognized effects on gastric glands.[40] Current evidence suggests that there is not only tissue-specific regulation of the expression of the gene coding for glucagon, but also a tissue-specific regulation of posttranslational processing of the synthesized glucagon precursor molecule. In the canine gastric and mammalian pancreatic islet A cells, the gene coding for glucagon is expressed, and the glucagon precursor molecule is predominantly processed to produce the 29-amino-acid-residue peptide, pancreatic glucagon, which readily reacts with C-terminally directed "pancreatic glucagon-specific antisera". In the intestine, and probably in the central nervous system, the predominant molecular forms appear to be glicentin and oxyntomodulin, which are recognized by the N-terminally or mid-molecule-directed nonspecific glucagon antisera. The name glicentin was derived from the fact that this entity was discovered as a glucagon-like *i*mmunoreactivity (therefore, "gli") and that the peptide was originally considered to have 100 amino acid residues (therefore, "centin"), but was subsequently found to contain 69 amino acid residues.[41] The functional significance of these tissue-specific differences in posttranslational processing of the glucagon precursor molecule is currently under intense investigation. Of interest, the C-terminal octapeptide (eight amino acid residues) extension of glucagon in the precursor molecule (see Figure 2B) is well conserved between species. The sequence of the octapeptide in human,[42] hamster,[43] and probably guinea pig[44] is –Lys–Arg–Asn–*Arg*–Asn–Asn–Ile–Ala. The pig,[39] dog,[35] and bovine octapeptide sequences[45] differ solely by a single substitution of a *Lys* for *Arg* residue in the middle of the molecule: –Lys–Arg–Asn–*Lys*–Asn–Asn–Ile–Ala. This striking degree of conservation, even in the guinea pig pancreatic glucagon molecule, which differs strikingly in its C-terminal region to the structures of the other mammalian glucagons (*vide infra*), is consistent with the suggestion that oxyntomodulin plays a physiological role in the mammal.[40]

The structure of the human glucagon precursor, deduced from cDNA sequences, confirmed the structure of a glicentin-like peptide, but unexpectedly the cDNA sequences were found to encode also for two peptides which are distinct from, but structurally related to glucagon; they were designated glucagon-like peptides 1 and 2, or GLP-1 and GLP-2.[42] The structural relationships between human glucagon and human GLP-1 and GLP-2 are shown in Figure 2C. Alignment of GLP-2 and glucagon at their N-termini reveals the presence of considerable structural homology throughout the two molecules. If the histidine residue in position 7 of GLP-1 is aligned with the N-terminal histidine residue of glucagon, a striking structural homology is also seen between these two peptides throughout their molecules, but is particularly intense in the N-terminal and central regions of the molecules. Molecules analogous to GLP-1 are also encoded in hamster and anglerfish genes, and recently the sequences of porcine and other mammalian GLP-1 molecules have been found to be similar to that of the human molecule (see Reference 46 and references cited therein). GLP-1[47] and, more potently, the N-terminally truncated GLP-1_{7-36}-amide form[48-50] stimulate the release of insulin from pancreatic islets *in vitro*; GLP-1 and its N-terminally truncated form are processed from proglucagon in the intestine and, to a smaller extent, in pancreatic islets. Consequently, GLP-1 is an excellent candidate to be considered a full member of this family of peptides. Because the deduced structure of GLP-1 ends with the –Gly–Arg–Arg sequence, uncertainty exists as to whether GLP-1 occurs as a 37-amino-acid-residue sequence ending in Gly or whether it exists in the amide form. Evidence for the structure *in vivo* being GLP-1_{7-36} amide was obtained recently by its isolation from human and pig small intestine.[46] Whether or not GLP-2 has inherent biological activity is as yet uncertain.

As seen from Figures 2A and 3A, the structures of the mammalian forms of secretin and especially glucagon are well conserved between species. The structures of glucagons isolated and characterized from bovine,[51] rabbit,[52] rat,[53] human,[54] dog,[33-35] and camel[55] pancreatic tissues and the structure of hamster glucagon deduced from cDNA sequences[43] are identical to the primary structure of porcine glucagon.[20] This indicates a remarkable degree of conservation. The avian glucagons (chicken,[56] turkey,[57] and duck[58]), the amphibian

A)

Secretin

Porcine, Bovine	H S D G T F T S E L S R L R D S A R L Q R L L Q G L V*
Rat	- - - - - - - - - - - - - - - - - Q - - - - - - - -
Human	- - - - - - - - - - - - - - - E G - - - - - - - - -
Dog	- - - - - - - - - - - - - - - E - - - - - - - - - -
Chicken	- - - - - Y - K M - G N - Q V - K F I - - N - M*

B)

VIP

Porcine, Bovine	H S D A V F T D N Y T R L R K Q M A V K K Y L N S I L N*
Human, Rat	- - - - - L - - T - - - - - - - - - - - - - - - - - -
Guinea pig	- - - - - - - - - - - - - - - M - - - V - -
Chicken	- - - - - - - - S - F - - - - - - - - - - V - T*

C)

PHI/PHM

Porcine	H A D G V F T S D F S R L L G Q L S A K K Y L E S L I*
Human	- - - - - - - - - - K - - - - - - - - - - - - - M*
Bovine	- - - - - - - - - Y - - - - - - - - - - - - -
Rat	- - - - - - - - Y - - - - - I - - - - - - - -

D)

GIP

Porcine	Y A E G T F I S D Y S I A M D K I R Q Q D F V N W L L A Q K G K K S D W K H N I T Q
Human	- - - - - - - - - - - - - - - - H - - - - - - - - - N - - - - - - - - - - -
Bovine	- I - - - - -

FIGURE 3.   Amino acid sequence variations between species for (A) secretin, (B) VIP, (C) PHI/PHM, and (D) GIP are portrayed using the one-letter notation for amino acid residues. The complete sequence is given for (A) porcine/bovine secretin, (B)porcine/bovine/human/rat VIP, (C) porcine PHI, and (D) porcine GIP. Species differences in amino acid sequences are shown for each species; a dash indicates the presence of an amino acid residue in that position identical to that of the full sequence portrayed. See Table 1 for symbols denoting specific chemical modification of amino acid residues.

glucagons,[59-61] and opossum glucagon[62] differ from the common mammalian structure at only one to three positions (Figure 2A). An important exception to this remarkable degree of structural conservation is guinea pig glucagon,[44] which differs at five amino acid residues in the C-terminal region (Figure 2A). Structure-function studies[63,64] suggest that the N-terminal region is essential for glucagon-like bioactivity, but that the C-terminal residues may enhance affinity for binding to the receptor. Hence, studies to determine the potency of guinea pig glucagon are clearly of interest, particularly since it is well accepted that the alteration in the structure of guinea pig insulin, compared to other mammalian structures, results in a lowered biological potency.

Similarly, the known mammalian forms of secretin indicate a strong degree of conservation. As seen in Figure 3A, the porcine[4,65] and bovine[66] molecules are identical, and the human form[67] differs by only two residues: at position 15 there is Asp/Glu interchange (porcine/human) and at position 16 a Ser/Gly interchange (porcine/human). Rat secretin differs from porcine secretin solely at position 14, where a Gln/Arg (rat/porcine) interchange occurs;[68] similarly, canine secretin also differs at one position from pig secretin, where at position 15 a Glu/Asp (dog/porcine) interchange occurs.[69] In contrast to glucagon, there is a considerably lesser degree of conservation between the avian and mammalian forms of secretin. There are 13 nonidentities in the 27 amino acid residues which compose the mammalian and chicken molecules.[70] Despite the large number of amino acid differences, chicken secretin does stimulate exocrine pancreatic secretion in the cat, albeit with a lower potency.[70] Of interest, at position 15, chicken secretin has a neutral residue (Gly) compared to the acidic residues (Asp or Glu) present in the mammalian forms of secretin and the basic amino acid residue (Lys) in mammalian VIP. The differences in structure at this position may play a role in the weaker potency of the ability of chicken secretin to stimulate pancreatic exocrine secretion. Also of interest, VIP is more potent than secretin in stimulating the exocrine avian pancreas.[71] In contrast to the recognized critical importance of the N-terminal histidine residue and the aspartic acid residue in position 3 to the bioactivity of secretin on the exocrine pancreas, the presence of the C-terminal valine amide structure does not seem to be of import as two C-terminally extended forms of secretin, one by a residue of glycine and the other by the Gly–Lys–Arg sequence,[72] are of comparable potency to secretin in stimulating pancreatic bicarbonate secretion.

The accepted physiological role for secretin is the stimulation of water and bicarbonate secretion from the pancreas in response to acidification of the duodenum. In this role, the potentiating effect of postprandial elevations of plasma CCK levels on the action of secretin is probably of physiological import. There is now general agreement that sufficient amounts of immunoreactive secretin are released into the peripheral circulation to account for postprandial pancreatic bicarbonate secretion. Hence, secretin, predominantly present in endocrine-like cells of the upper small intestinal mucosa, acts via a classical humoral mechanism to stimulate pancreatic secretion of water and bicarbonate in response to duodenal acidification (for review see Reference 65 and references cited therein).

B. THE VASOACTIVE INTESTINAL POLYPEPTIDE AND THE GASTRIC INHIBITORY POLYPEPTIDE

Said and colleagues detected the presence of a vasoactive peptide in lung tissue extracts, but since there was a far greater accessibility to intestinal extracts in Mutt's laboratory, they screened for and detected the presence of the presumed identical peptide in Mutt's intestinal extracts. Said and Mutt[73] isolated an intestinal peptide based on its vasodilatory properties, which they named the vasoactive intestinal polypeptide or VIP; it has been presumed that the originally identified lung peptide is the same as that isolated from the intestine, but this has yet to be proved by isolation and characterization of the lung peptide. On chemical characterization, VIP was found to be composed of 28 amino acid residues and to have an

N-terminal His residue and a C-terminal Asn amide residue.[21] Porcine VIP and secretin have identical amino acid residues in nine positions, with eight of these nine identities occurring in the N-terminal halves of the molecules (see Figure 1B). At about the same time, Brown and colleagues[74] identified a peptide with potent gastric acid inhibitory properties which was present in Mutt's side fractions derived from the purification of porcine cholecystokinin (CCK). The originally determined sequence of this peptide,[75] designated the gastric inhibitory polypeptide or GIP, has been corrected by the deletion of one Gln residue in the interior of the molecule.[22] Porcine GIP is composed of 42 amino acid residues and has an N-terminal tyrosine residue and a C-terminal glutamine residue (see Figures 1C and 3D). An N-terminally truncated form, by two amino acid residues, has been characterized from porcine small intestinal extracts.[22] The structural relation of VIP to secretin is shown in Figure 1B, and the structural relationships between VIP and GIP with the other members of this family are portrayed in Figure 1C.

The structure of VIP has been strongly conserved between mammalian species (see Figure 3B). Bovine, human, rat, and canine sequences are identical to porcine VIP.[76-79] Guinea pig VIP is exceptional in that it differs at four positions from the structure of VIP in other mammals;[80] interchanges occur (guinea pig/porcine, respectively) at position 5 (Leu/Val), position 9 (Thr/Asn), position 19 (Met/Val), and position 26 (Val/Ile). Chicken VIP[81] differs from porcine VIP at four positions including a different C-terminal amide residue, being Thr–NH$_2$ in the chicken and Asn–NH$_2$ in the mammal form. In the mammal, VIP appears to be strictly confined to neural structures of the peripheral and central nervous systems; hence, VIP is a neuropeptide. An exception occurs in human pathology, where VIP is present in and released into the peripheral circulation from certain islet cells or neural tumors to cause a recognized clinical syndrome, the Verner-Morrison syndrome (also known as the VIPoma, WDHA, or pancreatic cholera syndrome). VIP has been demonstrated to possess a number of pharmacological effects, including the ability to alter blood flow and motility, related to its potent activity on vascular and nonvascular smooth muscle. Another prominent effect of VIP is the stimulation of exocrine secretion of water and electrolytes; this is of clinical import in the Verner-Morrison syndrome. Both effects appear to be mediated primarily via the ability of VIP to stimulate adenylate cyclase activity. Of note, VIP possesses a number of activities attributed to other members of the secretin-glucagon family, such as the inhibition of gastric acid secretion, lipolysis, and glycogenolysis and the stimulation of the release of insulin from the endocrine pancreas; these effects are consistent with the structural homology of VIP with other members of the family. The pharmacological and potential physiological effects of VIP, including those briefly discussed above, and structure-function studies have been reviewed recently.[83-85] VIP has been the subject of a recent international symposium.[86]

GIP has been isolated from porcine,[22] human,[87] and bovine[88] small intestinal extracts, and sequence analyses indicate a strong conservation of the structure of GIP in mammalian species (see Figure 3D). The bovine molecule differs from the porcine molecule solely at position 37, where a residue of Ile is present in the bovine and a residue of Lys is in the porcine molecule. The human molecule differs from the porcine molecule at two positions; at position 18 a His residue in the bovine is substituted for an Arg residue in the porcine molecule, and at position 34 Asn is present in the human molecule instead of the Ser in the porcine molecule (see Figure 3D). The highest concentrations of immunoreactive GIP occur in the duodenum and upper jejunum, and immunohistochemical studies demonstrate the presence of GIP in a distinct population of endocrine-type cells in this region.[89] Increased plasma levels of immunoreactive GIP are seen after the ingestion of a mixed meal, oral glucose, protein, and, in humans and rats, with the lowering of duodenal pH (for review see Reference 90). Hence, GIP appears to be a classical GI hormone, being present in endocrine-type cells and exerting its biological effects after being delivered to target cells via the peripheral circulation. Because GIP is released into the peripheral circulation by oral fat ingestion, and because it was initially detected via its ability to inhibit gastric acid secretion in denervated canine gastric pouches, it was initially suggested that GIP might be a physiological enterogastrone. Recent studies demonstrating that GIP is a weak inhibitor of gastric acid secretion in humans and dogs[91,92] have cast doubt on its role as a physiological inhibitor of gastric acid secretion. Subsequent to the detection of gastric acid inhibitory effects of GIP, Dupré and colleagues, in collaboration with Brown, demonstrated that GIP is a potent insulin secretagogue in humans[93] and that GIP is released into the peripheral circulation in humans following oral glucose,[94] leading to the suggestion that GIP plays a role in the physiological regulation of insulin secretion. These results have been confirmed by numerous studies (e.g., see References 90 and 95), and it is now generally accepted that GIP is a physiologically important incretin. Brown has suggested that GIP be renamed the "glucose-dependent insulinotropic polypeptide", which describes a presumed physiological role and retains the acronym, GIP.

C. PHI, GROWTH HORMONE RELEASING FACTOR (GHRF), AND NONMAMMALIAN MEMBERS OF THE SECRETIN-GLUCAGON FAMILY

Utilizing their assay system that detects in crude tissue extracts the presence of peptides containing the C-terminal α-amide structures,[6] Tatemoto and Mutt[96] detected in and isolated from porcine upper small intestinal tissue extracts a peptide containing the C-terminal iso-leucine amide structure. On chemical characterization, it was demonstrated to be a linear polypeptide composed of 27 amino acid residues and to have an N-terminal histidine residue and a C-terminal isoleucine amide residue. Since the peptide was isolated on a strictly chemical basis, it was given a chemical name based on its N-terminal and C-terminal residues using the one-letter notation form for amino acid residues; hence, it was designated *p*eptide *h*istidine *i*soleucine amide or PHI. As shown in Figure 1C, PHI has extensive sequence homologies with other members of the secretin-glucagon family of peptides spread throughout the length of the molecules. Particularly striking are the 13 identities between porcine PHI and VIP. Subsequent to its characterization, PHI was found to have a number of biological activities, but the spectrum of activity was very similar to that of VIP and, as with VIP, PHI stimulates adenylate cyclase activity; for this and other reasons, it has been suggested that PHI and VIP exhibit their biological activities through common or similar receptors (for review see Reference 97). In the mammal, immunohistochemical studies have demonstrated that PHI is strictly confined to neural structures and, hence, is a neuropeptide. Based on the findings that, in a variety of tissues, immunoreactive PHI occurred in approximately equimolar concentrations to that of VIP and that VIP-producing tumors also produced PHI, Christofides and colleagues[98] suggested that VIP and PHI might be cosynthesized in the same precursor. Proof for their prediction came with the report by Itoh and colleagues,[78] who deduced the structure of the VIP precursor from a cloned cDNA sequence complimentary to mRNA coding for human VIP in a neuroblastoma. The precursor molecule was found to code also for a 27-amino-acid residue peptide identical to PHI with the exception of two amino acid residue interchanges; at position 12 a lysine (human) was substituted for arginine (porcine), and the C-terminal residue was a methionine amide residue rather than the porcine isoleucine amide (see Figure 3C). The latter amino acid residue interchange raised a problem in terms of nomenclature, with the consequence that the human molecule has been designated PHM, the M standing for the C-terminal methionine amide residue. The structure of the human form of PHI, deduced via the molecular biology techniques, was confirmed by isolation from intestinal tissue.[99] Similarly, bovine PHI has been isolated from small intestinal tissue and has been found to differ from the porcine molecule at a single position;[100] a tyrosine residue (bovine) is substituted for a phenylalanine residue (porcine) at position 10 (see Figure 3C).

Studies based on chromatographic and RIA techniques suggested that the predominant form of PHI/PHM-like immunoreactivity occurred in a larger molecular weight form in human gastric, nasal, and genital-urinary tissues, but not in central nervous system, intestinal, or pulmonary tissues.[101] On the basis of immunologic evidence, it was predicted that the larger form was composed of PHM C-terminally extended by the "spacer peptide sequence" which occurs between PHM and VIP in the precursor molecule. Structural evidence for this hypothesis was obtained by the same group with the isolation of a peptide from a human pheochromocytoma which contained 42 amino acid residues and was composed of the PHM sequence and the predicted C-terminal extension. Based on the chemical nomenclature system devised by Tatemoto and Mutt and the number of amino acid residues, these authors designated the peptide as *p*eptide *h*istidine *v*aline-42 or PHV-42.[102] The C-terminal extension corresponds to the predicted connecting peptide between PHM and VIP in the precursor protein, but lacks the C-terminal pair of basic amino acid residues. Recently, Cauvin and colleagues[103] have reported the presence of three molecular forms of immunoreactive PHI in the rat intestine. On sequence analysis, the predominant peak proved to be rat PHI extended at the C-terminus by an additional glycine residue, whereas the other two peaks were tentatively identified as PHI C-terminally extended with the connecting peptide and the 27-amino-acid residue rat form of PHI.[103] Hence, as with secretin and VIP, PHI occurs in tissues in C-terminally extended forms and, as with secretin and VIP, the C-terminally extended forms of PHI are bioactive.[102] In contrast to secretin and VIP, the C-terminally extended forms of PHI appear to be present in considerably greater quantities, and there also appears to be a tissue-specific difference in the amount of posttranslational processing which occurs. The significance of this finding in functional terms is unclear, but it has been suggested that PHV-42 may have specific functions.[102] PHI has been the subject of a brief review.[104]

The structure of the long sought after but elusive growth hormone releasing factor (GRF) or somatocrinin was finally characterized after isolation from two separate human pancreatic islet cell tumors which were present in patients suffering from acromegaly.[24,25] Rivier and colleagues[24] characterized the growth hormone releasing factor (GRF) as being composed of 40 amino acid residues with an N-terminal tyrosine residue and a C-terminal alanine residue. Guillemin and colleagues,[25] from a different pancreatic islet cell tumor, isolated an identical peptide apart from a C-terminal extension of four amino acid residues, with the C-terminal leucine residue being in the α-amide form. This human form of GRF possesses a striking structural homology with members of the glucagon-secretin family (see Figure 1C); particular sequence homology is seen with porcine PHI, with which it shares 12 identical residues (see Figure 1C). Once the structure of the human molecule had been determined, the isolation of GRF from hypothalamic tissue of different species rapidly followed.[26-28] In relation to the human sequence, three amino acid residue nonidentities occur in pig and four nonidentities in the cow; these nonidentities occur in the C-terminal part of the molecule. Compared to the human molecule, rat GRF has 14 amino acid interchanges spread throughout the molecule. Similar to other members of the secretin-glucagon family of peptides, the amino-terminal section of the GRF molecule contains the residues essential for biological activity; specifically, the tyrosine residue at the amino terminus is essential, whereas the C-terminal portion of the molecule appears not to be of import.[25,105] GRF has been the subject of a recent review.[106]

At about the same time, two groups isolated closely related but distinct peptides from Gila monster venom, and on characterization these peptides were discovered to have sequence homology with the secretin-glucagon family. Parker and colleagues[29] isolated two peptides from Gila monster venom which they designated helospectin I and II. Helospectin I is composed of 38 amino acid residues, including an N-terminal histidine residue and a C-terminal serine residue; there was no evidence for a C-terminal amidated residue. Helospectin

II is identical to helospectin I, but is foreshortened at the C-terminus by one serine residue (see Figure 1D). Hoshino and colleagues[30] isolated a closely related peptide, also from Gila monster venom, which they designated helodermin. Helodermin is composed of 35 amino acid residues, including an N-terminal histidine residue and a C-terminal proline amide residue (see Figure 1C,D). Helodermin and helospectin have clear structural homology with each other and with the secretin-glucagon peptide family. In particular, VIP and helodermin share 15 identities spread throughout their molecules, with the structural homology being particularly striking in the N-terminal and C-terminal regions (see Figure 1C). The C-terminal proline amide structure of helodermin was established not by direct chemical identification, but via HPLC analysis comparing the elution profiles of C-terminal fragments of the natural peptide with synthetic fragments containing the proline residue present either in the free acid or in the C-terminal amide form.[30] In contrast, clear chemical evidence was obtained that the helospectrins do not end in a C-terminal α-amide structure.[29] Considering the structural homology with VIP, it is not surprising that helospectrin and helodermin have been reported to have VIP-like biological actions in mammals,[29,107-109] and it has been suggested that, in certain tissues, helodermin and VIP occupy the same or a similar receptor site.[110] Using immunologic techniques, a helodermin-like peptide appears to be present in mammalian tissues,[31,111] strongly suggesting that another mammalian member of this family of peptides remains to be characterized.

III. THE GASTRIN-CHOLECYSTOKININ FAMILY

A. GASTRIN

Definitive proof for Edkins' gastrin hypothesis[2] occurred with Gregory and Tracy's isolation of two acidic heptadecapeptide amides from extracts of porcine antra in 1964.[5] Both gastrins were composed of 17 amino acid residues (designated "little gastrin" or LG and gastrin-17 or G-17), including an N-terminal pyroglutamyl residue and a C-terminal phenylalanine amide residue (see Figure 4A). Both peptides had identical amino acid sequences and differed only in whether or not the tyrosine phenolic group was esterified with sulfuric acid. Structure-function studies[112,113] demonstrated that (1) the C-terminal tetrapeptide amide possessed all the actions of G-17, but at a lower potency; (2) little or no activity was demonstrable with oxidation of the methionine residue present in this C-terminal four-amino-acid residue sequence or in the absence of the C-terminal amide structure; and (3) the presence or absence of the sulfate group on the tyrosine residue did not significantly affect the ability of gastrin to stimulate gastric acid secretion. A large number of biological activities have been attributed to gastrin, but most are probably pharmacological rather than physiological in nature. Gastrin is present in endocrine-type cells in the antral mucosa of many mammalian species (and to a lesser extent in duodenal mucosa of many species) is many mammalian species (and to a lesser extent in duodenal mucosa of many species), is released into the systemic circulation following neural stimulation or ingestion of a meal, Other putative physiological actions of gastrin include pepsin secretion, alteration of gastric and mucosal blood flow, and trophic activities on the pancreas and the gastric and small intestinal mucosa. It clearly has been established that gastrin is the causative entity released from pancreatic (or, rarely, extrapancreatic) islet cell tumors to stimulate hypersecretion of gastric acid, resulting in severe gastrointestinal ulceration in the Zollinger-Ellison syndrome (gastrinoma syndrome). The functional aspects of gastrin have been reviewed extensively.[114] The ready availability of homogeneous synthetic and natural preparations of gastrin has greatly enhanced investigations into the physiological and pathophysiological roles of gastrin, and these studies have acted as a valuable learning base for investigations into the roles of other GI hormones.

The structures of a number of mammalian gastrins have been determined,[5,116-125] and

A)   Mammalian "Little Gastrins" (G-17)

		1			5					10					15			

SO₃H

Porcine   Pyr–Gly–Pro–Trp–Met–Glu–Glu–Glu–Glu–Glu–Ala–Tyr–Gly–Trp–Met–Asp–Phe–NH₂

	1				5					10					15			
Human	–	–	–	–	Leu	–	–	–	–	–	–	–	–	–	–	–	–	–
Rat	–	Arg	–	Pro	–	–	–	–	–	–	–	–	–	–	–	–	–	–
Canine	–	–	–	–	–	–	Ala	–	–	–	–	–	–	–	–	–	–	–
Cow,Sheep,Goat	–	–	–	–	Val	–	–	–	Ala	–	–	–	–	–	–	–	–	–
Feline	–	–	–	–	Leu	–	–	–	Ala	–	–	–	–	–	–	–	–	–
Rabbit	–	–	–	–	Leu	Gln	–	–	–	–	–	–	–	–	–	–	–	–

B)   Guinea Pig "Little Gastrin" (G-16) versus Porcine G-17

SO₃H

Porcine G-17 | Pyr–Gly–Pro–Trp | Met | Glu–Glu–Glu–Glu | Glu | Ala–Tyr–Gly–Trp–Met–Asp–Phe–NH₂ |

SO₃H

Guinea Pig G-16 | Pyr–Gly–Pro–Trp | Ala | (Glu–Glu–Glu) | Ala | Ala–Tyr–Gly–Trp–Met–Asp–Phe–NH₂ |

C)   Mammalian "Big Gastrins" (G-34)

Human Gastrin-34	Q█L G P Q G P P H L V A D P S K K Q G P █ L E E E E E A Y^OG █ █ D F*
Porcine Gastrin-34	– – – L – – – – – – – – – – L A – – – – – – █ – – – – – – – – – – – –
Rat Gastrin-34	– – – – – – – Q – F I – – L – – – – R – P █ – – – – – – – – – – – –
Guinea Pig Gastrin-33	– – – – – V – A – – R T – L – – – – – – – A (E–E–E) A – – – – – – –

D)   Gastrin Forms Characterized

Gastrin-4		█ █ D F*
Gastrin-5		G █ █ D F*
Gastrin-6		Y^OG █ █ D F*
Human Gastrin-14		█ L E E E E E A Y^OG █ █ D F*
Human Gastrin-17		Q█G P █ L E E E E E A Y^OG █ █ D F*
Human Gastrin-34	Q█L G P Q G P P H L V A D P S K K	Q G P █ L E E E E E A Y^OG █ █ D F*

FIGURE 4.   (A) Portrayed are the species differences in the amino acid sequences of various mammalian little gastrins (G-17) using the three-letter notation for amino acid residues. The full amino acid sequence of porcine G-17 is given, blank spaces indicate that residues in that position are identical to those of porcine G-17, and the nonidentities are shown in the three-letter notation form. (B) The amino acid sequence differences between guinea pig little gastrin (G-16) and porcine G-17 are portrayed using the three-letter notation for amino acid residues. A deletion of one glutamic acid residue in the guinea pig molecule is portrayed by enclosing the area of the assumed deletion in brackets. (C) Mammalian forms of "big" gastrin (G-34) are portrayed using the one-letter notation for amino acid residues. The full amino acid sequence of human G-34 is given, identical residues to that of human G-34 in each position are portrayed by a dash, and nonidentities are portrayed using the one-letter notation. An assumed deletion of a glutamic acid residue in guinea pig G-33 is shown in brackets. (D) The various sizes of the characterized gastrin forms are portrayed using the one-letter notation for amino acid residues. See Table 1 for definition of symbols denoting specific chemical modification of residues.

are shown in Figure 4A to D. The structures of both "little gastrin" (G-17) and "big gastrin" (G-34) are relatively well conserved between mammalian species. Differences in the amino acid sequences occur in the mid- and N-terminal regions of G-17, and the substitutions are generally attributable to conservative one-base changes in the codon triplets. The C-terminal heptapeptide amide structure is identical in all mammalian species characterized to date and contains the residues essential for potent bioactivity.

The development of an RIA for gastrin provided a powerful tool for the demonstration of what, at the time, was an unexpected heterogeneity of molecular forms of gastrin which occurred in both tissue extracts and the peripheral circulation. Regarding gel filtration chromatography of plasma taken from hypergastrinemic patients, Yalow and Berson[126] discovered the existence of a larger form of gastrin-like immunoreactivity than G-17; they postulated that this larger form of gastrin might be a precursor for the formation of G-17. In 1972, Gregory and Tracy[127] isolated two forms of big gastrin (sulfated and unsulfated forms of the same molecule) from a gastrinoma and from porcine antra, which were demonstrated to be little gastrin (G-17) extended by 17 amino acid residues at the N-terminus (see Figure 4B); the originally published sequence of porcine and human gastrin-34 was subsequently corrected.[128] The N-terminal residue of the gastrin-34 molecule is, like G-17, a pyroglutamyl residue. In all mammalian species, the C-terminal 17 residues of G-17 are linked to the N-terminal extensions forming G-34 (G-33 in the guinea pig) by a pair of lysine residues; treatment of the larger molecules with trypsin liberates the biologically active G-17 fragment. The characterized mammalian G-34 structures for human, porcine, rat, and guinea pig are shown in Figure 4C. Human G-34 was isolated from a gastrinoma and porcine gastrin G-34 from antral tissue. The rat G-34 sequence was deduced from the nucleotide sequence of mRNA coding for rat gastrin.[123] The sequence of rat G-17 had been isolated previously using classical peptide chemistry[122] and is in agreement with the sequence deduced using molecular biology. The structures of guinea pig little and large gastrin were elucidated by isolation of the peptides from guinea pig antra.[124,125] Porcine G-34 differs from human G-34 by amino acid residue interchanges at four positions: at positions 4 and 14 a Leu/Pro interchange occurs (porcine/human, respectively), at position 15 an Ala/Ser interchange occurs, and at position 22 a Met/Leu interchange occurs (see Figure 4C). Rat G-34 differs from the human molecules at seven positions in the mid-molecule region.

The structure of guinea pig gastrin has proved to be an exception to those of other mammalian species. Guinea pig little gastrin isolated from antral tissue[124] proved to be a hexadecapeptide amide (i.e., G-16) rather than the usual heptadecapeptide amide molecule (see Figure 4B). For ease of comparison of the structures, the N-terminal and C-terminal residues of porcine and guinea pig molecules have been aligned to demonstrate the structural homologies; this alignment leaves a gap in the middle of the molecule where four consecutive Glu residues appear in the porcine molecule, but only three are present in the guinea pig. Hence, a one-residue deletion appears to have occurred in the central portion of the guinea pig molecule (see Figure 4B, region enclosed in brackets). Further differences occur at position five, numbering conventionally from the N-terminal residue, where an Ala/Met interchange occurs (guinea pig/pig, respectively) and at the eighth residue, numbered unconventionally from the C-terminal residue, where an Ala/Glu interchange occurs (guinea pig/pig, respectively). Of functional importance, however, the bioactive C-terminal heptapeptide of guinea pig gastrin is identical to that of the other mammalian species. Guinea pig large gastrin is composed of 33 amino acids (i.e., G-33) and differs by an additional 5 amino acid residue interchanges which occur in the N-terminal half of the molecule (see Figure 4C).

The differing molecular sizes of biologically active forms of gastrin isolated and characterized from tissue extracts to date include the C-terminal hexapeptide amide (G-6) from porcine antral mucosa,[129] the tetradecapeptide amide (G-14) from gastrinoma tissue,[130] and, as previously described, G-17 and G-34. Two smaller structures, the C-terminal tetrapeptide (G-4) and pentapeptide (G-5), which are identical in gastrin and CCK (*vide infra*), have been isolated and characterized; G-5 was isolated from dog brain and intestine,[131,132] and both G-5 and G-4 were isolated from pig brain.[133] These different forms are shown in Figure 4D. As is readily seen, all the larger molecular forms are N-terminal extended forms of the smaller molecules. That is, G-17 is identical to G-14 except for a three-amino-acid-residue N-terminal extension and, similarly, G-34 is an N-terminally extended form of G-17. Employing the tools of molecular biology, the structures of the gastrin precursor molecules determined for pig, human, and rat[134-136] are composed of 104, 101, and 115 amino acid residues, respectively. All of the precursor molecules contain a Gly–Arg–Arg sequence immediately C-terminal to the final phenylalanine residue of the active molecule; this structure provides the site and presumably the signal for cleavage and subsequent amidation of the phenylalanine residue. Similarly, the N-terminal glutaminyl residue of G-34 is N-terminally extended by a pair of Lys–Lys residues which provide the signal for proteolytic cleavage. During the formation of the mature gastrin molecules from the precursor molecules, two further important posttranslational modifications occur. The C-terminal phenylalanine residue is amidated, and the tyrosine residue may be esterified with sulfuric acid. Evidence has been reported that, in the porcine antrum, tyrosine-*O*-sulfation may occur prior to the C-terminal proteolytic cleavages and therefore precedes carboxyamidation.[18] As with the secretin-glucagon family of peptides, partially processed C-terminally extended forms of gastrin have been isolated. Such unprocessed forms are present in relatively small amounts in GI tissue, but the anterior pituitary contains considerably greater amounts of these inactive precursors compared to the bioactive amidated gastrin forms.[137]

In mammals, the antral mucosa contains the bulk of the gastrin present and is predominantly in the G-17 form; other components such as G-34 and smaller components are present in relatively minor amounts. In humans, the total gastrin content of duodenal mucosa is similar to that present in the antrum, and the proportion of G-34 to G-17 is approximately equal. Gastrin appears to be present in endocrine cells in the fetal pancreas, but disappears in the first few days of postnatal life. In pathological cases, functional gastrin-containing G cells occur in non-β islet cell tumors. Gastrin is present in the pituitary, but appears to remain in a bioinactive form.[137] The tissue occurrence of gastrin, biological actions, control of release, etc. have been reviewed.[114]

B. CHOLECYSTOKININ

Experiments by Ivy and Oldberg and colleagues in 1927 and 1928 described the presence of an intestinal hormone which had cholecystokinetic activity, and in 1941 Harper, Raper, and Vass and colleagues reported the presence of an entity in the intestinal mucosa which possessed pancreozymic effects (see Chapter 1); both entities acted via secretion into the bloodstream. Because the biological actions and, hence, putative physiological roles of these two hormones were distinct, cholecystokinin or CCK, acting on the gallbladder, and pancreozymin or PZ, stimulating pancreatic enzyme secretion, were considered to be distinct entities. Jorpes and Mutt[138] noted that during the purification of CCK and PZ from small intestinal tissue extracts the biological activities of each entity always occurred in the same chromatographic fractions; this led them to hypothesize that CCK and PZ were one and the same hormone. The proof of this hypothesis was attained with the isolation of a peptide containing 33 amino acid residues which possessed both activities from porcine intestinal extracts.[139] The discovery that CCK- and PZ-like activities were exerted by the same peptide was totally unexpected, but an even greater surprise awaited the structural characterization of CCK-PZ. Somewhat earlier, Anastasi and co-workers[140] had isolated a decapeptide (a peptide composed of ten amino acid residues) from frog skin which they designated cerulein, named after the species from which the peptide was isolated. Cerulein had biological activities similar to those of CCK, but sequence analysis revealed that it had a C-terminal pentapeptide amide which was identical to that of gastrin and that it, too, contained a sulfated tyrosine residue.[140] Mutt and Jorpes subsequently reported that CCK-PZ also had a C-terminal pentapeptide amide identical to that of gastrin and contained a sulfated tyrosine residue. This was a disturbing finding because gastrin was clearly recognized as a potent gastric acid secretagogue, while CCK and cerulein were potent stimulators of pancreatic enzyme secretion and gallbladder contraction. Subsequent biological studies revealed that CCK does indeed

A) Sequences of the C-Terminal Decapeptides of Gastrin, CCK and Caerulein

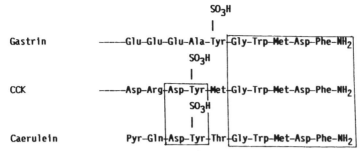

Gastrin — SO_3H over Tyr — -----Glu–Glu–Glu–Ala–Tyr–Gly–Trp–Met–Asp–Phe–NH$_2$

CCK — SO_3H over Tyr — -----Asp–Arg–Asp–Tyr–Met–Gly–Trp–Met–Asp–Phe–NH$_2$

Caerulein — SO_3H over Tyr — Pyr–Gln–Asp–Tyr–Thr–Gly–Trp–Met–Asp–Phe–NH$_2$

B) Characterized Porcine CCK Forms

CCK-58   A V Q K V D G E S R A H L G A L L A R –
            Y I Q Q A R K A P S G R V S M I K N L Q S L D P S H R I S D R D Y° M G W M D F*
CCK-39   Y I Q Q A R K A P S G R V S M I K N L Q S L D P S H R I S D R D Y° M G W M D F*
CCK-33   K A P S G R V S M I K N L Q S L D P S H R I S D R D Y° M G W M D F*
CCK-8   D Y° M G W M D F*
CCK-5   G W M D F*
CCK-4   W M D F*

C) Mammalian Forms of CCK-39

Porcine   Y I Q Q A R K A P S G R V S M I K N L Q S L D P S H R I S D R D Y° M G W M D F*
Human   – – – – – – – – – – – M – I V – – – – N –
Canine   – – – – – – – – – – – M – V – – – – – N –
Bovine   – – – – – – – – – – – M – V –
Mouse   – – – – V – – – – – – M – V L –
Rat   – – – – V – – – – – – M – V L – – – – G –
G-Pig CCK-22   S – – G – – – N – – – – – – – – – – V – – – – –

FIGURE 5.   (A) Portrayed are the C-terminal ten amino acid residues of gastrin and CCK and the full sequence of cerulein using the three-letter notation for amino acid residues. Amino acid residue identities between peptides are enclosed in boxes. (B) The characterized forms of porcine CCK are portrayed using the one-letter notation for amino acid residues. Identical residues are enclosed in a box structure. (C) Amino acid sequence differences in CCK-39 molecules occurring between mammalian species are portrayed using the one-letter notation for amino acid residues. The full sequence of porcine CCK-39 is given, amino acid residues identical with those in porcine CCK-39 are indicated by a dash, and nonidentities are shown in the one-letter code. The guinea pig CCK-39 sequence has not yet been reported, but the CCK-22 form is portrayed. See Table 1 for definition of symbols denoting specific chemical modifications to amino acid residues.

have gastric acid secretagogue activity, but is much less potent than gastrin in this regard; similarly, gastrin has weak CCK- and PZ-like activities.

In mammals, the important structure-function difference between gastrin and CCK appears to reside in the spatial relation of the tyrosine-O-sulfate group to the common C-terminal pentapeptide amide. In gastrin, this tyrosine residue (either in the sulfated or unsulfated form) is attached directly to the common pentapeptide amide, while in CCK and cerulein there is one amino acid residue interposed between the sulfated tyrosine and the common pentapeptide amide (see Figure 5A). Sulfation of the tyrosine residue in CCK is mandatory for potent CCK- or PZ-like activity, but the presence or absence of a sulfated

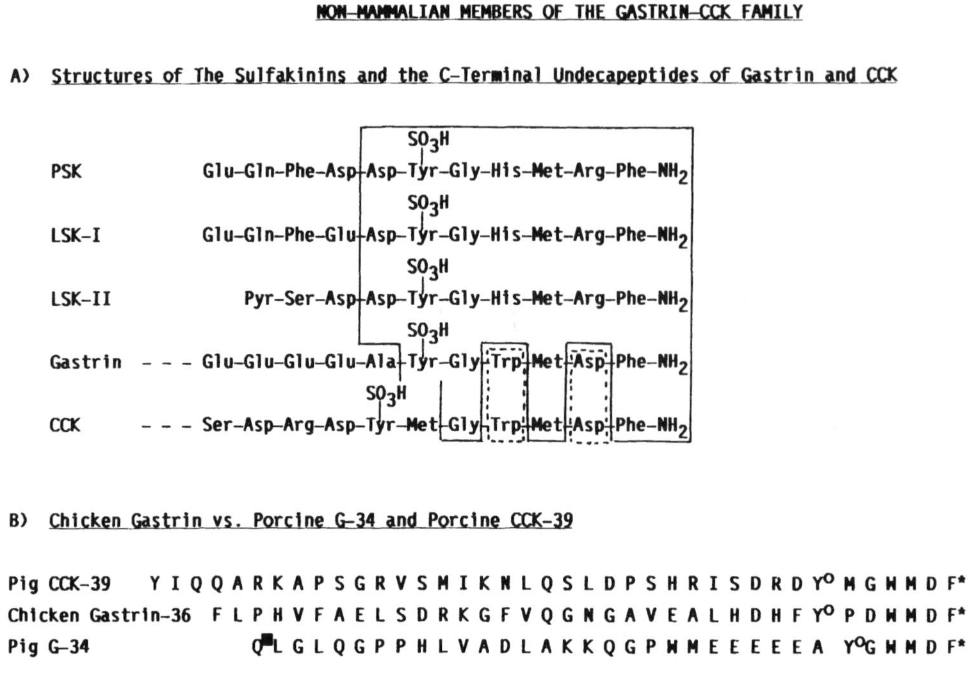

NON-MAMMALIAN MEMBERS OF THE GASTRIN-CCK FAMILY

A) **Structures of The Sulfakinins and the C-Terminal Undecapeptides of Gastrin and CCK**

```
                                         SO3H
                                          |
PSK           Glu-Gln-Phe-Asp─Asp-Tyr-Gly-His-Met-Arg-Phe-NH2
                                         SO3H
                                          |
LSK-I         Glu-Gln-Phe-Glu─Asp-Tyr-Gly-His-Met-Arg-Phe-NH2
                                         SO3H
                                          |
LSK-II            Pyr-Ser-Asp─Asp-Tyr-Gly-His-Met-Arg-Phe-NH2
                                         SO3H
                                          |
Gastrin  - - - Glu-Glu-Glu-Glu-Ala─Tyr-Gly─Trp┊Met┊Asp┊Phe-NH2
                              SO3H
                               |
CCK      - - -     Ser-Asp-Arg-Asp-Tyr-Met─Gly┊Trp┊Met┊Asp┊Phe-NH2
```

B) **Chicken Gastrin vs. Porcine G-34 and Porcine CCK-39**

```
Pig CCK-39         Y I Q Q A R K A P S G R V S M I K N L Q S L D P S H R I S D R D YO M G W M D F*
Chicken Gastrin-36     F L P H V F A E L S D R K G F V Q G N G A V E A L H D H F YO P D W M D F*
Pig G-34                   Q≡L G L Q G P P H L V A D L A K K Q G P W M E E E E A  YOG W M D F*
```

FIGURE 6. (A) The structures of the sulfakinins and the C-terminal undecapeptides of gastrin and CCK are portrayed using the three-letter notation for amino acid residues. Amino acid residue identities between the peptides are enclosed in boxes. (B) The structures of porcine CCK-39, porcine gastrin-34, and chicken gastrin-36 are portrayed using the one-letter identifier code for amino acid residues. See Table 1 for an explanation of symbols denoting specific modifications to amino acid residues.

tyrosine residue in gastrin is of little or no import in potent gastric acid secretagogue activity. In mammalian tissues, gastrin occurs in both the sulfated and unsulfated forms, but CCK occurs virtually entirely in the sulfated form. Oxidation of the methionine residue in the C-terminal pentapeptide amide structure completely abolishes the biological activity.[141] The view that peptides in this family have either potent gastrin-like or potent CCK-like activity due to the relation of the sulfated tyrosine residue to the common C-terminal pentapeptide amide appears to be correct in the mammal, but the situation is different in avian species. Dimaline and colleagues[142] isolated from chicken antra a novel peptide composed of 36 amino acid residues which, on pharmacological testing, potently stimulated avian gastric acid, but not pancreatic secretion and, hence, had gastrin-like activity. The structure of the C-terminal pentapeptide amide in this presumed chicken form of gastrin (see Figure 6B) differs from other members of the family; the glycine residue in the mammalian member of the family is replaced by an aspartic acid residue in chicken gastrin-36 (see Figure 6B). Furthermore, unlike gastrin and similar to CCK and cerulein, the sulfated tyrosine residue is not joined directly to the C-terminal pentapeptide amide, but has one interposed amino acid residue. Hence, the structure-function relations in avian species are clearly different from those in mammalian species and require further investigation. Specifically, isolation of peptides with CCK- and PZ-like activities from avian species is of interest. The late Professor Morton Grossman suggested that the term CCK-PZ be shortened to CCK due to the precedence of the discovery of the CCK-like effect; in general, this suggestion has been followed, and it is understood that the term CCK also implies pancreozymin activity.

   It is generally accepted that CCK, present in endocrine-type cells of mammalian upper small intestinal mucosa, is secreted into the systemic circulation following ingestion of a mixed meal to exert physiological functions on the biliary tract and exocrine pancreas.

Hence, it is generally accepted that CCK functions as a classical GI hormone. However, it has also been demonstrated that CCK is present in neural structures of the central nervous system, the peripheral autonomic nervous system, and the enteric nervous system. Hence, intense interest has been focused on CCK not only as a GI hormone, but also as a neuropeptide (for review see Reference 65 and references cited therein).

During the isolation of CCK-33 from porcine small intestine, Mutt[143] discovered the presence of a more basic form of CCK which consisted of CCK-33 N-terminally extended by six amino acids — hence, CCK-39. As discussed previously (*vide supra*), the biological activity resides in the far C-terminal region, the smallest entity giving rise to gastrin- or CCK-like activity being the C-terminal tetrapeptide amide or CCK-4. As seen in Figure 5B, a number of bioactive forms of CCK have been isolated from brain and intestine, and all consist of varying lengths of N-terminal extensions of CCK-4. To date, the smallest C-terminal fragment isolated and chemically characterized with fully potent CCK- or PZ-like bioactivity is the octapeptide of CCK-8; in certain assays, CCK-8 is more potent on a molar basis than any of the larger or smaller forms. CCK-8 was first isolated from sheep brain,[144] but subsequently identical molecules have been isolated from pig,[145] dog,[146] and human[147,148] brain and from dog[149] and rat[150] intestine. Recently, a unique form of CCK-8, along with a larger CCK-22 molecule, has been isolated from guinea pig.[151] The C-terminal pentapeptide amide of guinea pig CCK-8 is identical to that of the other mammalian species, but the adjacent residue (on the N-terminal side) is valine in the guinea pig and methionine in other mammalian species (see Figure 5C). Similarly, CCK-8 isolated from chinchilla brain also has a Val rather than a Met residue at this position.[152] CCK-8 isolated from the brains of two Australian marsupials is identical to that of the other mammalian species.[153] CCK-5 has been isolated from dog intestine[149] and pig brain.[145] The presence of CCK-4, the tetrapeptide amide common to gastrin and CCK, escaped isolation until recently, when it was characterized from pig intestine;[137] the presence of this, the smallest fragment of gastrin or CCK-containing bioactivity, had long been postulated as the N-terminal fragment resulting from the cleavage of CCK-4 from precursors that had been detected in tissue extracts. In the brain, CCK-4 appears to arise from the expression of the gene coding for the CCK precursor rather than the gastrin precursor.[137] The largest bioactive form of CCK isolated to date is CCK-58, which has been isolated from dog[149] and rat[155] intestine and from dog[146] and pig[154] brain. A number of intermediate molecular sizes of CCK have been isolated, but the significance of the demonstrated cleavage of these bioactive fragments into CCK forms of varying lengths is at present unknown. For example, CCK in the forms of 58, 39, 33, 25, 18, 8, 7, and 5 residues have been isolated from dog intestine;[149] in the forms of 39 and 33, from pig and cow;[157] and in the form of CCK-22, from rat[150] and guinea pig[151] intestine.

Sequence differences between species within the CCK-39 form are displayed in Figure 5C; also displayed is the longest known form for guinea pig CCK, CCK-22. The sequences of most of the different species of CCK-39 displayed were originally isolated and characterized from tissue extracts (*vide supra*), but the sequence for mouse CCK-39 was elucidated from cDNA encoding mouse brain CCK.[158] The bioactive C-terminal regions of all the mammalian CCK structures, except for the guinea pig, are identical (*vide supra*). Even the N-terminal region, which has not been demonstrated to possess bioactivity, is also well conserved. Pig CCK-39 differs from the other mammalian species in having a Val residue at position 13 rather than Met. Position 15 varies considerably between species, with the pig having a Met, the human an Ile, and the remaining species a Val residue. Mouse and rat have closely related structures and appear to differ only at position 21, where the mouse (as well as the pig) has a Ser compared to a Gly residue in the rat. The differences between species are minimal and, with one exception, absent in the bioactive C-terminal region; differences which do occur are generally conservative changes.

Using the techniques of molecular biology, the precursor molecules for CCK have been elucidated for pig,[159] human,[160] rat,[161-163] and mouse.[159] The rat and human precursor molecules consist of 115 amino acid residues, while that of the pig consists of 114. Each of the precursor molecules for the four species has a C-terminal extension of 12 amino acid residues attached to the C-terminal phenylalanine residue of CCK. These C-terminal extensions begin with the −Gly−Arg−Arg sequence, contain two tyrosine residues, and have minimal numbers of amino acid residue interchanges between species. This −Gly−Arg−Arg sequence is consistent with the formation of the C-terminal phenylalanine amide residue of CCK, and evidence has been presented indicating that not only is the tyrosine residue in the bioactive CCK-8 molecule fully sulfated, but so are the two tyrosine residues in the C-terminal extension.[164] It has been suggested that the sequence of posttranslational processing consists of sulfation of the three tyrosine residues followed by peptide cleavage at the pairs of basic amino acid residues and, finally, by the amidation reaction;[164] evidence in favor of this has been published recently.[137] Deschenes and colleagues[165] have compared the amino acid sequences of rat CCK and porcine gastrin precursors and have noted sequence similarities in the C-terminal extensions to the common C-terminal pentapeptide structures of gastrin and CCK. From this and other evidence, it has been suggested that genes coding for CCK and gastrin may have evolved from an identical ancestral gene;[165] however, it must be noted that the genes coding for human CCK and gastrin are located on different chromosomes.[166] Similar to other regulatory peptides, partially processed C-terminal extended fragments of CCK have been isolated, but, as with the other regulatory peptides, the physiological significance of this is unknown.

The characterization of the different forms of CCK from neural and intestinal tissue has demonstrated a preponderance of the larger CCK forms in the intestinal mucosa and of smaller forms, particularly CCK-8, in neural tissue. Hence, the processing of CCK to its smaller bioactive C-terminal fragments appears to occur much more rapidly in neural than in intestinal endocrine-type cells. This undoubtedly represents differences in posttranslational processing because the precursors for intestinal and brain CCK are identical.[159] In the intestine, the size of the predominant CCK molecule appears to be species dependent; CCK-58 appears predominant in man, dog, and cat, but CCK-39, CCK-33, and smaller molecules were predominant in pig, cow, and rat intestine.[167] A recent report has suggested that the failure of previous attempts to demonstrate larger molecular weight forms of CCK in the plasma of different species may have been due to rapid *in vitro* degradation; with prevention of such degradation, the major circulating form of cholecystokinin in canine blood was reported to be CCK-58.[168] Obviously, further work is needed to confirm this observation. Since the smaller CCK-8 form has at least the potency of any of the larger forms, the significance and necessity for the presence of larger forms of CCK are at present unknown; possibly they may provide some degree of protection against proteolytic degradation *in vivo*.

C. NONMAMMALIAN MEMBERS OF THE GASTRIN-CCK FAMILY

The recent characterizations of certain nonmammalian members of the gastrin-CCK family have posed interesting problems in both structure-function and evolutionary relationships between the species. The use of antisera directed toward the common C-terminal structure of the gastrin-CCK family of peptides demonstrated the presence of related peptides in nonmammalian species, but also suggested that the carboxy terminus was not identical to that of mammalian gastrin or CCK. Nachman and colleagues[169,170] detected the presence of an active agent(s) in head extracts of the Madeira cockroach, *Leucophaea maderae,* which on administration increased the frequency and amplitude of spontaneous contractions of the hindgut of the cockroach. Two closely homologous peptides were isolated and designated leucosulfakinin I (LSK-I)[169] and leucosulfakinin II (LSK-II).[170] LSK-I is an undecapeptide (composed of 11 amino acid residues) and has an N-terminal Glu residue, a C-terminal phenylalanine amide, and a sulfated tyrosine residue in position 6 (see Figure 6A). LSK-II is a decapeptide (composed of ten amino acid residues) with an N-terminal pyroglutamyl residue, a C-terminal phenylalanine amide residue, and a sulfated tyrosine residue. LSK-I and LSK-II have identical C-terminal heptapeptide amides. Recently, Veenstra[171] isolated two peptides from the corpora cardiaca of the American cockroach, *Periplaneta americana*. One peptide was identical with nonsulfated LSK-II, but the second, designated perisulfakinin (PSK), was identical to LSK-I with the exception of an Asp/Glu (PSK/LSK-I) amino acid residue interchange at position 4. As seen in Figure 6A, a comparison of PSK, LSK-I, and LSK-II with gastrin and CCK demonstrates interesting similarities and dissimilarities in their bioactive C-terminal structures. The C-terminal pentapeptide amides of the mammalian and nonmammalian peptides contain three identities and two nonidentities; the nonidentities are particularly interesting because the chemistries of the residue interchanges are remarkably different. The penultimate C-terminal residue is a basic amino acid residue, arginine, in the insects and is an acidic residue, aspartic acid, in the mammalian peptides. In the fourth position, numbering unconventionally from the C-terminal residue, the insect peptides have the mildly basic residue, histidine, whereas the mammalian peptides have an aromatic residue, tryptophan. The sulfated tyrosyl residue is directly linked to the C-terminal pentapeptide amide in the insect peptides and gastrin, but an additional amino acid residue is interposed in CCK (see Figure 6A). As with the mammalian peptides, all insect peptides require the C-terminal phenylalanine amide structure to possess bioactivity; as with CCK, but unlike gastrin, the sulfated tyrosyl residue is mandatory for activity.[169-171] Interestingly, CCK-8 and mammalian G-17s are not active in the cockroach hindgut preparation. There clearly are interesting evolutionary lessons to be learned from the isolation of such non-mammalian peptides. Similarly, the isolation of chicken gastrin-36, as discussed previously (*vide supra*), has raised interesting structure-function questions. The structure of chicken gastrin-36 is compared with porcine CCK-39 and porcine G-34 in Figure 6B.

IV. THE PANCREATIC POLYPEPTIDE FAMILY

The pancreatic polypeptide family was, until recently, unique in that all three recognized members were first isolated on a strictly chemical basis with their biological activities being demonstrated subsequent to their chemical characterizations. Avian pancreatic polypeptide (PP) was recognized by Kimmel and colleagues in an interesting chromatographic area obtained during the purification of insulin from chicken pancreatic extracts[172] and later was characterized chemically.[173] Chance and colleagues[174] isolated a series of mammalian pancreatic polypeptides from side fractions obtained during the purification of insulins and quickly realized that these mammalian pancreatic polypeptides were analogous to the avian pancreatic polypeptide molecule. Tatemoto, employing the chemical assay developed by himself and Mutt,[6] detected the presence of a peptide having the C-terminal tyrosine amide structure in crude extracts of porcine upper small intestinal tissue. This peptide was characterized chemically, and since no bioactivity had yet been demonstrated it was named PYY after the peptide with N-terminal tyrosine *(Y)* and C-terminal tyrosine amide *(Y)* residues.[175] During an attempt to characterize PYY from brain tissue, Tatemoto isolated a unique peptide with the C-terminal tyrosine amide structure[176] and, subsequently, the same peptide from intestinal tissue.[177] This peptide was designated *neuropeptide Y* or NPY.

The structural characterization of PYY and NPY demonstrated that they both had remarkable sequence similarities to the previously characterized porcine PP and an even more remarkable similarity to each other. As shown in Figure 7, all three peptides contain 36 amino acid residues and have N-terminal tyrosine residues and C-terminal tyrosine amide residues. The structural similarities between the molecules are distributed throughout their lengths; porcine PYY and PP have 18 identities spread throughout their molecules, with particular conservation of structure in the C-terminal region. In general, the nonidentities

THE PANCREATIC POLYPEPTIDE FAMILY – STRUCTURES OF PORCINE PP, NPY AND PYY

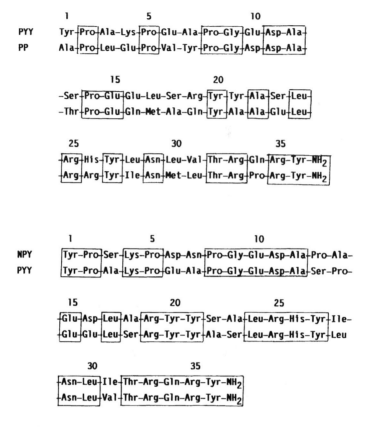

FIGURE 7.   The structures of porcine PYY and PP and the structures of porcine NPY and PYY are compared using the three-letter notation for amino acid residues. Amino acid residue identities between the peptides are enclosed within boxes.

are conservative changes. Even more remarkable is the structural similarity between porcine NPY and porcine PYY, which have no fewer than 25 identities in their 36 amino acid residues. Although the identities are spread throughout the molecule, particularly striking is the conservation of the structure of the C-terminal regions.

Recent studies examining the conformation of members of this family suggest that they maintain a specific three-dimensional structure which has been designated the "PP-fold".[178] X-ray crystallographic studies on turkey PP suggest the presence of two helical structures, an α-helix and a polyproline type II helix, which together with hydrophobic interactions produce a compact stable globular conformation.[178] Studies on NPY also noted the potential to form the two helical structures.[179] Because the amino acid residues essential for maintaining this so-called PP-fold are conserved in the primary structures of the three molecules, it has been suggested that the three-dimensional structure of the members of this family may be conserved to an even higher degree than suggested by their amino acid sequence homologies.[178] The C-terminal region of all members of this family is remarkably well conserved and is a region important for biological activity.[180] The proposed PP-fold structure allows a relatively free recognition of the C-terminal structure by receptors.

Despite the intense sequence homologies and perhaps even greater conservation of their tertiary structure, the functions, tissue distributions, and modes of action of the three members of the family are quite distinct. PP has been localized to a distinct endocrine-type cell in pancreatic islets, is released into the systemic circulation following ingestion of a mixed meal, and appears to be predominantly under the control of the vagus nerve (for review see Reference 181). Although PYY was first isolated from porcine upper small intestinal extracts, it is present in much higher concentrations in the lower small intestine and colon; in the distal ileum PYY has been localized to endocrine-type cells, and it has been suggested that it coexists with a probable glucagon precursor, glicentin.[182,183] Similar to PP, elevated plasma levels of PYY follow ingestion of a mixed meal, but unlike PP, PYY does not appear to be under dominant control by the vagus nerve.[184] In contrast, NPY appears to be strictly confined to neural structures in the mammal,[185-187] with the sole exception being its presence in certain cells of the adrenal medulla[188] and in tumors arising from adrenal medullary tissue, the pheochromocytoma.[189]

A. MAMMALIAN AND NONMAMMALIAN PANCREATIC POLYPEPTIDES

The primary structures of PP from a number of mammalian species have been elucidated;[174,190-192] these are displayed in Figure 8A. As is readily seen, there is a remarkable conservation of structure between the mammalian species, with the porcine, canine, bovine, human, and guinea pig structures differing from each other by only two or, at most, three positions. The rat appears to be exceptional in that it has eight nonidentities with human and other mammalian PP species, and the opossum has five nonidentities with the human molecule.[62] However, even in the rat the C-terminal hexapeptide is identical to that of other species; the C terminal 12 amino acid residues are identical in the other mammalian species. Such a high degree of conservation between species would be consistent with this region being important for biological activity.

Schwartz and colleagues have isolated and characterized an icosapeptide (a peptide composed of 20 amino acid residues) from canine[193] and human[194] pancreatic extracts and have suggested that it is a second stable product derived from the processing of mammalian PP precursors. The primary structures of human PP and the icosapeptide have been elucidated from cDNA sequences coding for the human PP precursor obtained from both normal and neoplastic islet tissue.[195-197] The human precursor contains 95 amino acid residues, and the primary structures of PP and the icosapeptide deduced from the cDNA sequences confirm those reported previously. The rat precursor structure has been recently deduced from cDNA sequences encoding for PP which was obtained from a rat islet cDNA library.[198] The rat precursor contains 98 amino acid residues, and the deduced rat PP sequence confirmed that reported previously. The rat precursor has significant structural homology in the signal and PP sequences to that of the PP precursor structures of other species, but, in contrast, the sequences of the icosapeptides have considerable sequence differences between species.[195,198] The functional significance of the lack of structure conservation of the icosapeptides between species and the potential evolutionary lessons to be learned from this striking difference are at present unclear. The elucidation of the gene structure coding for rat PP may add significantly to our knowledge.

Nonmammalian PP molecules are displayed in Figure 9. Similar to their mammalian counterparts, the structural conservation between the avian species is again remarkable with only two to four nonidentities being present; the C-terminal region again appears to be well conserved. However, there is considerable variation in the structure of the avian vs. the mammalian pancreatic polypeptides. A comparison of human PP to that of the avian species reveals nonidentities in more than one half of their molecules (Figure 9A). An amphibian pancreatic polypeptide, elucidated from the alligator, is more closely related to the avian species, but still differs at eight positions from the avian peptides.

Ingestion of protein or lipids elicits large rises in plasma PP levels, whereas carbohydrate ingestion results in relatively small elevations. These postprandial elevations of plasma PP

THE PANCREATIC POLYPEPTIDE FAMILY

A) Mammalian PP Structures

	1	10	20	30
Human	A P L E P V Y P G D N A T P E Q M A Q Y A A D L R R Y I N M L T R P R Y*
Porcine, Canine	- - - - - - - - - D - - - - - - - - - - - E - - - - - - - - - - - -
Bovine	- - - - E - - - - - - - - - - - - - - - - E - - - - - - - - - - - -
Ovine	- S - - - E - - - - - - - - - - - - - - - E - - - - - - - - - - - -
Rat	- - - - - M - - - - - Y - - H - - R - - - E T Q - - - - - - T - - - - - -
Guinea Pig	- - - - - - - - - - - - - Q - - - - - - - - M - - - - - - - - - - -
Opossum	- - Q - - - - - - - D - - - - - - - K - - - E - - - - - - R - - - - - -

B) Mammalian NPY Structures

	1	10	20	30
Human, Rabbit Guinea Pig, Rat	Y P S K P D N P G E D A P A E D M A R Y Y S A L R H Y I N L I T R Q R Y*
Porcine, Bovine	- - - - - - - - - - - - - - - - - L - - - - - - - - - - - - - - - - -
Ovine	- - - - - - - - - D - - - - - - L - - - - - - - - - - - - - - - - - -

C) Mammalian PYY Structures

	1	10	20	30
Porcine, Rat	Y P A K P E A P G E D A S P E E L S R Y Y A S L R H Y L N L V T R Q R Y*
Human	- - I - - - - - - - - - - - - - N - - - - - - - - - - - - - - - - - -

FIGURE 8. (A) The characterized mammalian PP structures are portrayed using the one-letter notation for amino acid residues. Human PP is given in its entirety; for the remaining species, residues identical to those in human PP are denoted by a dash, and nonidentities are given in the one-letter code. (B) The mammalian NPY structures characterized to date are portrayed in the one-letter notation for amino acid residues. The structure common to human, rabbit, guinea pig, and rat is given it its entirety; in the remaining species, residues identical to this common structure are indicated by a dash, and the nonidentities are given in the one-letter code. (C) The characterized mammalian PYY structures are portrayed in the one-letter notation for amino acid residues. The entire structure of porcine/rat PYY is portrayed, and nonidentitical residues in the human molecule are shown. See Table 1 for definition of symbols denoting specific chemical modifications of amino acid residues.

occur in a biphasic manner; the cephalic-phase plasma PP response occurs rapidly and is totally abolished by truncal vagotomy. The secondary or GI phase of PP release occurs later and may be mediated by enteropancreatic reflexes connecting the gut and pancreas and by humoral mechanisms. Although no clear physiological role for PP has yet been established, administration of PP in doses that result in plasma levels approximating those seen in the postprandial state produces inhibition of pancreatic exocrine secretion and biliary tract motility. The functional aspects of PP have been reviewed recently.[181] In summary, pancreatic polypeptide, contained in specific cells in the endocrine pancreas, appears to exert its activity in a classical hormonal fashion.

B. PYY AND NPY

The primary structure of PYY was determined first for the pig.[175] Subsequently, rat PYY, isolated from rat colonic extracts, was demonstrated to be identical to the porcine molecule.[199] More recently, human PYY has been characterized[200,201] and found to differ at

PANCREATIC POLYPEPTIDE FAMILY

A) Non-Mammalian PP Structures Compared with Human PP

```
             1          10           20          30
Chicken      G P S Q P T Y P G D D A P V E D L I R F Y D N L Q Q Y L N V V T R H R Y*
Turkey       - - - - - - - - - - - - - - - - - - - - N D - - - - - - - - - - - -
Goose        - - - - - - - - - N - - - - - - - ? - - - - - - - - - R L - - F - - - -
Alligator    T - L - - K - - - - G - - - - - - - Q - - N D - - - - - - - - - - P - F*
Frog         A - - E - H H - - - Q - T P D Q - A Q Y Y S D - Y - - I T F I - - P - F*
Human        A - L E - V - - - - N - T P - Q M A Q Y A A D - R R - I - M L - - P - -
```

B) Non-Mammalian NPY-like Structures

```
               1          10           20          30
Anglerfish PYG Y P P K P E T P G S N A S P E D N A S Y Q A A V R H Y V N L I T R Q R Y G
Salmon PP      - - - - - - N - - E D - P - - E L - K - Y T - L - - - I - - - - - - - Y*
Porcine NPY    - - S - - D N - - E D - P A - D L - R - Y S - L - - - I - - - - - - - Y*
Porcine PYY    - - A - - - A - - E D - - - - E L S R - Y - S L - - - L - - V - - - - Y*
```

FIGURE 9. (A) Certain nonmammalian PP structures are portrayed, using the one letter notation for amino acid residues, in comparison with the structure of human PP. The structure of chicken PP is given it its entirety; for the remaining species, residues identical to those of chicken PP are indicated by a dash, and nonidentities are given in the one-letter code. (B) Portrayed are certain characterized nonmammalian NPY-like structures in comparison with the structures of porcine NPY and porcine PYY. The entire sequence of anglerfish PYG is given using the one-letter notation for amino acid residues. In the remaining species, nonidentities are shown and residues identical with those in anglerfish PYG are indicated by a dash. See Table 1 for definition of symbols denoting specific modifications of amino acid residues.

two positions; an Ile/Ala (human/porcine) substitution occurs at position 3, and at position 18 an Asn/Ser (human/porcine) substitution is present (see Figure 8). Human PYY also occurs in an N-terminally truncated form PYY_{3-36}.[201] PYY is present in endocrine-type cells, predominantly in the distal intestine, and elevated plasma levels are seen postprandially. Hence, PYY also appears to act in a classical hormonal fashion, but, as with PP, its precise physiological role is uncertain. Because PYY inhibits pancreatic secretion[175] it has been suggested that PYY could be the "pancreatone" discovered by Harper and colleagues[202] as an inhibitor of pancreatic secretion present in ileal and colonic tissue extracts. However, PYY infusions into human did not result in significant inhibition of exocrine pancreas secretion;[203] hence, the precise physiological role for PYY remains to be determined.

NPY has been characterized in a number of mammalian species (see Figure 8B). The structure of human NPY was elucidated almost simultaneously by isolation and characterization from a pheochromocytoma[204] and by deducing its structure from a cDNA sequence encoding human NPY.[205] The sequences of guinea pig and rabbit NPY were elucidated via isolation and characterization from brain tissue,[206] and the primary structure of rat NPY was deduced from cDNA coding for rat NPY.[207] As shown in Figure 8B, the sequences of human, rat, guinea pig, and rabbit NPY are identical. The sequences of porcine NPY[176,177] and the recently elucidated bovine NPY structure characterized from brain tissue[208] are also identical, but differ from human, rabbit, guinea pig, and rat NPY at a single residue; at position 17 the porcine and bovine molecules have a leucine residue, whereas the other species have a methionine residue. Recently, NPY was elucidated from sheep brain and, like porcine and bovine NPY, was found to have a leucine residue in position 17, but it differed from all other species in having an aspartic acid rather than a glutamic acid residue at position 10 (see Figure 7B).[209]

In contrast to the other two members of the family, PP and PYY, NPY is confined to neural structures in both the central and peripheral nervous systems and in the central nervous system appears to be present in concentrations exceeding those of other known neuropeptides.[210] It has a widespread distribution in the peripheral nervous system; a number of studies suggest that NPY and noradrenaline coexist in perivascular nerves of the GI tract, heart, and other tissues and that, of functional importance considering the potent vasoconstrictor activity of NPY, it potentiates the activity of noradrenaline.[185-187] Evidence has also been reported that nerve fibers containing NPY, but not noradrenaline, arise from nerve cell bodies in the myenteric and submucosal plexuses of the GI tract.[211] A number of biological effects of NPY have been investigated recently; these include effects on vascular and nonvascular smooth muscle, on food intake following intracerebral administration, and indirectly and directly on anterior pituitary function. However, the effects of NPY to produce intense vasoconstriction in both systemic and local arterial circulations have received the most intense investigation. NPY has been the subject of a recent symposium.[208]

C. NONMAMMALIAN MEMBERS OF THE PANCREATIC POLYPEPTIDE FAMILY

The recent isolation from anglerfish pancreatic tissue of a 37-amino-acid-residue peptide with a carboxyl terminus ending in the Tyr–Gly sequence and having extensive structural identity with members of the mammalian PP family[212] has raised certain structure-function and evolutionary questions. Although it shares an extensive structural identity with family members, particularly striking are its homologies with porcine NPY and PYY and, to a lesser extent, with PP; this is despite the fact that it was isolated from the endocrine pancreas. This peptide also differs from the other members of the family in that it does not have the C-terminal tyrosine amide structure, but is extended by a residue of glycine; often the presence of a C-terminal glycine residue is a signal for conversion to the C-terminal amidated structure. At present it is unknown whether this is the final peptide product or if it represents a partially processed peptide. Because it is known that the C-terminal region is important for receptor binding in this family, questions concerning the absence of this C-terminal tyrosine amide structure are possibly of functional importance. Subsequently, Kimmel and colleagues[213] isolated a 36-amino-acid-residue peptide containing the C-terminal tyrosine amide structure from the pancreas of the Pacific salmon (*Oncorhynchus krsutch*). As with the anglerfish peptide, the salmon molecule had greater homology with NPY and PYY than with PP. In fact, the salmon peptide (designated salmon PP) had striking homology with porcine NPY, the two peptides being identical in 30 of their 36 amino acid residues (see Figure 9). The six amino acid residue interchanges can be explained by single base changes and are chemically conservative. Of note, the C-terminal 14 amino acid residues of both peptides are identical. Both groups reporting the structures of these peptides argue that the high concentrations in pancreatic tissue suggest that they are products of endocrine cells rather than of neural elements and, hence, are analogous to the mammalian PP molecules. However, the presence of related peptides has been detected in pancreatic tissue extracts of these fish, and further work is obviously necessary to clarify the evolutionary relationships of these peptides to the mammalian members of the family. Furthermore, functional studies of these peptides in fish are essential.

V. THE NEUROTENSIN FAMILY

During the process of isolating substance P from bovine hypothalamic extracts, the presence of a vasoactive peptide was discovered in chromatographic side fractions. Using its vasodilatory properties as the detection device, the peptide was isolated,[214] and sequence analysis revealed it to be composed of 13 amino acid residues, including an N-terminal

A) <u>The Neurotensin Family</u>

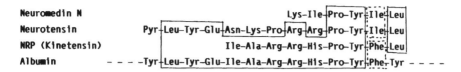

B) <u>Neurotensin Peptides</u>

C) <u>Neuromedin N Peptides</u>

FIGURE 10. (A) The structures of the characterized mammalian members of the neurotensin family are compared with a partial amino acid sequence occurring in the interior of the albumin molecule. The three-letter notation for amino acid residues is used, and identical residues between the peptides are enclosed within boxes. (B) The neurotensin-like peptides in mammals, chicken, and frog are given using the three-letter notation for amino acid residues. Identities between species are enclosed within boxes. (C) The structures of porcine and chicken neuromedin-N are given in the three letter notation for amino acid residues. Identities between the two peptides are enclosed within boxes.

pyroglutamyl residue and a C-terminal leucine residue.[215] The peptide was designated neurotensin after the tissue from which it was initially isolated and due to its vasoactive properties. Subsequently, an identical peptide was isolated from bovine intestinal extracts.[216] Following the synthesis of neurotensin, structure-function studies established that the C-terminal region of neurotensin, neurotensin (8—13) or NT_{8-13}, was essential for bioactivity,[217] a finding confirmed by a number of subsequent studies. Of particular importance are the positively charged arginine residues in positions 8 and 9 (see Figure 10A,B), the presence of a residue containing an aromatic structure in position 11, and the C-terminal Ile–Leu dipeptide structure.[218] It was immediately recognized that neurotensin had C-terminal sequence homology with the previously characterized amphibian skin octapeptide, xenopsin, named after the species from which it was isolated, *Xenopus laevis*.[219] Xenopsin and neurotensin have four identities in their C-terminal hexapeptides (see Figure 10B). The two amino acid interchanges (neurotensin/xenopsin), an Arg/Lys interchange at position 8 and a Tyr/Trp interchange at position 11 of neurotensin, are chemically conservative changes which leave a positively charged residue at position 8 and an aromatic structure in position 11. Hence, it is not surprising that xenopsin and neurotensin have a similar spectrum of bioactivity on administration to mammals.

Although a complete sequence analysis of the structure of human intestinal neurotensin was not obtained, the total composition of the isolated peptide, the partial sequence obtained, and a chromatographic analysis of fragments generated by proteolysis argue strongly that human neurotensin is identical to bovine neurotensin.[220] The primary structures of canine[221] and rat[222] neurotensins, deduced from cDNA sequences coding for their precursor molecules, are identical to bovine neurotensin. Hence, in all mammalian species from which they have been completely characterized to date (cow, dog, rat, and probably human), neurotensins are identical. An avian form of neurotensin, characterized from chicken intestine,[223] is also composed of 13 amino acids, is identical to the mammalian molecule in 10 of 13 positions, and is identical in the bioactive region, NT_{8-13}. The three amino acid interchanges which occur in the N-terminal halves of the molecules are (mammalian/avian) a Tyr/His interchange at position 3, a Glu/Val interchange at position 4, and a Pro/Ala interchange at position 7. These three amino acid interchanges can be accounted for by one-base changes in their codon triplets, but the resulting residue interchanges are not chemically conservative.

During studies on chicken intestinal extracts, Carraway and Ferris[224] noted the presence of a peptide which cross-reacted with a C-terminal-directed RIA for bovine neurotensin.[224] Using the RIA as the detection device, a hexapeptide (composed of six amino acids) was isolated and characterized; it contained an N-terminal residue of lysine and a C-terminal tetrapeptide identical to neurotensin (see Figure 10C). This peptide, although having clear sequence homology with both mammalian and chicken neurotensin, does have distinct differences in the C-terminal bioactive region of the molecule. Carraway and Ferris[224] proposed that this peptide be referred to as LANT-6, named unconventionally from the first letters of the N-terminal dipeptide Lys–Asn (hence, LA), NT referring to neurotensin and 6 referring to it being a hexapeptide. In contrast to chicken neurotensin, which demonstrated a spectrum of bioactivity similar to that of mammalian neurotensin, LANT-6 differed in certain important biological characteristics. For example, LANT-6 produced a hypertensive response, did not induce hyperglycemia or cyanosis, and did not increase vascular permeability. The authors raised the question whether or not a mammalian counterpart to chicken LANT-6 existed,[224] and an affirmative answer was provided shortly thereafter by Minamino and colleagues,[225] who isolated a hexapeptide from porcine spinal cord extracts. The mammalian peptide, designated neuromedin N, is identical to LANT-6 except for an Ile/Asn (neuromedin N/LANT-6) interchange at position 2 (see Figure 10C). Of interest, on smooth muscle tissue, neuromedin N appeared to have biological activity similar to but less potent than that of neurotensin.[225] Because LANT-6 and neuromedin N are distinct peptides, but have striking sequence homology to NT, they belong to the neurotensin family of peptides.

Carraway and colleagues[226] noted that treatment of mammalian plasma with the proteolytic enzyme pepsin generates large quantities of an entity which reacts with antisera recognizing the biologically active C-terminal region of neurotensin. Shortly thereafter, two groups, Mogard and colleagues[227] and Carraway and colleagues,[228] isolated and characterized a nonapeptide from pepsin-treated plasma. Mogard et al.[227] tentatively named the peptide they had isolated from human pepsin-treated plasma "kinetensin"; this peptide differed from neurotensin at two amino acid residues, Arg/His and Ile/Phe (neurotensin/kinetensin) interchanges in the biologically active C-terminal hexapeptide region (see Figure 10A). Carraway and colleagues[228] isolated the same peptide from pepsin-treated fractions of bovine, canine, human, and rat plasma and suggested that the peptide be designated the *neurotensin-related peptide* (NRP) until more evidence was obtained regarding its physiological significance. Both groups noted that NRP/kinetensin had intense sequence homology with at least three different domains in the structure of albumin, the most striking of which is portrayed in Figure 10A, and, to a lesser extent, with angiotensin I. It was suggested that the substrate for the generation of NRP is a minor isoalbumin molecule and that the generation of NRP from plasma or tissue substrates via proteolytic processing *in vivo* may represent a signaling process similar to the generation of angiotensin and bradykinin from inactive precursors.[228] Hence, the process by which NRP/kinetensin is generated to a potentially active form *in vivo* appears to differ greatly from that of the other two members of the family, neurotensin and neuromedin N. Rat, human, and canine NRP increased vascular permeability *in vivo* and released histamine from mass cells *in vitro*.[228] The elucidation of potential physiological role(s) for NRP/kinetensin awaits further studies.

The structure of the precursor molecule for neurotensin and neuromedin N has been deduced from clones isolated from a cDNA library derived from cultured canine enteric mucosal cells,[221] from bovine hypothalamic cDNA clones,[222] and from a rat genomic library.[222] The elucidated nucleotide sequences encode for a precursor containing both neurotensin and neuromedin N; the peptide-coding domains occur in tandem near the C-terminal region of the precursor, and the two peptides are separated and flanked by pairs of basic amino acid residues. An additional neuromedin N-like peptide also occurs in the central region of the precursor molecule. Comparisons of the neurotensin/neuromedin N precursor molecules from dog, cow, and rat demonstrate strong evolutionary conservation with an overall 76% identity for the three species.[222] The canine molecule from the intestine and the cow molecule from the hypothalamus are even more highly conserved (95%). Hence, neurotensin and neuromedin N are processed from similar precursors in both the gut and nervous system.

Neurotensin has been demonstrated to have a number of pharmacological properties on administration to mammals. For example, neurotensin has effects on vascular smooth muscle, gastrointestinal motility, inhibition of gastric acid secretion, and secretion of fluids and electrolytes, and hormones, etc. In the periphery, neurotensin is contained in a population of mucosal endocrine cells (designated N cells) which are present throughout the intestinal epithelium, but are particularly concentrated in the distal small intestine. Neurotensin is obviously present in neural structures of the central nervous system, but its presence in nerve fibers in the gastrointestinal tract is still a matter under investigation. The pharmacological effects and potential physiological roles of neurotensin have been reviewed recently.[229,230] Considerably less is known about potential physiological roles for neuromedin N and NRP/kinetensin.

VI. THE TACHYKININ FAMILY

In 1931, von Euler and Gaddum[231] described the presence of a factor in extracts of both equine brain and intestinal tissues which produced contractions of rodent upper small intestinal smooth muscle and lowered blood pressure. The significance of this finding was not fully appreciated initially, and it was thought possible that the activities could be due to a mixture of factors; the active factor was designated substance P. The peptidal nature of substance P was recognized in 1936, and biological activity attributable to substance P was demonstrated to be present in dorsal root ganglia and in the intestinal smooth muscle layers as well as in mucosal layers.[232] In the early 1960s, Erspamer and colleagues elucidated the structures of a number of peptides with a characteristic spectrum of smooth muscle activity which he designated tachykinin-like activity (for review see Reference 233). Two such peptides were eledoisin, isolated from extracts of molluscan salivary glands and possessing potent sialogogic activity, and physalaemin, isolated from amphibian skin extracts.[233] Eledoisin and physalaemin had striking C-terminal homology with each other and, as part of the definition, a tachykinin-like spectrum of smooth muscle activity. In 1970, Chang and Leeman[234] isolated an undecapeptide from bovine hypothalamic tissue extracts based on its ability to stimulate salivary gland secretion. They immediately recognized the similarity of the spectrum of biological activity of their peptide to that of substance P, including the previously demonstrated sialogogic effect of substance P.[235] Furthermore, they immediately recognized that the size and the amino acid composition of the substance P they had isolated were similar to that of eledoisin and physalaemin. The elucidation of the structure of substance P[236] demonstrated a striking C-terminal homology with these two nonmammalian peptides (see Figure 11). Shortly thereafter, Studer and colleagues[237] isolated an identical peptide from horse intestine, thereby providing the final evidence that substance P was the first of the so-called brain-gut peptides described.

Erspamer[233] is generally recognized as being the first to predict that additional mammalian

THE TACHYKININ FAMILY

A) **Mammalian Members**

Substance P	Arg–Pro–Lys–Pro–Gln–Gln⌐Phe⌐Phe⌐Gly–Leu–Met–NH$_2$
Neurokinin A	His–Lys–Thr–Asp–Ser⌐Phe–Val–Gly–Leu–Met–NH$_2$
Neurokinin B	Asp–Met–His–Asp–Phe⌐Phe–Val–Gly–Leu–Met–NH$_2$

B) **Substance P Subgroup**

Substance P	R	P	K	P	Q	Q	F F G L M*
Uperolein			Q*	D	P	N	A F Y G L M*
Physalaemin			Q*	A	D	P	N K F Y G L M*
Scyliorhinin I				A	K	F	D K F Y G L M*

C) **Kassinin/Eledoisin Subgroup**

NKA	H K T D S F V G L M*
NKB	D M H D F F V G L M*
Kassinin	D V P K S D Q F V G L M*
Scyliorhinin II	S P S N S K C P D G P D C F V G L M*
Eledoisin	Q* P S K D A F I G L M*

FIGURE 11. (A) The structures of the characterized mammalian members of the tachykinin family are portrayed using the three-letter notation for amino acid residues. Identical residues between peptides are enclosed within boxes. (B) The structures of certain members of the substance P subgroup are portrayed using the one-letter notation for amino acid residues. (C) The structures of certain members of the kasinin/eledoisin subgroup of peptides are portrayed using the one-letter notation for amino acid residues. See Table 1 for definition of symbols denoting specific chemical modifications to amino acid residues.

tachykinins remained to be discovered. Subsequently, based on numerous lines of investigation, including differential potencies of the biological effects of different amphibian peptides on administration into mammals, immunologic studies, and ligand-binding studies demonstrating the existence of different types of receptors for tachykinins, a number of researchers presented evidence suggesting the existence of additional mammalian tachykinins (for review see References 238 and 239). These predictions were fulfilled with the isolation of neurokinin-α and neurokinin-β from porcine spinal cord extracts by Kimura and colleagues.[240] Both peptides were composed of ten amino acid residues and had C-terminal pentapeptides identical to each other and to the amphibian skin peptide kassinin (see Figure 11A,C). Neurokinin-α and -β have considerable structural differences in the N-terminal halves of their molecules and, hence, are clearly distinct peptides. Independently and almost immediately thereafter, Nawa and colleagues[241] reported the deduced sequences of two types of bovine brain substance P precursors from cDNA clones coding for the so-called preprotachykinin A precursor (PPT-A). One PPT-A precursor contained only the sequence of substance P, but the second contained the sequence of a tachykinin-like structure which they termed substance K (named presumably for its structural similarity to kassinin) and which was identical to neurokinin-α.[240] Subsequently, Minamino and colleagues[242] isolated from porcine spinal cord extracts a peptide identical to neurokinin-α which they designated neuromedin L. Hence, an identical peptide had been identified independently, in rapid succession, by three groups and given three separate names: neurokinin-α, substance K, and neuromedin L. In accordance with current recommendations for the nomenclature of tachykinins,[243] in this review this decapeptide is designated neurokinin A or NKA, which is understood to be synonymous with neurokinin-α, substance K, and neuromedin L. A similar problem arose when Kangawa and colleagues[244] isolated from porcine spinal cord tissue extracts a peptide identical to neurokinin-β which they termed neuromedin K. Hence, this decapeptide was discovered independently by two groups about the same time, but was given different names. In accordance with the suggestions for tachykinin nomenclature,[243] in this review this decapeptide is designated neurokinin B or NKB, but is understood to be synonymous with neurokinin-β and neuromedin K.

We presently recognize three mammalian members of the tachykinin family: substance P, NKA, and NKB. The structures of these three tachykinins are given in Figure 11A and demonstrate their striking C-terminal homology. In all tachykinins identified to date, the C-terminal pentapeptide amide is composed of the structure Phe–X–Gly–Leu–Met-amide. Substance P is the progenitor of a subgroup composed of substance P and such nonmammalian peptides as physalaemin, uperolein,[233] and scyliorhinin-I[245] a peptide recently isolated from dogfish gastrointestinal tissue extracts. Members of this subgroup have an amino acid residue containing an aromatic structure (i.e., Tyr or Phe) at position X in the structure described earlier (see Figure 11B). The kassinin/eledoisin subgroup (see Figure 11C) have an amino acid residue containing an aliphatic structure (either Val or Ile) at this position. The differences in potencies between substance P, NKA, and NKB may be partially explained by their structural dissimilarities and by different binding affinities to receptor subgroups defined by functional studies (for review see References 238 and 239).

The structures of the mammalian tachykinins have been extraordinarily well conserved throughout evolution. Substance P has been isolated and sequenced from bovine brain,[234,236] horse[237] and guinea pig[246] intestinal tissue extracts, and its sequence has been deduced from cloned cDNAs obtained from cow,[241] human,[247,248] and rat.[249-251] In all cases, the determined or deduced structure for substance P is identical. Hence, the structure of mammalian substance P has been completely conserved. The structure for NKA has been determined by isolation and characterization from porcine spinal cord[240,242] and rabbit intestine,[252] and the NKA structure has been deduced from cloned cDNAs obtained from human,[247,248] rat[249-251] and bovine[241] species. In all cases, the NKA structure has been completely conserved. Similarly, the sequence of NKB as determined from peptides isolated from pig spinal cord extracts[240,244] and that deduced from cDNA clones obtained from a bovine intestinal cDNA library and from a bovine genomic library[253] are completely identical. Hence, to date, all mammalian structures determined or deduced for substance P, NKA, and NKB are identical.

Nawa et al.[241] identified two preprotachykinin A mRNAs which encoded for substance P precursors. The fact that one precursor molecule contained the substance P sequence alone while the second contained the sequences of both substance P and NKA raised certain questions about the organization of the PPT-A gene. Further work by this group[254,255] demonstrated that the bovine PPT-A gene contains seven exons; exon 3 encodes for the sequence of substance P, while exon 6 encodes for the NKA sequence. Since there appeared to be a single bovine PPT-A gene, they concluded that the previously identified two PPT-A mRNAs had been produced from a single gene as a consequence of alternate RNA splicing events.[254,255] The α-PPT-A mRNA, which codes only for substance P, is precisely missing the nucleotides present in exon 6 which encode for NKA; that is, the exon 6 nucleotides have been excluded. The β-PPT-A mRNA contains the peptide-encoding information from exons 2, 3, 4, 5, and 6 and encodes for both substance P and NKA. The substance P and NKA sequences are separated by pairs of basic amino acids on either side of a spacer peptide sequence (see Figure 12). In 1985, Tatemoto and colleagues[256] isolated a peptide from

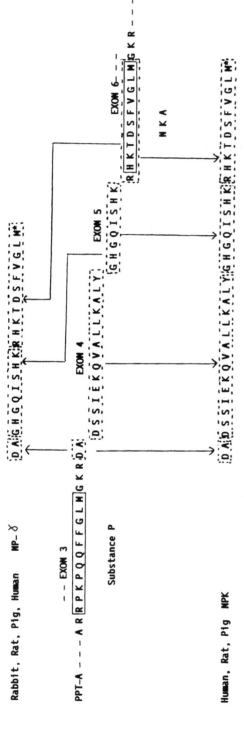

FIGURE 12.   The manner in which NP-α and NPK are formed from the PPT-A precursor molecule via alternate mRNA splicing events is portrayed schematically. Partial structures of the coding regions for exon 3 and exon 6 and the coding regions of exon 4 and exon 5 are portrayed. The one-letter notation for amino acid residues is used throughout.

porcine brain extracts which was composed of 36 amino acid residues and which contained the NKA sequence at its C-terminus. This 36-amino-acid-residue peptide, designated neuropeptide K or NPK, was precisely that predicted for the spacer peptide sequence minus the pair of basic amino acids and glycine residue flanking the C-terminus of the substance P sequence and included at its C-terminus the NKA sequence (see Figure 12). The sole structural difference between pig and cow NPK occurs at position 22, where a Leu residue is present in the bovine molecule and an Ile in the pig molecule. Since this time, the elucidation of the human and rat PPT-A precursors (*vide supra*) has demonstrated the identity of the human and rat NPK molecule with that of the pig. Hence, the NPK sequence, perfectly predicted by the encoding β-PPT-A mRNAs, appears to be almost as well conserved as the two previously recognized tachykinins, substance P and NKA. NPK was noted to have a similar spectrum of activity to NKA, but the time sequence of its activity and its potency differed, leading the authors to suggest that NPK might have different biological functions.[256] However, whether or not NPK has an expression and functional roles different from those of NKA is still under intense investigation. Of note, Deacon and colleagues[257] have demonstrated that, in neurons of the guinea pig small intestine, β-PPT-A is processed to substance P, NKA, and NPK in the cell bodies of the myenteric plexus, but that little NPK is present in synaptic vesicles, suggesting that NPK is converted to NKA during packaging into the storage vesicles for axonal transport.

Further studies have demonstrated the presence of a third rat PPT-A mRNA encoding substance P and NKA[249-251] which has been designated γ-PPT-A. β-PPT-A and γ-PPT-A are identical except for γ-PPT-A lacking a sequence of nucleotides coding for 15 consecutive amino acid residues located between substance P and NKA. This deleted portion corresponds exactly to the sequence specified by exon 4 of the PPT-A gene (see Figure 12). Kaga and colleagues[252] isolated two peptides from rabbit small intestine; one was identical to NKA and the second, termed neuropeptide-γ, was composed of 21 amino acids and was identical to NPK except for a deletion of 15 amino acids (residues 3 to 17 of NPK) in the interior of the molecule. These 15 deleted amino acid residues correspond exactly to exon 4 and arise from a third type of alternate RNA splicing which deletes exon 4. Since NPK was not present in the tissue extracts, the authors suggest that neuropeptide-γ represents a specific posttranslational product of γ-PPT-A and that the pathways of the processing of these peptides in rabbit gut differ from those in other species.[252] Recently, Takeda and colleagues[258] have demonstrated that neuropeptide-γ possesses potent sialogogic activity.

Hence, biologically active tachykinins can be produced in various tissues by differential processing at different molecular levels. Posttranslationally, precursor molecules can, in a tissue-specific manner, have different amounts or types of proteolytic processing, leading to different end products (i.e., NPK and/or NKA). The type of precursor molecule expressed in the tissue can be altered by tissue-specific generation of different PPT-A mRNAs which occur by alternate RNA splicing events. All three identified PPT-A precursor molecules encoded by α-, β-, and γ-PPT-A mRNAs include the potential for the production of substance P. α-PPT-A mRNA does not include exon 6 and, hence, does not code for NKA. β-PPT-A mRNA includes all exons and, hence, may produce substance P, NPK, and NKA; the relative amounts of NPK and NKA generated would be dependent upon posttranslational processing events. γ-PPT mRNA, which does not contain the coding section from exon 5, has the capability of producing substance P, neuropeptide-γ (an internally truncated form of NPK), and NKA; the amount of neuropeptide-γ and NKA produced will depend upon posttranslational processing events. Finally, regulation occurs via differing rates of expression of the gene in different tissues. Studies using combined HPLC and radioimmunoassay (RIA) analysis of tissue extracts have demonstrated considerable tissue-specific heterogeneity of tachykinins in central and peripheral tissues.[252,259-266] Similarly, tissue-specific expression of different mRNAs produced by alternate mRNA splicing events has been demonstrated.[254,267-269]

The distributions reported for substance P and NKA are understandably closely related because they are encoded by the same gene, the so-called PPT-A gene (*vide supra*). However, it has generally been reported that the amount of extractable substance P is either equal to or greater than that of NKA in all tissues, consistent with the fact that all PPT-A mRNAs encode substance P, whereas α-PPT-A mRNA does not encode NKA. A reassessment of the tissue distribution of substance P containing nerves and substance P plus NKA containing nerves at the microscopic level is indicated to determine whether there are differential projections for the two peptides. The gene encoding NKB (the so-called PPT-B gene) is clearly different from that encoding substance P and NKA. It is therefore not surprising that the distributions of NKB are different from those of substance P and NKA (see preceding references and Reference 270).

The tachykinins possess a similar spectrum of biological activities, but differ in their relative potencies.[230,233,238,239] Erspamer[233] pointed out that these relative potencies fall into two subgroups; one group is composed of substance P and physalaemin-like peptides, and the other is composed of kassinin-like peptides including NKA and NKB (see Figure 11B,C). At least part of the difference in potencies between the various compounds may reside in their interactions with functionally defined subgroups of receptors (for review see Reference 239). Of interest in this regard, studies are beginning to elucidate the structure of both the substance P receptor[271] and the NKA receptor.[272] A number of biological activities of the tachykinins, on both central and peripheral tissues of mammals, have been described. There is documentation on the potent effects of the tachykinins on gastrointestinal smooth muscle, thereby altering motility patterns; on vascular smooth muscle, resulting in lowered blood pressure; and on pancreatic and gastric exocrine secretion. Substance P and the neurokinins have been the subject of a recent international symposium[238] and recent reviews.[230,239]

VII. THE BOMBESIN FAMILY

The structure of ranatensin, a vasoactive undecapeptide isolated from extracts of amphibian skin, was described by Nakajima and colleagues,[273] and shortly thereafter Erspamer and colleagues[274,275] reported the structures of bombesin and alytesin, tetradecapeptides isolated from extracts of skin from two European frogs. Bombesin and alytesin differ in only 2 of their 14 amino acid residues, and both peptides have marked C-terminal sequence homology with ranatensin. The demonstrations by Erspamer and Melchiorri[276] that administration of bombesin to mammals resulted in a number of potent biological activities led them to predict the existence of a mammalian counterpart; subsequently, they were the first to report the existence of bombesin-like immunoreactivity in the mammalian gastrointestinal tract. This prediction was fulfilled with the isolation of the porcine gastrin-releasing polypeptide (GRP) from porcine upper gastric tissue extracts[277] and, subsequently, with the isolation of an avian GRP form from chicken proventiculus.[278] Both the mammalian and avian bombesin-like peptides, designated the gastrin-releasing polypeptides after the biological effect used to follow their purification, contained 27 amino acids and had striking C-terminal homology with bombesin and alytesin (see Figure 13A). The C-terminal ten amino acid residues of GRP and bombesin are identical except for a single nonidentity at the eighth position from the C-terminus, where a His residue is present in all the GRPs isolated to date, whereas a Gln residue is present in bombesin and alytesin (see Figure 13A). It had long been recognized that the C-terminal region of bombesin was the region of the molecule essential for bioactivity,[279,280] but this single interchange appears unimportant since a subsequent study demonstrated that GRP_{1-27} and bombesin were equipotent in stimulating plasma elevations of a number of gastroenteropancreatic hormones in the dog.[281] Of interest, the C-terminal seven amino acid residues, known to be the minimal fragment necessary for bombesin-like activity, are completely conserved in all known GRPs and in bombesin and alytesin.

A) <u>Bombesin, Alytesin and the Common C-terminal Tetradecapeptide Region of Mammalian GRPs</u>

GRP ——Met–Tyr–Pro–Arg┤Gly–Asn┤His┤Trp–Ala–Val–Gly–His–Leu–Met–NH$_2$

Bombesin Pyr┤Gln┤Arg–Leu–Gly–Asn–Gln–Trp–Ala–Val–Gly–His–Leu–Met–NH$_2$

Alytesin Pyr┤Gly┤Arg–Leu–Gly┤Thr┤Gln–Trp–Ala–Val–Gly–His–Leu–Met–NH$_2$

B) <u>Porcine GRP and Neuromedin B</u>

Porcine NMB$_{1-32}$ Ala–Pro–Leu–Ser–Trp–Asp–Leu–Pro–Glu–Pro–Arg–Ser–Arg–Ala–Gly–Lys–Ile–

Porcine GRP$_{1-27}$ Ala–Pro–Val–Ser–Val–Gly–Gly–Gly–Thr–Val–Leu–Ala–

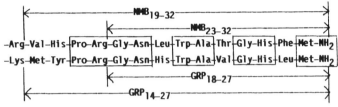

FIGURE 13. (A) The common C-terminal 14-amino acid residue region of the mammalian GRP is compared with the amphibian peptides bombesin and alytesin. The three-letter notation for amino acid residues is used, and identical residues between the peptides are enclosed within boxes. (B) The structures of porcine NMB$_{1-32}$ and porcine GRP$_{1-27}$ are compared after aligning the two peptides at their C-termini. The three-letter notation for amino acid residues is used, and identical residues between the two peptides are enclosed within boxes. The tryptic cleavage points which potentially may form shorter bioactive molecules (see text) are portrayed.

In accordance with the recommendations of the nomenclature committee on bombesin-like peptides,[282] this family of related peptides will be referred to as the bombesin family of peptides, and the name GRP will refer to mammalian and avian bombesin-like peptides. Molecular variants of GRP are identified by subscripts noting the number of the first and last amino acid residues of the variant peptides, with the total number of amino acid residue numbers based on the full length of the GRP molecule, i.e., the C-terminal decapeptide of the full molecule, GRP$_{1-27}$ is referred to as GRP$_{18-27}$.

GRPs have been isolated and characterized from intestinal tissue extracts of the pig,[277] chicken,[278] dog,[283] human,[284] guinea pig,[285] and a partial sequence has been obtained from dogfish.[286] The structure of human GRP was also deduced from cDNA clones obtained from the hepatic metastasis of a pulmonary carcinoid tumor[287] and, subsequently, from cDNA clones obtained from human small cell lung cancer.[288] The human GRP sequence deduced by molecular biology was unable to reveal whether or not human GRP was composed of 28 amino acids and began with an N-terminal Ala residue or whether the precursor was cleaved at the C-terminal side of the Ala residue to produce a 27-amino-acid-residue peptide with an N-terminal Val residue. The isolation of human GRP from extracts of a carcinoid tumor demonstrated that human GRP was, similar to the other GRPs, composed of 27 amino acids but began with an N-terminal Val residue.[284] Recently, the structure of rat GRP has been deduced from a cDNA clone obtained from a rat brain cDNA library.[289] It was unexpectedly reported that rat GRP is composed of 29 amino acids, in contrast to the other mammalian and avian heptacosapeptide GRPs.

A comparison of the structures of the various GRPs characterized to date reveals that the members of this family have remarkable C-terminal homology; with the exception of the rat, the C-terminal 15 amino acid residues of all mammalian GRPs are identical (see Figure 14A). Rat GRP, like chicken and dogfish GRP, has a Ser residue in place of an Asn residue (porcine, guinea pig, human, canine GRP) in the ninth position from the C-terminus, numbering unconventionally. This remarkable conservation of the entire C-terminal region of the GRP molecules is of significance in that numerous structure-function studies [279,280,290,291] have demonstrated this to be the region responsible for the recognized biological activities of bombesin-like peptides. It is also evident that a strong tendency to conserve the structure of the N-terminal region exists in the mammalian GRPs. Porcine and guinea pig GRP are identical, human GRP differs in only five positions, and canine GRP differs in four positions from the porcine molecule in the N-terminal halves of the molecules. The structure of rat GRP, deduced from cDNA clones encoding its structure, poses a problem. If rat GRP does consist of 29 amino acids and is aligned at its C-terminal with the other GRPs, there is still only one nonidentity with the structure of porcine GRP in the C-terminal 22 amino acid residues. If one assumes that a Thr–Gly dipeptide sequence (positions 5 and 6 in rat GRP) has been inserted in the rat, on deletion of this dipeptide and alignment of the porcine and rat peptides at their N- and C-termini only two nonidentities will occur in the entire 27 amino acid residue positions. The dogfish GRP sequence has only been partially elucidated, but with an assumed deletion of two amino acid residues in the interior of the molecule (marked with a box in Figure 14A), alignment of the dogfish sequence reveals a striking homology with the mammalian and avian GRPs. All the GRPs (mammalian, avian, and piscine) have a similar architecture; sites sensitive to trypsin-like cleavages occur at identical positions (positions 13 and 17 of porcine GRP).

A number of groups noted that large and small forms of bombesin-like immunoreactivity were present in mammalian tissue extracts and that the proportion of small to large forms varied between different tissues (for review see Reference 292). The identity of the large bombesin-like immunoreactive form was demonstrated to be GRP_{1-27}. Reeve and colleagues[283] first isolated the small bombesin-like immunoreactive form from canine intestinal extracts and demonstrated its identity to GRP_{18-27} (i.e., the C-terminal decapeptide of GRP). Subsequently, GRP_{18-27} was isolated from human,[284,293] dogfish,[286] and porcine spinal cord tissues,[294] the porcine spinal cord peptide was initially named neuromedin C, but, in accordance with recommended nomenclature,[282] is now designated GRP_{18-27}. Since GRP_{18-27} contains the residues essential for the known biological activities of this family of peptides, it has been suggested that, in certain tissues, a trypsin-like cleavage to form GRP_{18-27} may occur *in vivo*. Whether or not this cleavage is an important physiological event in certain tissues but not in others is uncertain; the relative proportions of each form are quite variable, and in certain bioassay systems the full molecule is equipotent to GRP_{18-27}. It appears clear that the cleavage of GRP_{1-27} to GRP_{18-27} does occur *in vivo* and is not simply an artifact of the extraction procedure.

Studies on the molecular biology of the gene encoding for the GRP precursor have proved interesting. Spindel and colleagues[287] cloned cDNAs from a hepatic metastasis of a pulmonary carcinoid tumor and demonstrated that the GRP precursor consisted of 148 amino acids and that the C-terminal Met residue of GRP was followed by the Gly–Lys–Lys sequence, consistent with cleavage of the precursor molecule and formation of the C-terminal methionine amide structure. The presence of a long C-terminal extension peptide in the precursor molecule was noted along with presence of two mRNA forms encoding the precursor. Blot hybridization analyses using human genomic DNA were consistent with the presence of a single GRP-encoding gene. Further work demonstrated the expression of the GRP gene in cultured human small cell lung cancer cell lines and in tumors,[288] and the presence of multiple types of prepro-GRP mRNAs.[288,295] These mRNAs all coded for GRP with the sequence differences occurring only in the C-terminal extension peptide-coding domain. The elucidation of the structure of the human GRP gene has provided a rational explanation for the occurrence of multiple mRNAs and pointed out certain unusual features of the GRP gene.[296] The human GRP gene consists of exon one, which encodes the 5'-untranslated region, the signal peptide, and the first 23 amino acid residues of GRP. The

A) Bombesin/GRP Subgroup

```
                              1           10          20
Porcine, Guinea Pig GRP    A P V S V G G G T V L A K M Y P R G N H H A V G H L M*
Human GRP                  V - L P A - - - - - - - T - - - - - - - - - - - - - -
Canine GRP                 - - - P G - Q - - - - D - - - - - - - - - - - - - -
Chicken GRP                - - - Q P - - S P A - T - I - - - - S - - - - - - - -
Rat GRP                    A P V S T G A - - - - - - - - - - - - - S - - - - - - - -
Dogfish GRP                - - - E N Q - S F P [   ] - - F - - - S - - - - - - - -
Bombesin                                           Q*Q R L - - Q - - - - - - - -
```

B) Neuromedin B Subgroup

```
Pig NMB        A P L S W D L P E P R S R A G K I R V H P R G N L W A T G H F M*
Human NMB      - - - - - - - - - - - - - S - - - - - S - - - - - - - - - -
Ranatensin R                           S N T A L R - Y - Q - - - - - - -
Ranatensin                                         Q*V P Q - - V - - - -
Litorin                                            Q*Q - - V - - - -
```

C) Phyllolitorin Subgroup

```
Phyllolitorin                                      Q*L W A V G S F M*
(Leu8) Phyllolitorin                               - - - - - - - L -
(Thr5, Leu8)                                       - - - - T - - L -
```

FIGURE 14. (A) The structural variation occurring between mammalian and nonmammalian forms of the GRP molecules are portrayed using the one-letter notation for amino acid residues. All peptides are aligned at their C-termini, and the full sequence of porcine/guinea pig GRP is given; in the remaining species, residues identical to that of porcine/guinea pig GRP are indicated by a dash, and nonidentities are portrayed using a one-letter identifier code. The assumed deletion of two amino acid residues in the interior of dogfish GRP is indicated by a box. (B) Certain of the molecules of the neuromedin B subgroup are portrayed using the one-letter identifier code for amino acid residues. All peptides are aligned at their C-termini. Porcine NMB is given in its entirety; in the remaining species, residues identical to that of porcine NMB are indicated by a dash, and nonidentities are given using the one-letter identifier code. (C) Certain members of the phyllolitorin subgroup are portrayed using the one-letter amino acid identifier code. The full sequence of phyllolitorin is given; in the remaining species, nonidentities are shown using the one-letter notation, and residues identical to those in phyllolitorin are indicated by a dash.

first intron interrupts the codons for amino acid residue 24 (a glycine residue) in GRP between the first and second nucleotides. Exon 2 provides the remaining three amino acids of GRP and the immediate C-terminal Gly–Lys–Lys sequence and also encodes most of the GRP C-terminal extension peptides. The remainder of the C-terminal extension peptide and the 3′-untranslated region are encoded by exon 3.[296] Spindel and colleagues[296] point out that intron 1, interrupting the coding region in the bioactive portion of GRP, is quite unusual; introns usually mark functional domains within a protein and rarely interrupt the coding for the biological active portion of peptides. The discovery of alternate donor sites and alternate acceptor sites was also an unusual finding, but provided an explanation for the formation of the three mRNAs encoding GRP precursors by alternate RNA splicing events. The most abundant mRNA (designated type I) is the full-sized mRNA, which encodes the GRP sequence and the full C-terminal extension peptide (termed GRP gene-associated peptide or GGAP). Type III mRNA is the next most abundant form and is produced by an alternate splicing event which deletes 19 nucleotides. Hence, the peptide encoded by Type III mRNA is missing 6 amino acid residues and, because the number of nucleotide deletions is not divisible by three, also results in a shift in reading frame; this results in a difference of 10 amino acids in size and an overall sequence difference of 27 amino acids compared to the peptide coded for by type I mRNA. Type II mRNA encodes for the six amino acids missing in type III clones and instead has a deletion of 21 nucleotides immediately 3' to the bases deleted in the type III clone; hence, no shift in reading frame results. The three mRNAs produced by the alternate RNA splicing events occur not only in neoplastic cells, but also in normal neural and endocrine tissue,[296] and the GGAPs predicted by the three mRNAs have been found in endocrine cells of human fetal lung and in human small cell lung cancer cells.[297] Although the presence of difference GGAPs has been described as a novel peptide family, the pathophysiological significance of the presence of these GGAPs and their expression in normal and neoplastic tissues are uncertain; they may provide markers for certain neoplastic processes.[297] Hence, studies on the GRP gene in humans have demonstrated some unusual features: (1) an intron interrupts the structure of GRP in a region of the molecule critical for its biological activity and (2) multiple precursor forms for GRP result from the use of alternate donor and acceptor splicing sites. Subsequently, the human GRP gene was found to be present on chromosome 18.[298,299] The recent elucidation of the structure of the GRP gene in rat has provided additional insights.[289] From rat brain cDNA libraries, a number of cDNA clones were isolated and found to have a similar overall organization to the human cDNA clone encoding prepro-GRP. However, it was found unexpectedly that a second initiation site was present in a rat brain transcript that was not present in duodenal transcripts. There was also no evidence for alternate RNA splicing events occurring in the rat mRNA encoding the GRP precursor; the single mRNA corresponded to the third human isoform.[289]

The bombesin-like family has been divided into three subgroups[300] based on amino acid residue differences occurring in the final C-terminal three amino acid residues, a region of the molecule critical for biological activity (see Figure 14). The bombesin/GRP subgroup contains the C-terminal tripeptide His–Leu–Met-amide sequence, whereas the second group, the ranatensin/neuromedin B subgroup, has the C-terminal tripeptide His–Phe–Met-amide structure. Hence, the differentiation between these subgroups is the His/Phe interchange, which occurs at the penultimate C-terminal residue. A third group, the phyllolitorin subgroup, differs from the other two subgroups in that it has a Ser rather than a His residue at the third position (numbering unconventionally) from the C-terminus. Certain members of the phyllolitorin subgroup, like the bombesin/GRP subgroup, have a Leu residue, and others, like the ranatensin/neuromedin B subgroup, have a Phe residue as the penultimate C-terminal residue. To date, no mammalian equivalent members of the phyllolitorin subgroup have been characterized. As seen in Figure 14, all three subgroups have well-conserved C-terminal heptapeptide structures which contain the region essential for biological activity.

Different lines of research had suggested the presence of a ranatensin-like peptide in the mammalian central nervous system. In 1983, Minamino and colleagues[301] isolated a decapeptide from mammalian spinal cord tissue extracts and demonstrated that this peptide, designated neuromedin B(NMB), had a C-terminal heptapeptide identical to that of ranatensin R. Subsequently, this group isolated two N-terminally extended forms of the original decapeptide which contained 32 and 30 amino acid residues.[302] Hence, assuming that the full molecule contains 32 amino acids, in keeping with the nomenclature committee's recommendations,[282] this molecule is designated NMB_{1-32}. The 30-amino-acid peptide is an N-terminally truncated form, neuromedin B_{3-32} and the originally isolated decapeptide becomes NMB_{23-32} (see Figure 13B). Since neuromedin B appears to be a major bombesin-like peptide present in the mammalian central nervous system[303-305] and has C-terminal structural homology with ranatensin, it is probable that neuromedin B is the mammalian equivalent of ranatensin (see Figure 14B). The human structure of neuromedin B was recently deduced from cDNA encoding neuromedin B obtained from a human hypothalamic cDNA library.[306] The human and porcine molecules differ in only two positions in their 32 amino acid residues;

human neuromedin B has Ser residues at positions 15 and 21, whereas the porcine molecule has residues of Gly and Pro, respectively. Hence, not only is the bioactive C-terminal region completely conserved in neuromedin B, but the N-terminal region is also almost completely conserved. Although no biological activity has as yet been attributed to the N-terminal region, such strong conservation suggests that further investigation should be undertaken.

The neuromedin B precursor contains 76 amino acid residues and, as with the GRP precursor, the sequence of NMB_{1-32} immediately follows the signal sequence.[306] The ranatensin precursor[306] has an N-terminally extended peptide interposed between the ranatensin coding region and the signal peptide sequence. Two neuromedin B mRNA species of slightly different sizes have widespread distributions in both the brain and GI tract. Further work is necessary to determine if the two mRNA species are formed by alternate splicing events similar to that described for the human GRP gene. The gene encoding neuromedin B is localized to the long arm of human chromosome 15.[306]

A comparison of the sequences of porcine NMB_{1-32} and porcine GRP_{1-27} (see Figure 13B) reveals considerable structure similarities. Of the final 12 C-terminal amino acid residues, 9 are identical in neuromedin B and GRP, and potential tryptic cleavage sites occur in identical positions of both molecules when they are aligned at their C-termini. The most C-terminal tryptic cleavage point would release the biologically active decapeptides NMB_{23-32} and GRP_{18-27}, both of which have been isolated and characterized from tissue extracts. Similarly, C-terminal tetradecapeptides NMB_{19-32} and GRP_{4-27} could be formed from a potential tryptic cleavage site, but neither of these peptides has been isolated as yet from tissue extracts. Neuromedin B has a clearly different tissue distribution from that of GRP in the mammal, and although they have a similar spectrum of biological activity, there are differences in relative potencies. Hence, neuromedin B probably serves different functions than GRP and additional investigation is necessary to determine the physiological and pathophysiological roles of these two peptides. Our current knowledge about the function of these peptides has recently been the subject of an international symposium.[307] The recognition of the phyllolitorin subgroup raises the obvious speculation that further mammalian peptides analogous to phyllolitorin remain to be detected. The differences in the bioactive spectrum and the structure-activity relationships of the phyllolitorin subgroup have been the subject of recent reviews.[300,308]

In the mammal, GRP appears to be confined strictly to neural elements of the central, peripheral, and enteric nervous systems, with the exceptions that it is also found in neuroendocrine cells of the peribronchial epithelium and in neuroendocrine cells present in small cell lung cancer, medullary carcinoma of the thyroid, and carcinoid tumors. Intracerebral administration of GRP to mammals results in potent and diverse biological effects, including the production of hypothermia, hyperglycemia, altered satiety mechanisms, and increased sympathetic tract outflow. Peripheral i.v. administration of GRP potently effects the release of a number of biologically active peptides (hormones, neuropeptides), alters gastrointestinal smooth muscle activity, and has potent effects on the exocrine secretion of the pancreas and stomach. Although no single biological function of GRP or bombesin has been demonstrated to be physiological, strong suggestive evidence has been reported for a role for GRP in vagal mechanisms resulting in the release of gastrin from endocrine cells in the gastric antrum. The bombesin-like peptides also have putative trophic roles in the lung, gastric antrum, and pancreas and have been implicated as autocrine growth factors in human small cell lung cancer.[309] The role of bombesin-like peptides in health and disease has recently been the subject of an international symposium[307] and has been the subject of recent reviews.[292,310]

VIII. THE ENDOTHELIN FAMILY

The story of endothelin, an endothelium-derived acidic 21-amino-acid-residue vasocon-

A) Comparison of Endothelin-1 with Sarafotoxin 6B

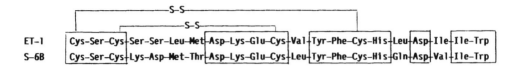

B) The Endothelin Family of Mammalian Peptides

		1	5	10	15	20
Human, Porcine, Mouse	ET-1	C S C S S L M D K E C V Y F C H L D I I M				
Human, Canine	ET-2	- - - - - W L - - - - - - - - - - - - -				
Mouse	VIC	- - - N - W L - - - - - - - - - - - - -				
Human, Porcine, Rat	ET-3	- T - F T Y K - - - - - Y - - - - - - -				

FIGURE 15.   (A) The full sequences of endothelin 1 and sarafotoxin 6B are given using the three-letter notation for amino acid residues. Identities between the two peptides are enclosed within boxes. The position of the disulfide bond is indicated. (B) The endothelin family of mammalian peptides is compared using the one-letter notation for amino acid residues. The entire sequence of ET-1 is given; in the remaining peptides, residues identical to those in ET-1 are indicated by a dash, and the nonidentities are given in the one-letter notation.

strictive peptide,[311] has unfolded with extraordinary rapidity, due both to the efficient use of "state-of-the-art" technology and to the high potential of the peptide for a significant role in physiology and pathophysiology.

Based on numerous lines of evidence that vascular endothelial cells produce various factors which relax or contract the underlying vascular smooth muscle, Yanagisawa and colleagues[311] screened for the presence of such active agents in supernatants from porcine aortic endothelial cell cultures. An agent was detected which caused a potent endothelium-independent, slow-onset but long-acting contraction of porcine coronary artery strips. After isolation, sequence analysis revealed the agent to be an acidic peptide composed of 21 amino acid residues, with an N-terminal Cys and C-terminal Trp residue. Via synthesis of possible disulfide analogues, it was found that the peptide, designated endothelin, contained two sets of intrachain disulfide bridges in which the Cys residues in positions 1 and 15 and those in positions 3 and 11 were joined[311,312] (see Figure 15A). The structure of preproendothelin was deduced from a cDNA clone obtained from a cDNA library constructed from porcine aortic endothelial cell mRNAs.[311] This precursor contained 203 amino acid residues, and the endothelin sequence was flanked on the N-terminal side by a pair of basic amino acids, but the C-terminal tryptophan residue was extended by a residue of valine; hence, it was postulated that proteolytic processing of the mature molecule from "big endothelin" involved a chymotrypsin-like endopeptidase which cleaved the bond between these two residues. Subsequent evidence has demonstrated that such a cleavage of big endothelium to produce the mature molecule is essential for potent bioactivity.[313,314] The endothelin precursor also contains a potentially interesting peptide, the so-called endothelin-like peptide, which has significant sequence homology to the N-terminal 15 amino acid residues of endothelin. This peptide structure has also been noted in subsequent studies, but its significance at present is unknown. The original report[311] also noted that (1) endothelin was the most potent of the known vasoconstrictor peptides and that its mode of action was direct on smooth muscle cells, (2) preproendothelin mRNA was expressed in porcine aortic intima *in vivo* as well as in cultured endothelial cells, (3) levels of the mRNA could be significantly altered by the addition of various agents, and (4) the action of endothelin was closely associated with calcium influx through dihydropyridine-sensitive calcium channels.

At virtually the same time, Takasaki and colleagues[315] isolated a series of highly homologous peptides from the lethal venom of the burrowing asp (*Atractaspis engaddensis*). Four such isotoxins were characterized, and one, S6B, is portrayed in Figure 15A. Previous studies[316,317] had demonstrated that the sarafotoxins were highly lethal and that a prominent feature of their cardiac toxicity was persistent coronary artery vasoconstriction. It was quickly noted[318,319] that the sarafotoxins and endothelin had extraordinary structural similarities. Both peptides were composed of 21 amino acid residues, the Cys residues were in identical positions, 14 of their 21 amino acid residues were identical (i.e., a 67% sequence homology), and the sequence homology was spread throughout the molecule; the most intensely variable region occurred in the N-terminal (4—7) residues (see Figure 15A). The functional similarities between the two peptides are not surprising given the intense structural homology. It was noted that since endothelin, the product of a mammalian endothelial cell, resembled an exocrine secretion from a reptile, it might possess functions other than vasoconstriction. Structure-function studies[314,320,321] have generally demonstrated the importance of the presence of the whole molecule for full potency, but particular importance is attached to the two disulfide bridge structures, the C-terminal hydrophobic moiety (in particular, the C-terminal Trp residue), and the sequence of charged residues in positions 8 to 10.

Shortly thereafter, Itoh and colleagues[322] deduced the sequence of the human endothelin precursor from cDNA clones isolated from a human placental cDNA library. The human precursor was composed of 212 amino acid residues and was 69% homologous with the porcine precursor. The primary structure of human endothelin, including its N-terminal and C-terminal flanking residues, was identical to that of the porcine molecule. Also present was the endothelin-related peptide. Yanagisawa and colleagues[323] cloned a cDNA from a rat genomic library which was considered to be a portion of the rat endothelin gene. A partial sequence of the "rat endothelin gene" was deduced, and the nucleotide sequence predicted a 21-amino-acid-residue peptide similar to but distinct from porcine or human endothelin. A total of 15 residues of rat endothelin were identical to the porcine/human molecule, and the Cys residues were in identical positions. Synthetic rat endothelin had potent vasoconstrictor activity, but certain pharmacological differences were clearly demonstrable between the rat and porcine molecules.[323]

The situation became more complex when Inoue and colleagues[324] discovered the presence of three distinct human endothelin-related genes which they cloned from a human genomic DNA library. These three genes predicted structurally distinct but related peptides. All peptides were composed of 21 amino acids and had remarkable sequence homology with each other (see Figure 15B). The first peptide was designated ET-1 and was identical to the previously described human and porcine molecule. The second peptide, designated ET-2, was identical to ET-1 except for two amino acid residue interchanges, a Leu/Trp (ET-1/ET-2) interchange at position 6 and a Met/Leu (ET-1/ET-2) interchange at position 7. The third peptide, designated ET-3, had six amino acid residue differences from ET-1; at position 2 a Ser/Thr interchange occurs (ET-1/ET-3, respectively), at position 4 a Ser/Phe interchange, at position 5 a Ser/Thr interchange, at position 6 a Leu/Tyr, and at position 14 a Phe/Tyr interchange (see Figure 15B). Of great interest, it was recognized that the previously described rat endothelin[323] was identical to ET-3, and the biological activities of the rat molecule were similar to those described for ET-3. Hence, rat endothelin is usually referred to as ET-3. Although all three peptides are potent constrictors of arterial smooth muscle, the pharmacological activities of each are quantitatively different, as are the time courses of their actions. In terms of maximal activity, the potency order of ET-2 and ET-1 > ET-3 has been reported. ET-3 produces a considerably shorter time course of vasopressor activity, but its initial depressor effect is significantly greater than that of ET-2 and ET-1. Hence, these peptides not only differ in structure, but also in their functional activities.

Based on RIA evidence suggesting that ET-2 was secreted from kidney cell lines, Itoh and colleagues[325] identified a cDNA clone from a canine kidney cDNA library which encoded ET-2; the deduced canine ET-2 structure is identical to that of human ET-2. Recently, Saida and colleagues[326] isolated two cDNA clones from a mouse genomic library. The first cDNA encoded for a peptide identical to ET-1. The second, however, encoded what appeared to be a novel peptide, designated the vasoactive intestinal contractor (VIC), which differed from ET-1 at three amino acid residue positions (see Figure 15B), but differed from ET-2 only at position four, where a Ser/Asn (ET-2/mouse VIC) interchange occurs.[326] Studies to determine the tissue-specific expression of the ET-1 and VIC genes demonstrated that ET-1 gene expression occurred in bovine and human endothelial cells, but not in other tissues. In contrast, the VIC gene appeared to be expressed in the intestine, but not in other tissues, and had potent contractile activity on the guinea pig ileum. Further studies are required to determine whether VIC is indeed a unique peptide or is the mouse equivalent to ET-2, containing one amino acid residue interchange.

The structure of the human ET-1 gene, assigned to human chromosome 6, has been elucidated recently[327,328] and reported to be organized into five exons. Exon 2 encodes the ET-1 sequence and the first four amino acids in the C-terminal extension region, while exon 3 encodes the remaining sequence of human "big ET-1" and also for the endothelin-like peptide. A number of potential regulatory sites for the expression of the ET-1 gene have been detected[327] which may be responsive to the various agents noted to increase the amount of ET-1 mRNA.

Figure 15A,B portrays the remarkably similar structural organization of the members of the endothelin family. The initial N-terminal residues are conserved, and there is a variable region between residues 4 and 7, followed by a conserved area of alternately charged residues (positions 8 to 10) and a highly conserved hydrophobic C-terminal region. The amino acid interchanges which occur between the family members are generally chemically conservative in nature. Further elucidation of the precursor structures demonstrated conservation of the N- and C-terminal regions flanking the mature peptides and also conservation of the endothelin-like peptides. Other regions of the genes coding for the precursors are not as well conserved, consistent with differences in evolution having occurred.[324]

The quantitative differences in the pharmacological activities of the different peptides suggest that they may serve different biological roles, and evidence is accumulating that they may be expressed differentially in various tissues. For example, ET-1 and ET-3 have been isolated and sequenced from porcine spinal cord and brain;[329,330] ET-2 was not detected in these studies. However, the presence of an ET-2-like peptide in kidney[325] and of the VIC peptide in intestine[326] suggests that they have different tissue distributions. Further work is required to determine the precise tissue distribution of these peptides and whether there is a matching expression of specific receptor subtypes. It is as yet unclear whether VIC is the mouse equivalent of ET-2 or is, indeed, a distinct peptide. The evidence to date has clearly established that the endothelin-like peptides are distinct and, hence, form a family of peptides.

Due to the obvious potentially important physiological role of the endothelins in cardiovascular regulation, the majority of pharmacological investigations have focused on their effects on vascular smooth muscle. However, a number of other biological effects have been noted for the endothelin-like peptides, including the ability to release eicosanoids and endothelial-derived relaxing factors from vascular beds, to inhibit (i.e., renin) or stimulate (atrial natriuretic peptide, catecholamines, etc.) hormone release, to modulate neuronal transmission, to contract nonvascular smooth muscle, and to stimulate proliferation of smooth muscle cells and fibroblasts. Although it has been only 2 years since the discovery of endothelin, a brief review has been published discussing the functional aspects of endothelin.[314]

IX. GEP REGULATORY PEPTIDES NOT YET DEMONSTRATED TO BE PART OF A FAMILY

A. GALANIN

Galanin was detected in and isolated from porcine upper small intestinal extracts by Tatemoto and colleagues[331] using a chemical assay that detected the presence of a peptide containing the C-terminal α-alanine structure. On sequence analysis, galanin was found to contain 29 amino acid residues and to have a unique amino acid sequence, an N-terminal residue of glycine, and a C-terminal alanine amide residue. Because it was isolated on a strictly chemical basis, galanin was named after its N- and C-terminal residues, glycine and *alanine* amide. Subsequent to its isolation, galanin was found to possess biological activity, exciting rodent smooth muscle and, on i.v. administration to dogs, lowering plasma insulin levels, resulting in mild plasma glucose level elevation.[331,332] Numerous studies have demonstrated that galanin in the mammal is confined to neural structures, and the detailed distribution of galanin-like immunoreactivity in the central and peripheral nervous systems has been reported (for review see Reference 333 and citations therein). The structure of the galanin precursor molecule, containing 123 amino acid residues, was deduced from cDNA clones obtained from a porcine adrenal medullary cDNA library.[334] Rökaeus and Carlquist[335] isolated and chemically characterized bovine galanin from intestinal extracts and also deduced the structure of the bovine molecule from cDNA clones obtained from a bovine adrenal medullary cDNA library. Bovine galanin differs from porcine galanin at only four amino acid residue positions; interchanges occur at position 16, Ile/Leu (porcine/bovine); at position 18, Asn/Ser; at position 23, His/Gln; and at position 26, Tyr/His (see Figure 16A). A structure for rat galanin was proposed recently by two groups. Vrontakis and colleagues[336] deduced the sequence of rat galanin from mRNA isolated from estrogen-induced rat pituitary tumors, and Kaplan and colleagues[337] deduced the sequence from cDNA samples obtained from a rat hypothalamic cDNA library. Rat galanin differs from pig galanin at three amino acid residue positions, all in the far C-terminal region; amino acid residue interchanges occur (porcine/rat, respectively) at position 23 (a His/Ser interchange) and at position 26 (a Tyr/His interchange), and the C-terminal residue in the pig is alanine amide and in the rat is threonine amide (see Figure 16A). Of note, all of the differences in amino acid sequences between the three molecules occur in the C-terminal halves of the molecule. Structure-function studies on the effect of galanin on smooth muscle,[338] on the endocrine pancreas,[339-341] and in the central nervous system[342] indicate that the region of the molecule of critical importance for biological activity resides in the N-terminal region of the molecule. Hence, the amino acid interchanges between species occur in a region not yet demonstrated to be important for bioactivity.

The specific distributions of galanin-like immunoreactive neurons, nerve fibers, and their projections have been described in the GI tracts of a number of species.[343-348] This widespread distribution in the enteric nervous system suggests a potential gastrointestinal regulatory role, and intense effects have indeed been found on intestinal smooth muscle. Curiously, in rodents the effect of galanin is stimulatory[349] while in the dog[350] and in humans[351] galanin inhibits motility. Galanin potently inhibits insulin responses to a number of physiological and pharmacological stimulatory agents in the dog,[352-354] rat,[355-357] and mouse[358] pancreas. Evidence has been reported that galanin exerts its pancreatic effects via specific β-cell receptors coupled to G regulatory proteins, it inhibits adenylate cyclase activity;[359-361] and activates ATP-sensitive potassium channels, coupled to pertussis toxin-sensitive G proteins, leading to hyperpolarization of β-cell membranes.[362,363] Perhaps similar to species-response differences on smooth muscles, the effect of galanin on the endocrine pancreas may be species specific because Bauer and colleagues[364] have reported little, if any, effect of galanin on insulin release in man despite its clear ability to elevate growth

A) Galanin

```
              1                                            10
Porcine    Gly-Trp-Thr-Leu-Asn-Ser-Ala-Gly-Tyr-Leu-Leu-Gly-Pro-His-Ala-
Bovine      -   -   -   -   -   -   -   -   -   -   -   -   -   -   -
Rat         -   -   -   -   -   -   -   -   -   -   -   -   -   -
```

```
                            20
Porcine    -Ile-Asp-Asn-His-Arg-Ser-Phe-His-Asp-Lys-Tyr-Gly-Leu-Ala-NH2
Bovine     -Leu   -  Ser  -   -   -   -  Gln  -   -  His  -   -   -
Rat         -   -   -   -   -   -   -  Ser  -   -  His  -   -  Thr-NH2
```

B) Motilin

```
                        1                                    10
Porcine, Human     Phe-Val-Pro-Ile-Phe-Thr-Tyr-Gly-Glu-Leu-Gln-Arg-
Canine              -   -   -   -   -   -  His-Ser  -   -   -  Lys-
```

```
                                        20
Porcine, Human     -Met-Gln-Glu-Lys-Glu-Arg-Asn-Lys-Gly-Gln
Canine             -Ile-Arg  -   -   -   -   -   -   -
```

C) Porcine Pancreastatin

```
             1                                    10
         Gly-Trp-Pro-Gln-Ala-Pro-Ala-Met-Asp-Gly-Ala-Gly-Lys-Thr-Gly-
                        20                                    30
         Ala-Glu-Glu-Ala-Gln-Pro-Pro-Glu-Gly-Lys-Gly-Ala-Arg-Glu-His-
                                    40
         Ser-Arg-Gln-Glu-Glu-Glu-Glu-Glu-Thr-Ala-Gly-Ala-Pro-Gln-Gly-

         Leu-Phe-Arg-Gly-NH2
```

FIGURE 16.   (A) The structures of porcine, bovine, the rat galanin are compared using the three-letter notation for amino acid residues. The full sequence of porcine galanin is given, residues in the remaining species identical to those in the porcine molecule are indicated by a dash, and nonidentities are given in the three-letter identifier code. (B) The structures of porcine, human, and canine motilin are portrayed using the three-letter identifier code for amino acid residues. The full structure of porcine/human motilin is given, nonidentities between the canine molecule and the porcine/human molecule are shown using the three-letter code, and identities between molecules are indicated by a dash. (C) The structure of porcine pancreastatin is portrayed using the three-letter identifier code.

hormone levels. The specific but widespread distribution of galanin in the central nervous system of mammals[343,344,365] and its anatomical and functional relationship with acetylcholine[342,365,367] have been the object of ongoing investigations due to its potential involvement in memory processes and dementia.[368] In summary, galanin is a recently identified peptide which has proved to be of great interest due to its potent and interesting biological activities; it has been the subject of brief reviews.[333,368,369]

B. MOTILIN

Stimulated by early observations during investigations into the mechanism of excitation of gastric motility (see Reference 370 for review), Brown and colleagues[371] detected the presence of a motility exciting agent in chromatographic side fractions of Mutt's porcine intestinal extracts. The isolated peptide was named motilin, and sequence analysis revealed it to be a unique linear polypeptide with 22 amino acid residues, including an N-terminal phenylalanine residue and a C-terminal residue of glutamine.[372] Motilin was demonstrated to be contained in a specific type of endocrine cell concentrated in the upper small intestinal mucosa and, on administration to fasted dogs, to stimulate fundic and antral motility.[370] Subsequently, Poitras and colleagues[373] isolated and characterized motilin from extracts of canine small intestinal tissue extracts. Canine motilin differs from the porcine molecule at five positions, all substitutions occurring within the N-terminal 14 amino acids (see Figure 16B).

The structure of the porcine motilin precursor molecule was deduced from cloned cDNAs isolated from a library derived from porcine intestinal mRNA.[374] The porcine precursor molecule consists of 119 amino acid residues, with the motilin sequence occurring directly after an alanine residue which ends the signal peptide sequence, an arrangement similar to that described for the GRP precursor (*vide supra* and Chapter 3). As with a number of other precursor molecules, Poitras and colleagues[373] identified a unique C-terminal sequence of amino acid residues, separated from the motilin sequence by a pair of basic amino acids, which they called the motilin-associated peptide; the functional significance of this C-terminal sequence is at present unknown. Northern blot analysis revealed that specific motilin mRNA occurred only in the GI tract.[374] The structure of the human motilin precursor molecule and the sequence of human motilin were deduced from cDNA clones encoding for the precursor molecule, obtained from libraries derived from human duodenal mRNA[375,376] and from human genomic libraries.[377,378] The organization of the human precursor is similar to that of the porcine precursor, with the motilin sequence occurring immediately after the signal peptide sequence. Human motilin has an identical sequence identical to that of porcine motilin. The human motilin gene is encoded by a single gene located on chromosome 6 and consists of five exons. The first 14 amino acids of the motilin sequence are coded by exon 2 and the remainder of the C-terminal amino acid residues by exon 3. This is a situation similar to that seen for GRP, in which the coding sequence for a small biologically active peptide is not encoded by a single exon, but is interrupted by an intron (*vide supra*). Expression of the motilin gene was found in human and monkey GI tissues, and a significant amount of motilin mRNA was also present in certain extra-gastrointestinal tissues (i.e., adrenal gland, liver). However, motilin mRNA was not detected in pituitary or cerebellum.[377] Despite the occurrence of several molecular forms of motilin in mucosal extracts and in human serum, Northern blot analysis reveals the presence of a single mRNA species.[376] Given the presence of a single gene coding for motilin and a single mRNA species resulting from expression of the gene, Yano et al.[377] conclude that the multiple forms of motilin-like immunoreactivity must be generated during the posttranslational processing of the precursor molecule.

A number of controversies surround the functional aspects and the physiological role of motilin in different mammalian species. Although there appears to be clear evidence that motilin may play an important physiological role regulating motility in the dog,[379,380] the evidence for a similar physiological significance in other mammalian species, including man, is lacking (for review see Reference 370). There have been discordant reports as to the qualitative and quantitative amounts of motilin-like immunoreactivity in various species, and there is still controversy as to the nature of the motilin-like immunoreactivity detected in the central nervous system. These and other controversies have been reviewed recently,[370] and motilin has been the subject of a recent monograph.[381]

```
                    1                                    10
                 ┌────────S-S────────┐
Human CGRP-α    Ala-Cys-Asp-Thr-Ala-Thr-Cys-Val-Thr-His-Arg-Leu-Ala-Gly-Leu-Leu-Ser-Arg-
Human CGRP-β     -   -  Asn  -   -   -   -   -   -   -   -   -   -   -   -   -   -
Rat CGRP-α      Ser  -  Asn  -   -   -   -   -   -   -   -   -   -   -   -   -   -
Rat CGRP-β      Ser  -  Asn  -   -   -   -   -   -   -   -   -   -   -   -   -   -


                    20                                   30
Human CGRP-α    Ser-Gly-Gly-Val-Val-Lys-Asn-Asn-Phe-Val-Pro-Thr-Asn-Val-Gly-Ser-Lys-Ala-Phe-NH₂
Human CGRP-β     -   -   -  Met  -   -  Ser  -   -   -   -   -   -   -   -   -   -   -
Rat CGRP-α       -   -   -   -   -   -  Asp  -   -   -   -   -   -   -  Glu  -   -
Rat CGRP-β       -   -   -   -   -   -  Asp  -   -   -   -   -   -   -  Lys  -   -
```

FIGURE 17.   The structures of human cGRP-α, human cGRP-β, rat cGRP-α, and rat cGRP-β are given using the three-letter notation for amino acid residues. The full structure of human cGRP-α is given; identical residues in the remaining peptides to those in the human cGRP-α molecule are indicated by a dash, and nonidentities are shown using the three-letter notation system. The positions of the disulfide bonds are indicated.

C. CALCITONIN GENE-RELATED PEPTIDES (CGRP)

In 1982 Amara and colleagues[382] noted a variable rate of calcitonin production in rat medullary thyroid carcinomas (MTC) which occurred spontaneously during serial transplantation of the MTC lines. This conversion to low calcitonin production was associated with the appearance of a larger calcitonin cDNA-reactive mRNA which did not direct the synthesis of calcitonin, but encoded the sequence of a potentially novel peptide as deduced from the nucleotide sequence.[382] This peptide is composed of 37 amino acids and in the precursor molecule is flanked on the N-terminal side by a pair of basic amino acids and on the C-terminal side by the Gly–Arg–Arg sequence; this suggests that this peptide, the so-called calcitonin gene-related peptide (CGRP), would contain the C-terminal phenylalanine amide structure. Evidence was presented that expression of the two mRNAs was tissue specific; the mRNA coding for calcitonin, but not CGRP, predominated in the thyroid, while the mRNA coding for CGRP, but not calcitonin, predominated in neural tissue. It was suggested that the production of the two mRNAs occurred via alternative splicing events.[382] Shortly thereafter, the same group elaborated on the tissue-specific production of CGRP as measured by the expression of specific mRNAs and the presence of CGRP immunoreactivity in tissue.[383] It was demonstrated that the brain CGRP mRNA sequence directed the synthesis of CGRP-like immunoreactivity in a cell-free translation system and that CGRP-like immunoreactive material extracted from rat hypothalami contained a component which comigrated chromatographically with synthetic CGRP.[383]

Morris and colleagues[384] isolated the human form of CGRP from MTC tissue and demonstrated that the structure of human CGRP differed at four positions in its amino acid sequence from that of the rat molecule (see Figure 17). Amino acid substitutions occurred (human/rat) at position 1 (an Ala/Sir interchange), at position 3 (an Asp/Asn interchange), at position 25 (an Asn/Asp interchange), and at position 35 (a Lys/Glu interchange). Interestingly, these four amino acid sustitutions are not chemically conservative changes. The presence of a disulfide bridge and the C-terminal phenylalamine α-amide structure, which had been predicted by the molecular biology study, was proved by the isolation of the human CGRP molecule.[384] Shortly thereafter, two groups[385,386] confirmed the sequence of human CGRP by deducing the structure using cloned cDNAs from libraries constructed from MTC tissue. The presence of CGRP-like peptides, identified by chromatography and specific RIAs, was demonstrated in human thyroid, pituitary, and brain tissues.[387]

The specific distribution of CGRP-producing cells and neural pathways led to the suggestion by Rosenfeld and colleagues[383] that CGRP played a role in nociception, ingestive behavior, and modulation of the autonomic nervous and endocrine systems. Early functional studies focused on the intense effects of CGRP on cardiac and vascular smooth muscle,[388,389] but since then CGRP has also been noted to have intense relaxant effects on intestinal smooth muscle[390] and a number of potent pharmacological effects following intracerebral or peripheral i.v. administration.[391] Similarly, the specific anatomical locations of CGRP-like immunoreactivity and the binding sites for CGRP, in both the peripheral and central nervous systems, have been documented.[383,387,391-393] In the mammal, CGRP appears to be predominantly a neuropeptide, but CGRP-like immunoreactivity is also present in adrenal medullary cells and in the lungs. It has been suggested that circulating levels of CGRP in normal subjects arise from perivascular nerves rather than from endocrine-type cells.[394]

The CGRP story has become more complex. Following the first suggestion of the possible presence of two genes coding for CGRP in the human genome,[395] strong evidence for the existence of a second CGRP gene has been reported. Steenbergh and colleagues[396] reported the existence of a second gene encoding CGRP in a human MTC metastasis. The structure of the second human CGRP precursor was deduced from cDNA clones and was demonstrated to differ from the first elucidated human CGRP structure at three amino acid positions. The originally elucidated human CGRP is usually referred to as CGRP-α and the more recently elucidated CGRP structure as CGRP-β. Human CGRP-α differs from CGRP-β (CGRP-α/CGRP-β) at three positions; at position 3 an Asp/Asn interchange occurs, at position 22 a Val/Met interchange occurs, and at position 25 an Asn/Ser interchange occurs (see Figure 17). Evidence was also presented that the second gene was not located immediately adjacent to the first noted human calcitonin gene on chromosome 11.[396] At about the same time, Amara and colleagues[397] identified a second rat gene encoding for CGRP in rat medullary thyroid carcinomas. The β-CGRP gene generated an mRNA which was considerably divergent from that derived from the CGRP-α gene, but the region encoding the CGRP sequence in CGRP-β differed by only a single amino acid residue; at position 35, CGRP-α contains a Glu residue while CGRP-β contains a Lys residue, similar to the human CGRPs (see Figure 17). Evidence was presented that the CGRP-α and -β genes were each single copy and that the CGRP-β gene appeared to contain no sequencce with close homology to the calcitonin-coding exon present in the CGRP-α gene. As the 5'- and 3'- noncoding regions of the CGRP-β and CGRP-α mRNAs diverge significantly, the generation of specific hybridization probes was possible. The pattern of expression of the mature transcripts in the brain was consistent with previously reported immunohistochemical data. However, hybridization studies revealed a variation in the relative levels of the mRNAs encoding CGRP-α and -β, particularly in cranial nerve nuclei.[397] Hence, the present evidence strongly suggests that CGRP-β is a novel peptide coded for by a separate gene. Considerably more investigation is necessary to determine the functional significance of these findings.

X. CONCLUDING REMARKS

The 1950s and 1960s saw the beginning of the elucidation of the primary structures of regulatory peptides; in this regard, the pioneering studies of Mutt and Jorpes in Stockholm, Erspamer and colleagues in Rome, and Gregory and Tracy in Liverpool deserve special recognition. During this period the first primary structures of mammalian and amphibian skin peptides were elucidated, and it was realized that these peptides formed families of peptides based on structural homology. Also during this period, structure-function studies began which elucidated the regions of the peptide molecules important for biological activity. The 1970s saw an unprecedented number of "new" biologically active peptides "discovered" (see Chapter 1). This was partly due to increased technological development, but was also due to the use of novel methods of detection that investigators employed to determine the presence of such peptides, a prominent example being the use of strictly chemical techniques to discover and purify peptides. Of particular importance was the development of an assay to detect the presence in tissue extracts of peptides containing the C-terminal α-amide structure, which led to the characterization of a number of novel biologically active peptides. Knowledge of the primary structures of regulatory peptides, elucidated by classical peptide chemical techniques, is now routinely used to construct probes which identify mRNAs encoding precursor molecules for these peptides, and, in turn, the gene transcribing the mRNA can be identified. An unexpected result of this molecular biology approach was the recognition of certain amino acid residue sequences within these precursor molecules, apart from the known peptide, which appeared to be previously unrecognized regulatory peptides. The synthesis of some of these peptides was followed by the demonstration that they possessed potent biological activities and had characteristic tissue-specific distributions. In this manner, the calcitonin gene-related peptide, the glucagon-like peptides 1 and 2, and additional members of the endothelian family were "discovered". However, the presence of a novel peptide structure encoded by deduced nucleotide sequences remains a theoretical possibility until the putative peptide is isolated and characterized from tissues. Due to our incomplete understanding of the rules governing the production of bioactive peptides at both the transcriptional and translational levels, the precise beginning or end of a novel amino acid sequence recognized at the nucleotide level may remain uncertain until a peptide is characterized by classical peptide chemistry techniques. For example, the final assignment of the N-terminal Val residue for human GRP required the isolation of human GRP from tissue extracts. The tools of molecular biology have greatly expanded our ability to "discover" novel peptide structures and have enhanced our ability to study the tissue-specific expression of such peptides. However, peptide chemistry is eventually required to determine whether or not the final product is produced in tissues and in what specific form. Hence, both the molecular biology approach and classical peptide chemistry are required; the two approaches are complimentary and not mutually exclusive.

Since McDonald's last review of this topic approximately 5 years ago[398] a number of novel peptides have been recognized, and it is probable that this process will continue. The endothelin story, which began just over 2 years ago, has developed with great rapidity and is the subject of a continuing intense investigation (*vide supra*). Minamino and colleagues[399] isolated from porcine spinal cord two apparently novel pepties, designated neuromedin U or NMU. One peptide form was composed of 25 amino acid residues and contained the C-terminal Asn amide structure, and the second form, NMU_{18-25} was the C-terminal octapeptide amide of the larger molecule. Rat NMU_{1-25} has also been isolated from the small intestine and the primary structure determined;[400,401] there is a striking C-terminal structural conservation between rat and pig NMU, an area recognized as being of import for bioactivity. The significance of the tissue-specific distribution and the biological effects of neuromedin U are presently unknown, but are under investigation.[402] In 1977, Price and Greenberg[403] isolated a molluscan cardioexcitatory neuropeptide, the so-called FMRF-amide, which was named for its structure using the one-letter identifying codes for amino acids. $FRMF-NH_2$ was demonstrated to have biological activity on administration to mammals, and immunologic evidence has been reported that a similar peptide exists in mammals. Recently Yang and colleagues[404] isolated an octadecapeptide and an octapeptide from bovine brain which have C-terminal homology with each other and the same C-terminal dipeptide amide as the molluscan peptide. The physiological significance of these peptides and whether they are the mammalian equivalent to the molluscan FMRF-amide peptide are uncertain. The distribution of the mammalian peptides in the central nervous system has been described,[405] and the actions of $FMRF-NH_2$ and related peptides in mammals have been reviewed recently.[406] Schaller and Bodenmüller[407] isolated a peptide from the coelenterate *Hydra* (designated the *Hydra* head activator) and determined its sequence to be Pyr–Pro–Pro–Gly–Gly–Ser–Lys–Val–Ile–Leu–Phe; recently, they have isolated an apparently identical peptide from rat intestine and human hypothalamus.[408] The *Hydra* head activator peptide has been demonstrated to have bioactivity in mammals, but its physiological significance is unknown (for review see Reference 409 and citations therein). A recently isolated peptide, the monitor peptide, is composed of 61 amino acid residues and was first isolated from rat pancreatic juice; it possesses the capability of stimulating the release of cholecystokinin.[410,411]

Mutt's laboratory at the Karolinska Institute in Stockholm has continued to play a leading role in the detection of "new" peptides. Agerberth and colleagues[412] have recently isolated a peptide from porcine small intestinal tissue extracts which is composed of 60 amino acid residues and is capable of inhibiting glucose-induced insulin secretion from the isolated perfused pancreas preparation; the amino acid sequence of this peptide contains a 42% homology with the previously described pancreatic secretory trypsin inhibitor. The significance of this recently characterized peptide is at present unknown. Based on previous work which demonstrated that peptides isolated from insects, designated cecropins,[413] and the structurally unrelated amphibian skin peptides, designated magainins,[414] may play important roles in protection against infections, a series of studies was initiated which demonstrated the existence of endogenous mammalian antibacterial peptides. Lee et al.[415] isolated a 31-amino-acid-residue peptide from pig intestinal extracts which possessed intense antibacterial activity. Sequence analysis revealed structural homologies with the previously described insect peptides cecropin 1A and cecropin B; on the basis of the structural homologies, the porcine peptide was designated porcine cecropin P1.[415] An additional antibacterial fraction was evidently identified, but remains to be characterized. The finding of an endogenous mammalian peptide with intense antibacterial activity opens up a new and exciting field of investigation. The presence of large amounts of a peptide with the C-terminal glycine amide structure in porcine pancreatic extracts was detected by Tatemoto and colleagues.[416] The isolated peptide was found to be composed of 49 amino acid residues and was designated pancreastatin based on its ability to inhibit the first phase of insulin secretion.[416] The structure of porcine pancreastatin is shown in Figure 16C. Almost immediately thereafter, the striking structural similarity between pancreastatin and an internal sequence of bovine chromogranin A was noted,[417] and later the sequence of pancreastatin was found to be contained within the structure of porcine chromogranin A.[418] Pancreastatin was demonstrated to inhibit the first phase of rat insulin secretion[416] and in another report the second phase of rat insulin secretion,[419] but did not affect insulin secretion in the pig.[420] Recently, it has been reported that pancreastatin has a potent effect to inhibit secretion of parathyroid hormone from porcine parathyroid cells in culture.[421] Because chromogranin A has a widespread distribution in secretory granules of many endocrine and neuroendocrine cells and, in certain systems, is secreted from such cells, its role in endocrine and neuroendocrine physiology has been a subject of intense speculation and investigation. The isolation of pancreastatin has led to the question of whether or not chromogranin A acts as a prohormone for the production of pancreastatin.[417] This is an exciting line of investigation that obviously requires considerably more investigation to answer the above question. Schmidt et al.,[422] using a chemical approach to screen for regulatory peptides, isolated a novel peptide composed of 25 amino acid residues which has been designated valosin. Recently, valosin has been shown to stimulate gastric and exocrine pancreatic secretions and to alter intestinal myoelectric activity in the dog.[423]

The development of immunologic (RIA, immunohistochemistry) and molecular biology techniques made possible our recognition of the complexity of the regulation of the tissue-specific expression of regulatory peptides. RIA using antibodies which recognize specific regions of peptide molecules (i.e., region-specific antibodies), combined with high-resolution chromatographic separation techniques (particularly reversed-phase HPLC), led to the recognition that these regulatory peptides were virtually universally cleaved from larger precursor molecules and that the degree of such proteolytic processing varied between tissues in which the gene coding for the peptide was expressed. Furthermore, the degree of post-translational processing involving modification of specific amino acid residues (i.e., sulfation of tyrosine residues, the formation of the C-terminal amide) also varies between tissues, and such variability may alter the functional state of the peptide. For example, there is clear evidence of differential proteolytic precursor processing for the glucagon family of peptides. The glucagon precursor, depending on the tissue site, may be processed by differential proteolytic cleavages to produce glucagon, glicentin, or the glucagon-like peptides. From the earliest times, it was recognized that these biologically active peptides were present in specific tissues; hence, it was implicitly understood that tissue-specific expression of the genes coding for these peptides occurred. What had not been recognized until recently was the diversity of mRNA production which may occur during transcription, resulting in the production of varying amounts of different precursor molecules and, hence, the production of different mature peptides in tissues in which the gene was expressed. This became evident with elucidation of the tissue-specific expression of different precursor molecules encoded by the calcitonin gene, via alternate mRNA splicing events, calcitonin being predominantly expressed in C cells of the thyroid and CGRP in neural tissue. The multiple forms of precursor molecules for the tachykinins is another example of such alternate mRNA splicing events leading to differing precursor molecules.

We now recognize that the production of a biologically active peptide can be regulated at multiple levels: (1) at the level of gene expression;(2) by alternate splicing events occurring during transcription, producing mRNAs encoding for different precursor molecules; (3) by tissue-specific differences in posttranslational processing events, such as differences in both the degree of proteolytic processing and in posttranslational modification of certain specific residues; and (4) in certain cases, by variation in the rate of catabolism of a biologically active peptide to inactive fragments which also may occur within secretory granules as a regulated process (i.e., the level of serum calcium regulates the degree of intrasecretory granule catabolism of the parathyroid hormone).

This chapter provides an overview of the amino acid sequences of some of the recognized regulatory peptides and the variation of their sequences between species and a comparison of the structural relationships between different peptides. It must be emphasized that the elucidation of the primary structure of a bioactive peptide is only a beginning. We are now starting to elucidate the structures of the specific receptors which interact with these peptides and to determine the tertiary structures formed by the peptides in solutions. A greater understanding of the structural basis for ligand-receptor interaction will greatly enhance our ability to understand the physiological roles played by individual members of peptide families (i.e., the specific roles played by each of the tachykinins) and to design specific and potent agonists and antagonists to their actions. The development of long-acting somatostatin agonists has led to an important practical application in the treatment of inoperable malignant neuroendocrine tumors. The development of specific competitive antagonists to the actions of locally acting peptides, such as neuropeptides, will greatly enhance the elucidation of their physiological roles. Studies are beginning to explain certain events which are attendant upon the occupation of a specific receptor by its ligand, specifically, the relationship between receptor activation and the so-called second messenger systems and specific membrane channel structures which alter the ion flux across cell membranes, thereby altering the functional status of target cells. Beyond second messenger systems, the linkage between the activation of such systems and the resultant specific alterations in cell function are still shrouded in mystery. Recent technological developments are enhancing our understanding of the events linking the activation of specific membrane channels by ligand-receptor interactions with an altered functional state of cells. It is probable that the future of this field will also include the identification of further previously unrecognized biologically active peptides.

REFERENCES

1. **Bayliss, W. M. and Starling, E. H.** On the causation of the so-called "peripheral reflex secretion" of the pancreas, *Proc. R. Soc. London*, 69, 352, 1902.
2. **Edkins, J. S.,** On the chemical mechanism of gastric secretion, *Proc. R. Soc. London Ser. B*, 76, 376, 1905.
3. **Jorpes, J. E. and Mutt, V.,** On the biological activity and amino acid composition of secretin, *Acta Chem. Scand.*, 15, 1790, 1961.
4. **Mutt, V. and Jorpes, J. E.,** Secretin: isolation and determination of structure, *Proc. I.U.P.A.C. 4th Int. Congr. Natural Products*, Stockholm, Sweden, 1966, Section 2C-3.
5. **Gregory, R. A. and Tracy, H. J.** The constitution and properties of two gastrins extracted from hog antral mucosa, *Gut.* 5, 103, 1964.
6. **Tatemoto, K. and Mutt, V.,** Chemical determination of polypeptide hormones, *Proc. Natl. Acad. Sci. U.S.A.*, 75, 4115, 1978.
7. **Li, C. H.,** Beta-lipotropin, a new pituitary hormone, *Arch. Biol. Med. Exp.*, 5, 55, 1968.
8. **Kimmel, J. R., Hayden, L. J. and Pollock, H. G.,** Isolation and characterization of a new pancreatic polypeptide hormone, *J. Biol. Chem.*, 250, 9369, 1975.
9. **Chance, R. E. and Jones, W. E.,** Polypeptide from Bovine, Ovine, Human, and Porcine Pancreas, U.S. Patent 3,842,063, 1974.
10. **Steiner, D. F., Clark, J. L., Nolan, C., Rubenstein, A. H., Margoliash, E., Aten. B., and Oyer, P. E.,** Proinsulin and the biosynthesis of insulin, *Recent Prog. Horm. Res.*, 25, 207, 1969.
11. **Takabatake, Y. and Sachs, H.,** Vasopressin biosynthesis. III. In vitro studies. *Endocrinology*, 75, 934, 1964.
12. **Chrétien, M. and Li, C. H.,** Isolation, purification, and characterization of β-lipotropic hormone from sheep pituitary glands, *Can. Journal Biochem.*, 45, 1163, 1967.
13. **Rholam, M., Nicolas, P., and Cohen, P.,** Precursors for peptide hormones share common secondary structures forming features at the proteolytic processing sites, *FEBS Lett.*, 207, 1, 1986.
14. **Gluschankof, P., Gomez, S., Lepage, A., Créminon, C., Nyberg, F., Terenius, L., and Cohen, P.,** Role of peptide substrate structure in the selective processing of peptide prohormones at basic amino acid pairs by endoproteases, *FEBS Lett.*, 234, 149, 1988.
15. **Bradbury, A. F., Finnie, M. D. A., and Smyth, D. G.,** Mechanism of C-terminal amide formation by pituitary enzymes, *Nature (London)*, 298, 686, 1982.
16. **Eipper, B. A., Mains, R. E., and Glembotski, C. C.,** Identification in pituitary tissue of a peptide α-amidation activity that acts on glycine-extended peptides and requires molecular oxygen, copper, and ascorbic acid, *Proc. Natl. Acad. Sci. U.S.A.*, 80, 5144, 1983.
17. **Bleakman, A. and Smyth, D. G.,** Sequential processing reactions in the formation of hormone amides, *Eur. J. Biochem.*, 167, 161, 1987.
18. **Hilsted, L. and Rehfeld, J. F.,** α-Carboxyamidation of antral progastrin, *J. Biol. Chem.*, 262, 16953, 1987.
19. **Huttner, W. B.,** Protein tyrosine sulfation, *Trends Biochem. Sci.*, 12, 361, 1987.
20. **Bromer, W. W., Sinn, L. G., and Behrens, O. K.,** The amino acid sequence of glucagon. V. Location of the amide groups, acid degradation studies and summary of sequential evidence, *J. Am. Chem. Soc.*, 79, 2807, 1957.
21. **Said, S. I. and Mutt, V.,** Isolation from porcine intestinal wall of a vasoactive octacosapeptide related to secretin and to glucagon, *Eur. J. Biochem.*, 28, 199, 1972.
22. **Jörnvall, H., Carlquist, M., Kwauk, S., Otte, S., McIntosh, C., Brown, J., and Mutt, V.,** Amino acid sequence and heterogeneity of gastric inhibitory polypeptide (GIP), *FEBS Lett.*, 123, 205, 1981.
23. **Tatemoto, K. and Mutt, V.,** Isolation and characterization of the intestinal peptide porcine PHI (PHI-27), a new member of the glucagon-secretin family, *Proc. Natl. Acad. Sci. U.S.A.*, 78, 6603, 1981.
24. **Rivier, J, Spiess, J., Thorner, M., and Vale, W.,** Characterization of a growth hormone-releasing factor from a human pancreatic islet tumour, *Nature (London)*, 300, 276, 1982.
25. **Guillemin, R., Brazeau, P., Böhlen, P., Esch, F., Ling, N., and Wehrenberg, W. B.,** Growth hormone-releasing factor from a human pancreatic tumor that caused acromegaly, *Science*, 218, 585, 1982.
26. **Böhlen, P., Esch, F., Brazeau, P., Ling, N., and Guillemin, R.,** Isolation and characterization of the porcine hypothalamic growth hormone releasing factor, *Biochem. Biophys. Res. Commun.*, 116, 726, 1983.
27. **Spiess, J., Rivier, J., and Vale, W.,** Characterization of rat hypothalamic growth hormone-releasing factor, *Nature (London)*, 303, 532, 1983.
28. **Esch, F., Böhlen, P., Ling, N., Brazeau, P., and Guillemin, R.,** Isolation and characterization of the bovine hypothalamic growth hormone releasing factor., *Biochem. Biophys. Res. Commun.*, 117, 772, 1983.
29. **Parker, D. S., Raufman, J.-P, O'Donohue, T. L., Bledsoe, M., Yoshida, H., and Pisano, J. J.,** Amino acid sequences of helospectins, new members of the glucagon superfamily, found in gila monster venom, *J. Biol. Chem.*, 259, 11751, 1984.

30. **Hoshino, M., Yanaihara, C., Hong. Y.-M, Kishida, S., Katsumaru, Y., Vanermeers, A., Vanermeers-Piret, M.-C., Robberecht, P., Christophe, J., and Yanaihara, N.,** Primary structure of helodermin, a VIP-secretin-like peptide isolated from Gila monster venom, *FEBS Lett.,* 178, 233, 1984.
31. **Robberecht, P., De Graef, J., Woussen, M.-C, Vandermeers-Piret, M.-C. Vanermeers, A., De Neef, P., Cauvin, A., Yanaihara, C., Yanaihara, N., and Christophe, J.,** Immunoreactive helodermin-like peptides in the rat; a new class of mammalian neuropeptides related to secretin and VIP, *Biochem. Biophys. Res. Commun.,* 130, 333, 1985.
32. **Sutherland, E. W. and DeDuve, C.,** Origin and distribution of the hyperglycemic-glycogenolytic factor of the pancreas, *J. Biol. Chem.,* 75, 663, 1948.
33. **Doi, K., Prentki, M., Yip, C., Muller, W. A., Jeanrenaud, B., and Vranic, M.,** Identical biological effects of pancreatic glucagon and a purified moiety of canine gastric immunoreactive glucagon, *J. Clin. Invest.,* 63, 525, 1979.
34. **Hatton, T. W., Yip, C. C., and Vranic, M.,** Biosynthesis of glucagon (IRG³⁵⁰⁰) in canine gastric mucosa, *Diabetes,* 34, 38, 1985.
35. **Shinomura, Y., Eng, J., and Yalow, R. S.,** Immunoreactive glucagons purified from dog pancreas, stomach and ileum, *Regul. Peptides,* 23, 299, 1988.
36. **Unger, R. H. and Orci, L.,** Physiology and pathophysiology of glucagon, *Physiol. Rev.,* 56, 778, 1976.
37. **Sasaki, H., Ed..,** Many faces of glucagon, *Biomed. Res.,* 6 (Suppl.), 1, 1985.
38. **Tager, H. S. and Steiner, D. F.,** Isolation of a glucagon-containing peptide: primary structure of a possible fragment of proglucagon, *Proc. Natl. Acad. Sci. U.S.A.,* 70, 2321, 1973.
39. **Bataille, D., Coudray, A. M., Carlquist, M., Rosselin, G., and Mutt, V.,** Isolation of glucagon-37 (bioactive enteroglucagon/oxyntomodulin) from porcine jejunal-ileum; isolation of the peptide, *FEBS Lett.,* 146, 73, 1982.
40. **Bataille, D., Gespach, C., Coudray, A. M., and Rosselin, G.,** "Enteroglucagon": a specific effect on gastric glands isolated from the rat fundus: evidence for an "oxyntomodulin" action, *Biosci. Rep.,* 1, 151, 1981.
41. **Thim, L. and Moody, A. J.,** The primary structure of porcine glicentin (proglucagon). *Regul. Peptides,* 2, 139, 1981.
42. **Bell, G. I., Sanchez-Pescador, R., Laybourn, P. J., and Najarian, R. C.,** Exon duplication and divergence in the human preproglucagon gene, *Nature (London),* 304, 368, 1983.
43. **Bell, G. I., Santerre, R. F., and Mullenbach, G. T.,** Hamster preproglucagon contains the sequence of glucagon and two related peptides, *Nature (London),* 302, 716, 1983.
44. **Conlon, J. M., Hansen, H. F., and Schwartz, T. W.** Primary structure of glucagon and a partial sequence of oxyntomodulin (glucagon-37) from the guinea pig, *Regul. Peptides,* 11, 309, 1985.
45. **Lopez, L. C., Frazier, M. L., Su, C.-J., Kumar, A., and Saunders, G. F.,** Mammalian pancreatic proglucagon contains three glucagon-related peptides, *Proc. Natl. Acad. Sci. U.S.A.,* 80, 5485, 1983.
46. **Ørskov, C., Bersani, M., Johnsen, A. H., Højrup, P., and Holst, J. J.,** Complete sequences of glucagon-like peptide-1 from human and pig small intestine, *J. Biol. Chem.,* 264, 12826, 1989.
47. **Schmidt, W. E., Siegel, E. G., and Creutzfeldt, W.,** Glucagon-like peptide-1 but not glucagon-like peptide-2 stimulates insulin release from isolated rat pancreatic islets, *Diabetologia,* 28, 704, 1985.
48. **Drucker, D. J., Philippe, J., Mojsov, S., Chick, W. L., and Habener, J. F.,** Glucagon-like peptide I stimulates insulin gene expression and increases cyclic AMP levels in a rat islet cell line, *Proc. Natl. Acad. Sci. U.S.A.,* 84, 3434, 1987.
49. **Holst, J. J., Ørskov, C., Nielsen, O. V., and Schwartz, T. W.,** Truncated glucagon-like peptide I, an insulin-releasing hormone from the distal gut, *FEBS Lett.,* 211, 169, 1987.
50. **Suzuki, S., Kawai, K., Ohashi, S., Mukai, H., and Yamashita, K.,** Comparison of the effects of various C-terminal and N-terminal fragment peptides of glucagon-like peptide-1 on insulin and glucagon release from the isolated perfused rat pancreas, *Endocrinology,* 125, 3109, 1989.
51. **Bromer, W. W., Boucher, M, E., and Koffenbeiger, J. E., Jr.,** Amino acid sequence of bovine glucagon, *J. Biol. Chem.,* 246, 2822, 1971.
52. **Sundby, F. and Markussen, J.,** Rabbit glucagon: isolation, crystallisation and amino acid compositon, *Horm. Metab. Res.,* 4, 56, 1971.
53. **Sundby, F. and Markussen, J.,** Isolation, crystallisation and amino acid composition of rat glucagon, *Horm. Metab. Res.,* 3, 184, 1971.
54. **Thomsen, J., Kristiansen, K., Brunfeldt, K., and Sundby, F.,** The amino acid sequence of human glucagon, *FEBS Lett.,* 21, 315, 1972.
55. **Sundby, F., Markussen, J., and Danbo, W.,** Camel glucagon: isolation, crystallisation and amino acid composition, *Horm. Metab. Res.,* 6, 425, 1974.
56. **Pollock, H. G. and Kimmel, J. R.,** Chicken glucagon: isolation and amino acid sequence studies, *J. Biol. Chem.,* 250, 9377, 1975.
57. **Markussen J., Frandsen, E. K., Heding, L. G., and Sundby, F.,** Turkey glucagon: crystallisation, amino acid composition and immunology, *Horm. Metab. Res.,* 4, 360, 1972.

58. **Sundby, F., Frandsen, E. K., Thomsen, J., Kristiansen, K., and Brunfeldt, K.,** Crystallisation and amino acid sequence of duck glucagon, *FEBS Lett.*, 26, 289, 1972.

59. **Lance, V., Hamilton, J. W., Rouse, J. B., Kimmel, J. R., and Pollock, H. G.,** Isolation and characterization of reptilian insulin, glucagon and pancreatic polypeptide: complete amino acid sequence of alligator (*Alligator mississippiensis*) insulin and pancreatic polypeptide, *Gen. Comp. Endocrinol.*, 55, 112, 1984.

60. **Pollock, H. G., Kimmel, J. R., Ebner, K. E., Hamilton, J. W., Rouse, J. B., Lance, V., and Rawitch, A. B.,** Isolation of alligator Gar (*Lepisosteus spatua*) glucagon, oxyntomodulin, and glucagon-like peptide: amino acid sequences of oxyntomodulin and glucagon-like peptide, *Gen. Comp. Endocrinol.*, 69, 133, 1988.

61. **Pollock, H. G., Hamilton, J. W., Rouse, J. B., Ebner, K. E., and Rawitch, A. B.,** Isolation of peptide hormones from the pancreas of the bullfrog (*Rana catesbeiana*), *J. Biol. Chem.*, 263, 9746, 1988.

62. **Yu, J.-H., Eng, J., Rattan, S., and Yalow, R. S.,** Opossum insulin, glucagon and pancreatic polypeptide: amino acid sequences, *Peptides*, 10, 1195, 1989.

63. **Frandsen, E. K., Gronvald, F. C., Heding, L. G., Johansen, N. L., Lundt, B. F., Moody, A. J., Markussen, J., and Volund, A.,** Glucagon: structure-function relationships investigated by sequence deletions, *Hoppe-Seyler's Z. Physiol. Chem.*, 362, 665, 1981.

64. **Unson, C. G., Gurzenda, E. M., Iwasa, K., and Merrifield, R. B.,** Glucagon antagonists: contribution to binding and activity of the amino-terminal sequence 1—5, position 12, and the putative alpha-helical segment 19—27, *J. Biol. Chem.*, 264, 789, 1989.

65. **Mutt, V.,** Secretin and Cholecystokinin, in *Gastrointestinal Hormones*, Mutt, V., Ed., Advances in Metabolic Disorders, Vol. II, Academic Press, New York, 1988, 251.

66. **Carlquist, M., Jörnvall, H., and Mutt, V.,** Isolation and amino acid sequence of bovine secretin, *FEBS Lett.*, 127, 71, 1981.

67. **Carlquist, M., Jörnvall, H., Forssmann, W.-G, Thulin, L., Johansson, C., and Mutt, V.,** Human secretin is not identical to the porcine/bovine hormone, *IRCS Med. Sci. Biochem.*, 13, 217, 1985.

68. **Gossen, D., Vandermeers, A., Vandermeers-Piret, M.-C., Rathé, J., Cauvin, A., Robberecht, P., and Christophe, J.,** Isolation and primary structure of rat secretin, *Biochem. Biophys. Res. Commun.*, 160, 862, 1989.

69. **Shinomura, Y., Eng, J., and Yalow, R. S.,** Dog secretin: sequence and biologic activity, *Life Sci.*, 41, 1243, 1987.

70. **Nilsson, A., Carlquist, M., Jörnvall, H., and Mutt, V.,** Isolation and characterization of chicken secretin, *Eur. J. Biochem.*, 112, 383, 1980.

71. **Dimaline, R. and Dockray, G. J.,** Potent stimulation of the avian exocrine pancreas by porcine and chicken vasoactive intestinal peptide, *J. Physiol. (London)*, 294, 153, 1979.

72. **Gafvelin, G., Carlquist, M., and Mutt, V.,** A proform of secretin with high secretin-like bioactivity, *FEBS Lett.*, 184, 347, 1985.

73. **Said, S. I. and Mutt, V.,** Polypeptide with broad biological activity: isolation from small intestine, *Science*, 169, 1217, 1970.

74. **Brown, J. C., Mutt, V., and Pederson, R. A.,** Further purification of a polypeptide demonstrating enterogastrone activity, *J. Physiol. (London)*, 209, 57, 1970.

75. **Brown, J. C. and Dryburgh, J. R.,** A gastric inhibitory polypeptide II: the complete amino acid sequence, *Can. J. Biochem.*, 49, 867, 1971.

76. **Carlquist, M., Mutt, V., and Jörnvall, H.,** Isolation and characterization of bovine vasoactive intestinal peptide (VIP), *FEBS Lett.*, 108, 457, 1979.

77. **Carlquist, M., McDonald., T. J., Go, V. L. M., Bataille, D., Johansson, C., and Mutt, V,.** Isolation and amino acid composition of human vasoactive intestinal polypeptide (VIP), *Horm. Metab. Res.*, 14, 28, 1982.

78. **Itoh, N., Obata, K., Yanaihara, N., and Okamoto, H.,** Human preprovasoactive intestinal polypeptide contains a novel PHI-27-like peptide, PHM-27, *Nature (London)*, 304, 547, 1983.

79. **Dimaline, R., Reeve, Jr., J. R., Shively, J. E., and Hawke, D.,** Isolation and characterization of rat vasoactive intestinal peptide, *Peptides*, 5. 183, 1984.

80. **Du, B.-H, Eng, J., Humes, J. D., Chang, M., Pan, Y.-C. E., and Yalow, R. S.,** Guinea pig has a unique mammalian VIP, *Biochem. Biophys. Res. Commun.*, 128, 1093, 1985.

81. **Nilsson, A.,** Structure of the vasoactive intestinal octacosapeptide from chicken intestine. The amino acid sequence, *FEBS Lett.*, 60, 322, 1975.

82. **Dimaline, R., Young, J., and Gregory, H.,** Isolation from chicken antrum, and primary amino acid sequence of a novel 36-residue peptide of the gastrin/CCK family, *FEBS Lett.*, 205, 318, 1986.

83. **Said, S. I.,** "Vasoactive Intestinal Peptide", in *Gastrointestinal Hormones*, Mutt, V., Ed., Advances in Metabolic Disorders, Vol. 11, Academic Press, New York, 1988, 369.

84. **Said, S. I.,** Vasoactive intestinal polypeptide (VIP): current status, *Peptides*, 5, 143, 1984.

85. **Christophe, J., Svoboda, M., Waelbroeck, M., Winand, J., and Robberecht, P.,** Vasoactive intestinal peptide receptors in pancreas and liver. Structure-function relationship, *Ann. N.Y. Acad. Sci.,* 527, 238, 1988.

86. **Christophe, J., Rosselin, G., Said, S. I., and Yanaihara, N., Eds.,** First international symposium on VIP and related peptides, *Peptides,* 5, 1, 1984.

87. **Moody, A. J., Thim, L., and Valverde, I.,** The isolation and sequencing of human gastric inhibitory peptide (GIP), *FEBS Lett.,* 172, 142, 1984.

88. **Carlquist, M., Maletti, M., Jörnvall, H., and Mutt, V.,** A novel form of gastric inhibitory polypeptide (GIP) isolated from bovine intestine using a radioreceptor assay, *Eur. J. Biochem.,* 145, 573, 1984.

89. **Polak, J. M., Bloom, S. R., Kuzio, M., Brown, J. C., and Pearse, A. G. E.,** Cellular localization of gastric inhibitory polypeptide in the duodenum and jejunum, *Gut,* 14, 284, 1973.

90. **Creutzfeldt, W.,** The incretin concept today, *Diabetologia,* 16, 75, 1979.

91. **Maxwell, V., Shulkes, A., Brown, J. C., Solomon, R., Walsh, J., and Grossman, M.,** Effect of gastric inhibitory polypeptide on pentagastrin-stimulated acid secretion in man, *Dig. Dis. Sci.,* 25, 113, 1980.

92. **Yamagishi, T. and Debas, H. T.,** Gastric inhibitory polypeptide (GIP) is not the primary mediator of the enterogastrone action of fat in the dog, *Gastroenterology,* 78, 931, 1980.

93. **Dupré, J., Ross, S. A., Watson, D., and Brown, J. C.** Stimulation of insulin secretion by gastric inhibitory polypeptide in man, *J. Clin. Endocrinol. Metab.,* 37, 826, 1973.

94. **Brown, J. C., Dryburgh, J. R., Ross, S. A., and Dupré, J.,** Identification and actions of gastric inhibitory polypeptide, *Recent Prog. Horm. Res.,* 31, 487, 1975.

95. **Andersen, D. K.,** "Physiological Effects of GIP in Man", in *Gut Hormones,* 2nd ed., Bloom, S. R. and Polak, J. M., Eds., Churchill Livingstone, Edinburgh, 1981, 256.

96. **Tatemoto, K. and Mutt, V.,** Isolation and characterization of the intestinal peptide porcine PHI (PHI-27), a new member of the glucagon-secretin family, *Proc. Natl. Acad. Sci. U.S.A.* 78, 6603, 1981.

97. **Tatemoto, K.,** PHI— a new brain-gut peptide, *Peptides,* 5, 151, 1982.

98. **Christofides, N. D., Yiangou, Y., Blank, M. A., Tatemoto, K., Polak, J. M., and Bloom, S. R.,** Are peptide histidine isoleucine and vasoactive intestinal peptide co-synthesised in the same pro-hormone?, *Lancet,* 2, 1398, 1982.

99. **Tatemoto, K., Jörnvall, H., McDonald, T. J., Carlquist, M., Go, V. L. W., Johansson, C., and Mutt, V.,** Isolation and primary structure of human PHI (peptide HI), *FEBS Lett.,* 174, 258, 1984.

100. **Carlquist, M., Kaiser, R., Tatemoto, K., Jörnvall, H., and Mutt, V.,** A novel form of PHI isolated in high yield from bovine upper intestine, *Eur. J. Biochem.,* 144, 243, 1984.

101. **Yiangou, Y., Requejo, F., Polak, J. M., and Bloom, S. R.,** Characterization of a novel prepro VIP derived peptide, *Biochem. Biophys. Res. Commun.,* 139, 1142, 1986.

102. **Yiangou, Y., Di Marzo, V., Spokes, R. A., Panico, M., Morris, H. R., and Bloom, S. R.,** Isolation, characterization, and pharmacological actions of peptide histidine valine 42, a novel prepro-vasoactive intestinal peptide-derived peptide, *J. Biol. Chem.,* 262, 14010, 1987.

103. **Cauvin, A., Vandermeers, A., Vandermeers-Piret, M.-C., Rathe, J., Robberecht, P., and Christophe, J.,** Peptide histidine isoleucinamide (PHI)-(1—27)-Gly as a new major form of PHI in the rat small intestine, *Endocrinology,* 125, 1296, 1989.

104. **Tatemoto, K.,** PHI— a new brain-gut peptide, *Peptides,* 5, 151, 1984.

105. **Wehrenberg, W. B. and Ling, N.,** In vivo biological potency of rat and human growth hormone-releasing factor, *Biochem. Biophys. Res. Commun.,* 115, 525, 1983.

106. **Guillemin, R., Brazeau, P., Böhlen, P., Esch, F., Ling, N., Wehrenberg, W. B., Bloch, B., Mougin, C., Zeytin, F., and Baird, A.,** Somatocrinin, the growth hormone releasing factor, *Recent Prog. Horm. Res.,* 40, 233, 1984.

107. **Vandermeers, A., Vandermeers-Piret, M.-C, Robberecht, P., Waelbroeck, M., Dehaye, J.-P., Winand, J., and Christophe, J.,** Purification of a novel pancreatic secretory factor (PSF) and a novel peptide with VIP- and secretin-like properties (helodermin) from Gila monster venom, *FEBS Lett.,* 166, 273, 1984.

108. **Robberecht, P., Waelbroeck, M., Dehaye, J.-P., Winand, J., Vandermeers, A., Vandermeers-Piret, M.-C., and Christophe, J.,** Evidence that helodermin, a newly extracted peptide from Gila monster venom, is a member of the secretin/VIP/PHI family of peptides with an original pattern of biological properties, *FEBS Lett.,* 166, 277, 1984.

109. **Cox, H. M. and Cuthbert, A. W.,** Secretory actions of vasoactive intestinal polypeptide, peptide histidine isoleucine and helodermin in rat small intestine: the effects of putative VIP antagonists upon VIP-induced ion secretion, *Regul. Peptides,* 26, 127, 1989.

110. **Robberecht, P., De Neef, P., Gourlet, P., Cauvin, A., Coy, D. H., and Christophe, J.,** Pharmacological charaterization of the novel helodermine/VIP receptor present in human SUP-T$_1$ lymphoma cell membranes, *Regul. Peptides,* 26, 117, 1989.

111. **Bjartell, A., Persson, P., Absood, A., Sundler, F., and Håkanson, R.,** Helodermin-like peptides in noradrenaline cells of adrenal medulla, *Regul. Peptides,* 26, 27, 1989.

112. **Tracy, H. J. and Gregory, R. A.** Physiological properties of a series of synthetic peptides structurally related to gastrin I, *Nature (London)*, 204, 935, 1964.

113. **Morley, J. S.,** Structure-function relationships in gastrin-like peptides, *Proc. R. Soc. London Ser. B*, 170, 97, 1968.

114. **Walsh, J. H. and Grossman, M. I.,** Gastrin. I, *N. Engl. J. Med.*, 292, 1324, 1975.

114a. **Walsh, J. H. and Grossman, M. I.,** Gastrin. II, *N. Engl. J. Med.*, 292, 1377, 1975.

115. **Gregory, R. A., Tracy, H. J., and Grossman, M. I.,** Isolation of two gastrins from human antral mucosa, *Nature (London)*, 209, 583, 1966.

116. **Beacham, J., Bentley, P., Gregory, R. A., Kenner, G. W., MacLeod, J. K., and Sheppard, R. C.,** Synthesis of human gastrin I, *Nature (London)*, 209, 585, 1966.

117. **Agarwal, K. L., Beacham, J., Bentley, P. H., Gregory, R. A., Kenner, G. W., Sheppard, R. C., and Tracy, H. J.,** Isolation, structure, and synthesis of ovine and bovine gastrin, *Nature (London)*, 219, 614, 1968.

118. **Gregory, R. A., Tracy, H. J., Grossman, M. I., de Valois, M., and Lichter, R.,** Isolation of canine gastrin, *Experientia*, 25, 345, 1969.

119. **Agarwal, K. L., Kenner, G. W., and Sheppard, R. C.,** Structure and synthesis of canine gastrin. *Experientia*, 25, 346, 1969.

120. **Agarwal, K. L., Kenner, G. W., and Sheppard, R. C.,** Feline gastrin. An example of peptide sequence analysis by mass spectrometry. *J. Am. Chem. Soc.*, 91, 3096, 1969.

121. **Jiang, R., Huebner, V. D., Lee, T. D., Chew, P., Ho, F. J., Shively, J. E., Walsh, J. H., and Reeve, J. E., Jr.,** Isolation and characterization of rabbit gastrin, *Peptides*, 9, 763, 1988.

122. **Reeve, J. R., Jr., Dimaline, R,. Shiveley, J. E., Hawke, D., Chew, P., and Walsh, J. H.,** Unique amino terminal structure of rat little gastrin, *Peptides*, 2, 453, 1981.

123. **Schaffer, M. H., Agarwal, K. L. and Noyes, B. E.,** Rat gastrin's amino acid sequence determined from the nucleotide sequence of the mRNA, *Peptides*, 3, 693, 1982.

124. **Bonato, C., Eng. J., Pan, Y.-C. E., Chang, M., Hulmes, J. D., and Yalow, R. S.,** Guinea pig "little" gastrin is a hexadecapeptide, *Life Sci.*, 37, 2563, 1986.

125. **Bonato, C., Eng. J., Pan, Y.-C. E., Miedel, M., Hulmes, J. D., and Yalow, R. S.,** Guinea pig 33-amino acid gastrin, *Life Sci.*, 39, 959, 1986.

126. **Yalow, R. S. and Berson, S. A.,** Size and charge distinctions between endogenous human plasma gastrin in peripheral blood and heptadecapeptide gastrins, *Gastroenterology*, 58, 609. 1970.

127. **Gregory, R. A. and Tracy, H. J.,** Isolation of two "big gastrins" from Zollinger-Ellison tumour tissue, *Lancet*, 2, 797, 1972.

128. **Dockray, G. J., Gregory, R. A., and Kenner, G. W.,** Immunochemical differences between natural and synthetic big gastrins, *Gastroenterology*, 75, 556, 1978.

129. **Gregory, R. A., Dockray, G. J., Reeve, J. R., Jr., Shiveley, J. E., and Miller, C.,** Isolation from porcine antral mucosa of a hexapeptide corresponding to the C-terminal sequence of gastrin, *Peptides*, 4, 319, 1983.

130. **Gregory, R. A. and Tracy, H. J.,** Isolation of two minigastrins from Zollinger-Ellison tumour tissue, *Gut*, 15, 683, 1974.

131. **Eysselein, V. E., Reeve., J. R., Jr., Shively, J. E., Miller, C. and Walsh, J. H.,** Isolation of a large cholecystokinin precursor from canine brain, *Proc. Natl. Acad. Sci. U.S.A.*, 81, 6565, 1984.

132. **Reeve, J. R., Jr., Eysselein, V., Walsh, J. H., Ben-Avram, C. M., and Shively, J. E.,** New molecular forms of cholecystokinin, *J. Biol. Chem.*, 261, 16392, 1986.

133. **Rehfeld, J. F. and Hansen, H. F.,** Characterization of preprocholecystokinin products in the porcine cerebral cortex, *J. Biol. Chem.*, 261, 5832, 1986.

134. **Yoo, O. J., Powell, C. T., and Agarwal, K. L.,** Molecular cloning and nucleotide sequence of full-length cDNA coding for porcine gastrin, *Proc. Natl. Acad. Sci. U.S.A.*, 79, 1049, 1982.

135. **Kato, K., Hayashizaki, Y., Takahashi, Y., Himenu, S., and Matsubara, K.,** Molecular cloning of the human gastrin gene, *Nucleic Acids Res.*, 11, 8197, 1983.

136. **Wiborg, O., Bergland, L., Boel, E., Norris, F., Norris, K., Rehfeld, J. F., Marcker, K. A., and Vuust, J.,** Structure of a human gastrin gene, *Proc. Natl, Acad. Sci. U.S.A.*, 81, 1067, 1984.

137. **Rehfeld, J. F.,** Accumulation of nonamidated preprogastrin and preprocholecystokinin products in porcine pituitary corticotrophs, *J. Biol. Chem.*, 261, 5841, 1986.

138. **Jorpes, J. E. and Mutt, V.,** Cholecystokinin and pancreozymin, one single hormone?, *Acta Physiol. Scand.*, 66, 196, 1966.

139. **Mutt, V. and Jorpes, J. E.,** Structure of porcine cholecystokinin-pancreozymin. I. Cleavage with throbin and with trypsin, *Eur. J. Biochem.*, 6, 156, 1968.

140. **Anastasi, A., Erspamer, V., and Endean, R.,** Isolation and structure of caerulein, an active decapeptide from the skin of *Hyla caerulea*, *Experientia*, 23, 699, 1967.

141. **Mutt, V.,** Behaviour of secretin, cholecystokinin and pancreozymin to oxidation with hydrogen peroxide, *Acta Chem. Scand.* 18, 2185, 1964.

142. **Dimaline, R., Young, J., and Gregory, H.,** Isolation from chicken antrum, and primary amino acid sequence of a novel 36-residue peptide of the gastrin/CCK family, *FEBS Lett.* 205, 318, 1986.
143. **Mutt, V.,** Further investigations on intestinal hormonal polypeptides, *Clin. Endocrinol.,* 5 (Suppl.), 175s, 1976.
144. **Dockray, G. J., Gregory, R. A., Hutchison, J. B., Harris, J. I., and Runswick, M. J.,** Isolation, structure and biological activity of two cholecystokinin octapeptides from sheep brain, *Nature (London),* 274, 711, 1978.
145. **Eng, J., Shiina, Y., Pan, Y.-C. E., Blacher, R., Chang, M., Stein, S., and Yalow, R. S.,** Pig brain contains cholecystokinin octapeptide and several cholecystokinin desoctapeptides, *Proc. Natl. Acad. Sci. U.S.A.,* 80, 6381, 1983.
146. **Eysselein, V. E., Reeve, J. R., Jr., Shively, J. E., Miller, C., and Walsh, J. H.,** Isolation of a large cholecystokinin precursor from canine brain, *Proc. Natl. Acad. Sci. U.S.A.,* 81, 6565, 1984.
147. **Miller, L. J., Jardine, I., Weissman, E., Go, V. L. M., and Speicher, D.,** Characterization of cholecystokinin from the human brain, *J. Neurochem.,* 43, 835, 1984.
148. **Reeve, J. R., Jr., Eysselein, V. E., Walsh, J. H., Sankaran, H., Deveney, C. W., Tourtellotte, W. W., Miller, C., and Shively, J. E.,** Isolation and characterization of biologically active and inactive cholecystokinin-octapeptides from human brain, *Peptides,* 5, 959, 1984.
149. **Reeve, J. R., Jr., Eysselein, V., Walsh, J. H., Ben-Avram, C. M., and Shively, J. E.** New molecular forms of cholecystokinin, *J. Biol. Chem.,* 261, 16392, 1986.
150. **Eng, J., Du, B.-H., Pan, Y.-C. E., Chang, M., Humes, J. D., and Yalow, R. S.,** Purification and sequencing of a rat intestinal 22 amino acid C-terminal CCK fragment, *Peptides,* 5, 1203, 1984.
151. **Zhou, Z.-Z., Eng, J., Pan, Y.-C. E., Chang, M., Hulmes, J. D., Raufman, J.-P, and Yalow, R. S.,** Unique cholecystokinin peptides isolated from guinea pig intestine, *Peptides,* 6, 337, 1985.
152. **Fan, Z.-W., Eng, J., Miedel, M., Hulmes, J. D., Pan Y.-C. E., and Yalow, R. S.,** Cholecystokinin octapeptides purified from chinchilla and chicken brains, *Brain Res. Bull.,* 18, 757, 1987.
153. **Fan, Z.-W., Eng, J,. Shaw, G., and Yalow, R. S.,** Cholecystokinin octapeptide purified from brains of Australian marsupials, *Peptides,* 9, 429, 1987.
154. **Tatemoto, K., Jörnvall, H., Siimesmaa, S., Halldén, G., and Mutt, V.,** Isolation and characterization of cholecystokinin-58 (CCK-58) from porcine brain, *FEBS Lett.,* 174, 289, 1984.
155. **Turkelson, C. M., Solomon, T. E., Bussjaeger, L., Turkelson, J., Ronk, M., Shively, J. E., Ho, F. J., and Reeve, J. R., Jr.,** Chemical characterization of rat cholecystokinin-58, *Peptides,* 9, 1255, 1989.
156. **Eysselein, V. E., Reeve, J. R., Jr., Shively, J. E., Hawke, D., and Walsh, J. H.,** Partial structure of a large canine cholecystokinin (CCK$_{58}$): amino acid sequence, *Peptides,* 3, 687, 1982.
157. **Carlquist, M., Mutt, V., and Jörnvall, H.,** Characterization of two novel forms of cholecystokinin isolated from bovine upper intestine. *Regul. Peptides,* 11, 27, 1985.
158. **Friedman, J., Schneider, B. S., and Powell, D.,** Differential expression of the mouse cholecystokinin gene during brain and gut development, *Proc. Natl. Acad. Sci. U.S.A.,* 82, 5593, 1985.
159. **Gubler, U., Chua, A. O., Hoffman, B. J., Collier, K. J., and Eng. J.,** Cloned cDNA to cholecystokinin mRNA predicts an identical preprocholecystokinin in pig brain and gut, *Proc. Natl. Acad. Sci. U.S.A.,* 81, 4307, 1984.
160. **Takahashi, Y., Kato, K., Hayashizaki, Y., Wakabayashi, T., Ohtsuka, E., Matsuki, S., Ikehara, M., and Matsubara, K.,** Molecular cloning of the human cholecystokinin gene by use of a synthetic probe containing deoxyinosine, *Proc. Natl. Acad. Sci. U.S.A.,* 82, 1931, 1985.
161. **Deschenes, R. J., Lorenz, L. J., Haun, R. S., Roos, B. A., Collier, K. J., and Dixon J. E.,** Cloning and sequence analysis of a cDNA encoding rat preprocholecystokinin, *Proc. Natl. Acad. Sci. U.S.A.,* 81, 726, 1984.
162. **Kuwano, R., Araki, K., Usui, H., Fukui, T., Ohtsuka, E., Ikenhara, M., and Takahashi, Y.,** Molecular cloning and nucleotide sequence of cDNA coding for rat brain cholecystokinin precursor, *J. Biochem.,* 96, 923, 1984.
163. **Deschenes, R. J., Narayana, S. V. L., Argos, P., and Dixon, J. E.,** Primary structural comparison of the preprohormones cholecystokinin and gastrin, *FEBS Lett.,* 182, 135, 1985.
164. **Eng, J., Gubler, U., Raufman, J.-P., Chang, M., Hulmes, J. D., Pan, Y.-C. E., and Yalow, R. S.,** Cholecystokinin-associated COOH-terminal peptides are fully sulfated in pig brain. *Proc. Natl. Acad. Sci. U.S.A ,* 83, 2832, 1986.
165. **Deschenes, R. J., Narayana, S. V. L., Argos, P., and Dixon, J. E.,** Primary structural comparison of the preprohormones cholecystokinin and gastrin, *FEBS Lett.,* 182, 135, 1985.
166. **Lund, T., van Kessel, A. H. M. G., Haun, S., and Dixon, J. E.,** The genes for human gastrin and cholecystokinin are located on different chromosomes, *Hum. Genet.,* 73, 77, 1986.
167. **Eberlein, G. A., Eysselein, V. E., and Goebell, H.,** Cholecystokinin-58 is the major molecular form in man, dog and cat but not in pig, beef and rat intestine, *Peptides,* 9, 993, 1988.

168. **Eysselein, V. E., Eberlein, G. A., Hesse, W. H., Singer, M. V., Goebell, H., and Reeve, J. R., Jr.,** Cholecystokinin-58 is the major circulating form of cholecystokinin in canine blood, *J. Biol. Chem.*, 262, 214, 1987.

169. **Nachman, R. J., Holman, G. M., Haddon, W. F., and Ling, N.,** Leucosulfakinin, a sulfated insect neuropeptide with homology to gastrin and cholecystokinin, *Science*, 234, 71, 1986.

170. **Nachman, R. J., Holman, G. M., Cook, B. J., Haddon, W. F., and Ling, N.,** Leucosulfakinin-II, a blocked sulfated insect neuropeptide with homology to cholecystokinin and gastrin, *Biochem. Biophys. Res. Commun.*, 140, 357, 1986.

171. **Veenstra, J. A.,** Isolation and structure of two gastrin/CCK-like neuropeptides from the American cockroach homologous to the leucosulfakinins, *Neuropeptides*, 14, 145, 1989.

172. **Kimmel, J. R., Hayden, L. J., and Pollock, H. G.,** Isolation and characterization of a new pancreatic polypeptide hormone, *J. Biol. Chem.*, 250, 9369, 1975.

173. **Kimmel, J. R., Pollock, H. G., and Hazelwood, R. L.** Isolation and characterization of chicken insulin, *Endocrinology*, 83, 1323, 1968.

174. **Chance, R. E., Moon, N. E., and Johnson, M. G.** Human pancreatic polypeptide (HPP) and bovine pancreatic polypeptide (BBP), in *Methods of Hormone Radioimmunoassay*, Jaffe, B. M. and Behrman, H. R., Eds., Academic Press, New York, 1979, 657.

175. **Tatemoto, K.,** Isolation and characterization of peptide YY (PYY), a candidate gut hormone that inhibits pancreatic exocrine secretion, *Proc. Natl. Acad. Sci. U.S.A.*, 79, 2514, 1982.

176. **Tatemoto, K.,** Neuropeptide Y: complete amino acid sequence of the brain peptide, *Proc. Natl. Acad. Sci. U.S.A.*, 79, 5485, 1982.

177. **Tatemoto, K., Siimesmaa, S., Jornvall, H., Allen, J. M., Polak, J. M., Bloom, S. R., and Mutt, V.,** Isolation and characterization of neuropeptide Y from porcine intestine, *FEBS Lett.*, 179, 181, 1985.

178. **Glover, I. D., Barlow, D. J., Pitts, J. E., Wood, S. P., Tickle, I. J., Blundell, T. L., Tatemoto, K., Kimmel, J. R., Wollmer, A., Strassburger, W., and Zhang, Y.-S.,** Conformational studies of the pancreatic polypeptide hormone family, *Eur. J. Biochem.*, 142, 379, 1985.

179. **Minakata, J., Taylor, J. W., Walker, M. W., Miller, R. J., and Kaiser, E. T.,** Characterization of amphiphilic secondary structure in neuropeptide Y through the design, synthesis, and study of model peptides, *J. Biol. Chem.*, 264, 7907, 1989.

180. **Servin, A. L., Rouyer-Fessard, C., Balasubramaniam, A., Saint Pierre, S., and Laburthe, M.,** Peptide-YY and neuropeptide-Y inhibit vasoactive intestinal peptide-stimulated adenosine $3',5'$-monophosphate production in rat small intestine: structural requirements of peptides for interacting with peptide-YY-preferring receptors. *Endocrinology*, 124, 692, 1989.

181. **Schwartz, T. W.,** Pancreatic polypeptide: a hormone under vagal control, *Gastroenterology*, 85, 1411, 1983.

182. **Lundberg, J. M., Tatemoto, K., Terenius, L., Hellström, P. M., Mutt, V., Hökfelt, T., and Hamberger, B.,** Localization of peptide YY (PYY) in gastrointestinal endocrine cells and effects on intestinal blood flow and motility, *Proc. Natl. Acad. Sci. U.S.A.* 79, 4471, 1982.

183. **Böttcher, G., Sjölund, K., Ekblad, E., Håkanson, R., Schwartz, T. W., and Sundler, F.,** Coexistence of peptide YY and glicentin immunoreactivity in endocrine cells of the gut, *Regul. Peptides*, 8, 261, 1984.

184. **Taylor, I. L.,** Distribution and release of peptide YY in dog measured by specific radioimmunoassay, *Gastroenterology*, 88, 731, 1985.

185. **Lundberg, J. M., Terenius, L., Hökfelt, T., Martling, C. R., Tatemoto, K., Mutt, V., Polak, J., Bloom, S. R., and Goldstein, M.,** Neuropeptide Y (NPY)-like immunoreactivity in peripheral noradrenergic neurons and effects of NPY on sympathetic function, *Acta Physiol. Scand.*, 116, 477, 1982.

186. **Furness, J. B., Costa, M., Emson, P. C., Håkanson, R., Moghimzadeh, E., Sundler, F., Taylor, I. L., and Chance, R. E.,** Distribution, pathways and reactions to drug treatment of nerves with neuropeptide Y- and pancreatic polypeptide-like immunoreactivity in the guinea-pig digestive tract, *Cell Tissue Res.*, 234, 71, 1983.

187. **Ekblad, E., Edvinsson, L., Wahlestedt, C., Uddman, R., Håkanson, R., and Sundler, F.,** Neuropeptide Y co-exists and co-operates with noradrenaline in perivascular nerve fibres, *Reg. Peptides*, 8, 225, 1984.

188. **Varndell, I. M., Polak, J. M., Allen, J. M., Terenghi, G., and Bloom, S. R.,** Neuropeptide tyrosine (NPY) immunoreactivity in norepinephrine-containing cells and nerves of the mammalian adrenal gland, *Endocrinology*, 114, 1460, 1984.

189. **Adrian, T. E., Allen, J. M., Terenghi, G., Bacarese-Hamilton, A. J., Brown, M. J., Polak, J. M., and Bloom, S. R.,** Neuropeptide Y in phaeochromocytomas and ganglioneuroblastomas, *Lancet*, 2, 540, 1983.

190. **Lin, T. M. and Chance, R. E.,** Candidate hormones of the gut. VI. Bovine pancreatic polypeptide (BPP) and avian pancreatic polypeptide (APP), *Gastroenterology*, 67, 737, 1974.

191. **Chance, R. E. and Jones, W. E.,** Polypeptides from Bovine, Ovine and Porcine Pancreas, U.S. Patent Office publ. X-3097 A, 22:3, 824, 063.

192. **Kimmel, J. R., Pollock, H. G., Chance, R. E., Johnson, M. G., Reeve, J. R., Jr., Taylor, I. L., Miller, C., and Shively, J. E.**, Pancreatic polypeptide from rat pancreas, *Endocrinology*, 114, 1725, 1984.
193. **Schwartz, T. W. and Tager, H. S.**, Isolation and biogenesis of a new peptide from pancreatic islets, *Nature (London)*, 294, 589, 1981.
194. **Schwartz, T. W., Hansen, H. F., Håkanson, R., Sundler, F., and Tager, H. S.**, Human pancreatic icosapeptide: isolation, sequence, and immunocytochemical localization of the COOH-terminal fragment of the pancreatic polypeptide precursor, *Proc. Natl. Acad. Sci. U.S.A.*, 81, 708, 1984.
195. **Boel, E., Schwartz, T. W., Norris, K. E., and Fiil, N. P.**, A cDNA encoding a small common precursor for human pancreatic polypeptide and pancreatic icosapeptide, *EMBO J.*, 3, 909, 1984.
196. **Leiter, A. B., Keutmann, H. T., and Goodman, R. H.**, Structure of a precursor to human pancreatic polypeptide, *J. Biol. Chem.*, 259, 14702, 1984.
197. **Takeuchi, T. and Yamada, T.**, Isolation of a cDNA clone encoding pancreatic polypeptide, *Proc. Natl. Acad. Sci. U.S.A.*, 82, 1536, 1985.
198. **Yamamoto, H., Nata, K., and Okamoto, H.**, Mosaic evolution of prepropancreatic polypeptide, *J. Biol. Chem.*, 261, 6156, 1986.
199. **Corder, R., Gaillard, R. C., and Böhlen, P.**, Isolation and sequence of rat peptide YY and neuropeptide Y, *Regul. Peptides*, 21, 253, 1988.
200. **Tatemoto, K., Nakano, I., Makk, G., Angwin, P., Mann, M., Schilling, J., and Go, V. L. W.**, Isolation and primary structure of human peptide YY, *Biochem. Biophys. Res. Commun.*, 167, 713, 1988.
201. **Eberlein, G. A., Eysselein, V. E., Schaeffer, M., Layer, P., Grandt, D., Goebell, H., Niebel, W., Davis, M., Lee, T. D., Shively, J. E., and Reeve, J. R., Jr.**, A new molecular form of PYY: structural characterization of human PYY(3-36) and PYY(1-36), *Peptides*, 10, 797, 1989.
202. **Harper, A. A., Hood, A. J. C., Mushens, J., and Smy, J. R.**, Pancreotone, an inhibitor of pancreatic secretion in extracts of ileal and colonic mucosa, *J. Physiol.*, 292, 455, 1979.
203. **Adrian, T. E., Savage, A. P., Sagor, G. R., Allen, J. M., Bacarese-Hamilton, A. J., Tatemoto, K., Polak, J. M., and Bloom, S. R.**, Effect of peptide YY on gastric, pancreatic, and biliary function in humans, *Gastroenterology*, 89, 494, 1985.
204. **Corder, R., Emson, P. C., and Lowry, P. J.**, Purification and characterization of human neuropeptide Y from adrenal-medullary phaeochromocytoma tissue, *Biochem. J.*, 219, 699, 1984.
205. **Minth, C. D., Bloom, S. R., Polak, J. M., and Dixon, J. E.**, Cloning, characterization, and DNA sequence of a human cDNA encoding neuropeptide tyrosine, *Proc. Nat. Acad. Sci. U.S.A.*, 81, 4577, 1984.
206. **O'Hare, M. M. T., Tenmoku, S., Aakerlund, L., Hilsted, L., Johnsen, A., and Schwartz, T. W.**, Neuropeptide Y in guinea pig, rabbit, rat and man. Identical amino acid sequence and oxidation of methionine-17, *Regul. Peptides*, 20, 293, 1988.
207. **Larhammar, D., Ericsson, A., and Håkan, P.**, Structure and expression of the rat neuropeptide Y gene, *Proc. Natl. Acad. Sci. U.S.A.*, 84, 2068, 1987.
208. **Tatemoto, K.**, Neuropeptide Y: isolation, structure, and function, in *Neuropeptide Y*, Mutt, V., Hökfelt, T., Fuxe, K., and Lundberg, J. M., Eds., Raven Press, New York, 1989, 13.
209. **Sillard, R., Agerberth, B., Mutt, V., and Jörnvall, H.**, Sheep neuropeptide Y. A third structural type of a highly conserved peptide, *FEBS Lett.*, 258, 263, 1989.
210. **Allen, Y. S., Adrian, T. E., Allen, J. M., Tatemoto, K., Crow, T. J., and Bloom, S. R.**, Neuropeptide Y distribution in the rat brain, *Science*, 221, 877, 1983.
211. **Furness, J. B., Costa, M., and Keast, J. R.**, Choline acetyltransferase and peptide immunoreactivity of submucous neurons in the small intestine of the guinea-pig, *Cell Tissue Res.*, 237, 239, 1984.
212. **Andrews, P. C., Hawke, D., Shively, J. E., and Dixon, J. E.**, A nonamidated peptide homologous to porcine peptide YY and neuropeptide YY, *Endocrinology*, 116, 2677, 1985.
213. **Kimmel, J. R., Plisetskaya, E. M., Pollock, H. G., Hamilton, J. W., Rouse, J. B., Ebner, K. E., and Rawitch, A. B.**, Structure of a peptide from Coho salmon endocrine pancreas with homology to neuropeptide Y, *Biochem. Biophys. Res. Commun.*, 141, 1084, 1986.
214. **Carraway, R. E. and Leeman, S. E.**, The isolation of a new hypotensive peptide, neurotensin, from bovine hypothalami, *J. Biol. Chem.*, 19, 6854, 1973.
215. **Carraway, R. and Leeman, S. E.**, The amino acid sequence of a hypothalamic peptide, neurotensin, *J. Biol. Chem.*, 250, 1907, 1975.
216. **Kitabgi, P., Carraway, R., and Leeman, S. E.**, Isolation of a tridecapeptide from bovine intestine tissue and its partial characterization as neurotensin, *J. Biol. Chem.*, 251, 7053, 1976.
217. **Carraway, R. E., and Leeman, S. E.**, Structural requirements for the biological activity of neurotensin, a new vasoactive peptide, in *Peptides: Chemistry, Structure and Biology*, Walter, R., and Meienhofer, J., Eds., Ann Arbor Science, Ann Arbor, MI, 1975, 679.
218. **Kitabgi, P., Poustis, C., Granier, C., Van Reitschoten, J., Rivier, J., Morgat, J. L., and Freychet, P,** Neurotensin binding to extraneural and neural receptors: comparison with biological activity and structure-activity relationships, *Mol. Pharmacol.*, 18, 11, 1980.

219. **Araki, K., Tachibana, S., Uchiyama, M., Nakajima, T., and Yasuhara, T.,** Isolation and structure of a new active peptide "Xenopsin" on the smooth muscle, especially on a strip of fundus from a rat stomach, from the skin of *Xenopus laevis, Chem. Pharm. Bull. (Tokyo),* 21, 2801, 1973.
220. **Hammer, R. A., Leeman, S. E., Carraway, R., and Williams, R. H.,** Isolation of human intestinal neurotensin, *J. Biol. Chem.,* 255, 2476, 1980.
221. **Dobner, P. R., Barber, D. L., Villa-Komaroff, L., and McKiernan, C.,** Cloning and sequence analysis of cDNA for the canine neurotensin/neuromedin N precursor, *Proc. Natl. Acad. Sci. U.S.A.,* 84, 3516, 1987.
222. **Kislauskis, E., Bullock, B., McNeil, S., and Dobner, P. R.,** The rat gene encoding neurotensin and Neuromedin N. *J. Biol. Chem.,* 263, 4963, 1988.
223. **Carraway, R. and Bhatnagar, Y. M.,** Isolation, structure and biologic activity of chicken intestinal neurotensin, *Peptides,* 1, 167, 1980.
224. **Carraway, R. E. and Ferris, C. F.,** Isolation, biological and chemical characterization, and synthesis of a neurotensin-related hexapeptide from chicken intestine, *J. Biol. Chem.,* 258, 2475, 1983.
225. **Minamino, N., Kangawa, K., and Matsuo, H.,** Neuromedin N: a novel neurotensin-like peptide identified in porcine spinal cord, *Biochem. Biophys. Res. Commun.,* 122, 542, 1984.
226. **Carraway, R. E., Mitra, S. P., and Ferris, C. F.,** Pepsin treatment of mammaliam plasma generates immunoreactive and biologically active neurotensin-related peptides in micromolar concentrations, *Endocrinology,* 119, 1519, 1986.
227. **Mogard, M. H., Kobayashi, R., Chen, C. -F., Lee, T. D., Reeve, J. R., Jr., Shively, J. E., and Walsh, J. J.,** The amino acid sequence of kinetensin, a novel peptide isolated from pepsin-treated human plasma: homology with human serum albumin, neurotensin and angiotensin, *Biochem. Biophys. Res. Commun.,* 136, 983, 1986.
228. **Carraway, R. E., Mitra, S. P., and Cochrane, D. E.,** Structure of a biologically active neurotensin-related peptide obtained from pepsin-treated albumin(s), *J. Biol. Chel.,* 262, 5968, 1987.
229. **Carraway, R. E., and Reinecke, M.** Neurotensin-like peptides and a novel model of the evolution of signaling systems, in *Evolution and Tumour Pathology of the Neuroendocrine System,* Falkmer, S., Håkanson, R., and Sundler, F., Eds., Elsevier, Amsterdam, 1984, 245.
230. **Armstrong, M. J. and Leeman, S. E.,** Neurotensin and substance P, in Gastrointestinal Hormones, Mutt, V., Ed., Academic Press, San Diego, Vol. 11, *Adv. Metab. Disord.,* 1988, 469.
231. **von Euler, U. S. and Gaddum, J. H.,** An unidentified depressor substance in certain tissue extracts *J. Physiol.* 72, 74, 1931.
232. **Pernow, B.,** Studies on substance P. Purification, occurrence and biological actions, *Acta Physiol. Scand.,* 29 (Suppl. 105), 1, 1953.
233. **Erspamer, V.,** The tachykinin peptide family, *Trends Neurol. Sci.,* 4, 267, 1981.
234. **Chang, M. M. and Leeman, S. E.,** Isolation of a sialogogic peptide from bovine hypothalamic tissue and its characterization as substance P, *J. Biol. Chem.,* 245, 4784, 1970.
235. **Lembeck, F. and Starke, K.,** Substanz P und Speichelsekretion, *Naunyn-Schmiedebergs Arch. Pharmakol. Exp. Pathol.,* 259, 375, 1968.
236. **Chang, M. M., Leeman, S. E., and Niall, H. D.,** Amino-acid sequence of substance P, *Nature New Biol.,* 232, 86, 1971.
237. **Studer, R. O., Trzeciak, A., and Lergier, W.,** Isolierung und Aminosauresequenz von Substanz P aus Pferdearm., *Helv. Chim. Acta,* 56, 860, 1973.
238. **Henry, J. L., Couture, R., Cuello, A. C., Pelletier, G., Quirion, R., and Regoli, D.,** *Substance P and neurokinins,* in *Proc. Substance P and Neurokinins — Montréal '86, Satellite Symp. 30th Int. Congr. Int. Union Physiological Sciences,* Springer-Verlag, Berlin, 1987.
239. **Daniel, E. E., Collins, S. M., Fox, J. -A., E. T., and Huizinga, J. D.,** Pharmacology of neuroendocrine peptides, in *Handbook of Physiology,* Vol. 20, The Gastrointestinal System I, Williams & Wilkins, Baltimore, 1989, 759.
240. **Kimura, S., Okada, M., Sugita, Y., Kanazawa, I., and Munekata, E.,** Novel neuropeptides, neurokinin-α and β, isolated from porcine spinal cord, *Proc. Jpn. Acad., Ser. B,* 59, 101, 1983.
241. **Nawa, H., Hirose, T., Takashima, H., Inayama, S., and Nakanishi, S.,** Nucleotide sequences of cloned cDNAs for two types of bovine brain substance P precursor, *Nature (London),* 306, 32, 1983.
242. **Minamino, N., Kangawa, K., Fukuda, A., and Matsuo, H.,** Neuromedin L: a novel mammalian tachykinin identified in porcine spinal cord, *Neuropeptides,* 4, 157, 1984.
243. **Henry, J. L.,** Discussions of nomenclature for tachykinins and tachykinin receptors, in *Substance P and Neurokinins, Proc. Substance P and Neurokinins — Montréal '86, Satellite Symp. 30th Int. Cong. Int. Union Physiological Sciences,* Springer-Verlag, Berlin, 1987, 1.
244. **Kangawa, K., Minamino, N., Fukuda, A., and Matuso, H.,** Neuromedin K: a novel mammalian tachykinin identified in porcine spinal cord, *Biochem. Biophys. Res. Commun.,* 114, 533, 1983.
245. **Conlon, J. M., Deacon, C. F., O'Toole, L., and Thim, L.,** Scyliorhinin I and II: two novel tachykinins from dogfish gut, *FEBS Lett.,* 200, 111, 1986.

246. **Murphy, R.,** Primary amino acid sequence of guinea-pig substance P, *Neuropeptides,* 14, 105, 1989.
247. **Bonner, T. I., Young, A. C., and Affolter, H. -H.,** Cloning and expression of rat and human tachykinin genes, in *Substance P and Neurokinins, Proc. Substance P and Neurokinins — Montréal '86, Satellite Symp. 30th Int. Congr. Int. Union Physiological Sciences,* Springer-Verlag, Berlin, 1987, 17.
248. **Harmar, A. J., Armstrong, A., Pascall, J. C., Chapman, K., Rosie, R., Curtis, A., Going, J., Edwards, C. R. W. and Fink, G.,** cDNA sequence of human B-preprotachykinin, the common precursor to substance P and neurokinin A, *FEBS Lett.,* 208, 67, 1986.
249. **Kawaguchi, Y., Hoshimaru, M., Nawa, H., and Nakanishi, S.,** Sequence analysis of cloned cDNA for rat substance P precursor: existence of a third substance P precursor, *Biochem. Biophys. Res. Commun.,* 139, 1040, 1986.
250. **Krause, J. E., Chirgwin, J. M., Carter, M.S., Xu, Z. S., and Hershey, A. D.,** Three rat preprotachykinin mRNAs encode the neuropeptides substance P and neurokinin A, *Proc. Natl. Acad. Sci. U.S.A.,* 84, 881, 1987.
251. **MacDonald, M. R., McCourt, D. W., and Krause, J. E.,** Posttranslational processing of α-, β-, and γ- preprotachykinins, *J. Biol. Chem.,* 263, 15176, 1988.
252. **Kaga, R., McGregor, G. P., Thim, L., and Conlon, J. M.,** Neuropeptide-γ: a peptide isolated from rabbit intestine that is derived from γ-preprotachykinin, *J. Neurochem.,* 50, 1412, 1988.
253. **Kotani, H., Hoshimaru, M., Nawa, H., and Nakanishi, S.,** Structure and gene organization of bovine neuromedin K precursor, *Proc. Natl. Acad. Sci. U.S.A.,* 83, 7074, 1986.
254. **Nawa, H. Kotani, H., and Nakanishi, S.,** Tissue-specific generation of two preprotachykinin mRNAs from one gene by alternative RNA splicing, *Nature (London),* 312, 729, 1984.
255. **Nakanishi, S.,** Structure and regulation of the preprotachykinin gene, *Trends Neurol. Sci.,* 9, 41, 1986.
256. **Tatemoto, K., Lundberg, J. M., Jörnvall, H., and Mutt, V.,** Neuropeptide K: isolation, structure and biological activities of a novel brain tachykinin, *Biochem. Biophys, Res. Commun.,* 128, 947, 1985.
257. **Deacon, C. F., Agoston, D. V., Nau, R., and Conlon, J. M.,** Conversion of neuropeptide K to neurokinin A and vesicular colocalization of neurokinin A and substance P in neurons of the guinea pig small intestine, *J. Neurochem.,* 48, 141, 1987.
258. **Takeda, Y. and Krause, J. E.,** γ-Preprotachykinin-(72—92)-peptide amide potentiates substance P-induced salivation, *Eur. J. Pharmacol.,* 161, 267, 1989.
259. **Arai, and Emson, P. C.,** Regional distribution of neuropeptide K and other tachykinins (neurokinin A, neurokinin B and substance P) in rat central nervous system, *Brain Res.,* 399, 240, 1986.
260. **McDonald, T. J., Christofi, F. L., Brooks, B. D., Barnett, W., and Cook, M. A.,** Characterization of content and chromatographic forms of neuropeptides in purified nerve varicosities prepared from guinea pig myenteric plexus, *Regul. Peptides,* 21, 69, 1988.
261. **Hua, X. -Y., Theodorsson-Norheim, E., Brodin, E., Lundberg, J. M., and Hökfelt, T.,** Multiple tachykinins (neurokinin A, neuropeptide K and substance P) in capsaicin-sensitive sensory neurons in the guinea-pig, *Regul. Peptides,* 12, 1, 1985.
262. **McDonald, T. J., Ahmad, S., Allescher, H. -D., Kostka, P., Daniel, E. E., Barnett, W., and Brodin, E.,** Canine myenteric, deep muscular, and submucosal plexus preparations of purified nerve varicosities: content and chromatographic forms of certain neuropeptides, *Peptides,* 11, 95, 1990.
263. **Brodin, E., Lindefors, N., Dalsgaard, C. -J., Theodorsson-Norheim, E., and Rossell, S.,** Tachykinin multiplicity in rat central nervous system as studied using antisera raised against substance P and neurokinin A, *Regul. Peptides,* 13, 253, 1986.
264. **Buscics, A., Holzer, P., Lippe, I. T., Pabst, M. A., and Lembeck, F.,** Density distribution of guinea pig myenteric plexus nerve endings containing immunoreactive substance P, *Peptides,* 7, 761, 1986.
265. **Toresson, G., Brodin, E., Wahlstrom, A., and Bertilsson, L.,** Detection of N-terminally extended substance P but not of substance P in human cerebrospinal fluid: quantitation with HPLC-radioimmunoassay, *J. Neurochem.,* 50, 1701, 1988.
266. **Too, H. -P., Cordova, J. L., and Maggio, J. E.,** A novel radioimmunoassay for neuromedin K. I. Absence of neuromedin K-like immunoreactivity in guinea pig ileum and urinary bladder. II. Heterogeneity of tachykinins in guinea pig tissues, *Regul. Peptides,* 26, 93, 1989.
267. **Goedert, M. and Hunt, S. P.,** The cellular localization of preprotachykinin A messenger RNA in the bovine nervous system, *Neuroscience,* 22, 983, 1987.
268. **Warden, M. K. and Young, W. S., II,** Distribution of cells containing mRNAs encoding substance P and neurokinin B in the rat central nervous system, *J. Comp. Neurol.,* 272, 90, 1988.
269. **Sternini, C., Anderson, K., Frantz, G., Krause, J. E., and Brecha, N.,** Expression of substance P/ neurokinin A-encoding preprotachykinin messenger ribonucleic acids in the rat enteric nervous system, *Gastroenterology,* 97, 348, 1989.
270. **Minamino, N., Masuda, H., Kangawa, K., and Matsuo, H.,** Regional distribution of neuromedin K and neuromedin L in rat brain and spinal cord, *Biochem. Biophys. Res. Commun.,* 124, 731, 1984.
271. **Yokota, T., Sasai, T., Tanaka, K., Fujiwara, T., Tsuchida, K., Shigemoto, R., Kakizuka, A., Ohkubo, H., and Nakanishi, S.,** Molecular characterization of a functional cDNA for rat substance P receptor, *J. Biol. Chem.,* 264, 17649, 1989.

272. **Sasai, Y. and Nakanishi, S.** Molecular characterization of rat substance K receptor and its mRNAs, *Biochem. Biophys. Res. Commun.* 165, 695, 1989.
273. **Nakajima, T., Tanimura, T., and Pisano, J. J.,** Isolation and structure of a new vasoactive peptide, *Fed. Proc.* 29, 282, 1970.
274. **Erspamer, V., Falconieri Erspamer, G., and Inselvini, M.,** Some pharmacological actions of alytesin and bombesin, *J. Pharm. Pharmacol.,* 22, 275, 1970.
275. **Anastasi, A., Erspamer, V., Bucci, M.,** Isolation and structure of bombesin and alytesin, two analogous active peptides from the skin of the European amphibians *Bombina* and *Alytes, Experientia,* 27, 166, 1971.
276. **Erspamer, V. and Melchiorri, P.,** Actions of bombesin on secretions and motility of the gastrointestinal tract, in: *Gastrointestinal Hormones,* Thompson, J. C., Ed., University of Texas Press, Austin, London 1975, 575.
277. **McDonald, T. J., Jörnvall, H., Nilsson, G., Vagne, M., Ghatei, M., Bloom. S. R., and Mutt, V.,** Characterization of a gastrin-releasing peptide from porcine non-antral gastric tissue, *Biochem. Biophys. Res. Commun.* 90, 227, 1979.
278. **McDonald, T. J., Jörnvall, H., Ghatei, M., Bloom, S. R., and Mutt, V.,** Characterization of an avian gastric (proventricular) peptide having sequence homology with the porcine gastrin releasing peptide and the amphibian peptides bombesin and alytesin, *FEBS Lett.,* 122, 45, 1980.
279. **Broccardo, M., Falconieri Erspamer, G., Melchiorri, P., Negri, L., and De Castiglione, R.,** Relative potency of bombesin-like peptides, *Br. J. Pharmacol.,* 55, 221, 1975.
280. **Marki, W., Brown, M., and Rivier, J. E.,** Bombesin analogs: effects on thermoregulation and glucose metabolism, *Peptides,* 2, 169, 1981.
281. **McDonald, T. J., Ghatei, M. A., Bloom, S. R., Adrian, T. E., Mochizuki, T., Yanaihara, C., and Yanaihara, N.,** Dose-response comparisons of canine plasma gastroenteropancreatic hormone responses to bombesin and the porcine gastrin-releasing peptide (GRP), *Regul. Peptides,* 5, 125, 1983.
282. **Erspamer, V., Go, V. L. W., Battey, J. F., Bloom, S. R., McDonald, T. J., Makhlouf, G. M., Minamino, N., Negri, L., Reeve, J. R., Jr., Rivier, J., Spindel, E. R. Taché, Y., and Walsh, J. H.,** Nomenclature meeting, *Ann. N.Y. Acad. Sci.,* 547, 1, 1988.
283. **Reeve, J. R., Jr., Walsh, J. H., Chew, P., Clark, B., Hawke, D., and Shively, J. E.,** Amino acid sequences of three bombesin-like peptides from canine intestine extracts, *J. Biol. Chem.,* 258, 5582, 1983.
284. **Orloff, M. S., Reeve, J. R., Jr., Miller Ben-Avram, C., Shively, J. E., and Walsh, J. H.,** Isolation and sequence analysis of human bombesin-like peptides, *Peptides,* 5, 865, 1984.
285. **Shaw, C., Thim, L., and Conlon, J. M.,** Primary structure and tissue distribution of guinea pig gastrin-releasing peptide, *J. Neurochem.,* 49, 1348, 1987.
286. **Conlon, J. M., Henderson, I. W., and Thim, L.,** Gastrin-releasing peptide from the intestine of the elasmobranch fish, *Scyliorhinus canicula* (common dogfish), *Gen. Comp. Endocrinol.,* 68, 415, 1987.
287. **Spindel, E. R., Chin, W. W., Price, J., Rees, L. H., Besser, G. M., and Habener, J. F.,** Cloning and characterization of cDNAs encoding human gastrin-releasing peptide, *Proc. Natl. Acad. Sci. U.S.A.,* 81, 5699, 1984.
288. **Sausville, E. A., Lebacq-Verheyden, A. -M., Spindel, E. R., Cuttitta, F., Gazdar, A. F., and Battey, J. F.,** Expression of the gastrin-releasing peptide gene in human small cell lung cancer, *J. Biol. Chemi.,* 261, 2451, 1986.
289. **Lebacq-Verheyden, A. -M., Krystal, G., Sartor, O., Way, J., and Battey, J. F.,** The rat preprogastrin releasing peptide gene is transcribed from two initiation sites in the brain, *Mol. Endocrinol.,* 2, 556, 1988.
290. **Mazzanti, G., Falconieri Erspamer, G., and Piccinelli, D.,** Relative potencies of porcine bombesin-like heptacosapeptide (PB-27), amphibian bombesin (B-14) and litorin, and bombesin C-terminal nonapeptide (B-9) on *in vitro* and *in vivo* smooth muscle preparations, *J. Pharm. Pharmacol.,* 34, 120, 1982.
291. **Girard, F., Aubé, C., St.-Pierre, S., and Jolicoeur, F. B.,** Structure-activity studies on neurobehavioral effects of bombesin (BB) and gastrin releasing peptide (GRP), *Neuropeptides,* 3, 443, 1983.
292. **McDonald, T. J.,** The gastrin releasing polypeptide, in Gastrointestinal Hormones, Mutt, V., Ed., Academic Press, San Diego, Vol. 11, *Adv. Metab. Disord.,* 1988, 199.
293. **Yoshizaki, D., de Bock, V., and Solomon, S.,** Origin of bombesin-like peptides in human fetal lung, *Life Sci.,* 34, 835, 1984.
294. **Minamino, N., Kangawa, K., and Matsuo, H.,** Neuromedin C: a bombesin-like peptide identified in porcine spinal cord, *Biochem. Biophys. Res. Commun.,* 119, 14, 1984.
295. **Spindel, E. R., Zilberberg, M. D., Habener, J. F., and Chin, W. W.,** Two prohormones for gastrin-releasing peptide are encoded by two mRNAs differing by 19 nucleotides, *Proc. Natl. Acad. Sci. U.S.A.,* 83, 19, 1986.
296. **Spindel, E. R., Zilberberg, M. D., and Chin. W. W.,** Analysis of the gene and multiple messenger ribonucleic acids (mRNAs) encoding human gastrin-releasing peptide: alternate RNA splicing occurs in neural and endocrine tissue, *Mol. Endocrinol.,* 1, 224, 1987.
297. **Cuttitta, F., Fedorko, J., Gu, J., Lebacq-Verheyden, A. -M., Linnoila, R. I., and Battey, J. F.,** Gastrin-releasing peptide gene-associated peptides are expressed in normal human fetal lung and small cell lung cancer: a novel peptide family found in man, *J. Clin. Endocrinol. Metab.,* 67, 576, 1988.

298. **Naylor, S. I., Sakaguchi, A. Y., Spindel, E., and Chin, W. W.,** Human gastrin-releasing peptide gene is located on chromosome 18, *Somatic Cell. Mol. Gene.,* 13, 87, 1987.

299. **Lebacq-Verheyden, A.-M., Bertness, V., Kirsch, H., Hollis, G. F., McBride, O. W., and Battey, J.,** Human gastrin-releasing peptide gene maps to chromosome band 18q21, *Somatic Cell Mol. Genet.,* 13, 81, 1987.

300. **Erspamer, V.,** Discovery, isolation, and characterization of bombesin-like peptides, *Ann. N. Y. Acad. Sci.,* 547, 3, 1988.

301. **Minamino, N., Kangawa, K., and Matsuo, H.,** Neuromedin B: a novel bombesin-like peptide identified in porcine spinal cord, *Biochem. Biophys. Res. Commun.,* 114, 541, 1983.

302. **Minamino, N., Sudoh, T., Kangawa, K., and Matsuo, H.,** Neuromedin B-32 and B-30: two ''big'' neuromedin B identified in porcine brain and spinal cord, *Biochem. Biophys. Res. Commun.,* 130, 685, 1985.

303. **Minamino, N., Kangawa, K., and Matsuo, H.,** Neuromedin B is a major bombesin-like peptide in rat brain: regional distribution of neuromedin B and neuromedin C in rat brain, pituitary and spinal cord, *Biochem. Biophys. Res. Commun.,* 124, 925, 1984.

304. **Namba, M., Ghatei, M. A., Bishop, A. E., Gibson, S. J., Mann, D. J., Polak, J. M., and Bloom, S. R.,** Presence of neuromedin B-like immunoreactivity in the brain and gut of rat and guinea pig, *Peptides,* 6, 257, 1986.

305. **Domin, J., Polak, J. M., and Bloom, S. R.,** The distribution and biological effects of neuromedins B and U, *Ann. N. Y. Acad. Sci.,* 547, 391, 1988.

306. **Krane, I. M., Naylor, S. L., Helin-Davis, D., Chin, W. W., and Spindel, E. R.,** Molecular cloning of cDNAs encoding the human bombesin-like peptide neuromedin B, *J. Biol. Chem.,* 263, 13317, 1988.

307. **Taché, Y., Melchiorri, P., and Negri, L.,** Bombesin-like peptides in health and disease, *Ann. N. Y. Sci.,* 547, 1, 1988.

308. **Negri, L., Improta, G., Broccardo, M., and Melchiorri, P.,** Phyllolitorins: a new family of bombesin-like peptides, *Ann. N.Y. Acad. Sci.,* 547, 415, 1988.

309. **Cuttitta, F., Carney, D. N., Mulshin, J., Moody, T. W., Fedorko, J., Fischler, A., and Minna, J. D.,** Bombesin-like peptides can function as autocrine growth factors in human small-cell lung cancer, *Nature (London),* 316, 823, 1985.

310. **Tache, Y. and Brown, M.,** On the role of bombesin in homeostasis, *Trends Neurosci.,* 5, 431, 1982.

311. **Yanagisawa, M., Kurihara, H., Kimura, S., Tomobe, Y., Kobayashi, M., Mitsui, Y., Yazaki, Y., Goto, K., and Masaki, T.,** A novel potent vasoconstrictor peptide produced by vascular endothelial cells, *Nature (London),* 332, 411, 1988.

312. **Kumagaye, S. -I., Kuroda, H., Nakajima, K., Watanabe, T. X., Kimura, T., Masaki, T., and Sakakibara, S.,** Synthesis and disulfide structure determination of porcine endothelin: an endothelium-derived vasoconstricting peptide, *Peptide Protein Res.,* 32, 519, 1988.

313. **McMahon, E. G., Fok, K. F., Moore, W. M., Smith, C. E., Siegel, N. R., and Trapani, A. J.,** *In vitro* and *in vivo* activity of chymotrypsin-activated big endothelin (porcine 1 — 40), *Biochem. Biophys. Res. Commun.,* 161, 406, 1989.

314. **Yanagisawa, M. and Masaki, T.,** Endothelin, a novel endothelium-derived peptide. Pharmacological activities, regulation and possible roles in cardiovascular control, *Biochem. Pharmacol.* 38, 1877, 1989.

315. **Takasaki, C., Tamiya, N., Bdolah, A., Wollberg, Z., and Kochva, E.,** Sarafotoxins S6: several isotoxins from atractaspis engaddensis (burrowing ASP) venom that affect the heart, *Toxicon,* 26, 543, 1988.

316. **Lee, S. -Y., Lee, C. Y., Chen, Y. M., and Kochva, E.,** Coronary vasospasm as the primary cause of death due to the venom of the burrowing ASP, atractaspis engaddensis, *Toxicon,* 24, 285, 1986.

317. **Kloog, Y., Ambar, I., Sokolovsky, M., Kochva, E., Wollberg, Z., and Bdolah, A.,** Sarafotoxin, a novel vasoconstrictor peptide: phosphoinositide hydrolysis in rat heart and brain, *Science,* 242, 268, 1988.

318. **Lee, C. Y. and Chiappinelli, V. A.,** Similarity of endothelin to snake venom toxin, *Nature (London),* 335, 303, 1988.

319. **Takasaki, C., Yanagisawa, M., Kimura, S., Goto, K., and Masaki, T.,** Similarity of endothelin to snake venom toxin, *Nature (London),* 335, 303, 1988.

320. **Kimura, S., Kasuya, Y., Sawamura, T., Shinmi, O., Sugita, Y., Yanagisawa, M., Goto, K., and Masaki, T.,** Structure-activity relationships of endothelin: importance of the C-terminal moiety, *Biochem. Biophys. Res. Commun.,* 156, 1182, 1988.

321. **Nakajima, K., Kumagaye, S. -I., Nishio, H., Kuroda, H., Watanabe, T. X., Kobayashi, Y., Tamaoki, H., Kimura, T., and Sakakibara, S.,** Synthesis of endothelin-1 analogues, endothelin-3, and sarafotoxin S6b: structure-activity relationships, *J. Cardiovas. Pharmacol.,* 13, S8, 1989.

322. **Itoh, Y., Yanagisawa, M., Ohkubo, S., Kimura, C., Kosaka, T., Inoue, A., Ishida, N., Mitsui, Y., Onda, H., Fujino, M., and Masaki, T.,** Cloning and sequence analysis of cDNA encoding the precursor of a human endothelium-derived vasoconstrictor peptide, endothelin: identity of human and porcine endothelin, *FEBS Lett.,* 213, 440, 1988.

323. **Yanagisawa, M., Inoue, A., Ishikawa, T., Kasuya, Y., Kimura, S., Kumagaye, S. -I., Nakajima, K., Watanabe, T. X., Sakakibara, S., Goto, K., and Masaki, T.**, Primary structure, synthesis, and biological activity of rat endothelin, and endothelium-derived vasoconstrictor peptide, *Proc. Natl. Acad. Sci. U.S.A.*, 85, 6964, 1988.

324. **Inoue, A., Yanagisawa, M., Kimura, S., Kasuya, Y., Miyauchi, T., Goto, K., and Masaki, T.**, The human endothelin family: three structurally and pharmacologically distinct isopeptides predicted by three separate genes, *Proc. Natl. Acad. Sci. U.S.A.*, 86, 2863, 1989.

325. **Itoh, Y., Kimura, C., Onda, H., and Fujino, M.**, Canine endothelin-2: cDNA sequence for the mature peptide, *Nucleic Acids Res.*, 17, 5389, 1989.

326. **Saida, K., Mitsui, Y., and Ishida, N.**, A novel peptide, vasoactive intestinal contractor, of a new (endothelin) peptide family, *J. Biol. Chem.*, 264, 14613, 1989.

327. **Inoue, A., Yanagisawa, M., Takuwa, Y., Mitsui, Y., Kobayashi, M., and Masaki, T.**, The human preproendothelin-1 gene. Complete nucleotide sequence and regulation of expression, *J. Biol. Chem.*, 264, 14954, 1989.

328. **Bloch, K. D., Friedrich, S. P., Lee, M. -E., Eddy, R. L., Shows, T. B., and Quertermous, T.**, Structural organization and chromosomal assignment of the gene encoding endothelin, *J. Biol. Chem.*, 264, 10851, 1989.

329. **Shinmi, O., Kimura, S., Yoshizawa, T., Sawamura, T., Uchiyama, Y., Sugita, Y., Kanazawa, I., Yanagisawa, M., Goto, K., and Masaki, T.**, Presence of endothelin-1 in porcine spinal cord: isolation and sequence determination, *Biochem. Biophys. Res. Commun.*, 162, 340, 1989.

330. **Shinmi, O., Kimura, S., Sawamura, T., Sugita, Y., Yoshizawa, T., Uchiyama, Y., Yanagisawa, M., Goto, K., Masaki, T., and Kanazawa, I.**, Endothelin-3 is a novel neuropeptide: isolation and sequence determination of endothelin-1 and endothelin-3 in porcine brain, *Biochem. Biophys. Res. Commun.*, 164, 587, 1989.

331. **Tatemoto, K., Rökaeus, Å., Jörnvall, H., McDonald, T. J., and Mutt, V.**, Galanin — a novel biologically active peptide from porcine intestine, *FEBS Lett.*, 164, 124, 1983.

332. **McDonald, T. J., Dupré, J., Tatemoto, K., Greenberg, G. R., Radziuk, J., and Mutt, V.**, Galanin inhibits insulin secretion and induces hyperglycemia in dogs, *Diabetes*, 34, 192, 1985.

333. **Rökaeus, Å.**, Galanin: a newly isolated biologically active neuropeptide, *Trends Neurosci.*, 10, 158, 1987.

334. **Rökaeus, Å. and Brownstein, M.**, Construction of a porcine adrenal medullary cDNA library and nucleotide sequence analysis of two clones encoding a galanin precursor, *Proc. Natl. Acad. Sci. U.S.A.*, 83, 6287, 1986.

335. **Rökaeus, Å. and Carlquist, M.**, Nucleotide sequence analysis of cDNAs encoding a bovine galanin precursor protein in the adrenal medulla and chemical isolation of bovine gut galanin. *FEBS Lett.*, 234, 400, 1988.

336. **Vrontakis, M. E., Peden, L. M., Duckworth, M. L., and Friesen, H. G.**, Isolation and characterization of a complementary DNA (galanin) clone from estrogen-induced pituitary tumor messenger RNA, *J. Biol. Chem.*, 262, 16755, 1987.

337. **Kaplan, L. M., Spindel, E. R., Isselbacher, K. J., and Chin, W. W.**, Tissue-specific expression oof the rat galanin gene, *Proc. Natl. Acad. Sci. U.S.A.*, 85, 1065, 1988.

338. **Fox, J. E. T., Brooks, B. D., McDonald, T. J., Barnett, W., Rökaeus, Å., Kostolanska, F., Yanaihara, C., and Yanaihara, N.**, Actions of galanin fragments on rat, guinea-pig and canine intestinal motility, *Peptides*, 9, 1183, 1988.

339. **Rökaeus, Å., McDonald, T. J., Brooks, B. D., and St.-Pierre, S.**, Galanin: structure-function relations and pancreatic forms,. *Diabetes*, 37 (Suppl. 1), 183, 1988.

340. **Hermansen, K., Yanaihara, N., and Ahren, B.**, On the nature of the galanin action on the endocrine pancreas: studies with six galanin fragments in the perfused dog pancreas, *Acta Endocrinol. (Copenhagen)*, 121, 545, 1989.

341. **Amiranoff, B., Lorinet, A. M., Yanaihara, N., and Laburthe, M.**, Structural requirement for galanin action in the pancreatic beta cell line Rin m5F, *Eur. J. Pharm.*, 163, 205, 1989.

342. **Fisone, G., Berthold, M., Bedecs, K., Unden, A., Bartafai, T., Bertorelli, R., Consolo, S., Crawley, J., Martin, B., Nilsson, S., and Hökfelt, T.**, N-terminal galanin-(1 — 16) fragment is an agonist at the hippocampal galanin receptor, *Proc. Natl. Acad. Sci. U.S.A.*, 86, 9588, 1989.

343. **Rökaeus, Å., Melander, T., Hökfelt, T., Lundberg, J.M., Tatemoto, K., Carlquist, M., and Mutt, V.**, A galanin-like peptide in the central nervous system and intestine of the rat, *Neurosci. Lett.*, 47, 161, 1984.

344. **Melander, T., Hökfelt, T., Rökaeus, A., Fahrenkrug, J., Tatemoto, K., and Mutt, V.**, Distribution of galanin-like immunoreactivity in the gastro-intestinal tract of several mammalian species, *Cell Tissue Res.*, 239, 253, 1985.

345. **Ekblad, E., Rökaeus, Å., Håkanson, R., Sundler, F.**, Galanin nerve fibres in the rat gut: distribution, origin and projections, *Neuroscience*, 16, 355, 1985.

346. **Bauer, F. E., Adrian, T. E., Christofides, N. D., Ferri, G. -L., Yanaihara, N., Polak, J. M., and Bloom, S. R.**, Distribution and molecular heterogeneity of galanin in human, pig, guinea-pig and rat gastrointestinal tracts, *Gastroenterology*, 91, 833, 1986.

347. **Furness, J. B., Costa, M., Rökaeus, Å., McDonald, T. J., and Brooks, B. D.**, Galanin-immunoreactive neurons in the guinea-pig small intestine: their projections and relationships to other enteric neurons, *Cell Tissue Res.*, 250, 607, 1987.

348. **Gonda, T., Daniel, E. E., McDonald, T. J., Fox, J. E. T., Brooks, B. D., and Oki, M.**, Distribution and function of enteric GAL-IR nerves in dogs: comparison with VIP, *Am. J. Physiol.*, 256, G884, 1989.

349. **Ekblad, E., Håkanson, R., Sundler, F., and Wahlestedt, C.**, Galanin: neuromodulatory and direct contractile effects on smooth muscle preparations, *J. Pharmacol.*, 86, 241, 1985.

350. **Fox, J. E. T., McDonald, T. J., Kostolanska, F., and Tatemoto, K.**, Galanin: an inhibitory neural peptide of the canine small intestine, *Life Sci.*, 39, 103, 1986.

351. **Bauer, F. E., Zintel, A., Kenny, M. J., Calder, D., Ghatei, M. A., and Bloom, S. R.**, Inhibitory effect of galanin on postprandial gastrointestinal motility and gut hormone release in humans, *Gastroenterology*, 97, 260, 1989.

352. **McDonald, T. J., Dupré, J., Greenberg, G. R., Tepperman, F., Brooks, B., Tatemoto, K., and Mutt, V.**, The effect of galanin on canine plasma glucose and gastroenteropancreatic hormone responses to oral nutrients and intravenous arginine, *Endocrinology*, 119, 2340, 1986.

353. **Hramiak, I. M., Dupré, J., and McDonald, T. J.**, Effects of galanin on insulin responses to hormonal, neuropeptidal and pharmacological stimuli in conscious dogs, *Endocrinology*, 122, 2486, 1988.

354. **Dunning, B. E., Ahrén, B., Veith, R. C., Böttcher, G., Sundler, F., and Taborsky, G. J., Jr.** Galanin: a novel pancreatic neuropeptide, *Am. J. Physiol.*, 251, E127, 1986.

355. **Silvestre, R. A., Miralles, P., Monge, L., Moreno, P., Villanueva, M. L. and Marco, J.**, Effects of galanin on hormone secretion from the *in situ* perfused rat pancreas and on glucose production in rat hepatocytes in vitro, *Endocrinology*, 121, 378, 1987.

356. **Schnuerer, E. M., McDonald, T. J., and Dupré, J.**, Inhibition of insulin release by galanin and gastrin-releasing peptide in the anaesthetized rat, *Reg. Peptides* 18, 307, 1987.

357. **Yoshimura, T., Ishizuka, J., Greeley, G. H., Jr., and Thompson, J. C.**, Effect of galanin on glucose-, arginine-, or potassium-stimulated insulin release, *Am. J. Physiol.*, 256, E619, 1989.

358. **Ahrén, B., Arkhammar, P., Berggren, P. -O., and Nilsson, T.**, Galanin inhibits glucose-stimulated insulin release by a mechanism involving hyperpolarization and lowering of cytoplasmic free Ca^{2+} concentration, *Biochem. Biophys. Res. Commun.*, 140, 1059, 1986.

359. **Amiranoff, B., Servin, A. L., Rouyer-Fessard, C., Couvineau, A., Tatemoto, K., and Laburthe, M.**, Galanin-receptors in a hamster pancreatic β-cell tumor: identification and molecular characterization, *Endocrinology*, 121, 284, 1987.

360. **Lagny-Pourmir, I., Amiranoff, B., Lorinet, A. M., Tatemoto, K., and Laburthe, M.**, Characterization of galanin receptors in the insulin-secreting cell line Rin m 5F: evidence for coupling with a pertussis toxin-sensitive guanosine triphosphate regulatory protein, *Endocrinology*, 124, 2635, 1989.

361. **Amiranoff, B., Lorinet, A. M., Lagny-Pourmir, I., and Laburthe, M.**, Mechanism of galanin-inhibited insulin release: occurrence of a pertussis toxin-sensitive inhibition of adenylate cyclase, *Eur. J. Biochem.*, 177, 147, 1989.

362. **De Weille, J., Schmid-Antomarchi, H., Fosset, M., and Lazdunski, M.**, ATP-sensitive K^+ channels that are blocked by hypoglycemia-inducing sulfonylureas in insulin-secreting cells are activated by galanin, a hyperglycemia-inducing hormone, *Proc. Natl. Acad. Sci. U.S.A.*, 85, 1312, 1988.

363. **Dunne, M. J., Bullett, M. J., Li, G., Wollheim, C. B., and Petersen, O. H.**, Galanin activates nucleotide-dependent K^+ channels in insulin-secreting cells via a pertussis toxin-sensitive G-protein, *EMBO J.*, 8, 413, 1989.

364. **Bauer, F. E., Ginsberg, L., Venetikou, M., MacKay, D. J., Burrin, J. M., and Bloom, S. R.**, Growth hormone release in man induced by galanin, a new hypothalamic peptide, *Lancet*, 2, 192, 1986.

365. **Skofitsch, G. and Jacobowitz, D. M.**, Immunohistochemical mapping of galanin-like neurons in the rat central nervous system, *Peptides*, 6, 509, 1985.

366. **Mastropaolo, J., Nadi, N. S., Ostrowski, N. L., and Crawley, J. N.**, Galanin antagonizes acetylcholine on a memory task in basal forebrain-lesioned rats, *Proc. Natl. Acad. Sci. U.S.A.*, 85, 9841, 1988.

367. **Chan-Palay, V.**, Neurons with galanin innervate cholinergic cells in the human basal forebrain and galanin and acetylcholine coexist, *Brain Res. Bull.*, 21, 465, 1988.

368. **Crawley, J. N. and Wenk, G. L.**, Co-existence of galanin and acetylcholine: is galanin involved in memory processes and dementia?, *Trends Neurosci.* 12, 278, 1989.

369. **Ahrén, B., Rorsman, P., and Berggren, P. -O.**, Galanin and the endocrine pancreas, *FEBS Lett.*, 229, 233, 1988.

370. **McIntosh, C. H. S. and Brown, J. C.**, Motilin, *Adv. Metab. Disord.*, 11, 439, 1988.

371. **Brown, J. C., Mutt, V., and Dryburgh, J. R.,** Further purification of motilin, a gastric motor activity stimulating polypeptide from the mucosa of small intestine of hogs, *Can. J. Physiol. Pharmacol.,* 49, 399, 1971.
372. **Schubert, N. and Brown, J. C.,** Correction to the amino acid sequence of porcine motilin, *Can. J. Biochem.,* 52, 7, 1974.
373. **Poitras, P., Reeve, J. R., Jr., Hunkapiller, M. W., Hood, L. E., and Walsh, J. H.,** Purification and characterization of canine intestinal motilin, *Regul. Peptides,* 5, 197, 1983.
374. **Bond, C. T., Nilaver, G., Godfrey, B., Zimmerman, E. A., and Adelman, J. P.,** Characterization of complementary deoxyribonucleic acid for precursor of porcine motilin, *Mol. Endocrinol.,* 2, 175, 1988.
375. **Seino, Y., Tanaka, K., Takeda, J., Takahashi, H., Mitani, T., Kurono, M., Kayano, T., Koh, G., Fukumoto, H., Yano, H., Fujita, J., Inagaki, N., Yamada, Y., and Imura, H.,** Sequence of an intestinal cDNA encoding human motilin precursor, *FEBS Lett.,* 223, 74, 1987.
376. **Dea, D., Boileau, G., Poitras, P., and Lahaie, R. G.,** Molecular heterogeneity of human motilinlike immunoreactivity explained by the processing of prepromotilin, *Gastroenterology,* 96, 695, 1989.
377. **Yano, H., Seino, Y., Fujita, J., Yamada, Y., Inagaki, N., Takeda, J., Gell, G. I., Eddy, R. L., Fan, Y.-S., Byers, M. G., Shows, T. B., and Imura, H.,** Exon-intron organization, expression and chromosomal localization of the human motilin gene, *FEBS Lett.,* 249, 248, 1989.
378. **Daikh, D. I., Douglass, J. O., and Adelman, J. P.,** Structure and expression of the human motilin gene, *DNA,* 8, 615, 1989.
379. **Poitras, P.,** Motilin is a digestive hormone in the dog, *Gastroenterology,* 87, 903, 1984.
380. **Lee, K. Y., Chang, T. M., and Chey, W. Y.,** Effect of rabbit antimotilin serum on myoelectric activity and plasma motilin concentration in fasting dog, *Am. J. Physiol.,* 245, G547, 1983.
381. **Itoh, Z, Ed.,** *Motilin,* Academic Press, San Diego, 1990.
382. **Amara, S. G., Jonas, V., and Rosenfeld, M. G.,** Alternative RNA processing in calcitonin gene expression generates mRNAs encoding different polypeptide products, *Nature (London),* 298, 240, 1982.
383. **Rosenfeld, M. G., Mermod, J. -J., Amara, S. G., Swanson, L. W., Sawchenko, P. E., Rivier, J., Vale, R. W., and Evans, R. M.,** Production of a novel neuropeptide encoded by the calcitonin gene via tissue-specific RNA processing, *Nature (London),* 304, 129, 1983.
384. **Morris, H. R., Panico, M., Etienne, T., Tippins, J., Girgis, S. I., and MacIntyre, I.,** Isolation and characterization of human calcitonin gene-related peptide, *Nature (London),* 308, 746, 1984.
385. **Steenbergh, P. H., Hoppener, J. W. M., Zandberg, J., Van de Ven, W. J. M., Jansz, H. S., and Lips, C. J. M.,** Calcitonin gene related peptide coding sequence is conserved in the human genome and is expressed in medullary thyroid carcinoma, *J. Clin. Endocrinol. Metab.,* 59, 358, 1984.
386. **Nelkin, B. D., Rosenfeld, K. I., de Bustros, A., Leong, S. S., Roos, B. A., and Baylin, S. B.,** Structure and expression of a gene encoding human calcitonin and calcitonin gene related peptide, *Biochem. Biophys. Res. Commun.,* 123, 648, 1984.
387. **Tschopp, F. A., Tobler, P. H., and Fischer, J. A.,** Calcitonin gene-related peptide in the human thyroid, pituitary and brain, *Mol. Cell Endocrinol.,* 36, 53, 1984.
388. **Etienne T., Girgis, S., MacIntyre, I., Morris, H. R., Manico, M., and Tippins, J. R.** An investigation of the activity of human and rat calcitonin gene-related peptide (CGRP) on isolated tissues from the rat and guinea-pig, *J. Physiol. (London),* 351, 48P, 1984.
389. **Brain, S. D., Williams, T. J., Tippins, J. R., Morris, H. R., and MacIntyre, I.,** Calcitonin gene-related peptide is a potent vasodilator, *Nature (London),* 313, 54, 1985.
390. **Barthó, L., Lembeck, F., and Holzer, P.,** Calcitonin gene-related peptide is a potent relaxant of intestinal muscle, *Eur. J. Pharmacol.,* 135, 449, 1987.
391. **Fischer, J. A. and Born, W.,** Novel peptides from the calcitonin gene: expression, receptors and biological function, *Peptides,* 6, 265, 1985.
392. **Gates, T. S., Zimmerman, R. P., Mantyh, C. R., Vigna, S. R., and Mantyh, P. W.,** Calcitonin gene-related peptide-alpha receptor binding sites in the gastrointestinal tract, *Neuroscience,* 31, 757, 1989.
393. **Sternini, C., Reeve, J. R., Jr., and Brecha, N.,** Distribution and characterization of calcitonin gene-related peptide immunoreactivity in the digestive system of normal and capsaicin-treated rats, *Gastroenterology,* 93, 852, 1987.
394. **Zaidi, M., Bevis, P. J. R., Girgis, S. I., Lynch, C., Stevenson, J. C., and MacIntyre, I.,** Circulating CGRP comes from the perivascular nerves, *Eur. J. Pharmacol.,* 117, 283, 1985.
395. **Jonas, V., Lin, C. R., Kawashima, E., Semon, D., Swanson, L. W., Mermod, J. -J., Evans, R. M., and Rosenfeld, M. G.,** Alternative RNA processing events in human calcitonin/calcitonin gene-related peptide gene expression, *Proc. Natl. Acad. Sci. U.S.A.,* 82, 1994, 1985.
396. **Steenbergh, P. H., Höppener, J. W. M., Zandberg, J., Lips, C. J. M., and Jansz, H. S.,** A second human calcitonin/CGRP gene, *FEBS Lett.,* 183, 403, 1985.
397. **Amara, S. G., Arriza, J. L., Leff, S. E., Swanson, L. W., Evans, R. M., and Rosenfeld, M. G.,** Expression in brain of a messenger RNA encoding a novel neuropeptide homologous to calcitonin gene-related peptide, *Science,* 229, 1094, 1985.

398. **McDonald, T. J.**, GI hormones, in: *Clinical Medicine*, Vol. 8, Spittell, Jr., J. A., Ed., 1986, 1.
399. **Minamino, N. K., Kangawa, K., and Matsuo, H.**, Neuromedin U-8 and U-25: novel uterus stimulating and hypertensive peptides identified in porcine spinal cord, *Biochem. Biophys. Res. Commu.*, 130, 1078, 1985.
400. **Conlon, J. M., Domin, J., Thim, L., DiMarzo, V., Morris, H. R., and Bloom, S. R.**, Primary structure of Neuromedin U from the rat, *J. Neurochem.*, 51, 988, 1988.
401. **Minamino, N., Kangawa, K., Honzawa, M., and Matsuo, H.**, Isolation and structural determination of rat neuromedin U, *Biochem Biophys. Res. Commun.* 156, 355, 1988.
402. **Domin, J., Polak, J. M., and Bloom, S. R.**, The distribution and biological effects of neuromedins B and U., *Ann. N.Y. Acad. Sci.*, 547, 391, 1988.
403. **Price, D. A. and Greenberg, M. J.**, The structure of a molluscan cardioexcitatory neuropeptide, *Science*, 197, 670, 1977.
404. **Yang, H. -Y. T., Fratta, W., Majane, E. A., and Costa, E.**, Isolation, sequencing, synthesis, and pharmacological characterization of two brain neuropeptides that modulate the action of morphine, *Proc. Natl. Acad. U.S.A.*, 82, 7757, 1985.
405. **Kivipelto, L., Majane, E. A., Yang, H. -Y. T., and Panula, P.**, Immunohistochemical distribution and partial characterization of FLFQPQRF amidelike peptides in the central nervous system of rats, *J. Comp. Neurol.*, 286, 269, 1989.
406. **Raffa, R. B.**, The action of FMRFamide (Phe-Met-Arg-Phe-NH$_2$) and related peptides on mammals, *Peptides*, 9, 915, 1988.
407. **Schaller, H. C. and Bodenmüller, H.**, Isolation and amino acid sequence of a morphogenetic peptide from hydra, *Proc. Natl. Acad. Sci. U.S.A.*, 78, 7000, 1981.
408. **Bodenmüller, H. and Schaller, H. C.**, Conserved amino acid sequence of a neuropeptide, the head activator, from coelenterates to humans, *Nature (London)*, 293, 579, 1981.
409. **Schaller, H. C. and Bodenmüller, H.**, Structure and function of the head activator in hydra and in mammals, *Adv. Metab. Disord.*, 11, 519, 1988.
410. **Iwai, K., Fukuoka, S. -I., Fushiki, T., Tsujikawa, M., Hirose, M., Tsunasawa, S., and Sakiyama, F.**, Purification and sequencing of a trypsin-sensitive cholocystokinin releasing peptide from rat pancreatic juice. Its homology with pancreatic secretory trypsin inhibitor, *J. Biol. Chem.*, 262, 8956, 1987.
411. **Iwai, K., Fukuoka, S. -I., Fushiki, T., Kido, K., Sengoku, Y., and Semba, T.**, Preparation of a verifiable peptide-protein immunogen: direction controlled conjugation of a synthetic fragment of the monitor peptide with myoglobin and application for sequences analysis, *Anal. Biochem.*, 171, 27, 1988.
412. **Agerberth, B., Soderling-Barros, J., Jörnvall, H., Chen, Z. -W., Ostenson, G. -G., Efendić, S., and Mutt, V.**, Isolation and characterization of a 60-residue intestinal peptide structurally related to the pancreatic secretory type of trypsin inhibitor: influence on insulin secretion, *Proc. Natl. Acad. Sci. U.S.A.*, 86, 8590, 1989.
413. **Bowan, H. G. and Hultmark, D.**, Cell-free immunity in insects, *Annu. Rev. Microbiol.*, 41, 103, 1987.
414. **Zasloff, M.**, Magainins, a class of antimicrobial peptides from Xenopus skin: isolation, characterization of two active forms, and partial cDNA sequence of a precursor, *Proc. Natl. Acad. Sci. U.S.A.*, 84, 5449, 1987.
415. **Lee, J. -Y., Boman, A., Chuanxin, S., Andersson, M., Jörnvall, H., Mutt, V., and Boman, H. G.**, Antibacterial peptides from pig intestine: isolation of a mammalian cecropin, *Proc. Natl. Acad. Sci. U.S.A.*, 86, 9159, 1989.
416. **Tatemoto, K., Efendić, S., Mutt, V., Makk, G., Feistner, G., and Barchas, J. D.**, Pancreastatin, a novel pancreatic peptide that inhibits insulin secretion, *Nature (London)*, 324, 476, 1986.
417. **Eiden, L. E.**, Is chromogranin a prohormone?, *Nature (London)*, 325, 301, 1987.
418. **Iacangelo, A. L., Fischer-Colbrie, R., Koller, K. J., Brownstein, M. J., and Eiden, L. E.**, The sequence of porcine chromogranin A messenger RNA demonstrates chromogranin A can serve as the precursor for the biologically active hormone, pancreastatin, *Endocrinology*, 122, 2339, 1988.
419. **Greeley, G. H., Cohn, D. V., Cooper, C. W., Levine, M. A., Gorr, S. -U., Ishizuka, J., and Thompson, J. C.**, Inhibition of glucose-stimulated insulin release in perfused rat pancreas by parathyroid secretory protein-I (chromogranin A), *Endocrinology*, 124, 1235, 1988.
420. **Bretherton-Watt, D., Ghatei, M. A., Bishop, A. E., Facer, P., Fahey, M., Hedges, M., Williams, G., Valentino, K. L., Tatemoto, K., Roth, K., Polak, J. M., and Bloom, S. R.**, Pancreastatin distribution and plasma levels in the pig, *Peptides*, 9, 1005, 1988.
421. **Fasciotto, B. H., Sven-Ulrik, G., DeFranco, D. J., Levine, M. A., and Cohn, D. V.**, Pancreastatin, a presumed product of chromogranin-A (secretory protein-I) processing, inhibits secretion from porcine parathyroid cells in culture, *Endocrinology*, 125, 1617, 1989.
422. **Schmidt, W. E., Mutt, V., Carlquist, M., Kratzin, H., Conlon, J. M., and Creutzfeldt, W.**, Valosin: isolation and characterization of a novel peptide from porcine intestine, *FEBS Lett.*, 191, 264, 1985.
423. **Konturek, S. J., Schmidt, W. E., Mutt, V., Konturek, J. W., and Creutzfeldt, W.**, Valosin stimulates gastric and exocrine pancreatic secretion and inhibits fasting small intestinal myoelectric activity in the dog, *Gastroenterology*, 92, 1181, 1987.

Chapter 3

REGULATION OF GASTROINTESTINAL NEUROPEPTIDE GENE EXPRESSION AND PROCESSING

Åke Rökaeus

TABLE OF CONTENTS

I.   Introduction 89
        A.   Methods and Technical Considerations 89
        B.   The Eukaryotic Gene and Its Regulation 91
                1.   Regulation of Tissue-Specific and Time-Dependent
                        Gene Expression 91
                2.   Regulatory Elements 93
                        a.   Enhancer and Blocking Elements 93
                        b.   TATA, CAAT, and GC "Boxes" 93
                        c.   Cyclic AMP-and Protein Kinase C-Responsive
                                Elements 94
                        d.   Steroid Responsive Elements 94
                        e.   Miscellaneous Elements 94
                3.   Posttranscriptional Processing 95
                        a.   Capping and Methylation 95
                        b.   Polyadenylation 95
                        c.   Splicing 96
                        d.   Nuclear-Cytoplasmic Transport and Cytoplasmic
                                Stability 96
                4.   Translation and Posttranslational Processing and
                        Modifications 96
                        a.   The Start and Stop Sites 97
                        b.   The Signal Peptide 97
                        c.   Further Enzymatic Processing of the Proprotein 97
                        d.   Amidation and Other Modifications 98

II.   Vasoactive Intestinal Polypeptide (VIP) — PHI/PHM-27 98
        A.   Gene and mRNA Organization 98
        B.   Gene and mRNA Regulation 100
                1.   Regulation of Gene Expression 100
                        a.   Cyclic AMP and Protein Kinase C 100
                        b.   Depolarization 100
                        c.   Hormonal Modulation 100
                        d.   Growth Factors and Glucocorticoids 101
                2.   Posttranscriptional Regulation and Maturation 101
        C.   Structure and Processing of the Precursor Protein 102

III.   Neurotensin (NT) and Neuromedin N 103
        A.   Gene and mRNA Organization 103
        B.   Gene and mRNA Regulation 103
                1.   Regulation of Gene Expression 103
                        a.   Cyclic AMP, Glucocorticoids, and Growth
                                Factors 103

    b. Lithium..105
  C. Preprohormone Structure and Processing.............................105
    1. Evolutionary Comparison...105
    2. Structure...106
    3. Molecular Forms..106

IV. Neuropeptide Y (NPY)...107
  A. Gene and mRNA Organization..107
  B. Gene and mRNA Regulation..108
    1. Regulation of Gene Expression...................................108
    2. mRNA Regulation...108
      a. Glucocorticoids and Growth Factors.....................108
      b. Cyclic AMP, Protein Kinase C, and Intracellular
        Ca^{2+}..108
      c. Miscellaneous Factors..................................109
  C. Structure and Processing of the Precursor Protein......................109

V. Gastrin-Releasing Peptide (GRP) and Neuromedin B.......................110
  A. Organization of the GRP Gene and mRNA.............................110
  B. Regulation of the GRP mRNA.......................................112
  C. Structure and Processing of the GRP Precursor Protein.................112
  D. Structure of the Neuromedin B mRNA and Precursor Protein...........113

VI. Galanin..114
  A. Gene and mRNA Organization..114
  B. Gene and mRNA Regulation..115
    1. mRNA Regulation and Biosynthesis..............................115
      a. Cyclic AMP and Protein Kinase C.......................115
      b. Depolarization...115
      c. Steroid Hormones......................................115
      d. Miscellaneous Factors..................................116
  C. Structure and Processing of the Precursor Protein......................116

VII. Motilin...117
  A. Gene and mRNA Organization..118
  B. Structure of the Precursor Protein....................................118
  C. Processing of the Precursor Protein and Molecular Forms..............119

VIII. Summary and Concluding Remarks....................................119

Acknowledgment..121

References...121

I. INTRODUCTION

During the past quarter century a sizeable number of bioactive peptides have been isolated from the gastrointestinal tract and/or brain as well as from other tissues.[1,2] A fair number of these have been demonstrated, by immunologic methods, to be present in nerves of the gastrointestinal tract.[3,4] During the last decade the progress in recombinant DNA technology has led to the elucidation of the gene and messenger RNA (mRNA) structure for several of these neuropeptides (for methodologies and terminology see Reference 5). The elucidation of the mRNA structure through cloning and sequence analysis of their complementary DNA (cDNA) has enabled deduction of their respective precursor protein structure, thus identifying new potentially biologically active peptides. However, one should bear in mind that cloning, i.e., elucidation of the cDNA structure, only allows identification of the primary amino acid structure of the precursor protein and does not reveal processing events,[6] such as where cleavage of signal peptides occurs and where other processing sites are located. Furthermore, cloning does not reveal whether the processed peptides are amidated, glycosylated, phosphorylated or sulfated and whether they have cysteine bridges, etc; to obtain this information, peptides or proteins still must be isolated and characterized. However, certain patterns governing some of these posttranslational modifications have evolved and certain ''rules'' and suggestive generalizations can be made from the primary structure, but they indeed must be proved.

In this chapter no attempts have been made to fully cover the vast body of literature in the area of gene regulation in general and of gastrointestinal neuropeptide genes in particular or to completely cover the corresponding literature concerning gastrointestinal neuropeptide processing and the processes involved therein. However, a brief introduction to the area of gene regulation and expression will be given, and a set of more recently characterized ''peptide'' genes (vasoactive intestinal polypeptide [VIP] — the peptide with an NH_2-terminal histidine and a COOH-terminal isoleucine/methionine residue [PHI/PHM]$_3$, neurotensin [NT]-neuromedin N_g neuropeptide Y [NPY], gastrin-releasing peptide [GRP], neuromedin B_g galanin [GAL] and motilin) which have been demonstrated to be present in nerves of the gastrointestinal tract, which illustrate different aspects of gene regulation and processing, and which have not been reviewed extensively elsewhere have been chosen for this chapter, whereas cholecystokinin,[7,8] substance P,[9,10] enkephalin,[11,12] calcitonin gene-related peptide,[13] and somatostatin[14,15] have not been included. General information about processes involved in gene regulation, transcription, RNA processing, translation, and precursor protein processing are of importance to fully understand the problems and limitations of the studies reviewed in this chapter. In addition, it should be noted that gene regulation and processing have in several cases been proved and in other cases suggested to be tissue specific, a fact to remember since most of the studies covering the individual gastrointestinal neuropeptides have not been performed in tissues originating in the gastrointestinal tract. Furthermore, some of these peptides, such as motilin and NT, are mainly localized in so-called endocrine cells bordering on the intestinal tract, and it is extremely difficult at present to discriminate whether regulatory events and processes for a particular peptide gene are the same in the neuronal cells and endocrine cells of the gut, which further complicated the evaluation of the data reported.

A. METHODS AND TECHNICAL CONSIDERATIONS

Studies of gene expression and mRNA regulation have become possible by new techniques which allow introduction of foreign genes, mutated genes, or gene constructs (i.e., by coupling the gene of interest to a reporter gene) into the pronucleus of fertilized eggs by direct injection or by retrovirus infection,[16,17] and more recently also into sperm cells,[18*] and developing chimeric/transgenic animals, or by stable or transient introduction into cells in culture. The estimation of the reporter gene activity or the isolation of RNA and analysis of mRNA levels using combinations of exon probes, intron probes, or synthetic oligonucleotides thereof has allowed discrimination between changes in mRNA turnover and increased transcriptional activity. Other techniques that have provided major impacts in this respect are DNA footprinting[19] and gel retardation (gel shift) assays,[20,21] which allow analysis and identification of protein-DNA interaction by regulatory factors (*trans*-acting factors) and specific DNA sequences (*cis*-acting elements).

The methods of studying gene regulation and gene expression or biosynthesis as well as the processing of precursor protein have in many cases been crude due to improper tools and systems, thus hampering the interpretation of the results, a fact which has not always been recognized. For example, the analysis of mRNA levels with different techniques (Northern blot, slot blots, liquid hybridization methods, *in situ* hybridization histochemistry)[5,22] using exon-related probes has been implicated to reflect gene transcription, which probably is true in many cases, but may instead reflect changes in mRNA turnover, such as in the processing of the primary RNA transcript or in mRNA stability. The use of intron and exon probes in combination helps clarify the picture and adds in strength to a suggestive change in gene transcription, since intron-containing probes tend to hybridize preferentially to nuclear unspliced RNA and are thus not detecting changes in mRNA stability and turnover to the same degree as exon-enriched probes, which hybridize preferentially to the mature mRNA forms.

In studies where the promoter region of the gene of interest has been fused to a reporter gene, such as the gene encoding for chloramphenicol acetyltransferase (CAT)[23,24] the β-lactamase gene,[25] or, more recently, the gene encoding firefly luciferase,[26,27] the concomitant measurement of the product encoded by the reporter gene does not have the limitations indicated earlier and presumably provides a fairly good measurement of gene regulation. However, the method relies heavily on the assumption that a cell line which expresses the gene endogenously has the correct set of regulatory factors and that the control cell line does not have these factors. Furthermore, the technique also demands several controls, such as cotransfection of another plasmid checking for transfection efficiency (different constructs may not transfect equally well). In addition, some cell lines may express endogenously the same type of activity as the reporter gene, which also needs the proper controls.

The analysis of the size of a particular mRNA by the Northern blot technique frequently results in a confusing picture which may be due to methods rather that facts. Thus, DNA markers are sometimes used which do not always run as corresponding single-stranded RNA. The ribosomal bands are also commonly used as the sole identification of size, attributing the wrong size to them; finally, mRNA may be degraded differentially during extraction by different protocols and from different tissues or isolated cells.

Similarly, the use of radioimmunoassay (RIA) to study biosynthesis-measuring changes of peptide/protein immunoreactivity content in cells and media, implicating gene activity, clearly may be an overinterpretation of the results, since not only are such changes effected by processes affecting mRNA turnover, but also by factors affecting precursor protein turnover, i.e., processing events, peptide degradation, and the ability of the antisera to measure precursor forms as well as the mature peptide correctly. In terms of processing, the use of RIA to detect elongated forms may be limited by the ability to antisera to detect them for immunologic reasons, which could be both steric as well as structural due to species differences. Furthermore, there is a lack of antibodies to flanking and/or spacer peptides, and only in a few cases as yet have such antibodies been developed, i.e., toward a spacer

---

*   The technique of introducing foreign genes into sperm cells[18] has recently been questioned. For discussion see the following references: Barinaga, M., Gene-transfer method fails test, *Science,* 246, 446, 1989; and Maddox, J., Transgenic route runs into sand, *Nature,* 341, 686, 1989.

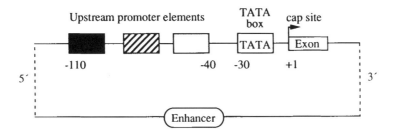

FIGURE 1. Schematic drawing illustrating the organization of a typical eukaryotic (polymerase II) promoter. The diagram shows the transcription initiation site (cap site), a TATA box located approximately 30 bases upstream from the cap site, and various other putative upstream (5′) promoter elements including an enhancer; the latter element also can be located downstream (3′) from the cap site.

peptide between PHI and VIP (see Section II), a C-terminal flanking peptide of NPY (see Section IV), and a GRP carboxy-terminal extension peptide (see Section V). In addition, even though the proper product may be detected, half-lives and the way they are detected (nonparallel serial dilution curves) may lead to false conclusions in terms of differential processing.

B. THE EUKARYOTIC GENE AND ITS REGULATION

A model eukaryotic gene (see Figures 1 and 2) consists of several principal components which are (starting from the 5′-end) enhancer elements and promoter elements that may mediate constitutive or regulated expression, including the "TATA" box (abbreviations for the different bases in DNA and RNA: A, adenine; C, cytosine; G, guanine; T, thymine; and U, uracil) which is found in most eukaryotic genes; enhancer elements also may be found within introns and downstream from the transcription termination site of the gene. Downstream, located approximately 30 bases from the TATA box, is the transcription initiation site (cap site; see Figure 2) followed by a number of exons (elements that are coding for sequences appearing in the mature mRNA) and introns (sequences which separate the exons and which are not found in the mature mRNA). In the last exon there are one or more signals for polyadenylation (see Section I.B.3.b), and further downstream from the last exon there is a site or sequence, of unknown nature, signaling transcriptional termination of the gene (for general information and reviews in this field see References 28 to 33).

The major control of gene expression is exerted at the transcriptional level, involving the specific interaction of regulatory proteins with specific DNA sequences in the promoter region. However, regulation of gene expression theoretically can be exerted at any step involved in the processing of the primary RNA transcript to the fully processed peptide/protein product (see Figure 2), i.e., in the processes involved in the modification of the 5′ and 3′ ends of the primary RNA transcript, in the processing of the RNA transcript leading to the correct splicing out of the introns in a differential and/or tissue-specific manner, in transport of the mRNAs into the cytoplasm, where steady-state levels are maintained, and finally at the translation of the mRNA and the posttranslational modification processes.[34,35]

1. Regulation of Tissue-Specific and Time-Dependent Gene Expression

Not all genes are expressed within a certain cell and within a given cell at all times; hence, a number of mechanisms must regulate these features, and it is obvious from the number of processes listed previously that there are several levels where such a control could take place, but tissue specificity seems to be regulated primarily at the transcriptional level.[29,36] There are several features which are characteristic for active vs. inactive genes, such

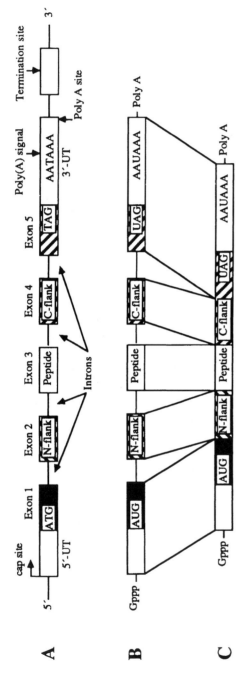

FIGURE 2. Schematic drawing illustrating the organization of: (A) a typical eukaryotic gene downstream from the cap site with a number of exons and introns; (B) its primary RNA transcript with cap (Gppp) and poly(A) tail (Poly A); and (C) the mature mRNA. Indicated in A, B, and C are a putative signal peptide encoded by exon 1(■); an N-terminal flanking peptide (N-flank) encoded by exon 2(▨); a single mature peptide, i.e., gastrointestinal neuropeptide, encoded by exon 3(☐); and a C-terminally located flanking peptide (C-flank) encoded by exon 4 and exon 5(▨) in the mature precursor protein. Abbreviations and explanations: 5'-untranslated region (5'-UT, ☐); 3'-untranslated region (3'-UT, ☐); ATG base triplet encoding the initiating methionine in the signal peptide; TAG — stop codon, poly(A) signal (AATAAA), and termination site of unknown nature.

as sensitivity to nuclease digestion (DNase I) (as shown, for example, by the gene encoding GRP [see Section V]), changes in chromatin structure, and extent of DNA methylation; the pattern and degree of DNA methylation seem to be both tissue- and species-specific. However, it is not known whether undermethylation is the cause or effect occurring as a result of transcription inhibiting methylation.[32,33,36-38] Furthermore, enhancer elements have been found to regulate cell specificity, probably by enhancer-activating proteins, as illustrated by insulin and chymotrypsin, which are only expressed in the endocrine and exocrine pancreas, respectively; *cis*-acting elements for such a control have been indentified within 300 base pairs (bp) from the transcriptional initiation site.[24,39]

2. Regulatory Elements

a. Enhancer and Blocking Elements

Although there is no strict difference between promoter and enhancer elements, the enhancers (see Figure 1) differ from other regulatory elements in that

1. They need not be located in the vicinity of the TATA box or the other promoter elements; in fact, their main feature is that they can exert the enhancement of transcription from the correct cap site from long distances ≥ 1 kilobases (kb) upstream or downstream from the cap site.
2. They work regardless of orientation.

The enhancer frequently consists of repeats in the nucleotide structure which occasionally enhance weakly by themselves, but usually more than one enhancer motif is needed for full enhancement. Although enhancers are usually found between 100 and 300 bp upstream from the transcription unit, enhancers have also been demonstrated within introns downstream from the cap site. The enhancers show both species- and tissue-specificity, but viral enhancers seem less discriminating and work in most combinations with foreign genes. The mechanism by which an enhancer exerts its effect is not known, but its function is restricted to the specific DNA molecule to which it is attached. However, there are a number of theories about the mechanism of action, such as that it may organize the chromatin structure and create regions of increased transcriptional activity, and that it may contain sequences which function as binding sites for factors that enhance transcription; enhancers have also been shown to be located in regions of increased DNase sensitivity corresponding to control regions of actively transcribing genes. The long distance function of the enhancer may be through anchoring to some nuclear compartment, and this effect may be transmitted to the upstream promoter region by means of folding the DNA molecule.[33,39-42]

Similarily, as in prokaryotes, blocking (silencing) elements have begun to be identified in eukaryotic genes,[38] although as yet not any of the gastrointestinal neuropeptide genes. Such blockade could be exerted either through the interaction of a *trans*-acting factor, with a special DNA blocking sequence in *cis,* but could theoretically also be obtained by competitive binding to an enhancer sequence or to any of the promoter elements listed below.[33,38]

b. TATA, CAAT, and GC "Boxes"

The TATA box and possibly also the CAAT and GC-rich boxes or other similar promoter elements probably aid in binding and direct the RNA polymerase II to the correct cap site.[28,29,38,42] The TATA box (consensus sequence 5′-TATA[A/T]A[A/T]-3′),[28,30] with minor discrepencies, is a promoter element found in almost all eukaryotic genes identified so far. The CAAT (consensus sequence ′5′-GG[C/T]CAATCT-3′)[33,42,43] and GC-rich boxes[38,42,44] are also commonly found in different promoters, but not to the extent of the TATA box. Occasionally genes may have several TATA motifs, such as the VIP gene (see Section II.A), which may be used in a regulated fashion. The CCAAT motif serves as the binding site for a CAAT-binding transcription factor/CAAT-binding protein (CTF/CP-1; see References 45 and 46), whereas the GC boxes which are usually found in multiple copies, serve as binding sites for the transcription factor Sp1, which activates transcription in several genes.[44,47] For review of recent studies that define structural domains for DNA-binding and transcriptional factors including the preceding and succeeding ones, see Reference 48.

c. Cyclic AMP- and Protein Kinase C-Responsive Elements

Several genes, including some of the reviewed neuropeptide genes, have been shown to be regulated by, or respond with, increased levels of mRNA following elevation of cyclic adenosine monophosphate (cAMP) and/or protein kinase C levels, i.e., with forskolin (an activator of adenylate cyclase) and a phorbol ester such as 12-*O*-tetradecanoyl(phorbol-13-acetate/phorbol, 12-myristate, 13-acetate (TPA/PMA), respectively. Enough genes and regulatory elements have been identified to allow compilation of consensus sequences on the DNA for these elements as well as identification of potential regulatory factors. The sequences are very similar to each other: 5′-TGACGTCA-3′, forskolin, factor CREB[24,49] and 5′-TGACTCA-3′, TPA, factor AP-1,[45,47,50] respectively. CREB (cAMP response element [CRE] binding protein) is a 43-kDa protein which seems to mediate transcriptional activation through the cAMP pathway when phosphorylated.[49] On the other hand, when the CRE is methylated, both binding and transcriptional activation are abolished in a tissue-specific manner.[51] Activator protein-1 (AP-1), a 47-kDa polypeptide isolated from HeLa cells, mediates both basal activation of transcription and specific responses to TPA.[45,50] The similarities in binding site and size suggest that CREB and AP-1 may be closely related or possibly identical transcription factors that sometimes may bind to both sites. However, data in the literature that demonstrate synergistic actions of cAMP and protein kinase C-activating drugs on certain genes or their mRNA speak against a complete identity between the binding sites and factors involved. However, as pointed out above, regulation of gene expression may take place at several different places other than directly on the gene; thus, synergism and/or additivity may be obtained through a combination of two such principally different mechanisms.

d. Steroid-Responsive Elements

In addition to responding to cAMP and TPA, several of the gastrointestinal neuropeptide genes and/or their mRNA also have been shown to respond to steroids such as glucocorticoids (dexamethasone) and estrogen; similarly, DNA-responsive elements have been identified: the consensus sequence for the glucocorticoid responsive element is 5′-GGT[A/T]CANNNTGTTCT-3′, and for the estrogen it is a 13-bp palindromic sequence (5′-GGTCANNNTGACC-3′), but an imperfect palindrome may be active in certain cases.[52-54] Interestingly, a single mutation in the estrogen element is sufficient to abolish estrogen inducibility, whereas a two-point mutation may convert the estrogen-responsive element into a glucocorticoid-responsive element.[55] Binding sites for steroids, i.e., the steroid-receptor complex,[54] are frequently found intermingled with binding sites for other transcription factors, and it has been shown that steroid binding units, if positioned correctly in relation to binding sites for other factors, may act synergistically with other nonrelated transcription factors to affect gene expression.[46] In this context it should again be pointed out that both dexamethasone and estrogen may affect mRNA turnover and stability in certain genes,[35] complicating the interpretation of data during an *in vivo* situation or when analyses of mRNA levels are only performed with exon-related probes (see Section I.A).

e. Miscellaneous Elements

In addition to the elements indicated above, several other more specific promoter elements and binding factors have been identified, i.e., elements for heavy metal ions and tryptophan oxygenase (a CACCC box) and a CACCC-binding factor,[31,33,42,44,46,50] as well as two tissue-specific transcription factors, Pit-1 and Oct-2 (POU domain), that activate genes specifying pituitary and lymphocyte phenotypes such as in the case of the prolactin gene.[56,57]

3. Posttranscriptional Processing
a. Capping and Methylation

RNA polymerase II, which is the enzyme responsible for transcription, is correctly oriented for its purpose by binding to promoter sites and initiates the primary RNA synthesis at the transcription initiation site, in 75% of the cases a purine (50% A and 25% C); the consensus structure around the cap site is 5'-Py-A/C-Py-Py-Py-Py-3' (Py — pyrimidine). Capping is a modification of the first 5' nucleotide in the primary RNA transcript which occurs instantaneously, before the entire gene has been transcribed. The modification consists of the addition of a guanosine residue, which is usually methylated, to the 5'-terminal nucleotide in the primary RNA transcript. This modification has been proposed to serve several purposes, i.e., to protect against 5'-exonuclease degradation, to be of importance in the processing of the RNA to mRNA, as in the splicing process, and to help bind the mRNA to the ribosomal RNA or its protein components during initiation of the translation process.[28,30,34,58,59]

It is not only the 5'-cap guanosine residue which is methylated, but also internal adenine residues. Although the purpose of this infrequent process is not clear, it has been suggested to play a role in RNA processing.[29,34]

b. Polyadenylation

Polyadenylation is a rapid event occurring within minutes, but it takes place after capping and immediately following trimming of the 3' end of the primary RNA transcript. However, it precedes splicing, although it is not a prerequisite for this later event.[30,34,60,61] Polyadenylation is believed to be signaled by the consensus poly(A) signal AATAAA[60,61] located in the last exon of the gene, but the 10 to 30 nucleotides downstream from the poly(A) signal also help determine the correct poly(A) site, i.e., the cleavage site of the primary transcript to which the poly(A) tail is added. Within a given gene there may be one or several poly(A) signals which are used in a regulated but not fully understood fashion, resulting in the formation of different length mRNAs (e.g., for NT; see Section III.A), thus being an additional way of regulating gene expression. In addition, there may be microheterogeneity in the site of poly(A) addition due to apparent inexactness in the poly(A) process leading to polyadenylation at several different sites downstream from the poly(A) signal, although within a given RNA molecule only at one place. It should be pointed out that the poly(A) is not copied from the DNA, but is actually synthesized in a very determined manner by a poly(A) polymerase. The poly(A) tail consists of approximately 200 to 250 bases added to the primary RNA transcript. It is gradually shortened during the processing of the RNA, and the size of the poly(A) tail in the mature mRNA, once it reaches the cytoplasm, varies considerably, between 30 and 250 bp. However, it has been shown that a length of only 30 adenylate residues is sufficient to confer full stability to the mRNA, whereas if the poly(A) tail is shortened to 16 bases it is ineffectual.[34,61] Moreover, the lack of a poly(A) tail does not render the mRNA untranslatable, but reduces its lifetime considerably. Although the length of the poly(A) tail is very homogeneous in size within a given tissue, it has been shown that the length of the poly(A) tail for a given gene transcript may vary in size between different organs, as shown for oxytocin and vasopressin;[62] this recently has been suggested to be the case also for NPY mRNA (see Section IV.A). The reason for this apparent tissue-specific polyadenylation is not known, and its possible functional importance as a regulatory event in gene expression remains to be elucidated.

c. Splicing

Splicing, which is one of the last steps in the posttranscriptional regulation of eukaryotic RNA, involves removal of intron sequences at defined consensus sites, thus being a link to the production of "stable" mature mRNA, and consequently may be used in the control of gene expression.[29,30,34,63] Indeed, a big VIP-related RNA species that contains a large intron has been identified and has been suggested to serve as a pool of immature VIP RNA (see Section II.B.2) which could be further processed and utilized upon demand. The consensus boundaries for the structure around the splice sites have been compiled, being [A/C]AG:GT[A/G] (exon:intron) and [C/T]AG:G (intron:exon) at the 5'- and 3'-splice junctions, respectively (underlined bases are most notably conserved). Alternative splicing has been demonstrated in several cases[34] as for the substance P-substance K gene,[9,10] the calcitonin-calcitonin gene-related peptide,[13] but more recently also for the neuropeptide GRP resulting in the formation of three different types of mRNA, all encoding GRP, but different carboxy-terminal extension peptides (see Section V.A).

In this context, it has been suggested that exons encode functional domains[64] in the precursor protein, as in the case of the exons in the VIP gene, also encoding PHI/PHM, where the individual exons encode these two biologically active peptides (see Section II.A). However, although this theory holds true in many cases, it is not always true, as exemplified by the human NPY and GRP genes, in which NPY and GRP, respectively, are encoded by two different exons, preliminary experiments also indicate that this is the case for bovine GAL (see Section IV.A). In addition, it has been suggested that the location of introns should be conserved in gene families believed to derive from a common ancestral gene, but in several cases this has been shown to be untrue; in these cases introns may have been inserted into nuclear genes after the divergency into gene families, which also may speak against the exon-functional domain theory.[64,65]

d. Nuclear-Cytoplasmic Transport and Cytoplasmic Stability

An additional place for regulation of gene expression could possibly be found in the transport step of the mature intact mRNA from the nucleus into the cytoplasm, since the portion of the primary RNA transcript which is removed during RNA processing is not transported to the cytoplasm, nor are prematurely finished transcripts transported.[34] Hence, there has to be a discriminating mechanism that only allows the transport of the mature fully processed mRNA to the cytoplasm, suggesting that such a potential mechanism may be regulated in some manner that could affect gene expression.[30,34,35]

Once in the cytoplasm, the mRNA is susceptible to regulation of its turnover, which varies considerably between mRNAs, whose half-lives vary between hours and days.[35] Thus, alterations in mRNA turnover have been shown to occur in response to several pharmacological and physiological stimuli, leading to increased to decreased turnover; e.g., estrogen and a growth factor have been shown to increase the half-life of some mRNAs, whereas a glucocorticoid such as dexamethasone has been shown to reduce the half life of one mRNA, suggesting again an additional way of affecting gene expression. Such an additional regulatory effect may further complicate the final outcome of a gene expression response, since several of these drugs or substances have been identified to interact positively with the promoter region of several genes and to affect neuropeptide biosynthesis.

4. Translation and Posttranslational Processing and Modifications

The translation of the mRNA and the further processing and modifications made to precursor protein (preproprotein) have been covered extensively in the literature; only a few points will be made here which are of importance for the further discussion of the individual neuropeptides.[15,66-72]

a. The Start and Stop Sites

The 5'-untranslated regions in the mRNA vary considerably in length, from 3 to 572 nucleotides, with 70% of them being in the range of 20 to 80 nucleotides. The translational start site is usually the most 5'-proximal AUG triplet in the mRNA (encoding the amino acid methionine [Met]), but in a minority of cases a more distal triplet is the functional one, suggesting that the nucleotides surrounding the AUG triplet also seem to play a role when choosing the correct start site.[73,74] Based on this assumption, a consensus sequence has been derived which covers almost all cases of different start sites, e.g., 5'-CC[A/G]CCAUGG-3'.[74,75] However, there are still a few cases where the first AUG has the correct surrounding, but still a second AUG triplet distal from the first has been shown to be the functional site. This has tentatively been explained by a "reinitiation" process occurring after an early stop, at a stop proximal to the functional second start site assuming that the ribosomal subunit "scans" the mRNA and remains on the mRNA after the first early stop.[73-75]

The stop signal sequence for the translational process is degenerate and could be any of three different types — UGA, UAA, and UAG. Commonly there are several stop sequences following the first, either in frame, such as in the GRP mRNA, or out of frame, such as in pig GAL mRNA, probably there to ensure translational stopping in case of a mutation. However, in the case of the vasopressin gene in the Brattleboro (diabetes insipidus) rat, there is a deletion mutation in the vasopressin coding exon, but no such backup stop sequence out of frame, leading to an unstopped translation and continuing into the poly(A) tail to create a mutated prohormone with a polylysine tail.[62]

b. The Signal Peptide

The signal (leader) peptide is a common principal feature in preproteins that are to be secreted and for that reason should be transported and inserted into the endoplasmatic reticulum for further processing and transport; i.e., it signals translocation.[66-70] The signal peptides for proteins to be secreted are typically 15 to 30 amino acid residues in length, starting with a Met residue. They usually contain a charged residue in the first five amino acids, followed by a hydrophobic core of at least nine amino acid residues presumably responsible for binding to the signal recognition protein. Following the hydrophobic core there is a second region, commonly with a high nonrandom and localized distribution of alanine residues, conferring the processing specificity.[76-78] After the signal sequence has served its apparent role, i.e., directing and inserting the prohormone into the space of the rough endoplasmic reticulum, the signal peptide is cleaved off by an endopeptidase, i.e., a metalloenzyme, liberating the proprotein/peptide.[70] The cleavage site per se is mainly characterized by the presence of amino acids with small, uncharged side chains, i.e., alanine and glycine for approximately 70% of the cases and serine, cysteine, and threonine making up an additional 24%.[78] In some cases the removal of the signal sequence seems to serve as the N-terminal processing site for the peptide/protein (e.g., for NPY and GRP).

c. Further Enzymatic Processing of the Proprotein

Additional enzymatic cleavage of the proprotein occurs in the Golgi complex and is usually needed to release the peptide or protein. This processing frequently involves cleavage at double basic residues, i.e., lysine (Lys) and arginine (Arg) and all combinations thereof, but occasionally at single bases[69-73] like NH_2-(N-)N-terminally of PHI (see Section II.C). However, the conformation of the proprotein and the peptide may also help in specifying where cleavage should occur, since cleavage does not always occur at single or double basis residue(s); i.e., neurotensin contains two internal arginine residues which are not processed (see Section III.C.2). The cleavage occurs at the COOH-(C-) terminal side of the basic amino acid residue by a trypsin-like enzyme, liberating a C-terminally extended peptide which is further processed by a carboxypeptidase-like enzyme that removes the basic residue(s).[69,70,79,80] However, it is not known whether it is the same set of enzymes involved in this principal process or whether they are different for each proprotein.[15] In a few cases, processing has been demonstrated at sites containing no basic amino acid residues, such as for endothelin (between tryptophan and valine)[81] and angiotensinogen (between leucine and valine/leucine by the enzyme renin).[82]

d. Amidation and Other Modifications

Amidation is a posttranslational modification process which is essential for bioactivity of several peptides occurring in the mature secretory granule. It occurs after the proteolytic processing at double basic residues and involves an enzyme that converts a C-terminal glycine (Gly) residue and the peptide bond between the ultimate residue preceding Gly to—$CO-NH_2$; the enzyme is dependent upon copper and ascorbate and has recently been cloned.[15,73,83-86]

In addition, there are a number of other modifications which could occur on the proprotein and its cleavage product(s): disulfide bridge formation, a tissue-specific acetylation of the NH_2-terminal; cyclization of an N-terminal glutamate residue; N-/(G-) and O-linked glycosylation partially signaled by the necessary sequence −asparagine−X−serine/threonine−, and −serine/threonine−, respectively; phosphorylation partially signaled by the sequence −serine−X−acidic−(X, any amino acid); methylation; fatty acid attachment; and *O*-sulfation of tyrosine residues catalyzed by a tyrosyl protein sulfotransferase in the *trans*-Golgi.[15,69,70,87-89] However, of the gastrointestinal neuropeptides reviewed in this chapter, only NT has been demonstrated to have one of these modifications, i.e., a blocked N-terminal pyroglutamyl residue but pro-VIP contains a possible signal for N-linked glycosylation.

II. VASOACTIVE INTESTINAL POLYPEPTIDE (VIP) — PHI/ PHM-27

A. GENE AND mRNA ORGANIZATION

The human VIP—PHM-27 gene, which is approximately 9 kb long,[90-95] is located on chromosone 6 (6q24)[96,97] and consists of seven exons interrupted by six introns (see Figure 3). The exons seem to correspond to functional domains encoded by the mature mRNA; it is approximately 1.8 to 2 kb long when encoding both peptides (for discussion about the size and individual references see Section II.B.2).[93,98-101] The functional domains transcribed from the exons are as follows: exon 1 — the 5′-untranslated region; exon 2 — the signal peptide; exon 3 — an amino-terminal peptide; exon 4 — PHI (rat)/PHM (man) and a part of a short intervening peptide; exon 5 — the C-terminal part of the intervening peptide and VIP; exon 6 — the remainder of the translated prohormone, the stop codons, and a small part of the 3′-untranslated region; and exon 7 — the large remainder of the 3′-untranslated region. The homology between exons 4 and 5, i.e., the exons encoding PHI/PHM-27 and VIP, respectively, and the surrounding intron boundaries suggests that an exon duplication of a common ancestral exon has occured. The VIP-PHI/PHM gene has four TATA boxes upstream from its cap site which potentially may be involved and used in connection with transcription activation, but at present only the TATA box located closest to the cap site 28 bases upstream has been identified as an active promoter element (see Section V.A; rat GRP mRNA).[94,95] No CAAT or GC boxes have been identified within 1.9 kb upstream from the cap site.[94,95] Downstream in the seventh exon, i.e., in the 3′-untranslated region of the mature mRNA, there are three signals for polyadenylation (AATAAA) in man and two in the rat.[94,95,100]

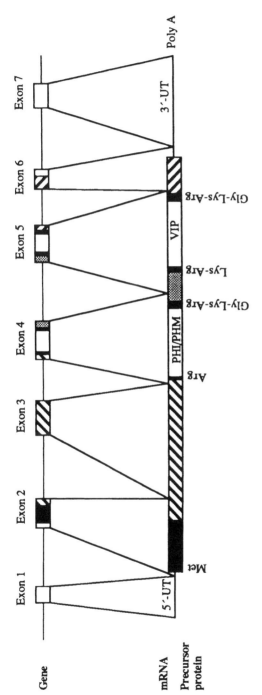

FIGURE 3. Schematic drawing illustrating the organization of the VIP-PHI/PHM-27 gene with seven exons, the mRNA, and a precursor protein. Also indicated is the initiating methionine (Met), the known sites for processing (■), and a spacer peptide between PHM and VIP (also identified to be the C-terminal part of PHV-42)(▨). Other symbols and abbreviations as in Figure 2.

B. GENE AND mRNA REGULATION
1. Regulation of Gene Expression
a. Cyclic AMP and Protein Kinase C

Data have shown that human VIP—PHM-27 gene in some human neuroblastoma cells in culture seems to be responsive to dibutyryl adenosine monophosphate ([Bt]$_2$cAMP) and/ or forskolin as well as to phorbol esters (TPA and MPA) as detected by elevated mRNA levels or by levels of VIP-like immunoreactivity (VIP-LI) in the cells and/or the cell culture medium, by either of the drugs or following combinations of the drugs — sometimes acting synergistically; the phorbol esters appear to induce gene expression independently from the cAMP-mediated pathway.[102-106] Similarly, in mouse neuroblastoma and in primary cultures of bovine adrenal chromaffin cells, VIP biosynthesis and secretion are increased by cyclic nucleotides and/or TPA; TPA was only tested in the latter, and an additive effect was obtained when it was combined with forskolin.[107-109]

Indeed, an apparently common cAMP/TPA element has been identified in the human VIP/PHM gene by CAT assay analysis and by sequence comparison to other genes; it is localized within 86 to 70 bp upstream from the cap site.[95,105] It was also demonstrated that two other genes responsive to cAMP elevation could compete for binding of nuclear proteins.[105] Furthermore, it was demonstrated that the element behaved as an enhancer in the sense that it had two sequence motif (CGTCA) and it was functional in both orientations.[105] In addition, it was recently shown that in one human neuroblastoma cell line (SK-N-SH) there seemed to be cell-type-specific transcription regulatory sequence(s) residing somewhere between 5.2 and 2.5 kb upstream from the cap site, since removal of this region resulted in an 80 to 90% reduction in basal CAT activity in the SH-IN subclone and in a 80% reduction of PMA-stimulated CAT activity in the SH-SY-5Y subclone.[106]

b. Depolarization

The effect on VIP-biosynthesis has only been studied in primary cultures of chromaffin cells obtained from bovine adrenal glands, and it has been found, in addition to the findings above, that after prolonged treatment with depolarizing agents (nicotine, potassium, veratridine, and barium) there is an increase in biosynthesis of VIP-LI (cell plus medium content), suggesting that the VIP-PHI gene, or the posttranscriptional RNA processing, may be stimulated by depolarizing agents.[109]

c. Hormonal Modulation

In the paraventricular nucleus in hypothalamus of lactating female rats a twofold increase in VIP-mRNA levels has been reported, suggesting that the VIP gene can be regulated by hormonal events associated with suckling and elevated prolactin secretion.[110] In addition, VIP-mRNA levels were increased in the hypothalamus at the time of sexual maturation.[110]

Within the rat ovaries, VIP-mRNA levels were 12-fold higher than in the rat brain cortex, but they were not affected by treatment with human chorionic gonadotropin (hCG); however, a large high molecular weight VIP-RNA form (see Section II.B.2 below) was reduced by hCG, suggesting a possible involvement of hCG in RNA processing.[111]

It has also been suggested that injection of corticosterone and estrogen into immature rats induces an increase in the VIP-mRNA in the hypothalamus and that both hyper- and hypocorticism elevate levels of VIP-LI in the hypothalamus of the rat after 14 d of treatment.[112,113] Similarly, hyperthyroidism significantly elevated VIP-LI in the basal ganglia, whereas hypothyroidism decreased levels of VIP-LI and the number of VIP-stained neurons in the developing cortex and the hypothalamus,[114] suggesting a possible linkage between thyroid hormone as well as corticosteroids and estrogen and VIP-LI expression.

However, whether these various suggestive hormonal actions on the VIP-PHI/PHM gene expression are indirect effects or effects per se on gene transcription remains to be shown.

Furthermore, it remains to be elucidated whether these effects are tissue specific and whether they also occur in the gastrointestinal tract.

d. Growth Factors and Glucocorticoids

Nerve growth factor (NGF) enhances the outgrowth of neurite-like processes in cultured human pheocromocytoma cells and in chromaffin cells dissociated from normal human adrenal glands and increases VIP-LI production.[115,116]   Dexamethasone, on the other hand, inhibits both neurite extension and production of VIP-LI,[115,116] suggesting an involvement of these factors in VIP—PHI/PHM gene expression, but both factors and dexamethasone have been shown to affect mRNA stability.[35]

Ascorbic acid (thought to be a cofactor for the enzyme involved in the amidation of VIP)[84] increased levels of VIP-LI in the medium from murine neuroblastoma cells, and the addition of forskolin potentiated this effect.[107] However, the enhancing action of ascorbic acid is more likely to be on the level of posttranslational processing of the precursor protein than on biosynthesis.[107]

2. Posttranscriptional Regulation and Maturation

A long high molecular weight intron (7 kb) containing RNA seems to be the major VIP-related RNA species in a human buccal tumor (HEp3),[117] but also can be detected in RNA isolated from rat brain nuclei; this long immature RNA may represent a form of cellular information storage, suggesting that VIP-PHI/PHM biosynthesis may be regulated at the RNA processing level. The 7-kb RNA also represents the major rat VIP-related RNA species during early ontogeny, but the relative amounts compared with mature VIP/PHI mRNA ("2 to 2.1" kb) decrease gradually by age.[101] A mature 2-kb mRNA has been observed embryonically at day 12 in the brain and at day 16 in the intestine, but it decreases thereafter, suggesting that the VIP-PHI gene is positively regulated by some growth factor, as demonstrated by increased levels of VIP-LI after NGF treatment (see Section II.B.1.d), and that the VIP/PHI mRNA product plays a role during embryonic development.[118]

The major mature mRNA in the rat, which encodes both VIP and PHI, has been stated to be 1.8[93] and 2 to 2.1 kb[101,118] and within this range in various other human neuroblastoma tumors.[103,106] However, the apparent size of these mRNAs seems to be the same upon visual inspection of the individual Northern blots, being located slightly below the 18S ribosomal band, suggesting that it is the same size VIP-PHM/PHI mRNA detected in all studies, and the difference resides in the interpretation of the size of the 18S ribosomal marker or the use of DNA markers, etc. With cDNAs being 1279 and 1341 bases in man[98] and rat,[100] respectively, with a cap site located 175 bases upstream from the initiating Met and with a mean length of the poly(A) tail of 150 to 250 bases, the correct size of the mature mRNA is likely to be in the range of the lower figure; otherwise, either the detected mRNA is unlikely to be the final mature mRNA corresponding to the cloned structure or the mature VIP-PHM/PHI mRNA must have an extremely long poly(A) tail. Despite these conflicting sizes for the mRNA, which probably are the same (the author's comment), a smaller mRNA of 1.6 kb has also been demonstrated and is the main mRNA in the HEp3 tumor. However, in the rat this smaller mRNA species was also found in adult gut (90 d) and brain (14 to 21 d) in addition to the "bigger" mature mRNA.[101] It should, however, be pointed out that the apparently mature VIP-related mRNA in the human buccal HEp3 tumor is only approximately 1.6 kb and does not seem to encode the PHM-27, suggesting that differential processing of the primary RNA transcript occurs in this tumor, making it different from other VIP-producing tumors which show both VIP and PHM transcripts;[98,99,119] this fact may reflect either cell type-specific expression and/or tissue-specific RNA processing. Whether the 1.6-kb mRNA found in adult rat is the rat counterpart of this "truncated" RNA or corresponds to a differentially polyadenylated RNA species (there are several potential poly[A] signals in both man and rat) or some other modification remains to be established; a deletion of only the bases coding for PHM or, in the case of the rat, PHI would not be resolved with Northern blot analysis and could not alone explain the nature of this smaller mRNA.

C. STRUCTURE AND PROCESSING OF THE PRECURSOR PROTEIN

The full-length translated VIP-PHI/PHM preprohormone is 170 amino acid residues long in man[98,99] and probably also in rat,[100] with the main difference being the encoding of PHM_{1-27} and PHI_{1-27} in man and rat, respectively, differing at 4 amino acid positions (see Figure 3).[2] The preprohormone contains a possible N-glycosylation site,[87,120] located at amino acid residues 68 to 70.[98] The signal peptide is assigned to the amino acids 1 through 21 and is followed by an amino-terminal peptide of 58 amino acids (residues 22 to 79). This amino-terminal peptide is followed by a single basic Arg residue on the N-terminal side of PHM/PHI-27 and the presumptive amide donor Gly and the double base Lys-Arg (residue [109—110]) C-terminally. This processing site is followed by a 12-amino-acid-residue spacer peptide and the VIP sequence, surrounded N-terminally by the double bases Lys-Arg, contrary to only Arg in front of PHI/PHM-27, and C-terminally (similar to PHI/PHM-27) by Gly–Lys–Arg. The latter processing site is followed by a 15-amino-acid-residue flanking peptide C-terminally.

Early cell-free translation studies of mRNA isolated from a human neuroblastoma suggested that prepro-VIP-PHM would be approximately 20 kDa whereas pro-VIP would be approximately 17.5 kDa,[121] findings which are in good agreement with the predicted weight of human prepro-VIP-PHM as deduced and calculated from its cloning, i.e., approximately 19.1 kDa.[98] Furthermore, labeling experiments of human neuroblastoma cells in culture in combination with VIP immunopurification demonstrated a 30-kDa species which could not be accounted for, but three other species (18, 11, and 8 kDa) in the cells, as well as three species (18, 8, and 3 kDa) in the surrounding cellular medium. These data are consistent with pro-VIP-PHM being the 18-kDa species, VIP-28 being the 3-kDa species, and the intermediate two species being transient intermediates formed during processing. Thus, the 11-kDa species may correspond to a C-terminal species obtained following processing N-terminally from PHM, and the 8-kDa species may be generated upon additional processing of the former species C-terminally from PHM, but this remains to be elucidated. However, when the neuroblastoma cells were extracted for their content, antibodies toward VIP and PHI demonstrated not only VIP-28 and PHM-27, but also an additional component which was slightly different from PHM since it cross-reacted only with N-terminally and not C-terminally directed PHI antisera, suggesting one of two things: either it was lacking the C-terminal of PHM, or it was C-terminally elongated and could not be detected by the latter antisera.[122] Such a C-terminally extended form of PHI/PHM, i.e., "big PHM", has been identified in pig and human stomach, but also in the nasal mucosa and urogenital tract in pigs, whereas "small PHM" is formed in the intestine, central nervous system, and lung.[123-125] This big PHM was shown to be C-terminally extended using an antibody toward the spacer peptide between PHI and VIP,[124] and such an elongated big PHM species, PHV-42, was recently isolated from a human adrenal pheochromocytoma.[126] PHV-42 corresponds exactly to prepro-VIP_{81-122} and consists of PHM-27, Gly–Lys–Arg (the amide donor and the dibasic processing site), and the 12-amino-acid spacer peptide between PHM and VIP. Although PHV-42 was not N-terminally extended in the human pheochromocytoma, another big PHM species that is N-terminally extended may very well exist in other tissues. The fact that big PHM[123-125] eluted just distally from cytochrome *c* (12.3 kDa) upon sizing chromatography, and could not be detected with VIP antibodies suggests that this may indeed be the case, although it remains to be demonstrated definitively. Another explanation may be that this species corresponds to the 11-kDa species described above and could not be detected by this VIP antisera. In any case, the data indicate differential, tissue-specific processing, based on the formation of big PHM and/or PHV-42 in only certain organs. Another such case, albeit at a different level, is demonstrated in the human buccal tumor (HEp3) described above, where the main mRNA species (1.6 kb) does not seem to encode PHM and cell-free translation of mRNA from this tumor gives rise to a protein of only 11 kDa.[127]

Furthermore, when several VIPomas were analyzed for PHM and VIP immunoreactivities and forms, it was found that levels of PHM-LI and VIP-LI did not correlate;[119] however, two forms of immunoreactivities were always found, a smaller form consistent with each peptide, respectively, as well as a bigger form. The relative amounts of the big and small forms were the same for VIP-LI, but they were highly variable for PHM-LI.[119] One explanation for the lack of correlation between PHM-LI and VIP-LI may be a lack of exact cross-reactivity for the bigger, variable PHM form which may obscure such an analysis, but if this is not the problem, these data indeed speak in favor of some differential processing. However, further analysis is definitely required, using more specific antibodies and antibodies directed toward other regions of pro-VIP-PHM/PHI, to fully prove or disprove the hypothesis of differential processing.

III. NEUROTENSIN (NT) and NEUROMEDIN N

Although neurotensin-like immunoreactivity (NT-LI) is mainly confined within endocrine-like cells in the gastrointestinal tract and only rarely, in some species as demonstrated so far (review and references[128]), within nerves, it will also be described in this chapter for the sake of completeness because of its possible paracrine action and because of some interesting aspects on gene expression and processing.

A. GENE AND mRNA ORGANIZATION
The neuromedin N-NT gene has been characterized only in the rat so far; it is approximately 10.2 kilobases (kb) in size and consists of four exons, all of which do not encode apparent functional domains (see Figure 4).[129] Exon 1 encodes the 5'-untranslated region and the putative signal peptide, exons 2 and 3 encode an intervening part of the precursor protein, and exon 4 encodes neuromedin N and NT in tandem. The processing of RNA to mature mRNA in rat and cow results in two different sizes of mRNA (1.5 and 1.0 kb) due to the presence of two signals for polyadenylation at two different sites in the 3'-untranslated region; this is different from the dog mRNA, which has only one such site and, consequently, only one size of mRNA (ca. 1 kb).[129,130] The localization of neuromedin N and NT within the same exon indicates that distinct mRNAs encoding only one or the other cannot be generated by alternative splicing.[130] A possible TATA box is located between 29 and 22 bp upstream from the assigned cap site, which is found approximately 100 bases upstream from the nucleotides encoding the initiating methionine in the precursor protein.

B. GENE AND mRNA REGULATION
1. Regulation of Gene Expression
In the rat, the two different sizes of mRNAs are detected in all neuronal tissues examined except the cerebellum and also in the gastrointestinal tract, where the 1.0-kb mRNA is clearly the dominating species, suggesting a tissue-specific regulation of their expression; whether lack of part of the 3'-untranslated region of the neuromedin N-NT mRNA is of any importance in terms of, e.g., stability, remains to be determined.[129]

a. Cyclic AMP, Glucocorticoids, and Growth Factors
NT-LI is not normally present in human adrenal medulla, but has been reported to be present occasionally in human pheochromocytomas[115] and in the adrenals from various other

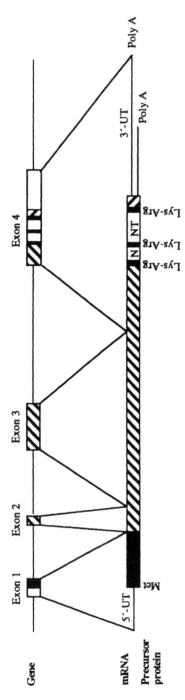

FIGURE 4. Schematic drawing illustrating the organization of the NT-neuromedin N (N) gene with four exons, the two different mRNAs, and a precursor protein. Symbols and abbreviations as in Figures 2 and 3.

species.[131] The rat pheochromocytoma cell line (PC-12) can be induced to increase intracellular levels and production of NT-LI up to 600-fold by a combination of forskolin, dexamethasone, and nerve growth factor (NGF), with the latter exerting a permissive effect.[132,133] In addition, in the rat medullary thyroid carcinoma cell line (rMTC-44), intracellular levels of NT-LI and release of NT-LI are also stimulated by dexamethasone.[134] Taken together, these studies suggest that the neuromedin N/NT gene could be positively regulated by glucocorticoid and cAMP elevation. The cloning of the rat gene has indeed identified (by sequence comparison with other genes) one and one half such putative cAMP responsive elements located approximately 25 bases upstream from the TATA box.[129] Along these lines, recent studies have shown that dexamethasone alone increases levels of mRNA, and in the presence of NGF the effect is potentiated.[135] On the other hand, the drug forskolin, which is without effect alone, increases levels of neuromedin N-NT mRNA in the presence of NGF; this effect is further increased when all three drugs are combined.[135]

b. Lithium

Lithium (Li), which is used to treat manic disorders,[136] was recently shown to potentiate the effect of dexamethasone, the combination of NGF and forskolin, and the combination of all three drugs, on neuromedin N-NT mRNA levels and cellular NT-LI levels in the rat pheochromocytoma PC-12. This was unexpected, since anatomically Li prevented neurite extension induced by NGF.[135] In the clinic, the target range for plasma Li is between 0.5 and 1.2 mM, with toxic effects observed if levels exceed 2 mM.[136] However, a concentration of Li $\geqslant$5 mM was needed to obtain the potentiating effect on neuromedin N-NT mRNA. There was a tendency for a potentiation of NT-LI levels with a 1 mM dose, but, if anything, mRNA levels tended to decrease, suggesting that if Li has any effect on the neuromedin N-NT gene expression during clinical conditions, it is probably a side effect. In addition, the rat adrenal gland has been shown to be particularly sensitive to Li in terms of its effect on inositol turnover, which may mediate the effect of Li,[136] further casting some doubt as to the general physiological significance of this finding, especially in terms of gastrointestinal regulation during normal conditions.

C. PREPROHORMONE STRUCTURE AND PROCESSING

In general, there is a good correlation between the relative amounts of neuromedin N-NT mRNA and NT-LI in different parts of the brain. However, in the hypothalamus the relative amounts of peptide, compared with the cortex, are much higher than the relative amounts of mRNA, which has been suggested to be due to differential processing of the precursor protein in these tissues.[129]

1. Evolutionary Comparison

The neuromedin N-NT precursor protein is 170 amino acid residues long in cow and dog, counting both Met residues at positions 1 and 2, but only 169 amino acids long in rat due to the presence of only one initiating Met.[129,130] The precursor protein is highly conserved between dog and cow (95%), but only 77 and 78%, respectively, when compared with the rat;[129] the region encoding neuromedin N and NT shows 100% conservation in these three species. In addition, the chemical isolation of human NT from the small intestine[137] and neuromedin N[138] from pig spinal cords demonstrates that, when compared with these two species, these peptides are also fully conserved. However, the apparent guinea pig form of NT is different in position 7.[139] The chicken form of NT is different in three amino acid positions (3, 4, and 7),[140] and the probable chicken neuromedin N counterpart, LANT-6, is different in amino acid position 7 as well as being one amino acid longer.[141] However, in principle, NT and neuromedin N seem to be very well conserved among these species. The peptide xenopsin, which has been isolated from the skin of the frog *Xenopus laevis*, also shares four of five C-terminal amino acids with NT and has been suggested to be the frog counterpart, but its precursor protein (consisting of only 80 amino acids)[142] bears little resemblance to the neuromedin N-NT precursor[129,130] and contains no obvious counterpart to neuromedin N.[142]

2. Structure

Within the rat precursor protein, amino acids 1 to 23 (1 to 24 for dog and cow) correspond to the putative signal peptide, and amino acids 24 (25) through 139 (140) correspond to an intervening peptide with unknown function encoded by exons 2, 3, and a part of 4; then neuromedin N_{1-6} and NT_{1-13} follows in tandem, flanked on both sides (as well as between) by double basic processing signals (Lys–Arg) and, ultimately, a pentapeptide of unknown function (see Figure 4).[129,130] It should be noted that within the 116-amino-acid peptide there is another potential site for processing, i.e., at positions 86 and 87 (87 and 88 in the cow and dog), where the amino acid sequence Arg–Lys is found, which (if serving as a target for processing enzymes) may liberate two additional peptides from this precursor protein. In addition, within NT itself there is another potential site for processing at amino acid residues 8 and 9 (Arg–Arg), which seem to be protected from such an endopeptidase cleavage, possibly by the two Pro residues bordering this site.[143] However, NT_{8-13} seems to possess sufficient information for full biological activity[128] as well as for binding, and NT_{9-13} is also biologically active, but by less than two magnitudes.[128] The complementary N-terminal peptide, NT_{1-8}, seems to be the inactive main (see Section 3) degradation product of NT_{1-13} rather than a peptide fragment liberated due to a final truly processing event of NT_{1-13} into a biologically active end product, NT_{9-13}.

The glutamine residue in the precursor protein corresponding to the first N-terminal residue in NT_{1-13} is also modified during the posttranslational processing and maturation, since NT_{1-13} has a pyroglutamyl residue in its N terminal, as originally determined by Caraway and Leeman (see Reference 128).

3. Molecular Forms

The major molecular form identified by RIA using NT_{1-13}-specific antibodies, in combination with chromatography, is consistent with NT_{1-13} regardless of tissue and mammalian species investigated and also in tissue extracts from tumors producing NT-LI.[128,144] However, with certain antibodies detected toward the N-terminal part of the NT molecule, a larger apparent form of NT-LI elutes in the void volume from some tumors of the gastrointestinal tract and particularly from plasma.[128,144] This material may represent a possible precursor form, but equally plausibly a void volume artifact in the RIA due to high protein concentrations. In addition, antibodies directed toward the N-terminus are prone to detect a smaller component NT_{1-8}, considered to be the major N-terminal metabolite of NT_{1-13} in plasma, but also albeit to a lesser extent, present in tissue extracts.[128,144] Using C-terminally directed antibodies, Carraway and Leeman described the presence of large amounts of smaller C-terminal forms of NT-LI in the upper gastrointestinal tract, whereas considerable less NT_{1-13} was found (see Reference 128). Retrospectively, this material may represent the detection of preferentially expressed neuromedin N, C-terminal degradation products of NT_{1-13}, and also kinetensin-like peptides, which can be generated from human plasma[145] and albumin(s) by pepsin[146] and have four out of six C-terminal amino acids in common with NT and sequence homologies to albumin(s).[145,146] Pepsin treatment of cat brain and ileal mucosa also generate neuromedin N-LI.[147]

Clarification of whether the neuromedin N-NT gene primary RNA transcript is processed differentially, resulting in different mRNAs producing either of the peptides alone, must await studies using peptide-specific mRNA probes. In addition, antibodies toward the N-terminal part of the precursor protein, preceding the neuromedin N-NT region, need to be

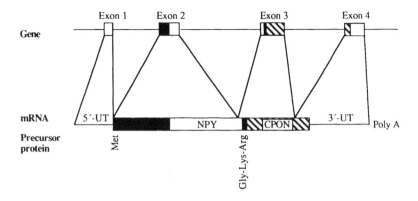

FIGURE 5. Schematic drawing illustrating the organization of the NPY gene with four exons, the mRNA, and a precursor protein. Abbreviation: C-terminal flanking peptide (CPON). Other symbols and abbreviations as in Figures 2 and 3.

developed to better study any tissue- and/or cell-specific processing of the neuromedin N-NT precursor protein. The limited data available at present may, however, indicate such a differential processing, since the ratios of NT-LI and neuromedin N-LI vary between tissues in the cat, and NT and LANT-6 immunoreactivities have similar, but not identical, distributions in the chicken.[148,149]

IV. NEUROPEPTIDE Y (NPY)

A. GENE AND mRNA ORGANIZATION

The human and rat neuropeptide Y (NPY) genes, the former located on chromosome 7 (pter-q22) in contrast to the related peptide gene for pancreatic polypeptide which is located on chromosome 17,[150] are approximately 8 and 7.2 kb long, respectively.[151,152] The gene consists of four exons (see Figure 5) which do not completely correspond to functional domains. Thus, exon 1 encodes the 5'-untranslated region, whereas exon 2 encodes the signal peptide and NPY, with the exception of the last two amino acid residues. The third base in the codon encoding the penultimate amino acid and the ultimate tyrosine (Y) residue in NPY, the amidation-signaling amino acid GLY, together with most of the remaining 3'-translated region are encoded by exon 3. Exon 4 encodes the last seven amino acids of the 3'-translated and -untranslated regions (approximately 170 bases), with the signal for polyadenylation located 19 bases from the poly(A) tail.[151-155]

The cap site in the human and rat NPY genes is located at a consensus A residue located 79 bases upstream from the bases encoding the initiating methionine in the precursor protein; a variant TATA box and a sequence resembling the CAAT box (human only) are found starting 30 and 79 bp, respectively, 5' upstream from the cap site.[151,152] Several apparently conservative mutations have been observed in the NPY gene, including one 80 bp upstream to the rat TATA-like box which may influence level of expression since it is within the upstream promoter region where regulatory elements are usually found.[151,152] The mRNAs transcribed from the NPY gene in both man and rat are approximately 0.8 kb in size, in reasonable agreement with the predicted transcription size of 552 bases and a poly(A) tail with 250 bases. However, careful analysis of the rat mRNA has demonstrated a tissue-dependent presence of two NPY mRNA species — approximately 0.82 and 0.76 kb.[4-6] The larger mRNA species is found in thymus, kidney, and adrenal gland, whereas the smaller is present in heart, ovary, and spleen; in the brain both types of mRNAs are found, but the relative amounts vary from region to region.[152] Similarly, mouse spleen NPY mRNA is smaller than the predominating brain mRNA.[152] There are no obvious explanations for the existence of two NPY mRNAs differing in size, nor is it known whether there are any functional implications in terms of stability etc.; a simple explanation for the differences in size may be tissue-specific addition of different length poly(A) tails, as has been described previously for vasopressin and oxytocin.[62]

B. GENE AND mRNA REGULATION
1. Regulation of Gene Expression

It has been shown in transient expression experiments performed with the promoter region of the NPY gene, 530 bases 5' to the transcription initiating site, fused with a promoterless reporter gene that this region of the NPY gene is sufficient for expression in two neuronal cell lines — rat pheochromocytoma (PC12) and thyroid medullary carcinoma (CA-77) cells.[151] Furthermore, it has been demonstrated that when the cDNA for human NPY (i.e., lacking its own promoter) is placed under the control of the mouse metallothionein promoter and stably transfected into the mouse anterior pituitary corticotrope cell line (AtT-20), NPY mRNA levels and pro-NPY synthesis can be induced artificially by cadmium and glucocorticoids, agents known to stimulate the metallothionein promoter.[156]

2. mRNA Regulation

Although relatively low levels of NPY mRNA have been demonstrated in the rat stomach and even lower levels in the small intestine,[155] other tissues have mainly been used to study NPY mRNA regulation.

Embryonically there is a developmental regulation, and detectable levels of NPY mRNA are observed in the rat brain at embryonic day 16, with levels increasing drastically by day 18 and 20 and remaining high in adult rats.[152]

a. Glucocorticoids and Growth Factors

NPY-LI has been demonstrated in adrenal glands (see Section IV.C) and treatment of rat PC-12 cells with NGF, but not with the glucocorticoid dexamethasone, increased NPY mRNA levels fivefold (apparent maximum after 3 d)[157] and cellular levels of NPY-LI.[158] In addition, there was a smaller (twofold) increase of mRNA levels after treatment with epidermal growth factor.[157] In another study, however, also using PC-12 cells and the same dose of dexamethazone (1 μM), there was a time-dependent increase in NPY mRNA after treatment with dexamethasone; the reason for this discrepancy was not discussed, but a similar effect by dexamethasone was also demonstrated in a mouse neuroblastoma cell line (N18TG-2).[155] The latter finding is in agreement with the fact that when another mouse neuroblastoma cell line (NG108-15) was treated with either dexamethasone or NGF, both drugs caused a twofold increase in cellular levels of NPY-LI, and when administered together they were acting synergistically.[159] Taken together the data suggest that, at least in rodent pheochromocytomas and neuroblastomas, NPY gene expression (mRNA levels) seems to be stimulated by both glucocorticoids and NGF.

b. Cyclic AMP, Protein Kinase C, and Intracellular Ca²⁺

The rat PC-12 cells and the mouse neuroblastoma cells (N18TG-2) have been used also to test the effect of raising cAMP levels (by forskolin) and of activation of the protein kinase C system (by phorbol esters — PMA) on NPY mRNA levels.[155] In PC-12 cells, forskolin caused a rapid but transient increase in NPY mRNA levels (maximally 2.7-fold at 8 h) with apparent desensitization upon prolonged exposure, but the responses at 48 and 72 h were potentiated by dexamethasone. In the mouse cells, weak responses were obtained with both forskolin and dexamethasone, but they had a dramatic synergistic action when combined.[155] The phorbol ester PMA and the calcium ionophore A23187, which can elevate intracellular Ca^{2+}, had little or no effect on NPY mRNA alone, but a moderate effect when combined.

However, when PMA was combined with forskolin, there was a 20 to 70-fold elevation which was further enhanced with dexamethasone or A23187 (200-fold). The magnitude of the NPY mRNA elevation by forskolin and PMA suggest transcriptional activation, and in the rat gene between bases −88 to −65 there is indeed a sequence resembling the consensus sequence for the phorbol ester element (see Section I.C.2.c) and, to a lesser degree, the cAMP element.[155]

c. Miscellaneous Factors

When fed but not fasted, rats are given an insulin injection to stimulate splanchnic nerve activation, and NPY-LI levels decrease to about 60% after 2 h.[160] On the other hand, NPY mRNA levels become 6.5-fold elevated 24 h after such a 2-h insulin stress, a rise which is prevented by bilateral transection of the splanchnic nerves.[161] The injection of reserpine into rats causes levels of NPY-LI to drop initially (at 24 h) in the adrenal gland, but for several days thereafter levels of NPY-LI become progressively elevated;[162] NPY mRNA levels are already elevated 1 d after drug treatment in the adrenal chromaffin cells as well as in the ganglion cells.[162]

In addition to being found in the gastrointestinal tract, NPY-LI and NPY mRNAs have been demonstrated in peripheral blood cells (probably due to circulating megakaryocytes which contain NPY mRNA), bone marrow, and spleen, among several places. Levels of NPY mRNA and NPY-LI are elevated in the spleens of mouse strains developing an autoimmune disease resembling human systemic lupus erythematosis.[163] However, the increased levels of NPY mRNA did not seem to be due to a general increase in body NPY mRNA, since brain levels were approximately the same in all mouse strains investigated, suggesting a tissue-specific elevation.

C. STRUCTURE AND PROCESSING OF THE PRECURSOR PROTEIN

The NPY precursor protein consists of 97 and 98 amino acids, respectively, in man and rat (see Figure 5). The rat has an "initiating" Met residue followed by an adjacent Met residue in its signal peptide of 29 amino acids. Human prepro-NPY has a three-amino-acid repeat in the signal peptide, i.e., amino acids 1 to 3 (Met–Leu–Gly) also occurring at residues 7 to 9, but an assignment of the former Met fits better with the size and structure of a typical signal peptide of 28 amino acids. The signal peptide is followed by NPY, an amide-donating Gly residue, two double basic residues Lys–Arg, and a C-terminal 30-amino-acid flanking peptide (C-PON).[151-155]

Cell-free translation experiments *in vitro* using mRNA isolated from postmortem human hypothalami or total RNA isolated from a human pheochromocytoma, followed by immunoprecipitation, have estimated human prepro-NPY to be approximately 12 kDa[164,165] and 10.5 kDa,[153] respectively, which is in reasonable accordance with the molecular weight of 10.8 kDa estimated from cloning studies.[153]

When AtT-20 mouse anterior pituitary cells, stably transfected with an artificial gene construct (inducible with Cd^{2+}) expressing human prepro-NPY, are treated with Cd^{2+}, there is an increase in immunoprecipitable material, with antibodies toward NPY and CPON. The precipatated material elutes as one peak just distal to cytochrome c (12.3 kDa), roughly corresponding to the predicted size of pro-NPY, and as an additional single smaller peak representing either NPY or CPON, respectively.[156] Characterization of the pro-NPY products by microsequencing confirmed the production of NPY and CPON, in agreement with the chromatographic data. The data from this artificial system are in agreement with results obtained by gel chromatography of adrenal gland extracts from various species[161,166] demonstrating single peaks of NPY-LI. However, several peaks of immunoreactivity are frequently demonstrated by high pressure liquid chromatography (HPLC), although there is usually a major peak corresponding to NPY in adrenal glands, in human pheochromocytomas

A

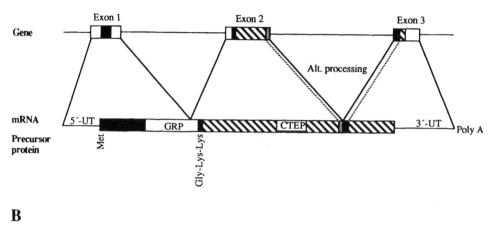

B

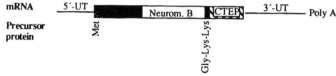

FIGURE 6.   Schematic drawing illustrating the organization of: (A) the GRP gene with three exons, the mRNA, and a precursor protein; (B) the organization for neuromedin B mRNA and a precursor protein. Also indicated in A are the sites for alternative processing resulting in three different types of GRP mRNAs and precursor proteins. Abbreviation: C-terminal extension peptide (CTEP). Other symbols and abbreviations as in Figures 2 and 3.

and ganglioneuroblastomas, as well as in rat PC-12 cells and in the gastrointestinal tract.[158,161,165-167] In addition, it has also been demonstrated by immunohistochemistry and RIA that NPY and CPON have qualitatively the same distribution in the brain and in peripheral neurons and that they are colocalized in cell bodies and neurons.[168-170]

V. GASTRIN-RELEASING PEPTIDE (GRP) AND NEUROMEDIN B

Gastrin-releasing peptide (GRP_{1-27}), which presumably is the mammalian counterpart to frog skin peptide bombesin, is widely distributed in neurons in the gastrointestinal tract, but it is also found in endocrine cells in the lung (for review and references see Reference 171). The peptides neuromedin C[172] and neuromedin B[173] have been isolated from pig spinal cord as biologically active entities. The former was identical to GRP_{18-27}, whereas the latter was a decapeptide having seven positional identities with the C-terminal part of the peptides in the GRP/bombesin family. The C-terminal heptapeptide was, however, identical to that of the frog peptides ranatensins R and C, suggesting that neuromedin B is the mammalian counterpart to frog ranatensin.[171,173] Similarly to GRP-LI, specific neuromedin C-(from here on GRP_{18-27})[174] and B-immunoreactivities have also been demonstrated to be present in neurons in the gastrointestinal tract.[175,176]

A. ORGANIZATION OF THE GRP GENE AND mRNA

The human GRP gene, located on chromosome 18 (18q21 band),[177,178] is approximately 10 kb in size and consists of three exons and two introns (4.8 and 3.9 kb)[179] the three exons do not encode apparent functional domains (see Figure 6A). Thus, exon 1 encodes the 5'-

untranslated region, the signal peptide, and the first 23 amino acids (and the first base in the codon corresponding to amino acid 24) of human GRP, whereas exon 2 encodes the remainder of amino acid 24 and the remaining 3 complete amino acids of GRP as well as the first 74 amino acids of a C-terminal extension peptide (CTEP). Exon 3 encodes the remainder of CTEP and the 3′-untranslated region. The primary cap site in the human GRP gene is located approximately 109 or 102 bases upstream from the bases encoding the initiating Met in prepro-GRP, as demonstrated by two different studies;[179,180] a second, weak possible cap site was located an additional 21 bases upstream.[179] A typical TATA box is found 132 bases upstream from the initiating Met, and a consensus CAAT box is found 140 bases upstream from the TATA box.[179] Between these two boxes 60 bases upstream from the latter box is a stretch of 10 nucleotides which are homologous to sequences in the somatostatin and enkephalin genes reported to be involved in their regulation by cAMP, but preliminary experiments have not supported such an influence on the human GRP gene in SCLC cell lines.[179] In addition, two GC-rich potential regulatory sequences are found in this region, one on each side of the potential cAMP element.[8] Furthermore, specific sites sensitive to DNase I (see Section I.C.1) have been demonstrated near the promoter region of the GRP gene in cell lines expressing GRP, but not in cell lines that do not express GRP.

Alternative splicings of the primary GRP RNA transcript have been demonstrated to lead to the formation of at least three different mRNA species. Thus, two mature human GRP mRNAs were demonstrated in a hepatic metastatic pulmonary carcinoid tumor; they were approximately 0.9 and 0.85 kb in size.[181] In addition, in a cDNA library made from a pulmonary carcinoid tumor (possibly the same as above), two different populations of cDNAs (the corresponding mRNA size was approximately 0.95 kb) were discovered differing by 19 nucleotides (causing a frame shift) in the translated region (types I and III, respectively), hence encoding different sized precursor proteins (see Section V.C).[181,182] The same types of cDNA species, together with a third species (type II) with a nucleotide deletion of 21 bases, were identified from another cDNA library made from a small cell lung cancer (SCLC) cell line; it was shown that these three mRNA species could be obtained through differential splicing of a single primary RNA transcript, around the second intron, being approximately 3.9 to 4.4 kb, by the use of different combinations of splice donor and acceptor sites encoding different CTEPs (see Figure 6A).[179,181] It was shown that the GRP mRNAs which were approximately ≤0.8 kb in size were present in some, but not all, SCLCs and metastatic SCLC tumors, in good agreement with the amount of bombesin-LI demonstrated by RIA measurements.[183] However, the immunoreactive material was not consistent with GRP, but with bombesin or $GRP_{18—27}$ upon chromatography (for further discussion see Section V.C).[184] The alternative RNA splicing into three different mRNAs occurs similarly in both neuronal tissues (gut and brain) and endocrine tissues as well as in normal and neoplastic tissue,[179] which is detected as various sized mRNAs or broad hybridizing bands of approximately 0.85 to 1.0 kb.[179,181-183] The full-sized form I^{174} mRNA predominates (63 to 85% of total), with form III (19-base deletion and frame shift) being the second most abundant (13 to 29%) and form II (21-base deletion) being the least abundant (<8%).[179] Further evidence that the three different mRNAs arise through differential splicing from a common primary GRP RNA transcript is the fact that with a short oligonucleotide probe, specifically detecting the 19-base deletion in form III, but not with a probe lacking these 19 bases, a big GRP RNA-related species (>5 kb) is detected, suggesting that the 19-base sequence was spliced out from a common, less mature GRP RNA.[183]. However, with the use of a much longer cDNA probe, covering a large region of the C-terminal flanking peptide but lacking the 19 bases, larger intermediate GRP RNA species (ca. 7.8 and 3.8 kb) were observed.[179,181]

The cloning of rat brain prepro GRP and the subsequent determination of the rat GRP mRNA size demonstrated it to be approximately 1.1 kb in the duodenum and brain, with an additional, ca. 1.5-kb GRP mRNA found in the rat brain.[184] The 5′-untranslated region in the cDNAs was found to be longer than 55 bases (man) and 512 (the longer mRNA in rat), whereas the 3'-untranslated region was approximately 289 (man) and 316 (rat) bases, with single polyadenylation signals located 19 and 28 bases from the 3' end, respectively.[181,185] The findings of two mRNA speices in the rat were tentatively explained by transcription initiation from one TATA box located 125 nucleotides upstream from the bases encoding the initiating Met; the cap site for this promoter was localized 29 bases downstream from the TATA box.[185] Transcription and concomitant polyadenylation of this mRNA species are in agreement with the finding of the 1.1-kb mRNA.[185] The finding of the larger 1.5-kb mRNA species in the brain is explained by transcription being driven from a promoter ≥ 500 bases upstream from the initiating Met and apparently used selectively in the rat brain; the confirmation of such a hypothesized promoter must await cloning of the rat gene. The functional implications for GRP transcription regulation by these two promoters in the rat GRP gene remain to be elucidated, but similar findings have been shown for the rat thyrotropin (TSH) β-subunit gene, where the upstream promoter is used constitutively and the downstream promoter is regulated by thyroidal status.[186] Whereas the sizes of the mRNAs found in the rat are in accord with the cDNA structure, the sizes of the human GRP mRNAs reported in the different studies (varying between ≤ 0.8 and 0.95 kb) are harder to reconcile, since the demonstrated human cDNA structure had a 5'-untranslated region ≥ 55 bases long, a coding sequence, and a 3'-untranslated region of 742 bases. By assuming a mean poly(A) tail of 100 to 200 bases, the size ought to be ≥ 0.9 to 1.0 bases, provided the mRNAs have not been degraded during isolation or the mRNA species detected have undergone differential processing other than those discussed above (authors comments),[182,183] which are only 19 and 21 nucleotides, respectively, and beyond the resolution of agarose gels (Northern blot technique).

B. REGULATION OF THE GRP mRNA

In the human fetal lung, GRP mRNA levels were first detected at gestational weeks 9 to 10. During the canalicular phase of pulmonary development (from approximately 16 to 30 weeks of gestation) the GRP mRNA levels became markedly elevated, 25-fold higher than in the adult lung; GRP mRNA levels fell to almost adult levels by week 34.[187] By contrast, levels of GRP-LI remained elevated until several months after birth. Similarly, GRP mRNAs and levels of GRP-LI are strongly expressed in neonatal thyroid, but almost undetectable in fetal or normal adult thyroid, although the relative amounts of GRP-LI vs. GRP mRNA were higher.[187] The findings suggest that pro GRP or one of its mature products play a role during normal lung and thyroid development.[187] However, whether similar changes in GRP mRNA levels and peptide levels occur in the gastrointestinal tract is not known, nor is the mechanism by which GRP gene transcription in the lung is stimulated during gestation.

C. STRUCTURE AND PROCESSING OF THE GRP PRECURSOR PROTEIN

The human and rat prepro GRP consist of 148 and 147 amino acids, respectively, as originally deduced from cloning studies (see Figure 6A).[182,185] The human GRP_{1-27} and rat GRP_{1-29} structures are preceded by a signal peptide of presumably 23 amino acid length, and followed by an amide-donating Gly residue, a double base (Lys–Lys) processing site, and CTEP, being 95 (residues 54 to 148) and 92 amino acids, respectively.[182,185] However, three other human prepro GRPs have been deduced from two other cloning studies; form I had a substitution at amino acid position 134 (Pro instead of Leu),[181,183] whereas the other two types had major nucleotide deletions (see Section IV.A), resulting in the deletion of amino acid residues 128 to 134 (form II) and the deletion of amino acid residues 122 to 148 together with an insertion of 17 new residues (form III), the latter two prepro-GRPs being 141 and 138 residues, respectively.[181,183] The rat CTEP in brain seems to be homologous to the human form III prepro GRP.[181]

Preliminary cell-free translation experiments support the existence of at least two GRP prohormones, with the predicted molecular weights of forms I and III.[183] In addition, preliminary immunohistochemistry studies in human fetal pulmonary endocrine cells have reported coexpression of GRP and of at least two of the three pro GRP forms (forms I and III).[181,188]

Concerning the processing of prepro GRP, no larger immunoreactive form than GRP_{1-27} or GRP_{1-29} (rat) have been demonstrated in the gastrointestinal tract,[189-191] brain[192], and in fetal lung or in lung carcinoid tumors[184,193-195] using antibodies raised against bombesin or against GRP,[192,193] which may or may not detect C-terminally extended peptides. However, in all studies a smaller immunoreactive form, apparently coeluting with frog bombesin,[189,190] but later found to correspond to the interspecies conserved (except for in the rat, which has a serine residue instead of aspartic acid at position 9 from the C-terminal end) C-terminal decapeptide of GRP, i.e., GRP_{18-27}[191-194] or possibly (but less likely) GRP_{19-27} was obtained.[194] GRP_{1-27} or GRP_{1-29} (rat) is presumably processed, similar to several other prohormones (i.e., prosomatostatin[14,15]), at the single basic Arg residue located at position 17 or 19 (rat), thus generating GRP_{1-16} or GRP_{1-18} (rat) and GRP_{18-27} or GRP_{20-29} (rat). The latter is identical to the peptide neuromedin C isolated from pig spinal cord[172] and an identical peptide isolated from dog intestine[191] and a human bronchial carcinoid.[195]

In conclusion, based on immunologic properties, the processing machinery for pro GRP seems to be apparently identical, at least in the gastrointestinal tract, lung, and brain, in the sense that no C-terminally extended GRP forms have been demonstrated, but only GRP and GRP_{18-27}-like/identical material. However, it should be pointed out that the relative amounts of GRP and GRP_{18-27}/neuromedin C-like peptide seem to vary between these tissues, a fact which tentatively could be explained by the ability of different antibodies to cross-react with the two species in an identical manner. However, there seems to be a real tissue-specific difference in either the processing machinery for GRP or the stability of GRP, since the GRP_{18-27}/neuromedin C-like component contributes 80 to 90% of the total GRP-LI found in the pylorus and the small intestine, whereas in the stomach and pancreas it contributes less than 40%.[189,190] Concerning the processing of the three different pro-GRPs, this has to be addressed when suitable specific antibodies have been raised toward all of the different CTEPs. However, the use of recombinant prepro GRP encoding form I demonstrated tissue- and/or species-specific processing of the CTEP (form I), identifying three, two, and zero potential processing sites when transfected into a moth ovary cell line (Sf9), an SCLC (NCI-H510), and into Swiss 3T3 mouse embryo fibroblasts;[196] in the two former cell lines the proper pro GRP form was generated after cleavage of the signal peptide and GRP was generated through processing at the double basic side C-terminal to GRP, whereas pro GRP was not further processed in the 3T3 cells.[196]

D. STRUCTURE OF THE NEUROMEDIN B mRNA AND PRECURSOR PROTEIN

The original chemical isolation and characterization of the neuromedin B (NB) decapeptide from pig spinal cord,[173] as well as the isolation of two N-terminally elongated forms (NB_{1-32} and NB_{3-32}) from the same organ,[197] have recently been extended by the cloning of human hypothalamic cDNA encoding prepro NB (see Figure 6B).[198]

It was found that human prepro NB was encoded by a single gene located on chromosome 15 (in contrast to the GRP gene, which was located on chromosome 18) and translated from a ca. 800-base mRNA species, in agreement with the size of the cDNA (639 bases) to yield a 76-amino-acid-residue NB precursor protein (see Figure 6B).[198] The deduced NB mRNA structure revealed a 5′-untranslated region of ≥36 bases, a 3′-untranslated region of 372 bases, and a single poly(A) signal located 14 bases from the 3′-end. Two NB mRNA species differing by 50 to 100 bases were identified by the Northern blot technique, similar to those

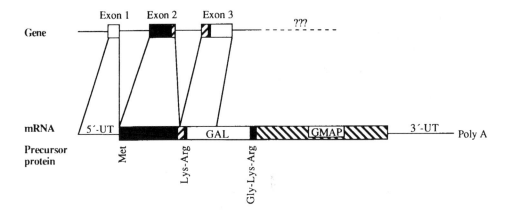

FIGURE 7.   Schematic drawing illustrating the partial, preliminary organization of the bovine GAL gene and its first three exons, as well as the organization of the bovine, porcine, and rat mRNAs and precursor proteins. Abbreviation: galanin message-associated peptide (GMAP). Other symbols and abbreviations as in Figures 2 and 3.

of human GRP mRNAs;[181] the mechanism by which they are produced is not known, but they were found both in brain and in the gastrointestinal tract (stomach and colon).[198]

The deduced structure of human prepro-NB (76 amino acids) consisted of a typical signal peptide (residues 1 to 24), NB_{1-32}, followed C-terminally by an amide-donating Gly residue and a double basic (Lys–Lys) processing site, and a C-terminal extension peptide (residues 60 to 76).[198] Human NB_{1-32} differed from pig NB_{1-32} at positions 15 and 21 (serine instead of glycine and serine instead of proline, respectively). Furthermore, as in GRP (see Section V.C), a potential single basic (Arg) processing site was located immediately before the C-terminal NB decapeptide,[197,198] suggesting that NB_{1-32} is further processed to NB_{23-32}, identical to the originally isolated form of NB[173] (i.e., the decapeptide) and similar to the proposed formation of GRP_{18-27}. This is further supported by chromatographic characterization of NB-LI in the gastrointestinal tract and in the brain, showing one peak consistent with NB_{1-32} and the second peak consistent with the NB decapeptide, the latter peak being the dominating component both in the gastrointestinal tract and brain.[175,176]

VI. GALANIN

Contrary to neurotensin and to motilin, galanin (GAL)-LI is widely distributed in the enteric system throughout the gastrointestinal tract and in all mammalian species investigated so far, including man.[199,200]

A. GENE AND mRNA ORGANIZATION

The author has seen no publication regarding the structure of the GAL gene; however, preliminary characterization of one bovine genomic clone so far has identified a putative promoter region with a possible TATA box as well as three exons (see Figure 7).[201] The two first exons seem to encode functional domains in the mature mRNA, i.e., the 5'-untranslated region and the signal sequence, respectively, whereas the third exon seems to encode only the N-terminal half of the GAL molecule, at least in cows (see Figure 7), i.e., the part which through cloning and isolation studies of GAL from pig,[202,203] rat,[204,205] and cow,[206] has been shown to be conserved. Immunologic and chromatographic evidence suggests that the C-terminal part of the GAL molecule is structurally different from that of pig in a number of other mammalian species in which the GAL structure has not yet been elucidated, including man.[199,200,207,208] Using the isolated cDNAs as hybridization probes, there is seemingly only one GAL gene in the rat[205] as well as in the pig.[201]

The mature mRNA in pig, rat, and cow are of similar size, e.g., approximately 900 bases, with the cow being possibly slightly smaller, which is in good agreement with the sizes of the cDNAs and the assumption of poly(A) tails of approximately 100 to 200 bases.[203-206] The 5'-untranslated region was longer than 140 nucleotides in all three species, and the 3'-untranslated varied between 155 and 180 nucleotides across the three species, with a single presumptive polyadenylation signal located at a position varying between 17 and 21 from the 3' end.

B. GENE AND mRNA REGULATION

In general, GAL-LI and GAL mRNA have been demonstrated to be present in the same tissues, including the gastrointestinal tract (for review and references see Reference 199).[203,205,206,209] Within the rat gastrointestinal tract, GAL mRNA seemed to be most abundant in the duodenum, lower in the stomach, least abundant in the small intestine, and not detectable in the colon and esophagus when identical amounts of RNA were analyzed for content of GAL mRNA.[205] However, GAL-LI has been demonstrated both in the colon and esophagus.[199]

1. mRNA Regulation and Biosynthesis

As for many other peptides, GAL-LI is present in the chromaffin cells of the adrenal medulla in many species, such as pig, cow, cat, and dog, whereas in others, such as rat and man, GAL-LI seems to be nonexistent or only rarely found, respectively.[199,207,208] Primary cultures of bovine adrenal medullary chromaffin cells also express and release GAL-LI and have been used to study GAL gene regulation, i.e., mRNA regulation, and biosynthesis as measured by changes in total (cellular + media) amounts of GAL-LI.[210,211] GAL mRNA has also been studied *in vivo* in pig intestine and rat brain,[209] with special reference to the rat hypothalamus[212] and the rat anterior pituitary.[204,213]

a. Cyclic AMP and Protein Kinase C

When bovine adrenal medullary chromaffin adrenal containing (epinephrine) cells, in which GAL-LI is present in a population of cells, are treated with the drug forskolin or with a protein kinase C activating phorbolester (TPA), there is an increase in GAL mRNA as well as in GAL biosynthesis as determined by RIA measurements of GAL-LI,[210,211] suggesting the presence of a cAMP and a TPA regulatory element in the GAL gene, e.g., the bovine GAL gene. This is supported by studies in progress[214] in which TPA increases GAL mRNA levels in human neuroblastoma cells (SK-N-SH-cells, see Section II.B.1.a). However, forskolin seemed to be without any effect in these cells, which may indicate species and/or organ differences in the GAL gene regulation, but possibly also that this tumor cell line has lost the factor mediating the forskolin response.

b. Depolarization

In addition, in the bovine chromaffin cells, treatment with depolarizing agents (veratridine, barium, and potassium) also increases GAL mRNA levels and stimulates biosynthesis of GAL-LI; the effects by veratridine and barium are blocked by inhibiting depolarization and blocking calcium entry by tetrodotoxin and D600, respectively, suggesting that the GAL gene, at least in bovine chromaffin cells, may respond to some factor mediating depolarization as well.[210,211]

c. Steroid Hormones

In Fisher rats, one established way of inducing tumors in the pituitary is treatment with high doses of estrogen. The estrogen treatment not only increases prolactin mRNA, but also GAL mRNA expression in the pituitary, where it is not normally detected.[204] The induction of GAL mRNA is time dependent and rapid in onset and occurs within 3 h after a single injection of estrogen.[204] In a recent, more detailed, study performed in Sprague Dawley rats, it was found that not only did levels of GAL mRNA change in relation to the estrous cycle but changes in levels of plasma GAl-LI could also be detected in ovariectomized females treated with estrogen. In addition, estrogen was demonstrated to induce GAL mRNA in a rat transplantable growth hormone- and prolactin-secreting pituitary tumor (MtTW$_{15}$).[205] However, it was also stated that estrogen-induced increase of GAL mRNA did not occur "in neuronal tissues expressing this gene" without any details as to which (and how) neuronal tissues had been examined.[205] Thus, GAL mRNA regulation by estrogen may be of limited relevance for the gastrointestinal tract and may be the first good demonstration of tissue-specific regulation of the GAL gene. However, the fact that (at least in cow and man) GAL mRNA may be induced through several other mechanisms, such as depolarization, cAMP elevation, and protein kinase C activation,[210,211,214] may suggest alternative indirect mechanisms for the estrogen-induced GAL mRNA elevation in the rat pituitary until an estrogen-inducible consensus element has been identified on at least the rat GAL gene.

d. Miscellaneous Factors

Interestingly, there is another aspect of GAL mRNA regulation provided by nature itself, e.g., in the Brattleboro (diabetes insipidus) rat, which has a genetic mutation in the gene encoding preprovasopressin, resulting in lack of correctly processed vasopressin.[62] In the Brattleboro rat there is an elevation in the GAL mRNA in the paraventricular and supraoptical nuclei (SON) of the hypothalamus; GAL-LI coexists with vasopressin in the SON.[212] This finding, by *in situ* hybridization histochemistry, was not accompanied by increased levels of GAL-LI in either nuclei nor in the median eminence or in the posterior pituitary (PP) to which these nuclei project; on the contrary, levels of GAL-LI were reduced in the PP.[212] Taken together, these data suggest that GAL biosynthesis was increased, and probably also turnover and release of immunologically detectable GAL-LI.[212] Whether the GAL mRNA levels were changed any other place, such as in the gastrointestinal tract of the Brattleboro rat, was not investigated.[212] It is, however, more likely that the increase in GAL mRNA levels probably represent a local nuclei compensation of deficient vasopressin production, i.e., a tissue-specific gene activation rather than a general overall GAL gene activation, but nothing is known about the nature of this activating mechanism.

C. Structure and Processing of the Precursor Protein

The mature mRNA encodes a 123 (pig and cow) or 124 (rat) amino acid residue precursor protein (see Figure 7) which has the same principal arrangement and is well conserved in the areas of the signal peptide and the galanin molecule, but is less well conserved in the C-terminal flanking peptide (GAL-message-associated peptide [GMAP]).[203-206] The exact size of the signal peptide has not been determined, but is likely to consist of 23 or 25 amino acids (positions with Gly residues). However, in the cow, exon 2 encodes the first 27 amino acids, the 27th being a proline (Pro).[201] This Pro residue is conserved in cow, pig, and rat, and in a few rare cases Pro has been shown to be the end point of a signal peptide.[78] Thus, depending upon the assignment of the signal peptide, a possible 7-, 5-, or 3-amino-acid peptide may follow which may or may not be liberated upon processing (pre-GAL message peptide), i.e., residues 24 to 30. Then follows a double basic cleavage site (Lys–Arg), the 29 amino acid GAL molecule, followed C-terminally by the amide-donating Gly residue and another set of two double basic residues (Lys–Arg). This latter double basic processing site is followed by a 59 (pig and cow) and 60 (rat) GAL message-associated peptide with unknown function and with less conservation across the three species, but with a couple of potential single basic processing sites conserved across all three species at positions 73/74 and 88/89 in the prepro-GAL molecule and at a few other places which are not fully conserved.[203-206]

There has been no study aiming at the elucidation of the processing of prepro-GAL, although there have been a number of studies describing different molecular forms using RIAs based on pig GAL antibodies, in combination with chromatography. The problem with these studies is that the C-terminus to the midportion of the GAL molecule, in most species, seems to vary from that of pig GAL, causing it to behave differently from standard pig GAL upon chromatography, or it remains undetected when antibodies directed toward the C-terminus of the pig GAL molecule are used.[199,208] However, in the pig adrenal gland, chromaffin cells will be enriched for all immunoreactive components of a certain peptide, since different components are not transported out in long neurons, such as in the gastrointestinal tract. Four immunoreactive components have been described: component 1 elutes before cytochrome c (approximately 12.3 kDa), close to the void volume; component 2, at $\geq$12.3 kDa, in agreement with the size of prepro-GAL; and a third component, approximately half way in between component 2 and 4, which is identical to GAL_{1-29} (approximately 3.2 kDa).[207] Knowing the size of pig prepro-GAL, component 1 is difficult to account for and may represent some aggregation, nonspecific immunoreactivity, or indeed some posttranslational modification, although pig prepro-GAL does not have any apparent signal for N-glycosylation or phosphorylation. Component 2 is in agreement with the expected size of pig prepro-GAL, whereas component 3 probably represents a possibly C-terminally (to one of the single basic amino acid sites; see above) and N-terminally (5 to 7 residues, see above) elongated form (molecular weight between 3.2 and 12.3 kDa). In the pig gastrointestinal tract only one component was observed after sizing chromatography corresponding to pig GAL_{1-29}, which on reverse-phase HPLC was resolved into two components[200] perhaps representing intact GAL_{1-29} and chemically rearranged GAL_{1-29} that may have undergone β-aspartic rearrangement; there are two potential places in pig GAL for this rearrangement.[206] We have observed similar dual components in bovine gastrointestinal tract after HPLC, and they did not differ in their amino acid composition.[206,215] These data in the pig may indicate differential processing in the gastrointestinal tract and the adrenal gland having the same end product GAL_{1-29}, but more likely represent a detection problem in the gastrointestinal tract. In the human gastrointestinal tract as well as in the adrenal gland, two components are being detected; one is larger, roughly equal, or slightly smaller than component 3 in the pig above and has a smaller component in good agreement with the size of pig GAL_{1-29}.[200] The reason for this difference in number of GAL immunoreactive components between human and pig gastrointestinal tracts is not known, and the true pattern of processing in either species, or for that matter in any other species, can not be addressed properly until the human GAL sequence has been elucidated and antibodies toward the flanking peptides in pro-GAL of the species investigated have been raised. Thus, the pattern in pig, suggested earlier, remains highly speculative at the present time.

VII. MOTILIN

Although pig motilin (M) was isolated in 1971 by Brown and co-workers (see Reference 216), since then there has been only one report demonstrating M-LI in the nervous system of the gastrointestinal tract in man, monkey, and dog by immunohistochemistry.[217] In addition, two other reports support this finding, at least in the dog, since selective separation of the muscularis and mucosal layers followed by the measurement by RIAs identified M-LI of the same type as in the former layer.[216,218] However, in general, M-LI seems to be found in an endocrine type of cell, and the literature lacks confirmation of the demonstration of immunoreactive nerves in the gut.[219]

Only recently the structures for the human gene,[220] the mRNA encoding human[221,222] and porcine[223] intestinal motilin (approximately 0.7 kb), and the structure of each precursor protein have been deduced (see Figure 8). However, Northern blot analysis using M cDNA

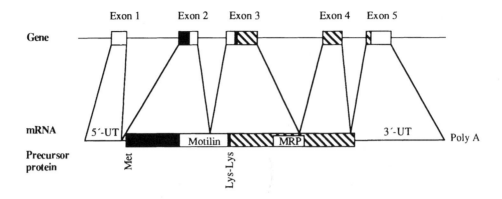

FIGURE 8.   Schematic drawing illustrating the organization of the motilin gene with five exons, the mRNA, and a precursor protein. Abbreviation: motilin-related peptide (MRP). Other symbols and abbreviations as in Figures 2 and 3.

probes identified M mRNA only in the antrum and duodenum, whereas in all other parts of the human gastrointestinal tract, in the human gallbladder, and in the porcine brain and adrenal gland, M mRNA levels were too low to be detected.[220,223]

A. GENE AND mRNA ORGANIZATION

There seems to be only one motilin gene that cross-reacts with the isolated motilin cDNAs. Thus, it has been found that the human M gene, which is approximately 9 kb long (see Figure 8),[220] is located on chromosome 6 (p21.2→p21.3 region) and consists of five exons interrupted by four introns (see Figure 8). The exons do not correspond to obvious functional domains encoded by the mature mRNA; it is approximately 0.7 kb long.[220-223] The domains transcribed from the exons are: exon 1 — the 5'-untranslated region; exon 2 — the signal peptide and M_{1-14}; exon 3 — the remaining part of M, i.e., residues 15 to 22, and one half of a carboxy-terminal extension peptide; exon 4 — the remainder of the translated prohormone except for the last two and two thirds of the amino acids in the flanking peptide; these acids are encoded by exon 5, which also encodes the 3'-untranslation region.[220] There is a single TATA box located 19 bases upstream from the cap site. No obvious CAAT or GC boxes have been identified within 0.3 kb upstream from the cap site.[220] Downstream in exon 5, i.e., in the 3'-untranslated region of the mature mRNA, there is a single signal for polyadenylation (AATAAA).[220]

B. STRUCTURE OF THE PRECURSOR PROTEIN

The M cDNA structures and the deduced structures for each precursor protein in the two cloning studies in man (115[220,221] and 114[222] amino acids, respectively) and pig (119 amino acids)[223] revealed several differences at the nucleotide level and some differences at the amino acid level, but in principle they all had the same organization (see Figure 8); a signal peptide of 25 amino acids (residues 1 to 25) was followed by the conserved 22 amino acids comprising M (residues 26 to 47). The M structure was followed by the two double basic residues Lys–Lys and a 66 or 65, respectively (man), and 70 amino acid C-terminal polypeptide "M-related peptide" (MRP) with unknown function, but with several conserved single basis residues which could serve as potential monobasic processing sites, e.g., at amino acid positions 70, 76, 88, 91, and 96 (see Section VII.C). The difference between the two deduced human structures consisted of one hydrophobic amino acid substitution in the signal sequence and the insertion of one amino acid at the C-terminal part of MRP. The porcine form of prepro-M, consisting of 119 amino acid residues, had three substitutions in the signal peptide and 29 substitutions in the MRP, mainly toward the C-terminal end, and was also elongated C-terminally. The 5'-untranslated region was 71 nucleotides in man and at least 39 nucleotides in pig, and the 3'-untranslated region was approximately the same in the two human studies, e.g., approximately 155 nucleotides, while being slightly longer (170 nucleotides) in the pig.[223] The poly(A) signal was located 17 to 21 nucleotides upstream from the start of the poly(A) tail.

C. PROCESSING OF THE PRECURSOR PROTEIN AND MOLECULAR FORMS

Early characterization of extracts from various regions of dog gastrointestinal tract, mucosa or muscularis, revealed that the majority of M-LI in principle coeluted with M_{1-22},[216] but two additional larger, possible precursor components of M of 14 and 6 kDa, were also demonstrated in another study.[218] An additional peak was also eluted in the void volume, representing either some nonspecific interference in the RIAs, some nonpeptide-related component (not degraded by trypsin), or M-related-LI associated with some protein component.[218] These findings in part agree with an older study by Christofides and co-workers[219] demonstrating two components in the human and porcine gut, but no M-immunoreactive nerves. The first peak eluted at a position before cytochrome c (≥ 12.3 kDa), and the second peak coeluted with M_{1-22} and constituted 50 and 5% of the total immunoreactivity in human and porcine gut, respectively.[219] However, in a more recent study, this two-component pattern was confirmed in man, but the first peak in this study eluted at a position slightly after cytochrome c, suggesting a size slightly smaller than 12.3 kDa;[222] this larger peak is absent from human plasma.[219,222] In addition, cell-free translation in vitro of human intestinal mRNA followed by immunoprecipitation with M antibodies resulted in the identification of a single band with an estimated molecular weight of 14 to 15 kDa.[222] However, the estimated molecular weight of the 119-amino-acid prepro-M has been calculated to be approximately 13 kDa. These data are not at all consistent, and they may indicate species differences in the processing, imprecision in the determination of sizes and weights, and/or additional modifications of the precursor protein structure, although pig prepro-M does not seem to be N-glycosylated.[223] However, one study has suggested a reasonable route for processing of human prepro-M (approximately 14 to 15 kDa based on cell-free translation, but not in accordance with the molecular weight deduced from the cDNA structure). This route of processing may in principle be valid also for prepro-M from other species in which the signal peptide is cleaved off in a first step generating pro-M (approximately 11 to 12 kDa), in reasonable agreement with the chromatographic data.[222] In a second step, pro-M seems to be further processed, first generating an immunoreactive fragment of 6 kDa, suggesting a processing step at one of the single conserved bases in the MRP (see Section VII.B) located approximately in the middle or slightly N-terminally in pro-M. In the third and final step, processing seems to take place at the double basic site immediately C terminally of M, thus liberating M_{1-22} and a nonimmunoreactive (with M antibodies) fragment.

Remaining to be elucidated are whether M indeed is located in the nerves of the gastrointestinal tract, and if so, whether the isolated cDNA also is representative for neuronal M and whether the suggested processing pattern is correct and also representative for the neuronal type of M. Furthermore, the recent cloning of the M gene and several M cDNAs may now allow the regulation of the M-gene expression and mRNA levels to be studied.

VIII. SUMMARY AND CONCLUDING REMARKS

This chapter reviews the current information available concerning some of the more recently cloned gastrointestinal neuropeptides in terms of their chromosomal localization, gene and mRNA organization, and the structure of their precursor protein, as well as their processing (see Table 1). Although these neuropeptides have action diversity, they show apparent similarities in the sense that several of them have genes with similar/identical cis-

TABLE 1
Comparison of Organizational and Regulatory Features in Some Gastrointestinal Neuropeptide Genes

	VIP-PHI/PHM	NT-N	NPY	GRP	Neuromedin-B	GAL	Motilin
Chromosomal localization	6	ND	7	18	15	ND	6
Size of the gene (kb)	9	10.2 (rat)	8	10	ND	≥3 (cow)	9
Number of exons	7	4	4	3	ND	>3 (cow)	5
Size of the encoded precursor protein (number of amino acids)	170		97	148	74		115/114
	170 (rat)	169 (rat)	98 (rat)	147 (rat)		124 (rat)	
		170 (cow)				123 (cow)	
		170 (dog)				123 (pig)	119 (pig)

Increased gene expression, mRNA-levels and/or biosynthesis

	VIP-PHI/PHM	NT-N	NPY	GRP	Neuromedin-B	GAL	Motilin
Protein kinase C	+		+			+	
cAMP	+	+				+	
Depolarization	+	+	+			+	
Glucocorticoids		+	+				
Estrogen	+						
Other hormones	+					+	
Growth factors	+	+	+				
Developmental	+		+	+			
Lithium		+					

Note: ND, not determined; data refer to man if not indicated otherwise (top part of the table).

acting elements and they respond with increased gene expression, increased mRNA levels, and/or peptide levels as a result of elevation of cAMP, protein kinase C, glucocorticoids, estrogen, and/or growth factors. However, although the final outcome may be the same for two different neuropeptides in response to the same stimuli, e.g., increased expression/ production of the neuropeptide, it is not known whether the effect is exerted through either increased transcriptional activity, stimulated RNA processing, changes in mRNA turnover, or stimulation of precursor protein processing. Furthermore, it is difficult to determine the relevance for much of the data reviewed here, in terms of regulation and processing in the gastrointestinal tract, since it has clearly been demonstrated that these processes in many cases are tissue- and/or species-specific events. The data reviewed here may, at present, be regarded as a panorama of possibilities to regulate expression and processing of each individual neuropeptide gene, and much remains to be studied in reference to regulation indicative of the gastrointestinal tract. Thus, the human GAL gene has not been characterized, and the human GAL and motilin genes have not yet been studied in terms of their regulation. Moreover, for several of the peptide genes, studies discriminating between regulation on the transcriptional level and RNA processing and/or mRNA turnover remains to be done. In addition, the function of spacer and flanking peptides needs elucidation, e.g., whether they may signal different steps in transport, processing, and modification, similar to the function of the signal peptide, or whether they serve some biological action per se.

ACKNOWLEDGMENT

The author would like to thank Professors Viktor Mutt and Thomas J. McDonald for their advice, ideas, and encouragement through the work of this chapter; Mrs. Delphi Post for reviewing and correcting the language; and the staff at the Karolinska Institute's Medical Library for their efficiency and speed in making publications available. In addition, the Swedish Medical Research Council (Projects 7906 and 8997) and Skandigen AB are to be acknowledged for their support during the writing of this chapter and the author's studies on galanin.

REFERENCES

1. **Mutt, V.**, Questions answered and raised by work on the chemistry of gastrointestinal and cerebrogastrointestinal hormonal polypeptides, *Chem. Scri.*, 26B, 191, 1986.
2. **McDonald, T. J.**, A historical overview of the gastroenteropancreatic regulatory peptides, in *Neuropeptide, Function in the Gastrointestinal Tract*, Daniel, E. E., Ed., CRC Press, Boca Raton, FL, 1990, 1.
3. **Furness, J. B. and Costa, M.**, Types of nerves in the enteric nervous system, *Neuroscience*, 5, 1, 1980.
4. **Schultzbeg, M., Hökfelt, T., Nilsson, G., Terenius, L., Rehfeld, J. F., Brown, M., Elde, R., Goldstein, M., and Said, S.**, Distribution of peptide- and catecholamine-containing neurons in the gastro-intestinal tract of rat and guinea-pig: immunohistochemical studies with antisera to substance P, vasoactive intestinal polypeptide, enkephalins, somatostatin, gastrin/cholecystokinin, neurotensin and dopamine β-hydroxylase, *Neuroscience*, 5, 689, 1980.
5. **Maniatis, T., Fritsch, E. F., and Sambrook, J.**, *Molecular Cloning: A Laboratory Manual*, Cold Spring Harbor Laboratory, Cold Spring Harbor, NY, 1982.
6. **Treston, A. M. and Mulshine, J. L.**, Beyond transcriptional events, *Nature (London)*, 337, 406, 1989.
7. **Deschenes, R. J., Haun, R. S., Sunkel, D., Roos, B. A., and Dixon, J. E.**, Modulation of cholecystokinin gene expression, *Ann. N.Y. Acad. Sci.*, 448, 53, 1985.
8. **Eysselein, V. E., Reeve, J. R., Jr., and Eberlein, G.**, Cholecystokinin — gene structure, and molecular forms in tissue and blood, *Z. Gastroenterol.*, 24, 645, 1986.
9. **Nakanishi, S.**, Structure and regulation of the preprotachykinin gene, *Trends Neurosci.*, 9, 41, 1986.
10. **Krause, J. E., MacDonald, M. R., and Takeda, Y.**, The polyprotein nature of substance P precursors, *BioEssays*, 10, 62, 1989.

11. **Gubler, U.**, Enkephalin genes, in *Molecular Cloning of Hormone Genes*, Habener, J. F., Ed., Humana Press, Clifton, NJ, 1987, chap. 10.

12. **Lynch, D. R. and Snyder, S. H.**, Enkephalins: focus on biosynthesis, in *Brain Peptides Update*, Vol. 1, Martin, J. B., Brownstein, M. J., and Krieger, D. T., Eds., John Wiley & Sons, New York, 1987, chap. 11.

13. **Rosenfeld, M. G., Leff, S., Amara, S. G., and Evans, R. M.**, Calcitonin and calcitonin-related peptide genes: tissue specific RNA processing, in *Molecular Cloning of Hormone Genes*, Habener, J. F., Ed., Humana Press, Clifton, NJ, 1987, chap. 11.

14. **Goodman, R. H., Montminy, M. R., Low, M. J., and Habener, J. F.**, The somatostatin genes, in *Molecular Cloning of Hormone Genes*, Habener, J. F., Ed., Humana Press, Clifton, NJ, 1987, chap. 5.

15. **Schwartz, T. W.**, Processing of peptide precursors, in *Molecular Biology of the Islet of Langerhans*, Okamoto, H., Ed., Cambridge University Press, London, 1990.

16. **Jaenisch, R.**, Transgenic animals, *Science*, 240, 1468, 1988.

17. **Capecchi, M. R.**, Altering the genome by homologous recombination, *Science*, 244, 1288, 1989.

18. **Lavitrano, M., Cammaioni, A., Fazio, V. M., Dolci, S., Farace, M. G., and Spadafora, C.**, Sperm cells as vectors for inducing foreign DNA into eggs: genetic transformation of mice, *Cell*, 57, 717, 1989.

19. **Galas, D. J. and Schmitz, A.**, DNAase footprinting: a simple method for detection of protein-DNA binding specificity, *Nucleic Acids Res.*, 5, 3157, 1978.

20. **Garner, M. M. and Revzin, A.**, The use of gel electrophoresis to detect and study nucleic acid-protein interactions, *Trends Biochem. Sci.*, 11, 395, 1986.

21. **Crothers, D. M.**, Gel electrophoresis of protein-DNA complexes, *Nature (London)*, 325, 464, 1987.

22. **Siegel, R. E.**, *In situ* hybridization histochemistry, in *Brain Peptides Update*, Vol. 1, Martin, J. B., Brownstein, M. J., and Krieger, D. T., Eds., John Wiley & Sons, New York, 1987, chap. 7.

23. **Gorman, C. M., Moffat, L. F., and Howard, B. H.**, Recombinant genomes that express chlorampenicol acetyltransferase in mammalian cells, *Mol. Cell. Biol.*, 2, 1044, 1982.

24. **Comb, M., Hyman, S. E., and Goodman, H. M.**, Mechanisms of trans-synaptic regulation of gene expression, *Trends Neurosci.*, 10, 473, 1987.

25. **Edlund, T., Walker, M. D., Barr, P. J., and Rutter, W. J.**, Cell-specific expression of the rat insulin gene: evidence for role of two distinct 5′ flanking elements, *Science*, 230, 912, 1985.

26. **De Wet, J. R., Wood, K. V., DeLuca, M., Helinski, D. R., and Subramani, S.**, Firefly luciferase gene: structure and expression in mammalian cells, *Mol. Cell. Biol.*, 7, 725, 1987.

27. **Nordeen, K.**, Luciferase reporter gene vectors for analysis of promoters and enhancers, *BioTechniques*, 6, 454, 1988.

28. **Corden, J., Wasylyk, B., Buchwalder, A., Sassone-Corsi, P., Kedinger, C., and Chambon, P.**, Promoter sequences of eukaryotic protein-coding genes, *Science*, 209, 1406, 1980.

29. **Darnell, J. E., Jr.**, Variety in the level of gene control in eukaryotic cells, *Nature (London)*, 297, 365, 1982.

30. **Breathnach, R. and Chambon, P.**, Organization and expression of eukaryotic split genes coding for proteins, *Annu. Rev. Biochem.*, 50, 349, 1981.

31. **Gluzman, Y., Ed.**, Eukaryotic transcription: the role of *cis*- and *trans*-acting elements in initiation, Current Communications in Molecular Biology, Cold Spring Harbor Laboratory, NY, 1985.

32. **Hayashi, H., Ed.**, Hormonal regulation of gene expression, in *Gunma Symposia on Endocrinology*, Vol. 24, Institute of Endocrinology, Gunma University Center for Academic Publications, Gunma, Japan, 1987.

33. **Maniatis, T., Goodbourn, S., and Fischer, J. A.**, Regulation of inducible and tissue-specific gene expression, *Science*, 236, 1237, 1987.

34. **Nevins, J. R.**, The pathway of eukaryotic mRNA formation, *Annu. Rev. Biochem.*, 52, 441, 1983.

35. **Raghow, R.**, Regulation of messenger RNA turnover in eukaryotes, *Trends Biochem. Sci.*, 12, 358, 1987.

36. **Lübbe, A. and Schaffner, W.**, Tissue-specific gene expression, *Trends Neurosci.*, 8, 100, 1985.

37. **Razin, A. and Riggs, A. D.**, DNA methylation and gene function, *Science*, 210, 604, 1980.

38. **Sassone-Corsi, P. and Borrelli, E.**, Transcriptional regulation by *trans*-acting factors, *Trends Genet.*, 2, 215, 1986.

39. **Walker, M. D., Edlund, T., Boulet, A. M., and Rutter, W. J.**, Cell-specific expression controlled by the 5′-flanking region of insulin and chymotrypsin genes, *Nature (London)*, 306, 557, 1983.

40. **Khoury, G. and Gruss, P.**, Enhancer elements, *Cell*, 33, 313, 1983.

41. **Gruss, P.**, Magic enhancers?, *DNA*, 3, 1, 1984.

42. **Serfling, E., Jasin, M., and Schaffner, W.**, Enhancers and eukaryotic gene transcription, *Trends Genet.*, 1, 224, 1985.

43. **Benoist, C., O'Hare, K., Breathnach, R., and Chambon, P.**, The ovalbumin gene — sequence of putative control regions, *Nucleic Acids Res.*, 8, 127, 1980.

44. **Dynan, W. S. and Tjian, R.**, Control of eukaryotic messenger RNA synthesis by sequence-specific DNA-binding proteins, *Nature (London)*, 316, 774, 1985.

45. **Lee, W., Mitchell, P., and Tjian, R.**, Purified transcription factor AP-1 interacts with TPA-inducible enhancer elements, *Cell*, 49, 752, 1987.

46. **Schüle, R., Muller, M., Kaltschmidt, C., and Renkawitz, R.,** Many transcription factors interact synergistically with steroid receptors, *Science,* 242, 1418, 1988.

47. **Lee, W., Haslinger, A., Karin, M., and Tjian, R.,** Activation of transcription by two factors that bind promoter and enhancer sequences of the human metallothionein gene and SV40, *Nature (London),* 325, 368, 1987.

48. **Mitchell, P. J. and Tjian, R.,** Transcriptional regulation in mammalian cells by sequence-specific DNA binding proteins, *Science,* 245, 371, 1989.

49. **Montminy, M. R. and Bilezikjian, L. M.,** Binding of a nuclear protein to the cyclic-AMP response element of the somatostatin gene, *Nature (London),* 328, 175, 1987.

50. **Angel, P., Imagawa, M., Chiu, R., Stein, B., Imbra, R. J., Rahmsdorf, H. J., Jonat, C., Herrlich, P., and Karin, M.,** Phorbol ester-inducible genes contain a common *cis* element recognized by a TPA-modulated *trans*-acting factor, *Cell,* 49, 729, 1987.

51. **Iguchi-Ariga, S. M. M. and Schaffner, W.,** CgG methylation of the cAMP-responsive enhancer/promoter sequence TGACGTCA abolishes specific factor binding as well as transcriptional activation, *Genes Dev.,* 3, 612, 1989.

52. **Karin, M., Haslinger, A., Holtgreve, H., Richards, R. I., Krauter, P., Westphal, H. M., and Beato, M.,** Characterization of DNA sequences through which cadmium and glucocorticoid hormones induce human metallothionein-IIA gene, *Nature (London),* 308, 513, 1984.

53. **Tora, L., Gaub, M.-P., Mader, S., Dierich, A., Bellard, M., and Chambon, P.,** Cell-specific activity of a GGTCA half-palindromic oestrogen-responsive element in the chicken ovalbumin gene promoter, *EMBO J.,* 7, 3771, 1988.

54. **Beato, M.,** Gene regulation by steroid hormones, *Cell,* 56, 335, 1989.

55. **Martinez, E., Givel, F., and Wahli, W.,** The estrogen-responsive element as an inducible enhancer: DNA sequence requirements and conversion to a glucocorticoid-responsive element, *EMBO J.,* 6, 3719, 1987.

56. **He, X., Treacy, M. N., Simmons, D. M., Ingraham, H. A., Swanson, L. W., and Rosenfeld, M. G.,** Expression of a large family of POU-domain regulatory genes in mammalian brain development, *Nature (London),* 340, 35, 1989.

57. **Crenshaw, E. B., III, Kalla, K., Simmons, D. M., Swanson, L. W., and Rosenfeld, M. G.,** Cell-specific expression of the prolactin gene in transgenic mice is controlled by synergistic interactions between promoter and enhancer elements, *Genes Dev.,* 3, 959, 1989.

58. **Banerjee, A. K.,** 5'-terminal cap structure in eukaryotic messenger ribonucleic acids, *Microbiol. Rev.,* 44, 175, 1980.

59. **Konarska, M. M., Padgett, R. A., and Sharp, P. A.,** Recognition of cap structure in splicing *in vitro* of mRNA precursors, *Cell,* 38, 731, 1984.

60. **Proudfoot, N. J. and Brownlee, G. G.,** 3' non-coding region sequences in eukaryotic messenger RNA, *Nature (London),* 263, 211, 1976.

61. **Brawerman, G.,** The role of the poly(A) sequence in mammalian messenger RNA, *Crit. Rev. Biochem.,* 10, 1, 1981.

62. **Ivell, R., Schmale, H., Krisch, B., Nahke, P., and Richter, D.,** Expression of a mutant vasopressin gene: differential polyadenylation and read-through of the mRNA 3' end in a frame shift mutant, *EMBO J.,* 5, 971, 1986.

63. **Sharp, P. A.,** Splicing of messenger RNA precursors, *Science,* 235, 1987.

64. **Gilbert, W.,** Genes-in-pieces revisited, *Science,* 228, 823, 1985.

65. **Rogers, J. H.,** How were introns inserted into nuclear genes?, *Trends Genet.,* 5, 213, 1989.

66. **Blobel, G. and Doberstein, B.,** Transfer of proteins across membranes. I. Presence of proteolytically processed and unprocessed nascent immunoglobulin light chains on membrane-bound ribosomes of murine myeloma, *J. Cell Biol.,* 67, 852, 1975.

67. **Steiner, D. F., Quinn, P. S., Chan, S. J., Marsh, J., and Tager, H. S.,** Processing mechanisms in the biosynthesis of proteins, *Ann. N.Y. Acad. Sci.,* 343, 1, 1980.

68. **Strauss, A. W. and Boime, I.,** Compartmentation of newly synthesized proteins, *Crit. Rev. Biochem.,* 12, 205, 1982.

69. **Mains, R. E., Eipper, B. A., Glembotski, C. C., and Dores, R. M.,** Strategies for the biosynthesis of bioactive peptides, *Trends Neurosci.,* 6, 229, 1983.

70. **Turner, A. J.,** Neuropeptide processing enzymes, *Trends Neurosci.,* 7, 258, 1984.

71. **Andrews, P. C., Brayton, K., and Dixon, J. E.,** Precursors to regulatory peptides: their proteolytic processing, *Experientia,* 43, 784, 1987.

72. **Gluschankof, P. and Cohen, P.,** Proteolytic enzymes in the post-translational processing of polypeptide hormone precursors, *Neurochem. Res.,* 12, 951, 1987.

73. **Bürger, E.,** Peptide hormones and neuropeptides, *Arzneim. Forsch. Drug Res.,* 38, 754, 1988.

74. **Kozak, M.,** Compilation and analysis of sequences upstream from the translational start site in eukaryotic mRNAs, *Nucleic Acids Res.,* 12, 857, 1984.

75. **Kozak, M.,** Point mutations define a sequence flanking the AUG initiator codon that modulates translation by eukaryotic ribosomes, *Cell,* 44, 283, 1986.
76. **von Heijne, G.,** Patterns of amino acids near signal sequences, *Eur. J. Biochem.,* 133, 17, 1983.
77. **Perlman, D. and Halvorson, H. O.,** A putative signal peptidase recognition site and sequence in eukaryotic and prokaryotic signal peptides, *J. Mol. Biol.,* 167, 391, 1983.
78. **Watson, M. E. E.,** Compilation of published signal sequences, *Nucleic Acids Res.,* 12, 5145, 1984.
79. **Skidgel, R. A.,** Basic carboxypeptidases: regulators of peptide hormone activity, *Trends Pharmacol. Sci.,* 9, 299, 1988.
80. **Fricker, L. D., Adelman, J. P., Douglass, J., Thompson, R. C., von Strandmann, R. P., and Hutton, J.,** Isolation and sequence analysis of cDNA for rat carboxypeptidase E [EC3.4.17.10], a neuropeptide processing enzyme, *Mol. Endocrinol.,* 3, 666, 1989.
81. **Yanagisawa, M., Kurihara, H., Kimura, S., Tomobe, Y., Kobayashi, M., Mitsui, Y., Yazaki, Y., Goto, K., and Masaki, T.,** A novel potent vasoconstrictor peptide produced by vascular endothelial cells, *Nature (London),* 332, 411, 1988.
82. **Kageyama, R., Ohkubo, H., and Nakanishi, S.,** Primary structure of human preangiotensinogen deduced from the cloned cDNA sequence, *Biochemistry,* 23, 3603, 1984.
83. **Bradbury, A. F. and Smyth, D. G.,** Biosynthesis of the C-terminal amide in peptide hormones, *Biosci. Rep.,* 7, 907, 1987.
84. **Eipper, B. A. and Mains, R. E.,** Peptide alpha-amidation, *Annu. Rev. Physiol.,* 50, 333, 1988.
85. **Eipper, B. A., Park, L. P., Dickerson, I. M., Keutmann, H. T., Thiele, E. A., Rodriguez, H., Schofield, P. R., and Mains, R. E.,** Structure of the precursor to an enzyme mediating COOH-terminal amidation in peptide biosynthesis, *Mol. Endocrinol.,* 1, 777, 1987.
86. **Busby, W. H., Quackenbush, G. E., Humm, J., Youngblood, W. W., and Kizer, J. S.,** An enzyme(s) that converts glutaminyl-peptides into pyroglutamyl-peptides, *J. Biol. Chem.,* 262, 8532, 1987.
87. **Sharon, N. and Lis, H.,** Glycoproteins: research booming on long-ignored ubiquitous compounds, *Mol. Cell. Biochem.,* 42, 167, 1982.
88. **Towler, D. A., Eubanks, S. R., Towery, D. S., Adams, S. P., and Glaser, L.,** Amino-terminal processing of proteins by *N*-myristoylation, *J. Biol. Chem.,* 262, 1030, 1987.
89. **Huttner, W. B.,** Tyrosine sulfation and the secretory pathway, *Annu. Rev. Physiol.,* 50, 363, 1988.
90. **Bodner, M., Fridkin, M., and Gozes, I.,** Coding sequences for vasoactive intestinal peptide and PHM-27 peptide are located on two adjacent exons in the human genome, *Proc. Natl. Acad. Sci. U.S.A.,* 82, 3548, 1985.
91. **Tsukuda, T., Horovitch, S. J., Montminy, M. R., Mandel, G., and Goodman, R. H.,** Structure of the human vasoactive intestinal polypeptide gene, *DNA,* 4, 293, 1985.
92. **Gozes, I., Bodner, M., Shani, Y., and Fridkin, M.,** Structure and expression of the vasoactive intestinal peptide (VIP) gene in a human tumor, *Peptides,* 7, 1, 1986.
93. **Linder, S., Barkhem, T., Norberg, A., Persson, H., Schalling, M., Hökfelt, T., and Magnusson, G.,** Structure and expression of the gene encoding the vasoactive intestinal peptide precursor, *Proc. Natl. Acad. Sci. U.S.A.,* 84, 605, 1987.
94. **Yamagami, T., Ohsawa, K., Nishizawa, M., Inoue, C., Gotoh, E., Yanaihara, N., Yamamoto, H., and Okamoto, H.,** Complete nucleotide sequence of human vasoactive intestinal peptide/PHM-27 gene and its inducible promoter, *Ann. N.Y. Acad. Sci.,* 527, 87, 1988.
95. **Ohsawa, K., Yamagami, T., Nishizawa, M., Inoue, C., Gotoh, E., Yamamoto, H., and Okamoto, H.,** Structure and expression of human vasoactive intestinal peptide/PHM-27 gene, *Gunma Symp. Endocrinol.,* 24, 169, 1987.
96. **Gozes, I., Avidor, R., Yahav., Y., Katznelson, D., Croce, C. M., and Huebner, K.,** The gene encoding vasoactive intestinal peptide is located on human chromosome 6p21→6qter, *Hum. Genet.,* 75, 41, 1987.
97. **Gozes, I., Nakai, H., Byers, M., Avidor, R., Weinstein, Y., Shani, Y., and Shows, T. B.,** Sequential expression in the nervous system of c-MYB and VIP genes, located in human chromosomal region 6 q24, *Somatic Cell Mol. Genet.,* 13, 305, 1987.
98. **Itoh, N., Obata, K.-I., Yanaihara, N., and Okamoto, H.,** Human preprovasoactive intestinal polypeptide contains a novel PHI-27-like peptide, PHM-27, *Nature (London),* 304, 547, 1983.
99. **Bloom, S. R., Christofides, N. D., Delamarter, J., Buell, G., Kawashima, E., and Polak, J.,** Diarrhoea in vipoma patients associated with cosecretion of a second active peptide (peptide histidine isoleucine) explained by single coding gene, *Lancet,* 2, 1163, 1983.
100. **Nishizawa, M., Hayakawa, Y., Yanaihara, N., and Okamoto, H.,** Nucleotide sequence divergence and functional constraint in VIP precursor mRNA evolution between human and rat, *FEBS Lett.,* 183, 55, 1985.
101. **Gozes, I., Shani, Y., and Rostène, W. H.,** Developmental expression of the VIP-gene in brain and intestine, *Mol. Brain Res.,* 2, 137, 1987.
102. **Yanaihara, N., Suzuki, T., Sato, H., Hoshino, M., Okaru, Y., and Yanaihara, C.,** Dibutyryl cAMP stimulation of production and release of VIP-like immunoreactivity in a human neuroblastoma cell line, *Biomed. Res.,* 2, 728, 1981.

103. **Hayakawa, Y., Obata, K.-I., Itoh, N., Yanaihara, N., and Okamoto, H.,** Cyclic AMP regulation of pro-vasoactive intestinal polypeptide/PHM-27 synthesis in human neuroblastoma cells, *J. Biol. Chem.*, 259, 9207, 1984.

104. **Ohsawa, K., Hayakawa, Y., Nishizawa, M., Yamagnami, T., Yamamoto, H., Yanaihara, N., and Okamoto, H.,** Synergistic stimulation of VIP/PHM-27 gene expression by cyclic AMP and phorbol esters in human neuroblastoma cells, *Biochem. Biophys. Res. Commun.*, 132, 885, 1985.

105. **Fink, J. S., Verhave, M., Kasper, S., Tsukada, T., Mandel, G., and Goodman, R. H.,** The CGTCA sequence motif is essential for biological activity of the vasoactive intestinal peptide gene cAMP-regulated enhancer, *Proc. Natl. Acad. Sci. U.S.A.*, 85, 6662, 1988.

106. **Waschek, J. A., Hsu, C.-M., and Eiden, L. E.,** Lineage-specific regulation of the vasoactive intestinal peptide gene in neuroblastoma cells in conferred by 5.2 kilobases of 5'-flanking sequence, *Proc. Natl. Acad. Sci. U.S.A.*, 85, 9547, 1988.

107. **Brick, P. L., Howlett, A. C., and Beinfeld, M. C.,** Synthesis and release of vasoactive intestinal polypeptide (VIP) by mouse neuroblastoma cells: modulation by cyclic nucleotides and ascorbic acid, *Peptides*, 6, 1075, 1985.

108. **Beinfeld, M. C., Brick, P. L., Howlett, A. C., Holt, I. L., Pruss, R. B., Moskal, J. R., and Eiden, L. E.,** The regulation of vasoactive intestinal peptide synthesis in neuroblastoma and chromaffin cells, *Ann. N.Y. Acad. Sci.*, 527, 68, 1988.

109. **Waschek, J. A., Pruss, R. M., Siegel, R. E., Eiden, L. E., Bader, M.-F., and Aunis, D.,** Regulation of enkephalin, VIP, and chromogranin biosynthesis in actively secreting chromaffin cells: multiple strategies for multiple peptides, *Ann. N.Y. Acad. Sci.*, 493, 308, 1988.

110. **Gozes, I. and Shani, Y.,** Hypothalamic vasoactive intestinal peptide messenger ribonucleic acid is increased in lactating rats, *Endocrinology*, 119, 2497, 1986.

111. **Gozes, I. and Tsafriri, A.,** Detection of vasoactive intestinal peptide-encoding messenger ribonucleic acid in the rat ovaries, *Endocrinology*, 119, 2606, 1986.

112. **Gozes, I.,** Biosynthesis and regulation of expression: the vasoactive intestinal peptide gene, *Ann. N.Y. Acad. Sci.*, 527, 77, 1988.

113. **Nobou, F., Besson, J., Rostene, W., and Rosselin, G.,** Ontogeny of vasoactive intestinal peptide and somatostatin in different structures of the rat brain: effects of hypo- and hypercorticism, *Dev. Brain Res.*, 20, 296, 1985.

114. **Woodhams, P. L., McGovern, J., McGregor, G. P., O'Shaughnessy, D. J., Ghatei, M. A., Adrian, T. H., Lee, Y., Polak, J. M., Bloom, S. R., and Balazs, R.,** Effects of changes in neonatal thyroid status on the development of neuropeptide systems in the rat brain, *Int. J. Dev. Neurosci.*, 1, 155, 1983.

115. **Tischler, A. S., Lee, Y. C., Perlman, R. L., Costopoulos, D., Slayton, V. W., and Bloom, S. R.,** Production of "ectopic" vasoactive intestinal peptide-like and neurotensin-like immunoreactivity in human pheochromocytoma cell cultures, *J. Neurosci.*, 4, 1398, 1984.

116. **Tischler, A. S., Lee, Y. C., Perlman, R. L., Costopoulos, D., and Bloom, S. R.,** Production of "ectopic" vasoactive intestinal peptide-like immunoreactivity in normal human chromaffin cell cultures, *Life Sci.*, 37, 1881, 1985.

117. **Gozes, I., Giladi, E., and Shani, Y.,** Vasoactive intestinal peptide gene: putative mechanism of information storage at the RNA level, *J. Neurochem.*, 48, 1136, 1987.

118. **Gozes, I., Schächter, P., Shani, Y., and Giladi, E.,** Vasoactive intestinal peptide gene expression from embryos to aging rats, *Neuroendocrinology*, 47, 27, 1988.

119. **Fahrenkrug, J.,** Evidence for common precursors but differential processing of VIP and PHM in VIP-producing tumors, *Peptides*, 6, 357, 1985.

120. **Pless, D. D. and Lennarz, W. J.,** Enzymatic conversion of proteins to glycoproteins, *Proc. Natl. Acad. Sci. U.S.A.*, 74, 134, 1977.

121. **Obata, K.-I., Itoh, N., Okamoto, H., Yanaihara, C., Yanaihara, N., and Suzuki, T.,** Identification and processing of biosynthetic precursors to vasoactive polypeptide in human neuroblastoma cells, *FEBS Lett.*, 136, 123, 1981.

122. **Hoshino, M., Yanaihara, C., Ogino, K., Iguchi, K., Sato, H., Suzuki, T., and Yanaihara, N.,** Production of VIP- and PHM (human PHI)-related peptides in human neuroblastoma cells, *Peptides*, 5, 155, 1984.

123. **Yiangou, Y., Christofides, N. D., Blank, M. A., Yanaihara, N., Tatemoto, K., Bishop, A. E., Polak, J. M., and Bloom, S. R.,** Molecular forms of peptide histidine isoleucine-like immunoreactivity in the gastrointestinal tract, *Gastroenterology*, 89, 516, 1985.

124. **Yiangou, Y., Requejo, F., Polak, J. M., and Bloom, S. R.,** Characterization of a novel prepro VIP derived peptide, *Biochem. Biophys. Res. Commun.*, 139, 1142, 1986.

125. **Yiangou, Y., Williams, S. J., Bishop, A. E., Polak, J. M., and Bloom, S. R.,** Peptide histidine-methionine immunoreactivity in plasma and tissue from patients with vasoactive intestinal peptide-secreting tumors and watery diarrhea syndrome, *J. Clin. Endocrinol. Metab.*, 64, 131, 1987.

126. **Yiangou, Y., Di Marzo, V., Spokes, R. A., Panicio, M., Morris, H. R., and Bloom, S. R.**, Isolation, characterization, and pharmacological actions of peptide histidine valine 42, a novel prepro-vasoactive intestinal peptide-derived peptide, *J. Biol. Chem.*, 262, 14010, 1987.

127. **Gozes, I., Bodner, M., Shwartz, H., Shani, Y., and Fridkin, M.**, Studies toward the biosynthesis of vasoactive intestinal peptide (VIP), *Peptides*, 5, 161, 1984.

128. **Rökaeus, Å.**, Studies on neurotensin as a hormone: assay and release of neurotensin-like immunoreactivity and effects of neurotensin, *Acta Physiol. Scand. Suppl.*, 501, 1981.

129. **Kislauskis, E., Bullock, B., McNeil, S., and Dobner, P. R.**, The rat gene encoding neurotensin and neuromedin N, *J. Biol. Chem.*, 263, 4963, 1988.

130. **Dobner, P. R., Barner, D. L., Villa-Komaroff, L., and McKiernan, C.**, Cloning and sequence analysis of cDNA for the canine neurotensin/neuromedin N precursor, *Proc. Natl. Acad. Sci. U.S.A.*, 84, 3516, 1987.

131. **Rökaeus, Å, Fried, G., and Lundberg, J. M.**, Occurrence, storage and release of neurotensin-like immunoreactivity from the adrenal gland, *Acta Physiol. Scand.*, 120, 373, 1984.

132. **Tischler, A. S., Lee, Y. C., Slayton, V. W., Jason, W. J., and Bloom, S. R.**, Content and release of neurotensin in PC12 pheochromocytoma cell cultures: modulation by dexamethasone and nerve growth factor, *Regul. Peptides*, 3, 415, 1982.

133. **Tischler, A. S., Lee, Y. C., Coustopoulos, D., Slayton, V. W., Jason, W. J., and Bloom, S. R.**, Cooperative regulation of neurotensin content in PC12 pheochromocytoma cell cultures: effects of nerve growth factor, dexamethasone, and activators of adenylate cyclase, *J. Neurosci.*, 6, 1719, 1986.

134. **Zeytinoglu, F. N., Brazeau, P., and Mougin, C.**, Regulation of neurotensin secretion in a mammalian C cell line: effect of dexamethasone, *Regul. Peptides*, 6, 147, 1983.

135. **Dobner, P. R., Tischler, A. S., Lee, Y. C., Bloom, S. R., and Donahue, S. R.**, Lithium dramatically potentiates neurotensin/neuromedin N gene expression, *J. Biol. Chem.*, 263, 13983, 1988.

136. **Drummond, A. H.**, Lithium and inositol lipid-linked signalling mechanisms, *Trends Pharmacol. Sci.*, 8, 129, 1987.

137. **Hammer, R. A., Leeman, S. R., Carraway, R., and Williams, R. H.**, Isolation of human intestinal neurotensin, *J. Biol. Chem.*, 255, 2476, 1980.

138. **Minamino, N., Kangawa, K., and Matsuo, H.**, Neuromedin N: a novel neurotensin-like peptide identified in porcine spinal cord, *Biochem. Biophys. Res. Commun.*, 122, 542, 1984.

139. **Shaw, C. and Conlon, M.**, Isolation and characterization of guinea pig neurotensin, *Can. J. Physiol. Pharmacol.*, 21 (Abstr.), 146, 1986.

140. **Carraway, R. and Bhatnagar, Y. M.**, Isolation, structure and biological activity of chicken intestinal neurotensin, *Peptides*, 1, 167, 1980.

141. **Carraway, R. E. and Ferris, C. F.**, Isolation, biological and chemical characterization, and synthesis of a neurotensin-related hexapeptide from chicken intestine, *J. Biol. Chem.*, 258, 2475, 1983.

142. **Sures, I. and Crippa, M.**, Xenopsin: the neurotensin-like octapeptide from *Xenopus* skin at the carboxyl terminus of its precursor, *Proc. Natl. Acad. Sci. U.S.A.*, 81, 380, 1984.

143. **Cotrait, M.**, Conformational study of the neurotensin and substance P fragments: Pro-Arg-Arg-Pro and Arg-Pro-Lys-Pro, *Int. J. Peptide Protein Res.*, 22, 110, 1983.

144. **Theodorsson-Norheim, E., Öberg, K., Rosell, S., and Boström, H.**, Neurotensinlike immunoreactivity in plasma and tumor tissue from patients with endocrine tumors of the pancreas and gut, *Gastroenterology*, 85, 881, 1983.

145. **Mogard, M. H., Kobayashi, R., Chen, C.-F., Lee, T. D., Reeve, J. R., Jr., Shively, J. E., and Walsh, J. H.**, The amino acid sequence of kinetensin, a novel peptide isolated from pepsin-treated human plasma: homology with human serum albumin, neurotensin and angiotensin, *Biochem. Biophys. Res. Commun.*, 136, 983, 1986.

146. **Carraway, R. E., Mitra, S. P., and Cochrane, D. E.**, Structure of a biologically active neurotensin-related peptide obtained from pepsin-treated albumin(s), *J. Biol. Chem.*, 262, 5968, 1987.

147. **Carraway, R. and Mitra, S. P.**, Potential precursors to neuromedin N in the cat and their processing by pepsin, *Endocrinology*, 120, 2101, 1987.

148. **Carraway, R. and Mitra, S. P.**, The use of radioimmunoassay to compare the tissue and subcellular distributions of neurotensin and neuromedin N in the cat, *Endocrinology*, 120, 2093, 1987.

149. **Carraway, R. E., Ruane, S. E., and Ritsema, R. S.**, Radioimmunoassay for Lys^8, Asn^9, Neurotensin 8-13: tissue and subcellular distribution of immunoreactivity in chickens, *Peptides*, 4, 111, 1982.

150. **Takeuchi, T., Gumucio, D., Meisler, M., Minth, C., Dixon, J., Eddy, R., Shows, T. B., and Yamada, T.**, Genes encoding pancreatic polypeptide and neuropetide Y are on human chromosomes 17 and 7, *J. Clin. Invest.*, 77, 1038, 1986.

151. **Minth, C. D., Andrews, P. C., and Dixon, J. E.**, Characterization, sequence, and expression of the cloned human neuropeptide Y gene, *J. Biol. Chem.*, 261, 11974, 1986.

152. **Larhammar, D., Ericsson, A., and Persson, H.**, Structure and expression of the rat neuropeptide Y gene, *Proc. Natl. Acad. Sci. U.S.A.*, 84, 2068, 1987.

153. **Minth, C. D., Bloom, S. R., Polak, J. M., and Dixon, J. E.,** Cloning, characterization, and DNA sequence of a human cDNA encoding neuropeptide tyrosine, *Proc. Natl. Acad. Sci. U.S.A.,* 81, 4577, 1984.

154. **Allen, J., Novotný, J., Martin, J., and Heinrich, G.,** Molecular structure of mammalian neuropeptide Y: analysis by molecular cloning and computer-aided comparison with crystal structure of avian homologue, *Proc. Natl. Acad. Sci. U.S.A.,* 84, 2532, 1987.

155. **Higuchi, H., Yang, H.-Y.T., and Sabol, S. L.,** Rat neuropeptide Y precursor gene expression: mRNA structure, tissue distribution, and regulation by glucocorticoids, cyclic AMP, and phorbol ester, *J. Biol. Chem.,* 263, 6288, 1988.

156. **Dickerson, I. M., Dixon, J. E., and Mains, R.,** Transfected human neuropeptide Y cDNA expression in mouse pituitary cells, *J. Biol. Chem.,* 262, 13646, 1987.

157. **Allen, J. M., Martin, J. B., and Heinrich, G.,** Neuropeptide Y gene expression in PC12 cells and its regulation be nerve growth factor: a model for developmental regulation, *Mol. Brain Res.,* 3, 39, 1987.

158. **Allen, J. M., Tischler, A. S., Lee, Y. C., and Bloom, S. R.,** Neuropeptide Y (NPY) in PC12 phaeo-chromocytoma cultures: responses to dexamethasone and nerve growth factor, *Neurosci. Lett.,* 46, 291, 1984.

159. **Yeats, J. C., Allen, J. M., Bloom, S. R., Leigh, P. J., and MacDermot, J.,** Neuropeptide Y in neuroblastoma x glioma hybrid cells, *FEBS Lett.,* 163, 57, 1983.

160. **de Quidt, M. E. and Emson, P. C.,** Neuropeptide Y in the adrenal gland: characterization, distribution, and drug effects, *Neuroscience,* 19, 1011, 1986.

161. **Fischer-Colbrie, R., Iacangelo, A., and Eiden, L.,** Neural and humoral factors separately regulate neuropeptide Y, enkephalin, and chromogranin A and B mRNA levels in rat adrenal medulla, *Proc. Natl. Acad. Sci. U.S.A.,* 85, 3240, 1988.

162. **Schalling, M., Ericsson, A., Persson, H., Chai, S.-Y., Dagerlind, Å, Brene, S., Seroogy, K., Lundberg, J. M., Larhammar, D., Terenius, L., Massoulie, J., Goldstein, M., and Hökfelt, T.,** Expression of neuropeptide Y mRNA in neuronal and non-neuronal tissue and its response to reserpine treatment and stress — *in situ* hybridization and immunohistochemistry, *Soc. Neurosci. Abstr.,* 13, 1286, 1987.

163. **Ericsson, A., Schalling, M., McIntyre, K. R., Lundberg, J. M., Larhammar, D., Seroogy, K., Hökfelt, T., and Persson, H.,** Detection of neuropeptide Y and its mRNA in megakaryocytes: enhanced levels in certain autoimmune mice, *Proc. Natl. Acad. Sci. U.S.A.,* 84, 5585, 1987.

164. **Ivell, R., Emson, P., and Richter, D.,** Human neuropeptide Y, somatostatin and vasopressin precursors identified in cell-free translations of hypothalamic mRNA, *FEBS Lett.,* 177, 175, 1984.

165. **Adrian, T. E., Allen, J. M., Terenghi, G., Bacarese-Hamilton, A. J., Brown, M. J., Pollak, J. M., and Bloom, S. R.,** Neuropeptide Y in phaeochromocytomas and ganglioneuroblastomas, *Lancet,* 2, 540, 1983.

166. **Majane, E. A., Alho, H., Kataoka, Y., Lee, C. H., and Yang, H.-Y.T.,** Neuropeptide Y in bovine adrenal glands: distribution and characterization, *Endocrinology,* 117, 1162, 1985.

167. **Allen, J. M., Hughes, J., and Bloom, S. R.,** Presence, distribution, and pharmacological effects of neuropeptide Y in mammalian gastrointestinal tract, *Dig. Dis. Sci.,* 32, 506, 1987.

168. **Allen, J. M., Polak, J. M., and Bloom, S. R.,** Presence of the predicted C-flanking peptide of neuropeptide Y (CPON) in tissue extracts, *Neuropeptides,* 6, 95, 1985.

169. **Gulbenkian, S., Wharton, J., Hacker, G. W., Varndell, I. M., Bloom, S. R., and Polak, J. M.,** Co-localization of neuropeptide tyrosine (NPY) and its C-terminal flanking peptide (C-PON), *Peptides,* 6, 1237, 1985.

170. **Cailliez, D., Danger, J.-M., Polak, J. M., Pelletier, G., Andersen, A. C., Leboulenger, F., and Vaundry, H.,** Co-distribution of neuropeptide Y and its C-terminal flanking peptide in the brain and pituitary of the frog, *Rana ridibunda, Neurosci. Lett.,* 74, 163, 1987.

171. **Spindel, E.,** Mammalian bombesin-like peptides, *Trends Neurosci.,* 9, 130, 1986.

172. **Minamino, N., Kangawa, K., and Matsuo, H.,** Neuromedin C: a bombesin-like peptide identified in porcine spinal cord, *Biochem. Biophys. Res. Commun.,* 119, 14, 1984.

173. **Minamino, N., Kangawa, K., and Matsuo, H.,** Neuromedin B: a novel bombesin-like peptide identified in porcine spinal cord, *Biochem. Biophys. Res. Commun.,* 119, 14, 1984.

174. **Erspamer, V., Go, V. L. W., Battey, J. F., Bloom, S. R., McDonald, T. J., Makhlouf, G. M., Minamino, N., Negri, L., Reeve, J. R., Rivier, J., Spindel, E. R., Taché, Y., and Walsh, J. H.,** Nomenclature meeting: report and recommendations 14 October 1987, *Ann. N.Y. Acad. Sci.,* 547, 1, 1988.

175. **Namba, M., Ghatei, M. A., Bishop, A. E., Gibson, S. J., Mann, D. J., Polak, J. M., and Bloom, S. R.,** Presence of neuromedin B-like immunoreactivity in the brain and gut of rat and guinea-pig, *Peptides,* 6, 257, 1985.

176. **Sakamato, A., Kitamura, K., Haraguchi, Y., Yoshida, T., and Tanaka, K.,** Immunoreactive neuro-medin B and neuromedin C: distribution and molecular heterogeneity in rat and human tissue extracts, *Am. J. Gastroenterol.,* 82, 1035, 1987.

177. **Lebacq-Verheyden, A. M., Bertness, V., Kirsch, I., Hollis, G. F., McBride, O. W., and Battey, J.,** Human gastrin-releasing peptide gene maps to chromosome band 18q21, *Somatic Cell Mol. Genet.,* 13, 81, 1987.

178. **Naylor, S. L., Sakaguchi, A. Y., Spindel, E., and Chin, W. W.,** Human gastrin-releasing peptide gene is located on chromosome 18, *Somat. Cell Mol. Genet.,* 13, 81, 1987.

179. **Spindel, E. R., Zilberberg, M. D., and Chin, W. W.,** Analysis of the gene and multiple messenger ribonucleic acids (mRNAs) encoding human gastrin-releasing peptide: alternate RNA splicing occurs in neural and endocrine tissue, *Mol. Endocrinol.,* 1, 224, 1987.

180. **Markowitz, S., Krystal, G., Lebacq-Verheyden, A.-M., Way, J., Sausville, E. A., and Battey, J. F.,** Transcriptional activation and DNase I hypersensitive sites are associated with selective expression of the gastrin-releasing peptide gene, *J. Clin. Invest.,* 82, 808, 1988.

181. **Sausville, E. A., Lebacq-Verheyden, A. M., Spindel, E. R., Cuttitta, F., Gazdar, A. F., and Battey, J. F.,** Expression of the gastrin-releasing peptide gene in human small cell lung cancer: evidence for alternative processing resulting in three distinct mRNAs, *J. Biol. Chem.,* 261, 2451, 1986.

182. **Spindel, E. R., Chin, W. W., Price, J., Rees, L. H., Besser, G. M., and Habener, J. F.,** Cloning and characterization of cDNAs encoding human gastrin-releasing peptide, *Proc. Natl. Acad. Sci. U.S.A.,* 81, 5699, 1984.

183. **Spindel, E. R., Zilberberg, M. D., Habener, J. F., and Chin, W. W.,** Two prohormones for gastrin-releasing peptide are encoding by two different mRNAs differing by 19 nucleotides, *Proc. Natl. Acad. Sci. U.S.A.,* 83, 19, 1986.

184. **Cuttitta, F., Carney, D. N., Mulshine, J., Moody, T. W., Fedorko, J., Fischler, A., and Minna, J. D.,** Bombesin-like peptides can function as autocrine growth factors in human small-cell lung cancer, *Nature (London),* 316, 823, 1985.

185. **Lebacq-Verheyden, A. M., Krystal, G., Sartor, O., Way, J., and Battey, J. F.,** The rat prepro gastrin releasing peptide gene is transcribed from two intiation sites in the brain, *Mol. Endocrinol.,* 2, 556, 1988.

186. **Carr, F. E., Need, L. R., and Chin, W. W.,** Isolation and characterization of the rat thyrotropin β-subunit gene, *J. Biol. Chem.,* 296, 981, 1987.

187. **Sunday, M.,** Tissue-specific expression of the mammalian bombesin gene, *Ann. N.Y. Acad. Sci.,* 547, 95, 1988.

188. **Cuttitta, F., Fedorko, J., Gu, J., Lebacq-Verheyden, A.-M., Linnoila, R. I., and Battey, J. F.,** Gastrin-releasing peptide gene-associated peptides are expressed in normal human fetal lung and small cell lung cancer: a novel peptide family found in man, *J. Clin. Endocrinol. Metab.,* 67, 576, 1988.

189. **Dockray, G. J., Vaillant, C., and Walsh, J. H.,** The neuronal origin of bombesin-like immunoreactivity in the rat gastrointestinal tract, *Neuroscience,* 4, 1561, 1979.

190. **Price, J., Penman, E., Wass, J. A. H., and Rees, L. H.,** Bombesin-like immunoreactivity in human gastrointestinal tract, *Regul. Peptides,* 9, 1, 1984.

191. **Reeve, J. R., Jr., Walsh, J. H., Chew, P., Clark, B., Hawke, D., and Shively, J. E.,** Amino acid sequences of three bombesin-like peptides from canine intestine extracts, *J. Biol. Chem.,* 258, 5582, 1983.

192. **Roth, K. A., Evans, C. J., Lorenz, R. G., Weber, E., Barchas, J. D., and Chang, J.-K.,** Identification of gastrin releasing peptide-related substances in guinea pig and rat brain, *Biochem. Biophys. Res. Commun.,* 112, 528, 1983.

193. **Roth, K. A., Evans, C. J., Weber, E., Barchas, J. D., Bostwick, D. G., and Bensch, K. G.,** Gastrin-releasing peptide-related peptides in a human malignant lung carcinoid tumor, *Cancer Res.,* 43, 5411, 1983.

194. **Yoshizaki, K., de Bock, V., and Slomon, S.,** Origin of bombesin-like peptides in human fetal lung, *Life Sci.,* 34, 835, 1984.

195. **Orloff, M. S., Reeve, Jr., J. R., Miller Ben-Avram, C., Shively, J. E., and Walsh, J. H.,** Isolation and sequence analysis of human bombesin-like peptides, *Peptides,* 5, 865, 1984.

196. **Battey, J. F., Lebacq-Verheyden, A.-M., Krystal, G., Markowitz, S., Sartor, O., and Way, J.,** Regulation of the expression of the human preprogastrin-releasing peptide gene and post-translational processing of its gene product, *Ann. N.Y. Acad. Sci.,* 547, 31, 1988.

197. **Minamino, N., Sudoh, T., Kangawa, K., and Matsuo, H.,** Neuromedin B-32 and B-30: two "big" neuromedin B identified in porcine brain and spinal cord, *Biochem. Biophys. Res. Commun.,* 130, 685, 1985.

198. **Krane, I. M., Naylor, S. L., Helin-Davis, D., Chin, W. W., and Spindel, E. R.,** Molecular cloning of cDNAs encoding human bombesin-like peptide neuromedin B, *J. Biol. Chem.,* 263, 13317, 1988.

199. **Rökaeus, Å.,** Galanin: a newly isolated biologically active neuropeptide, *Trends Neurosci.,* 10, 158, 1987.

200. **Bauer, F. E., Adrian, T. E., Christofides, N. D., Ferri, G. L., Yanaihara, N., Polak, J. M., and Bloom, S. R.,** Distribution and molecular heterogeneity of galanin in human, pig, guinea pig and rat gastrointestinal tracts, *Gastroenterology,* 91, 877, 1986.

201. **Rökaeus, Å.,** unpublished data, 1987.

202. **Tatemoto, K., Rökaeus, Å., Jörnvall, H., McDonald, T. J., and Mutt, V.,** Galanin — a novel biologically active peptide from porcine intestine, *FEBS Lett.,* 164, 124, 1983.

203. **Rökaeus, Å. and Brownstein, M.**, Construction of a porcine adrenal medullary cDNA library and nucleotide sequence analysis of two clones encoding a galanin precursor, *Proc. Natl. Acad. Sci. U.S.A.*, 83, 6287, 1986.

204. **Vrontakis, M. E., Peden, L. M., Duckworth, M. L., and Friesen, H. G.**, Isolation and characterization of a complementary DNA (galanin) clone from estrogen-induced pituitary tumor messenger RNA, *J. Biol. Chem.*, 262, 16755, 1987.

205. **Kaplan, L. M., Spindel, E. R., Isselbacher, K. J., and Chin, W. W.**, Tissue-specific expression of the rat galanin gene, *Proc. Natl. Acad. Sci. U.S.A.*, 85, 1065, 1988.

206. **Rökaeus, Å. and Carlquist, M.**, Nucleotide sequence analysis of cDNAs encoding a bovine galanin precursor protein in the adrenal medulla and chemical isolation of bovine gut galanin, *FEBS Lett.*, 234, 400, 1988.

207. **Bauer, F. E., Adrian, T. E., Yanaihara, N., Polak, J. M., and Bloom, S. R.**, Chromatographic evidence for high-molecular-mass galanin immunoreactivity in pig and cat adrenal glands, *FEBS Lett.*, 201, 327, 1986.

208. **Bauer, F. E., Hacker, G. W., Terenghi, G., Adrian, T. E., Polak, J. M., and Bloom, S. R.**, Localization and molecular forms of galanin in human adrenals: elevated levels in pheochromocytomas, *J. Clin. Endocrinol. Metab.*, 63, 1372, 1986.

209. **Rökaeus, Å. and Young, W. S., III**, Galanin cell bodies in the pig and rat located by *in situ* hybridization histochemistry, *Soc. Neurosci. (Abstr.)*, 157, 2, 1986.

210. **Eiden, L., Eskay, R., Fischer-Colbrie, R., Pruss, R., Rausch, D., Rökaeus, Å., and Waschek, J.**, First and second messenger control of neuropeptide synthesis and secretion in neuroendocrine cells, *Neurochem. Int.*, 13 (Suppl. 1), 39 (Abstr. W7; 2), 1988.

211. **Rökaeus, Å., Pruss, R. M., and Eiden, L. E.**, Galanin gene expression in chromaffin cells is controlled by calcium and protein kinase signalling pathways, *Endocrinol.*, 1990, submitted for publication.

212. **Rökaeus, Å., Young, W. S., III, and Mezey, É.**, Galanin coexists with vasopressin in the normal rat hypothalamus and galanin's synthesis is increased in the Brattleboro (diabetes insipidus) rat, *Neurosci. Lett.*, 90, 45, 1988.

213. **Kaplan, L. M., Gabriel, S. M., Koenig, J. I., Sunday, M. E., Spindel, E. R., Martin, J. B., and Chin, W. W.**, Galanin is an estrogen-inducible, secretory product of the rat anterior pituitary, *Proc. Natl. Acad. Sci. U.S.A.*, 85, 7408, 1988.

214. **Rökaeus, Å. and Waschek, J. A.**, unpublished data, 1989.

215. **Rökaeus, Å. and Carlquist, M.**, unpublished data, 1988.

216. **Chey, W. Y., Escoffery, R., Roth, F., Chang, T. M., and Yajima, H.**, Motilin-like immunoreactivity (MLI) in the gut and neurons of peripheral and central nervous system, *Regul. Peptides*, 1 (Suppl. 1), S19, 1980.

217. **Fox, J. E. T., Track, N. S., Daniel, E. D., and Yanaihara, N.**, Immunoreactive motilin is not exclusive to the gastrointestinal mucosa, *Biomed. Res.*, 2, 321, 1981.

218. **Poitras, P., Trudel, L., Lahaie, R. G., and Pomier-Layrargue, G.**, Motilin-like immunoreactivity in intestine and brain of dog, *Life Sci.*, 40, 1391, 1987.

219. **Christofides, N. D., Bryant, M. G., Ghatei, M. A., Kishimoto, S., Buchan, A. M. J., Polak, J. M., and Bloom, S. R.**, Molecular forms of motilin in the mammalian and human gut and human plasma, *Gastroenterology*, 80, 292, 1981.

220. **Yano, H., Seino, Y., Fujita, J., Yamada, Y., Inagaki, N., Takeda, J., Bell, G. I., Eddy, R. L., Fan, Y.-S., Byers, M. G., Shows, T. B., and Imura, H.**, Exon-intron organization, expression and chromosomal localization of the human motilin gene, *FEBS Lett.*, 249, 248, 1989.

221. **Seino, Y., Tanaka, K., Takeda, J., Takahashi, H., Mitani, T., Kurono, M., Kayano, T., Koh, G., Fukumoto, H., Yano, H., Fujita, J., Inagaki, N., Yamada, Y., and Imura, H.**, Sequence of an intestinal cDNA encoding human motilin precursor, *FEBS Lett.*, 223, 74, 1987.

222. **Dea, D., Boileau, G., Poitras, P., and Lahaie, R. G.**, Molecular heterogeneity of human motilinlike immunoreactivity explained by the processing of prepromotilin, *Gastroenterology*, 96, 695, 1989.

223. **Bond, C. T., Nilaver, G., Godfrey, B., Zimmerman, E. A., and Adelman, J. P.**, Characterization of complementary deoxyribonucleic acid precursor of porcine motilin, *Mol. Endocrinol.*, 2, 175, 1988.

Chapter 4

MICROANATOMY AND CHEMICAL CODING OF PEPTIDE-CONTAINING NEURONS IN THE DIGESTIVE TRACT

E. Ekblad, R. Håkanson, and F. Sundler

TABLE OF CONTENTS

I.   Introduction 132
     A.   Extrinsic Nervous Control 132
     B.   Intrinsic Nervous Control 132
           1.   Neuronal Location 133
           2.   Electrophysiological Properties 134
           3.   Morphology 134
           4.   Ultrastructure 134
           5.   Histochemical Properties 134

II.   Neuropeptides in the Gut 135
     A.   Calcitonin Gene-Related Peptide (CGRP) 137
     B.   Cholecystokinin (CCK) 143
     C.   Galanin 145
     D.   Gastrin-Releasing Peptide (GRP) 147
     E.   Neuromedin U (NMU) 149
     F.   Neuropeptide Y (NPY) 150
     G.   Neurotensin 152
     H.   Opioid Peptides 152
     I.   Somatostatin 154
     J.   Substance P (SP) 155
     K.   Vasoactive Intestinal Polypeptide (VIP) 156

III.   Enteric Neurons: Projection Patterns and Plasticity 159

IV.   Enteric Neurons as Multimessenger Systems 161

Acknowledgments 164

References 164

I. INTRODUCTION

The nervous control of visceral functions is exerted by "central" (parasympathetic, sympathetic, and sensory) as well as local neurons (See Figure 1). The digestive tract is conspicuously rich in local neurons, collectively referred to as the enteric nervous system. Processes such as absorption, secretion, and coordinated peristaltic activity are to a great extent regulated by intramural nerve reflexes (for reviews see References 1 to 4).

A. EXTRINSIC NERVOUS CONTROL

The degree of extrinsic control varies between different regions of the gastrointestinal tract. After cutting the extrinsic nerves, the striated muscle in the esophagus and external anal sphincter is more or less paralyzed and the esophageal and gastric motor activity is disturbed while the intestines maintain their various functions (for a review see Reference 5). Extrinsic nerve fibers in the gut are relatively few and terminate in the intramural ganglia and around blood vessels. They have their cell bodies in the jugular and nodose ganglia, the abdominal sympathetic ganglia, the thoracolumbar and sacral spinal ganglia, and the brain stem (dorsal vagal motor nucleus) and sacral spinal cord and reach the gut via the vagus, the splanchnic, and the sacral nerves (see Figure 1) (for reviews see References 5 to 8). The vagus carries sensory afferents and preganglionic parasympathetic fibers in the proportion 30:1.[9] Vagotomy results in the disappearance of a relatively small proportion of the nerve endings in the gastric myenteric ganglia.[10,11] The vagus is known to participate in the control of both motor and secretory functions; the vagal control of motor function seems to play a greater role in the proximal part of the gut (esophagus and stomach) than in the more distal part.[5,12] Cervical vagotomy impairs but does not abolish esophageal motility.[13] Similarly, gastric emptying is impaired but not abolished by vagotomy.[14] Vagotomy also impairs secretory functions, as exemplified by the pronounced reduction in gastric acid output (cf. References 15 and 16).

The sympathetic nerves inhibit intestinal secretion, gut motility, and blood flow. The effects are thought to be exerted at least partly via inhibition of intrinsic cholinergic motor neurons.[17,18]

Sensory afferents reach the gut via the vagal and splanchnic nerves. Peripheral nerve endings are thought to act as mechano-, thermo-, and chemoreceptors, activating intramural reflexes, but also conveying signals to the central nervous system (cf. References 1, 3 to 5, and 19). Much information is still lacking on the identity and precise location of receptor sites and pathways integrating the nervous control of gut activities.

B. INTRINSIC NERVOUS CONTROL

The intrinsic fibers, which represent the majority of the nervous elements in the gut, have their cell bodies in the intramural ganglia. In the small intestine of the cat the number of cell bodies is around 6 million.[20] The basic requirements for an integrative circuitry, i.e., functional interaction between local sensory neurons, interneurons, and secreto-motor neurons, are present in the enteric ganglia. Intramural reflexes are initiated by sensory receptors that respond to, e.g., distension or chemicals in the lumen. This information triggers action potentials that lead to release of sensory neurotransmitters. Interneurons within the ganglia receive the information, which is then processed and modulated before it is transmitted to the secreto-motor neurons. They in turn respond by the appropriate activation of epithelial cells and smooth muscle cells including the local vasculature.[1,3,4] Complex reflexes are involved in the propulsion of a bolus. One is referred to as the enteric ascending excitatory reflex and another one as the enteric descending inhibitory reflex (for reviews see References 2, 3, and 21). Reflexes maintaining mucosal functions are thought to be initiated by mechanical or chemical activation of mucosal sensory receptors, which excite interneurons and secreto-motor neurons in the submucous ganglia (cf. References 1 and 22).

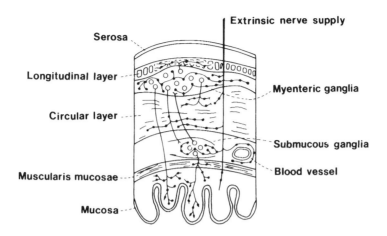

FIGURE 1. Schematic illustration of the organization of the extrinsic (parasympathetic, sympathetic, and sensory) and intrinsic gut nerve supply. The cell bodies of the enteric neurons are accumulated in the myenteric (Auerbach) ganglia and the submucous (Meissner) ganglia.

The intrinsic neurons can be described and classified on the basis of location, electrophysiological properties, shape, ultrastructure, and histochemical properties.

1. Neuronal Location

The collections of nerve cell bodies that constitute the myenteric ganglia are situated between the outer (longitudinal) and inner (circular) smooth muscle layers. The ganglia and the interconnecting nerve strands form a continuous meshwork with the nerve cell bodies accumulated at the nodes. The size of the ganglia and the shape of the meshwork vary from one part of the gut to another and between different species.[6,8] The number of myenteric nerve cell bodies in the small and large intestine ranges from 10,000 to 20,000 per square centimeter as estimated in the rat, guinea pig, and cat (for a review see Reference 23). The projections from myenteric neurons terminate mainly in the smooth muscle and within other myenteric ganglia and, to a minor extent, also in submucous ganglia and the mucosa.[24,25] In addition, some myenteric neurons project to extramural ganglia.[24,26] The submucous ganglia are more numerous than the myenteric ganglia, but each individual submucous ganglion contains fewer cell bodies. Together with the interconnecting nerve strands the submucous and myenteric ganglia form similarly patterned meshworks, although the submucous plexus is larger meshed and less regular.[6,8] In the guinea pig ileum and colon there are about 4000 submucous neurons per square centimeter.[27] Submucous neurons project preferentially to the mucosa and submucous ganglia. In addition to the myenteric and submucous plexuses a subserous nerve plexus occupying a narrow strip at the attachment of the mesentery was described by Auerbach.[28] The fibers were found to penetrate the longitudinal muscle layer to join the myenteric plexus. Later Schabadasch[29] showed that the subserous plexus in the cat extended around the circumference of the gut and that it harbored ganglion cells. Extrinsic nerve fibers are suggested to comprise the major part of this plexus.[30] Since vagal nerve bundles frequently are ganglionated, Schofield[6] suggested that the subserous plexus is an intermediary between the extramural nerve supplies and the intramural nerve plexuses. Recently, ganglia containing gastrin releasing peptide (GRP) and vasoactive intestinal peptide (VIP) immunoreactive nerve fibers and cell bodies have been described in the rat stomach serosa along the minor curvature.[31]

Scattered neurons in the lamina propria of the gastric mucosa were described by Vau.[32] Similarly, neuron-like cells, issuing dendritic processes, have been described in the ileal mucosa.[33]

2. Electrophysiological Properties

Many enteric neurons display spontaneous electrical activity. Extracellular recordings from single myenteric neurons show that their discharge pattern belongs to one of three types: (1) "burst" units (50% of all neurons), (2) "single-spike" units (10%), and (3) "mechanosensitive" units.[34-36] In the submucous plexus approximately 50% of the neurons are spontaneously active, and both burst and single-spike firings have been recorded.[37] By intracellular recordings two main categories of myenteric neurons can be identified on the basis of their excitability. They are type 1 or S cells, showing an excitatory postsynaptic potential, and type 2 or AH cells, showing a long-lasting after-hyperpolarization.[38,39] Type 2 (AH) cells have not been found in the stomach.[40,41] The submucous neurons also seem to fit into this classification in type 1 (S) cells and type 2 (AH) cells and a composite of these two.[1,42]

3. Morphology

A number of conventional histological staining techniques, such as methylene blue and various silver staining procedures, have been used to study the morphology of the enteric nervous system. Three types of neurons in the intramural ganglia were described by Dogiel in his classical studies.[43,44] Type I cells are characterized by numerous short dendrites and a single long axon. Type II cells have a short axon and fewer and longer dendrites. Type III cells have a long axon and dendrites of intermediate length. Dogiel's classification on morphological grounds has been suggested to be too rigid in that it does not take into account the possibility of intermediate forms.[45] Further attempts to classify enteric neurons were made by Stach,[46] who extended Dogiel's classification to eight different neuronal populations. Conceivably there are species differences. This is illustrated in Figure 2. Other factors, such as age, are also probably of importance. Cavazzana and Borsetto,[47] Greving,[48] and Gabella[49] suggested that the distinction between the morphologically different cell types is blunted by old age.

4. Ultrastructure

The neurons in the gut can be classified ultrastructurally by the appearance of the vesicles present in the terminals. The different vesicular populations have been described as small clear vesicles, small granular vesicles, large granular vesicles, large opaque vesicles, flattened vesicles, or ring vesicles.[8,50-53] Nerve profiles containing different types of vesicles are illustrated in Figure 3. Although these various types of vesicles usually occur in different nerve endings, two or more may coexist in the same terminal. It is generally held that dense-cored vesicles (granular vesicles) contain neuropeptides (as well as classical transmitters) and that small clear vesicles lack neuropeptides, but represent another storage site for classical transmitters.[50,54-56] Attempts are being made to identify the ultrastructural appearance of peptide-containing vesicles in order to distinguish individual neurons (for a review see Reference 57), so far with limited success.

5. Histochemical Properties

Specific histochemical techniques have been used to identify the individual neuronal populations on the basis of their content of transmitter or transmitter-related enzymes. These techniques include staining for cholinesterase activity, monoamine fluorescence histochemistry, and immunocytochemistry for demonstration of peptides, amino acids, and amines. Based on the results of such techniques, individual populations of enteric neurons have been claimed to store acetylcholine,[58,59] serotonin,[60] gamma aminobutyric acid (GABA),[61-63] histamine,[64,65] and/or numerous neuropeptides (cf. References 57, 66, and 67). The digestive tract is richer in neuropeptides than any other tissue outside the brain. For several of these peptides a tentative functional role in gut regulation has been proposed (for reviews see References 68, 69, and 70).

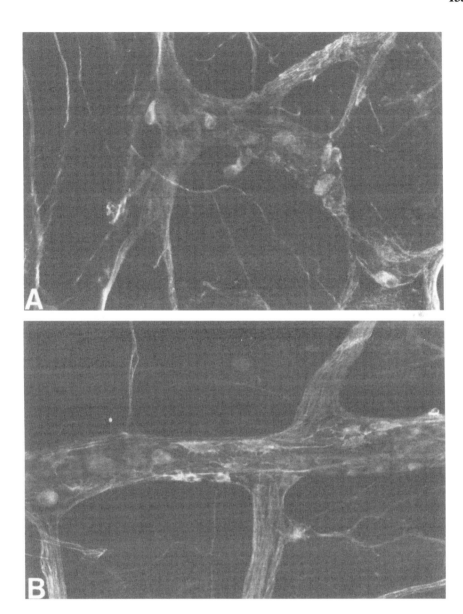

FIGURE 2. Whole mounts of myenteric ganglia and longitudinal muscle from (A) rat and (B) guinea pig small intestine immunostained with neurofilament antiserum. Note the difference between the two species in the shape of the myenteric neurons. In the rat the myenteric nerve cell bodies make up a fairly homogeneous population with an oval to rounded shape while in the guinea pig several different neuronal cell types (as described by Dogiel) can be distinguished. (Magnification × 175.)

In order to link a particular function to a particular neuron, attempts have been made to correlate morphology and electrophysiological characteristics of individual enteric neurons. Studies of single myenteric neurons in guinea pig small intestine have, for instance, revealed some S cells to be Dogiel type I cells and to contain enkephalin and/or VIP, whereas the majority of AH cells are Dogiel type II cells and lack enkephalin.[71-73]

II. NEUROPEPTIDES IN THE GUT

In the following section we will review the distribution, origin, projection patterns, and

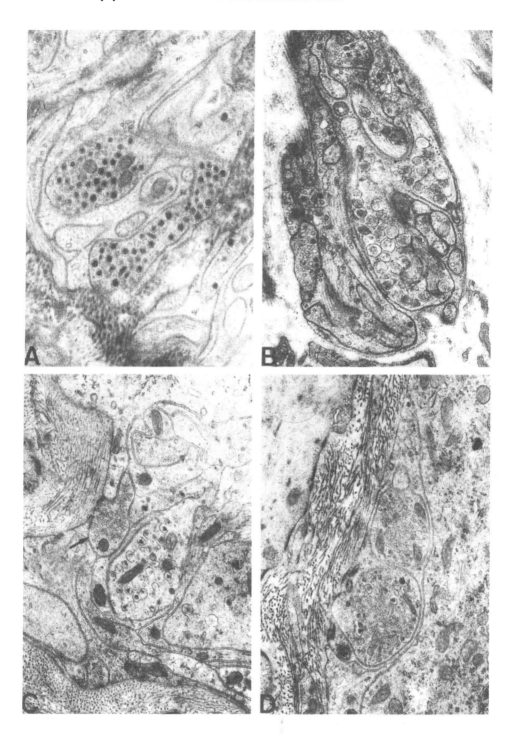

FIGURE 3.   Electron micrographs of nerve profiles in myenteric ganglia in (A) cat esophagus and (B — D) mouse colon. In (A) all nerve profiles are *p*-type, i.e., contain a predominance of large, dense core vesicles (approximately 100 nm). *P*-type nerve profiles containing dense core vesicles of somewhat different texture and electron density are illustrated in (B) and (C). Two nerve profiles containing a predominance of small clear vesicles are seen in (D). A few large granular vesicles are also seen. Those terminals are thought to be cholinergic. The presence of dense core vesicles, however, suggests that they also contain neuropeptides. In (C) a nerve profile containing exclusively small clear vesicles is also seen (arrow). (Magnification [A] and [B] × 20,200; [C] and [D] × 18,400.)

chemical coding of a variety of peptide-containing enteric neurons. A brief survey of the molecular biology and functional role of each peptide is also given. Included in the review are those peptides which are established as neuropeptides by immunocytochemistry and chemical analysis of gut extracts. In many cases a functional role has been tentatively ascribed to them. Some regulatory peptides, such as motilin,[74] CRF,[75] τ-MSH, and ACTH,[76] have been claimed to occur in enteric neurons. However, their presence has not been confirmed chemically, and they are therefore not included. The descriptions made are limited to mammals; only in a few cases are data from submammalian species included.

The various neuronal populations will be referred to by their chemical coding, i.e., their content of different messenger molecules. We will restrict the classification to neuropeptides serving as markers for the different precursors expressed by the neurons. An attempt is made to have each neuropeptide precursor represented by one well-characterized fragment. Thus, when we refer to the presence of VIP, for instance, we refer to the VIP precursor and all fragments thereof.

Neuronal cell counts are not included. This is mainly due to the difficulties in comparing and evaluating results from different investigators. The difficulty is mainly due to the fact that the number of immunocytochemically demonstrable neuronal cell bodies differs greatly between untreated and colchicine-treated animals (e.g., see Reference 57). The way colchicine is administered (systemic vs. local application) is also important in this respect. Furthermore, the general quality and specificity of the antisera used are decisive for the number of nerve cell bodies that will be detected. Different peptide-containing neurons differ greatly in number and distribution within one and the same location, as illustrated in Figures 4 to 7, which also illustrate differences in the innervation pattern between stomach and small and large intestines in the rat. When attempting to outline the organization of the enteric peptide-containing neurons, species differences will have to be kept in mind. The recently discovered neuropeptide neuromedin U (NMU) may serve as an example (see Figure 8). NMU-containing nerve fibers are numerous in the smooth muscle in some species and in the mucosa/submucosa in other species while being extremely rare altogether in still other species. The neuropeptides are described next in alphabetic order.

A. CALCITONIN GENE-RELATED PEPTIDE (CGRP)

CGRP is a 37-amino-acid peptide that arises either by alternative processing of the calcitonin gene mRNA transcript (α-CGRP)[77,78] or from the mRNA at a different gene (β-CGRP).[79] In the rat the two CGRPs differ only at position 35. Evidence has been presented that sensory neurons express α-CGRP preferentially while enteric neurons express β-CGRP.[80] In organs where CGRP occurs predominantly in sensory fibers, the concentration of α-CGRP was found to be three to six times higher than that of β-CGRP. In the intestine the concentration of β-CGRP was up to seven times higher than that of α-CGRP. In capsaicin-treated rats (capsaicin depletes C fibers of their substance P (SP) and CGRP content), no α-CGRP could be detected in the digestive tract, while β-CGRP was still present.[80] Immunoreactive CGRP occurs together with SP in primary sensory neurons in a number of organs (cf. References 81 to 83). Immunoreactive CGRP exists in the gut both in intramural neurons and in nerve fibers of extramural origin (often together with SP), probably of sensory origin.[84-87] Immunocytochemistry has revealed CGRP-containing nerve fibers throughout the mammalian gastrointestinal tract. In the esophagus such fibers are particularly numerous in the upper portions. CGRP-immunoreactive nerve fibers in the esophagus and esophageal sphincter form an abundant subepithelial plexus from which fibers penetrate into the epithelium. In the muscle layer, CGRP fibers occur in bundles and run as single varicose fibers in smooth muscle and around blood vessels.[84,86,88,89] In the striated muscle of the upper esophagus, CGRP occurs in the motor fibers and the peptide is demonstrable in the motor end plates.[88] No immunoreactive nerve cell bodies are present in the esophagus.[84,86,88]

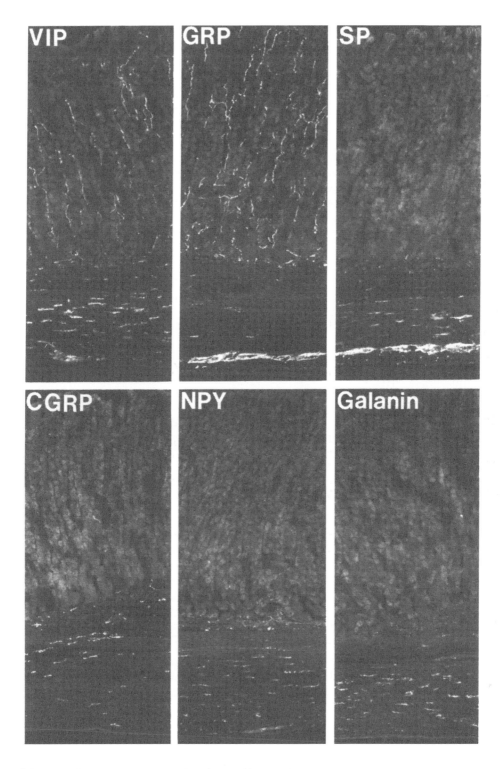

FIGURE 4.   Rat stomach, cryostat sections. The peptide-containing neuronal systems illustrated are arranged in order of overall nerve fiber frequency. VIP fibers are numerous throughout the wall. GRP fibers are numerous in the mucosa/submucosa and in the myenteric ganglia, but less numerous in the smooth muscle. SP, CGRP, NPY, and galanin fibers are few to moderate in number in the mucosa/submucosa, but quite numerous in the smooth muscle and intramural ganglia. (Magnification × 115.)

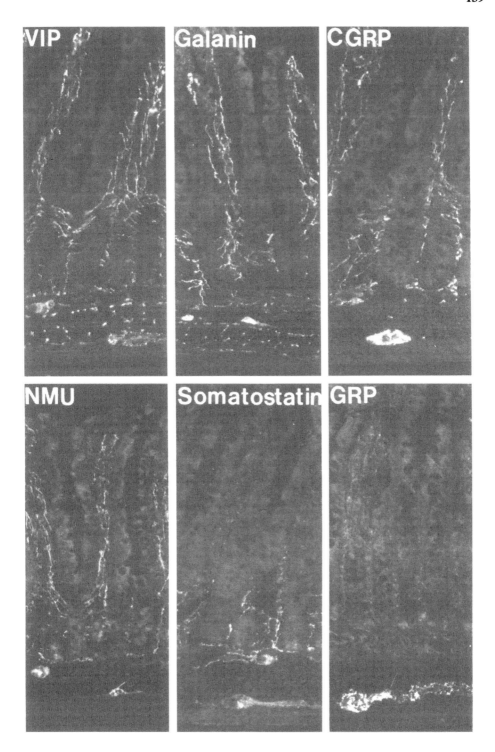

FIGURE 5.   Rat small intestine, cryostat sections. The peptide-containing neuronal systems illustrated are arranged in order of overall nerve fiber frequency. VIP and galanin fibers are numerous throughout the wall. CGRP, NMU, and somatostatin fibers are moderate in number in the mucosa/submucosa and within the myenteric ganglia and are rare in the smooth muscle. Somatostatin fibers predominate greatly in the basal portion of the mucosa and in the submucosa. GRP fibers are numerous in the myenteric ganglia, but rare in other layers of the wall. (Magnification × 115.)

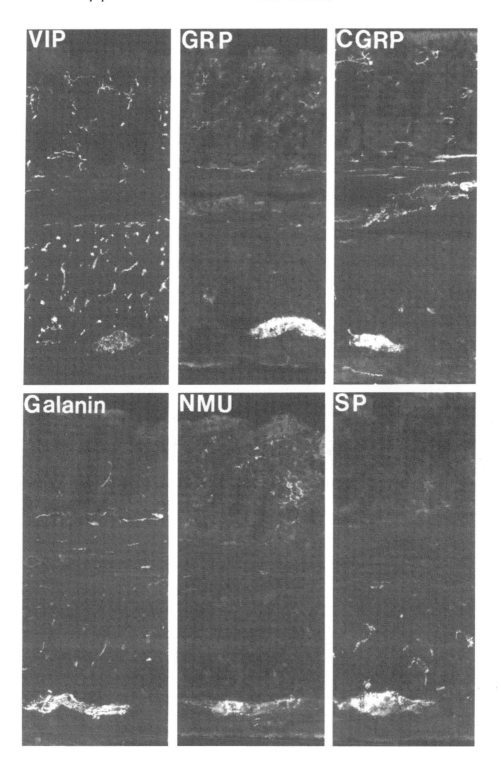

FIGURE 6.   Rat large intestine, cryostat sections. The peptide-containing neuronal systems illustrated are arranged in order of overall nerve fiber frequency. VIP fibers are numerous throughout the wall. GRP, CGRP, and galanin fibers are moderate in number in the mucosa/submucosa and smooth muscle, but numerous in the intramural ganglia. NMU fibers are moderate in number in the mucosa/submucosa and intramural ganglia, but rare in the smooth muscle. SP fibers are very few in the mucosa/submucosa and submucous ganglia, moderate in number in the smooth muscle, and numerous in the myenteric ganglia. (Magnification × 115.)

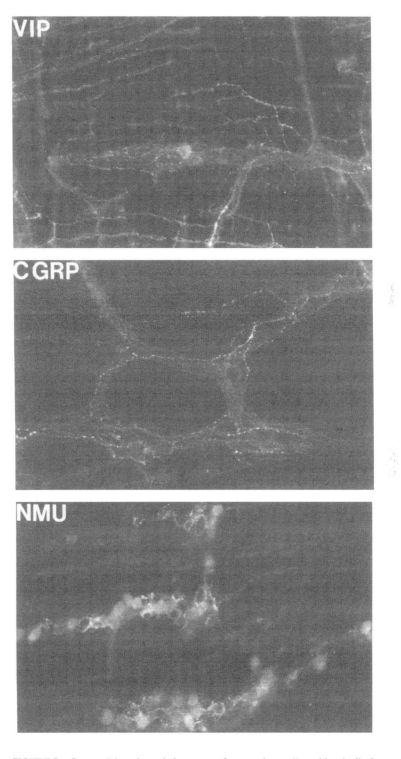

FIGURE 7. Rat small intestine, whole mounts of myenteric ganglia and longitudinal muscle. VIP fibers are numerous both within the ganglia and in the smooth muscle. A moderate number of VIP-immunoreactive nerve cell bodies can be detected. CGRP and NMU fibers are moderate in number within the ganglia, but rare in the smooth muscle. CGRP-immunoreactive nerve cell bodies are few (none are shown here), while NMU-immunoreactive nerve cell bodies are numerous. (Magnification × 140.)

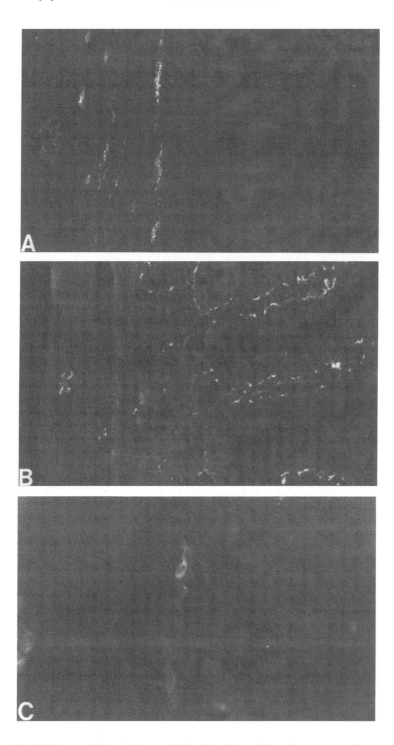

FIGURE 8.   (A) Chicken, (B) rat, and (C) guinea pig small intestine, cryostat sections immunostained for NMU. In chicken small intestine NMU fibers are numerous in the smooth muscle and myenteric ganglia, but rare in the mucosa/submucosa. In rat small intestine a moderate number of NMU fibers are found in the mucosa/submucosa and intramural ganglia. Very few NMU fibers occur in the smooth muscle. In guinea pig small intestine a few NMU immunoreactive nerve cell bodies, but no fibers, can be detected. (Magnification × 150.)

The CGRP fibers in the esophagus belong to two different populations; one population originates in the nodose ganglion (sensory fibers), and the other population originates in the nucleus ambiguus and is thought to control motor activity.[88]

In the rat stomach, CGRP-immunoreactive nerves are fairly numerous and are present in all layers as well as around blood vessels[19,84,86,90] (see Figure 4). In contrast, the stomach of pig and man harbors very few CGRP-containing fibers.[91] There is a notable lack of CGRP-immunoreactive nerve cell bodies in the stomach wall of the species examined. The bulk of CGRP fibers are eliminated by extrinsic denervation[90] or capsaicin treatment.[19,86,90] In the small and large intestine of a number of mammals, CGRP-immunoreactive fibers are quite numerous particularly in the mucosa/submucosa, intramural ganglia, and around blood vessels[84-87,90,92-94] (see Figures 5 to 7). Most of the intestinal CGRP-containing fibers have their nerve cell bodies in the submucous and myenteric ganglia. A minor portion is extrinsic in origin, innervating blood vessels and, to some extent, the myenteric ganglia. These fibers, which usually also contain SP, are sensory.[81,86,87] Perivascular CGRP-containing fibers lacking SP have been described.[87,92,95] In the rat colon they harbor VIP and are intrinsic in origin.[87] In the stomach and duodenum of the rat a substantial proportion of the CGRP fibers seem to be extrinsic[19,90,96,97] in that transection of extrinsic nerves[90] or capsaicin treatment[19,86,90] results in a marked decrease in the number of CGRP fibers.

Intrinsic CGRP-immunoreactive fibers in guinea pig small intestine also contain choline acetyltransferase (ChAT), neuropeptide Y (NPY), cholecystokinin (CCK), and somatostatin. They originate in both the submucous and myenteric ganglia and project to the mucosa.[92] In the rat small intestine a subpopulation of both myenteric and submucous CGRP neurons contain NMU as well.[98,99] In the rat small and large intestine, myenteric CGRP fibers terminate in the smooth muscle and in other myenteric ganglia. Myenteric CGRP neurons in the rat small intestine issue both ascending and descending projections for approximately 2 mm (See Figure 9); while in the large intestine they issue descending projections for approximately 5 mm. Submucous CGRP neurons project to the mucosa and other submucous ganglia for approximately 8 mm in the oral direction in the small intestine (see Figure 9) and for 6 mm in both the oral and anal directions in the large intestine.[85,87] In the rat large intestine a minor population of the submucous CGRP-containing neurons also stores somatostatin. The projections from these neurons are both ascending and descending.[87] CGRP-immunoreactive terminals were claimed to contain both electron-lucent and large granular vesicles; the CGRP-immunoreactive material is confined to the granular vesicles, 80 to 90 nm in diameter.[100]

CGRP is a potent vasodilator.[101-103] Few reports have dealt with the actions of CGRP on gut motor activities. It has been reported to cause contraction of the guinea pig ileum via an activation of cholinergic neurons.[104,105] In lower concentrations, however, CGRP has been reported to cause relaxation of longitudinal smooth muscle of guinea pig ileum[105] and rat duodenum.[97] The relaxation is suggested to be a direct effect on the smooth muscle. CGRP is a potent inhibitor of gastric acid secretion.[106] A stimulation of somatostatin release by CGRP has been found in the stomach and intestine.[107,108] In rat colonic epithelia both α- and β-CGRP were found to elicit indirect antisecretory (in low concentrations) and direct secretory (in higher concentrations) effects.[109]

B. CHOLECYSTOKININ (CCK)

CCK was first isolated as a 33-amino-acid peptide (CCK-33) from porcine small intestine.[110] This was followed by the demonstration of an N-terminally extended form, CCK-39.[111] An even larger form, CCK-58, has been isolated from canine intestine[112] and a smaller form, CCK-8, from sheep brain.[113] Thus, CCK occurs in several different molecular forms all arising from the same precursor, the most abundant being CCK-8.[114] Cloning of the CCK gene has revealed that the CCK precursor contains 115 amino acids in the rat[115] and 114

CGRP

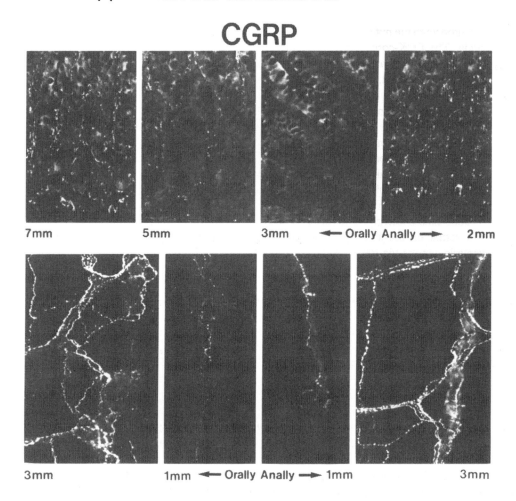

7mm   5mm   3mm   ← Orally Anally →   2mm

3mm   1mm ← Orally Anally → 1mm   3mm

FIGURE 9.   CGRP-immunoreactive nerve fibers in rat small intestine, after local denervation (intestinal clamping). Upper panel: cryostat sections, mucosa and submucosa visible. No loss of CGRP fibers anally to the lesion. CGRP fibers disappeared orally for approximately 2 mm in the submucous ganglia and for approximately 4 mm in the mucosa and submucosa. Beyond these points CGRP fibers increased gradually in number. At 3 to 4 mm in the submucous ganglia and at 7 to 8 mm in the mucosa and submucosa, the number of CGRP fibers was back to normal. Lower panel: whole mounts of longitudinal muscle with adherent myenteric ganglia. CGRP fibers are reduced in number for a distance of 1 to 2 mm in both oral and anal direction. Further orally and anally the nerve fiber density is normal. (Magnification × 90 [upper panel], × 130 [lower panel].)

amino acids in the pig.[116] All active forms of CCK share the same C-terminal pentapeptide, which is identical with the C-terminus of gastrin. Immunocytochemical studies[117] using C-terminally directed antisera to gastrin described "gastrin-like" immunoreactivity in brain. This was later shown to be CCK, predominantly in the form of CCK-8.[118,119] Authentic gastrin seems to occur in certain hypothalamo-hypophyseal neurons[117] and in the vagal nerve.[113,121] In the gut a few gastrin-containing nerves have been described in the muscular wall of the colon,[122] possibly originating from the vagal nerve. Enteric neurons, however, seem to lack gastrin.

CCK-immunoreactive neurons are found in low numbers throughout the gastrointestinal tract. The bulk of CCK-immunoreactive nerve fibers occurs within the submucous and myenteric ganglia. Here some immunoreactive nerve cell bodies are situated as well. The number of CCK-containing fibers is particularly low in the upper part of the gut; the fibers increase in number more distally, being most numerous in the colon.[95,123-128] They are thought to be intrinsic to the gut since immunoreactive nerve cell bodies can be detected in intramural ganglia. The CCK-containing submucous neurons in guinea pig small intestine also store ChAT (suggesting that they are cholinergic), CGRP, NPY, and somatostatin (Som)[24] and project to the mucosa. ChAT/CCK/CGRP/NPY/Som-containing neurons are found also in the myenteric ganglia of guinea pig small intestine.[24] Likewise, these neurons are reported to project to the mucosa. An additional population of CCK-containing myenteric neurons store CCK/Dynorphin (Dyn)/Enkephalin (Enk)/GRP/VIP and project to prevertebrate ganglia, while others contain CCK/Dyn/VIP/GRP/SP or CCK/SP/Som.[24] These latter CCK-containing neuronal populations probably project to other myenteric ganglia and to submucous ganglia. Schultzberg et al.,[124] in their studies on the guinea pig intestine, reported that the majority of submucous CCK neurons contained somatostatin as well, but that both CCK and somatostatin also occurred in separate populations.

CCK causes contraction of the intestinal smooth muscle. This effect reflects both a direct action on the smooth muscle cells and an indirect neuromodulatory effect in the myenteric ganglia via stimulation of acetylcholine release[127,129] and of SP release.[130]

C. GALANIN

This 29-amino-acid peptide was first isolated from porcine upper small intestine.[131] The name is derived from the N-terminal and C-terminal residues, glycine and alanine amide. Galanin has no significant sequence similarities with any other known peptide.[131] Studies on galanin mRNA from pig adrenal medulla revealed the galanin precursor to contain 123 amino acids and to be composed of an N-terminal portion followed by galanin and a C-terminal 59-amino-acid portion.[132] In the rat and mouse esophagus a few galanin-immunoreactive nerve fibers innervate the mucosa and circular muscle, and scattered galanin-immunoreactive cell bodies occur in the myenteric ganglia.[133] In the stomach the galanin-containing fibers are mainly located in the circular muscle and the intramural ganglia, as studied in the rat and pig[133-135] (see Figure 4). In the guinea pig stomach wall, galanin-containing nerves are rare; only a few fibers can be detected in the mucosa.[133] In the intestines the galanin innervation is notably extensive in a number of species. Numerous galanin-containing nerve fibers are found in all layers of the gut wall, and immunoreactive nerve cell bodies occur in both myenteric and submucous ganglia[87,93,133-138] (see Figures 5 and 6). Perivascular galanin-containing nerve fibers have been described in the gut wall of rat, guinea pig, and cat.[95,133,134,137] Galanin and VIP coexist in neurons in the guinea pig, dog, and pig intestines,[133,135,138] but they occur in separate neuronal populations in rat and human gut.[87,93,134] In the guinea pig small intestine about two thirds of the submucous CCK neurons store galanin; most of them also contain VIP, while the others contain NPY.[137] Since a number of neuropeptides are contained in these neurons, they have been designated Dyn/Gal/VIP and CGRP/CCK/ChAT/Gal/NPY/Som neurons.[137] In the guinea pig small intestine a small population of myenteric galanin neurons display Dogiel type III morphology and contain VIP and NPY. Most of the myenteric galanin neurons exhibit Dogiel type I morphology.[137] The galanin-immunoreactive fibers in the rat and guinea pig intestine are thought to be intrinsic to the gut, since extrinsic denervation does not visibly affect their number.[87,134,135,137] Local denervations within the gut wall have revealed that myenteric galanin neurons issue descending projections to the circular muscle and other myenteric ganglia. In guinea pig small intestine they are up to 4 mm long;[137] in rat small intestine, approximately 15 mm;[134,135] and in rat large intestine, approximately 6 mm[87] (see Figure 10). Submucous galanin-containing neurons supply the mucosa, submucosa, submucous arteries and arterioles, and other submucous ganglia in the rat small intestine. These neurons issue both ascending and descending projections for approximately 2 mm in both directions, while in the rat large intestine no oro-anal projections can be demonstrated[87] (see Figure 10). In the rat small intestine, galanin-immunoreactive nerve terminals contain both agranular vesicles

Galanin

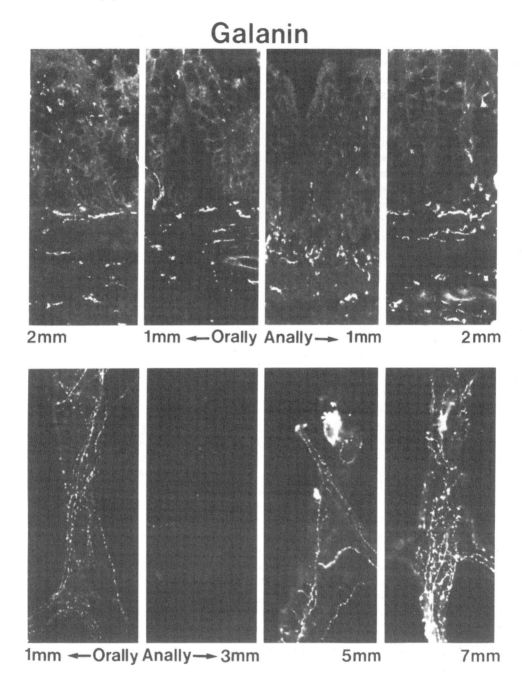

FIGURE 10.   Galanin-immunoreactive nerve fibers in rat colon after local denervation (intestinal clamping). Upper panel: cryostat sections, mucosa and submucosa visible. No obvious change in the number of fibers orally or anally to the lesion. Lower panel: whole mounts of longitudinal muscle with adherent myenteric ganglia. Orally to the lesion there was no loss of galanin fibers. There was a loss of fibers for 3 mm anally to the lesion. Beyond this point there was a gradual increase in the number of fibers, and a normal frequency could be seen at approximately 6 mm anally to the lesion. A moderately immunoreactive nerve cell body is marked by asterisk. (Magnification × 100 [upper panel], × 140 [lower panel].)

(30 to 40 nm) and large granular vesicles (80 to 120 nm).[139] Galanin has been shown to have a number of actions, including an inhibitory effect on neurons in the guinea pig taenia coli[140] and direct effects on rat jejunal longitudinal muscle (contractile)[131,140] and on dog small intestinal circular muscle (inhibitory).[141] Galanin markedly potentiates VIP-induced relaxation of isolated smooth muscle cells from guinea pig small intestine.[142]

D. GASTRIN-RELEASING PEPTIDE (GRP)

GRP, a 27-amino-acid peptide first isolated from the porcine nonantral gastric mucosa,[143] is thought to be the mammalian counterpart of bombesin, a 14-amino-acid peptide originally isolated from amphibian skin.[144] The C-terminal 7-amino-acid sequence of GRP is identical with that of bombesin. This region is the biologically active portion, are consequently bombesin and GRP have similar biological activity. Other bombesin-like peptides and neuromedin B and neuromedin C, both of which have been isolated from porcine spinal cord.[145,146] The human GRP and neuromedin B precursors have been identified.[147,148] The GRP precursor consists of 148 amino acids; the signal sequence is followed by GRP and a C-terminal extension peptide. The neuromedin B precursor is 74 amino acids long and contains neuromedin B (NMB-32) next to the signal peptide followed by an extension peptide. Alternative processing of the single GRP precursor RNA transcript may give rise to three different GRP precursors.[147] The precursors differ in the C-terminal extension peptide; the structure of GRP itself is unaffected. Two potential cleavage sites within human GRP could generate GRP_{14-27} (GRP-14), or GRP_{18-27} (GRP-10). GRP-10 is identical with neuromedin C.

In the mammalian gut, GRP is localized to neurons exclusively.[149-151] This is not the case in fish, amphibians, and birds, where bombesin/GRP-like peptides are present not only in neurons, but also in endocrine cells in the stomach.[152,153] Many antisera used in immunocytochemical and immunochemical studies recognize the common C-terminal portion of GRP and neuromedin B and therefore do not differentiate between the two peptides. An extensive network of bombesin/GRP-immunoreactive nerve fibers occurs throughout the digestive tract in a number of mammals. Information on the GRP innervation of the esophagus is sparse. The esophagus of rabbit and man harbors a few immunoreactive fibers in the myenteric ganglia and smooth muscle.[154,155] The stomach is densely innervated by GRP-containing nerve fibers in a number of mammals, including man[31,128,148,150,151,154,156-161] (see Figure 4). Mucosal GRP-immunoreactive fibers extend high up between the glands and are particularly numerous in the mouse, rat (see Figure 4), and pig, while there are fewer in the cat and guinea pig.[151] The density of the GRP nerve fibers differs only slightly between the oxyntic and pyloric mucosa in experimental animals. In man, however, the GRP fibers in the oxyntic area clearly outnumber those in the antral portion.[161] GRP fibers are also numerous in the smooth muscle and myenteric ganglia. GRP-immunoreactive nerve cell bodies are regularly seen in the myenteric ganglia.

In the mammalian intestine, GRP-immunoreactive nerves are mainly found in the myenteric ganglia and smooth muscle.[85,87,93,127,148,150,151,154,156,162] Mucosal GRP fibers are few in the small intestine (see Figure 5), while they are numerous in the large intestine (see Figure 6), as studied in the rat and mouse.[151] GRP-immunoreactive nerve cell bodies are few in both the myenteric and submucous ganglia.

GRP coexists with dynorphin and VIP in myenteric neurons in the guinea pig small intestine.[24] In the rat stomach, GRP coexists with SP in myenteric neurons.[163] In the oxyntic mucosa of the stomach of rat and man, the GRP-containing fibers also store VIP (see Figure 11).[163a] The GRP- (and VIP-) containing fibers in the rat oxyntic mucosa are suggested to be extrinsic in origin.[156,164] However, sympathectomy or vagotomy does not overtly affect the frequency of GRP fibers in the rat stomach wall.[128] Furthermore, the finding of nerve cell bodies containing both GRP and VIP in the myenteric ganglia of the rat stomach (see Figure 11) strongly suggests an intramural origin. In the intestines the bulk of GRP-containing

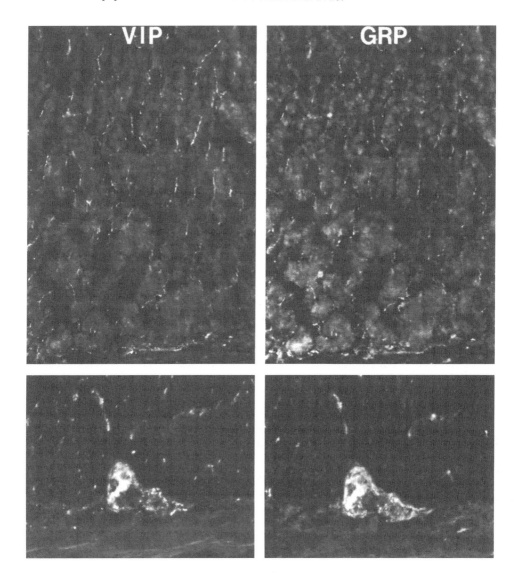

FIGURE 11.   Rat stomach, oxyntic gland area, cryostat sections. Upper panel: mucosa and submucosa. Lower panel: myenteric ganglia. Double immunostaining revealed coexistence of VIP and GRP in mucosal nerve fibers and in myenteric nerve cell bodies. VIP and GRP are found in separate populations in the smooth muscle. (Magnification × 185 [upper panel], × 205 [lower panel].)

nerve fibers is of intramural origin, since the number of such fibers does not change after extrinsic denervation.[87,162,165,166] The myenteric GRP-containing neurons issue comparatively long descending projections to other myenteric ganglia and to the smooth muscle. In the guinea pig small intestine they are 9 to 12 mm in length;[162] in the rat small intestine, approximately 20 mm;[165] in the rat large intestine, approximately 10 mm;[87] and in the canine small intestine, 15 to 30 mm in length.[166] Myenteric GRP neurons also project, to some extent, to the submucous ganglia and to the mucosa in the three species studied. Submucous GRP neurons are absent from the guinea pig small intestine,[162] but occur in the small and large intestine of the rat[87,165] and in the small intestine of the dog,[166] where they project to the mucosa and to other submucous ganglia. The direction and length of these projections have been established in the rat small and large intestine, where submucous GRP neurons issue both ascending and descending projections for 2 to 3 mm in both directions.[85,87] A population of myenteric GRP-containing neurons in the rat and guinea pig small intestine project to the celiac ganglion.[162,167,168] These neurons also contain CCK, dynorphin, enkephalin, and VIP in the guinea pig.[24] An electron microscopic study of GRP nerves (rat stomach)[156] shows that the GRP-immunoreactive material is stored in round granular vesicles with a solid-appearing dense core.

Bombesin/GRP produces a wide variety of effects, such as stimulation of gall bladder and intestinal smooth muscle contraction, gastric acid secretion, and pancreatic enzyme secretion (for reviews see References 169 to 171). Many of these actions are likely to be secondary to the release of other neurohormonal peptides, such as gastrin, somatostatin, pancreatic polypeptide, insulin, or glucagon.[172,173]

E. NEUROMEDIN U (NMU)

Two peptides capable of stimulating uterine motor activity have been isolated from porcine spinal cord.[174] The two peptides were found to be chemically related and are referred to as neuromedin U (U for uterus). One of them was found to consist of 8 amino acid residues (NMU-8) while the other was 25 residues long (NMU-25). NMU-25 contains NMU-8 at its C-terminal end, preceded by paired basic residues. Both peptides are C-terminally amidated.[174] NMU-like peptides in guinea pig and rat have been shown to differ from those of the pig.[175,176] Rat NMU consists of 23 amino acids and differs from porcine NMU-25 in 14 positions.[177] Porcine NMU-25 contains an Arg–Arg sequence at positions 16 and 17. In rat NMU-23 these residues are replaced with Gly–Gly. It is unlikely, therefore, that proteolytic processing can give rise to NMU-8 in the rat.

The distribution and frequency of NMU-immunoreactive nerve fibers in the gastrointestinal tract show remarkable species variation (see Figure 8). NMU-immunoreactive fibers and nerve cell bodies are numerous in the gut of toad, chicken, rat (see Figures 5 and 6), mouse, and pig, less numerous in the fish (roach), mole, hamster, and cat, and rare or absent in man.[98,99,176] Reports on the supply of NMU-immunoreactive nerves in the guinea pig gastrointestinal tract are conflicting. In two studies, NMU-containing fibers were found to be few or lacking in both small and large intestine,[99,176] while in a third study, NMU-containing fibers were reported to be numerous in the small intestine.[178] Since the amino acid sequence of guinea pig NMU is not yet known, it is possible that some of the NMU antibodies (all of which were raised against porcine NMU) do not recognize guinea pig NMU or recognize it poorly. In this context it should be noted that NMU-immunoreactive endocrine cells were described in the small intestine of the guinea pig,[178] but not in any of the other animals studied.[98,99,176] The topographic distribution of NMU-immunoreactive nerve fibers varies considerably between different species. In the chicken gut, NMU-containing nerve fibers predominate in the myenteric ganglia and smooth muscle, while in the pig they occur almost exclusively in the mucosa/submucosa and the submucous ganglia.[99] Intestinal NMU fibers are probably intramural in origin, since NMU-immunoreactive nerve cell bodies are present in the intramural ganglia (see Figure 7) and since extrinsic denervation does not affect the number of NMU-immunoreactive fibers in the rat and guinea pig intestine.[98,99,178] NMU fibers in the rat small intestine are both ascending and descending, but the projection distances have not been established.[98] In guinea pig small intestine myenteric NMU-containing neurons project to the celiac ganglia, circumferentially or anally within the myenteric plexus, or to the submucous ganglia, while submucous NMU-containing neurons project to the mucosa.[178] Most of the NMU-containing nerve cell bodies in both myenteric and submucous ganglia in rat small intestine also contain CGRP.[98,99] In the pig intestines all NMU-containing fibers in the mucosa also store VIP, whereas only approximately 50% of the NMU-immunoreactive nerve cell bodies in the submucous ganglia contain this peptide.[99] In the smooth muscle the NMU-containing fibers (which are few) store VIP as well. Somewhat surprisingly, however, VIP and NMU seem to be located in different nerve cell bodies in the myenteric ganglia.[99] In the chicken small intestine the majority of NMU fibers in the smooth muscle and myenteric ganglia and all NMU fibers in the mucosa contain VIP.[99] In guinea pig small intestine the majority of the NMU-containing myenteric neurons are Dogiel type II, some of which also store SP.[178] In the submucous ganglia NMU is contained within four different neuronal populations: (1) Dyn/Gal/NMU/VIP, (2) ChAT/CCK/CGRP/Gal/NMU/NPY/Som, (3) ChAT/CCK/CGRP/NMU/NPY/Som, or (4) ChAT/SP/NMU.[178] Electron microscopic studies of NMU-immunoreactive neurons in rat gut showed the NMU-immunoreactive material to be stored in large (60 to 70 nm) dense-cored vesicles.[98]

At present, little information is available on the effects of NMU on intestinal functions. In preparations from the porcine small intestine, NMU-8 failed to affect motor activity while stimulating intestinal ion transport (increase in short circuit current) via an effect on noncholinergic enteric neurons.[179]

F. NEUROPEPTIDE Y (NPY)

Neuropeptide Y, a 36-amino-acid peptide, was first isolated from porcine brain.[180] NPY has a high proportion of tyrosine (Y) residues including one at the N-terminus and another one (amidated) at the C-terminus and was therefore named neuropeptide Y. Because of sequence homology, NPY is a member of the pancreatic polypeptide family.

The NPY precursor is composed of 69 amino acids and has the NPY sequence at the N-terminal end followed by a 30-amino-acid C-terminal flanking peptide.[181] NPY is found exclusively in neuronal elements in the gastrointestinal tract.

NPY-immunoreactive nerve fibers are numerous throughout the gastrointestinal tract and in all layers of the wall. In the esophagus a dense network of NPY fibers supplies the muscularis mucosae, external muscle, myenteric ganglia, and blood vessels.[154,155,182] In the stomach, NPY-containing fibers occur mainly in the smooth muscle and myenteric ganglia, although they are present also in the mucosa and submucosa (around blood vessels and at the basal portion of the glands)[128,154,183-186] (see Figure 4). Immunoreactive nerve cell bodies are found in the myenteric ganglia of the esophagus and stomach.[128,155] In the small intestine, NPY-containing nerve fibers are numerous in all layers and around blood vessels. The dense mucosal innervation in this part of the gut is particularly notable, and immunoreactive nerve cell bodies are numerous in both submucous and myenteric ganglia[154,166,183,184,187-190] (see Figure 12). The topography of the NPY innervation of the large intestine resembles that of the small intestine, except that the NPY-containing fibers are less frequent.[87,183,184] The perivascular NPY innervation of the gut is generally dense and resembles that of adrenergic nerves. Treatment of rats or guinea pigs with the adrenergic neurotoxin 6-hydroxydopamine eliminates the NPY-containing perivascular nerves as well as some NPY fibers in the myenteric ganglia.[96,183,187,190]

The following observations are pertinent in defining the origin of NPY-containing fibers in the gut wall: (1) extrinsic denervation causes a loss of perivascular NPY-immunoreactive fibers, while the bulk of the NPY fibers in the gut wall remains;[85,87,96,183,187] (2) perivascular NPY fibers contain the catecholamine-forming enzymes dopamine-β-hydroxylase (DBH) and tyrosine hydroxylase (TH);[183,188] and (3) intramural ganglia are rich in NPY-immunoreactive nerve cell bodies. Together these observations suggest that NPY-containing nerve fibers in the gut derive from two different sources. One population is identical to the adrenergic (sympathetic) nerve fibers mainly supplying blood vessels. Another population is intrinsic to the gut and supplies the mucosa and smooth muscle. In guinea pig small intestine, the submucous as well as some of the myenteric NPY-immunoreactive neurons also contain ChAT, CCK, CGRP, and somatostatin,[92] while other myenteric NPY-storing nerve cell bodies contain dynorphin, enkephalin, and VIP.[24] In the pig, rat, and mouse small intestine the whole population of myenteric and submucous NPY-containing neurons stores

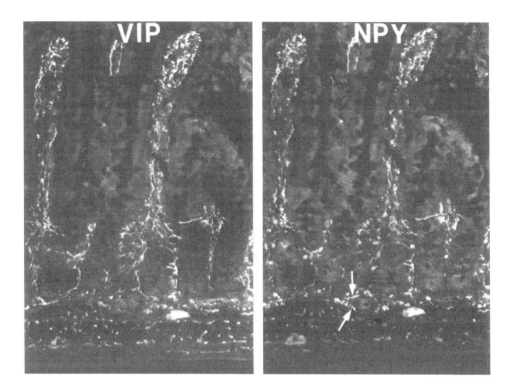

FIGURE 12.   Rat small intestine, cryostat sections. Double immunostaining revealed coexistence of VIP and NPY in the same nerve fibers and nerve cell bodies. Note that NPY does not coexist with VIP in perivascular (sympathetic) nerve fibers (arrows). (Magnification × 160.)

VIP[191] (see Figure 12). In rat stomach and large intestine as well as in human stomach and intestine, myenteric neurons contain both NPY and VIP, whereas submucous neurons harbor NPY and VIP in separate neuron populations.[87,93,128,186,191a] Microsurgical techniques have been used to outline the projections of NPY-containing nerves in the intestinal wall. In the guinea pig, most myenteric NPY-containing neurons project anally for approximately 2 mm to the circular muscle and other myenteric ganglia.[183] Some myenteric NPY-containing neurons (those storing NPY/ChAT/CCK/CGRP/Som) as well as the submucous NPY-containing neurons project to the mucosa.[92] In rat intestine NPY- (and VIP-) containing myenteric neurons project anally to other myenteric ganglia and to the smooth muscle. These projections are approximately 2 mm in length in the small intestine[85] and approximately 4 mm in the large intestine.[87] The submucous NPY- (and VIP-) containing neurons in rat small intestine issue ascending projections for approximately 4 mm,[85] while no oro-anal projections could be detected in the large intestine.[87] In both locations the submucous neurons supply the mucosa/submucosa and other submucous ganglia. By contrast, the canine small intestine harbors NPY-immunoreactive myenteric neurons which issue orally directed fibers up to 30 mm long to reach other myenteric ganglia and circular muscle.[166]

The effects of NPY that are best studied are those concerned with vascular regulation and the interaction with noradrenaline. NPY is a potent vasoconstrictor *in vivo*.[192-194] *In vitro* NPY has little or no effect on vascular tone, but it powerfully augments the vasoconstrictor effects of noradrenaline.[195-197] In guinea pig ileum, NPY is thought to inhibit motor activity through a neuromodulatory effect. This probably reflects a suppressed release of acetylcholine.[198,199] The effect of NPY on electrogenic ion transport has been studied in the rat intestinal epithelia.[200] NPY causes a reduction in short circuit current that is insensitive to the neurotoxin tetrodotoxin (TTX) or to α-adrenoreceptor antagonists. Net chloride absorption is increased while no effect on sodium movements has been detected.

G. NEUROTENSIN

Neurotensin is a 13-amino-acid peptide originally isolated from bovine hypothalamus.[201] Later it was also demonstrated in human and bovine intestine.[202,203] The amino acid sequence of neurotensin was, at the time of its discovery, unique.[204] In 1984 Minamino et al.[205] isolated and sequenced a six-amino-acid peptide from porcine spinal cord, with the four C-terminal amino acids identical to the C-terminal end of neurotensin. This peptide was named neuromedin N. Nucleotide sequencing of cloned cDNA encoding neurotensin revealed a 170-amino-acid precursor containing neurotensin and neuromedin N in tandem, separated by a Lys–Arg sequence near the C-terminus. An additional neuromedin N-like peptide was found to be located in the mid-portion of the precursor.[206] Immunocytochemical studies have shown that neurotensin in the gut has a dual localization to both endocrine cells and neurons (cf. Reference 207). The endocrine cells are numerous in the distal small intestine and represent the main source of gut neurotensin. In mammals, neuronally located neurotensin has been found in the intestinal wall of sheep,[159] dog,[208] rat,[124] and guinea pig[127,209] and in the stomach wall of cat.[210] On the whole, neurotensin-containing nerve fibers are few and are found in the smooth muscle and myenteric ganglia. Neurotensin-containing nerve fibers are numerous in the intestinal wall of chicken and in some bony fishes.[211-213] The origin of neurotensin fibers in the gastrointestinal tract is not yet known. Immunoreactive nerve cell bodies occur in the intramural ganglia in the chicken,[213] suggesting an intrinsic origin. There is no information available on the costorage of neurotensin and other neuropeptides or on the projection pattern of neurotensin-containing neurons.

In vitro studies on guinea pig small intestine have revealed that neurotensin evokes contraction of the longitudinal muscle via an indirect mechanism involving cholinergic and SP neurons,[214] while it has a direct inhibitory action on the circular smooth muscle.[215] However, in dog small intestine, neurotensin contracts the circular smooth muscle through both direct and indirect mechanisms and inhibits the longitudinal muscle via indirect mechanisms.[216,217] The results of *in vivo* studies on dog intestinal smooth muscle suggest that neurotensin inhibits circular muscle contractions by the release of noradrenaline.[217,218] Neurotensin stimulates secretion of anions in the guinea pig ileal mucosa.[219,220]

H. OPIOID PEPTIDES

Leu- and Met-enkephalin were isolated from brain extracts based on their ability to act as opiate receptor agonists.[221] Subsequently, many more opioid peptides have been isolated and characterized. Today there are three known opioid peptide precursors in the body: pro-opiomelanocortin, proenkephalin A, and proenkephalin B (cf. Reference 222). Neuronal opioids in the gut probably arise from either proenkephalin A or B. The evidence that opioids in the gut derive from pro-opiomelanocortin is much less convincing. Several fragments of proenkephalin A and B have been localized immunocytochemically to neuronal elements in the gut.[93,223-228] Among opioids thought to occur in the enteric nervous system are Met-enkephalin, Met-enkephalin–Arg–Phe, Met-enkephalin–Arg–Gly–Leu, bovine adrenal medulla docosapeptide (BAM-22), Leu-enkephalin, and dynorphin. Proenkephalin A gives rise to Met- and Leu-enkephalin and extended forms of Met-enkephalin.[229] Proenkephalin B gives rise to Leu-enkephalin and its extended forms dynorphin A and B.[230] β-Endorphin-like material (arising from pro-opiomelanocortin) has been reported to be present in neurons of the gastrointestinal tract.[124,127,231] However, chemical data confirming its identity are lacking (see also Reference 228).

Enkephalin-immunoreactive nerve fibers are particularly numerous in the myenteric ganglia and smooth muscle throughout the gastrointestinal tract. Leu- and Met-enkephalin have been claimed to occur in separate nerve populations,[232,233] but most studies have been carried out with antibodies that do not discriminate between the different enkephalin variants. The esophagus of many mammals contains numerous enkephalin-immunoreactive nerve fibers in the external muscle and in the muscularis mucosae as well as in the myenteric ganglia, where a few enkephalin-containing nerve cell bodies have been demonstrated.[154,155,184,234-237] In the stomach, enkephalin-immunoreactive nerve fibers are numerous in the myenteric ganglia and smooth muscle while they are few in the submucous ganglia and in the basal part of the mucosa and submucosa.[124,128,154,184,187,223,231,237,238] In the small and large intestine, enkephalin-containing nerves are found almost exclusively in the myenteric ganglia and smooth muscle. They are scarce or absent from the submucosa and mucosa in most species examined.[85,87,93,124,166,184,187,231,233,237-245] The enkephalin-containing fibers are probably intramural in origin; nerve cell bodies occur in the myenteric ganglia, and extrinsic denervation does not overtly change the number of enkephalin-immunoreactive nerve fibers in the gut wall.[85,87,240,245] Immunoreactive enkephalin occurs also in endocrine cells in the digestive tract;[223,238,239,246,247] they are particularly numerous in the pig.[238]

Dynorphin-immunoreactive nerve fibers are reported to form an extensive network in the gastrointestinal tract of the guinea pig, particularly in the myenteric ganglia and smooth muscle. Dynorphin-containing nerve cell bodies are detected in both submucous and myenteric ganglia.[24,225,248] In the rat,[225,248] dynorphin-immunoreactive fibers in the gut are few, while they seem to be lacking altogether in pig and man.[93,248a]

Met-enkephalin-immunoreactive neurons also seem to contain extended forms of Met-enkephalin, including BAM,[93,223,228] suggesting that the peptides derive from proenkephalin A. At least in the guinea pig gut, proenkephalin B also seems to contribute to neuronal enkephalin-like peptides; approximately 95% of the enkephalin-immunoreactive neurons seem to contain dynorphin as well.[226] It is not yet known whether proenkephalin A and B occur together or in separate neurons. The majority of the enkephalin/dynorphin-containing myenteric neurons in guinea pig small intestine contain VIP[226] and PHI.[249] Some Dyn/Enk/VIP-containing myenteric neurons contain NPY as well, while others contain CCK and GRP.[24] The latter neuronal population (CCK/Dyn/Enk/GRP/VIP) projects to the prevertebral ganglia.[24] Approximately 50% of the nerve cell bodies in the submucous ganglia of the guinea pig small intestine contains dynorphin and VIP, but not enkephalin.[24] Coexistence of enkephalin and SP has been reported in small intestinal myenteric neurons of the guinea pig[228,249] and cat[228,250] as well as in the entire gastrointestinal tract of man.[186] In nerve fibers in the circular muscle of the guinea pig small intestine, enkephalin invariably coexists with either VIP or SP.[251] By contrast, Uchida et al.[249] demonstrated nerve terminals with dense-cored vesicles in the guinea pig circular muscle that were immunoreactive for both Met–enkephalin–Arg–Gly–Leu and SP or PHI, but also terminals that were immunoreactive for Met–enkephalin–Arg–Gly–Leu only. The projections of intrinsic enkephalin neurons in the gut have been studied in the rat, guinea pig, and dog. In the rat, myenteric enkephalin-containing neurons issue ascending projections to other myenteric ganglia and to the smooth muscle. In the small intestine these projections are 6 to 7 mm long, and in the large intestine they are 5 mm long.[85,87] In the guinea pig small intestine, myenteric enkephalin neurons issue both ascending and descending projections.[240,252] The descending ones project to the smooth muscle and are 2 mm long; the ascending ones project to other myenteric ganglia[240] and to smooth muscle[24] and are 3.5 to 4 mm long. In guinea pig colon the great majority of the myenteric enkephalin-containing neurons are ascending.[252] In the canine small intestine, myenteric enkephalin-containing neurons issue ascending projections up to 30 mm long to other myenteric ganglia and to the circular muscle.[166] Myenteric enkephalin-containing neurons have been reported to project also to submucous ganglia.[24,166]

The population of dynorphin-containing neurons in the guinea pig small intestine seems to be identical to the one that contains VIP.[24] From available data it can be concluded that the myenteric dynorphin- (and VIP-) containing neurons issue descending projections to other myenteric ganglia and to the smooth muscle.[24] Thus, the approximate length of these projections is 6 mm.[253] The submucous dynorphin (and VIP) neurons project to the mucosa and submucosa.[24]

The ultrastructure of enkephalin-containing nerve fibers has been studied in the guinea pig and cat intestine.[224,243,244,249,254] The enkephalin-immunoreactive material is located in large dense-cored or opaque vesicles. In the guinea pig the diameter of these vesicles ranges between 80 and 100 nm and in the cat between 100 and 200 nm. The nerve terminals also contain small clear vesicles that are nonimmunoreactive.

Enkephalins have a morphine-like effect on the digestive tract, i.e., a presynaptic inhibition of the cholinergic neurons mediating contraction,[236,255,256] and in some cases, may have a direct contractile effect on the muscle cells.[236,257,258] The constipating effect of opiates is in part due to inhibition of intestinal motility and in part due to enhanced absorption of water, sodium, and chloride[259,260] as well as an inhibition of stimulated intestinal secretion.[261,262] The effects of opioids on mucosal function are abolished by TTX and are thus indirect, probably mediated by the enteric neurons. The lack of demonstrable opioid receptors on epithelial cells[263] seems to support this view.

I. SOMATOSTATIN

Somatostatin, a 14-amino-acid cyclic peptide, was originally isolated from the hypothalamus[264,265] by virtue of its ability to inhibit the release of growth hormone. In fact, somatostatin inhibits secretion in a number of exocrine and endocrine cells (cf. References 266 and 267). Somatostatin-28 contains somatostatin-14 at its C-terminal end preceded by two basic residues.[268] Human and rat preprosomatostatin consists of 116 amino acid residues, with somatostatin-28 at the C-terminus.[269,270] Somatostatin-14 is the predominant form in the upper gut, while there is a relative increase in the proportion of somatostatin-28 further distally.[271] In the gut, somatostatin has a dual localization: in endocrine/paracrine cells and in neurons.[272,273] When separating gut mucosa and musculature, somatostatin-28 was found in high concentrations in the mucosa (endocrine cells?), while somatostatin-14 predominated in the musculature (nerves).[274,275] Somatostatin-immunoreactive nerve fibers are found throughout the gastrointestinal tract, although they are few in comparison with many other neuropeptide-containing fibers. The esophagus contains only a few somatostatin fibers[124,155] located mainly in the smooth muscle and the myenteric plexus; nerve cell bodies are rare or absent. A similar frequency and distribution of somatostatin fibers with a lack of nerve cell bodies also were observed in the stomach wall.[124,128,154,155,184,186,237,276,277] In the intestines the somatostatin fibers are by comparison more numerous and occur predominantly in the myenteric ganglia and to some extent also in smooth muscle.

In addition, somatostatin-containing nerve fibers can be found in the submucous ganglia and in the submucosa and mucosa, predominantly at the base of the glands (see Figure 5). In the intestines, somatostatin-containing nerve cell bodies are regularly seen in both myenteric and submucous ganglia.[85,87,93,124,154,166,184,237,243,272,276-278] Most of the somatostatin-containing nerve fibers in the intestines are intramural in origin; their distribution and frequency are virtually unaffected by extrinsic denervation.[85,87,245,279] In guinea pig small intestine a few somatostatin-containing nerve fibers seem to be extramural in origin.[280,281] They originate in the celiac-mesenteric ganglia and display the features of adrenergic neurons; their terminals are found in the submucous ganglia and in the mucosa. Intrinsic somatostatin-containing neurons (most of them are submucous; some are myenteric) in guinea pig small intestine harbor ChAT, CCK, CGRP, and NPY and terminate in the mucosa.[92] Besides the population of som/ChAT/CCK/CGRP/NPY-containing neurons, another population of myenteric somatostatin-containing neurons harbors SP and/or CCK.[24] In the rat colonic mucosa, somatostatin has been found to occur in a subpopulation of the CGRP-immunoreactive neurons.[87]

Myenteric somatostatin-containing neurons give off descending projections that terminate in the smooth muscle and within the myenteric ganglia. The lengths of the projections are 8 to 12 mm in guinea pig small intestine,[279] 6 mm in rat small and large intestine,[85,87] and 15 to 25 mm in canine small intestine.[166] In the rat small and large intestine, submucous somatostatin-containing neurons project to the mucosa/submucosa and to other submucous ganglia. The length of these projections is 5 mm.[85,87] Ultrastructural studies of somatostatin-containing terminals in myenteric plexus in guinea pig colon have revealed somatostatin to be localized in large (approximately 90 nm) granular vesicles.[254,282]

The actions of somatostatin on intestinal motility are thought to be neuromodulatory. Thus, somatostatin reduces the amplitude of ileal contractions *in vitro* by inhibiting acetylcholine release from neurons of the myenteric plexus.[283-285] As studied in rat colon, somatostatin enhances both VIP release and the descending relaxation.[286] In human appendix, somatostatin failed to affect neuronally mediated contractions of the longitudinal muscle.[93] Somatostatin effects on mucosal functions include a decrease in short circuit current as a result of an increased net absorption of sodium and chloride. This is suggested to be a direct effect on the epithelium as studied on rabbit ileum mucosa,[287,288] while in guinea pig small intestine the effect of somatostatin is abolished by TTX and therefore thought to be indirect.[289]

J. SUBSTANCE P (SP)

SP is an 11-amino-acid peptide belonging to the tachykinin family (cf. Reference 290). Although it was discovered many years ago in extracts of intestine,[291] it was not isolated and sequenced until 1971.[292] SP and other tachykinins are widely distributed in the body, and SP is known to occur in a population of primary sensory (C fiber) neurons (cf. References 82 and 290). In bovine brain, two SP mRNAs seem to arise by alternative processing of a single primary RNA transcript encoding two different SP precursors (α- and β-preprotachykinin [PPT]).[293] Unlike α-PPT, β-PPT gives rise to a second tachykinin, neurokinin A (NKA), in addition to SP. Later, Krause et al.[294] described three different PPT mRNAs in rat brain, deriving from the same gene. Two of the three resulting precursors (β- and τ-PPT) give rise to both SP and NKA while differing from each other by an extra pentadecapeptide sequence in the intervening portion between the two tachykinins in β-PPT. There is good evidence that both SP and NKA are produced in gut neurons and that all SP-containing neurons also store NKA (cf. Reference 82). To what extent α-PPT (which generates SP, but not NKA) is expressed in individual gut neurons remains to be clarified. Analysis of RNA extracts from rat duodenum has revealed the presence of all three PPT-encoding transcripts in the following order of magnitude: τ-mRNA $>$ β-mRNA $\gg$ α mRNA.[295]

In the gut, SP has a dual localization in neuronal elements and in endocrine cells.[296-298] The endocrine SP-containing cells contain also serotonin and comprise a minor population of the enterochromaffin cells. However, the SP-containing neurons represent the main source of SP in the gut. Nerve cell bodies and fibers displaying SP immunoreactivity are present throughout the gut. In the esophagus, SP-containing fibers seem to occur mainly in the smooth muscle and intramural ganglia and around blood vessels. The esophageal striated muscle is devoid of SP-immunoreactive fibers. A few SP-containing neuronal cell bodies occur in myenteric ganglia in the cat, but not in other species.[124,154,186,299,300] In the stomach wall, SP fibers are numerous in the myenteric ganglia and smooth muscle (see Figure 4). The myenteric ganglia contain a few scattered SP-immunoreactive nerve cell bodies, while none are found in the submucous ganglia. Mucosal SP fibers occur in varying frequency in various species and are seen to extend high up between the oxyntic and pyloric glands.[124,128,154,159,184,186,241,299,301,302] They are notably numerous in the pig.[91] On the whole, SP-containing nerve fibers are more numerous in the small and large intestine than in the esophagus and stomach (see Figure 6). The SP-containing fibers occur in all layers, and immunoreactive nerve cell bodies are numerous in both myenteric and submucous ganglia. In the intestines the number of mucosal SP-containing nerve fibers varies between species, being high in man, pig, and rat and quite low in cat and rabbit.[85,87,93,124,154,159,166,184,186,237,243,245,297,303-308]

Perivascular SP-containing nerve fibers which are intrinsic in origin are found throughout the gastrointestinal tract in all species. They are numerous (particularly around submucosal arteries) in guinea pig and rat and few in pig and man. Extrinsic SP-containing nerve fibers often contain CGRP[81,82] and are thought to be sensory in origin.[302,306,309-314] They terminate not only around blood vessels, but also in intramural ganglia. This has been revealed by extrinsic denervation and capsaicin treatment.[85,87,128,302,303,306,315]

Subpopulations of myenteric SP-containing neurons also contain serotonin (guinea pig large intestine),[316] Met-enkephalin (guinea pig small intestine, cat ileum, and human gut)[24,186,250] or somatostatin and CCK (guinea pig small intestine).[24] Submucous SP neurons in guinea pig small intestine make up approximately one fifth of the ChAT-containing neurons.[317] In the guinea pig small intestine, populations of myenteric SP-containing neurons give rise to projections to underlying smooth muscle, submucous ganglia, and mucosa and within the myenteric ganglia (orally as well as anally directed).[318] However, the majority of the mucosal SP-containing fibers arise in the submucous ganglia.[318] Projections from myenteric SP neurons to the submucous ganglia have also been described in the guinea pig cecum.[319] In the rat small intestine, myenteric SP-containing neurons project anally for approximately 7 mm to other myenteric ganglia and to the smooth muscle;[85] in the large intestine they issue orally directed, approximately 5-mm-long projections.[87] Submucous SP neurons in the rat small intestine issue both ascending and descending projections for approximately 4 mm, supplying other submucous ganglia and the mucosa/submucosa.[85] In the dog small intestine, myenteric SP neurons project for approximately 2 mm orally and anally to the smooth muscle and to other myenteric ganglia.[166]

Ultrastructural observations indicate that the SP-immunoreactive material is present in both small and large vesicles. Small (up to 80 nm) SP-containing vesicles are found in nerve fibers in the human fetal stomach,[320] in the guinea pig stomach,[321] and also in the human small intestine.[322] Large (up to 185 nm) SP-containing vesicles are found in nerve fibers in the guinea pig taenia coli,[243] guinea pig colon,[254] human small intestine,[322] and guinea pig small intestine.[249]

SP dilates blood vessels and contracts nonvascular smooth muscle (cf. Reference 290). The contractile effect of SP on intestinal smooth muscle is partly indirect and mediated by myenteric cholinergic neurons and partly a direct action on the smooth muscle cells. SP-induced contraction of gut smooth muscle has been demonstrated in a number of mammals (pig, cat, rabbit, guinea pig, and rat; for reviews see References 69 and 323). However, in the human appendix, NKA, but not SP, is capable of contracting the longitudinal smooth muscle.[93,323a] The effect of SP on intestinal mucosal transport has been studied both *in vivo* and *in vitro*. When given intravenously, SP causes secretion in the dog[324] while it stimulates absorption in the rat.[325] When studied under *in vitro* conditions, SP causes a TTX-sensitive stimulation of chloride secretion.[326,327]

K. VASOACTIVE INTESTINAL POLYPEPTIDE (VIP)

VIP, a 28-amino-acid peptide, was isolated from porcine duodenum.[328,329] The human VIP precursor was identified in 1983 by Itoh et al.[330] It contains not only VIP, but also a VIP-like peptide referred to as PHM (PHI in pig). The name derives from the fact that the N-terminal amino acid is His and the C-terminal one is Met (Ile in pig).[331] VIP is a neuropeptide with a wide distribution in the body. The digestive tract is richly supplied with VIP-containing nerve fibers which are intramural in origin.[332]

VIP and PHI/PHM in the mammalian gut are present exclusively in neuronal elements,[191,333,334] with the exception of the cat antrum,[335,336] where a few VIP-containing endocrine cells are present. In the avian gastrointestinal tract and in some species of fish, VIP is present not only in nerves, but also in endocrine cells.[337-340] VIP-immunoreactive fibers are numerous in the esophageal wall, with a predominance in the smooth muscle. VIP-immunoreactive cell bodies occur in both the submucous and the myenteric ganglia.[124,155,235,341] VIP-immunoreactive fibers are numerous in all layers of the stomach wall (see Figure 4), both in the oxyntic and in the pyloric regions. The VIP fibers form a dense network in the smooth muscle and mucosa, where they extend high up between the glands. VIP-immunoreactive nerve cell bodies are found in both the submucous and the myenteric ganglia.[124,128,154,157-159,186,237,342] Sphincteric regions receive a remarkable dense supply of VIP-immunoreactive fibers.[343] In the small and large intestine, VIP-containing fibers are numerous in all of the layers, and the intramural ganglia are rich in VIP-immunoreactive nerve cell bodies (see Figures 5 to 7).[24,85,87,93,154,161,166,178,233,241,243,245,344,345] Blood vessels in the gut wall are richly supplied with VIP-containing fibers, particularly those in the submucosa. The large number of VIP-containing nerve cell bodies in the intramural ganglia, together with the fact that extrinsic denervation fails to affect the VIP nerve supply, indicates that the VIP-containing nerve fibers in the gut are intramural in origin.[85,87,126,166,245,253] In the rat intestine, VIP-containing neurons in the submucous ganglia issue ascending projections that are 4 mm long in the small intestine and 2 mm long in the large intestine (see Figure 13); they seem to terminate mainly in other submucous ganglia and in the mucosa.[85,87] Myenteric VIP-containing neurons send descending projections to other myenteric ganglia and to the smooth muscle layer; these fibers are approximately 6 mm long in guinea pig intestine,[253] approximately 2 mm in rat small intestine,[85] approximately 4 mm in rat large intestine[87] (see Figure 13), and up to 20 mm long in dog ileum.[166] Some myenteric VIP-containing neurons in the guinea pig small intestine project to submucous ganglia,[24] while some submucous VIP neurons in guinea pig cecum seem to project to myenteric ganglia.[319] Other myenteric VIP-containing fibers terminate in the celiac and inferior mesenteric ganglia.[253,346]

VIP-containing neurons in rat small intestine are thought to be cholinergic since they are acetylcholinesterase positive[347] and contain ChAT.[348] However, VIP-containing neurons in guinea pig small intestine seem to lack ChAT immunoreactivity.[316] The anticipated coexistence of VIP and PHI was once questioned by the finding that, in the pig stomach, VIP-immunoreactive fibers are more numerous than PHI-immunoreactive fibers.[349] Such a difference could not be confirmed in studies on the stomach of rat and mouse.[128] Recent studies have revealed C-terminally extended forms of PHI, mostly peptide histidine valine (PHV-42), in extracts of stomach.[350] Conceivably, such variants will not be recognized by all PHI antisera used, which may explain the relative shortage of PHI-like immunoreactivity in the pig stomach. VIP is also reported to coexist with other neuropeptides. One subpopulation of myenteric VIP-containing neurons in guinea pig small intestine contains CCK, dynorphin, and GRP plus enkephalin or SP, while another contains dynorphin and GRP plus enkephalin or NPY.[24] The myenteric VIP neurons in the rat stomach and the myenteric and submucous VIP neurons in the mouse stomach contain NPY as well.[128] The VIP-containing nerve fibers in the gastric mucosa in both rat and man seem to contain GRP rather than NPY (see Figure 11).[163a] In the small intestine of mouse, rat, and pig all VIP-immunoreactive neurons contain NPY (see Figure 12).[85] In the rat colon some perivascular VIP-containing nerve fibers harbor CGRP as well; these fibers do not disappear after extrinsic denervation and are therefore thought to be intramural in origin.[87] VIP and galanin are reported to coexist in myenteric neurons in the small intestine of guinea pig, pig, and dog and in the large intestine of guinea pig.[133,136,137] In the submucous neurons in guinea pig small intestine VIP coexists with dynorphin and galanin.[136]

Ultrastructural studies have revealed VIP-immunoreactive material in both large granular vesicles and small agranular vesicles in enteric neurons. Large (90 to 150 nm) granular VIP-containing vesicles were found in canine distal esophagus,[351] cat colon,[352] guinea pig small and large intestine,[243,254,353] and rat small intestine.[354-356] Small (35 to 50 nm) agranular vesicles containing VIP were claimed to be present in rat small intestine[354,355] and guinea pig small intestine.[353]

VIP

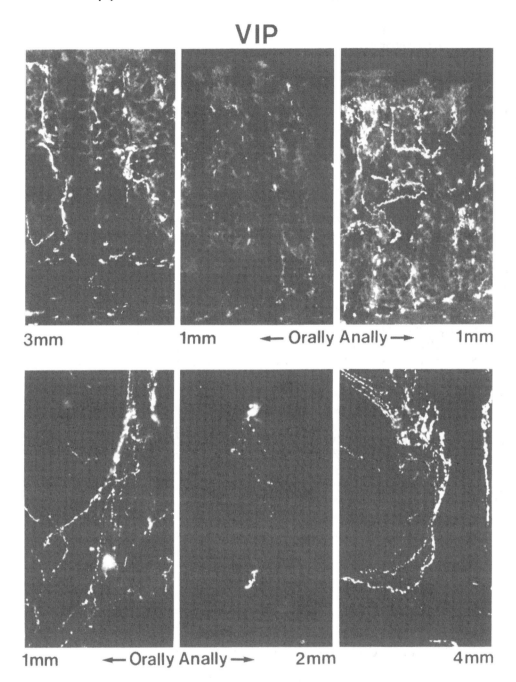

FIGURE 13.   VIP-immunoreactive nerve fibers in rat colon, after local denervation (intestinal clamping). Upper panel: cryostat sections, mucosa and submucosa visible. No loss of VIP fibers anally to the lesion. Orally there was a markedly reduced number of VIP fibers for approximately 2 mm, further orally the fiber frequency was normal. Lower panel: whole mounts of longitudinal muscle with adherent myenteric ganglia. No loss of VIP fibers orally to the lesion. Anally there was a loss of fibers for 1 to 2 mm, followed by a gradual increase in the number of fibers up to 4 mm; beyond this point the fiber frequency was normal. (Magnification × 120 [upper panel], × 140 [lower panel].)

The main effects of VIP (and PHI/PHM) on gastrointestinal functions are increased blood flow, relaxation of smooth muscle, and increased intestinal secretion (for reviews see References 1,69, 357, and 358). Relaxation is induced in gastric and intestinal smooth muscle and in sphincters, such as the lower esophageal sphincter and the internal anal sphincter. The effect is thought to be direct. VIP causes a marked secretion of water and electrolytes in both the small and large intestine. This reflects active chloride secretion and has been attributed to a direct action of VIP on the enterocytes.

III. ENTERIC NEURONS: PROJECTION PATTERNS AND PLASTICITY

Processes such as absorption, secretion, and peristaltic activity are controlled by intramural reflexes that are likely to involve peptide-containing neurons. Conceivably, the various neuropeptides play important roles as mediators and modulators in these processes (for recent reviews see References 1 to 3 and 69). In order to meet the demands for precise regulation and coordination, the enteric neurons have to interact in a highly ordered and well-integrated fashion. Therefore, it is to be expected that each population of enteric neurons displays a characteristic distribution and projection pattern. A knowledge of the neuroanatomy and microcircuitry of the nerves of the digestive tract is necessary before we can begin to understand the intricate mechanisms by which the enteric nervous system controls the physiology of the gut. In order to study the projection pattern of individual neuron populations, lesion experiments resulting in local interruption of neuronal pathways within the gut wall have been used. Such lesions, e.g., myectomy and myotomy, were first employed for this purpose by Furness and Costa.[359]

Generally, most myenteric neurons issue descending projections to other myenteric ganglia and to the smooth muscle, while most submucous neurons issue both ascending and descending projections to other submucous ganglia and the mucosa/submucosa. There is, however, evidence that the two sets of intramural (i.e., submucous and myenteric) ganglia are interconnected and that the connections are reciprocal. Inputs from submucous neurons to the myenteric ganglia have been demonstrated by neuronal tracing.[360] Beside these morphological findings, physiological observations support the view that submucous neurons innervate myenteric ganglia[361] as well as smooth muscle.[362] In guinea pig cecum, VIP-containing submucous neurons are reported to project to myenteric ganglia.[319] Projections from myenteric to submucous ganglia are represented by myenteric neurons containing CCK, dynorphin, enkephalin, GRP, somatostatin, SP, or VIP in guinea pig small intestine[24] and SP in guinea pig cecum.[319] Earlier morphological studies demonstrated processes from submucous ganglia to smooth muscle (for review see Reference 6). Wilson et al.,[363] on the other hand, concluded that in guinea pig small intestine the myenteric ganglia are the source of all nerve fibers of intramural origin in smooth muscle. The discrepancy may reflect species differences. In the rat some submucous neurons storing VIP/NPY and CGRP in the small intestine and submucous VIP-containing neurons in the large intestine seem to project also to subjacent circular muscle.[85,87] Projections from myenteric ganglia to the mucosa are also reported; e.g., neurons containing CGRP/CCK/NPY/Som/SP in guinea pig small intestine[24] and GRP in rat large intestine.[87]

Plate 1* outlines the projection patterns of the different subpopulations of enteric neurons in the rat small intestine. The polarities and projection distances of enteric neurons in rat small and large intestines are outlined in Figure 14.

So far projection studies utilizing local denervations have been performed in three species: guinea pig (cf. Reference 243), rat (cf. Reference 25), and dog.[166] A comparison of the projections of some peptide-containing myenteric neurons in the small intestine of the three

* Plate 1 follows page 160.

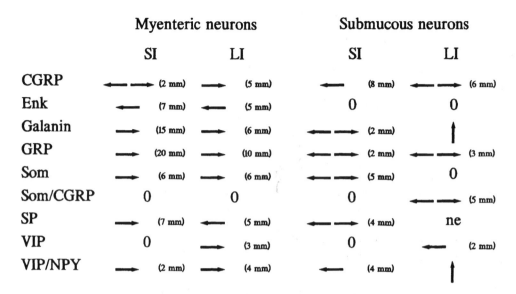

FIGURE 14. Polarities (marked with arrows) and projection distances (in millimeters) of various peptide-containing neuronal populations in the rat small (SI) and large (LI) intestine. → Indicates descending projections, ← indicates ascending projections, and ↑ indicates that no oro-anal projections can be demonstrated; o means absence of nerve fibers, while ne stands for not examined.

Myenteric neurons, SI

	Rat	Guinea-pig	Dog
Enk	← (7 mm)	← (6-8 mm)	← (20-30 mm)
GRP	→ (20 mm)	→ (9-12 mm)	→ (15-30 mm)
NPY	→ (2 mm)	→ (1-2 mm)	← (30 mm)
Som	→ (6 mm)	→ (8-12 mm)	→ (15-25 mm)
SP	→ (7 mm)	⇄ (1-2 mm)	⇄ (2 mm)
VIP	→ (2 mm)	→ (2-10 mm)	→ (5-20 mm)

FIGURE 15. Comparison of the projections of myenteric peptide-containing neurons in the small intestine of rat, guinea pig, and dog. There are many similarities in the projection pattern among the three species. → Indicates descending projections; ← indicates ascending projections. The projection lengths are given in millimeters. (Data are from References 24, 85, 87, 165, 166, 183, 240, 253, and 318.)

species is given in Figure 15 from which it is evident that the projection patterns are fairly similar. One characteristic that they have in common is that most myenteric neurons issue descending projections of varying length. In the rat small intestine they range from 2 mm (VIP/NPY-containing neurons) to 20 mm (GRP-containing neurons).[25] In the rat and guinea pig the projections of the corresponding neurons seem to be of fairly similar length, while in the dog there is a tendency for longer projections. Different populations of submucous neurons issue either ascending projections or ascending as well as descending projections for 2 to 8 mm, depending on the type of neuron. A comparison between species of submucous neurons cannot be made since the polarities and projection lengths of these neurons have only been described in the rat intestine.[85,87]

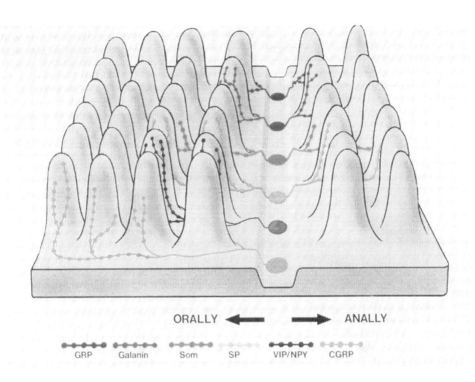

ORALLY ⟵   ⟶ ANALLY

GRP   Galanin   Som   SP   VIP/NPY   CGRP

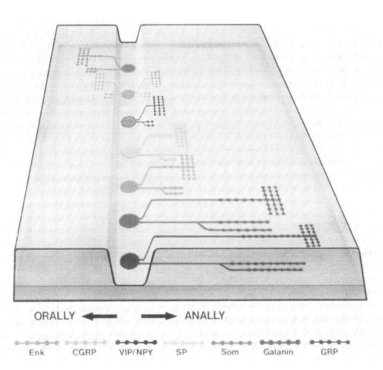

ORALLY ⟵   ⟶ ANALLY

Enk   CGRP   VIP/NPY   SP   Som   Galanin   GRP

PLATE 1. Schematic illustration attempting to outline the terminations and projections of the different peptide-containing neurons in rat small intestine. Upper panel: nerve fibers emanating from the submucous ganglia and terminating within the ganglia and in the mucosa and submucosa. Lower panel: nerve fibers emanating from the myenteric ganglia and terminating within the ganglia and longitudinal smooth muscle (straight course in figure) and in the circular smooth muscle (bent course in figure). The actual course of the fibers from the cell bodies to the termination area is unknown.

On the whole, the basic organization of the enteric neurons is quite similar both in different species and in the small and large intestine of the same species (rat), but certain differences may be noted. In order to illustrate the species differences, the myenteric NPY- and VIP-containing neurons in the small intestine may serve as an example. In the rat the two peptides coexist in neurons issuing descending pathways.[85] In the guinea pig[24] and dog[166] most VIP- and NPY-containing neurons are separate populations. In the guinea pig both of them project anally, while in the dog VIP-containing neurons project anally and NPY-containing neurons project orally. Myenteric SP-containing neurons in the rat may be used to exemplify differences between the small and large intestine. These neurons issue descending projections in the small intestine[85] and ascending projections in the large intestine.[87] Although the differences mentioned here are relatively minor in the context of the overall organization of the enteric innervation, they emphasize the need to secure a precise mapping of each neuronal population as a basis for the understanding of the neuronal circuitry involved in the regulation of gut function.

Local denervation results in axonal degeneration of the damaged enteric neurons. Little is known about the plasticity, i.e., the regeneration capacity of the damaged enteric neurons. Neuronal outgrowth has been described in guinea pig small intestine as soon as 2 to 4 days after lesioning of VIP-, enkephalin-, GRP-, and somatostatin-containing enteric neurons.[240,252,253] GRP-containing neurons in the rat small intestine have been examined at various times after myectomy, and a reinnervation as well as a remarkable hyperinnervation was observed[364] (see Figure 16). Myectomy which involves removal of the longitudinal muscle with adherent myenteric ganglia results in transient depletion of GRP fibers in the circular muscle underlying the myectomized segment and from the myenteric ganglia and smooth muscle up to 10 mm distally to the lesion. As examined at different time intervals up to 60 weeks postoperatively, fine varicose GRP fibers of a normal appearance were found to gradually reappear in the portion anally to the lesion beginning at 20 weeks, first in the more distal portions and then (after 40 to 60 weeks) also in the more adjacent portions. The circular muscle in the myectomized segment also became reinnervated during this time period. These fibers were notably coarse, more numerous than in control circular muscle, and arranged in thick bundles (hyperinnervation). Such nerve fibers were particularly frequent 40 weeks after the operation. Whether or not the regrowth of nerve fibers means a functional reinnervation is yet not known. Studies on the migration of the myoelectric complex after local denervation (transection) have recently been performed in the guinea pig small intestine. Recovery occurred about 2 months after the operation.[365] At that time nerve fibers containing VIP, GRP, or somatostatin were seen traversing the lesion.

IV. ENTERIC NEURONS AS MULTIMESSENGER SYSTEMS

The number of neuropeptides detected in the enteric nervous system has increased rapidly over the last few years. Many of these neuropeptides are found to be colocalized with other neuropeptides in the same neuron. A list illustrating the chemical coding, i.e., the different combinations of coexisting neuropeptides, is shown in Table 1. However, peptides represent only one type of putative messenger in enteric neurons. Enteric neurons probably also contain transmitters of a more conventional type. Thus, neuropeptides may coexist with, for instance, acetylcholine, noradrenaline, histamine, serotonin, or GABA. Hence, it is to be expected that many (perhaps all) enteric neurons represent multimessenger systems. The coexistence of neuropeptides may be anticipated (when the two peptides arise from the same precursor). This is exemplified by VIP/PHI[191,330,333,334] and SP/NKA.[82,93,293] Besides the anticipated coexistence there are several examples of unexpected coexistence, i.e., coexistence of peptides that arise from different precursors, e.g., VIP/NPY,[191] VIP/galanin,[133] or CCK/CGRP/ NPY/somatostatin.[24]

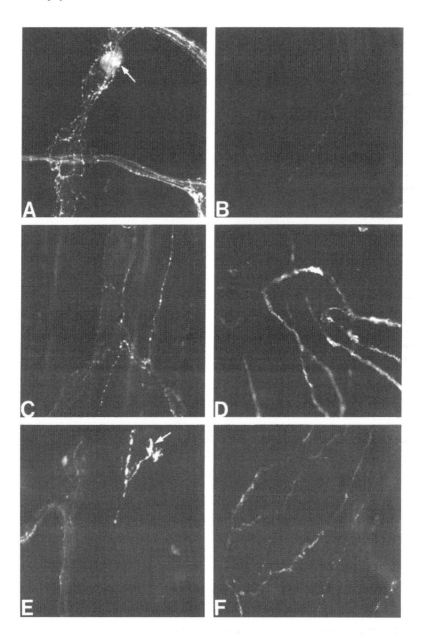

FIGURE 16.   Whole mounts of (A, C, E) longitudinal muscle with adherent myenteric ganglia and (B,D,F) circular muscle, immunostained for GRP. Specimens from (A,B) intact jejunum, (C,D) 40 weeks after myectomy, and (E,F) 60 weeks after myectomy. Myectomy (circumferential removal of 10 mm of the outer longitudinal smooth muscle with adherent myenteric ganglia) resulted in disappearance of GRP fibers from the myectomized circular muscle and from myenteric ganglia and both muscle layers for approximately 10 mm anally to the lesion (not shown). (A) Numerous GRP fibers and a nerve cell body (arrow) in myenteric ganglia. (B) Few, delicate GRP fibers in the circular muscle. (C) At 40 weeks postoperatively GRP fibers can be detected 6 mm anally to the lesion. (Futher anally the fibers gradually increase in number. Approximately 15 mm from the lesion the number of GRP fibers is back to normal). (D) At 40 weeks postoperatively, numerous GRP fibers are seen to run in thick bundles in the myectomized circular muscle. (E) At 60 weeks postoperatively, GRP fibers can be detected 3 mm anally to the lesion, sometimes markedly thickened (arrow). (Further anally the fibers gradually increase in number). From 6 to 8 mm anally to the lesion the GRP innervation is back to normal. (F) At 60 weeks the fiber bundles in the myectomized circular muscle are somewhat less numerous than at 40 weeks. In addition, single varicose GRP fibers can be seen. (Magnification × 180.)

TABLE 1
Chemical Coding of Recognized Enteric Neurons

Coexisting peptides	Species	Location	Ref.
CGRP/CCK/Gal/NPY/Som	Guinea pig	Small intestine	24, 137
CGRP/CCK/Gal/NMU/NPY/Som	Guinea pig	Small intestine	178
CGRP/NMU	Rat	Small intestine	98, 99
CGRP/Som	Rat	Large intestine	87
CGRP/VIP	Rat	Large intestine	87
CCK/Dyn/Enk/GRP/VIP	Guinea pig	Small intestine	24
CCK/Dyn/GRP/SP/VIP	Guinea pig	Small intestine	24
CCK/Som	Guinea pig	Large intestine	124
CCK/Som/SP	Guinea pig	Small intestine	24
Dyn/Enk/NPY/VIP	Guinea pig	Small intestine	24
Dyn/Gal/NMU/VIP	Guinea pig	Small intestine	178
Enk/SP	Cat	Small intestine	228, 250
	Man	Stomach, small and large intestine	186
	Guinea pig	Small intestine	24, 228, 249
Gal/VIP	Dog, pig	Small intestine	133, 135, 138
	Guinea pig	Large intestine	133
Gal/NPY/VIP	Guinea pig	Small intestine	137
GRP/SP	Rat	Stomach	163
GRP/VIP	Rat, man	Stomach	163a
NMU/NPY/VIP	Rat, pig	Small intestine	99
NPY/VIP	Rat, man	Stomach, small and large intestine	87, 93, 128, 186, 191
	Mouse, pig	Small intestine	191
SP/NMU	Guinea pig	Small intestine	178

The characteristic set of messengers in a certain population of neurons has been referred to as the chemical coding of those neurons.[366] The chemical coding reflects the programming of the neuron. Conceivably the release of multiple messengers may lead to an increased capacity for precise communication with other cell types and to an enhanced autocontrol by feedback mechanisms. A number of questions still need to be answered. Is the messenger programming of the neuron variable? Are the messengers costored in the same cytoplasmic vesicles and, thus, coreleased? Do the different messengers act on the same or on different target cells? Do the messengers cooperate after their release? Is there an "optimal" number of coexisting messengers? Studies on these subjects on the enteric nervous system are still in an initial stage, but some observations made with other neuronal systems may be worth mentioning. The fact that the expression of messengers may change is illustrated by the findings of Kessler and Black,[367] who showed that the sympathetic superior cervical ganglia in the rat greatly increased their content of SP after preganglionic denervation. Interestingly, these decentralized neurons express both tyrosine hydroxylase and SP.[368]

These findings, together with those of an increased density of CGRP-containing nerves in the iris stroma and peripheral blood vessels after sympathectomy and an increase of TH-immunoreactive nerves after capsaicin treatment, suggest interaction between sympathetic and sensory nerves.[369] Denervation of the rat adrenal medulla results in an increase in Leu-enkephalin content with no change in the catecholamine content.[370] This increase is preceded by an increase in the amount of preproenkephalin mRNA.[371] Sympathetic neurons grown in culture produce both noradrenaline and acetylcholine and display a gradual transition from adrenergic to cholinergic features.[372] Altered neurotransmitter expression in cholinergic neurons has also been shown.[373] Cholinergic neurons from ciliary ganglion, when injected into chick embryos, begin to manufacture catecholamines. Also the local environment may influence the expression of messengers. Serotonin-containing neurons from mesencephalic raphe nucleus start expressing SP when transplanted into hippocampus and striatum, but not when they are transplanted to the spinal cord.[374] A recent study of the rat small and large intestines[375] touches upon these phenomena. Sympathectomy by the long-term use of guanethidine was found to elevate the intestinal VIP and neurotensin content, while the content of Met-enkephalin was lowered. Whether these changes are due to altered release or synthesis is unknown. It is also not known whether the increased levels of VIP reflect accumulation of the peptide in the "normal" VIP neurons or whether VIP is being expressed by other neurons.

The question of whether costored messengers are coreleased has no definite answer. Conceivably, peptides arising from the same precursor are to be found in the same vesicles, although they may show a compartmentalization within the vesicles. This can be exemplified by the glicentin and glucagon immunoreactants in the pancreatic islets.[376] Also peptides arising from different precursors are generally found to be colocalized within the same granules. Examples are Met-enkephalin–Arg–Gly–Leu and SP or PHI in enteric neurons,[249] CGRP and SP, CGRP and somatostatin, SP and somatostatin in neurons in dorsal root ganglia;[377] and glicentin and PYY in gut endocrine cells.[378,379] Such colocalized peptides are probably released together. When it comes to coexistence of a neuropeptide and a classical transmitter, the principle seems to be that the neuropeptide is stored in large dense-core vesicles, while the classical transmitter occurs in small "synaptic" vesicles as well. This has been shown by subcellular fractionation for VIP and acetylcholine[380] and for NPY and noradrenaline.[381] Differential storage indicates that differential release may occur. There is evidence that differential release exists and that it is dependent on the mode and intensity of neuronal stimulation.[382-385] Future studies of these mechanisms will hopefully provide us with information on the possible interactions betweeen the various messengers and the possible advantages of multimessenger systems.

ACKNOWLEDGMENTS

Grant support was provided by the Swedish Medical Research Council (project numbers 4499 and 1007), The Påhlsson Foundation, and the Swedish Medical Society.

REFERENCES

1. **Cooke, H. J.,** Neural and humoral regulation of small intestinal electrolyte transport, in *Physiology of the Gastrointestinal Tract,* 2nd ed., Johnson, L. R., Ed., Raven Press, New York, 1987, 1307.
2. **Furness, J. B. and Costa, M.,** Influence of the enteric system on motility, in *The Enteric Nervous System,* Churchill Livingstone, London, 1987, 137.
3. **Wood, J. D.,** Neural basis of enteric motor function, in *Nerves and the Gastrointestinal Tract,* Singer, M. and Goebell, H., Eds., Falk Symp. No. 50, Kluwer Academic, Dordrecht, The Netherlands, 1989, 227.
4. **Lundgren, O., Svanvik, J., and Jivegård, L.,** Enteric nervous system. I. Physiology and pathophysiology of the intestinal tract, *Dig. Dis. Sci.,* 34, 264, 1989.
5. **Roman, C. and Gonella, J.,** Extrinsic control of digestive tract motility, in *Physiology of the Gastrointestinal Tract,* 2nd ed., Johnson, L. R., Ed., Raven Press, New York, 1987, 507.
6. **Schofield, G. C.,** Anatomy of muscular and neuronal tissues in the alimentary canal, in *Handbook of Physiology, Vol. 4, The Alimentary Canal,* Code, C. F., Ed., Waverly Press, Maryland, 1968, 1579.
7. **Gershon, M. D.,** The enteric nervous system, *Annu. Rev. Neurosci.,* 4, 227, 1981.
8. **Gabella, G.,** Structure of muscles and nerves in the gastrointestinal tract, in *Physiology of the Gastrointestinal Tract,* 2nd ed., Johnson, L. R., Ed., Raven Press, New York, 1987, 335.
9. **Hoffman, H. H. and Schnitzlein, H. N.,** The number of nerve fibers in the vagus nerve of man, *Anat. Rec.,* 139, 429, 1961.
10. **Oki, M. and Daniel, E. E.,** Effects of vagotomy on the ultrastructure of the dog stomach, *Gastroenterology,* 73, 1029, 1977.

11. **Vaithilingam, U. D., Wong, W. C., and Ling, E. A.,** Transneural changes in the myenteric ganglia of the monkey following vagotomy, *Neuroscience,* 17, 829, 1986.

12. **Grundy, D.,** Vagal control of gastrointestinal functions, *Baillieres Clin. Gastroenterol.,* 2, 23, 1988.

13. **Mukhopadhyay, A. K. and Weisbrodt, N. W.,** Neural organization of esophageal peristalsis: role of vagus nerve, *Gastroenterology,* 68, 444, 1975.

14. **Kalbasi, H., Hudsen, F. R., Herring, A., Moss, S., Glass, H. I. and Spencer, J.,** Gastric emptying following vagotomy and antrectomy and proximal gastric vagotomy, *Gut,* 16, 501, 1975.

15. **Håkanson, R., Hedenbro, J., Liedberg, G., and Vallgren, S.,** Effect of vagotomy on gastric acid secretion in the rat, *Acta Physiol. Scand.,* 115, 135, 1982.

16. **Hirschowitz, B. I.,** Central and peripheral nervous control of gastric secretion, in *Nerves and the Gastrointestinal Tract,* Singer, M. and Goebell, H., Eds., Falk Symp. No. 50, Kluwer Academic, Dordrecht, The Netherlands, 1989, 127.

17. **Furness, J. B. and Costa, M.,** The adrenergic innervation of the gastrointestinal tract, *Ergeb. Physiol.,* 69, 1, 1974.

18. **Sjövall, H., Jodal, M., and Lundgren, O.,** Sympathetic control of intestinal fluid and electrolyte transport, *News Physiol. Sci.,* 2, 214, 1987.

19. **Dockray, G. J., Green, T., and Varro, A.,** The afferent peptidergic innervation of the upper gastrointestinal tract, in *Nerves and the Gastrointestinal Tract,* Singer, M. and Goebell, H., Eds., Falk Symp. No. 50, Kluwer Academic, Dordrecht, The Netherlands, 1989, 105.

20. **Sauer, M. E. and Rumble, C. T.,** The number of nerve cells in the myenteric and submucous plexus of the small intestine of the cat, *Anat. Rec.,* 96, 373, 1946.

21. **Kosterlitz, H. W. and Watt, A. J.,** The peristaltic reflex, in *Methods in Pharmacology,* Vol. 3, Daniel, E. E. and Paton, D. M., Eds., Plenum Press, New York, 1975, 391.

22. **Lundgren, O.,** Enteric nervous control of mucosal functions of the small intestine *in vivo,* in *Nerves and the Gastrointestinal Tract,* Singer, M. and Goebell, H., Eds., Falk Symp. No. 50, Kluwer Academic, Dordrecht, The Netherlands, 1989, 275.

23. **Gabella, G.,** Innervation of the gastrointestinal tract, *Int. Rev. Cytol.,* 59, 129, 1979.

24. **Furness, J. B. and Costa, M.,** Studies of neuronal circuitry of the enteric nervous system, in *The Enteric Nervous System,* Churchill Livingstone, London, 1987, 111.

25. **Ekblad, E., Håkanson, R., and Sundler, F.,** Projections of enteric peptide-containing neurons in the rat, in *Nerves and the Gastrointestinal Tract,* Singer, M. V. and Goebell, H., Eds., Falk Symp. No. 50, Kluwer Academic, Dordrecht, The Netherlands, 1989, 47.

26. **Schultzberg, M. and Dalsgaard, C.-J.,** Enteric origin of bombesin immunoreactive fibers in the rat coeliac-superior mesenteric ganglion, *Brain Res.,* 269, 190, 1983.

27. **Ohkubo, K.,** Studies on the intrinsic nervous system of the digestive tract. I. The sub-mucous plexus of guinea-pig, *Jpn. J. Med. Sci.,* 6, 1, 1936.

28. **Auerbach, L.,** Fernere vorläufige Mitteilung über den Nervenapparat des Darmes, *Virchows Arch.,* 30, 457, 1864.

29. **Schabadasch, A.,** Die Nerven des Magens der Katze, *Z. Zellforsch.,* 110, 254, 1930.

30. **Rash, R. M., and Thomas, M. D.,** The intrinsic innervation of the gastro-oesophageal and pyloro-duodenal junctions, *J. Anat.,* 96, 389, 1962.

31. **Skak-Nielsen, T., Poulsen, S. S., and Holst, J. J.,** Immunohistochemical detection of ganglia in the rat stomach serosa, containing neurons immunoreactive for gastrin-releasing peptide and vasoactive intestinal peptide, *Histochemistry,* 87, 47, 1987.

32. **Vau, E.,** Über die subglandulären Ganglienzellen in der Magenvand einiger Haussäugetier, *Anat. Anz.,* 73, 380, 1937.

33. **Newson, B., Ahlman, H., Dahlström, A., and Nyhus, L. M.,** Ultrastructural observations in the rat ileal mucosa of possible epithelial "taste cells" and submucosal sensory neurons, *Acta Physiol. Scand.,* 114, 161, 1982.

34. **Wood, J. D.,** Neurophysiology of parasympathetic and enteric ganglia, in *Autonomic Ganglia,* Elvin, L. G., Ed., John Wiley & Sons, Chichester, England, 1983, 367.

35. **Wood, J. D.,** Physiology of the enteric nervous system in, *Physiology of the Gastrointestinal Tract,* 2nd Ed., Johnson, L. R., Ed., Raven Press, New York, 1987, 67.

36. **Hodgkiss, J. P. and Lees, G. M.,** Transmission in enteric ganglia, in *Autonomic and Enteric Ganglia,* Karczmar, A. G., Koketsu, K., and Nishi, S., Eds., Plenum Press, New York, 1986, 405.

37. **Nozdrachew, A. D. and Vataev, S. I.,** Neuronal electrical activity in the submucosal plexus of the cat small intestine, *J. Autonom. Nerv. Syst.,* 3, 45, 1981.

38. **Nishi, S. and North, R. A.,** Intracellular recording from the myenteric plexus of the guinea pig ileum, *J. Physiol. (London),* 231, 471, 1973.

39. **Hirst, G. D. S., Holman, M. E., and Spence, I.,** Two types of neurons in the myenteric plexus of duodenum in the guinea pig, *J. Physiol. (London),* 231, 303, 1974.

40. **King, B. F. and Szurszewski, J. H.,** Intracellular recordings from vagally innervated intramural neurons in opossum stomach, *Am. J. Physiol.,* 246, G209, 1984.
41. **Schemann, M. and Wood, J. D.,** Neurophysiology of the gastric myenteric plexus, in *Nerves and the Gastrointestinal Tract,* Singer, M. V. and Goebell, H., Eds., Falk Symp. No. 50, Kluwer Academic, Dordrecht, The Netherlands, 1989, 97.
42. **Hirst, G. D. S. and McKirdy, H. C.,** Synaptic potentials recorded from neurons of the submucous plexus of guinea-pig small intestine, *J. Physiol. (London),* 249, 369, 1975.
43. **Dogiel, A. S.,** Zur Frage über die Ganglien der Darmgeflechte bei den Säugetieren, *Anat. Anz.,* 10, 517, 1895.
44. **Dogiel, A. S.,** Über den Bau der Ganglien in den Geflechten des Darmen und der Gallenblase des Menschen und der Säugetieren, *Arch. Anat. Physiol. Anat. Abt.,* 1899, 130, 1899.
45. **Kuntz, A.,** On the occurrence of reflex arcs in the myenteric and submucous plexuses, *Anat. Rec.,* 24, 193, 1922.
46. **Stach, W.,** A revised morphological classification of neurons in the enteric nervous system, in *Nerves and the Gastrointestinal Tract,* Singer, M. V. and Goebell, H., Eds., Falk Symp. No. 50, Kluwer Academic, Dordrecht, The Netherlands, 1989, 29.
47. **Cavazzana, P. and Borsetto, P. L.,** Recherches sur l'aspect microscopic des plexus nerveux intramuraux et sur les modifications morphologies de leurs neurones dans les divers traits de l'intestin humain, *Acta Anat.,* 5, 17, 1948.
48. **Greving, R.,** Histologische Studien um Plexus myentericus des Magens. I. Der Plexus myentericus und seine Zelltypen *Dsch. Z. Nervenheilk.,* 165, 622, 1951.
49. **Gabella, G.,** Fall in the number of myenteric neurons in aging guinea pigs, *Gastroenterology,* 96, 1487, 1989.
50. **Baumgarten, H. G., Holstein, A.-F., and Owman, Ch.,** Auerbach's plexus of mammals and man: electron microscopic identification of three different types of neuronal processes in myenteric ganglia of the large intestine from Rhesus monkeys, guinea-pigs and man, *Z. Zellforsch. Mikrosk. Anat.,* 106, 376, 1970.
51. **Gabella, G.,** Fine structure of the myenteric plexus in the guinea-pig ileum, *J. Anat.,* 111, 69, 1972.
52. **Cook, R. D. and Burnstock, G.,** The ultrastructure of Auerbach's plexus in the guinea-pig. I. Neuronal elements, *J. Neurocytol.,* 5, 171, 1976.
53. **Llewellyn-Smith, I. J., Furness, J. B., Wilson, A. J., and Costa, M.,** Organization and fine structure of enteric ganglia, in *Autonomic Ganglia,* Elfvin, L. G., Ed., John Wiley & Sons, Chichester, England, 1983, 145.
54. **Sundler, F., Håkanson, R., Leander, S. and Uddman, R.,** Neuropeptides in the gut wall: cellular and subcellular localization, topographic distribution and possible physiological significance, in *Cytochemical Methods in Neuroanatomy,* Palay, S. and Chan-Palay, V., Eds., Alan R. Liss, New York, 1982, 341.
55. **Gordon-Weeks, P. R.,** The ultrastructure of noradrenergic and cholinergic neurons in the autonomic nervous system, in *Handbook of Chemical Neuroanatomy,* Vol. 6, the Peripheral Nervous System, Björklund, A., Hökfelt, T., and Owman, Ch., Eds., Elsevier, Amsterdam, 1988, 117.
56. **Varndell, I. M. and Polak, J. M.,** The ultrastructure of peptide-containing neurons, in *Handbook of Chemical Neuroanatomy,* Vol. 6, The Peripheral Nervous System, Björklund, A., Hökfelt, T., and Owman, Ch., Eds., Elsevier, Amsterdam, 1988, 143.
57. **Costa, M., Furness, J. B., and Llewellyn-Smith, I. J.,** Histochemistry of the enteric nervous system, in *Physiology of the Gastrointestinal Tract,* Johnson, L. R., Eds., Raven Press, New York, 1987, 1.
58. **Coupland, R. E. and Holmes, R. L.,** The use of cholinesterase techniques for the demonstration of peripheral nervous structures, *Q.J. Microsc. Sci.,* 98, 327, 1957.
59. **Furness, J. B., Costa, M., and Eckenstein, F.,** Neurons localized with antibodies against choline acetyltransferase in the enteric nervous system, *Neurosci. Lett.,* 40, 105, 1983.
60. **Costa, M., Furness, J. B., Cuello, A. C., Verhofstad, A. A. J., Steinbusch, H. W. M., and Elde, R. P.,** Neurons with 5-hydroxytryptamine-like immunoreactivity in the enteric nervous system: their visualization and reactions to drug treatment, *Neuroscience,* 7, 351, 1982.
61. **Baetge, G. and Gershon, M. D.,** GABA in the PNS: demonstration in enteric neurons, *Brain Res. Bull.,* 16, 421, 1986.
62. **Jessen, K. R., Mirsky, R., and Hills, J. M.,** GABAergic neurons in the vertebrate peripheral nervous system, in *GABAergic Mechanisms in the Mammalian Periphery,* Erdö, S. L. and Bowery, N. G., Eds., Raven Press, New York, 1986, 117.
63. **Furness, J. B., Trussell, D. C., Pompolo, S., Bornstein, J. C., Maley, B. E., and Storm-Mathisen, J.,** Shapes and projections of neurons with immunoreactivity for gamma-aminobutyric acid in the guinea-pig small intestine, *Cell Tissue Res.,* 256, 293, 1989.
64. **Ekblad, E., Wahlestedt, C., Håkanson, R., Sundler, F., Watanabe, T., and Wada, F.,** Is histamine a neurotransmitter in the gut? Evidence from histidine decarboxylase immunocytochemistry, *Acta Physiol. Scand.,* 123, 225, 1985.

65. **Panula, P., Kaartinen, M., Mäckling, M., and Costa, E.,** Histamine-containing peripheral neuronal and endocrine systems, *J. Histochem. Cytochem.,* 33, 933, 1985.

66. **Polak, J. M. and Bloom, S. R.,** Peptidergic nerves in the gastrointestinal tract, *Invest. Cell Pathol.,* 1, 301, 1978.

67. **Sundler, F., Böttcher, G., Ekblad, E., and Håkanson, R.,** The neuroendocrine system of the gut, *Acta Oncol.,* 28, 303, 1989.

68. **Makhlouf, G. M.,** Enteric neuropeptides: role in neuromuscular activity of the gut, *Trends Pharmacol. Sci.,* 6, 214, 1985.

69. **Dockray, G. J.,** Physiology of enteric neuropeptides, in *Physiology of the Gastrointestinal Tract,* 2nd ed., Johnson, L. R., Eds., Raven Press, New York, 1987, 41.

70. **Taylor, G. S. and Bywater, R. A. R.,** Novel autonomic neurotransmitters and intestinal function, *Pharmacol. Ther.,* 40, 401, 1989.

71. **Hodgkiss, J. P. and Lees, G. M.,** Morphological studies of electrophysiologically identified myenteric plexus neurons of the guinea-pig ileum, *Neuroscience,* 8, 593, 1984.

72. **Bornstein, J. C., Costa, M., Furness, J. B., and Lees, G. M.,** Electrophysiology and enkephalin immunoreactivity of identified myenteric plexus neurones of guinea-pig small intestine, *J. Physiol. (London),* 351, 313, 1984.

73. **Katayama, Y., Lees, G. M., and Pearson, G. T.,** Electrophysiology and morphology of vasoactive-intestinal-peptide-immunoreactive neurons of the guinea pig ileum, *J. Physiol. (London),* 378, 1, 1986.

74. **Lee, K. Y., Chang, T., and Chey, W. Y.,** Effect of electrical stimulation of the vagus on plasma motilin concentration in dog, *Life Sci.,* 29, 1093, 1981.

75. **Wolter, H. J.,** Corticotropin-releasing factor is contained within perikarya and nerve fibres of rat duodenum, *Biochem. Biophys. Res. Commun.,* 122, 381, 1984.

76. **Wolter, H. J.,** α-Melanotropin, β-endorphin and adrenocorticotropin-like immunoreactivities are colocalized within duodenal myenteric plexus perikarya, *Brain Res.,* 325, 290, 1985.

77. **Amara, S. G., Jonas, V., Rosenfeld, M. G., Ong, E. S., and Evans, R. M.,** Alternative RNA processing in calcitonin gene expression generates mRNAs encoding different polypeptide products, *Nature (London),* 298, 240, 1982.

78. **Rosenfeld, M. G., Mermod, J. J., Amara, S. G., Swanson, L. W., and Evans, R. M.,** Production of a novel neuropeptide encoded by the calcitonin gene via tissue specific RNA processing. *Nature (London),* 304, 129, 1983.

79. **Amara, S. G., Arriza, J. L., Leff, S. E., Swanson, L. W., Evans, R. M., and Rosenfeld, M. G.,** Expression in brain of a messenger RNA encoding a novel neuropeptide homologous to calcitonin gene-related peptide, *Science,* 229, 1094, 1985.

80. **Mulderry, P. K., Ghatei, M. A., Spokes, R. A., Jones, P. M., Pierson, A. M., Hamid, Q. A., Kanse, S., Amara, S. G., Burrin, J. M., Legon, S., Polak, J. M., and Bloom, S. R.,** Differential expression of α-CGRP and β-CGRP by primary sensory neurons and enteric autonomic neurons of the rat, *Neuroscience,* 25, 195, 1988.

81. **Gibbins, I. L., Furness, J. B., Costa, M., McIntyre, I., Hillyard, S. J., and Girgis, S.,** Co-localization of calcitonin gene-related peptide-like immunoreactivity with substance P in cutaneous, vascular and visceral sensory neurons of guinea-pigs, *Neurosci. Lett.,* 57, 125, 1985.

82. **Sundler, F., Brodin, E., Ekblad, E., Håkanson, R., and Uddman, R.,** Sensory nerve fibers: distribution of substance P, neurokinin A and calcitonin gene-related peptide, in *Tachykinin Antagonists,* Proc. Fernström Symp., Håkanson, R. and Sundler, F., Eds., Elsevier, Amsterdam, The Netherlands, 1985, 213.

83. **Lundberg, J. M. and Hökfelt, T.,** Multiple co-existence of peptides and classical transmitters in peripheral autonomic and sensory neurons — functional and pharmacological implications, in *Progress in Brain Research,* Vol. 68, Hökfelt, T., Fuxe, K., and Pernow, B., Eds., Elsevier, Amsterdam, 1986, 241.

84. **Clague, J. R., Sternini, C., and Brecha, N. C.,** Localization of calcitonin gene-related peptide-like immunoreactivity in neurons of the rat gastrointestinal tract, *Neurosci. Lett.,* 56, 63, 1985.

85. **Ekblad, E., Winther, C., Ekman, R., Håkanson, R., and Sundler, F.,** Projections of peptide-containing neurons in rat small intestine, *Neuroscience,* 20, 169, 1987.

86. **Sternini, C., Reeve, J. R., and Brecha, N.,** Distribution and characterization of calcitonin gene-related peptide immunoreactivity in the digestive system of normal and capsaicin-treated rats, *Gastroenterology,* 93, 852, 1987.

87. **Ekblad, E., Ekman, R., Håkanson, R., and Sundler, F.,** Projections of peptide-containing neurons in rat colon, *Neuroscience,* 27, 655, 1988.

88. **Rodrigo, J., Polak, J. M., Fernandez, L., Ghatei, M. A., Mulderry, P., and Bloom, S. R.,** Calcitonin gene-related peptide immunoreactive sensory and motor nerves of the rat, cat and monkey esophagus, *Gastroenterology,* 88, 444, 1985.

89. **Parkman, H. P., Reynolds, J. C., Elfman, K. S., and Ogorek, C. P.,** Calcitonin gene-related peptide: a sensory and motor neurotransmitter in the feline lower esophageal sphincter, *Regul. Peptides,* 25, 131, 1989.

90. **Lee, Y., Shiotani, Y., Hayashi, N., Kamada, T., Hillyard, C. J., Girgis, S. I., MacIntyre, I., and Tohyama, M.,** Distribution and origin of calcitonin gene-related peptide in the rat stomach and duodenum: an immunocytochemical analysis, *J. Neural Transm.,* 68, 11, 1987.

91. **Sundler, F., Ekblad, E., and Håkanson, R.,** Occurrence and distribution of SP- and CGRP-containing nerve fibers in gastric mucosa. Species differences, in *Sensory Nerves and Neuropeptides in Gastroenterology: From Basic Science to Clinical Perspectives,* Costa, M., Surrenti, C., Gorini, S., Maggi, C. A., and Meli, A., Eds., Plenum Press, NY, in press.

92. **Furness, J. B., Costa, M., Gibbins, I. L., Llewellyn-Smith, I. J., and Oliver, J. R.,** Neurochemically similar myenteric and submucous neurons directly traced to the mucosa of the small intestine, *Cell Tissue Res.,* 241, 155, 1985.

93. **Ekblad, E., Arnbjörnsson, E., Ekman, R., Håkanson, R., and Sundler, F.,** Neuropeptides in the human appendix: distribution and motor effects, *Dig. Dis. Sci.,* 34, 1217, 1989.

94. **Ohtani, R., Kaneko, T., Kline, L. W., Labedz, T., Tang, Y. and Pang, P. K. T.,** Localization of calcitonin gene-related peptide in the small intestine of various vertebrate species, *Cell Tissue Res.,* 258, 35, 1989.

95. **Dahlström, A., Nilsson, O., Lundgren, O., and Ahlman, H.,** Nonadrenergic, noncholinergic innervation of gastrointestinal vessels: morphological and physiological aspects, in *Nonadrenergic Innervation of Blood Vessels,* Vol. 2, Regional Innervation, Burnstock, G. and Griffith, S. G., Eds., CRC Press, Boca Raton, FL, 1988, 143.

96. **Su, H. C., Bishop, A. E., Power, R. F., Hamada, Y., and Polak, J. M.,** Dual intrinsic and extrinsic origin of CGRP- and NPY-immunoreactive nerves of rat gut and pancreas, *J. Neurosci.,* 7, 2674, 1987.

97. **Maggi, C. A., Manzini, S., Giuliani, S., Santicioli, P., and Meli, A.,** Extrinsic origin of the capsaicin-sensitive innervation of rat duodenum: possible involvement of calcitonin gene-related peptide (CGRP) in the capsaicin-induced activation of intramural non-adrenergic non-cholinergic neurons, *Naunyn-Schmiedeberg's Arch. Pharmacol.,* 334, 172, 1986.

98. **Ballestra, J., Carlei, F., Bishop, A. E., Steel, J. H., Gibson, S. J., Fehey, M., Hennessey, R., Domin, J., Bloom, S. R., and Polak, J. M.,** Occurrence and developmental pattern of neuromedin U-immunoreactive nerves in the gastrointestinal tract and brain of the rat, *Neuroscience,* 25, 797, 1988.

99. **Ekblad, E., Persson, P., Håkanson, R., and Sundler, F.,** Neuronal neuromedin U in mammalian and submammalian species, An immunochemical and immunocytochemical study, manuscript in preparation.

100. **Fehér, E., Burnstock, G., Varndell, I. M., and Polak, J. M.,** Calcitonin gene-related peptide-immunoreactive nerve fibres in the small intestine of the guinea-pig: electron-microscopic immunocytochemistry, *Cell Tissue Res.,* 245, 353, 1986.

101. **Brain, S. D., Williams, T. J., Tippins, H. R., and MacIntyre, I.,** Calcitonin gene-related peptide is a potent vasodilator, *Nature (London),* 313, 54, 1985.

102. **Uddman, R., Edvinsson, L., Ekblad, E., Håkanson, R., and Sundler, F.,** Calcitonin gene-related peptide (CGRP): perivascular distribution and vasodilatory effects, *Regul. Peptides,* 15, 1, 1986.

103. **Kawasaki, H., Takasaki, K., Saito, A., and Goto, H.,** Calcitonin gene-related peptide acts as a novel vasocilator neurotransmitter in mesenteric resistance vessels of the rat, *Nature (London),* 335, 164, 1988.

104. **Tippins, J. R., Morris, H. R., Panico, M., Etienne, T., Bevis, P., Girgis, S., MacIntyre, I., Azria, M., and Attinger, M.,** The myotropic and plasma-calcium modulating effects of calcitonin gene-related peptide, *Neuropeptides,* 4, 425, 1984.

105. **Barthó, L., Lembeck, F., and Holzer, P.,** Calcitonin gene-related peptide is a potent relaxant of intestinal muscle, *Eur. J. Pharmacol.,* 135, 449, 1987.

106. **Taché, Y., Pappas, T., Lauffenburger, M., Goto, Y., Walsh, J. H., and Debas, H.,** Calcitonin gene-related peptide: potent peripheral inhibitor of gastric acid secretion in rats and dogs, *Gastroenterology,* 87, 344, 1984.

107. **Yamatani, T., Kadowaki, S., Shiba, R., Abe, H., Chihara, K., Fukase, M., and Fujita, T.,** Calcitonin gene-related peptide stimulates somatostatin release from isolated perfused rat stomach, *Endocrinology,* 118, 2144, 1986.

108. **Dunning, B. E. and Taborsky, G. J., Jr.,** Calcitonin gene-related peptide: a potent and selective stimulator of gastrointestinal somatostatin secretion, *Endocrinology,* 120, 1774, 1987.

109. **Cox, H. M., Ferrar, J. A., and Cuthbert, A. W.,** Effects of α- and β-calcitonin gene-related peptides upon ion transport in rat descending colon, *Br. J. Pharmacol.,* 97, 996, 1989.

110. **Mutt, V. and Jorpes, J. E.,** Hormonal polypeptides of the upper intestine, *Biochem. J.,* 125, 57P, 1971.

111. **Mutt, V.,** Further investigations on intestinal hormonal polypeptides, *Clin. Endocrinol.,* 5 (Suppl.), 175, 1976.

112. **Eysselein, V. E., Reeve, J. R., Jr., Shively, J. E., Hawke, D., and Walsh, J. H.,** Partial structure of a large canine cholecystokinin (CCK.58): amino acid sequence, *Peptides,* 3, 687, 1982.

113. **Dockray, G. J., Gregory, R., and Hutchison, J.,** Isolation, structure and biological activity of two cholecystokinin octapeptides from sheep brain, *Nature (London),* 274, 711, 1978.

114. **Dockray, G. J., Desmond, H., Gayton, R. J., Jönsson, A.-C., Raybould, H., Sharkey, K. A., Varro, A., and Williams, R. G.**, Cholecystokinin and gastrin forms in the nervous system, *Ann. N.Y. Acad. Sci.*, 448, 32, 1985.

115. **Deschenes, R. J., Lorenz, L. J., Haun, R. S., Roos, B. A., Collier, K. J., and Dixon, J. E.**, Cloning and sequence analysis of a cDNA encoding rat preprocholecystokinin, *Proc. Natl. Acad. Sci. U.S.A.*, 81, 726, 1984.

116. **Gubler, U., Chua, A. O., Hoffman, B. I., Collier, K. J., and Eng, J.**, Cloned cDNA to cholecystokinin mRNA predicts an identical preprocholecystokinin in pig brain and gut, *Proc. Natl. Acad. Sci. U.S.A.*, 81, 4307, 1984.

117. **Vanderhaeghen, J.-J., Signeau, J. C., and Gepts, W.**, New peptide in the vertebrate CNS reacting with gastrin antibodies, *Nature (London)*, 257, 604, 1975.

118. **Dockray, G. J.**, Immunochemical evidence of cholecystokinin-like peptides in brain, *Nature (London)*, 264, 568, 1976.

119. **Lorén, I., Alumets, J., Håkanson, R., and Sundler, F.**, Distribution of gastrin and CCK-like peptides in rat brain, *Histochemistry*, 59, 249, 1979.

120. **Rehfeld, J. F.**, Localization of gastrin to neuro- and adeno-hypophysis, *Nature (London)*, 271, 771, 1978.

121. **Uvnäs-Wallensten, K., Rehfeld, J. F., Larsson, L.-I., and Uvnäs, B.**, Heptadeca-peptide gastrin in the vagal nerve, *Proc. Natl. Acad. Sci. U.S.A.*, 74, 5709, 1977.

122. **Rehfeld, J. F., Goltermann, N., Larsson, L.-I., Emson, P. C., and Lee, C. M.**, Gastrin and cholecystokinin in central and peripheral neurons, *Fed. Proc.*, 38, 2325, 1979.

123. **Larsson, L.-I. and Rehfeld, J. F.**, Localization and molecular heterogeneity of cholecystokinin in the central and peripheral nervous system, *Brain Res.*, 165, 201, 1979.

124. **Schultzberg, M., Hökfelt, T., Nilsson, G., Terenius, L., Rehfeld, F. F., Brown, M., Elde, R., Goldstein, M., and Said, S.**, Distribution of peptide- and catecholamine-containing neurons in the gastrointestinal tract of rat and guinea-pig: immunohistochemical studies with antisera to substance P, vasoactive intestinal polypeptide, enkephalins, somatostatin, gastrin/cholecystokinin, neurotensin and dopamine β-hydroxylase, *Neuroscience*, 5, 689, 1980.

125. **Dockray, G. J., Vaillant, C., and Hutchison, J. B.**, Immunochemical characterization of peptides in endocrine cells and nerves with particular reference to gastrin and cholecystokinin, in *Cellular Basis of Chemical Messengers in the Digestive System*, Grossman, M. I., Brazier, M. A. B., and Lechago, J., Eds., Academic Press, New York, 1981, 215.

126. **Keast, J. R., Furness, J. B., and Costa, M.**, Origins of peptide and norepinephrine nerves in the mucosa of the guinea pig small intestine, *Gastroenterology*, 86, 637, 1984.

127. **Leander, S., Ekman, R., Uddman, R., Sundler, F., and Håkanson, R.**, Neuronal cholecystokinin, gastrin-releasing peptide, neurotensin, and β-endorphin in the intestine of the guinea-pig. Distribution and possible motor functions, *Cell Tissue Res.*, 235, 521, 1984.

128. **Ekblad, E., Ekelund, M., Graffner, H., Håkanson, R., and Sundler, F.**, Peptide-containing nerve fibers in the stomach wall of rat and mouse, *Gastroenterology*, 89, 73, 1985.

129. **Vizi, E. S., Bertaccini, G., Impicciatore, M., and Knoll, J.**, Evidence that acetylcholine released by gastrin and related polypeptides contributes to their effect on gastrointestinal motility, *Gastroenterology*, 64, 268, 1973.

130. **Hutchison, J. B. and Dockray, G. J.**, Evidence that the action of cholecystokinin octapeptide on the guinea-pig ileum longitudinal muscle is mediated in part by substance P release from the myenteric plexus, *Eur. J. Pharmacol.*, 69, 87, 1981.

131. **Tatemoto, K., Rökaeus, Å., Jörnvall, H., MacDonald, T. J., and Mutt, V.**, Galanin — a novel biologically active peptide from porcine intestine, *FEBS Lett.*, 164, 124, 1983.

132. **Rökaeus, Å. and Brownstein, M. J.**, Construction of a porcine adrenal medullary cDNA library and nucleotide sequence analysis of two clones encoding a galanin precursor, *Proc. Natl. Acad. Sci. U.S.A.*, 83, 6287, 1986.

133. **Melander, T., Hökfelt, T., Rökaeus, Å., Fahrenkrug, J., Tatemoto, K., and Mutt, V.**, Distribution of galanin-like immunoreactivity in the gastrointestinal tract of several mammalian species, *Cell Tissue Res.*, 239, 253, 1985.

134. **Ekblad, E., Håkanson, R., Rökaeus, Å., and Sundler, F.**, Galanin nerve fibers in the rat gut: distribution, origin and projections, *Neuroscience*, 16, 355, 1985.

135. **Bishop, A. E., Polak, J. M., Bauer, F. E., Christofides, N. D., Carlei, F., and Bloom, S. R.**, Occurrence and distribution of a newly discovered peptide, galanin, in the mammalian enteric nervous system, *Gut*, 27, 849, 1986.

136. **Bauer, F. E., Adrian, T. E., Christofides, N. D., Ferri, G.-L., Yanaihara, N., Polak, J. M., and Bloom, S. R.**, Distribution and molecular heterogeneity of galanin in human, pig, guinea-pig and rat gastrointestinal tract, *Gastroenterology*, 91, 833, 1986.

137. **Furness, J. B., Costa, M., Rökaeus, Å., McDonald, T. J., and Brooks, B.**, Galanin-immunoreactive neurons in the guinea-pig small intestine: their projections and relationships to other enteric neurons, *Cell Tissue Res.*, 250, 607, 1987.

138. **Gonda, T., Daniel, E. E., McDonald, T. J., Fox, J. E. T., Brooks, B. D., and Oki, M.,** Distribution and function of enteric GAL-IR nerves in dogs: comparison with VIP, *Am. J. Physiol.,* 256, G884, 1989.

139. **Fehér, E. and Burnstock, G.,** Galanin-like immunoreactive nerve elements in the small intestine of the rat. An electron microscopic immunocytochemical study, *Neurosci. Lett.,* 92, 137, 1988.

140. **Ekblad, E., Håkanson, R., Sundler, F., and Wahlestedt, C.,** Galanin: neuromodulatory and direct contractile effects on smooth muscle preparations, *Br. J. Pharmacol.,* 86, 241, 1985.

141. **Fox, J. E. T., McDonald, T. J., Kostolanska, F., and Tatemoto, K.,** Galanin: an inhibitory neural peptide of the canine small intestine, *Life Sci.,* 39, 103, 1986.

142. **Grider, J. R. and Makhlouf, G. M.,** The modulatory function of galanin: potentiation of VIP-induced relaxation in isolated smooth muscle cells, *Gastroenterology,* 94, A157, 1988.

143. **McDonald, T. J., Nilsson, G., Vagne, M., Ghatei, M., Bloom, S. R., and Mutt, V.,** A gastrin releasing peptide from nonantral gastric tissue, *Gut,* 19, 767, 1978.

144. **Anastasi, A., Erspamer, V., and Bucci, V.,** Isolation and structure of bombesin and alytesin, two analogous active peptides from skin of the European amphibians *Bombina* and *Alytes, Experientia,* 27, 166, 1971.

145. **Minamino, N., Kanagawa, K., and Matsuo, H.,** Neuromedin B: novel bombesin-like peptide identified in porcine spinal cord, *Biochem. Biophys. Res. Commun.,* 114, 541, 1983.

146. **Minamino, N., Kanagawa, K., and Matsuo, H.,** Neuromedin C: a bombesin-like peptide identified in porcine spinal cord, *Biochem. Biophys. Res. Commun.,* 119, 14, 1984.

147. **Spindel, E. R., Chin, W. W., Price, J., Rees, L. H., Besser, G. M., and Habener, J. F.,** Cloning and characterization of cDNAs encoding human gastrin-releasing peptide, *Proc. Natl. Acad. Sci. U.S.A.,* 81, 5699, 1984.

148. **Spindel, E. R. and Krane, I. M.,** Molecular biology of bombesin-like peptides: comparison of cDNAs encoding human gastrin-releasing peptide, human neuromedin B and amphibian ranatensin, *Ann. N.Y. Acad. Sci.,* 547, 10, 1988.

149. **Dockray, G. J., Vaillant, C., and Walsh, J. H.,** The neuronal origin of bombesin-like immunoreactivity in the rat gastrointestinal tract, *Neuroscience,* 4, 1561, 1979.

150. **Yanaihara, N., Yanaihara, C., Mochizuki, T., Imura, K., Fujita, T., and Iwanaga, T.,** Immunoreactive GRP, *Peptides,* 2 (Suppl. 2), 185, 1981.

151. **Moghimzadeh, E., Ekman, R., Håkanson, R., Yanaihara, N., and Sundler, F.,** Neuronal gastrin-releasing peptide in the mammalian gut and pancreas, *Neuroscience,* 10, 553, 1983.

152. **Timson, C. M., Polak, J. M., Wharton, J., Ghatei, M. A., Bloom, S. R., Usellini, L., Capella, C., Solcia, E., Brown, M. R., and Pearse, A. G. E.,** Bombesin-like immunoreactivity in the avian gut and its localization to a distinct cell type, *Histochemistry,* 61, 213, 1979.

153. **Vaillant, C., Dockray, G., and Walsh, J. H.,** The avian proventriculus is an abundant source of endocrine cells with bombesin-like immunoreactivity, *Histochemistry,* 64, 307, 1979.

154. **Keast, J. R., Furness, J. B., and Costa, M.,** Distribution of peptide-containing neurons and endocrine cells in the rabbit gastrointestinal tract, with particular reference to the mucosa, *Cell Tissue Res.,* 248, 565, 1987.

155. **Wattchow, D. A., Furness, J. B., Costa, M., O'Brien, P. E., and Peacock, M.,** Distribution of neuropeptides in the human esophagus, *Gastroenterology,* 93, 1363, 1987.

156. **Iwanaga, T.,** Gastrin-releasing peptide (GRP)/bombesin-like immunoreactivity in the neurons and para-neurons of the gut and lung, *Biomed. Res.,* 4, 93, 1983.

157. **Forssmann, W. G. and Reinecke, M.,** Peptidergic innervation of the stomach, *Hepatogastroenterology,* 29, 87, 1982.

158. **Kuwahara, A., Ishikawa, T., Mikami, S.-I., and Yanaihara, N.,** Distribution of neurons containing immunoreactivity for gastrin-releasing peptide (GRP), substance P and vasoactive intestinal polypeptide (VIP) in the rat gastric wall, *Biomed. Res.,* 4, 473, 1983.

159. **Wathuta, E. M.,** The distribution of vasoactive intestinal polypeptide-like, substance P-like and bombesin-like immunoreactivity in the digestive system of the sheep, *Q. J. Exp. Physiol.,* 71, 615, 1986.

160. **Miller, A. S., Furness, J. B., and Costa, M.,** The relationship between gastrin cells and bombesin-like immunoreactive nerve fibres in the gastric antral mucosa of guinea-pig, rat, dog and man, *Cell Tissue Res.,* 257, 171, 1989.

161. **Sjövall, M., Ekblad, E., Lundell, L., and Sundler, F.,** Gastrin-releasing peptide: neuronal distribution and spatial relation to endocrine cells in the human upper gut, *Regul. Peptides,* 28, 47, 1990.

162. **Costa, M., Furness, J. B., Yanaihara, N., Yanaihara, C., and Moody, T. W.,** Distribution and projections of neurons with immunoreactivity for both gastrin-releasing peptide and bombesin in the guinea-pig small intestine, *Cell Tissue Res.,* 235, 285, 1984.

163. **Kuwahara, A., Mikami, S., and Yanaihara, N.,** Coexistence of immunoreactive gastrin-releasing peptide and substance P in the myenteric plexus of rat stomach, *Biomed. Res.,* 6, 443, 1985.

163a. **Ekblad, E., Håkanson, R. and Sundler, F.,** Innervation of the stomach of rat and man with special reference to the endocrine cells, in *The Stomach as an Endocrine Organ,* Proc. Fernström Symp., Håkanson, R. and Sundler, F., Eds., Elsevier, Amsterdam, The Netherlands, 1990, in press.

164. **Bosshard, A., Protat, C., Dechelette, M. A., and Chery-Croze, S.,** Gastrin-releasing peptide (GRP) immunoreactivity in the gastrointestinal tract of normal and capsaicin-treated rats, *Regul. Peptides,* 22, 391, 1988.

165. **Ekblad, E., Ekman, R., Håkanson, R., and Sundler, F.,** GRP neurones in the rat small intestine issue long anal projections, *Regul. Peptides,* 9, 279, 1984.

166. **Daniel, E. E., Furness, J. B., Costa, M., and Belbeck, L.,** The projections of chemically identified nerve fibres in canine ileum, *Cell Tissue Res.,* 247, 377, 1987.

167. **Schultzberg, M. and Dalsgaard, C.-J.,** Enteric origin of bombesin immunoreactive fibres in the rat coelica-superior mesenteric ganglion, *Brain Res.,* 269, 190, 1983.

168. **Hanaji, M., Kawai, Y., Kawashima, Y., and Tohyama, M.,** Projections of bombesin-like immunoreactive fibers from the rat stomach to the celiac ganglion revealed by a double-labeling technique, *Brain Res.,* 416, 192, 1987.

169. **Bertaccini, G., Impicciatore, M., Molina, E., and Zappia, L.,** Action of bombesin on human gastrointestinal motility, *R. Gastroenterol.,* 6, 45, 1974.

170. **Erspamer, V. and Melchiorri, P.,** Actions of bombesin on secretion and motility of the gastrointestinal tract, in *Gastrointestinal Hormones,* Thompson, J. C., Ed., University of Texas Press, Austin, 1975, 575.

171. **Porreca, R., Burks, T. F., and Sheldon, R. J.,** Central and peripheral visceral effects of bombesin, *Ann. N.Y. Acad. Sci.,* 547, 194, 1988.

172. **Du Val, J. W., Saffouri, B., Weir, G. C., Walsh, J. H., Arimura, A., and Maklouf, G. M.,** Stimulation of gastrin and somatostatin secretion from the isolated rat stomach by bombesin, *Am. J. Physiol.,* 241, G242, 1981.

173. **McDonald, T. J., Ghatei, M. A., Bloom, S. R., Track, N. S., Radziuk, J., Dupre, J., and Mutt, V.,** A qualitative comparison of canine plasma gastroenteropancreatic hormone responses to bombesin and the porcine gastrin-releasing peptide (GRP), *Regul. Peptides,* 2, 293, 1981.

174. **Minamino, N., Kangawa, K., and Matsuo, H.,** Neuromedin U-8 and U-25: novel uterus stimulating and hypertensive peptides identified in porcine spinal cord, *Biochem. Biophys. Res. Commun.,* 130, 1070, 1985.

175. **Domin, J., Ghatei, M. A., Cohen, P., and Bloom, S. R.,** Neuromedin U — a study of its distribution in the rat, *Peptides,* 8, 779, 1987.

176. **Augood, S. J., Keast, J. R., and Emson, P. C.,** Distribution and characterization of neuromedin U-like immunoreactivity in rat brain and intestine and in guinea pig intestine, *Regul. Peptides,* 20, 281, 1988.

177. **Conlon, J. M., Domin, J., Thim, L., DiMarzo, V., Morris, H. R., and Bloom, S. R.,** Primary structure of neuromedin U from the rat, *J. Neurochem.,* 51, 988, 1988.

178. **Furness, J. B., Pompolo, S., Murphy, R., and Giraud, A.,** Projections of neurons with neuromedin U-like immunoreactivity in the small intestine of the guinea pig, *Cell Tissue Res.,* 257, 415, 1989.

179. **Brown, D. R. and Quito, F. L.,** Neuromedin U octapeptide alters ion transport in porcine jejunum, *Eur. J. Pharmacol.,* 155, 159, 1988.

180. **Tatemoto, K., Carlqvist, M., and Mutt, V.,** Neuropeptide Y — a novel brain peptide with structural similarities to peptide YY and pancreatic polypeptide, *Nature (London),* 296, 659, 1982.

181. **Minth, C. D., Bloom, S. R., Polak, J. M., and Dixon, J. E.,** Cloning, characterization and DNA sequence of a human cDNA encoding neuropeptide tyrosine, *Proc. Natl. Acad. Sci. U.S.A.,* 81, 4577, 1984.

182. **Aggestrup, S., Emson, P., Uddman, R., Sundler, F., Landkaer Jensen, S., and Rahbek Sörensen, H.,** Distribution and content of neuropeptide Y in the human lower esophageal sphincter, *Digestion,* 36, 68, 1987.

183. **Furness, J. B., Costa, M., Emson, P. C., Håkanson, R., Moghimzadeh, E., Sundler, F., Taylor, I. L., and Chance, R. E.,** Distribution, pathways and reactions to drug treatment of nerves with neuropeptide Y- and pancreatic polypeptide-like immunoreactivity in the guinea-pig digestive tract, *Cell Tissue Res.,* 234, 71, 1983.

184. **Keast, J. R., Furness, J. B., and Costa, M.,** Distribution of certain peptide-containing nerve fibres and endocrine cells in the gastrointestinal mucosa in five mammalian species, *J. Comp. Neurol.,* 236, 403, 1985.

185. **Lee, Y., Shiosaka, S., Emson, P. C., Powell, K. F., Smith, A. D., and Tohyama, M.,** Neuropeptide Y-like immunoreactive structures in the rat stomach with special reference to the noradrenaline neuron system, *Gastroenterology,* 89, 110, 1985.

186. **Wattchow, D. A., Furness, J. B., and Costa, M.,** Distribution and coexistence of peptides in nerve fibers of the external muscle of the human gastrointestinal tract, *Gastroenterology,* 95, 32, 1988.

187. **Sundler, F., Moghimzadeh, E., Håkanson, R., Ekelund, M., and Emson, P. C.,** Nerve fibers in the gut and pancreas of the rat displaying neuropeptide Y immunoreactivity. Intrinsic and extrinsic origin, *Cell Tissue Res.,* 230, 487, 1983.

188. **Ekblad, E., Wahlestedt, C., Ekelund, M., Håkanson, R., and Sundler, F.,** Neuropeptide Y in the gut and pancreas: distribution and possible vasomotor function, *Front. Horm. Res.,* 12, 85, 1984.

189. **Ferri, G. L., Ali-Rachedi, A., Tatemoto, K., Bloom, S. R., and Polak, J. M.,** Immunocytochemical localization of neuropeptide Y-like immunoreactivity in extrinsic noradrenergic and intrinsic gut neurons, *Front. Horm. Res.,* 12, 81, 1984.

190. **Lundberg, J. M., Terenius, L., Hökfelt, T., and Tatemoto, K.,** Comparative immunohistochemical and biochemical analysis of pancreatic polypeptide-like peptides with special reference to the presence of neuropeptide Y in central and peripheral neurons, *J. Neurosci.,* 4, 2376, 1984.

191. **Ekblad, E., Håkanson, R., and Sundler, F.,** VIP and PHI coexist with an NPY-like peptide in intramural neurones of the small intestine, *Regul. Peptides,* 10, 47, 1984.

191a. **Ekblad, E.,** unpublished data, 1990.

192. **Lundberg, J. M. and Tatemoto, K.,** Pancreatic polypeptide family (APP, BPP, NPY and PYY) in relation to sympathetic vasoconstriction resistant to α-adrenoreceptor blockade, *Acta Physiol. Scand.,* 116, 393, 1982.

193. **Dahlöf, C., Dahlöf, P., and Lundberg, J. M.,** Neuropeptide Y (NPY): enhancement of blood pressure increase upon alpha-adrenoceptor activation and direct pressor effects in pithed rats, *Eur. J. Pharmacol.,* 109, 289, 1985.

194. **Aubert, J. F., Waeber, B., Rossier, B., Geering, K., Nussberger, J., and Brunner, H. R.,** Effects of neuropeptide Y on the blood pressure response to various vasoconstrictor agents, *J. Pharmacol Exp. Ther.,* 246, 1088, 1988.

195. **Ekblad, E., Edvinsson, L., Wahlestedt, C., Uddman, R., Håkanson, R., and Sundler, F.,** Neuropeptide Y co-exists and co-operates with noradrenaline in perivascular nerve fibers, *Regul. Peptides,* 8, 225, 1984.

196. **Hellström, P. M., Olerup, O., and Tatemoto, K.,** Neuropeptide Y may mediate effects of sympathetic nerve stimulations on colonic motility and blood flow in the cat, *Acta Physiol. Scand.,* 124, 613, 1985.

197. **Wahlestedt, C., Edvinsson, L., Ekblad, E., and Håkanson, R.,** Neuropeptide Y potentiates noradrenaline-evoked vasoconstriction: mode of action, *J. Pharmacol. Exp. Ther.,* 234, 735, 1985.

198. **Garzón, J., Höllt, V., and Sánches-Blázquez, P.,** Neuropeptide Y is an inhibitor of neural function in the myenteric plexus of the guinea-pig ileum, *Peptides,* 7, 623, 1096.

199. **Allen, J. M., Hughes, J., and Bloom, S. R.,** Presence, distribution and pharmacological effects of neuropeptide Y in mammalian gastrointestinal tract, *Dig. Dis. Sci.,* 32, 506, 1987.

200. **Cox, H. M., Cuthbert, A. W., Håkanson, R., and Wahlestedt, C.,** The effect of neuropeptide Y and peptide YY on electrogenic ion transport in rat intestinal epithelia, *J. Physiol.,* 398, 65, 1988.

201. **Carraaway, R. and Leeman, S. E.,** The isolation of a new hypotensive peptide, neurotensin, from bovine hypothalami, *J. Biol. Chem.,* 248, 6854, 1973.

202. **Kitabgi, P., Carraway, R., and Leeman, S. E.,** Isolation of a trideca peptide from bovine intestinal tissue and its partial characterization as neurotensin, *J. Biol. Chem.,* 251, 7053, 1976.

203. **Hammer, R. A., Leeman, S. E., Carraway, R., and Williams, R. H.,** Isolation of human intestinal neurotensin, *J. Biol. Chem.,* 255, 2476, 1980.

204. **Carraway, R. and Leeman, S. E.,** The amino acid sequence of a hypothalamic peptide neurotensin, *J. Biol. Chem.,* 250, 1907, 1975.

205. **Minamino, N., Kangawa, K., and Matsuo, H.,** Neuromedin N: a novel neurotensin-like peptide identified in porcine spinal cord, *Biochem. Biophys. Res. Commun.,* 122, 542, 1984.

206. **Dobner, P. R., Barber, D. L., Villa-Komaroff, L., and McKiernan, C.,** Cloning and sequence analysis of cDNA for the canine neurotensin/neuromedin N precursor, *Proc. Natl. Acad. Sci. U.S.A.,* 84, 3516, 1987.

207. **Sundler, F., Håkanson, R., Leander, S., and Uddman, R.,** Light and electron microscopic localization of neurotensin in the gastrointestinal tract, *Ann. N.Y. Acad. Sci.,* 400, 94, 1982.

208. **Buchan, A. M. J. and Barber, D. L.,** Neurotensin-containing neurons in the canine enteric innervation, *Neurosci. Lett.,* 76, 13, 1987.

209. **Reinecke, M., Forssmann, W. G., Thickotter, G., and Triepel, J.,** Localization of neurotensin-immunoreactivity in the spinal cord and peripheral nervous system of the guinea pig, *Neurosci. Lett.,* 37, 37, 1983.

210. **Lundberg, J. M., Hökfelt, T., Änggård, A., Uvnäs-Wallensten, K., Brimijoin, S., Brodin, E., and Fahrenkrug, J.,** Peripheral peptide neurones: distribution, axonal transport, and some aspects on possible functions, in *Neural Peptides and Neural Communications,* Costa, E. and Trabucchi, M., Eds., Raven Press, New York, 1980, 25.

211. **Sundler, F., Håkanson, R., Hammer, R. A., Alumets, J., Carraway, R., Leeman, S. E., and Zimmerman, E. A.,** Immunohistochemical localization of neurotensin in endocrine cells of the gut, *Cell Tissue Res.,* 178, 313, 1977.

212. **Langer, M., Van Noorden, S., Polak, J. M., and Pearse, A. G. E.,** Peptide hormone-like immunoreactivity in the gastrointestinal tract and endocrine pancreas of eleven teleost species, *Cell Tissue Res.,* 199, 493, 1979.

213. **Saffrey, M. J., Polak, J. M., and Burnstock, G.,** Distribution of vasoactive intestinal polypeptide-, substance P-, enkephalin-, and neurotensin-like immunoreactive nerves in the chicken gut during development, *Neuroscience,* 7, 279, 1982.

214. **Monier, S. and Kitabgi, P.**, Substance P-induced autodesensitization inhibits atropine-resistant neurotensin-stimulated contractions in guinea pig ileum, *Eur. J. Pharmacol.*, 65, 461, 1980.

215. **Goedert, M., Hunter, J. C., and Ninkovic, M.**, Evidence for neurotensin as a non-adrenergic, non-cholinergic neurotransmitter in guinea pig ileum, *Nature (Lond)*, 311, 59, 1984.

216. **Karaus, M., Prasad, K. R., Sarna, S. K., and Lang, I. M.**, Differential effects of [Gln^4] neurotensin on circular and longitudinal muscle of dog ileum *in vitro*, *Am. J. Physiol.*, 16 (4), G566, 1987.

217. **Fox, J. E. T., Ahmed, S., Kostka, P., Checler, F., Allescher, H. D., Vincent, J. P., Kostolanska, F., Kwan, C. Y., and Daniel, E. E.**, Neurotensin and the canine small intestine, motility *in vivo* and *in vitro*, muscle and nerve receptors and sites of degradation, in *Nerves and the Gastrointestinal Tract*, Singer, M. and Goebell, H., Eds., Falk Symp. No. 50, Kluwer Academic, Dordrecht, The Netherlands, 1989, 680.

218. **Sakai, Y., Daniel, E. E., Jury, J., and Fox, J. E. T.**, Neurotensin inhibition of canine intestinal motility *in vivo* via α-adrenoceptors, *Can. J. Physiol. Pharmacol.*, 62, 403, 1984.

219. **Kachur, J. F., Miller, R., Field, M., and Rivier, J.**, Neurohumoral control of ileal electrolyte transport. II. Neurotensin and substance P, *J. Pharmacol. Exp. Ther.*, 220, 456, 1982.

220. **Tapper, E. J.**, Local modulation of intestinal ion transport by enteric neurons, *Am. J. Physiol.*, 244, 456, 1983.

221. **Hughes, J., Smith, T. W., Kosterlitz, H. M., Fothergill, L. A., Morgan, B. A., and Morris, R.**, Identification of two related pentapeptides from the brain with potent opiate agonist activity, *Nature (London)*, 258, 577, 1975.

222. **Numa, S.**, Precursors of opioid peptides and corticotropin-releasing factor and their genes, in *Biogenetics of Neurohormonal Peptides*, Håkanson, R. and Thorell, J., Eds., Academic Press, London, 1985, 29.

223. **Giraud, A., Jönsson, A.-C., and Dockray, G. J.**, Pro-enkephalin gene derived peptides in the porcine stomach: cellular distribution and molecular forms, *Peptides*, 5, 757, 1984.

224. **Kobayashi, S., Suzuki, M., Uchida, T., and Yanaihara, N.**, Enkephalin neurons in the guinea pig duodenum: a light and electron microscopic immunocytochemical study using an antiserum to methionine-enkephalin-arg^6-gly^7-leu^8, *Biomed. Res.*, 5, 489, 1984.

225. **Vincent, S. R., Dalsgaard, C.-J., Schultzberg, M., Hökfelt, T., Christensson, I., and Terenius, L.**, Dynorphin-immunoreactive neurons in the autonomic nervous system, *Neuroscience*, 11, 973, 1984.

226. **Costa, M., Furness, J. B., and Cuello, A. C.**, Separate populations of opioid containing neurons in the guinea-pig intestine, *Neuropeptides*, 5, 445, 1985.

227. **Wang, Y.-N. and Lindberg, I.**, Distribution and characterization of the opioid octapeptide met^5-enkephalin-arg^6-gly^7-leu^8 in the gastrointestinal tract of the rat, *Cell Tissue Res.*, 244, 77, 1986.

228. **Sundler, F., Bjartell, A., Böttcher, G., Ekblad, E., and Håkanson, R.**, Localization of enkephalins and other endogenous opioids in the digestive tract, *Gastroenterol. Clin. Biol.*, 11, 14B, 1987.

229. **Gubler, U., Seeburg, P., Hoffman, B. J., Gage, L. P., and Udenfriend, S.**, Molecular cloning establishes proenkephalin as precursor of enkephalin-containing peptides, *Nature (London)*, 295, 206, 1982.

230. **Horikawa, S., Takai, T., Totosato, M., Takahashi, H., Noda, M., Kakidani, H., Kubo, T., Hirose, T., Inayama, S., Hayashida, H., Miyata, T., and Numa, S.**, Isolation and structural organization of the human preproenkephalin B gene, *Nature (london)*, 306, 611, 1983.

231. **Wolter, H. J.**, α-Melanotropin, β-endorphin and adrenocorticotropin-like immunoreactivities are colocalized within duodenal myenteric plexus perikarya, *Brain Res.*, 325, 290, 1985.

232. **Linnoila, R. I., DiAugustine, R. P., Miller, R. J., Chang, K. J., and Cuatrecasas, P.**, An immuno-histochemical and radioimmunological study of the distribution of [Met^5]- and [Leu^5]-enkephalin in the gastrointestinal tract, *Neuroscience*, 3, 1187, 1979.

233. **Larsson, L.-I. and Stengaard-Pederson, K.**, Immunocytochemical and ultrastructural differentiation between met-enkephalin-, leu-enkephalin-, and met/leu-enkephalin-immunoreactive neurons of feline gut, *J. Neurosci.*, 2, 861, 1982.

234. **Uddman, R., Alumets, J., Håkanson, R., Sundler, F., and Walles, B.**, Peptidergic (enkephalin) innervation of the mammalian esophagus, *Gastroenterology*, 78, 732, 1980.

235. **Aggestrup, S., Uddman, R., Lindkaer-Jensen, S., Sundler, F., Schaffalitzky de Muckadell, O., Holst, J. J., Håkanson, R., Ekman, R., and Rahbek-Sörensen, H.**, Regulatory peptides in the lower esophageal sphincter of man, *Regul. Peptides*, 10, 167, 1985.

236. **Daniel, E. E., Fox, J. E. T., Allescher, H.-D., Ahmad, S., and Kostolanska, F.**, Peripheral actions of opiates in canine gastrointestinal tract: actions on nerves and muscles, *Gastroenterol. Clin. Biol.*, 11, 35B, 1987.

237. **Tange, A.**, Distribution of peptide-containing endocrine cells and neurons in the gastrointestinal tract of the dog: immunohistochemical studies using antisera to somatostatin, substance P, vasoactive intestinal polypeptide, met-enkephalin and neurotensin, *Biomed. Res.*, 4, 9, 1983.

238. **Alumets, J., Håkanson, R., Sundler, F., and Chang, K. J.**, Leu-enkephalin-like material in nerves and enterochromaffin cells in the gut, *Histochemistry*, 56, 187, 1978.

239. **Bu'Lock, A. J., Vaillant, C., and Dockray, G. J.,** Immunohistochemical studies on the gastrointestinal tract using antisera to met-enkephalin and met-enkephalin Arg[6] Phe[7], *J. Histochem. Cytochem.,* 31, 1356, 1983.

240. **Furness, J. B., Costa, M., and Miller, R. J.,** Distribution and projections of nerves with enkephalin-like immunoreactivity in the guinea-pig small intestine, *Neuroscience,* 8, 653, 1983.

241. **Ferri, G.-L., Botti, P., Biliotti, G., Rebecchi, L., Bloom, S. R., Tonelli, L., Labo, G., and Polak, J. M.,** VIP-, substance P-, and met-enkephalin-immunoreactive innervation of the human gastroduodenal mucosa and Brunner's glands, *Gut,* 25, 948, 1984.

242. **Jessen, K. R., Saffrey, M. J., Van Noorden, S., Bloom, S. R., Polak, J. M., and Burnstock, G.,** Immunohistochemical studies on the enteric nervous system in tissue culture and in situ: localization of vasoactive intestinal polypeptide (VIP), substance P and enkephalin immunoreactive nerves in the guinea pig, *Neuroscience,* 5, 1717, 1980.

243. **Leander, S., Håkanson, R., and Sundler, F.,** Nerves containing substance P, vasoactive intestinal polypeptide, enkephalin or somatostatin in the guinea pig taenia coli, *Cell Tissue Res.,* 215, 21, 1981.

244. **Kobayashi, S., Suzuki, M., and Yanaihara, N.,** Enkephalin neurons in the guinea pig proximal colon: immunocytochemical study using an antiserum to methionine-enkephalin-Arg[6]-Gly[7]-Leu[8], *Arch. Histol. Jpn.,* 48, 27, 1985.

245. **Malmfors, G., Leander, S., Brodin, E., Håkanson, R., Holmin, T., and Sundler, F.,** Peptide containing neurones intrinsic to the gut wall. An experimental study in the pig, *Cell Tissue Res.,* 214, 225, 1981.

246. **Larsson, L.-I. and Stengaard-Pedersen, K.,** Enkephalin/endorphin-related peptides in antropyloric gastrin cells, *J. Histochem. Cytochem.,* 29, 1088, 1981.

247. **Nihei, K., Iwanaga, T., Yanaihara, N., Mochizuki, T., and Fujita, T.,** Proenkephalin A occurs in the enterochromaffin (EC) cells of the porcine intestine: an immunocytochemical study using antisera to met-enkephalin-Arg[6]-Gly[7]-Leu[8] and to serotonin, *Biomed. Res.,* 4, 393, 1983.

248. **Lees, G. M., Leishman, D. J., and Pearson, G.T.,** Dynorphin A (1—8)-immunoreactivity in the guinea-pig and rat isolated small intestine, *J. Physiol.,* 403, 65, 1988.

248a. **Ekblad, E.,** unpublished data, 1990.

249. **Uchida, T., Kobayashi, S., and Yanaihara, N.,** Occurrence and projections of three subclasses of met-enkephalin-Arg[6]-Gly[7]-Leu[8] neurons in the guinea pig duodenum: immunoelectron microscopic study on the co-storage of met-enkephalin-Arg[6]-Gly[7]-Leu[8] with substance P or PHI (1—15), *Biomed. Res.,* 6, 415, 1985.

250. **Domoto, A., Gonda, T., Oki, M., and Yanaihara, N.,** Coexistence of substance P- and methionine[5]-enkephalin-like immunoreactivity in nerve cells of the myenteric ganglia in the cat ileum, *Neurosci. Lett.,* 47, 9, 1984.

251. **Llewellyn-Smith, I. J., Furness, J. B., Gibbins, I. L., and Costa, M.,** Quantitative ultrastructural analysis of enkephalin-, substance P-, and VIP-immunoreactive nerve fibers in the circular muscle of the guinea pig small intestine, *J. Comp. Neurol.,* 272, 139, 1988.

252. **Kobayashi, S. and Nishisaka, T.,** Myenteric enkephalin neurons around the laser-photocoagulation necrosis: an immunocytochemical investigation in the guinea pig jejunum and proximal colon, *Arch. Histol. Jpn,* 48, 239, 1985.

253. **Costa, M. and Furness, J. B.,** The origins, pathways and terminations of neurons with VIP-like immunoreactivity in the guinea-pig small intestine, *Neuroscience,* 8, 665, 1983.

254. **Probert, L., De Mey, J., and Polak, J. M,.** Ultrastructural localization of four different neuropeptides within separate populations of p-type nerves in the guinea pig colon, *Gastroenterology,* 85, 1094, 1983.

255. **Van Neuten, J. M., Van Ree, J. M., and Vanhoutte, P. M.,** Inhibition by met-enkephalin of peristaltic activity in the guinea pig ileum, and its reversal by naloxone, *Eur. J. Pharmacol.,* 41, 341, 1977.

256. **Waterfield, A. A., Smokcum, R. W. J., Hughes, J., Kosterlitz, H. W., and Henderson, G.,** *In vitro* pharmacology of the opioid peptides, enkephalins and endorphins, *Eur. J. Pharmacol.,* 43, 107, 1977.

257. **Bitar, K. N. and Makhlouf, G. M.,** Specific opiate receptors on isolated mammalian gastric smooth muscle cells, *Nature (London),* 297, 72, 1982.

258. **Hellström, P. M.,** Pharmacological analysis of the mechanisms of of action for colonic contraction induced by neurotensin, substance P and methionine-enkephalin, *Acta Physiol. Scand.,* 125, 13, 1985.

259. **Kachur, J. F., Miller, R. J., and Field, M.,** Control of guinea pig intestinal electrolyte secretion by a δ-opiate receptor, *Proc. Natl. Acad. Sci. U.S.A.,* 77, 2753, 1980.

260. **Dobbins, J. W., Dharmsathaphorn, K., Racuscen, L., and Binder, H.J.,** The effect of somatostatin and enkephalin on ion transport in the intestine, *Ann. N.Y. Acad. Sci.,* 372, 594, 1981.

261. **Beubler, E. and Lembeck, F.,** Inhibition of stimulated fluid secretion in the rat small and large intestine by opiate antagonists, *Naunyn-Schmiedeberg's Arch. Pharmacol.,* 306, 113, 1979.

262. **McKay, J. S., Linaker, B. D., and Turnberg, L. A.,** Influence of opiates on ion transport across rabbit ileal mucosa, *Gastroenterology,* 80, 279, 1981.

263. **Gaginella, T. S., Rimele, T. J., and Wietecha, M.,** Studies on rat intestinal epithelial cell receptors for serotonin and opiates, *J. Physiol. (London),* 335, 101, 1983.

264. **Brazeau, P., Vale, W., Burgus, R., Ling, N., Butcher, M., Rivier, J., and Guillemin, R.,** Hypothalamic polypeptide that inhibit the secretion of immunoreactive pituitary growth hormone, *Science,* 179, 77, 1973.

265. **Burgus, R., Ling, N., Butcher, M., and Guillemin, R.,** Primary structure of somatostatin, a hypothalamic peptide that inhibits the secretion of pituitary growth hormone, *Proc. Natl. Acad. Sci. U.S.A.,* 70, 684, 1973.

266. **Raptis, S., Schlege, W., and Pfeiffer, E. F.,** Effects of somatostatin on gut and pancreas, in *Gut Hormones,* Bloom, S. R., Ed., Churchill Livingstone, London, 1978, 446.

267. **Walsh, J. H.,** Gastrointestinal hormones, in *Physiology of the Gastrointestinal Tract,* Johnson, L. R., Ed., Raven Press, New York, 1987, 181.

268. **Pradayrol, L., Jörnvall, H., Mutt, V., and Ribet, A.,** N-Terminally extended somatostatin: the primary structure of somatostatin-28, *FEBS Lett.,* 109, 55, 1980.

269. **Goodman, R. H., Aron, D. C., and Roos, B. A.,** Rat pre-prosomatostatin: structure and processing by microsomal membranes, *J. Biol. Chem.,* 258, 5570, 1983.

270. **Shen, L.-P., Pictet, R., and Rutter, W.,** Human somatostatin I: sequence of the cDNA, *Proc. Natl. Acad. Sci. U.S.A.,* 79, 4575, 1982.

271. **Penman, E., Wass, J., Butler, M., Penny, E., Price, J., Wu, P., and Rees, L.,** Distribution and characterization of immunoreactive somatostatin in human gastrointestinal tract, *Regul. Peptides,* 7, 53, 1983.

272. **Hökfelt, T., Johansson, O., Efendic, S., Luft, R., and Arimura, A.,** Are there somatostatin-containing nerves in the rat gut? Immunohistochemical evidence for a new type of peripheral nerves, *Experientia,* 31, 852, 1975.

273. **Alumets, J., Sundler, F., and Håkanson, R.,** Distribution, ontogeny and ultrastructure of somatostatin immunoreactive cells in the pancreas and gut, *Cell Tissue Res.,* 185, 465, 1977.

274. **Baskin, D. G. and Ensinck, J. W.,** Somatostatin in epithelial cells of intestinal mucosa is present primarily as somatostatin 28, *Peptides,* 5, 615, 1984.

275. **Baldissera, F. G. A., Holst, J. J., Jensen, S. L., and Krarup, T.,** Distribution and molecular forms of peptides containing somatostatin immunodeterminants in extracts from the entire gastrointestinal tract of man and pig, *Biomed. Biophys. Acta,* 838, 132, 1985.

276. **Keast, J. R., Furness, J. B., and Costa, M.,** Somatostatin in human enteric nerves. Distribution and characterization. *Cell Tissue Res.,* 237, 199, 1984.

277. **Skak-Nielsen, T., Holst, J. J., Baldissera, F. G. A., and Seier Poulsen, S.,** Localization in the gastrointestinal tract of immunoreactive prosomatostatin, *Regul. Peptides,* 19, 183, 1987.

278. **Costa, M., Patel, Y., Furness, J. B., and Arimura, A.,** Evidence that some intrinsic neurons of the intestine contain somatostatin, *Neurosci. Lett.,* 6, 215, 1977.

279. **Costa, M., Furness, J. B., Llewellyn-Smith, I. J., Davies, B., and Oliver, J.,** An immunohistochemical study of the projections of somatostatin-containing neurons in the guinea-pig intestine, *Neuroscience,* 5, 841, 1980.

280. **Hökfelt, T., Elfvin, L. G., Elde, R., Schultzberg, M., Goldstein, M., and Luft, R.,** Occurrence of somatostatin-like immunoreactivity in some peripheral sympathetic noradrenergic neurons, *Proc. Natl. Acad. Sci. U.S.A.,* 74, 3587, 1977.

281. **Costa, M. and Furness, J. B.,** Somatostatin is present in a subpopulation of noradrenergic nerve fibres supplying the intestine, *Neuroscience,* 13, 911, 1984.

282. **Endo, Y., Uchida, T., and Kobayashi, S.,** Somatostatin neurons in the small intestine of the guinea pig: a light and electron microscopic immunocytochemical study combined with nerve lesion experiments by laser irradiation, *J. Neurocytol.,* 15, 725, 1986.

283. **Guillemin, R.,** Somatostatin inhibits the release of acetylcholine induced electrically in the myenteric plexus, *Endocrinology,* 99, 1653, 1976.

284. **Furness, J. B. and Costa, M.,** Actions of somatostatin on excitatory and inhibitory nerves in the intestine, *Eur. J. Pharmacol.,* 55, 69, 1979.

285. **Teitelbaum, D. H., O'Dorisio, T. M., Perkins, W. E., and Gaginella, T. S.,** Somatostatin modulation of peptide-induced acetylcholine release in guinea pig ileum, *Am. J. Physiol.,* 246, 6509, 1984.

286. **Grider, J. R., Arimura, A., and Makhlouf, G. M.,** Role of somatostatin neurons in intestinal peristalsis: facilitatory interneurons in descending pathways, *Am. J. Physiol.,* 16, G434, 1987.

287. **Dharmsathaphorn, K., Binder, H. J., and Dobbins, J. W.,** Somatostatin stimulates sodium and chloride absorption on the rabbit ileum, *Gastroenterology,* 78, 1559, 1980.

288. **Guadalini, S., Kachur, J. F., Smith, P. L., Miller, R. J., and Field, M.,** In vitro effects of somatostatin on ion transport in rabbit intestine, *Am. J. Physiol.,* 238, G67, 1980.

289. **Keast, J. R., Furness, J. B., and Costa, M.,** Effects of noradrenaline and somatostatin on basal and stimulated mucosal ion transport in the guinea-pig small intestine, *Naunyn-Schmiedeberg's Arch. Pharmacol.,* 337, 393, 1986.

290. **Pernow, B.,** Substance P, *Pharmacol. Rev.,* 35, 85, 1983.

291. **Von Euler, U. S. and Gaddum, J. H.,** An unidentified depressor substance in certain tissue extracts, *J. Physiol. (London),* 72, 74, 1931.

292. **Chang, M. M., Leeman, S. E., and Niall, H. D.,** Amino-acid sequence of substance P, *Nature New Biol.,* 232, 86, 1971.

293. **Nawa, H., Hirose, T., Takashima, H., Inayama, S., and Nakanishi, S.,** Nucleotide sequences of cloned cDNAs for two types of bovine brain substance P precursor, *Nature (London),* 306, 31, 1983.

294. **Krause, J. E., Chirgwin, J. M., Carter, M. S., Xu, Z. S., and Hershey, A. D.,** Three rat preprotachykinin mRNAs encode the neuropeptides substance P and neurokinin A, *Proc. Natl. Acad. Sci. U.S.A.,* 84, 881, 1987.

295. **Sternini, C., Anderson, K., Frantz, G., Krause, J. E., and Brecha, N.,** Expression of substance P/neurokinin A-encoding preprotachykinin messenger ribonucleic acids in the rat enteric nervous system, *Gastroenterology,* 97, 348, 1989.

296. **Nilsson, G., Larsson, L.-I., Håkanson, R., Brodin, E., Pernow, B., and Sundler, F.,** Localization of substance P-like immunoreactivity in mouse gut, *Histochemistry,* 43, 97, 1975.

297. **Pearse, A. G. E. and Polak, J. M.,** Immunocytochemical localization of substance P in mammalian intestine, *Histochemistry,* 41, 373, 1975.

298. **Sundler, F., Håkanson, R., Larsson, L.-I., Brodin, E., and Nilsson, G.,** Substance P in the gut: an immunochemical and immunohistochemical study of its distribution and development, in *Substance P,* von Euler, U. S. and Pernow, B., Eds., Raven Press, New York, 1977, 59.

299. **Hayashi, H., Ohsumi, K., Fujiwara, M., Mizuno, N., Kanazawa, I., and Yajima, H.,** Immunohistochemical studies on enteric substance P of extrinsic origin in the cat, *J. Autonom. Nerv. Syst.,* 5, 207, 1982.

300. **Leander, S., Brodin, E., Håkanson, R., Sundler, F., and Uddman, R.,** Neuronal substance P in the esophagus. Distribution and effects on motor activity, *Acta Physiol. Scand.,* 115, 427, 1982.

301. **Minagawa, H., Shiosaka, S., Inoue, H., Harashi, N., Kasahara, A., Kamata, T., Tohyama, M., and Shiotani, Y.,** Origins and three-dimensional distribution of substance P-containing structures in the rat stomach using whole-mount tissue, *Gastroenterology,* 86, 51, 1984.

302. **Sharkey, K. A., Williams, R. G., and Dockray, G. J.,** Sensory substance P innervation of the stomach and pancreas, *Gastroenterology,* 87, 914, 1984.

303. **Costa, M., Cuello, A. C., Furness, J. B., and Franco, R.,** Distribution of enteric neurons showing immunoreactivity for substance P in the guinea-pig ileum, *Neuroscience,* 5, 323, 1980.

304. **Ferri, G. L., Botti, P. L., Vezzadini, P., Biliotti, G., Bloom, S. R., and Polak, J. M.,** Peptide-containing innervation of the human intestinal mucosa, *Histochemistry,* 76, 413, 1982.

305. **Brodin, E., Sjölund, K., Håkanson, R., and Sundler, F.,** Substance P-containing nerve fibers are numerous in human intestinal mucosa, *Gastroenterology,* 85, 557, 1983.

306. **Domoto, T., Daniel, E. E., Oki, M., and Yanaihara, N.,** Substance P-like immunoreactivity in nerves of dog ileum, with special reference to its depletion by vagotomy, intestinal distension and scorpion venom, *Biomed. Res.,* 4, 479, 1983.

307. **Sjölund, K., Schaffalitzky de Muckadell, O., Fahrenkrug, J., Håkanson, R., Peterson, B. G., and Sundler, F.,** Peptide-containing nerve fibers in the gut wall in Crohn's disease, *Gut,* 24, 724, 1983.

308. **Llewellyn-Smith, I. J., Furness, J. B., Murphy, R., O'Brien, P. E., and Costa, M.,** Substance P-containing nerves in the human small intestine, *Gastroenterology,* 86, 421, 1984.

309. **Lundberg, J. M., Hökfelt, T., Nilsson, G., Terenius, L., Rehfeld, J. F., Elde, R., and Said, S.,** Peptide neurons in the vagus, splanchnic and sciatic nerves, *Acta Physiol. Scand.,* 104, 499, 1978.

310. **Gamse, R., Lembeck, F., and Cuello, A. C.,** Substance P in the vagus nerve. Immunochemical and immunohistochemical evidence for axoplasmic transport, *Naunyn-Schmiedeberg's Arch. Pharmacol.,* 306, 37, 1979.

311. **Brimijoin, S., Lundberg, J. M., Brodin, E., Hökfelt, T., and Nilsson, G.,** Axonal transport of substance P in the vagus and sciatic nerves of the guinea pig, *Brain Res.,* 191, 443, 1980.

312. **Matthews, M. R. and Cuello, A. C.,** Substance P immunoreactive peripheral branches of sensory neurons innervate guinea-pig sympathetic neurons, *Proc. Natl. Acad. Sci. U.S.A.,* 79, 1668, 1982.

313. **De Groat, W. C., Kawatani, M., Hisamitsu, T., Lowe, I., Morgan, C., Roppolo, J., Booth, A. M., Nedelhaft, U., Kuo, D., and Thor, K.,** The role of neuropeptides in the sacral autonomic reflex pathways of the cat, *J. Autonom. Nerv. Syst.,* 7, 339, 1983.

314. **Lindh, B., Dalsgaard, C. J., Elfvin, L.-G., Hökfelt, T., and Cuello, A. C.,** Evidence of substance P immunoreactive neurons in dorsal root ganglia and vagal ganglia projecting to the guinea-pig pylorus, *Brain Res.,* 269, 365, 1983.

315. **Furness, J. B., Papka, R. E., Della, N. G., Costa, M., and Eskay, R. L.,** Substance P-like immunoreactivity in nerves associated with the vascular system in guinea-pigs, *Neuroscience,* 7, 447, 1982.

316. **Legay, C., Saffrey, M. J., and Burnstock, G.,** Coexistence of immunoreactive substance P and serotonin in neurones of the gut, *Brain Res.,* 302, 379, 1984.

317. **Furness, J. B., Costa, M., and Keast, J. R.,** Choline acetyltransferase and peptide immunoreactivity of submucous neurons in the small intestine of guinea-pig, *Cell Tissue Res.,* 237, 328, 1984.

318. **Costa, M., Furness, J. B., Llewellyn-Smith, I. J., and Cuello, A. C.,** Projections of substance P-containing neurons within the guinea-pig small intestine, *Neuroscience,* 6, 411, 1981.

319. **Jessen, K. R., Polak, J. M., Van Noorden, S., Bloom, S. R., and Burnstock, G.,** Peptide-containing neurons connect the two ganglionated plexuses of the enteric nervous system, *Nature (London),* 283, 391, 1980.

320. **Kapadia, S. E. and Kapadia, C. R.,** Ultrastructure and localization of substance P and met-enkephalin immunoreactivity in the human fetal gastric antrum, *Cell Tissue Res.,* 243, 284, 1986.

321. **Hoyes, A. D., Sikri, K. L., and Barber, P.,** Localization of substance P-like immunoreactivity in the intramural nerve plexuses of the guinea pig stomach using immunofluorescence and immunoperoxidase techniques, *J. Anat.,* 135, 319, 1982.

322. **Llewellyn-Smith, I. J., Furness, J. B., Murphy, R., O'Brien, P. E., and Costa, M.,** Substance P-containing nerves in the human small intestine. Distribution, ultrastructure and characterization of the immunorective peptide, *Gastroenterology,* 86, 421, 1984.

323. **Barthó, L. and Holzer, G.,** Search for a physiological role of substance P in gastrointestinal motility, *Neuroscience,* 16, 1, 1985.

323a. **Ekblad, E., Arnbjörnsson, E., Sundler, F., and Håkanson, R.,** Does the human appendix contain a unique tachykinin receptor? Manuscript in preparation, 1990.

324. **McFadden, D., Zinner, M. J., and Jaffe, B. M.,** Substance P-induced intestinal secretion of water and electrolytes, *Gut,* 27, 267, 1986.

325. **Mitchenere, P., Adrian, T. E., Kirk, R. M., and Bloom, S. R.,** Effect of gut regulatory peptides on intestinal luminal fluid in the rat, *Life Sci.,* 29, 1563, 1981.

326. **Hubel, K. A., Renquist, K. S., and Shirazi, S.,** Neural control of ileal ion transport: role of substance P (SP) in rabbit and man, *Gastroenterology,* 86, 1118, 1984.

327. **Keast, J. R., Furness, J. B., and Costa, M.,** Different substance P receptors are found on mucosal epithelial cells and submucous neurons of the guinea pig small intestine, *Naunyn-Schmiedeberg's Arch. Pharmacol.,* 329, 382, 1985.

328. **Said, S. I. and Mutt, V.,** Polypeptide with broad biological activity: isolation from small intestine, *Science,* 169, 1217, 1970.

329. **Mutt, V. and Said, S. I.,** Structure of the porcine vasoactive intestinal octacosa peptide, *Eur. J. Biochem.,* 42, 581, 1974.

330. **Itoh, N., Obata, K., Yanaihara, N., and Okamoto, H.,** Human preprovasoactive intestinal polypeptide contains a novel PHI-27-like peptide, PHM-27, *Nature (London),* 304, 547, 1983.

331. **Tatemoto, K. and Mutt, V.,** Isolation and characterization of the intestinal peptide porcine PHI (PHI-27), a new member of the glucagon-secretin family, *Proc. Natl. Acad. Sci. U.S.A.,* 78, 6603, 1981.

332. **Sundler, F., Ekblad, E., Grunditz, T., Håkanson, R., and Uddman, R.,** Vasoactive intestinal peptide in the peripheral nervous system, *Ann. N.Y. Acad. Sci.,* 527, 143, 1988.

333. **Yanaihara, N., Nokihara, K., Yanaihara, C., Iwanaga, T., and Fujita, T.,** Immunocytochemical demonstration of PHI and its co-existence with VIP in intestinal nerves of the rat and pig, *Arch. Histol. Jpn.,* 46, 575, 1983.

334. **Fahrenkrug, J., Bek, T., Lundberg, J. M., and Hökfelt, T.,** VIP and PHI in cat neurons: co-localization but variable tissue content possibly due to differential processing, *Regul. Peptides,* 12, 21, 1985.

335. **Larsson, L.-I., Fahrenkrug, J., Schaffalitzky de Muckadell, O., Sundler, F., Håkanson, R., and Rehfeld, J. F.,** Localization of vasoactive intestinal polypeptide (VIP) to central and peripheral neurons, *Proc. Natl. Acad. Sci. U.S.A.,* 73, 3097, 1976.

336. **Yamada, J., Kitamura, N., Yamashita, T., Misu, M., and Yanaihara, N.,** Vasoactive intestinal poly-peptide (VIP) immunoreactivity of endocrine-like cells in the feline pyloric mucosa, *Cell Tissue Res.,* 226, 113, 1982.

337. **Sundler, F., Alumets, J., Fahrenkrug, J., Håkanson, R., and Schaffalitzky de Muckadell, O. B.,** Cellular localization and ontogeny of immunoreactive vasoactive intestinal polypeptide (VIP) in the chicken gut, *Cell Tissue Res.,* 196, 193, 1979.

338. **Falkmer, S., Fahrenkrug, J., Alumets, J., Håkanson, R., and Sundler, F.,** Vasoactive intestinal polypeptide (VIP) in epithelial cells of the gut mucosa of an elasmobranchian cartilaginous fish, the ray, *Endocrinol. Jpn.,* 1, 31, 1980.

339. **Reinecke, M., Schlüter, P., Yanaihara, N., and Forssmann, W. G.,** VIP immunoreactivity in enteric nerves and endocrine cells of the vertebrate gut, *Peptides,* 12, 149, 1981.

340. **Ekblad, E., Alumets, J., Ekman, R., Falkmer, S., Håkanson, R., and Sundler, F.,** Two distinct VIP precursors in cartilaginous fish?, *Peptides,* 6, 383, 1985.

341. **Uddman, R., Alumets, J., Edvinsson, L., Håkanson, R., and Sundler, F.,** Peptidergic (VIP) innervation of the esophagus, *Gastroenterology,* 75, 5, 1978.

342. **Edin, R., Lundberg, J. M., Ahlman, H., Dahlström, A., Fahrenkrug, J., Hökfelt, T., and Kewenter, J.,** On the VIP-ergic innervation of the feline pylorus, *Acta Physiol. Scand.,* 107, 185, 1979.

343. **Alumets, J., Fahrenkrug, J., Håkanson, R., Schaffalitzky de Muckadell, O., Sundler, F., and Uddman, R.,** A rich VIP nerve supply is characteristic of sphincters, *Nature (London),* 280, 155, 1979.

344. **Larsson, L.-I., Polak, J. M., Buffa, R., Sundler, F., and Solcia, E.,** On the immunocytochemical localization of the vasoactive intestinal polypeptide, *J. Histochem. Cytochem.,* 27, 936, 1979.

345. **Costa, M., Furness, J. B., Buffa, R., and Said, S.,** Distribution of enteric nerve cell bodies and axons showing immunoreactivity for vasoactive intestinal polypeptide (VIP) in the guinea pig intestine, *Neuroscience,* 5, 587, 1980.

346. **Dalsgaard, C.-J., Hökfelt, T., Schultzberg, M., Lundberg, J. M., Terenius, L., Dockray, G. J., and Goldstein, M.,** Origin of peptide-containing fibers in the inferior mesenteric ganglion of the guinea pig: immunohistochemical studies with antisera to substance P, enkephalin, vasoactive intestinal polypeptide, cholecystokinin and bombesin, *Neuroscience,* 9, 191, 1983.

347. **Iwanaga, T., Fujita, T., and Yanaihara, N.,** Occurrence of gastrin-releasing peptide (GRP)-like and vasoactive intestinal polypeptide (VIP)-like immunoreactivities in cholinergic neurons in the digestive tract of the rat, *Biomed. Res.,* 4, 167, 1983.

348. **Willard, A. L. and Nishi, R.,** Neuropeptides mark functionally distinguishable cholinergic enteric neurons, *Brain Res.,* 422, 163, 1987.

349. **Bishop, A. E., Polak, J. M., Yiangou, Y., Christofides, N. D., and Bloom, S. R.,** The distribution of PHI and VIP in porcine gut and their co-localization to a proportion of intrinsic ganglion cells, *Peptides,* 5, 255, 1984.

350. **Yiangou, Y., Requejo, F., Polak, J. M., and Bloom, S. R.,** Characterization of a novel prepro-VIP derived peptide, *Biochem. Biophys. Res. Commun.,* 139, 1142, 1986.

351. **Berezin, I., Allescher, H.-D., and Daniel, E. E.,** Ultrastructural localization of VIP-immunoreactivity in canine distal oesophagus, *J. Neurocytol.,* 16, 749, 1987.

352. **Larsson, L.-I.,** Ultrastructural localization of a new neuronal peptide (VIP), *Histochemistry,* 54, 173, 1977.

353. **Llewellyn-Smith, I. J., Costa, M., and Furness, J. B.,** Light and electron microscopic immunocytochemistry of the same nerves from whole mount preparations, *J. Histochem. Cytochem.,* 33, 857, 1985.

354. **Feher, E. and Léránth, C.,** Light and electron microscopic immunocytochemical localization of vasoactive intestinal polypeptide (VIP)-like activity in the rat small intestine, *Neuroscience,* 10, 97, 1983.

355. **Loesch, A. and Burnstock, G.,** Ultrastructural identification of VIP-containing nerve fibres in the myenteric plexus of the rat ileum, *J. Neurocytol.,* 14, 327, 1985.

356. **Maeda, M., Takagi, H., Kubota, Y., Morishima, Y., Akai, F., Hashimoto, S., and Mori, S.,** The synaptic relationship between vasoactive intestinal polypeptide (VIP)-like immunoreactive neurons and their axon terminals in the rat small intestine: light and electron microscopic study, *Brain Res.,* 329, 356, 1985.

357. **Grider, J. R. and Makhlouf, G. M.,** Vasoactive intestinal peptide: transmitter of inhibitory motor neurons of the gut, *Ann. N.Y. Acad. Sci.,* 527, 368, 1988.

358. **McCabe, R. D. and Dharmsathaphorn, K.,** Mechanism of VIP-stimulated chloride secretion by intestinal epithelial cells, *Ann. N.Y. Acad. Sci.,* 527, 326, 1988.

359. **Furness, J. B. and Costa, M.,** Projections of intestinal neurons showing immunoreactivity for vasoactive intestinal peptide are consistent with these neurons being the enteric inhibitory neurons, *Neurosci. Lett.,* 15, 199, 1979.

360. **Kirchgessner, A. L. and Gershon, M. D.,** Identification of vagal and submucosal inputs to the myenteric plexus by retrograde and anterograde transport, in *Nerves and the Gastrointestinal Tract,* Singer, M. V. and Goebell, H., Eds., Falk Symp. No. 50, Kluwer Academic, Dordrecht, The Netherlands, 1989, 69.

361. **Jin, J.-G., Neya, T., and Nakayama, S.,** Contractions of the guinea-pig ileum evoked by stimulation of the submucous plexus, *Eur. J. Pharmacol.,* 161, 73, 1989.

362. **Sanders, K. M. and Smith, T. K.,** Motorneurones of the submucous plexus regulate electrical activity of the circular muscle of canine proximal colon, *J. Physiol. (London),* 380, 292, 1986.

363. **Wilson, A. J., Llewellyn-Smith, I. J., Furness, J. B., and Costa, M.,** The source of the nerve fibres forming the deep muscular and circular muscle plexuses in the small intestine of the guinea pig, *Cell Tissue Res.,* 247, 497, 1987.

364. **Ekblad, E., Ekman, R., Håkanson, R., and Sundler, F.,** Return of nerve fibers containing gastrin-releasing peptide in rat small intestine after local removal of myenteric ganglia, *Neuroscience,* 24, 309, 1988.

365. **Galligan, J. J., Furness, J. B., and Costa, M.,** Migration of the myoelectric complex after interruption of the myenteric plexus: intestinal transection and regeneration of enteric nerves in the guinea pig, *Gastroenterology,* 97, 1135, 1989.

366. **Furness, J. B., Morris, J. L., Gibbins, I. L., and Costa, M.,** Chemical coding of neurons and plurichemical transmission, *Annu. Rev. Pharmacol. Toxicol.,* 29, 289, 1989.

367. **Kessler, J. A. and Black, I. B.,** Regulation of substance P in adult rat sympathetic ganglia, *Brain Res.,* 234, 182, 1982.

368. **Bohn, M. C., Kessler, J. A., Adler, J. E., Markey, K., Goldstein, M., and Black, I. B.,** Simultaneous expression of the SP-peptidergic and noradrenergic phenotypes in rat sympathetic neurons, *Brain Res.,* 298, 378, 1984.

369. **Terenghi, G., Zhang, S.-O., Unger, W. G., and Polak, J. M.**, Morphological changes of sensory CGRP-immunoreactive and sympathetic nerves in peripheral tissues following chronic denervation, *Histochemistry*, 86, 89, 1986.

370. **La Gamma, E. F., Adler, J. E., and Black, I. B.**, Impulse activity differentially regulates leu-enkephalin and catecholamine characteristics in the adrenal medulla, *Science*, 224, 1102, 1984.

371. **Black, I. B., Adler, J. E., and La Gamma, E. F.**, Impulse activity differentially regulates co-localized transmitters by altering messenger RNA levels, in *Coexistence of Neuronal Messengers: A New Principle in Chemical Transmission*, Hökfelt, T., Fuxe, K., and Pernow, B., Eds., Progress in Brain Research, Vol. 68, Elsevier, Amsterdam, 1986, 121.

372. **Potter, D. D., Matsumoto, S. G., Landis, S. C., Sah, D. W. Y., and Furshpan, E. J.**, Transmitter status in cultured sympathetic principal neurons: plasticity, graded expression and diversity, in *Coexistence of Neuronal Messengers: A New Principle in Chemical Transmission*, Progress in Brain Research, Vol. 68, Hökfelt, T., Fuxe, K., and Pernow, B., Eds., Elsevier, Amsterdam, 1986, 103.

373. **Coulombe, J. N. and Bronner-Fraser**, Cholinergic neurones acquire adrenergic neurotransmitters when transplanted into an embryo, *Nature (London)*, 324, 569, 1986.

374. **Schultzberg, M., Foster, G. A., Gage, F. A., Björklund, A., and Hökfelt, T.**, Coexistence during ontogeny and transplantation, in *Coexistence of Neuronal Messengers: A New Principle in Chemical Transmission*, Progress in Brain Research, Vol. 68, Hökfelt, T., Fuxe, K. and Pernow, B., Eds., Elsevier, Amsterdam, 1986, 129.

375. **Nelson, D. K., Service, J. E., Studelska, D. R., Brimijoin, S., and Go, V. L. W.**, Gastrointestinal neuropeptide concentrations following guanethidine sympathectomy, *J. Autonom. Nerv. Syst.*, 22, 203, 1988.

376. **Ravazzola, M. and Orci, L.**, Glucagon and glicentin immunoreactivity are topologically segregated in the granule of the human pancreatic A cell, *Nature (London)*, 284, 66, 1980.

377. **Merighi, A., Polak, J. M., Gibson, S. J., Gulbenkian, S., Valentino, K. L., and Peirone, S. M.**, Ultrastructural studies on calcitonin gene-related peptide-, tachykinins- and somatostatin-immunoreactive neurones in rat dorsal root ganglia: evidence for the colocalization of different peptides in single secretory granules, *Cell Tissue Res.*, 254, 101, 1988.

378. **Ali-Rachedi, A., Varndell, I. M., Adrian, T. E., Gapp, D. A., Van Noorden, S., Bloom, S. R., and Polak, J. M.**, Peptide YY (PYY) immunoreactivity is co-stored with glucagon-related immunoreactants in endocrine cells of the gut and pancreas, *Histochemistry*, 80, 487, 1984.

379. **Böttcher, G., Alumets, J., Håkanson, R., and Sundler, F.**, Co-existence of glicentin and peptide YY in colorectal L-cells in cat and man. An electron microscopic study, *Regul. Peptides*, 13, 283, 1986.

380. **Lundberg, J. M.**, Evidence for coexistence of vasoactive intestinal polypeptide (VIP) and acetylcholine in neurons of cat exocrine glands. Morphological, biochemical and functional studies, *Acta Physiol. Scand.*, 112, (Suppl. 496), 1, 1981.

381. **Fried, G., Terenius, L., Hökfelt, T., and Goldstein, M.**, Evidence for differential localization of noradrenaline and neuropeptide Y (NPY) in neuronal storage vesicles isolated from rat vas deferens, *J. Neurosci.*, 5, 540, 1985.

382. **Edwards, A. V., Järhult, J., Andersson, P.-O., and Bloom, S. R.**, The importance of the pattern of stimulation in relation to the response of autonomic effectors, in *Systemic Role of Regulatory Peptides*, Bloom, S. R., Polak, J. M., and Lindenlaub, E., Eds., Schattauer, Stuttgart, 1982, 145.

383. **Lundberg, J. M. and Hökfelt, T.**, Multiple co-existence of peptides and classical transmitters in peripheral autonomic and sensory neurons — functional and pharmacological implications, in *Coexistence of Neuronal Messengers: A New Principle in Chemical Transmission*, Progress in Brain Research, Vol. 68, Hökfelt, T., Fuxe, K., and Pernow, B., Eds., Elsevier, Amsterdam, 1986, 241.

384. **Agoston, D. V., Conlon, J. M., and Whittaker, V. P.**, Selective depletion of the acetylcholine and vasoactive intestinal polypeptide of the guinea-pig myenteric plexus by differential mobilization of distinct transmitter pools, *Exp. Brain Res.*, 72, 535, 1988.

385. **Lundberg, J. M., Rudehill, A., Sollevi, A., Fried, G., and Wallin, G.**, Co-release of neuropeptide Y and noradrenaline from pig spleen *in vivo*: importance of subcellular storage, nerve impulse frequency and pattern, feedback regulation and resupply by axonal transport, *Neuroscience*, 28, 475, 1989.

Chapter 5

NEUROPEPTIDE MOTOR ACTIONS VARY BETWEEN *IN VIVO* AND *IN VITRO* EXPERIMENTAL CONDITIONS

Jo-Ann E.T. Fox-Threlkeld

TABLE OF CONTENTS

I.   Introduction 182

II.   Characteristics of the Experimental Preparations 183
        A.   Isolated Cells 183
        B.   *In vitro* Strips 184
        C.   *In situ* 184

III.   Peptide Families, Variable Responses in the Same Tissues 184
        A.   Gastrin/Cholecystokinin (CCK) Family 184
        B.   Gastrin-Releasing peptide (GRP)/Bombesin 185
        C.   Motilin 185
        D.   Substance P, Tachykinin/Neurokinin Family 185
        E.   Neurotensin 186
        F.   Vasoactive Intestinal Polypeptide (VIP) 187
        G.   Opiates 188

IV.   Summary, Conclusions, and Cautions 189

References 190

I. INTRODUCTION

Recently, the list of peptides, amines, amino acids, and other compounds purported to be neurotransmitters or neuromodulators in the gastrointestinal tract has expanded rapidly. Studies to determine the locus and mechanism of action of motor function have led to the development of a number of *in vivo*, *ex vivo*, and *in vitro* experimental preparations in a variety of species. Demonstration of similarities in responses in these varying situations leads readily to generalizations of this particular mode of action in all situations.

Differences in responses require different interpretations. Most researchers have little difficulty in accepting variability in the responses of the musculature of the gastrointestinal tract to any given peptide among different species, among different segments, and among different layers of the tract. However, rarely is the response of the same muscle layer in the same species studied systematically in a number of experimental conditions which vary in the complexity and intactness of organization of the biological system. This may happen because of technical limitations such as the size of the species. Close intra-arterial injections into the gastrointestinal tract are difficult in a mouse, so certain *in vivo* studies are precluded. A preparation may be used by so many workers that comparison of results between compounds and techniques becomes relatively easy, and it becomes tempting to generalize them to other experimental conditions. The guinea pig longitudinal muscle myenteric plexus preparation is one of these which has served the scientific community well, particularly as the bioassay for opiates. However, demonstration of the inhibition of acetylcholine release by stimulation of nerves in the myenteric plexus became generalized or extrapolated to the proposal that this action accounted for the constipating effect of these agents. In fact, it is unclear what the mode of action of opioids might be on the intact or *in vivo* guinea pig intestine. It may, as in other species, involve initiation of a nonpropulsive spasm reduced by atropine or tetrodotoxin and operate through control systems unavailable *in vitro*. The enhanced absorption, inhibited secretion, and increased mucosal blood flow initiated by opioids are also known to be contributing factors to their constipating effects present *in vivo*, but not *in vitro*. (See Chapters 11 and 12 and review in Reference 1 for a detailed discussion.)

Another illustrative case has been the attempt to establish that vasoactive intestinal polypeptide (VIP) is the elusive nonadrenergic noncholinergic (NANC) inhibitory transmitter to circular muscle. The positive evidence from sphincteric regions is that VIP antibodies selectively reduce both VIP and NANC nerve-induced relaxations. The negative evidence is a lack of correspondence between the cellular modes of action of VIP and the mediator (see Reference 2). In most systems studied so far, the activation of adenylate cyclase and increase cyclic adenosine monophosphate (cAMP) levels are frequently associated with relaxation of gastrointestinal sphincters; this has been assumed to support its identification with a relaxing mediator. Such an extrapolation is difficult to maintain in the face of evidence that excitation results from addition to VIP to some tissues, but in others it causes a rise in cAMP levels in muscle, while the NANC mediator increases cyclic guanosine monophosphate (cGMP); both cause relaxation[3] (see References 2 and 4 for detailed discussions).

The action of a peptide in a given preparation is the result of the summation of its interactions with a variety of control systems, neural, hormonal, and myogenic, which may individually cause excitatory and inhibitory effects by both intra- and intracellular mechanisms. The observed outcome will depend greatly, as suggested above, on which of these control systems are allowed to operate. It will also depend on the cellular transducer used to demonstrate the action. Consider measurement of intracellular Ca^{2+} levels, which regulate contraction and secretion. This can be influenced by processes which regulate Ca^{2+} extrusion, Ca^{2+} sequestration, and Ca^{2+} entrance or release from stores. Stimulation of neuropeptide receptors in a membrane may increase Ca^{2+} conductance alone, resulting in increased Ca^{2+} entry and increased contraction. If, however, this increase in Ca^{2+} is linked to increased

K^+ conductance, membrane hyperpolarization and concomitant relaxation may occur. (See discussion of neurotensin to follow.)

The use of the isolated cell and isolated cell organelles facilitates examination of the individual intracellular and membrane mechanisms which control contractility, but may not take into account the alterations in the state of other components in control systems by the preparations for the measurements; e.g., release by a peptide of Ca^{2+} from endoplasmic reticulum in a skinned preparation must necessarily fail to demonstrate changes in membrane conductance of Ca^{2+} or K^+, in $Na^+ — Ca^{2+}$ exchange, in the plasmalemmal Ca^{2+} pump, or in the production and utilization of adenosine triphosphate (ATP). The more complex the biological system used as a measure of response, the greater the number of mechanisms which can influence outcomes and the greater the extent of their interactions. The observed net effect will then provide little mechanistic insight as to cellular and subcellular causation. Major difficulties in mechanistic interpretations will arise when the end point of change is contractility in an intact tissue.

II. CHARACTERISTICS OF THE EXPERIMENTAL PREPARATIONS

A. ISOLATED CELLS

A homogeneous population of isolated cells represents a simple preparation available for measurement of the action of an agonist in altering cell length. This preparation has the advantage of eliminating diffusion barriers, allowing measurement of the direct action of the agonist unaffected by other control systems and providing the opportunity for studying simultaneously several aspects of cell function such as contractility, membrane potential, and intracellular Ca^{2+}. It also has a number of major disadvantages. One is difficulty in interpreting the extremely low range of range (10^{-12} to 10^{-15} M) of concentrations at which many agonists are effective in inducing shortening in these cells. This is in contrast to the strips of muscle from which the cells were derived which require 3 to 5 log units higher concentrations to produce contraction.[5] Since similar concentrations of K^+ contract the isolated cells and the strips by a mechanism bypassing the activation of a receptor, this increased sensitivity to receptor-mediated actions suggests that the isolation procedure has in some way unregulated the receptor population.[6] Furthermore, the broad-spanned shape of the dose-response curve for contraction determined from the isolated cell suggests that the isolation procedure may induce the expression of new receptors or inhibition of responses to partial receptor occupation in the isolated cell. Alternatively, the steep dose-response curve for contracting agonists in the strip preparation could be the result of down-regulation of receptors by the release of local factors such as neuropeptides or arachidonic acid metabolites. There is also a large discrepancy between the concentrations of agonists required to induce contraction (picomolar range) and agonists required to displace antagonist binding (micromolar range). Some of these differences were suggested to result from the duration of exposure required to study these phenomena <5 for contraction and >15 min for binding studies. The initial exposure to the agonist may result in rapid alteration of down-regulation of the receptor with a resulting decrease in affinity in the equilibrium binding assay (for a detailed discussion see Reference 7).

For relaxation the situation is different. Equal concentrations appear to relax both isolated cells and strip preparations.[7] In the isolated cell, relaxation is likely to be an energetically demanding process since it involves a re-elongation of a previously contracted cell now lacking connections to the extracellular elastic elements present in tissues which store energy for elongation during contraction (see Reference 8 for review). If these energetic differences were somehow eliminated, perhaps relaxation of isolated cells would be similar to contraction in its concentration dependence.

B. *IN VITRO* STRIPS

In an isolated strip, the level of complexity of control systems is clearly increased compared to the single cell. A strip of gut may be prepared in a number of ways so that it includes both layers of the muscularis externa and the nerve plexuses within as well as the associated interstitial cells of Cajal (ICC), any associated inflammatory cells, such as neutrophils and mast cells, and the connective tissue. Depending on dissection and on which layers remain undamaged, the preparation, in our experience, may be quiescent or spontaneously active. When these isolated strip preparations are exposed to similar agonists as the isolated cell, one frequent finding is that a number of additional responses to the agonists appear.[4,5] These frequently are attributed to the heterogeneity of the cell population. Mast cells, macrophages, interstitial cells of Cajal, or nerves are the cells to which these additional actions are most commonly attributed. In addition, different levels of production of arachidonic acid products by the muscle or other cells themselves must be considered.

C. *IN SITU*

When the level of complexity of the system is further increased by studying the system *in situ*, *in vivo*, or even *ex vivo*, there appears to be a major increment in the complexity of control systems responding to peptides. The receptors for peptides on nerves appear to predominate in their control functions over receptors which have been found previously on muscle cells. Consider one simple case, the actions of substance P in the canine antrum.[9] *In vivo* the response and presumably the receptor can be demonstrated both on the nerves and muscle (i.e., responses present before and after elimination of nerve stimulated contraction, but those "after" required higher peptide doses), but only the muscle response could be demonstrated *in vitro*. Here the explanation can be given that nerve function was lost with dissection. The situation becomes much less clear when a direct smooth muscle response is not apparent *in vivo*, but appears *in vitro* e.g., with gastrin on canine antrum.[9] In this case all responses to the peptide were eliminated *in vivo* by the prior application of either atropine to block cholinergic receptors or tetrodotoxin to block nerve conduction, yet the strip was contracted in response to the peptide in the presence of these blocking agents and the isolated cell shortened in response to picomolar concentrations of the peptide. In these cases the simplest explanation appears to be that *in vivo* there was a down-regulation of the receptor or its coupling to response in the smooth muscle membrane. This may occur by internalization of the peptide receptor complex, perhaps resulting in a decreased production of the receptor or uncoupling of the receptor from the response, both of which have been shown to occur in the same system for VIP.[10] Whether the continued presence of the peptide extracellularly by release from nerves or endocrine cells is the triggering factor for this down-regulation *in vivo* is unknown. Such findings as these prohibit extrapolation of findings about site and mechanism of peptide action from one situation to another without direct studies of each. It should be noted in passing that these observations, although more striking for peptides, also apply to classical mediators.[5,9]

III. PEPTIDE FAMILIES, VARIABLE RESPONSES IN THE SAME TISSUES

A. GASTRIN/CHOLECYSTOKININ (CCK) FAMILY

Receptors (sensitive in the picomolar range) for this family of peptides have been identified on isolated smooth muscle cells of the guinea pig gastric fundus,[7,11] canine gastric antrum,[6] and canine gallbladder[12] (see References 1 and 7 for summary). In the *in vitro* strip preparations from these sites, the responses (nanomolar range) of the unstimulated strips were contractions which were unchanged in the presence of cholinergic antagonists and tetrodotoxin (see Reference 1 for summary). Additional responses were attributed to release of histamine presumably from mast cells[13] (guinea pig stomach) and amplification of field-stimulated release of acetylcholine (canine gallbladder).[14] These results are indicative of the presence of a receptor on the gastrointestinal smooth muscle for which the numbers or coupling to the response appear to be either amplified in the isolated cell conditions or, conversely, reduced in the strip conditions. Receptors also appear to be present on the cholinergic nerves in stimulated strip preparations. *In vivo* in the dog (*in vivo* guinea pig studies are rarely conducted) the situation is reversed: gastrin[9] and CCK[15] act only to release acetylcholine in the antrum and in the gallbladder.[15] No responses to these agonists remain after the administration of atropine and/or hexamethonium or tetrodotoxin. Thus, there are receptors on both nerves and muscle, but their expression depends upon the experimental conditions.

B. GASTRIN-RELEASING PEPTIDE (GRP)/BOMBESIN

Similar to gastrin, GRP stimulates antral motor activity in the anesthetized dog after intra-arterial injection, but at concentrations 20 times more than required for secretion of gastrin.[16,17] Motor changes, but not gastrin release, were eliminated by tetrodotoxin (TTX). In the small intestine the response was inhibition of field-stimulated motor activity. Neural responses were not evident in the strip preparations, but as for the stomach, excitation occurred at a presumably smooth muscle receptor[18] at nanomolar concentrations. Isolated cells of the human antrum and small intestine contracted to bombesin and GRP analogues in the femto- to picomolar range, but concentration-response data from human tissue strip studies are not available.[19] Thus, this family represents another group of peptides whose neural receptors become ineffective after the dissection procedure and whose smooth muscle receptors are unregulated by isolation of single cells.

C. MOTILIN

Motilin excites the dog antrum *in vivo* by releasing acetylcholine and the small intestine by releasing both acetylcholine and a naloxone-sensitive neurotransmitter (see Reference 1 for review and Section III.G). A smooth muscle receptor is not evident in strip preparations or in binding studies from the site in the dog[9] (see Chapter XI). Only rabbit and human tissues show smooth muscle responses *in vitro*, and these have been confirmed by binding studies (see Reference 20 for review). Unfortunately, studies in these species *in vivo* before and after administration atropine are not available, so differences associated with experimental procedure have not been explored.

D. SUBSTANCE P, TACHYKININ/NEUROKININ FAMILY

Receptors for substance P have been demonstrated on most gastrointestinal smooth muscles (see Reference 1 for review) and are apparent under all three experimental conditions in the dog stomach[9] and small intestine.[5,21] Similar to the gastrin findings, receptors in the isolated cell preparation are activated at a picomolar concentration, while the quiescent and field-stimulated isolated strip preparation demonstrates no neural receptors, only smooth muscle receptors in the nanomolar range. *In vivo* in the dog, neural receptors are stimulated by lower concentrations of intra-arterial substance P than are the muscle receptors.[17] In the canine small intestine, two separate neural receptors could be identified beside a muscle receptor. Release of a transmitter which inhibited excitation induced by field stimulation of nerves occurred at a lower concentration than initiation of release of acetylcholine to stimulate motor activity. Neither of the neural receptors were apparent in a quiescent nor in a field-stimulated strip preparation. Only postjunctional (smooth muscle) responses could be elicited *in vitro*. Full thickness strips of canine small intestine only contract to nanomolar range concentrations of substance P.[17] If, however, the longitudinal muscle and myenteric plexus is removed, phasic activity is initiated by picomolar range concentrations.[53] This could be interpreted as removal of a suppressing factor from the myenteric plexus which down-regulates the remaining muscle cells, elimination of neural inhibition from the myenteric plexus, or allowance of expression of a population of cells such as the interstitial cells of Cajal in the deep muscular plexus area. It also poses the question of how complex the system must be before down-regulation of receptors occurs.

In a further analysis, highly selective analogues for the subtypes of neurokinin receptors were used to study responses of the canine antrum, pylorus, and duodenum *in vivo* and *in vitro*. The NK-1 (substance P preferring) excitatory responses were suggested to be present on both neural and muscle (post-receptor) sites in all three *in vivo* and only muscle sites *in vitro*. The NK-2 (neurokinin A preferring) excitatory responses appeared to be confined to post-receptor sites under both conditions. Both NK-1 and NK-2 receptors appear to be present on the intestinal circular muscle by binding studies.[22] The NK-3 (neurokinin B preferring) excitatory receptors were confined to neural (TTX-sensitive) structures and were apparent only *in vivo* in the antrum and duodenum.[23] The neural receptors for inhibition were of the NK-1 and NK-3 types in both the pylorus[18] and the small intestine.[24] These studies illustrate that some differences in the responses in varying experimental conditions are resolvable when the different receptor subtypes are examined. Explanations of other differences are still incomplete. Others appear to require a multiplicity of *in vitro* preparations for clarification.

In the cat stomach *in vivo*, substance P produces excitation at neural (atropine-sensitive) sites, but acts only at atropine-insensitive sites in the *in vitro* strip preparation. This illustrates again the generality of the apparent true down-regulation of muscle receptors *in vivo*[25] and the loss of functional neural receptors *in vitro*.

In the guinea pig, the longitudinal muscle myenteric plexus preparation responds to low-frequency field stimulation with release of acetylcholine predominately. Substance P stimulation of acetylcholine release also can be demonstrated by monitoring contractile response.[26] However, inhibition of acetylcholine release by substance P was demonstrable only by studying [^{3}H]-choline output because the ED_{50} for muscle contraction was the same as the ED_{50} for inhibition of acetylcholine release.[27] These results suggest that the presence of sites of peptide action may require simultaneous study of neurotransmitter release and motor activity. However, due to the lack of *in vivo* studies in the guinea pig, so far there is no indication of the sites of substance P receptors in this experimental condition.

E. NEUROTENSIN

The most easily demonstrable location of neurotensin in the gastrointestinal tract is endocrine cells concentrated in the ileum, although there are occasional reports of enteric neural neurotensin immunoreactivity (see Reference 1 for review). This peptide produces both excitation and inhibition in many gastrointestinal muscles. The characteristics of the responses and the receptors responsible for these responses in the canine gastrointestinal tract have been studied in detail. Four types of responses are apparent in the canine small intestinal circular muscle.

In the anesthetized animal, neurotensin administered to the small intestine by close intra-arterial injection produces inhibition of field-stimulated circular muscle activity. Both a TTX-sensitive and an α_2-adrenoceptor antagonist-sensitive site (presumably involving adrenergic nerves) and a TTX-insensitive site (presumably smooth muscle) are present.[28] The later site requires higher doses for activation. Inhibition at a smooth muscle receptor is also apparent at nanomolar concentrations *in vitro* on phasic contractions of isolated gastric[29] and small intestinal[30,31] circular muscle strips whether the myenteric plexus and the longitudinal muscle is present or not. Neurally mediated inhibition was not found *in vitro*. The inhibition of the smooth muscle activity of the circular muscle strip by neurotensin was not altered by reduction of disulfide bonds by dithiothreitol (DDT),[32] which abolished phasic excitation to low concentrations.

In the sucrose gap, inhibition is associated with a hyperpolarization of the muscle membrane in the nanomolar range. The hyperpolarization is reversed by blockade of the apamin-sensitive K^+ conductance.[33] In the presence of apamin, neurotensin induces a contraction of the muscle at the time when inhibition occurs during hyperpolarization. Even in the absence of apamin and despite the initial hyperpolization, a tonic contraction was produced by higher (100 nm) concentrations of neurotensin.

A further TTX-insensitive, Ca^{2+} channel antagonist-sensitive response involving stimulation of phasic activity at picomolar concentrations was initially identified in the muscle strip studies. [26,27] This response requires the presence of the inner circular muscle layer and the deep muscular plexus. The experimental conditions of the sucrose gap preparations (probably due to membrane hyperpolarization) appear to inhibit expression of this receptor; i.e., slow waves are absent or small, and phasic excitation at low concentrations usually does not occur in these conditions. *In vivo* excitation to neurotensin injected as a bolus also does not occur during resting or field-stimulated contractions. It is only apparent as a delayed amplification of the excitation induced by intra-arterial bolus injections of acetylcholine when low concentrations are given.[26] Thus, expression of the receptor for neurotensin on the smooth muscle-inducing excitation is more labile than expression of receptor-inducing inhibition.

A fourth type of response found only *in vitro* is a tonic contraction which is maximal at micromolar concentrations. A histamine (H_1) antagonist-sensitive contraction was identified in the canine corpus[25] as being due to degranulation of mast cells. A similar contraction, not yet evaluated as to sensitivity to H_1 antagonists, occurs in response to high neurotensin concentrations in the small intestine[26] in the sucrose gap preparation or muscle bath. It is not affected by reduction of disulfide bonds,[33] but is dependent on the presence of Ca^{2+} in the extracellular medium and blocked by nifedipine. Release of histamine from mast cells has been documented for neurotensin in the rat,[34] but binding studies are not available for canine mast cells.

High- and low-affinity receptors for neurotensin are present on both the deep muscular and the submucous plexuses.[35,54] No ligand sites could be identified on longitudinal muscle or on myenteric plexus synaptosomes. It is tempting to associate these binding sites with the neural and smooth muscle sites of secretion described previously, but so far no definite association can be made. The binding is affected by the ionic strength of the binding medium, and the possibility exists that high- and low-affinity sites reflect different states of the same receptor. Neurotensin binding sites on both muscle membranes and synaptosomes are similarly affected by DTT.

The studies on neurotensin are the most complete for any peptide so far and clearly illustrate the necessity for a multiplicity of studies to determine the range of receptors and coupling mechanisms available to a peptide in any one tissue layer. They also illustrate that the coupling of the receptor to the membrane and intracellular pathways ultimately determines the response to the peptide. Neurotensin requires the presence of extracellular Ca^{2+} to produce contractions and intracellular Ca^{2+} stores to hyperpolarize the membrane, but the coupling of the receptor to the apamin-sensitive Ca^{2+}-dependent K^+ conductance results in hypolarization of the membrane and a relaxation response. Excitation results from neurotensin stimulation of Ca^{2+} entry through L-Ca channels which open despite hyperpolarization and overwhelm the effects of increased K^+ conductance.

F. VASOACTIVE INTESTINAL POLYPEPTIDE (VIP)

VIP is one of the most abundant peptides in the enteric nervous system (see Reference 36). Its role as a stimulant of enterocyte secretion by raising intracellular adenylate levels in enterocytes is well established.[37] The VIP receptors on these cells have been identified by binding studies[38] and localized *in situ* by autoradiographic studies.[39] The role of VIP in control of motor function of the gastrointestinal is less clear-cut because of the numerous discrepancies between the *in vivo* and *in vitro* experimental situation and the lack of an adequate antagonist for VIP. Precontracted isolated cells of the human[40] and guinea pig[41] antrum relax to nanomolar-range VIP concentrations. The lower esophageal sphincters of the opossum (see Reference 42 for extensive review), the cat,[43] and the dog[44] relax to VIP both *in vivo* and *in vitro* in the presence and absence of blockade of neural transmission by tetrodotoxin, a criteria often used to define smooth muscle receptors. VIP receptors have not been demonstrated on the smooth muscle by binding in this tissue, and the TTX insensitivity may reflect an action of VIP at the interstitial cells of Cajal (see Reference 1 for a review). In the absence of adequate antagonists, the finding that immunoneutralization of VIP reduces the neural responses in this tissue has been used to argue for VIP as the sole and direct nonadrenergic noncholinergic transmitter to smooth muscle in this area. There are multiple lines of evidence to contradict this claim, the strongest of which is that relaxation of the sphincter *in vitro* by field stimulation is associated with rises in intracellular cGMP, not cAMP, which VIP increases[3] and that VIP decreases membrane conductance (apparently to chloride) while the mediator increases K^+ conductance.[2]

In the dog, VIP relaxed the pylorus by a TTX-insensitive mechanism *in vivo*, but only produced a weak, inconsistent relaxation of the same tissue *in vitro*, even though activation of intrinsic nerves and adenylate cyclase were capable of relaxing these tissues fully.[45] The authors suggested that the dissection procedure rendered nonfunctional the structures on which VIP acts *in vivo* to produce TTX-sensitive relaxation and that these were therefore unlikely to be the smooth muscles, but might be the interstitial cells of Cajal.

In the small intestine the role of VIP in controlling circular motor activity is not clear. Responses to intra-arterial VIP are not always consistent. Both excitation and inhibition occur in the dog.[46] This may be a consequence of the high ambient level of immunoreactive VIP *in vivo*. VIP output in an *ex vivo* segment is maintained at a level 100 times that of substance P (a peptide with a similar distribution) by tonic neural activity blocked by hexamethonium. High ambient VIP could down-regulate its receptors.[47] Tonic VIP output was reduced by α_2-adrenoceptor agonists, μ- and δ-opiate agonists, motilin, and TTX, accompanied by activated circular motor contractions. This motor activity apparently results from removal of a tonic inhibitory input from VIP release. Whether VIP causes inhibition by acting at the muscle or at another site, such as the interstitial cells of Cajal, is unknown. Functional biochemical studies of VIP receptors on the interstitial cells of Cajal have not been reported. The binding of VIP to the circular muscle has not been demonstrated either by autoradiography[48] (receptor density extremely low as compared to the mucosa) or by binding to purified plasma membranes (no specific binding in preliminary studies,[55] strengthening the suggestion that VIP receptors are localized on other non-muscle structures such as nerves and the interstitial cells of Cajal. There are no *in vitro* strip motility or electrophysiological studies of VIP action on the dog small intestine.

The variations in the VIP responses and the uncertainty about the location of receptors illustrate the problems of making sweeping generalizations about the peptide as the NANC mediator without a multiplicity of analytic studies in the same tissues under a variety of conditions.

G. OPIATES

Opiate receptors of the μ-, δ, and κ-subtypes have been identified as presynaptic to cholinergic enteric motor neurons (see Reference 1 for detailed review). When cholinergic nerves are stimulated *in vitro*, opiates inhibit the contractions induced by this acetylcholine release. Inhibition of motor activity induced by stimulation of cholinergic nerves after intra-arterial injection of opioids to the gastrointestinal tract of the anesthetized dog could be interpreted as a similar response.[49,50] The interpretation of the excitation induced by μ- and δ-agonists in the small intestine is not so simple. Excitation was reduced by atropine and or by tetrodotoxin when the opiate was given at a time when the TTX-induced phasic activity had died away, but field-stimulated motor activity was still absent. Such a finding is usually interpreted as indicative of an action on smooth muscle receptors. However, isolated circular muscle strips were not excited by any class of opioid agonists: field-stimulated contractions were inhibited by opiates of all classes acting to inhibit release of acetylcholine. Binding studies to smooth muscle plasma membrane vesicles and synaptosomes revealed binding sites only to the neural (synaptosomal) sites.[51] Contractions of isolated circular muscle cells to opiates have been reported to guinea pig,[52] but the interpretation of this may be complicated by the occurrence of amplification of muscle receptors during cell isolation, as previously described.

The discovery that the opiates produce motor activity by inhibiting release of VIP in the canine small intestine provided an explanation of some of these discrepancies. When the release of VIP was examined over time, it was shown that this release recovered after TTX suppression, accompanied by the disappearance of TTX-induced contractions and occurring before motor responses to field stimulation. In this experimental circumstance, opiates could both depress VIP output and induce motor activity. Thus, the TTX sensitivity of enteric nervous systems may vary, possibly due to their varied levels of tonic neural activity. In any case there was no need to postulate the presence of opiate receptors on smooth muscles to explain TTX-insensitive responses to opioids in this tissue. The presence or absence of field-stimulated activity clearly is not a complete verification of the absence or presence of total neural blockade.

IV. SUMMARY, CONCLUSIONS, AND CAUTIONS

When measuring motor activity, receptors for peptides on nerves appear to respond to the lowest concentrations under *in vivo* conditions, but receptors at nerves are less apparent or nonfunctional in the strip preparation. Their expression or demonstration requires optimizing conditions for their expression or measurement of release of the transmitter which the receptor modulates. If receptors on smooth muscles are apparent *in vivo*, they are activated by higher doses of the agonist. Such receptors are generally robust and can be demonstrated under most experimental conditions. However, smooth muscle receptors or their expression can be down-regulated *in vivo*, and they may be apparent only under *in vitro* conditions. Expression of smooth muscle receptors can be further amplified in the isolated cell condition until, as for bombesin, contractions occur at femtomolar concentrations. If all the cells responded to such low concentrations of agnists *in vivo*, a state of constant receptor activation would probably result. The decreased expression of receptors *in vivo* may protect the cells from a state of constant activation and allow institution of controls by neurotransmitters and hormones.

The major differences in sites and mechanisms of peptide actions outlined in this chapter strongly sugest that conclusions and extrapolation require a multiplicity of experimental preparations, an open mind as to the probable finding of aberrant, different results, and an awareness of the dangers of extrapolation at all times.

REFERENCES

1. **Daniel, E. E., Collins, S. M., Fox, J. E. T., and Huizinga, J. D.,** Pharmacology of neuroendocrine peptides, in *Handbook of Physiology, Section 6, The Gastrointestinal System,* Vol. 1, Motility and Circulation, Part 1, Schultz, S. G., Wood, J. D., and Rauner, B. B., Eds., American Physiological Society, Bethesda, MD, 1989, chap. 20.
2. **Daniel, E. E.,** Nonadrenergic, noncholinergic (NANC) neuronal inhibitory interactions with smooth muscle, in *Calcium and Contractility,* Grover, A. K. and Daniel, E. E., Eds., Humana Press, Clifton, NJ, 1985, chap. 15.
3. **Torphy, T. J., Fine, C. F., Burman, M., Barnette, M. S., and Ormsbee, H. S., III,** Lower esophageal sphincter relaxation is associated with increased cyclic nucleotide content, *Am. J. Physiol.,* 251, G786, 1986.
4. **Fox, J. E. T.,** Control of gastrointestinal motility by peptides: old peptides, new tricks-new peptides, old tricks, *Gastroenterol. Clin. North Am.,* 18, 163, 1989.
5. **Fox, J. E. T., Collins, S. M., and Daniel, E. E.,** Expression of peptide responses depends upon study environment, in *Regulatory Peptides in Digestive, Nervous and Endocrine Systems,* INSERM Symp. no. 25, Lewin, M. J. M. and Bonfils, S., Eds., Elsevier, Amsterdam, 1985, 265.
6. **Collins, S. M.,** Calcium utilization by dispersed canine gastric smooth muscle cells, *Am. J. Physiol.,* 251, G181, 1986.
7. **Daniel, E. E., Collins, S. M., Fox, J. E. T., and Huizinga, J. D.,** Pharmacology of drugs acting on gastrointestinal motility, in *Handbook of Physiology, Section 6, The Gastrointestinal System,* Vol. 1, Motility and Circulation, Part 1, Schultz, S. G., Wood, J. D., and Rauner, B., Eds., American Physiological Society, Bethesda, MD, 1989, chap. 19.
8. **Sanders, K. M.,** Electrophysiology of dissociated gastrointestinal muscle cells, in *Handbook of Physiology, Section 6, The Gastrointestinal System,* Vol. 1, Motility and Circulation, Part 1, Schultz, S. G., Wood, J. D., and Rauner, B. B., Eds., American Physiological Society, Bethesda, MD, 1989, chap. 4.
9. **Fox, J. E. T., Daniel, E. E., Jury, J., Fox, A. E., and Collins, S. M.,** Sites and mechanisms of action of neuropeptides on canine gastric motility differ *in vivo* and *in vitro, Life Sci.,* 33, 817, 1983.
10. **Robberecht, P., De Neef, P., Waelbroeck, M., Tastenoy, M., and Christophe, J.,** VIP and related peptides induce rapid homologous desensitization in the humanlymphoma SUP T1 cell line, *Peptides,* 10, 441, 1989.
11. **Collins, S. M. and Gardener, J. D.,** Cholecystokinin-induced contraction of dispersed smooth muscle cells, *Am. J. Physiol.,* 243, G497, 1982.
12. **Bitar, K. N. and Makhlouf, G. M.,** Cholinergic and CCK receptors on canine isolated gall bladder smooth muscle cells, *Gastroenterology,* 80 (Abstr.), 111, 1981.
13. **Gerner, T., Haffner, J. F. W., and Norstein, J.,** The effects of mepyramine and cimetidine on the motor responses to histamine, cholecystokinin and gastrin in the fundus and antrum in isolated guinea-pig stomachs, *Scand. J. Gastroenterol.,* 14, 65, 1979.
14. **Milenov, K., Rakovska, A., and Kalfin, R.,** Cholinergic mechanism in cholecystokinin action on gall bladder motility, *Methods Find. Exp. Clin. Pharmacol.,* 10, 741, 1988.
15. **Kuwahara, A., Ozawa, K., and Yaniahara, N.,** Effects of cholecystokinin-octapeptide on gastric motility of anesthetized dogs, *Am. J. Physiol.,* 251, G678, 1986.
16. **Fox, J. E. T. and McDonald, T. J.,** Motor effects of gastrin releasing peptide (GRP) and bombesin in the canine small intestine, *Life Sci.,* 35, 1667, 1986.
17. **McDonald, T. J. and Fox, J. E. T.,** Effects of porcine gastrin releasing peptide (GRP) in canine antral motility and gastrin release *in vivo, Life Sci.,* 35, 1415, 1984.
18. **Mayer, E. A., Elashoff, J., and Walsh, J.,** Characterization of bombesin effects on canine gastric muscle, *Am. J. Physiol.,* 243, G141, 1982.
19. **Micheletti, R., Grider, J. R., and Makhlouf, G. M.,** Identification of bombesin receptors on isolated muscle cells from human intestine, *Regul. Peptides,* 21, 219, 1988.
20. **Fox, J. E. T.,** Mechanisms of motilin excitation as determined by *in situ* and *in vitro* studies, in *Motilin,* Itoh, Z., Ed., Academic Press, 1990, 73.
21. **Fox, J. E. T. and Daniel, E. E.,** Substance P: a potent inhibitor of the canine small intestine *in vivo, Am. J. Physiol.,* 253, G21, 1986.
22. **Müller, M. J., Sato, H., Bowker, P., and Daniel, E. E.,** Receptors for tachykinins in canine intestinal circular muscle, *J. Pharmacol. Exp. Ther.,* 246, 739, 1988.
23. **Allescher, H. - D., Kostalanska, F., Tougas, G., Fox, J. E. T., Regoli, D., Drapeau, D., and Daniel, E. E.,** The action of neurokinin and substance P receptors in canine pylorus, antrum and duodenum, *Peptides,* 10, 671, 1989.
24. **Fox, J. E. T., McDonald, T. J., Alford, L., and Kostalanska, F.,** Tachykinin activation of muscarinic inhibition in canine small intestine is SPP in nature, *Life Sci.,* 39, 1123, 1986.

25. **Edin, R., Lundberg, J. M., Lidberg, P., Dählstrom, A. and Ahlman, H.,** Atropine sensitive contractile motor effects of substance on the feline pylorus and stomach *in vivo, Acta Physio. Scand.,* 110, 207, 1980.

26. **Holzer, P. and Lembeck, F.,** Neurally mediated contractionof ileal longitudinal muscle by substance P, *Neurosci. Lett.,* 17, 101, 1980.

27. **Kilbinger, H., Sharp, S., Erlhof, I., and Holzer, P.,** Antagonist discrimination between subtypes of tachykinin receptors in the guinea-pig ileum, *Naunyn-Schmiedeberg's Arch. Pharmacol.,* 334, 181, 1986.

28. **Sakai, Y., Daniel, E. E., Jury, J., and Fox, J. E. T.,** Neurotensin inhibition of canine intestinal motility *in vivo* via α-adrenoceptors, *Can. J. Physiol. Pharmacol.,* 62, 403, 1984.

29. **Mclean, J. and Fox, J. E. T.,** Mechanisms of action of neurotensin on motility of canine gastric corpus *in vitro, Can. J. Physiol. Pharmacol.,* 61, 29, 1983.

30. **Fox, J. E. T., Kostalanska, F., Daniel, E. E., Allescher, H. D., and Hanke T.,** Mechanism of excitatory actions of neurotensin on canine small intestinal circular muscle *in vivo and in vitro, Can. J. Physiol. Pharmacol.,* 65, 2254, 1987.

31. **Fox, J. E. T., Kostalanska, F., Allescher, H. D., and Daniel, E. E.,** Neurotensin and the canine small intestine. I. Circular muscle motility *in vivo* and *in vitro, Dig. Dis. Sci.,* 32, 910, 1987.

32. **Ahmad, S. and Daniel, E. E.,** Neurotensin receptors on circular smooth muscle of canine small intestine — the role of disulfide bridges, *Biophys. Biochem. Res. Commun.,* 165, 422, 1989.

33. **Christinck, F., Daniel, E. E., and Fox, J. E. T.,** Electrophysiological responses of canine ileal circular muscle to electrical stimulation and neurotensin (NT), *Gastroenterology,* 96, A680, 1989.

34. **Shanahan, F., Denberg, J. A., Fox, J. E. T., Bienenstock, J., and Befus, A. D.,** Effects of neuroenteric peptides on histamine release, *J. Immunol.,* 135, 1131, 1985.

35. **Ahmad, S., Berezin, I., Vincent, J. P., and Daniel, E. E.,** Neurotensin receptors in canine intestinal smooth muscle: preparation of plasma membranes and characterization of ^{125}I-Tyr^3-neurotensin binding, *Biochim. Biophys. Acta,* 896, 224, 1987.

36. **Furness, J. B. and Costa, M.,** *The Enteric Nervous System,* Churchill Livingstone, Edinburgh, 1987.

37. **Donowitz, M. and Welsh, M. J.,** Regulation of mammalian small intestinal motility, in *Physiology of the Gastrointestinal Tract,* 2nd ed., Johnson, L. R., Ed., Raven Press, New York, 1987, 1351.

38. **Laburthe, M., Breant, B. and Rouyer-Fesard, C.,** Molecular identification of receptors for vasoactive intestinal peptide in rat intestinal epithelium by covalent cross-linking, *Eur. J. Biochem.,* 139, 181, 1984.

39. **Sayadi, H., Harmon, J. W., Moody, T. W., and Korman, L. Y.,** Autoradiographic distribution of vasoactive intestinal polypeptide receptors in rabbit and rat small intestine, *Peptides,* 9, 23, 1988.

40. **Bitar, K. N., Saffouri, B., and Makhlouf, G. M.,** Cholinergic and peptidergic receptors on isolated human antral smooth muscle cells, *Gastroenterology,* 82, 832, 1982.

41. **Bitar, K. N. and Makhlouf, G. M.,** Relaxation of isolated gastric smooth muscle cells by vasoactive intestinal peptide, *Science,* 216, 531, 1982.

42. **Goyal, R. K. and Patterson, W. G.,** Esophageal motility, in *The Handbook of Physiology, Section 6, The Gastrointestinal System,* Vol. 1, Motility and Circulation, Part 2, Schultz, S. G., Wood, J. D., and Rauner, b. B., Eds., American Physiological Society, Bethesda, MD, 1989, chap. 22.

43. **Biancani, P., Walsh, J. H., and Behar, J.,** Vasoactive intestinal polypeptide: a neurotransmitter for lower esophageal sphincter relaxation, *J. Clin. Invest.,* 73, 378, 1984.

44. **Allescher, H. D., Berezin, I., Jury, J., and Daniel, E. E.,** Characteristics of canine lower esophageal sphincter: a new electrophysiological tool, *Am. J. Physiol.,* 255, G441, 1988.

45. **Allescher, H. D., Daniel, E. E., Dent, J., and Fox, J. E. T.,** Inhibitory function of VIP-PHI and galanin in canine pylorus, *Am. J. Physiol.,* 256, G789, 1989.

46. **Gonda, T., Daniel, E. E., McDonald, T. J., Fox, J. E. T., Brooks, B. D., and Oki, M.,** Distribution and function of enteric galanin-immunoreactive nerves in dog: comparison to VIP, *Am. J. Physiol.,* 256, G884, 1989.

47. **Manaka, H., Manaka, Y., Kostalanska, F., Fox, J. E., T., and Daniel, E. E.,** Release of VIP and substance P from isolated perfused canine ileum, *Am. J. Physiol.,* 257, G180, 1989.

48. **Zimmerman, R. P., Gates, T. S., Mantyh, C. R., Vigna, S. R., Boehmer, C. G., and Mantyh, P. W.,** Vasoactive intestinal peptide (VIP) receptors in the canine gastrointestinal tract, *Peptides,* 9, 1241, 1989.

49. **Fox, J. E. T. and Daniel, E. E.,** Exogenous opiates: their local mechanism of action in the canine small intestine and stomach, *Am. J. Physiol.,* 253, G179, 1987.

50. **Fox, J. E. T. and Daniel, E. E.,** Activation of endogenous excitatory opiate pathways in the canine small intestine by field stimulation and motilin, *Am. J. Physiol.,* 253, G189, 1987.

51. **Allescher, H. D., Ahmad, S., Kostka, P., Kwan, C. Y., and Daniel, E. E.,** Distribution of opioid receptors in canine small intestine: implications for function, *Am. J. Physiol.,* 256, G966, 1989.

52. **Bitar, K. N. and Makhlouf, G. M.,** Selective presence of opiate receptors on intestinal circular muscle cells, *Life Sci.,* 37, 531, 1982.

53. **Müller, M. J., Daniel, E. E., and Bowker, P.,** unpublished observations.

54. **Ahmad, S. and Daniel, E. E.,** unpublished observations.

55. **Daniel, E. E.,** preliminary studies, 1989.

Chapter 6

RECEPTORS FOR NEUROPEPTIDES (NEUROKININS): FUNCTIONAL STUDIES

D. Regoli, S. Dion, and G. Drapeau

TABLE OF CONTENTS

I.   Introduction 194

II.   Neurokinin Receptors: Early Studies...194

III.   Neurokinin Receptors in Intestinal Tissues.....................................195

IV.   Neurokinin Receptors in Isolated Vessels: New Pharmacological
        Preparations..196

V.   Structure-Activity Studies: Naturally Occurring Peptides, Fragments,
        and Selective Agonists...198
        A.   Characterization of the NK-1 Receptor 198
        B.   Characterization of the NK-2 Receptor 198
        C.   Characterization of the NK-3 Receptor 200

VI.   Some Results Obtained with Antagonists 202

VII.   Neurokinin Receptors in the Gastrointestinal Tract............................203

VIII.   Conclusions 204

Acknowledgments 204

References..204

I. INTRODUCTION

The hypothesis that substance P and related neurokinins may play an important role in the regulation of gastrointestinal functions was suggested shortly after the discovery of the first peptide, substance P,[1] and has since been reaffirmed repeatedly.[2,3] The peptides, substance P in particular, have been shown to activate intestinal motility *in vivo*,[4,5] possibly through direct (on smooth muscle) or indirect (through acetylcholine or other agents) mechanisms, as well as to stimulate exocrine secretions, especially in the salivary glands and the pancreas,[6,7] and to modulate intestinal blood flow.[8] Indeed, substance P and neurokinin A appear to be present in afferent (sensory) and possibly also in efferent nerve terminals in the intestinal wall, around the vessels, and in exocrine glands.[9] When released by physiological stimuli, neurokinins are able to exert a variety of local effects, such as direct stimulation of intestinal or vascular smooth muscles, activation of local reflexes, facilitation of the release of other transmitters (acetylcholine, 5-hydroxytryptamine) or autacoids (histamine, prostanoids), and stimulation of exocrine secretions such as in the salivary gland.[10] When released in the vessel walls, they may produce vasodilation by activating the generation of the endothelium-derived relaxing factor (EDRF)[11] or induce venoconstriction (especially NKB) by acting directly on vascular smooth muscles.[12]

II. NEUROKININ RECEPTORS:EARLY STUDIES

These various effects are brought about by the activation of multiple receptors whose existence was suggested from early findings by the Erspamer group.[13] In 1982, a comparison of data obtained in different isolated organs (intestinal, extraintestinal) brought Lee et al.[14] to suggest the existence of two different receptor types named SP-P and SP-E. Further progress was made when the other two mammalian neurokinins (neurokinin A and neurokinin B) were identified[15] and when it was shown that each natural neurokinin may have its own receptor.[16] Shortly after, three receptor types were identified with biological[17] and biochemical assays.[18] The three-receptor hypothesis was further confirmed and definitely demonstrated by the use of selective agonists.[19,20] The three receptors have been named NK-1, NK-2, and NK-3 to indicate the receptors for substance P (SP), neurokinin A (NKA), and neurokinin B (NKB), respectively.[21]

For decades, the pharmacology of substance P was studied by measuring the hypotensive effects of partially purified preparations in various laboratory animals (for a review see Reference 22) or the myotropic effect on the guinea pig ileum (G.P.I.).[22] Organs other than the G.P.I. were used by Erspamer and Melchiorri[23] namely, the hamster and dog urinary bladders for the characterization of eledoisin, physalaemin, kassinin, and other tachykinins. The guinea pig trachea was studied extensively by Mizrahi[24] in 1982 and since then has become one of the preparations most frequently used because of the important implications of neurokinins in the non-adrenergic noncholinergic (NANC) component of bronchial functions.

The various organs mentioned above respond to SP and related peptides with concentration-dependent contractions which were thought to be due primarily to direct activation of smooth muscle receptors; indirect effects through release of other neurotransmitters (acetylcholine, noradrenaline) were known from the studies of Hedqvist and von Euler[25] on the electrically stimulated guinea pig ileum and the data of Lee et al.[14] on the rat vas deferens. Some of the peripheral organs (e.g., the hamster urinary bladder) were used also for biochemical studies on receptor characterizations,[18] although the majority of binding assays have been performed on mouse mesencephalic cells in primary culture[26] or rat cerebral membranes.[18,27,28]

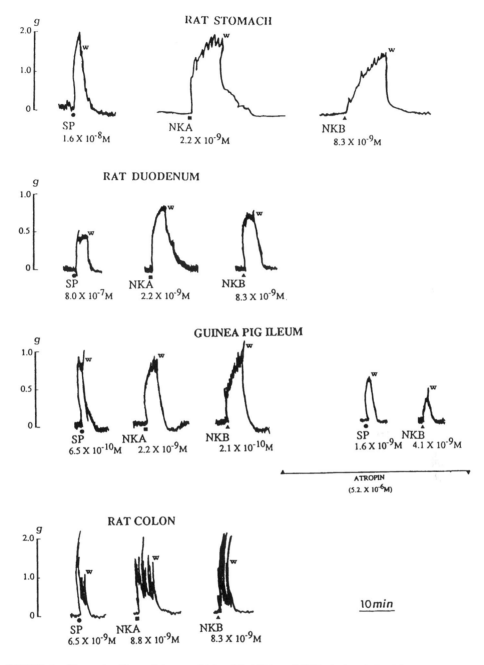

FIGURE 1. Myotropic effects of the neurokinins, SP, NKA, and NKB, in four isolated intestinal tissues; w indicates washing of the contractile agent. Abscissa, time in minutes. Ordinate, tension in grams.

III. NEUROKININ RECEPTORS IN INTESTINAL TISSUES

Substance P and related peptides have been found to stimulate the smooth muscles of the gastrointestinal tract of practically all animals, even the goldfish.[29] The majority of the studies on neurokinins in intestinal tissues have been performed *in vitro*, using isolated smooth muscle preparations, some of which have been chosen to illustrate (see Figure 1) the myotropic effects of neurokinins. Tracings recorded on four isolated tissues, the rat stomach, (prepared as a strip of the fundus according to Vane),[30] the rat duodenum, the guinea pig ileum, and the rat colon, are shown in Figure 1. NKA is the most active of the three neurokinins in the rat stomach; it produces a rapid initial contraction (phasic) followed by a tonic sustained increase of tension, whereas SP is weaker and it produces only the phasic contraction, which is not maintained. The response of the stomach to NKB develops rather slowly and is sustained.

Similar responses have been observed with the three neurokinins in the rat duodenum; again, NKA is the most active of the three natural peptides, followed by NKB and SP. The response of this tissue to NKA is also to be separated into a rapid initial phasic contraction followed by a sustained tonic response. SP only induces the rapid initial phasic response which, in some tissues (for instance, that shown in Figure 1) can be maintained for some time at a stable plateau. NKB shows an effect similar to that of NKA.

The guinea pig ileum is the preparation most frequently used in the pharmacology of tachykinins and neurokinins. It is very sensitive to SP and to NKB and less sensitive to NKA. The guinea pig ileum responds to SP with a rapid increase of tension which is rapidly reversible, while the contractions provoked by NKB and NKA are composed of an initial phasic and a late prolonged tonic response, as shown in Figure 1. The effects of neurokinins, especially those of NKB, are markedly inhibited by atropine, suggesting that, in parallel with their action on the smooth muscle, the neurokinins are able to activate the release of acetylcholine from the myenteric plexus.

Receptors for neurokinins appear to be present also in the last portion of the gut, the colon. Again, substance P induces a rapid increase of tension, quickly reversible, while NKA and particularly NKB are less active as stimulants of the initial phasic response, but they are more active than SP in increasing colon motility.

IV. NEUROKININ RECEPTORS IN ISOLATED VESSELS: NEW PHARMACOLOGICAL PREPARATIONS

Characterization of neurokinin receptors with pharmacological assays has raised difficulties and controversies because of the complexity of some organs such as the intestine (the guinea pig ileum, the rat duodenum) or the respiratory and urinary tracts (guinea pig trachea, hamster urinary bladder, guinea pig urinary bladder). The situation has improved somewhat since the introduction of some isolated vessels in the pharmacology of substance P; the first isolated vessel, the rabbit mesenteric vein, was characterized extensively by Bérubé et al.,[31] but did not correspond to the expectation and, hence, was rapidly abandoned; in 1980, Couture et al.[32] carried out experiments on the dog carotid artery (D.C.A.), a preparation whose intrinsic tone has to be raised by noradrenaline in order to facilitate the recording of the inhibitory (relaxing) effects of neurokinins. The D.C.A. has now become the most sensitive and selective preparation for SP and has been shown to contain only receptors of the NK-1 type.[33] A few years later, D'Orléans-Juste et al.[34] identified the rabbit pulmonary artery (R.P.A.) as a pharmacological preparation particularly sensitive to NKA and later as an NK-2 monoreceptor system.[17] After an intensive search, an isolated vessel sensitive to NKB, the rat portal vein, was discovered by Mastrangelo et al.[35] and since then has become the selective NK-3 preparation.[17,20,33]

Tracings obtained in the three vascular preparations are shown in Figure 2. When contracted with noradrenaline (2.0×10^{-8} M) to a stable plateau of approximately 1 to 1.5 g, the dog carotid artery responds to substance P and related peptides with a rapid relaxation which is maintained. This effect is observed only in the presence of intact endothelium and has been shown to be due to the release of the endothelium-dependent relaxing factor.[11] In the absence of endothelium, the tissue is insensitive to neurokinins.

When utilized in the same conditions, contracted with noradrenaline and with intact

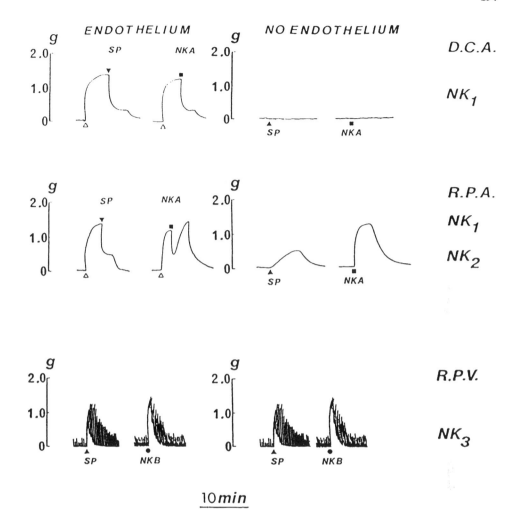

FIGURE 2. Tracings recorded with neurokinins in three isolated vascular preparations. The dog carotid artery (D.C.A., the NK-1 receptor system) and the rabbit pulmonary arter (R.P.A., containing NK-1 in the endothelium and NK-2 in the smooth muscle) with endothelium were stimulated with noradrenaline (2.0×10^{-8} M) to improve the recording of the inhibitory effects of the neurokinins. The rabbit pulmonary artery without endothelium and the rat portal vein (R.P.V., the NK-3 receptor system) were used without treatment. Abscissa, time in minutes. Ordinate, tension in grams. Concentrations of peptides on D.C.A.: SP (3.3×10^{-10} M), NKA (2.2×10^{-9} M); R.P.A.: SP (3.3×10^{-10} M), NKA (2.2×10^{-9} M). R.P.A. without endothelium: SP (3.3×10^{-7} M), NKA (2.2×10^{-9} M); R.P.V.: SP (1.6×10^{-7} M), NKB (8.3×10^{-9} M).

endothelium, the rabbit pulmonary artery shows a different response to SP and NKA (see Figure 2). While the relaxation produced by SP remains stable, the relaxation induced by NKA is rapidly reversible and is followed by a contraction. When the endothelium is removed, the rabbit pulmonary artery responds to neurokinins with a contraction and is much more sensitive to NKA than to SP. D'Orléans-Juste et al.[34] have shown that this effect is due to the activation of NK-2 receptors.

The rat portal vein responds to neurokinins with a rapid contraction which is not maintained. Spontaneous activity is also strongly accentuated by these peptides, the most active of which is NKB, followed by NKA and SP. Identical responses are observed in this tissue in the presence or in the absence of endothelium. The rat portal vein is a NK-3 monoreceptor system.

V. STRUCTURE-ACTIVITY STUDIES: NATURALLY OCCURRING PEPTIDES, FRAGMENTS, AND SELECTIVE AGONISTS

In Table 1, the preparations most commonly used in pharmacological assays of different neurokinin receptors are analyzed by showing the apparent affinities (pD_2) and relative activities (in percent of SP of the three mammalian neurokinins, three tachykinins, three neurokinin fragments, and the most selective agonists for each receptor type.[17,19,20,33] The preparations are the guinea pig ileum not treated (G.P.I. n.t.) or treated (G.P.I. t.) with atropine ($10^{-6} M$), the dog carotid artery (D.C.A.) for the NK-1 receptors, the rat duodenum (R.D.) and the rabbit pulmonary artery (R.P.A.) for the analysis of the NK-2 receptor system, and the hamster urinary bladder (H.U.B.) and rat portal vein (R.P.V.) for the NK-3 receptor (see Table 1). Indeed, the muscarinic receptor antagonist reduces the activities of NKB, NKA, and particularly that of the NK-3-selective agonist [MePhe^7]NKB (see also Figure 1).

A. CHARACTERIZATION OF THE NK-1 RECEPTOR

The two NK-1 receptor preparations of the guinea pig ileum (G.P.I.) show some important differences in the order of potency of tachykinins and neurokinin fragments as well as in the activities of the selective agonists. To eliminate the interference of the prejunctional receptor which promotes the release of acetylcholine and which has been shown to be of the NK-3 type,[36] the G.P.I. was treated with atropine (Table 1). Even after treatment with atropine the two preparations show differences when compared to the D.C.A., particularly with neurokinin fragments and the selective agonists. For instance, SP$_{4-11}$ is only half as active as SP in the D.C.A., but is 2.4 to 3.3 times more active in the G.P.I. The NK-2-selective agonist [Nle^{10}]NKA$_{4-10}$ is inactive on the D.C.A., while maintaining activity in the intestinal preparation. The pattern of biological activities found on the D.C.A. was taken as representative of the NK-1 receptor because it is similar (1) to that measured with the neurokinins and the selective peptides in two *in vivo* assays, the rat blood pressure and rat salivary secretion, according to Maggi et al.;[37] and (2) to the abilities of the same peptides to displace Bolton-Hunter substance P labeled with iodine ([^{125}]BH-SP) from rat brain synaptosomes in binding assays[38] (see Figure 3). NK-1 receptors can therefore be described by the order of potency of agonists as follows:

$$SP > NKA > NKB$$

$$PHY > ELE > KAS$$

$$[Sar^9, Met(O_2)^{11}]SP >>> [MePhe^7]NKB > [Nle^{10}]NKA_{4-10}$$

The data presented in Table 1 are indicative of the complexity of the gastrointestinal tissues; even after elimination of the mucosa, the submucosa, and the circular smooth muscle, the longitudinal strip of the guinea pig ileum contains the three neurokinin receptor types which all contribute, directly or indirectly, to contraction of the intestinal smooth muscle.[39,40] In the entire ileum *in vitro* and in ordinary life, the system is even more complex, considering the various tissues or cell types on which neurokinins may exert other biological effects (e.g., mastocytes, blood vessels, intestinal glands, etc.).

B. CHARACTERIZATION OF THE NK-2 RECEPTOR

A similar situation is found in the other gastrointestinal tissue, for instance, in the rat duodenum analyzed in Table 1, in comparison with the rabbit pulmonary artery.

TABLE 1
Summary of Pharmacological Data Obtained in Isolated Organs with a Variety of Neurokinin-Related Peptides

Pharmacological preparations

Peptide	NK-1						NK-2				NK-3			
	G.P.I. n.t.		G.P.I. t.		D.C.A.		R.D.		R.P.A.		H.U.B.		R.P.V.	
	pD_2	R.A.	pD_2	R.A.	pD_2	R.A.	pD_2	R.A.	pD_2	R.A.	pD_2	R.A.	pD_2	R.A.
Substance P (SP)	8.78	100	8.48	100	10.00	100	6.50	100	6.13	100	5.57	100	5.82	100
Neurokinin A (NKA)	8.40	42	7.90	26	9.40	25	8.22	5,248	8.22	12,303	7.40	6,761	6.45	427
Neurokinin B (NKB)	8.64	72	7.94	29	8.90	8	8.15	4,467	7.45	2,089	7.20	4,266	7.68	7.244
Physalaemin	9.04	182	8.80	209	10.00	100	6.52	106	6.02	78	5.57	100	6.18	229
Eledoisin	9.21	269	8.60	132	9.20	16	7.70	1,700	8.22	12,303	7.06	3,090	7.11	1,950
Kassinin	9.35	372	8.76	191	9.50	32	8.00	3,200	7.66	3,388	7.90	21,400	7.20	2,399
SP_{4-11}	9.30	331	8.86	240	9.67	47	6.30	64	6.23	126	5.26	49	5.87	112
NKA_{4-10}	7.77	10	7.63	14	8.62	4.2	8.19	4,898	8.52	24,547	7.65	12,023	6.79	933
NKB_{4-10}	—		7.80	21	8.70	5	7.50	1,000	7.95	6,607	7.24	4,677	6.67	708
[Sar^9,Met(O$_2$)^{11}]SP	8.91	135	8.51	107	10.45	282	6.67	148	Inactive			<0.1	Inactive	
[Nle^{10}]NKA$_{4-10}$	7.99	16	7.43	9	7.00	0.1	8.02	3,311	7.90	5,888	7.09	3,311		<1
[MePhe^7]NKB	8.76	96	6.59	1.3	7.15	0.14	8.66	14,454	5.24	13	6.14	372	8.30	30,200

Note: 1 pD_2, R.A., NK-1, NK-2, NK-3, G.P.I., D.C.A., R.D., R.P.A., H.U.B., and R.P.V. — see text. R.D., R.P.A., H.U.B., and R.P.V. were used without treatment as in the majority of studies. D.C.A. with intact endothelium and contracted with noradrenaline (2.0 × 10^{-8} M). R.P.A., without endothelium. G.P.I. t., the tissues were treated (1) with atropine (10^{-6} M).

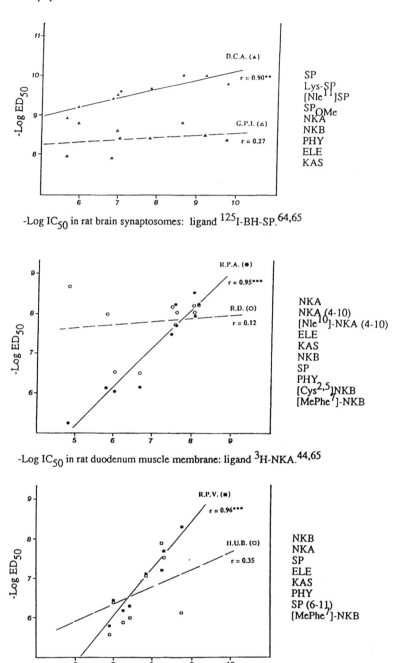

FIGURE 3. Log-log plots of the binding affinities and the biological activities of neurokinins and their congeners. Top panel: binding affinities of a series of SP-related peptides evaluated on rat brain synaptosomes and their biological activities measured on the dog carotid artery (D.C.A., ▲) and the guinea pig ileum (G.P.I., △). Central panel: binding affinities of NKA-related peptides to smooth muscle membranes of the rat duodenum and their biological activities on the rabbit pulmonary artery (R.P.A., ●) or the rat duodenum (R.D., ○). Bottom panel: binding of NKB-related peptides to rat brain synaptosomes and biological activities on the rat portal vein (R.P.V., ■) or the hamster urinary bladder (H.U.B., ▢). Abscissa, $-$log of the peptide concentration utilized to displace the respective ligand. Ordinate, $-$log of the peptide ED_{50}. Each ED_{50} was calculated from at least eight determinations. ** $p < 0.01$, *** $p < 0.001$. For other explanations, see text.

In spite of a certain correspondence in the order of potency of the neuro- and tachykinins, NKB and kassinin show high affinity on the R.D., especially kassinin, which is more active than eledoisin, contrary to the results obtained in the R.P.A. The potencies of the fragments, particularly those of the selective agonists, differ markedly in the two preparations; in fact, all three selective agonists are able to contract the R.D., which has therefore been considered as a multiple receptor system.[36] The R.P.A. responds to [Nle^{10}]NKA$_{4-10}$ and only slightly to [MePhe^7]NKB and has therefore been considered to be representative of the NK-2 receptor, since the same pattern of agonist activities that has been obtained in this tissue was also found in a variety of other NK-2 preparations, for instance, the rat and rabbit vas deferens,[36] the human urinary bladder,[41] the human bronchus,[42] and the human intestine.[43] Moreover, a good correlation was demonstrated (see Figure 3) between the myotropic effects of agonists on the R.P.A. and their abilities to displace [^3H]NKA from rat duodenal membranes[44] (see Figure 3). NK-2 receptors can therefore be described by the other of potency of agonist recorded on the R.P.A. as follows:

$$NKA > NKB > SP$$

$$ELE > KAS > PHY$$

$$[Nle^{10}]NKA_{4-10} >>> [MePhe^7]NKB > [Sar^9, Met(O_2)^{11}]SP$$

Again, the intestinal preparation appears to be a complex one in which the three neurokinin receptors contribute to the duodenal contraction; this is true both for the whole tissue, which was used in the majority of the experiments, as well as for the longitudinal muscle strip of the R.D., which was assayed in a few experiments by Dion et al.[45] and gave the same results as the whole organ. The presence of multiple receptors in the rat duodenum was also demonstrated in *in vivo* experiments by Maggi et al.,[46] who suggested that NK-1 receptors may be involved in the acute phasic initial response while NK-2 receptors are responsible for the late, protracted tonic contraction observed after i.v. applications of NKA in the rat.

C. CHARACTERIZATION OF THE NK-3 RECEPTOR

As for the NK-3 receptor, the data shown in Table 1 indicate that NKB and KAS are the most potent agents in this system (see data of the R.P.V.). Among the fragments the most active are NKB$_{4-10}$ and [MePhe^7]NKB, a selective agonist which is four times more active than NKB. The typical NK-3 system is the rat portal vein (R.P.V.), while the other organ, the hamster urinary bladder (H.U.B.), appears to contain a fairly large number of functionally active NK-2 and some NK-3 receptors (see data obtained with NKA and the selective agonists, Table 1). The correlation between biological data obtained in the R.P.V. and the binding affinities evaluated with [^{125}I]BH-ELE in rat brain homogenates[38] is fairly good (see Figure 3), while no correlation was found between the binding and the biological activities of the same peptides in the H.U.B.

Based on the results of the biological assays performed on the R.P.V., the NK-3 receptor can be described by the following orders of potency of agonists:

$$NKB > NKA > SP$$

$$KAS > ELE > PHY$$

$$[MePhe^7]NKB >>> [Nle^{10}]NKA_{4-10} > [Sar^9, Met(O_2)^{11}]SP$$

TABLE 2
Neurokinin Antagonists: Apparent Affinities Measured against the Natural Neurokinins and Some Selective Receptor Agonists

Peptide	NK-1 D.C.A.		NK-2 R.P.A.		NK-3 R.P.V.	
	SP	NK-1S.A.[a]	NKA	NK-2S.A.[b]	NKB	NK-3S.A.[c]
[D-Pro⁴,D-Trp⁷˒⁹,Nle¹¹]SP₄₋₁	6.85	6.85	5.07	5.40	5.37	5.37
[D-Pro⁴,D-Trp⁷˒⁹,Phe¹¹]SP₄₋₁₁	6.54	6.39	5.69	5.78	5.78	5.09
[D-Tyr⁴,D-Trp⁷˒⁹,Nle¹¹]SP₄₋₁₁	7.02	6.44	5.39	5.39	5.79	5.42
[D-Tyr⁴,D-Trp⁷˒⁹˒¹⁰,Phe¹¹]SP₄₋₁₁	6.13	6.20	5.57	5.71	5.51	5.16

[a] NK-1 selective agonist: $[Sar^9,Met(O_2)^{11}]SP$
[b] NK-2 selective agonist: $[Nle^{10}]NKA_{4-10}$
[c] NK-3 selective agonist: $Succ[Asp^5,MePhe^8]SP_{6-11}$

Note: D.C.A., dog carotid artery; R.P.A., rabbit pulmonary artery; R.P.V., rat portal vein.

VI. SOME RESULTS OBTAINED WITH ANTAGONISTS

Antagonists for substance P were discovered before 1981 by Folkers et al.,[47] who found that [D-Pro²,D-Trp⁷˒⁹]SP was able to reduce the effect of SP in several tests *in vitro* and *in vivo*[48,49] (for a review see Reference 49). In an attempt to reduce the histamine-releasing activity and possibly the neurotoxicity of the antagonists, Mizrahi et al.[50] and Regoli et al.[51] prepared and tested a series of analogues of the sequence SP_{4-11}, containing D-Trp in positions 7 and 9 and eventually in position 10. A large number of analogues of both the undeca- and octapeptide antagonists were prepared in order to find compounds more potent and possibly more selective than the initial prototypes. The results of these studies have been summarized in three review papers.[16,52,53]

Undecapeptide and octapeptide antagonists were found to be competitive against not only SP, but also other neurokinins in both *in vitro* and *in vivo* assays.[16,50]

The majority of the studies reported with SP antagonists have been performed with a compound published by Folkers et al.[54] and called spantide ([D-Arg¹,D-Trp⁷˒⁹,Leu¹¹]SP). Spantide has been shown to reduce salivary secretion induced by substnace P in anesthetized rats and to prevent the peripheral effects of substance P in the skin, such as flare and wheal in humans and plasma extravasation in animals.[49] SP antagonists have also been found to antagonize the algesic effect of SP in rat and mice when given intrathecally;[57] this use of the antagonists in the rat is, however, being strongly limited by their neurotoxicity and the complexity and unreliability of the pharmacological tests, as recently discussed by Vaught.[55]

More success has been obtained with the use of antagonists to block or reduce peripheral effects of SP and related neurokinins. Spantide and some of the octapeptide antagonists have been found to reduce or eliminate peripheral effects that have been attributed to local release of neurokinins — for instance, the contraction of the sphincter pupilla in the rabbit, the contraction of the stomach evoked by vagal stimulation in the cat, and vasodilation and plasma extravasation as well as the vasodilation in the dental pulp which is brought about by stimulation of cranial nerves in the rat.[49] It is conceivable that these various effects are brought about by the activation of different receptor types, but the antagonists appear to be unable to discriminate between NK-1, NK-2, and NK-3 receptors; in other words, the majority of the undeca- and octapeptide antagonists do not appear to be selective for one or the other receptor type. Recent studies performed in our laboratory following the discovery of monoreceptor systems and of selective agonists for the NK-1, NK-2, and NK-3 receptors have given results that support the above conclusion. This is shown in Table 2, where three octapeptide antagonist have been tested in the dog carotid artery (D.C.A.), rabbit pulmonary artery (R.P.A.), and the rat portal vein (R.P.V.) against the three natural neurokinins and some selective agonists. The data in Table 2 indicate that all three compounds are more active on the NK-1 than on the other receptors by one log unit or more.

The results summarized in Table 2 are representative of the data obtained in numerous studies with SP antagonists and indicate that these types of compounds are of little value for receptor classification. In fact, these compounds are even too weak on the NK-1 receptor; their affinity is two to three log units lower than that of the natural neurokinins in practically all pharmacological tests, including the three tissues utilized for the experiments summarized in Table 2. Neurokinin antagonists are more active on the NK-1 receptor by at least one log unit than on the other receptors. They do not discriminate at all between NK-2 and NK-3 receptors.

VII. NEUROKININ RECEPTORS IN THE GASTROINTESTINAL TRACT

Substance P and neurokinins appear to be present in every part of the gastrointestinal tract, including the exocrine glands. According to Pernow,[2] "SP is almost entirely connected to the nervous structures...of both the myenteric and submucous plexus from all intestinal segments." Neurokinins subserve practically all relevant intestinal functions, and their receptors are involved in the regulation of intestinal blood flow, exocrine secretion, and motility. All receptors are represented. NK-1 receptors appear to be predominant in the salivary glands to promote vasodilation and salivary secretion in the rat[37] and other species. NK-1 and NK-2 receptors are present in the rat stomach and appear to subserve different functions; both NK-1 and NK-2 are in the muscle and mediate contraction of the fundus and antrum, while NK-2 receptors are localized in the secretory glands where they reduce acid secretion and mediate the protective effect of neurokinins on the stomach mucosa.[56] NK-1 and NK-2 receptors have been demonstrated in the rat duodenum both *in vitro*[37] and *in vivo*.[46] According to Maggi et al.,[46] NK-1 receptors mediate the rapid initial contraction of the rat duodenum in response to intravenously applied substance P, while NK-2 functional sites are responsible for the tonic prolonged contraction that follows the initial spike. NK-2 receptors are also present in the biliary duct and the gallbladder of several mammals.[2]

Substance P and related neurokinins are present in the pancreas and stimulate exocrine secretion when applied intravenously to anesthetized dogs. Bicarbonates, amylase, total protein content, and the volume of the pancreatic flow are markedly increased by substance P,[57] which appears to act through activation of guanylate cyclase.[58] Neurokinins also act on endocrine pancreatic functions by inhibiting both the release of insulin induced by glucose or arginine[59] and the release of glucagon.[60] On the choleresis in the dog, SP exerts an inhibitory effect similar to somatostatin.[61]

The pharmacological effect and the possible physiological role of neurokinins on the intestine has mainly been studied in the guinea pig. *In vitro*, the guinea pig ileum has been used extensively by practically all investigators in the field of substance P. The three neurokinin receptors are all present in the tissue and contribute to activation of intestinal movements. Thus, NK-1 receptors are present in the smooth muscle membrane and are responsible for the contraction of the ileum *in vitro* and possibly for the peristaltic movements *in vivo*. NK-3 receptors are localized in the parasympathetic nerve endings to promote the release of acetylcholine.[39,40] Both NK-1 and NK-3 receptors have been demonstrated by the use of selective agonists.[19,20,36] The presence of NK-2 receptors, which appear to contribute to contraction, has been suggested on the basis of positive results obtained with NK-2 selective agonists, such as $[Nle^{10}]NKA_{4-10}$ and $[\beta\text{-Ala}^8]\text{-}NKA_{4-10}$.[36,45] It still has not been established whether SP and related neurokinins have receptors on the sensory nerves from which the reflexogenic arc is initiated for peristaltic movements. It has not been excluded that the neurokinin receptor may be localized in the sensory nerves, but it has not yet been identified.

Neurokinin receptors are present in the colon of various species, including men.[43,62] SP and related neurokinins may be involved in the pathogenesis of colonic diseases such as the Hirschsprung's disease[63] and other diseases.[2] Whether the lesion is due to changes in the content of the neurokinins or to variations in their receptor number remains to be established.

VIII. CONCLUSIONS

A large number of neuropeptides have been found in the intestine. In mammals, these neuropeptides are contained in nervous structures and presumably act as neurotransmitters and neuromodulators. The neurokinin SP and its homologue NKA are well represented and have been shown to participate in the regulation of gastrointestinal functions by the intermediacy of specific receptors.

Neurokinins find three different receptor types in the gut. NK-1 receptors particularly sensitive to substance P are found in the intestinal smooth muscles and in a variety of cells (mastocytes, macrophages, secretory cells of the glands, etc.) within the intestinal wall. NK-1 receptors are to be found also in vascular endothelia, where they mediate the release of EDRF, which is a potent vascular smooth muscle relaxant. NK-2 receptors are to be found on the smooth muscles, where they act as stimulants, and on the secretory glands (for instance, in the stomach), where they inhibit gastric secretion. NK-3 receptors have been identified in myenteric neurons, where they activate the release of neurotransmitters, particularly acetylcholine, and in this way promote intestinal motility and possibly intestinal secretions.

Because of the complexity of intestinal smooth muscles, these preparations, which were used for decades as assay organs for neurokinins, are now being progressively abandoned, and they are being replaced by more selective monoreceptor systems such as the dog carotid artery (NK-1), the rabbit pulmonary artery (NK-2) and the rat portal vein (NK-3). These organs have been used successfully to identify selective neurokinin agonists which provide new tools for the elucidation of the physiology and pharmacology of substance P and related peptides.

ACKNOWLEDGMENTS

The secretarial work of Mrs. Cécile Théberge is gratefully acknowledged. Experimental work quoted in this paper has been performed by the authors in collaboration with graduate students (N. Rouissi, C. Tousignant, D. Jukic, and S. Télémaque), summer students (F. Nantel and A. Madarnas), and technicians (M. Boussougou, R. Laprise, and M. Battistini), with the financial support of the Medical Research Council of Canada (MRCC) and the Quebec Heart Foundation. D.R. is a Career Investigator of the MRCC and S.D. is a student of the Fonds de la Recherche en Santé du Québec.

REFERENCES

1. **Von Euler, U. S.,** Untersuchugen über Substanz P, die atropinfeste, darmerregende und gefasserweiternde Substanz aus Darm und Gehirn, *Naunyn-Schmiedeberg's Arch. Exp. Pathol. Pharmakol.*, 181, 181, 1936.
2. **Pernow, B.,** Substance P, *Pharmacol. Rev.*, 35, 85, 1983.
3. **Bartho, L. and Holzer, P.,** Commentary. Search for a physiological role of substance P in gastrointestinal motility, *Neuroscience*, 16, 1, 1985.

4. **Pernow, B.,** Pharmacology of substance P, *Ann. N.Y. Acad. Sci.,* 104, 393, 1963.
5. **Milenov, K., Oehme, P., Bienert, M., and Bergmann, J.,** Effect of SP on mechanical and myoelectrical activities of stomach and small intestines in conscious dog, *Arch. Int. Pharmacodyn. Ther.,* 233, 251, 1978.
6. **Vogler, K., Haefely, W., Hürlimann, A., Studer, R. O., Lergier, W., Strassle, R., and Berneis, K. H.,** A new purification procedure and biological properties of substance P, *Ann. N.Y. Acad. Sci.,* 104, 378, 1963.
7. **Starke, K., Lembeck, F., Lorenz, W., and Weiss, U.,** Gallen- und Pankreas-Sekretion unter Substanz P und einem Physalamin-Derivat, *Naunyn-Schmiedeberg's Arch. Exp. Pathol. Pharmacol.,* 260, 269, 1968.
8. **Hallberg, D. and Pernow, B.,** Effect of substance P on various vascular beds in the dog, *Acta Physiol. Scand.,* 93, 277, 1975.
9. **Furness, J. B. and Costa, M.,** *The Enteric Nervous System,* Churchill Livingstone, Edinburgh, 1987.
10. **Liang, T. and Cascieri, M. A.,** Substance P receptor on parotid cell membranes, *J. Neurosci.,* 1, 1133, 1981.
11. **D'Orléans-Juste, P., Dion, S., Mizrahi, J., and Regoli, D.,** Effects of peptides and non-peptides on isolated arterial smooth muscles: role of endothelium, *Eur. J. Pharmacol.,* 114, 9, 1985.
12. **D'Orléans-Juste, P.,** personal communication.
13. **Erspamer, V.,** The tachykinin peptide family, *Trends Neurosci.,* 4, 267, 1981.
14. **Lee, C. M., Iversen, L. L., Hanley, M. R., and Sandberg, B. E. B.,** The possible existence of multiple receptors for substance P, *Naunyn-Schmiedeberg's Arch. Pharmacol.,* 318, 281, 1982.
15. **Nawa, H., Kotani, H., and Nakanishi, S.,** Tissue-specific generation of two preprotachykinin mRNAs from one gene by alternative RNA splicing, *Nature, (London),* 312, 729, 1984.
16. **Regoli, D., D'Orléans-Juste, P., Drapeau, G., Dion, S., and Escher, E.,** Pharmacological characterization of substance P antagonists, in *Tachykinin Antagonists,* Hakanson, R. and Sundler, F., Eds., Elsevier, Amsterdam, 1985, 277.
17. **Regoli, D., Drapeau, G., Dion, S., and D'Orléans-Juste, P.,** Minireview, pharmacological receptors for substance P and neurokinins, *Life Sci.,* 40, 109, 1987.
18. **Buck, S. H., Burcher, E., Shults, C. W., Lovenberg, W., and O'Donahue, T. L.,** Novel pharmacology of substance K binding sites: a third type of neurokinin receptor, *Science,* 226, 987, 1984.
19. **Drapeau, G., D'Orléans-Juste, P., Dion, S., Rhaleb, N. - E., Rouissi, N., and Regoli, D.,** Selective agonists for substance P and neurokinins, *Neuropeptides,* 10, 43, 1987.
20. **Regoli, D., Drapeau, G., Dion, S., and Couture, R.,** New selective agonists for neurokinin receptors: pharmacological tools for receptor characterization, *Trends Pharmacol. Sci.,* 9, 290, 1988.
21. **Henry, J. L., Couture, R., Cuello, A. C., Pelletier, G., Quirion, R., and Regoli, D., Eds.,** *Substance P and Neurokinins, Montreal 1986,* Springer-Verlag, New York, 1987.
22. **Gaddum, J. H. and Schild, H.,** Depressor substances in extracts of intestine, *J. Physiol. (London),* 83, 1, 1934.
23. **Erspamer, V. and Melchiorri, P.,** Active polypeptides of the amphibian skin and their synthetic analogues, *Pure Appl. Chem.,* 35, 463, 1973.
24. **Mizrahi, J., Couture, R., Caranikas, S., and Regoli, D.,** Pharmacological effects of peptides on tracheal smooth muscle, *Pharmacology,* 25, 39, 1982.
25. **Hedqvist, P. and Euler, U. S.,** Influence of substance P on the response of guinea pig ileum to transmural nerve stimulation, *Acta Physiol. Scand.,* 95, 341, 1975.
26. **Beaujouan, J. C., Torrens, Y., Herbet, A., Daguet, M. C., Glowinski, J., and Prochiantz, A.,** Specific binding of an immunoreactive and biologically active ^{125}I-labeled substance P derivative to mouse mesencephalic cells in primary culture, *Mol. Pharmacol.,* 22, 48, 1982.
27. **Beaujouan, J. C., Torrens, Y., Viger, A., and Glowinski, J.,** A new type of tachykinin binding site in the rat brain characterized by specific binding of a labelled eledoisin derivative, *Mol. Pharmacol.,* 26, 248, 1984.
28. **Cascieri, M. A. and Liang, T.,** Characterization of the substance P receptor in rat brain cortex membranes and the inhibition of radio-ligand binding by guanine nucleotides, *J. Biol. Chem.,* 258, 5158, 1983.
29. **Gaddum, J. H. and Szerb, J. C.,** Assay of substance P on goldfish intestine in a microbath, *Br. J. Pharmacol.,* 17, 451, 1961.
30. **Vane, J. R.,** A sensitive method for the assay of 5-hydroxytryptamine, *Br. J. Pharmacol.,* 12, 344, 1957.
31. **Bérubé, A., Marceau, F., Rioux, F., and Regoli, D.,** The rabbit mesenteric vein: a specific bioassay for substance P, *Can. J. Physiol. Pharmacol.,* 56, 603, 1978.
32. **Couture, R., Gaudreau, P., St.-Pierre, S., and Regoli, D.,** The dog common carotid artery: a sensitive bioassay for studying vasodilator effects of substance P and of kinins, *Can, J. Physiol. Pharmacol.,* 58, 1234, 1980.
33. **Regoli, D., Drapeau, G., Dion, S., and D'Orléans-Juste, P.,** Receptors for neurokinins in peripheral organs, in *Substance P and Neurokinins,* Henry, J. L., Couture, R., Cuello, A. C., Pelletier, G., Quirion, R., and Regoli, D., Eds., Springer-Verlag, New York, 1987, 99.

34. **D'Orléans-Juste, P., Dion, S., Drapeau, G., and Regoli, D.,** Different receptors are involved in the endothelium-mediated relaxation and the smooth muscle contraction of the rabbit pulmonary artery in response to substance P and related neurokinins, *Eur. J. Pharmacol.,* 125, 37, 1986.

35. **Mastrangelo, D., Mathison, R., Huggel, H. J., Dion, S., D'Orléans-Juste, P., Rhaleb, N. E., Drapeau, G., Rovero, P., and Regoli, D.,** The rat isolated portal vein: a preparation sensitive to neurokinins, particularly neurokinin B, *Eur. J. Pharmacol.,* 134, 321, 1986.

36. **Dion, S., D'Orléans-Juste, P., Drapeau, G., Rhaleb, N. E., Rouissi, N., Tousignant, C., And Regoli, D.,** Characterization of neurokinin receptors in various isolated organs by the use of selective agonists, *Life Sci.,* 41, 2269, 1987.

37. **Maggi, C. A., Guiliani, S., Santicioli, P., Regoli, D., and Meli, A.,** Mini-review. Peripheral effects of neurokinins: functional evidence for the existence ofmultiple receptors, *J. Autonom. Pharmacol.,* 7, 11, 1987.

38. **Dion, S., Drapeau, G., Rhaleb, N. E., D'Orléans-Juste, P., and Regoli, D.,** Receptors for substance P and neurokinins, correlation between binding and biological activities, *Eur. J. Pharmacol.,* 138, 125, 1987.

39. **Regoli, D., Mizrahi, J., D'Orléans-Juste, P., and Escher, E.,** Receptors(s) for substance P: single or multiple type?, in *Substance P,* Skrabanek, P. and Powell, D., Eds., Boole Press, Dublin, 1983, 10.

40. **Laufer, R., Wormser, U., Friedman, Z. Y., Gilon, C., Chorev, M., and Selinger, Z.,** Neurokinin B is a preferred agonist for a neuronal substance P receptor and its action is antagonized by enkephalin, *Proc. Natl. Acad. Sci. U.S.A.,* 82, 7444, 1985.

41. **Dion, S., Corcos, J., Carmel, M., Drapeau, G., and Regoli, D.,** Substance P and neurokinins as stimulants of the human isolated urinary bladder, *Neuropeptides,* 11, 83, 1988.

42. **Advenier, C., Naline, E., Drapeau, G., and Regoli, D.,** Relative potencies of neurokinins in guinea pig trachea and human bronchus, *Eur. J. Pharmacol.,* 139, 133, 1987.

43. **Rouissi, N., Dion, S., Drapeau, G., and Regoli, D.,** unpublished.

44. **Bergstrom, L., Beaujouan, J. C., Torrens, Y., Saffroy, M., Glowinski, J., Lavielle, S., Chassaing, G,. Marquet, A., D'Orléans-Juste, P., Dion, S., and Regoli, D.,** ³H-neurokinin A labels a specific tachykinin binding site in the rat duodenal smooth muscle, *Mol. Pharmacol.,* 32, 764, 1987.

45. **Dion, D., Rhaleb, N. E., Drapeau, G., and Regoli,D.,** unpublished results.

46. **Maggi, C. A., Giuliani, S., Manzini, S., Santicioli, P., and Meli, A.,** Motor effect of neurokinins on the rat duodenum: evidence for the involvement of substance K and substance P receptors, *J. Pharmacol. Exp. Ther.,* 238, 341, 1986.

47. **Folkers, K., Horig, J., Rosell, S., and Bjorkroth, U.,** Chemical design of antagonists of substance P, *Acta Physiol. Scand.,* 111, 505, 1981.

48. **Leander, S., Hakanson, R., Rosell, S., Folkers, K., Sundler, F., and Tornqvist, K.,** A specific substance P antagonist blocks smooth muscle contractions induced by non-cholinergic, non-adrenergic nerve stimulation, *Nature (London),* 294, 467, 1981.

49. **Rosell, S. and Folkers,** Substance P antagonists: a new type of pharmacological tool, *Trends Pharmacol. Sci.,* 3, 211. 1982.

50. **Mizrahi, J., Escher, E., Caranikas, S., D'Orléans-Juste, P., and Regoli, D.,** Substance P antagonists active *in vitro* and *in vivo, Eur. J. Pharmacol.,* 82, 101, 1982.

51. **Regoli, D., Escher, E., Drapeau, G., D'Orléans-Juste, P., and Mizrahi, J.,** Receptors for substance P. III. Classification by competitive antagonists, *Eur. J. Pharmacol.,* 97, 179, 1984.

52. **Folkers, K., Rosell, S., Hakanson, R., Chu, J. Y., Lu, L. A., Leander, S., Tang, P. F. L., and Ljungqvist, A.,** Chemical design of antagonists of substance P, in *Tachykinin antagonists,* Hakanson, R. and Sundler, F., Eds., Elsevier, Amsterdam, 1985, 259.

53. **Escher, E., Parent, P., Mizrahi, J., D'Orléans-Juste, P., Dion, S., Drapeau, G., and Regoli, D.,** Synthesis and structure-activity studies of novel substance P antagonists, an exercise in medicinal chemistry, in *Tachykinin Antagonists,* Hakanson, R. and Sundler, F., Eds., Elsevier, Amsterdam, 1985, 267.

54. **Folkers, K., Rosell, S., Xu, J. C., Bjorkroth, U., Lu, Y. A., and Liu, Y. Z.,** Antagonists of substance P from emphasis on position 11, *Acta Chem. Scand., Ser. B,* 37, 623, 1983.

55. **Vaught, J. L.,** Substance P antagonists and analgesia: a review of the hypothesis, *Life Sci.,* 43, 1419, 1988.

56. **Evangelista, S., Lippe, I. Th., Rovero, P., Maggi, C. A., and Meli, A.,** Tachykinins protect against ethanol-induced gastric ulcers in rats, *Regul. Peptides,* 22, 66, 1988.

57. **Thulin, L. and Holm, I.,** Effect of substance P on the flow of hepatic bile and pancreatic juice, in *Substance P, Nobel Symp.* 37, von Euler, U. S. and Pernow, B., Eds., Raven Press, New York, 1977, 247.

58. **Sjodin, L., Brodin, E., Nilsson, G., and Conlon, T. P.,** Interaction of substance P with dispersed pancreatic acinar cells from the guinea pig. Bindig of radioiodinated peptide, *Acta Physio. Scand.,* 109, 97, 1980.

59. **Lundqvist, I., Sundler, F., Ahren, B., Alumets, J., and Hakanson, R.,** Somatostatin, pancreatic polypeptide, substance P, and neurotensin: cellular distribution and effects on stimulated insulin secretion in the mouse, *Endocrinology,* 104, 832, 1979.

60. **Efendic, S., Luft, R., and Pernow, B.,** Effect of substance P on arginine-induced insulin and glucagon release from the isolated perfused rat pancreas, in *Substance P, Nobel Symp. 37,* von Euler, U. S. and Pernow, B., Eds., Raven Press, New York, 1977, 241.

61. **Holm, I., Thulin, L., and Hellgren, M.,** Anticholeretic effect of substance P in anesthetized dogs, *Acta Physiol. Scand.,* 102, 274, 1978.

62. **Couture, R., Mizrahi, J., Regoli, D., and Devroede, G.,** Peptides and the human colon: an *in vitro* pharmacological study, *Can. J. Physiol. Pharmacol.,* 59, 957, 1981.

63. **Ehrenpreis, T. and Pernow, B.,** On the occurrence of substance P in the recto sigmoid in Hirschprung's disease, *Acta Physiol. Scand.,* 17, 380, 1952.

64. **Lavielle, S., Chassaing, G., Ploux, O., Loeuillet, D., Besseyre, J., Julien, S., Marquet, A., Convert, O., Beaujouan, J. C., Torrens, Y., Bergstrom, L., Saffroy, M., and Glowinski, J.,** Analysis of tachykinin binding site interactions using constrained analogues of tachykinins, *Biochem. Pharmacol.,* 37, 41, 1988.

65. **Beaujouan, J. C.,** personal communication.

Chapter 7

RECEPTORS FOR NEUROPEPTIDES: LIGAND BINDING STUDIES

Sultan Ahmad, H.-D. Allescher, and C.-Y. Kwan

TABLE OF CONTENTS

I.   Introduction 210

II.   Purification of Biological Membranes 210

III.   Peptide Receptors in Gastrointestinal Tract 212
        A.   General Considerations 212
        B.   Neurotensin 214
        C.   Opioids 215
        D.   Neurokinins 218
        E.   Vasoactive Intestinal Polypeptide 219
        F.   Cholecystokinins 221
        G.   Somatostatin 222

IV.   Conclusion 223

References 223

I. INTRODUCTION

Perhaps the most important factor in rendering ligand binding studies so popular and common is the simplicity of their execution. A radioactive tag (usually ^{3}H or ^{125}I) is attached to the ligand, which is then incubated with tissue slices, isolated cells, purified biological membranes, or solubilized preparations under appropriate conditions. Bound ligand is then separated from unbound ligand by centrifugation, filtration, simple washing (in the case of tissue slices), or column chromatography, etc. and radioactivity of the bound ligand is enumerated in a counter. Specificity, reversibility, saturability, and other kinetic parameters of the binding sites for the ligand, in general, can easily be obtained using this simple technique.

The present chapter is not aimed at elucidating the advantages/disadvantages of ligand binding studies, moreover, a detailed description of the methodology of these studies is also beyond the scope of this article. However, since most of the binding studies are performed using biological membranes preparations, we feel that it is of relevance to discuss the importance of the care that should be exercised in preparation and use of such membranes before any interpretation of the data is attempted.

II. PURIFICATION OF BIOLOGICAL MEMBRANES

Why does one need a purified preparation of membranes for binding studies? There are two obvious reasons. One is to remove contaminating macromolecules and organelles (nucleus, endoplasmic reticulum, mitochondria, etc.) from the preparation so that a better signal-to-noise ratio is obtained, yielding reproducible and more trustworthy results. The second reason is to obtain the preparation of the relevant membranes from a complex homogenate which, even after careful dissection, may contain membranes from several different cell types (e.g., smooth muscle cells, neuronal elements, glial cells, interstitial cells, etc.) so that the appropriate conclusion regarding the actual locus and density of the binding sites can be reached.

In so far as possible, it is desirable to begin by reducing the number of cell types present (or enhancing the proportion of the desirable cell types) by dissection. Another approach is to obtain a single cell type by culture techniques, but culturing some cell types has not as yet been possible. Furthermore, cultured cells may have altered properties, not to mention the problems associated with the culture technique itself. Once the cellular heterogeneity has been simplified as much as feasible, the residual tissue is homogenized. The time and force of homogenization may vary depending upon the tissue, and this also must be standardized for every tissue.

The membrane purification procedure involves a series of centrifugation steps to separate membranes and particles according to their weight. This is usually followed by further purification according to the buoyant density of individual membranes with density gradient centrifugation. Membranes of similar densities but diverse origins may not be separated over the density gradient. In such cases one must be content with the quantification of the contaminating membranes. Various fractions obtained during this procedure are therefore evaluated as to the content of the specific markers for different cell and membrane types. Recently, Kwan[1] has critically evaluated the current methods of membrane fractionation for various smooth muscle types.

Markers for membranes from smooth muscle (5'-nucleotidase, EC 3.1.3.5), mitochondria (cytochrome C oxidase, EC 1.9.3.1), and endoplasmic reticulum (NADPH cytochrome C reductase, EC 1.6.99.3) are relatively good, so identification and reasonable separation of these membranes is possible. However, since the starting homogenate may contain membranes from several cell types, it is essential to assess and separate these membranes. Until recently, not much effort has been devoted to addressing this problem. Since many smooth muscles along the GI tract are highly innervated, one major hinderance to progress in this aspect has been the meager knowledge of contamination by neuronal membranes due to the lack of specific markers for neuronal membranes.

Recently, specific binding of the radioactively labeled sodium channel blocker [3H]saxitoxin has been used to identify neuronal membranes in the homogenates of gastrointestinal tissue and to assess neuronal contamination in isolated smooth muscle membrane preparations.[2] Using this approach, it became possible to separate the circular smooth muscle plasma membranes and deep muscular plexus neuronal membranes from the circular muscle homogenate,[2] longitudinal smooth muscle plasma membranes and myenteric plexus neuronal membranes from the homogenates of longitudinal muscle/myenteric plexus preparation,[3] and neuronal membranes of the submucous plexus from homogenates of submucosa and muscularis mucosa of canine small intestine.[4] The procedure first required extensive dissection to separate various layers of the intestinal wall, i.e., myenteric plexus longitudinal muscle layer, circular muscle containing deep muscular plexus, and submucosa containing submucosal plexus.[2-4] Except for submucous plexus membranes, which are relatively well purified on differential centrifugation, all other preparations required sucrose density gradient centrifugation[5] (see also Figure 1). Smooth muscle plasma membranes sedimented between 14 and 33% (w/v) sucrose on the gradient, while neuronal membranes concentrated at a higher density, i.e., >35% sucrose. This observation was further validated by the measurement of immunoreactive vasoactive intestinal polypeptide (VIP-IR) as a measure of intact synaptosomes. An excellent correlation was obtained between saxitoxin binding and VIP-IR content, while the correlation between 5'-nucleotidase and saxitoxin binding and 5'-nucleotidase and VIP-IR was extremely poor (see Figure 2). Using these highly purified and well-defined membranes, a number of peptide and nonpeptide receptors as well as ion channels in these tissues have been characterized recently.[2-6] Furthermore, this approach has been extended to vascular[7] and myometrial[8] tissues. In rat myometrium, the specific binding of saxitoxin decreased by about 85% at day 22 of pregnancy, when natural denervation of the tissue is known to take place.[9] Shi et al.[10] have also shown that the dog aorta, which has very few nerves, contains few [3H]saxitoxin-binding sites compared to mesenteric arteries and veins and saphenous arteries. Dog mesenteric nerves had a high density of saxitoxin binding.[10] These observations further reinforce the validity of saxitoxin as a neuronal marker. Thus, it is clear that not only can saxitoxin be used as an excellent marker for neuronal membranes, but also it is possible to separate these membranes from those of smooth muscle by construction of appropriate differential centrifugation and sucrose density gradient schemes. So far it is uncertain that the density of saxitoxin binding is uniform over all components of neuron or that no other cell types (e.g., glia) possess saxitoxin binding sites.

Two more classes of cells in the gastrointestinal tract, for which little biophysical and biochemical information is available, are the glial cells and the interstitial cells of Cajal (ICCs). With increasing evidence for their possible role in gastrointestinal physiology, especially that of ICCs,[11] it becomes even more important to identify them in the homogenate. A good candidate marker for glial cells is the glial fibrillary acidic protein (GFAP).[12] Antibodies against GFAP are now available which could be used to label the glial cells in the tissue. However, since GFAP is not a membrane-bound protein, this marker has limitations for use in membrane purification. ICCs still remain the mysterious members of unidentified cell types. The only information available about them is based on some electrophysiological and morphological experiments;[11] practically nothing is known regarding their biophysical and biochemical properties. Nevertheless, with the use of available markers for other cell types, it may be possible to identify them on the basis of a probability elimination process. At the moment we are actively pursuing this aspect of identifying ICCs of the canine colon.

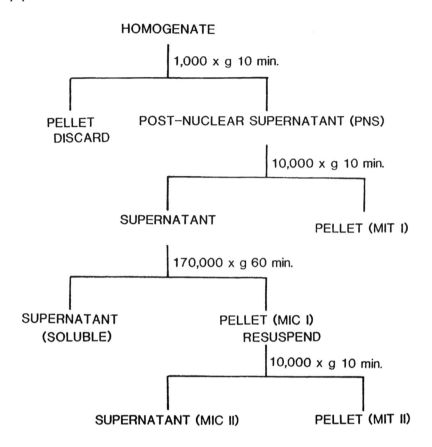

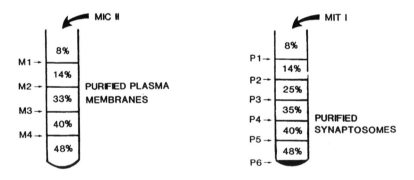

FIGURE 1. Scheme for the processing of the circular smooth muscle of canine small intestine to obtain purified plasma membranes from circular smooth muscle and purified synaptosomes from the deep muscular plexus. (From Ahmad, S., Rausa, J., Jang, E., and Daniel, E. E., *Biochem. Biophys. Res. Commun.*, 159, 119, 1989. With permission.)

III. PEPTIDE RECEPTORS IN GASTROINTESTINAL TRACT

A. GENERAL CONSIDERATIONS

Receptors are macromolecules on cell membranes or inside the cell that, upon binding the agonist, initiate a chain of events culminating in the biological response. Ligand binding studies give us information regarding only the first part of the receptors definition, i.e., binding and binding properties. Therefore, in the strict sense, one should defer from assuming

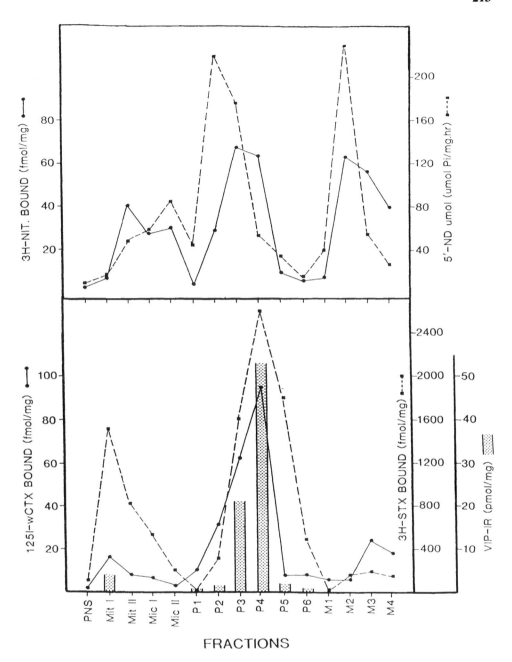

FIGURE 2. The distribution of the smooth muscle plasma membrane marker enzyme 5'-nucleotidase (5'-ND), neuronal marker [³H]-saxitoxin binding ([³H]STX), vasoactive intestinal polypeptide immunoreactivity (VIP-IR) as a measure of intact synaptosomes, and the binding of [³H]nitrendipine ([³H]NIT) and [¹²⁵I]ω-conotoxin ([¹²⁵I]ω-CTX) to the fractions obtained after differential and sucrose density gradient centrifugation of the homogenate of circular smooth muscle of canine small intestine (see Figure 1). (From Ahmad, S., Rausa, J., Jang, E., and Daniel, E. E., *Biochem. Biophys. Res. Commun.*, 159, (1), 119, 1989. With permission.)

these "binding sites" to be receptors until their physiological significance is established. For the sake of simplicity, in the following sections we have used the term "receptor" for all the binding sites whether or not their physiological significance is established.

B. NEUROTENSIN

In agreement with the direct action of neurotensin (NT) on smooth muscle and the indirect action via nerves, NT receptors have been recognized on smooth muscles and nerves of the gastrointestinal tract of rats, guinea pigs, and dogs.[13-16] Autoradiographic studies with [³H]NT have demonstrated the presence of NT receptors on circular smooth muscle and to a lesser extent on longitudinal smooth muscle of guinea pig ileum.[14] Using [³H]NT and the crude membrane preparation[14] or the homogenate of isolated longitudinal smooth muscle cells,[13] the equilibrium constant (K_d) of NT was measured to be 3.4 and 4.0 nM, respectively. The obvious experimental difficulty in these studies was the use of tritium-labeled neurotensin with low specific activity (56 to 77 Ci/mmol). Such a low specific activity combined with the low capacity (B_{max}) of binding sites may have precluded the detection of high-affinity binding sites in these preparations.

Mazella et al.[17] reported iodination of a neurotensin analogue, Trp¹¹-NT. Since iodination at Tyr¹¹ causes a drastic reduction in the biological potency of the peptide, selective iodination of Trp¹¹-NT or at Tyr³ yielded radiolabeled peptide with high potency. Using this high-specific-activity iodinated derivative of Trp¹¹-NT, Kitabgi et al.[15] characterized NT receptors on the purified membrane preparation from rat gastric fundus. Clearly, high-affinity and low-affinity sites were recognized.

Iodinated Trp¹¹-NT served as a good label for NT receptors in murine species; however, guinea-pig[18] and human brain[19] NT receptors exhibited low affinity for this label. Sadoul et al.[19] reported the iodination and chromatographic separation of native NT selectively iodinated at the Tyr³ residue. Using [¹²⁵I]Tyr³-NT, NT receptors on canine small intestinal circular smooth muscle plasma membranes have been characterized.[16] NT receptors have been observed[16] and characterized on the deep muscular plexus neuronal membranes and submucous plexus membranes[150] as well. Little or no binding is present in the myenteric plexus and the longitudinal smooth muscle plasma membranes.

In all studies performed in the gastrointestinal tract with iodinated neurotensin, NT receptors are present as a heterogeneous population of binding sites demonstrating high affinity/low capacity and low affinity/high capacity. The guanosine 5′-triphosphate (GTP) analogue Gpp(NH)p modulated binding,[16] as did sodium ions,[15,16] indicating the involvement of GTP-binding proteins in the signal transduction mechanism. Structure-activity studies have revealed that carboxy-terminal residues (8 to 13) of the peptide are essential for binding activity.[15,16] In fact NT_{8-13} had approximately three times higher affinity than NT itself. Replacement of Tyr¹¹ by Trp¹¹ did not affect the affinity in membranes from rat fundus,[15] while in canine small intestine it reduced affinity to 10% of native NT.[16] Phe¹¹-NT had an affinity about 10% that of NT in canine small intestine.[16] The affinity of D-Trp¹¹-NT was also drastically low (0.34% as compared to NT) in rat fundus tissue[15] and even more so in the canine intestinal smooth muscle (0.04% as compared to NT).[16]

Covalent labeling with a photoreactive NT analogue or by cross-linking reagent and subsequent sodium dodecyl sulfate polyacrylamide electrophoresis (SDS-PAGE) of rat fundus NT receptors revealed a specifically labeled receptor protein band of molecular weight 110,000 Da.[20] This molecular weight corresponds well with that determined by target size analysis by radiation inactivation,[21] which was calculated to be 103,000 Da. The molecular weight of rat brain NT receptors obtained by radiation inactivation was 108,000 Da.[21] However, rat brain NT receptors after covalent labeling moved on SDS-PAGE as two species of M_r 49,000 and 51,000 Da.[22]

Our unpublished observations suggest there are differences in the target size of NT receptors on canine small intestine smooth muscle plasma membranes and deep muscular plexus membranes, as calculated by the radiation inactivation technique.[151] Smooth muscle plasma membrane receptors were inactivated at lower radiation doses, yielding a target size of approximately 200,000 Da, while deep muscular plexus NT receptors were inactivated with a target size of 100,000 Da. We further studied the role of disulfide bridges in such observed differences in molecular organization. Reduced sulfhydryl groups are required for the binding activity, since sulfhydryl reducing agents (dithiothreitol [DTT] and cysteine) augmented and the oxidant reduced specific binding.[152] These receptors appear to be different from those in rat mast cells, where DTT failed to influence binding.[23]

The antihistamine levocabastine is reported to inhibit NT binding to rat[24] and mouse brain membranes.[25] Levocabastine selectively inhibited neurotensin binding to low-affinity sites without affecting high-affinity sites.[25] We found no effect of levocabastine on NT binding to canine small intestine circular smooth muscle plasma membranes or to neuronal membranes from deep muscular plexus or submucous plexus.[153] Several other nonmurine tissues are also known to be insensitive to levocabastine.[26] More extensive screening of several tissues from different species with regard to the action of levocabastine would provide interesting information about the biological evolution of the levocabastine-sensitive sites of NT receptors.

In summary, neurotensin receptors in the gastrointestinal tract demonstrate heterogeneity in their recognition properties, affinities, densities, and structural molecular organization. Solubilized and purified NT receptors from mouse[25] and bovine brain[27] also differ in their electrophoretic mobility. Extensive work at the molecular and genetic level needs to be done to account for these differences more clearly.

C. OPIOIDS

Morphine and morphine-related compounds have been known to affect gastrointestinal function since antiquity either by a direct pharmacological maneuver or as the result of disturbing side effects. In this chapter, the term "opioids" will be used in a broad sense to encompass this group of peptides and nonpeptides, endogenous or exogenous.

Much of the information regarding the molecular properties of opioid receptors is from the studies on the central nervous system and from neuronal cell lines. However, gastrointestinal bioassay systems have also provided some valuable information (see Chapter 11), and ligand binding studies have provided detailed insights into these macromolecules.

The concept of opioid receptor multiplicity was first suggested by Portoghese.[28] Martin et al.[29] and Gilbert et al.[30] obtained the first pharmacological evidence for the existence of opioid receptor subtypes in the central nervous system using chronic spinal dogs. The criteria for identification and characterization of opioid receptors and subtypes thereof have recently been reviewed by several workers.[31-35] The concept of multiple opioid receptors has been further strengthened by the development of opioid drugs highly selective for each subtype.[33] Some of the opioid drugs and their selectivity for receptor subtypes are listed in Table 1.

In the gastrointestinal tract, three major subtypes of opioid receptors — namely, mu (μ), delta (δ), and kappa (κ) — have been recognized. Subclassification of μ-receptors has also been suggested such that μ_1-sites bind both enkephalins and morphine with affinities higher than their respective preferred sites.[36] In guinea pig ileum, the μ-opioid receptors were recognized as the μ_2-subtypes, i.e., lower affinities for enkephalins and morphine.[37]

The presence of opioid receptors in the gastrointestinal tract using the ligand binding approach was first suggested by Terenius,[38] who reported the "specific uptake" of ^{3}H-labeled dihydromorphine by a crude membrane fraction from the guinea pig ileum longitudinal muscle (perhaps containing myenteric plexus) preparation. With a refinement of binding techniques, better characterization followed. Pert and Snyder,[39] using a crude membrane preparation, reported the presence of opioid receptors in guinea pig ileum myenteric

TABLE 1

Affinities of Various Opioids for μ, δ, and κ Subtypes of Opioid Receptors

Opioids	Apparent Ki (nM)		
	μ	δ	κ
Morphiceptin	20.00	30,000.00	>20,000.00
[N-Me-Phe^3, D-Pro] Morphiceptin	5.50	10,000.00	20,000.00
DAGO	3.60	700.00	5,000.00
Fentanyl	0.30	200.00	170.00
Syndiphalin	0.29	1,250.00	13,000.00
DTLET	25.30	1.35	
DSLET	31.00	4.80	
DTE$_{12}$	96.30	1.06	
ICI-154129	17,000.00	840.00	
Dynorphin A$_{1-17}$	0.73	2.38	10.12
	27.00	175.00	10.00
Dynorphin A$_{1-13}$	0.22	0.49	0.05
Ethylketocyclazocine	1.00	3.20	0.37
Etorphine	0.15	0.30	0.48
Diprenorphine	0.20	0.18	0.50

Constructed from Chang, K. J., in *The Receptors*, Vol. 1, Conn, P. M., Ed., Academic Press, Orlando, FL, 1985, 1. With permission.

TABLE 2

Opioid Binding to Canine Small Intestine[4,6]

	LM	CM	DMP	MP	SMP
Diprenorphine	ND	ND	0.18	0.12	0.09
K$_d$ (NM)					
B$_{max}$ (fmol/mg)			492	400	683
% Subtype					
μ (PLO-17)			40 — 45	40 — 45	64
δ (DPDPE)			40 — 45	40 — 45	24
κ (U-50488H)			10 — 15	10 — 15	12

Note: LM, longitudinal muscle; CM, circular muscle; DMP, deep muscular plexus; MP, myenteric plexus; SMP, submucous plexus; ND, not detected.

plexus, no binding to the longitudinal muscle could be observed. In 1975, Terenius[40] and Creese and Snyder[41] separately reported the detailed characterization of opioid receptors in guinea pig ileum longitudinal muscle/myenteric plexus. Good correlation was found between receptor binding and pharmacological potency of both agonists and antagonists.[41] Since then, specific opioid binding has been reported in several GI systems, including parietal cells from guinea pig ileum[42] and rat small intestine.[43] No binding, however, was found on rat intestinal epithelial cells.[44] Opioid receptors have even been demonstrated in the midgut of the insect *Leucophaea madierae*.[45]

Recently, some highly selective opioid drugs have been used to characterize opioid receptors in various highly purified membranes isolated from canine small intestine.[4,6] [^{3}H]-diprenorphine, a nonselective opioid antagonist ligand, bound with similar affinities to the deep muscular plexus, myenteric plexus,[6] and submucous plexus[4] (also see Table 2). Competition with subtype-selective opioid compounds revealed that both deep muscular and myenteric plexus contain about 40 to 45% μ- and δ-sites and 10 to 15% κ-sites. Submucous plexus had a higher proportion of μ-sites (64%) and fewer δ-sites (24%), and κ-sites again constituted a small proportion of binding sites (see Table 2). However, κ-sites were actually present since a relatively selective κ-ligand, [³H]ethylketocyclazosin, bound to both deep muscular plexus (DMP), and myenteric plexus (MP) at a high- and low-affinity site (curvilinear Scatchard plot). In the presence of high concentrations (0.5 μM each) of an unlabeled cyclic penicillin analogue of enkephalin, DPDPE, and a morphiceptin analogue, PLO-17, together to shield μ- and δ-subtypes, high-affinity sites could still be demonstrated.[6]

In autoradiographic studies, binding has been detected only in mucosa, deep muscular plexus, and submucous plexus of rat stomach[46] or in villi and crypts of rat small intestine.[47] Studies using isolated membranes have provided no evidence of high-affinity binding to either longitudinal or circular muscle,[6] nor to the muscularis mucosa[4] of canine small intestine. Nishimura et al.,[48] in their autoradiographic studies on rats and guinea pig, reported the presence of both μ- and δ-receptors in the circular muscle and the muscularis mucosa of gastric fundus. In rat ileum and duodenum, no binding sites could be detected in the muscle layers, while in guinea pigs only μ-receptors were diffusely present over muscular layers.[48] However, due to the limited resolution of light microscopic studies, the binding sites detected in muscular layers could be present on the axons/varicosities of myenteric, deep muscular, or submucous plexus. Autoradiographic studies with high resolution at the electron microscopic level need to be performed to resolve this issue. So far, ligand binding studies have presented no convincing evidence concerning the presence of opioid receptors on any muscle layer of the GI tract. Most of the opioid receptors found were primarily present on neuronal membranes. Such a lack of receptors on smooth muscle plasma membranes contrasts with some *in vivo* and *in vitro* functional studies suggesting a direct action of opioids on smooth muscles (see Chapter 5).

Opioid agonist and antagonist binding is differentially regulated by sodium ions. Agonist binding is decreased while antagonist binding is enhanced or not altered in the presence of sodium.[6,43,45,49,50] Among other alkali metal ions, lithium (although to a much lesser extent) also distinguished the antagonist vs. agonist binding.[50] The discrimination of agonist/antagonist binding correlates impressively with their agonist/antagonist properties.[50] It has been suggested that the two conformations of opioid receptors, i.e., agonist binding and antagonist binding, are at any given time in equilibrium and that sodium ions shift the equilibrium farther toward the antagonist-binding conformation.[50]

Various sulfhydryl active agents alter the binding properties of opioid agonists and antagonists in some non-GI systems,[51-56] and intact disulfide bridges are implicated for binding activity. The sulfhydryl groups are apparently located at or near the opioid binding sites, since preincubation of membranes with *N*-ethyl-maleimide (NEM) in the presence of opioid agonist or antagonist protects the receptors from inactivation by NEM.[56] Sodium ions also protect against inactivation, probably by rendering the thiol groups less accessible to the action of NEM. A difference in the sensitivity of opioid receptor subtypes to inactivation by NEM has also been observed.[56]

Opioid receptors from various nongastrointestinal sources have been purified and reconstituted.[57,58] When purified to homogeneity, both μ and δ opioid receptors were noted to have a molecular weight of 58,000 Da as observed over SDS-PAGE.[57-59] Differences in molecular weight have also been observed.[60-62] Sucrose density gradients of detergent solubilized preparation identified κ-binding activity at a sedimentation coefficient 19S, while μ and δ combined peak was identified at 34 to 39S.[63] δ-Receptors also appear to differe immunogenically, since μ- and δ but not binding was inhibited by a monoclonal antibody raised against opioid receptors.[64]

To summarize, the opioid receptors have been clearly demonstrated both functionally as well as biochemically in GI systems, at least on neuronal membranes. The status of opioid receptors on the smooth muscle cells of the gastrointestinal tract still remains controversial because most binding studies have been unable to demonstrate binding sites while some functional studies suggest the presence of opioid receptors on the smooth muscle cells of the gastrointestinal tract.

D. NEUROKININS

Substance P (SP) has received much attention due to a number of biological actions and its presence in both central and peripheral neurons.[65] However, only recently has it been realized that SP is one member of a family of structurally related neuropeptides, the neurokinins.

Several peptides structurally related and attributed to the family of neurokinins have been identified in mammalian and non-mammalian systems.[66] The biological activity of these peptides resides in C-terminal pentapeptide residues, which are conserved throughout this family of peptides. The canonical tachykinin sequence is Phe–X–Gly–Leu–Met–NH$_2$, where X is an aromatic or branched aliphatic amino acid.[66] Three known mammalian neurokinins (substance P, neurokinin A and neurokinin B [SP, NK-A, and NK-B, respectively]) are encoded by two distinct but remarkably similar genes.[66-67] The preprotachykinin A gene encodes for SP and NKA. Alternative splicing of this gene generates three mRNAs encoding three precursor proteins. Of the three precursor proteins, the minor species α-preprotachykinin gives rise to SP. Two other precursor proteins, β-and τ-preprotachykinins, contain the sequences for both SP and NK-A. The other gene preprotachykinin B encodes NK-B.[66,67]

At least a dozen mammalian neurokinins have also been identified.[66] The major non-mammalian neurokinins have also been identified.[66] The major nonmammalian neurokinins include physalaemin,[68] kassinin,[69] and eledoisin.[70] These nonmammalian neurokinins also have C-terminal structural homology with mammalian neurokinins.[66]

Initially, two subtypes of receptors, i.e., SP-P and SP-E, were suggested.[71] The SP-P subtype was considered as the one where relative potencies of SP, eledoisin, and physalaemin were similar. The SP-E subtype was described as the one where elesoisin and physalaemin were more potent than SP. As additional neurokinins and their analogues were discovered, the relative potency picture could not fit into this classification scheme. The problem was (and to a great extent still is) compounded by the fact that selective, high-potency antagonists were (and still are) not available.[72-74] However, designation of three classes of neurokinin receptors has become widely acceptable in the past few years. SP is most potent at the NK-1 receptor subtype. The preferred ligand for the NK-2 subtype is NK-A, and NK-B is most effective at the NK-3 subtype of neurokinin receptor.[72] Three peripheral bioassays provide good assay systems for the three subclasses of neurokinin receptors. Dog carotid artery, rabbit pulmonary artery, and rat portal vein are the selective assay systems used for NK-1, NK-2, and NK-3 receptor subtypes, respectively.[74] (see Chapter 6). Interestingly, the potencies of NKA and NKB in dog carotid artery, an SP-preferring site, are higher than those in their own respective preferred sites.[74] The presence of NK-4 receptor has also been suggested for guinea pig trachea.[74]

Using various radioactively labeled neurokinins, several groups have reported the presence of these receptors in the gastrointestinal tract of rat, guinea pig, and dog.

Manyth et al.,[75] using [^{3}H]SP, localized neurokinin receptors in the guinea pig gastrointestinal tract by autoradiographic studies. Heavy receptor densities were observed in muscularis mucosa of duodenum and circular muscle of ileum and colon. Lower densities were observed in other layers of the wall. Burcher et al.[76] described Bolton-Hunter (BH)-conjugated SP- and SK (NKA)-binding sites (NK-1 and NK-2 receptors) in the rat GI tract. In most regions, the binding of BHSP and BHSK were similar. Some differences in regional distribution were observed. Manyth et al.,[77] using several radioligands to label NK-1, NK-2, and NK-3 receptors, exensively studied the distribution of NK receptor subtypes in all the regions of canine GI tract. Differenetial distribution of NK-1 and NK-2 receptors was observed throughout the length of the GI tract. NK-1 receptors were more widely distributed than NK-2, while no NK-3 receptors were observed in any area.

Using classical ligand-binding assays, receptors for SP have been demonstrated in the guinea pig ileum longitudinal muscle/myenteric plexus preparation.[78-90] Buck et al.[78] showed a curvilinear Scatchard plot strongly suggestive of multiple binding sites (or negative cooperativity), while Watson et al.[79] found no evidence for multiplicity in the binding sites. The rank order of potency for neurokinins and analogues was quite similar in these two studies and reflected the displacement for the SP-P (NK-1) site. The displacement profile, however, did not match the functional potency ratio of these peptides. The possibility for the presence of a second site was suggested. Using three radioactively labeled ligands specific for different subclasses, Burcher et al.[80] showed the displacement pattern of all three NK receptor subclasses in guinea pig ileum. Rat duodenum, however, contained only NK-1 and NK-2 (according to the present categorization) sites. Binding of BH-conjugated eledoisin (BHE) to rat duodenum was also observed, but this binding site did not show typical SP-E (NK-3) displacement characteristics. BHE was likely binding to the NK-2 site.[80] Bergstrom et al.,[81] using [^{3}H]NKA, demonstrated the presence of specific binding in rat duodenum, clearly reminiscent of NK-2 sites. The possiblity of multiple neurokinin binding sites in canine small intestine has also been discussed.[82] In this study, although the rank order of potency in displacing [^{125}I]Tyr^8-SP was similar to that recognized for the NK-1 sits, the *in vitro* contractility data did not match this rank order of potency. The tissue used for the membrane preparation as well as the contractility experiments contained both smooth muscle and deep muscular plexus neurons.[16] During the membrane preparation, a majority of the contamination by neuronal membranes was removed.[6,16] Binding studies may, therefore, have reflected only the binding to smooth muscle plasma membranes (NK-1 receptors), while functional responses may have been the reflection of the response at NK-1 (smooth muscle) and the NK-2 (neuronal receptors).

Where is the locus of neurokinin receptors? Functional and electrophysiological studies suggest that receptors in the GI tract should be present both on smooth muscle cells and on neurons as well as on mucosal cells. Autoradiographic studies by Manyth et al.[77] demonstrated the presence of both NK-1 and NK-2 receptors in the circular muscle layer in many parts of the GI tract. NK-2 binding sites were present only on the myenteric plexus of jejunum while NK-1 binding was present only on the myenteric plexus in ileum and colon. No binding on the myenteric plexus in any other region was demonstrated. Differential distribution of NK-1 and NK-2 receptors in mucosa and muscularis mucosa along the length of the GI tract was also evident. Using the purified membranes from the circular muscle of canine small intestine, Muller et al.[82] reported the presence of NK-1 receptors in smooth muscle plasma membranes. Other reported binding studies have been performed in heterogeneous preparations of membranes, and therefore the locus cannot be ascribed to a particular cell type. The expression of certain subtypes of receptors by a given cell type has extreme physiological significance, especially in the case of SP and NK-A, where these peptides are encoded by the same gene, colocalized, and probably released together. In such a case, the presynaptic structure fails to confer the specificity of the response, and the specificity will therefore have to be determined at the postsynaptic site. That is to say, the presence or absence and the relative density of the receptor(s) located post-synaptically would determine the functional response. Therefore, it is important to determie the locus of binding sites as to a particular cell type. More careful binding studies are required to define the exact locus of neurokinin binding sites.

E. VASOACTIVE INTESTINAL POLYPEPTIDE

Vasoactive intestinal polypeptide (VIP) is a widely distributed neuropeptide, and VIP-immunoreactive nerves have been localized throughout the GI tract.[65] These nerves supply the circular smooth muscle and also the myenteric and submucous plexuses. The mucosa is also richly innervated in the small and large intestines. The gastric mucosa is, however, sparse in VIP-immunoreactive nerves.[65] VIP-immunoreactive nerves are also associated with small blood vessels of the GI tract, especially in mucosal and submucosal layers, throughout the GI tract.[65] Consistent with such a wide distribution of VIP in the GI tract, several studies have elaborated the functional role of VIP with regard to motility and secretion.[65]

Wide projection of VIP-containing nerves and a variety of functional responses where VIP has been implicated within the GI tract suggest multiple target cells for the action of VIP. The major emphasis of the binding studies with VIP has been secretory or absorptive cells in the GI tract. Autoradiographic distribution of VIP receptors in rabbit and rat small intestine revealed a high density in mucosal layers, while low density was noted in the smooth muscle layers.[83] Other binding studies elaborated the receptors on pancreatic acinar cells, intestinal epithelial cells, liver, human colonic adenocarcinoma cell lines, or gastric mucosa.[83-95]

Many structurally related, biologically active peptides interact with VIP at its binding sites. These peptides include secretin, peptide histidine isoleucine (PHI), peptide histidine methionine (PHM), rat and human growth hormone-releasing factors (rGHRF and hGHRF, respectively), and three peptides isolated from Gila monster venom — namely, helodermin, helospectin I, and helospectin II. These peptides have structural homology with VIP at its N-terminus.

Evidence has been presented for the presence of two subtypes of receptors for this family or peptides; one is VIP preferring and the other is secretin preferring (see below). Other subtypes with intermediate affinity for VIP or secretin have also been proposed.[87,89,90] VIP is several orders of magnitude more potent than secretin in competing for the binding of labeled VIP. Conversely, secretin is more effective in displacing labeled secretin than VIP.[87,89 90 92] The Gila venom peptides, however, displace VIP and secretin with similar potency.[90] A good peptide to distinguish VIP and secretin receptors is D-Phe^4-PHI, which potently displaced VIP from rat pancreatic membranes, while it had no effect on secretin binding up to a concentration of 10 μM.[92] D-Phe^4-PHI was an agonist (cAMP accumulation) with efficacy close to that of VIP albeit about 20 times less potent than VIP in binding as well as in an adenylate cyclase assay.[92] VIP fragment VIP$_{10-28}$ antagonized the action of VIP on accumulation of cAMP in rat pancreatic acini, while it had no effect up to a concentration of 100 μM at secretin-stimulated cAMP accumulation.[87]

There appear to be some species differences in the structural requirements for binding as well as for adenylate cyclase activation. Two VIP-related peptides from rat and human, PHI and PHM, respectively, are only about five times less potent than VIP in rat intestinal epithelial cells.[96] In human epithelia, however, these peptides are about three orders of magnitude less potent than VIP.[96] It appears, therefore, that PHI in rats may act on VIP receptors to produce a physiological response. The role of its human counterpart, PHM in human epithelia, is still uncertain. Several alternative explanations can be given, namely, the action of PHM through a distinct but as yet unidentified receptor in human or the difference in degradation pattern of PHM in human as compared to the rat epithelia. In epithelia of rabbit and rat intestine, VIP receptors are present on the basolateral membranes, and no binding could be observed on the brush border side.[97]

Studies with VIP fragments have revealed some important information. A whole VIP sequence is required for the agonistic activity and, to a lesser extent, for the binding activity both in human and in rat eptithelia.[98] Removal of the first 13 residues caused an elimination of the agonistic property. Binding potency is reduced but not abolished,[98] conferring the fragment VIP$_{14-28}$ with antagonistic properties. The fragment VIP$_{10-28}$ was also found to be a specific antagonist for VIP-preferring receptors.[87] The removal of only one amino acid from the N-terminus causes a shift in displacement potency by about 80-fold both in rat and human epithelia. In the rat this peptide (VIP_{2-28}) acted as a partial agonist, while in human it was a full agonist.[98]

The analysis of molecular size by either covalent cross-linking or photoaffinity labeling from different tissues yielded one, two, or more bands on sodium dodecyl sulfate poly-acrylamide gel electrophoresis (SDS-PAGE) in molecular radium ranging from 33,000 to 250,000 Da.[93-100] In many studies a 60,000- to 70,000-Da species appears to be the high-affinity receptor (i.e., the labeling was sensitive to low concentrations of VIP and to mod-ulation by guanine nucleotides).[85,95,99,100] A molecular entity of M_r = 33,000 Da appears to be the low-affinity receptor (low affinity for VIP, no sensitivity for guanine nucleotides) and may represent either the receptors that are uncoupled from guanine nucleotide binding regulatory proteins or neo- or nascent receptors not yet coupled to the G proteins. VIP binding is sensitive to disulfide reducing agents, and the presence of intramolecular as well intermolecular disulfide bonds is indicated.[101]

Binding studies have established the presence of VIP receptors on the secretory cells of the gastrointestinal tract. As mentioned previously, VIP is not only important in the secretory function of the GI tract, but it also plays an important role in motility. Actions of VIP on nerves, interstitial cells of Cajal, and smooth muscle cells have been proposed. More thorough binding studies on preparations of these cell types need to be done to establish clearly the site of action of VIP for its role in motility.

F. CHOLECYSTOKININS

Cholecystokinin (CCK) activity was first described by Ivy and Oldberg[102] in 1928 as an agent from the small intestine which caused strong contraction of the gallbladder.[102-104] CCK is present throughout the digestive tract, primarily in endocrine cells in the proximal gut and in nerves in the ileum and colon.[103] CCK occurs in several molecular forms as defined by the number of amino acid residues (i.e., CCK-58, CCK-39, CCK-33, CCK-8, CCK-5, CCK-4, and probably CCK-21 and CCK-12 as well[103]) and may be present in a sulfated or unsulfated form. Two major forms are CCK-33 and CCK-8, both present in similar concentrations in proximal intestine. In distal intestine, however, CCK-8 predomi-nates.[105] CCK-4 may be the major molecular form in the nerve terminals of the endocrine pancreas.[105]

CCK receptors have ben identified on the pancreatic acinar cells, smooth muscle tumors, mucosal cells, and gallbladder in several species. Initial studies were performed using [³H]CCK-8 and [³H]caerulein,[105] a CCK-related peptide, but the low specific activity of the label did not permit detection of high-affinity sites. More recently, [¹²⁵I]labeled Bolton-Hunter conjugated CCK-33 and CCK-8 have been prepared[106,107] and appear to be the labels of choice. Of these two labeled derivatives, [¹²⁵I]BH-CCK-8 is preferred because it is less likely to generate the unlabeled biologically active and labeled biologically inactive fragments due to peptidase cleavage of the label.[105,107] Furthermore, the affinity of CCK-8 is higher than that of CCK-33.[105]

In pancreatic acinar cells, CCK receptors are present as a heterogeneous population with high-affinity (K_d 0.026 to 2.0 nM) and low-affinity sites (K_d 2.0 to 20 nM).[108-110] It is interesting to note that in pancreatic membranes only one class of binding sites has been observed.[111-112] The reason for this discrepancy is unclear.

CCK receptors have been identified on bovine[113,114] and rabbit[115] gallbladder membranes either as homogeneous or heterogeneous populations of binding sites,[113-115] possibly reflecting the species differences. The receptors on gallbladder were confined to the membranes of the smooth muscle of muscularis,[113-115] while no binding to the serosal or mucosal membranes was reported.

In relation to its role in gastric inhibitory and satiety effects, CCK receptors have been identified in the smooth muscle layer of the pyloric sphincter.[116,117] CCK receptors are also present in human gastric smooth muscle tumors as identified by the binding of [^{125}I]BH-CCK-8.[118] A single, high-affinity site (K_d 0.075 nM) has been identified.

Various CCK and gastrin-related peptides inhibit the binding of radiolabeled CCK in pylorus membranes, usually in the following order; caerulein > CCK-8 > CCK-7 = CCK-33 > tyrosine-Hnl (SO$_3$H)CCK-6 > des (SO$_3$H)CCK-7 > gastrin I.[115] Cholecystokinins and gastrins have their C-terminal residue amidated and have a common C-terminal pentapeptide sequence. Due to this sequence homology there is some degree of cross-reactivity of CCKs and gastrins for the receptors.[119]

CCK receptors in pancreas appear to be different from those in brain. In particular, central CCK receptors demonstrate higher affinity for gastrin as compared to those in pancreas.[104,107,110,120] Moreover, dibutyryl cGMP and proglumide, two CCK receptor antagonists, were effective in inhibiting [^{125}I]CCK binding in pancreas, while in guinea pig cortical membranes these agents were ineffective.[104,121,122]

Determination of molecular weight by affinity linking or by cross-linking agents also reveals the heterogeneity in CCK receptors. In rat pancreatic plasma membranes, a major band of 80,000 to 95,000 Da has been recognized on SDS-PAGE.[123,124] Mouse pancreatic plasma membranes contained a CCK-binding protein of 76,000 Da linked by a disulfide bond to a 40,000 nonbinding protein.[125] The major band labeled in bovine gallbladder muscularis was 70,000 to 85,000 Da,[126] while human gastric smooth muscle tumor contained a 75,000-Da labeling protein.[118]

G. SOMATOSTATIN

Somatostatin, a tetradecapeptide, was discovered by Guillemin et al.[127] and is present both in endocrine cells and nerve fibers of the GI tract.[65,127-129] Initially, somatostatin was recognized as an inhibitor of growth hormone secretion, but it has become increasingly clear that this is only one of a multitude of actions of this peptide, including several inhibitory actions on the gastrointestinal tract.[127-130] Somatostatin-like immunoreactivity is present throughout the gastrointestinal tract of many species and is present in both endocrine cells and in nerve fibers of the myenteric and submucous plexuses, innervating circular muscle, mucosa, and submucosa.[65,128-130] Two major forms of somatostatin have been recognized: a 14-amino-acid peptide, somatostatin 14 (SS-14), and a larger, 28-amino-acid peptide, SS-28, containing SS-14 at its carboxy terminus. Other forms, SS-25 and SS-20, have also been isolated from porcine intestinal extracts.[131]

Receptors for somatostatin have been identified in several tissues. In mucosal cells, many studies have described the high-affinity binding of radioactively labeled somatostatin ([^{125}I]Tyr^{11}-somatostatin and [^{125}I]Tyr^1-somatostatin) to a cytosolic protein.[132-144] Intracellular somatostatin receptors have also been recognized in other tissues such as pancreas, skeletal muscle, liver, anterior pituitary, brain, spleen, submandibular gland, and heart.[145,146] That is not to say, however, that somatostatin receptors are confined to the cytosol because membrane bound receptors have also been recognized in some instance,[132,136,147,148] such as in pig small intestinal epithelial cells.[132] High-affinity/low-capacity and low-affinity/high-capacity binding sites have been recognized in many tissues examined, with high-affinity sites having dissociation constants (K_d) between 10 and 30 nM and low-affinity K_d of around 200 nM.[140-143] Rat jejuno-ileal mucosal somatostatin receptors had atypical affinity profiles because high- and low-affinity K_d values were 0.07 and 1.05 μM.[137] This particular study contrasts with another study from the same group[142] in which the affinities of the rat jejunal mucosal somatostatin receptors were in the nanomolar range (18.2 and 224 nM for high- and low-affinity sites, respectively), although the number of binding sites was approximately half as compared to that from the jejuno-ileal mucosa.[137]

An intracellular location of somatostatin receptors requires an internalization of somatostatin by these cells. In fact, evidence in favor of such a possibility has been presented for isolated gastric cells[148] and for anterior pituitary cells.[139] Presumably active intracellularly, these receptors activate phosphoprotein phosphatase and decrease the phosphorylation state of some intracellular protein(s).[133,138] Reyl and Lewin[133] have recognized a 130,000-Da somatostatin-binding protein and two non-binding proteins of M_r 64,000 and 13,000 in rat gastric membranes which were able to dephosphorylate histone. When purified somatostatin-binding protein was added to either of the purified phosphorylating proteins, a concentration-dependent inhibition of dephosphorylation was observed. This inhibition of dephosphorylation was reversed in a concentration-dependent manner by the addition of somatostatin.[133] The molecular weights of somatostatin receptors in rat pancreatin plasma membranes were determined to be 90,000[147] and 92,000 Da.[149]

As is evident from the above discussion, many of the studies regarding the localization and characterization of the somatostatin receptors have been performed on the secretory/absorptive cells of the gastrointestinal tract. However, somatostatin does have inhibitory effects on gastrointestinal motility as well.[128-130] The localization and the properties of these receptors involved in motility remain to be elucidated.

IV. CONCLUSION

Functional and biochemical data support the existence of receptors for a number of neuropeptides in diverse cell types within the GI tract. It becomes imperative, therefore, to obtain reasonably pure preparations of membranes from homogeneous cell types to establish the loci of these receptors. So far, identification and separation of membranes of intracellular organelles, neurons, and smooth muscle have been possible. More effort needs to be expended on the identification and separation of membranes from other cell types such as glial and interstitial cells.

Within the limits of the space allotted, it has not been possible to cover all the neuropeptide ligand binding studies in the GI tract. We have discussed some of the cases where more studies have been performed and more data are available. In many cases there is a dearth of available binding data. The space allotted to each peptide in this chapter does not necessarily reflect its relative importance in the GI tract; rather it reflects the quantum of relevant data at hand. Therefore, more attention should be given to those peptides which have been shown to alter the GI function, but for which relatively little binding data is available, namely, galanin, neuropeptide Y, bombesin, and calcitonin gene-related peptide.

REFERENCES

1. **Kwan, C. Y.,** Preparation of smooth muscle plasma membranes: a critical evaluation, in, *Sarcolemmal Biochemistry,* Kidwai, A. M., Ed., CRC Press, Boca Raton, FL, 1987, 59.
2. **Ahmad, S., Allescher, H. D., Manaka, H., Manaka, Y., and Daniel, E. E.,** ³H-Saxitoxin as a marker for deep muscular plexus neurons, *Am. J. Physiol.,* 255, G462, 1988.
3. **Kostka, P., Ahmad, S. Allescher, H, D., Berezin, I., Kwan, C. Y., and Daniel, E. E.,** Subcellular fractionation of longitudinal smooth muscle myenteric plexus (LSM/MP) of dog ileum: dissociation of the distribution of two plasma membrane marker enzymes, *J. Neurochem.,* 49, 1124, 1987.
4. **Ahmad, S., Allescher, H. D., and Daniel, E. E.,** Biochemical studies on alpha-2 adrenoceptors and opioid receptors on canine submucosal neurons, *Am. J. Physiol.,* 256, G957, 1989.
5. **Ahmad, S., Rausa, J., Jang, E., and Daniel, E. E.,** Calcium channel binding in nerves and muscle of canine small intestine, *Biochem. Biophys. Res. Commun.,* 159(1), 119, 1989.
6. **Allescher, H. D., Ahmad, S., Kostka, P., Kwan, C. Y., and Daniel, E. E.,** The distribution of opioid receptors in canine small intestine: implications for the function, *Am. J. Physio.,* 256, G966, 1989.
7. **Shi, a. G., Ahmad, S., Kwan, C. Y., and Daniel, E. E.,** Alpha adrenoceptors in dog mesenteric vessels — subcellular distribution and the number of ³H-prazosin and ³H-rauwolscine binding sites, *J. Cardiovasc. Pharmacol.,* 15, 515, 1990.

8. **Kyozuka, M., Crankshaw, D. J., Crankshaw, J., Kwan, C. Y., and Daniel, E. E.**, Alpha-2 adreno-ceptors on nerves and muscles of rat uterus, *J. Pharmacol. Exp. Ther.*, 244(3), 1128, 1988.

9. **Garfield, R. E.**, Cell to cell communication in smooth muscle, in *Calcium and Contractility*, A. K. Grover and E. E. Daniel, Eds., Humana Press, Clifton, NJ, 1985, 142.

10. **Shi, A. G., Kwan, C. Y., and Daniel, E. E.**, Characterization of alpha-adrenoreceptor subtypes by ^{3}H-Prazosin and ^{3}H-rauwolseine binding to canine venous smooth muscle membranes, *Can. J. Physiol. Pharmacol.*, 67, 1067, 1989.

11. **Berezin, I., Huizinga, J. D., and Daniel, E. E.**, Interstitial cells of Cajal in the canine colon: a special communication network at the inner border of the circular muscle, *J. Comp. Neurol.*, 273, 42, 1988.

12. **McCarthy, K. D., Salm, A., and Lorea, L. S.**, Astroglial receptors and their regulation of intermediate filament protein phosphorylation, in *Glial Cell Receptors*, Kimelberg, H. K., Ed., Raven Press, New York, 1988, 1.

13. **Kitabgi, P. and Freychet, P.**, Neurotensin: contractile activity, specific binding, and lack of effect on cyclic nucleotides in intestinal smooth muscle, *Eur. J. Pharmacol.*, 55, 35, 1979.

14. **Goedert, M., Hunter, J., and Ninkovic, M.**, Evidence for neurotensin as non-adrenergic, non-cholinergic neurotransmitter in guine apig ileum, *Nature (London)*, 311, 59, 1984.

15. **Kitabgi, P., Kwan, C. Y., Fox, J. E. T., and Vincent, J. P.**, Characterization of neurotensin binding to rat gastric smooth muscle receptor sites, *Peptides*, 5, 917, 1984.

16. **Ahmad, S., Berezin, I., Vincent, J. P., and Daniel, E. E.**, Neurotensin receptors in canine intestinal smooth muscle: preparation of plasma membranes and characterization of (tyr^{3}-^{125}I)-labelled neuroensin binding, *Biochim. Biophys. Acta*, 896, 195, 1987.

17. **Mazella, J., Poustis, C., Labbe, C., Checler, F., Kitabgi, P., Granier, C., Van Rietschoten, J., and Vincent, J. P.**, Monoiodo Trp^{11}-neurotensin, a highly radioactive ligand of neurotensin receptors. Preparation, biological activity and binding properties to rat brain synaptic membranes, *J. Biol. Chem.*, 258, 3476, 1983.

18. **Checler, F., Labbe, C., Granier, C., Van Rietschoten, J., Kitabgi, P., and Vincent, J. P.**, [Trp^{11}]Neurotensin and xenopsin discriminate between rat and guinea pig neurotensin receptors, *Life Sci.*, 31, 1145, 1982.

19. **Sadoul, J. L., Mazella, J., Amar, S., Kitabgi, P., and Vincent, J. P.**, Preparation of neurotensin selectively iodinated on tyrosine 3 residue. Biological activity and binding properties on mammalian neurotensin receptors, *Biochem. Biophys. Res. Commun.*, 120, 812, 1984.

20. **Mazella, J., Kwan, C. Y., Kitabgi, P., and Vincent, J. P.**, Covalent labelling of neurotensin receptors in rat gastric fundus plasma membranes, *Peptides*, 6, 1137, 1985.

21. **Ahmad, S., Kwan, C. Y., Vincent, J. P., Grover, A. K., Jung, C. Y., and Daniel, E. E.**, Target size analysis of neurotensin receptors, *Peptides*, 8, 195, 1987.

22. **Mazella, J., Kitabgi, P., and Vincent, J. P.**, Molecular properties of neurotensin receptors in rat brain. Identification of subunits by covalent labelling, *J. Biol. Chem.*, 260, 508, 1985.

23. **Lazarur, L. H., Perrin, M. H., and Brown, M. R.**, Mast cell binding of neurotensin. I. Iodination of neurotensin and characterizatin of the interaction of neurotensin with mast cell receptor sites, *J. Biol. Chem.*, 252, 7174, 1977.

24. **Kitabgi, P., Rostene, W., Dussailant, M., Schote, A., Laduron, P. M., and Vincent, J. P.**, Two populations of neurotensin binding sites in murine brain; discrimination by antihistamine levocabastine reveals markedly different radioautographic distribution, *Eur. J. Pharmacol.*, 140, 285, 1987.

25. **Mazella, J., Chabry, J., Kitabgi, P., and Vincent, J. P.**, Solubilization and characterization of active neurotensin receptors from mouse brain, *J. Biol. Chem.*, 263, 144, 1988.

26. **Schote, A. and Laduron, P. M.**, Differential postnatal ontogeny of two [^{3}H]neurotensin binding sites in rat brain, *Brain Res.*, 408, 326, 1987.

27. **Mills, A., Demoliou-Mason, C. D., and Bernard, E. A.**, Characterization of neurotensin binding sites in intact and solubilized bovine brain membranes, *J. Neurochem.*, 50, 904, 1988.

28. **Portoghese, P. S.**, A new concept on the mode of interaction of narcotic analgesics with receptors, *J. Med. Chem.*, 8, 609, 1965.

29. **Martin, W. R., Eades, C. G., Thompson, J. A., Huppler, R. E., and Gilbert, P. E.**, The effects of morphine- and nalorphine-like drugs in the nondependent and morphine-dependent chronic spinal dog, *J. Pharmacol. Exp. Ther.*, 197, 517, 1976.

30. **Gilbert, P. E. and Martin, W. R.**, The effects of morphine- and nalorphine-like drugs in nondependent, morphine-dependent and cyclazocine-dependent chronic spinal dogs, *J. Pharmacol. Exp. Ther.*, 198, 66, 1976.

31. **Goldstein, A. and James, I. F.**, Multiple opioid receptors, criteria for identification and classification, *Trends Pharmacol. Sci.*, 5, 503, 1984.

32. **Leslie, F.**, Methods used for the study of opioid receptors, *Pharmacol. Re.*, 39(3), 197, 1987.

33. **Chang, K. J.**, Opioid receptors: multiplicity and sequelae of ligand-receptor interactions, in *The Receptors*, Vol. 1, Conn, P. M., Ed., Academic Press, Orlando, FL, 1985, 1.

34. **Simonds, W. F.**, The molecular basis of opioid receptor function, *Endocrinol. Rev.*, 9(2), 200, 1988.
35. **Pasternak, G. W.**, Multiple mu opioid receptors: biochemical and pharmacological evidence for multiplicity, *Biochem. Pharmacol.*, 35(3), 361, 1986.
36. **Wolozin, B. L. and Pasternak, G. W.**, Classification of multiple morphine and enkephalin binding sites in the central nervous system, *Proc. Natl. Acad. Sci. U.S.A.*, 78, 6181, 1981.
37. **Gintzer, A. R. and Pasternak, G. W.**, Multiple mu receptors: evidence for mu_2 sites in the guinea pig ileum, *Neurosci. Lett.*, 39, 51, 1983.
38. **Terenius, L.**, Specific uptake of narcotic analgesics by subcellular fractions of the guinea pig ileum, *Acta Pharmacol. Toxicol.*, 31 (Suppl. 1), 50, 1972.
39. **Pert, C. B. and Snyder, S. H.**, Opiate receptor: demonstration in nervous tissue, *Science*, 179, 1011, 1973.
40. **Terenius, L.**, Comparison between narcotic "receptors" in the guinea pig ileum and the rat brain, *Acta Pharmacol. Toxicol.*, 37, 211, 1975.
41. **Creese, I. and Snyder, S. H.**, Receptor binding and pharmacological activity of opiates in the guinea pig intestine, *J. Pharmacol. Exp. Ther.*, 194(1), 205, 1975.
42. **Kromer, W., Skowronek, B., Stark, H., and Netz, H.**, Modulation of acid secretion from enriched guinea pig parietal cells by opioid receptors, *Pharmacology*, 27, 298, 1983.
43. **Monferini, E., Strada, D., and Manara, L.**, Factor from rat small intestine potently affects opiate receptor binding, *Life Sci.*, 29, 603, 1981.
44. **Gaginella, T. S., Rimele, T. J., and Wietecha, M.**, Studies on rat intestinal epithelial cell receptors for serotonin and opiates, *J. Physiol.*, 335, 101, 1983.
45. **Stefano, G. B., Scharrer, B., and Assanah, P.**, Demonstration, characterization and localization of opioid binding sites in the midgut of the insect *Leucophaea madierae* (Blattaria), *Brain Res.*, 253, 205, 1982.
46. **Nishimura, E., Buchan, A. M. J., and McIntosh, C. H. S.**, Autoradiographic localization of opioid receptors in rat stomach, *Neurosci. Lett.*, 50, 73, 1984.
47. **Dashwood, M. R., Debnam, E. S., Bagnall, J., and Thompson, C. S.**, Autoradiographic localisation of opiate receptors in rat small intestine, *Eur. J. Pharmacol.*, 107, 267, 1985.
48. **Nishimura, E., Buchan, A. M. J., and McIntosh, C. H. S.**, Autoradiographic localization of mu- and delta-type opioid receptors in the gastrointestinal tract of the rat and guinea pig, *Gastroenterology*, 91, 1084, 1986.
49. **Simon, E. J., Hiller, J. M., Grith, J., and Edelman, I.**, *J. Pharmacol. Exp. Ther.*, 192, 531, 1975.
50. **Pert, C. B. and Snyder, S. H.**, Opiate receptor binding of agonists and antagonists affected differentially by sodium, *Mol. Pharmacol.*, 10, 868, 1974.
51. **Simon, E. J. and Groth, J.**, Kinetics of opiate receptor inactivation by sulfhydryl reagents: evidence for the conformational change in presence of sodium ions, *Proc. Natl. Acad. Sci. U.S.A.*, 72, 2404, 1975.
52. **Wilson, H. A., Pasternak, G. W., and Snyder, S. H.**, Differentiation of opiate agonist and antagonist receptor binding by protein modifying agents, *Nature (London)*, 253, 448, 1975.
53. **Pasternak, G. W., Wilson, H. A., and Snyder, S. H.**, Differential effects of protein modifying agents on receptor binding of opiate agonists and antagonists, *Mol. Pharmacol.*, 11, 340, 1975.
54. **Nozaki, M. and Cho, T. M.**, Effect of 5,5'-dithiobis-(2-nitrobenzoic acid) on opiate binding to both the membrane bound receptor and partially purified opiate receptor, *J. Neurochem.*, 45, 461, 1985.
55. **Kamikubo, K., Murase, H., Murayama, M., Matsuda, M., and Miura, K.**, Evidence for disulfide bonds in membrane-bound and solubilized opioid receptors, *J. Neurochem.*, 50, 503, 1988.
56. **Simon, E. J. and Hiller, J. M.**, The opiate receptors, *Ann. Rev. Pharmacol. Toxicol.*, 18, 371, 1978.
57. **Simonds, W. F.**, The molecular basis of opioid receptor function, *Endocrinol. Rev.*, 9, 200, 1988.
58. **Ueda, H., Harada, H., Nozaka, M., Katada, T., Ui, M., Satoh, M., and Takagi, H.**, Reconstitution of rat brain opioid receptors with purified guanine nucleotide binding regulatory proteins, G_i and G_o, *Proc. Natl. Acad. Sci. U.S.A.*, 85, 7013, 1988.
59. **Roy, S., Zhu, Y. X., Lee, N. M., and Loh, H. H.**, Different molecular weight forms of opioid receptors revealed by polyclonal antibodies, *Biochem. Biophys. Res. Commun.*, 150, 237, 1988.
60. **Szucs, M., Belcheva, M., Simon, J., Benyhe, S., Toth, G., Hepp, J., Wollemann, M., and Medzihradszky, K.**, Covalent labelling of opioid receptors with ^{3}H-D-Ala^2-Leu^5-enkephalin chloromethyl ketone. I. Binding characteristics in rat brain, *Life Sci.*, 41, 177, 1987.
61. **Yeung, C. W.**, Photoaffinity labelling of opioid receptors of rat brain membranes with ^{125}I(D-Ala^2-p-N$_3$-Phe^4-Met^5) enkephalin, *Arch. Biochem. Biophys.*, 254, 81, 1987.
62. **Ueda, H., Harada, H., Misawa, H., Nozaki, M., and Tagaki, H.**, Purified opioid mu-receptor is of a different molecular size than delta- and kappa-receptors, *Neurosci. Lett.*, 75, 339, 1987.
63. **Itzhak, Y., Hiller, J. M., and Simon, E. J.**, Solubilization and characterization of mu, delta and kappa opioid binding sites from guinea pig brain: physical separation of kappa receptors, *Proc. Natl. Acad. Sci. U.S.A.*, 81, 4217, 1984.
64. **Bidlack, J. M. and O'Malley, W. E.**, Inhibition of and but not opioid binding to membranes by Fab fragments from a monoclonal antibody directed against the opioid receptor, *J. Biol. Chem.*, 261, 15844, 1986.

65. **Furness, J. B. and Costa, M.,** *The Enteric Nervous System,* Churchill Livingstone, New York, 1987.
66. **Maggio, J. E.,** Thachykinins, *Ann. Rev. Neurosci.,* 11, 13, 1988.
67. **Kotani, H., Hoshimura, M., Nawa, H., and Nakanishi, S.,** Structure and gene organization of bovine neuromedin K precursor, *Proc. Natl. Acad. Sci. U.S.A.,* 83, 7074, 1986.
68. **Erspamer, V., Anastasi, A., Bertaccini, G., and Cei, J. M.,** Structure and pharmacological actions of physalaemin, the main active polypeptide of the skin of *Physalaemus fuscumusculatus, Experientia,* 20, 489, 1964.
69. **Anastasi, A., Montecucchi, P., Erspamer, V., and Visser, J.,** Amino acid composition and sequence of kassinin, a tachykinin dodecapeptide from the skin of the african frog *Kassina senegalensis, Experientia,* 33, 857, 1977.
70. **Erspamer, V. and Anastasi, A.,** Structure and pharmacological actions of eledoisin, the active endeca-peptide of the posterior salivary glands of *Eledone, Experientia,* 18, 58, 1962.
71. **Lee, C. M., Iverson, L. L., Hanley, M. R., and Sandberg, B. E. B.,** The possible existence of multiple receptors for substance P, *Naunyn-Schiedeberg's Arch. Pharmacol.,* 318, 281, 1982.
72. **Quirion, R. and Damm, T. V.,** Multiple neurokinin receptors: recent developments, *Regul. Peptides,* 22, 1, 1988.
73. **Buck, S. H. and Burcher, E.,** The tachykinins: a family of peptides with a brood of receptors, *Trends Pharmacol. Sci.,* 7, 65, 1986.
74. **Regoli, D., Drapeau, G., Dion, S., and Do'Orléans-Juste, P.,** Pharmacological receptors for substance P and neurokinins, *Life Sci.,* 40, 109, 1987.
75. **Manyth, P., Goedert, M., and Hunt, S. P.,** Autoradiographic visualization of receptor binding sites for substance P in the gastrointestinal tract of the guinea pig, *Eur. J. Pharmacol.,* 100, 133, 1984.
76. **Burcher, E., Shults, C. W., Buck, S. H., Chase, T. N., and O'Donohue, T. L.,** Autoradiographic distribution of substance K binding sites in rat gastrointestinal tract: a comparison with substance P, *Eur. J. Pharmacol.,* 102, 561, 1984.
77. **Manyth, P. W., Manyth, C. R., Gates, T., Vigna, S. R., and Maggio, J. E.,** Receptor binding sites for substance P and substance K in canine gastrointestinal tract and their possible role in inflammatory bowel disease, *Neuroscience,* 25, 817, 1988.
78. **Buck, S. H., Maurin, Y., Burks, T. F., and Yamamura, H. I.,** High affinity ^{3}H-substance P binding to longitudinal muscle membranes of the guinea pig small intestine, *Life Sci.,* 34, 497, 1984.
79. **Watson, S. P. and Iverson, L. L.,** ^{3}H-substance P binding to the guinea pig ileum longitudinal smooth muscle membranes, *Regul. Peptides,* 8, 273, 1984.
80. **Burcher, E., Buck, S. H., Lowenberg, W., and O'Donohue, T. L.,** Characterization and autoradiographic localization of multiple tachykinin binding sites in gastrointestinal tract and bladder, *J. Pharmacol. Exp. Ther.,* 236, 819, 1986.*
81. **Bergstrom, L., Beaujouan, J. C., Torrens, Y., Saffroy, M., Glowinski, J., Lavielle, S., Chassaing, G., Marquet, A., D'Orléans, P., Dion, S., and Regoli, D.,** ^{3}H-neurokinin A labels a specific tachykinin binding site in the rat duodenal smooth muscle, *Mol. Pharmacol.,* 32, 764, 1987.
82. **Muller, M. J., Sato, H., Bowker, P., and Daniel, E. E.,** Receptors for tachykinins in canine intestine circular muscle, *J. Pharmacol. Exp. Ther.,* 246, 739, 1988.
83. **Sayadi, H., Harmon, J. W., Moody, T. W., and Korman, L. Y.,** Autoradiographic distribution of vasoactive intestinal polypeptide receptors in rabbit and rat small intestine, *Peptides,* 9, 23, 1988.
84. **Binder, H. J., Lemp, G. G., and Gardner, J. D.,** Receptors for vasoactive intestinal peptide and secretin on small intestinal epithelial cells, *Am. J. Physiol.,* 238, G190, 1980.
85. **Gesapach, C., Bawab, W., Chastre, E., Emami, S., Yanaihara, N., and Rosselin, G.,** Pharamcology and molecular identification of vasoactive intestinal polypeptide (VIP) receptors in normal and cancerous gastric mucosa in man, *Biochem. Biophys. Res. Commun.,* 151 (2), 929, 1988.
86. **Amiranoff, B., Laburthe, M., and Rosselin, G.,** Characterization of specific binding sites for vasoactive intestinal polypeptide in rat intestinal epithelial cells, *Biochim. Biophys. Acta,* 627, 215, 1980.
87. **Bissionet, B. M., Collen, M. J., Adachi, H., Jensen, R. T., and Gardner, J. D.,** Receptors for vasoactive intestinal peptide on rat pancreatic acini, *Am. J. Physiol.,* 246, G710, 1984.
88. **Robichon, a. and Marie, J. C.,** Selective photolablelling of high and low affinity binding sites for vasoactive intestinal peptide (VIP): evidence for two classes of covalent VIP receptor complexes in intestinal cell membranes, *Endocrinology,* 120 (3), 978, 1987.
89. **Zhou, Z., Gardner, J. D., and Jensen, R. T.,** Receptors for vasoactive intestinal peptide and secretin on guinea pig pancreatic acini, *Peptides,* 8, 633, 1987.
90. **Zhou, Z., Gardner, J. D., and Jensen, R. T.,** Interaction of peptides related to VIP and secretin with guinea pig pancreatic acini, *Am. J. Physiol.,* 256, G283, 1989.
91. **Luis, J., Martin, J., El Battari, A., Fantini, J., Giannellini, F., Marvaldi, J., and Pichon, J.,** Cyclohexamide induces accumulation of vasoactive intestinal peptide (VIP) binding sites at the cell surface of a human colonic adenocarcinoma cell line (HT29-D4); evidence for the presence of an intracellular pool of VIP receptors, *Eur. J. Biochem.* 167, 391, 1987.

92. **Robberecht, P., Coy, D. H., De Neef, P., Camus, J., Cauvin, A., Waelbroeck, M., and Christophe, J.,** [D-Phe⁴]Peptide histidine-isoleucinamide ([D-Phe⁴]PHI), a highly selective vasoactive intestinal peptide agonist, discriminates VIP-preferring from secretin-preferring receptors in rat pancreatic membranes, *Eur. J. Biochem.,* 165, 243, 1987.

93. **Robichan, A., Kuks, P. F. M., and Beson, J.,** Characterization of vasoactive intestinal peptide receptors by a photoaffinity label; site-specific modification of vasoactive intestinal peptide by derivatization of the receptor-bound peptide, *J. Biol. Chem.,* 262 (24), 11539, 1987.

94. **Couvineau, A., Rousset, M., and Laburthe, M.,** Molecular identification and structural requirement of vasoactive intestinal peptide (VIP) receptors in human colon adenocarcinoma cell line, HT-29, *Biochem. J.,* 231, 139, 1985.

95. **Martin, J., Darbon, H., Luis, J., El Battari, A., Marvaldi, J., and Pichon, J.,** Photoaffinity labelling of vasoactive intestinal peptide binding sites on intact human colonic adenocarcinoma cell line HT29-D4; synthesis and use of photosensitive vasoactive intestinal peptide derivatives, *Biochem. J.,* 250, 679, 1988.

96. **Laburthe, M., Couvineau, A., Fessard, C. R., and Moroder, L.,** Interaction of PHM, PHI and 24-glutamine PHI with human VIP receptors from colonic epithelium: comparison with rat intestinal receptors, *Life Sci.,* 36, 991, 1985.

97. **Dharmsthaphorn, K., Harms, V., Yamashiro, D. J., Hughes, R. J., Binder, H. J., and Wright, E. M.,** Preferential binding of vasoactive intestinal polypeptide to basolateral membrane of rat and rabbit enterocytes, *J. Clin. Invest.,* 71, 27, 1983.

98. **Couvineau, A., Fessard, R. C., Fournier, A., St. Pierre, S., Pipcorn, R., and Laburthe, M.,** Structural requirements for VIP interaction with specific receptors in human and rat intestinal membranes: effect of nine partial sequences, *Biochem. Biophys. Res. Commun.,* 121 (2) 493, 1984.

99. **Couvineau, A. and Laburthe, M.,** The human vasoactive intestinal peptide receptor: molecular identification by covalent cross-linking in colonic epithelium, *J. Clin. Endocrinol. Metab.,* 61, 50, 1985.

100. **Laburthe, M., Breant, B., and Fessard, C. R.,** Molecular identification of receptors for vasoactive intestinal peptide in rat intestinal epithelium by covalent cross-linking; evidence for two classes of binding sites with different structural and functional properties, *Eur. J. Biochem.,* 139, 181, 1984.

101. **Robbercht, P., Waelbroeck, M., Camus, J., and De Neef, P.,** Importance of disulfide bonds in receptors for vasoactive intestinal polypeptide and secretin in rat pancreatic plasma membranes, *Biochim. Biophys. Acta,* 773, 271, 1984.

102. **Ivy, A. C. and Oldberg, E.,** A hormone mechanism for gall-bladder contraction and evacuation, *Am. J. Physiol.,* 86, 599, 1928.

103. **Marx, M. ,Guillermo, G., Lonovics, J., and Thompson, J. C.,** Cholecystokinin, in *Gastointestinal Endocrinology,* Thompson, J. C., Greenley, G. H., Jr., Rayford, P. L., and Townsend, C. M., Jr., Eds., McGraw-Hill, New York, 1987, 213.

104. **Rosenweig, S. A. and Jamison, J. D.,** The cholecystokinin receptor, in *The Receptors,* Vol. 4, Conn, M. P., Ed., Academic Press, London, 1986, 213.

105. **Desbuquois, B.,** Receptors for gastrointestinal polypeptides, in *Polypeptide Hormone Receptors,* Posner, B. I., Ed., Marcel Dekker, New York, 1985, 419.

106. **Sankaran, H., Deveney, C. W., Goldfine, I. D., and Williams, J. A.,** Preparation of biologically active radioiodinated cholecystokinin for radioreceptor assay and readioimmuno assay, *J. Biol. Chem.,* 256, 12417, 1981.

107. **Miller, L. J., Roxenzweig, S. A., and Jamieson, J. D.,** Preparation and characterization of a probe for the cholecystokinin octapeptide receptor, N*(^{125}I-desaminotyrosyl)CCK-8, and its interaction with pancreatic acini, *J. Biol. Chem.,* 256, 12417, 1981.

108. **Sankaran, H., Goldfine, I. D., Deveney, C. W., Wong, K. Y., and Williams, J. A.,** Binding of cholecystokinin to high affinity receptors on isolated rat pancreatic acini, *J. Biol. Chem.,* 255, 1849, 1980.

109. **Sankaran, H., Goldfine, I. D., Bailey, A., Licko, V., and Williams, J. A.,** Relationship of cholecystokinin receptor binding to regulation of biological functionin pancreatic acini, *Am. J. Physiol.,* 242, G250, 1982.

110. **Jensen, R. T., Lemp, G. F., and Gardner, J. D.,** Interaction on cholecystokinin with specific membrane receptors on pancreatic acinar cells, *Proc. Natl. Acad. Sci. U.S.A.,* 77, 2079, 1980.

111. **Steigerwalt, R. W. and Williams, J. A.,** Characterization of cholecystokinin receptors on rat pancreatic membranes, *Endocrinology,* 109, 1746, 1981.

112. **Innis, R. B. and Snyder, S. H.,** Distinct cholecystokinin receptors in brain and pancreas, *Proc. Natl. Acad. Sci. U.S.A.,* 77, 6917, 1980.

113. **Schjoldager, B., Shaw, M. J., Powers, S. P., Schmalz, P. F., Szurszewski, J., and Miller, L. J.,** Bovine gallbladder muscularis: source of a myogenic receptor for cholecystokinin, *Am. J. Physiol.,* 254, G294, 1988.

114. **Steigerwalt, R. W., Goldfine, I. D., and Williams, J. A.,** Characterization of cholecystokinin receptors on bovine gall-bladder membranes, *Am. J. Physiol.,* 247 G709, 1984.

115. **Singh, P. and Thompson, J. C.,** Receptors for gastrin and cholecystokinin, in *Gastrointestinal Endocrinology,* Thompson, J. C., Greenley, G. H., Jr., Rayford, P. L., and Townsend, C. M., Jr., Eds., McGraw-Hill, New York, 1987, 69.

116. **Robinson, P. H., Moran, T. H., Goldrich, M., and McHugh, P. R.,** Development of cholecystokinin binding sites in rat upper gastrointestinal tract, *Am. J. Physiol.,* 252, G529, 1987.

117. **Smith, G. T., Moran, T. H., Coyle, J. T., Kuhar, M. J., O'Donahue, T. L., and McHugh, P. R.,** Anatomic localization of cholecystokinin receptors to the pyloric sphincter, *Am. J. Physiol.,* 246, R127, 1984.

118. **Miller, L. J.,** Characterization of cholecystokinin receptors on human gastric smooth muscle tumors, *Am. J. Physiol.,* 247, G402, 1984.

119. **Cherner, J. A., Sutliff, V. E., Grybowski, D. M., Jensen, D. M., and Gardner, J. D.,** Functionally distinct receptors for cholecystokinin and gastrin on dispersed chief cells from guinea pig stomach, *Am. J. Physiol.,* 254, G151, 1988.

120. **Innis, R. B. and Snyder, S. H.,** Distinct cholecystokinin receptors in brain and pancreas, *Proc. Natl. Acad. Sci. U.S.A.,* 77, 6917, 1980.

121. **Gardner, J. D. and Jensen, R. T.,** Cholecystokinin receptor antagonists, *Am. J. Physiol.,* 246, G471, 1984.

122. **Lin, C. W. and Miller, T.,** Characterization of cholecystokinin receptor sites in guinea pig cortical membranes using [^{125}I]Bolton Hunter-cholecystokinin octapeptide, *J. Pharmacol. Exp. Ther.,* 232, 775, 1985.

123. **Pearson, R. K. and Miller, L. J.,** Affinity labelling of a novel cholecystokinin-binding protein in rat pancreatic plasmalemma using new short probes for the receptor, *J. Biol. Chem.,* 262, 869, 1987.

124. **Madison, L. D., Rosenzweig, S. A., and Jamieson, J. D.,** Use of heterobifunctional cross-linker *m*-maleimidobenzoyl *N*-hydroxysuccinimide ester to affinity label cholecystokinin binding proteins on rat pancreatin plasma membranes, *J. Biol. Chem.,* 259, 14818, 1984.

125. **Sakamoto, C., Goldfine, I. D., and Williams, J. A.,** Characterization of cholecystokinin receptor subunits on pancreatic plasma membranes, *J. Biol. Chem.,* 258, 12707, 1983.

126. **Shaw, M. J., Hadac, E. M., and Miller, L. J.,** Preparation of enriched plasma membranes from bovine gall bladder muscularis for characterization of cholecystokinin receptors, *J. Biol. Chem.,* 262, 14313, 1987.

127. **Guillemin, R. and Gerich, J. E.,** Somatostatin: physiological and clinical significance, *Annu. Rev. Med.,* 27, 379, 1976.

128. **Newman, J. B., Lluis, F., and Townsend, T. M., Jr.,** Somatostatin, in *Gastrointestinal Endocrinology,* Thompson, J. C., Greeley, G. H., Rayford, P. L., and Townsend, T. M., Jr., Eds., McGraw-Hill, New York, 1987, 286.

129. **Krejs, G. J.,** Physiological role of somatostatin in the digestive tract: Gastric acid secretion, intestinal absorption and motility, *Scand. J. Gastroenterol.,* 21 (Suppl 119), 47, 1986.

130. **McIntosh, C. H. S.,** Gastrointestinal somatostatin: distribution, secretion and physiological significance, *Life Sci.,* 37, 2043, 1985.

131. **Arakawa, Y. and Tichibana, S.,** Somatostatin-20, a novel NK$_2$-terminally extended form of somatostatin isolated from porcine duodenum together with somatostatin-28 and somatostatin-25, *Life Sci.,* 35, 2529, 1984.

132. **Weber, W. Cole, T., and Conlon, J. M.,** Specific binding and degradation of somatostatin by membrane vesicles from pig gut, *Am. J. Physiol.,* G679, 1986.

133. **Reyl, F. J. and Lewin, M. J. M.,** Intracellular receptors for somatostatin in gastric mucosal cells: decomposition and reconstitution of somatostatin-stimulated phosphoprotein phosphatases, *Proc. Natl. Acad. Sci. U.S.A.,* 79, 978, 1982.

134. **Grossman, A., Richardson, S. B., Moloshok, T., and Frangione, B.,** Evidence that somatostatin (SRIF^{14}) is the primary coligand in pancreas required for specific binding of [^{3}H]estradiol in pancreatic tissue: demonstration that [^{3}H]estradiol and [^{125}I]SRIF^{14} form complexes of varying size with a specific binding protein, *J. Steroid. Biochem.,* 21 (3), 279, 1984.

135. **Reyl, F. and Levin, M. J. M.,** Evidence for an intracellular somatostatin receptor in pancreas: a comparative study with reference to gastric mucosa, *Biochem. Biophys. Res. Commun.,* 109 (4), 1324, 1982.

136. **Susini, C., Esteve, J. P., Vaysse, M., and Ribet, A.,** Calcium dependence of somatostatin binding to receptors, *Peptides,* 6, 831, 1985.

137. **Arilla, E., Lopez-Ruiz, M. P. Guijarro, L., Prieto, J. C., Gomez-Pan, A., and Hirst, B.,** Characterization of somatostatin binding sites in cytosolic fraction of rat intestinal mucosa, *Biochim. Biophys. Acta,* 802, 203, 1984.

138. **Reyl, F. and Lewin, M. J. M.,** Somatostatin is a potent activator of phosphoprotein phosphatases in the digestive tract, *Biochim. Biophys. Acta,* 675, 297, 1981.

139. **Draznin, B., Sherman, N., Sussman, K., Dahl, R., and Vatter, A.,** Internatlization and cellular processing of somatostatin in primary culture of rat anterior pituitary cells, *Endocrinology,* 117 (2), 960, 1985.

140. **Arilla, E., Ruiz, M. P., Guijarro, L., and Prieto, J. C.,** Somatostatin binding sites in cytosolic fractions of parietal and non-parietal cells from rabbit fundic mucosa, *Biosci. Rep.,* 5, 321, 1985.

141. **Ruiz, M. P., Arilla, E., Guijarro, L., and Prieto, J. C.,** Somatostatin binding sites in cytosolic fraction of rabbit intestinal mucosa: distribution throughout the gastrointestinal tract, *Comp. Biochem. Physiol.,* 81B (4), 1041, 1985.

142. **Ruiz, M. P., Guijarro, L. G., and Arrilla, E.,** Species variations of somatostatin concentrations and binding sites in jejunal mucosa, *Comp. Biochem. Physiol.,* 89A (2), 237, 1988.

143. **Arilla, E., Colas, B., Ruiz, M. P., and Prieto, J. C.,** Effect of small intestinal resection on somatostatin binding to cytosol of rabbit gastric mucosa, *Life Sci.,* 39, 1395, 1986.

144. **Colas, B., Bodegas, G., Sanz, M., Prieto, J. C., and Arilla, E.,** Partial enterectomy decreases somatostatin binding sites in residual intestine of rabbits, *Clin. Sci.,* 74, 499, 1988.

145. **Reyl, F. and Lewin, M. J. M.,** Evidence for an intracellular receptor in pancreas: a comparative study with reference to gastric mucosa, *Biochem. Biophys. Res. Commun.,* 109, 1324, 1982.

146. **Ogawa, N., Thompson, T., and Friesnen, H. G.,** Characteristics of somatostatin binding protein, *Can. J. Physiol. Pharmacol.,* 56, 48, 1977.

147. **Sakamoto, C., Goldfine, I. D., and Williams, J. A.,** The somatostatin receptor on isolated pancreatic acinar cell plasma membranes; identification of subunit structure and direct regulation by cholecystokinin, *J. Biol. Chem.,* 259 (15), 9623, 1984.

148. **Reyl, F., Silve, C., and Levin, M. J. M.,** in *Hormonal Receptors in Digestion and Nutrition,* Rosselin, G., Fromageot, P., and Bonfils, S., Eds., Elsevier/North-Holland, Amsterdam, 1979, 391.

149. **Susini, c. Bailey, A., Jaroslaw, S., and Williams, J. A.,** Characterization of covalently cross-linked pancreatic somatostatin receptors, *J. Biol., Chem.,* 261 (35), 16738, 1986.

150. Unpublished, 1989.

151. Unpublished, 1988.

152. Unpublished, 1989.

153. Unpublished, 1989.

Chapter 8

RECEPTORS FOR NEUROPEPTIDES: RECEPTOR ISOLATION STUDIES AND MOLECULAR BIOLOGY

Jean-Pierre Vincent and Patrick Kitabgi

TABLE OF CONTENTS

I.   Introduction 232

II.   Isolation of Peptide Receptors by Direct Solubilization and Purification 232
      A.   Preliminary Choices 232
      B.   Receptor Solubilization 233
            1.   Binding Assay to Soluble Receptor 233
            2.   Detergent 235
            3.   Stabilizing Agent 235
            4.   Peptidase Inhibitors 235
      C.   Characterization of the Crude Soluble Receptor 236
            1.   Binding Properties 236
            2.   Molecular Structure 236
      D.   Purification 237
            1.   Prepurification Step 237
            2.   Affinity Chromatography 237
                  a.   Preparation of Affinity Gels 237
                  b.   Properties of Affinity Columns 239
                  c.   Chromatography of Soluble Receptors 239
            3.   Postaffinity Step 240
            4.   Quantitative Results 241
            5.   Structural and Functional Properties of Purified
                  Neuropeptide Receptors 241
                  a.   Binding Parameters and Specificity 241
                  b.   Molecular Properties 241
      E.   Conclusion 242

III. Receptor Expression in the *Xenopus* Oocyte: A Promising Alternative
      Approach for Receptor cDNA Cloning 243

Addendum 245

Acknowledgments 245

References 245

I. INTRODUCTION

Neuropeptides generally fulfill a dual function of neurotransmitter and of hormone in the brain and the periphery. Both modes of action imply as a first step the selective association of the neuropeptide with a specific receptor located on the plasma membrane of the target cell. The peptide-receptor interaction triggers changes of membrane ionic conductance and/ or variations of intracellular enzymatic activity. These transduction mechanisms often involve guanyl nucleotide binding proteins as intermediates. Finally, the biological response associated to receptor occupancy is terminated by proteolytic degradation of the neuropeptide.

Because of the central role played by receptors in the mode of action of hormones and neurotransmitters, a number of studies have been devoted to the elucidation of their structural and functional properties. In particular, several laboratories are currently trying to solubilize and to purify neuropeptide receptors from various sources. In spite of significant advances in the biochemical characterization of membrane-bound proteins, purification to homogeneity of a neuropeptide receptor remains a difficult task. At first sight, the only property of the receptor that represents an advantage in view of its purification is its high affinity for its specific ligand. All other characteristics of the receptor, including the fact that it is a membrane-bound protein, should be considered as major drawbacks. The most severe limitation to solubilization and purification studies is that there exists no rich source of neuropeptide receptors. For this reason, it has been impossible until now to purify such a receptor from the gastrointestinal tract. The only neuropeptide receptors purified to date (the μ- and δ-opioid receptors and the somatostatin and neurotensin receptors) were obtained from mammalian brain homogenates that contained a few hundred femtomoles of neuropeptide binding sites per milligram of protein.[1-6]

This chapter will describe the various approaches and methodologies that have been used to purify neuropeptide receptors from brain, with the purpose being to underline and to discuss the peculiar advantages and drawbacks of each method. It is hoped that these data will be useful to those who will undertake the purification of neuropeptide receptors from gastrointestinal tract preparations. Recently, the cDNA cloning of the bovine neurokinin A (substance K) receptor through the oocyte expression system has been described.[7] This alternative method for the isolation of pure neuropeptide receptors will be presented in the last part of this chapter.

II. ISOLATION OF PEPTIDE RECEPTORS BY DIRECT SOLUBILIZATION AND PURIFICATION

These studies are always conducted according to the general strategy summarized in Table 1. However, a number of various techniques can be used to progress along this common scheme.

A. PRELIMINARY CHOICES

The very first choice that should be made concerns the type of receptor to purify. Almost all neuropeptide receptors are now known to exist under various forms that differ by their pharmacological and binding properties. Thus, there exist at least three different types of opiate receptors, termed μ, δ, and κ that are unevenly distributed in the central nervous system and in peripheral tissues.[8,9] Functionally distinct somatostatin receptor subtypes also can be distinguished in brain, pancreas, and pituitary by their different affinities for somatostatin and its analogues.[10-14] Although there seems to exist a single functional receptor for neurotensin in brain and peripheral membranes, binding sites of different affinites and properties have been described.[15-17] Once a given receptor subtype has been chosen, the selection of the preparation that will be used as starting material for solubilization and

TABLE 1
General Scheme for Purification of a Functional and Stable Receptor

Solubilization

Choice of starting material
Choice of solubilization conditions
    Detergent
    Stabilizing agent
    Binding assay
Properties of the crude soluble receptor
    Binding parameters (affinity, binding capacity, selectivity)
    Molecular weight (gel filtration)
    Subunit structure (covalent labeling)

Purification

Prepurification step
Affinity chromatography
    Preparation and properties of the affinity gel
    Chromatography of the prepurified receptor
Final purification step
Characterization of the purified receptor
    Binding properties
    SDS-PAGE analysis (silver staining, autoradiography after radioiodination, covalent labeling)

purification of the receptor is made essentially on the basis of its number of specific binding sites. The availability of a preparation containing a single receptor subtype is obviously a great advantage. The receptor source should also be readily available fresh and in quantity. These considerations led to the purification of the δ-opiate receptor from neuroblastoma-glioma NG108-15 hybrid cells, a relatively rich source of this single type of opiate binding sites.[3] Neurotensin receptors were purified from brain homogenates of newborn mouse because this preparation is the richest known source of functional neurotensin binding sites and because it does not contain low-affinity sites (see Table 2). Brain membranes were used to purify the somatostatin receptor because their binding capacity is higher than those of the pituitary and pancreas (see Table 2).

B. RECEPTOR SOLUBILIZATION

Neuropeptide receptors can be purified according to two different approaches. In the first, the specific radiolabeled neuropeptide is covalently attached to is membrane-bound receptor by photoaffinity labeling or by means of chemical cross-linking. The ligand-receptor complex is then solubilized and can be followed through later purification steps by monitoring radioactivity. The main advantage of this approach is that there is no need to keep the receptor in an active binding state during solubilization and purification. However, since the binding site is definitely blocked, it will be impossible to test the function of the purified receptor in reconstitution experiments. This strategy has been used successfully to purify the δ-opiate receptor from NG108-15 cells after covalent labeling with [³H]methylfentanylisothiocyanate.[3]

In the second approach, experimental conditions are devised which do not destroy the binding properties of the receptor. It is possible in that case to estimate the effect of each purification step on the affinity and the binding capacity of the receptor, which will be obtained in a free form at the end of the purification.

1. Binding Assay to Soluble Receptor

When the receptor is purified in a free and active form, a convenient assay to measure the binding of a radiolabeled peptide analogue to the soluble receptor should be available.

TABLE 2

Binding Properties of Somatostatin and Neurotensin Receptors in Central and Peripheral Membrane Preparations

Peptide	Receptor preparation	Radioligand	Binding parameters		Ref.
			K_d (nM)	B_{max} (fmol/mg protein)	
Somatostatin	Cortical brain membranes	[125I]-CGP 23996[a]	2.4	450	10
	Anterior pituitary membranes	[125I]SS-28 analogue[b]	0.46	262	11
		[125I-Tyr11]SS-14[c]	2.33	100	12
	Pancreatic acinar cell membranes	[125I]SS-28 analogue[b]	1.73	101	12
		[125I-Tyr11]SS-14[c]	0.5	266	13
Neurotensin	Brain synaptic membranes (adult rat)	[125I-Tyr3]neurotensin			
	High affinity		0.15	15.5	15
	Low affinity		7.2	135.7	15
	Fundus smooth muscle membranes (adult rat)	[125I-Tyr]-Trp11-neurotensin			
	High affinity		0.056	6.6	16
	Low affinity		1.92	11.4	16
	Brain homogenate (8-d-old mouse)[d]	[125I-Tyr3]neurotensin	0.13	250	5

a Des-Ala1,Gly2-desamino-Cys3[125I-Tyr11]-dicarba3,14-somatostatin-14.
b 125I[Leu8,D-Trp22,Tyr25]somatostatin-28.
c [125I-Tyr11]somatostatin-14.
d This preparation only contains high-affinity sites.

One of the main initial difficulties of this strategy is the necessity to find simultaneously suitable conditions for solubilization of the receptor in an active form and for measurement of the soluble binding activity. Indeed, failure to detect neuropeptide binding in a soluble extract can be due either to the fact that the receptor has lost its activity upon solubilization or to the inability of the assay to detect receptor binding. As long as the binding assay has not been validated, solubilization experiments cannot be interpreted properly. The best way to overcome this difficulty is probably to use a binding assay as simple and rapid as possible. The separation of radiolabeled neurotensin bound to the soluble receptor from free ligand was carried out either by gel filtration on small columns of Sephadex® G-50 or by filtration on GF/B filters pretreated with polyethylenimine.[5,6] For the soluble μ-opioid receptor, both the polyethylene glycol precipitation and the charcoal absorption methods were used.[1,2] Purification of the somatostatin receptor was followed by means of a covalent binding assay.[4]

2. Detergent

Receptor solubilization can be carried out with a variety of detergents including Triton® X-100 (one of the most frequently used), Lubrol® PX, Nonidet P-40, 3-[(3-cholamidopropyl) dimethylammonio]-1-propanesulfonic acid (CHAPS), and digitonin. There exists no general rule or law which would permit selection of the most convenient detergent to solubilize a given receptor. Thus, the μ-opioid receptor has been solubilized in an active form with either digitonin,[1] Triton® X-100,[2] glycodeoxycholate,[18] or CHAPS.[19] Only CHAPS was found to solubilize the high-affinity neurotensin receptor without loss of binding activity,[20] whereas digitonin seems to be selective for the solubilization of active low-affinity neurotensin binding sites.[6]

3. Stabilizing Agent

The various opiate receptors are generally stable after detergent solubilization. However, in the absence of any additive, the half-life of the CHAPS-solubilized mouse brain neurotensin receptor was only 3 h at 0° C, which strongly limited the possibility of purifying this receptor in an active form. Fortunately, the presence of cholesteryl hemisuccinate in the solubilization buffer largely improved the stability of the receptor ($t^1/_2$ = 31 h). Addition of 1 nM neurotensin to mouse brain homogenate before solubilization further increased the half-life to 65 h.[20] The lower-affinity neurotensin receptors from bovine brain were found to be stabilized selectively by azolectin.[6]

Although addition of specific stabilizing agents is not always required, receptors are systematically solubilized in the presence of glycerol concentrations comprised between 5 and 25% (w/v). In our hands, this additive did not increase the half-life of the neurotensin receptor at 0° C, but protected it against losses of binding activity that occurred upon freezing and thawing.[20]

4. Peptidase Inhibitors

In order to purify receptor proteins in their native form, it is necessary to work in the presence of protease inhibitors. A great variety of natural or synthetic inhibitors have been used in receptor purification studies. Since only four different families of proteolytic enzymes exist, the presence of four specific inhibitors in the purification buffer should be sufficient to completely protect the receptor against degradation by proteases. A list of protease inhibitors currently used in receptor purification is given in Table 3. For example, high-affinity neurotensin receptors were solubilized and purifed in a buffer containing 0.1 mM phenylmethylsulfonyl fluoride (PMSF) (inhibitor of serine proteases), 1 mM iodoacetamide (inhibitor of thiol proteases), 5 mM ethylenediaminetetracetic acid (EDTA) (inhibitor of metalloproteases), and 1 μM pepstatin (inhibitor of acidic proteases).[5,20]

Taking into account all the parameters discussed earlier, receptor solubilization is gen-

TABLE 3
Protease Inhibitors Used in Receptor Purification

Proteolytic activity	Inhibitor[a]	
	General	Specific[b]
Serine proteases	DFP[c] (10^{-3} *M*) PMSF[d] (10^{-4} *M*)	Benzamidine (10^{-3} *M*) Aprotinin (10^{-7} *M*) STI[e] (10^{-7} *M*)
Thiol proteases	SH reagents (ex: iodoacetamide 10^{-3} *M*)	Leupeptin (10^{-4} *M*)[f] Antipain (10^{-4} *M*)[f] Chymostatin (10^{-4} *M*)[f]
Metalloproteases	Metal chelators EGTA (5×10^{-3} *M*) EDTA (5×10^{-3} *M*) *o*-Phenanthroline (10^{-3} *M*) Dithiothreitol (10^{-3} *M*)	Bestatin (10^{-4} *M*) Amastatin (10^{-5} *M*)
Acid proteases	Neutral pH	Pepstatin (10^{-6} *M*)

[a]   Useful concentrations are given in parentheses,
[b]   These inhibitors are specific for one or a limited number of peptidases in the corresponding family,
[c]   Diisopropyl fluorophosphate,
[d]   Phenylmethylsulfonyl fluoride,
[e]   Soybean trypsin inhibitor,
[f]   Also inhibits serine proteases.

erally carried out as follows. Membranes or cells used as the source of the receptor are incubated at a concentration of 1 to 10 mg of protein per milliliter in a buffered solution (at neutral pH) containing glycerol, a cocktail of protease inhibitors, eventually a stabilizing lipid, and the detergent at a final concentration of about 1% (w/v). After 30 to 60 min at 0° C, the incubation medium is centrifuged at high speed for 15 to 60 min. The supernatant which contains the soluble receptor is removed and either used immediately or stored at low temperature.

C. CHARACTERIZATION OF THE CRUDE SOLUBLE RECEPTOR

The purpose of these experiments is to evaluate the consequence of detergent solubilization on the structural and functional properties of the receptor before undertaking its purification.

1. Binding Properties

When neuropeptide receptors are solubilized under the free form, their binding parameters can be evaluated immediately after solubilization and compared to those of the membrane-bound receptor. The CHAPS-solubilized neurotensin receptor from mouse brain exhibited the same binding parameters and the same selectivity for neurotensin and its analogues as the high-affinity neurotensin binding sites that were initially present in brain homogenates.[20] Similarly, the affinities of various opioid ligands for membrane-bound and soluble μ-opiate receptors from bovine striatum were found to be strongly correlated.[18] These data indicate that both the neurotensin and the μ-opiate receptors have not been altered extensively by their removal from the membrane.

2. Molecular Structure

The apparent molecular weight of solubilized receptors can be obtained by gel filtration on a calibrated column. The glycodeoxycholate-solubilized μ-opiate receptor was eluted from a Sepharose® CL-6B column as a symmetrical peak at a position corresponding to a molecular weight of approximately 300 to 400 kDa.[18] Chromatography of the CHAPS-solubilized neurotensin receptor on Ultrogel® AcA34 gave a molecular weight value of about 100 kDa.[20]

Covalent labeling of the crude soluble receptor with a radiolabeled analogue provides a direct means not only to detect and to identify the protein which corresponds to the receptor in a complex mixture, but also to analyze the subunit structure of the receptor protein. Immediately after its solubilization by CHAPS, the mouse brain neurotensin receptor could be covalently labeled with radioactive neurotensin derivatives either by photoaffinity or by chemical cross-linking. Both techniques led to the specific labeling of a single protein band of about 100 kDa as shown by sodium dodecyl sulfate-polyacrylamide gel electrophoresis (SDS-PAGE) analysis and autoradiography.[5,20] Comparison of these results with the gel filtration data indicates that the neurotensin receptor is made of a single 100-kDa polypeptide chain. The structure of μ-opioid receptors seems to be complex, since subunit molecular weights of 58 or 65 kDa were obtained by covalent labeling of bovine striatum or rat brain, as compared to more than 300 kDa for the native receptor.[1,2] The δ-opiate receptor of NG108-15 cells is oligomeric, and the molecular weight of the binding subunit is 58 kDa.[3] The size of the brain somatostatin receptor as determined by photo-cross-linking is about 60 kDa.[4]

D. PURIFICATION

Brain neuropeptide receptors were purified essentially in a single step of affinity chromatography. However, in order to minimize the total amount of protein loaded on the affinity column, it may be of interest to prepurify the crude soluble receptor.

1. Prepurification Step

A 30-fold purification of the prelabeled δ-opioid receptor was achieved by chromatography of the crude soluble NG108-15 extract on WGA-agarose with a yield of more than 90%.[3] CHAPS-solubilized neurotensin receptors were prepurified by passage through two columns of SP-Sephadex® C-25 and hydroxylapatite connected in series. About 50% of total protein, including several brain proteases,[21] was retained on the gels, whereas the neurotensin binding activity was recovered quantitatively in the flowthrough.[5]

2. Affinity Chromatography

This powerful technique is the only one that can be considered as suitable for purification of membrane proteins as rare as neuropeptide receptors. Two different versions of this technique exist. In ligand affinity chromatography a ligand specifically recognized by the receptor is covalently coupled to an inert gel, whereas in immunoaffinity chromatography the specific ligand is replaced by an antibody directed against the receptor. Alternatively, when a covalent ligand-receptor complex is purified, the antibody can be specific for the ligand moiety, which remains accessible in the complex.

a. Preparation of the Affinity Gels

Ligand affinity gels used to purify the μ-opiate receptors were prepared by coupling nonpeptidic μ-agonists to various supports equipped with spacer arms which prevent steric hindrance (see Table 4). By contrast, immobilization of antibodies is generally carried out directly on CNBr-activated gels, as illustrated by the anti-Fit IgG-Sepharose® support, which specifically recognized the Fit-δ-opiate receptor complex. The somatostatin and neurotensin receptors were purified on supports in which the native peptide or one of its analogues was covalently attached to activated spacer arms (see Table 4).

The most important quality of an affinity gel is its receptor binding capacity. It is therefore necessary to achieve maximum efficiency in the coupling reaction. For this reason, neurotensin$_{2-13}$ (which contains two reactive amine functions) was preferred over neurotensin$_{1-13}$ (whose N-terminal residue is pGlu) for the preparation of an affinity gel designed for purification of the neurotensin receptor. The yield of covalent binding to

TABLE 4
Structure and Properties of Affinity Gels Used to Purify Neuropeptide Receptors

Receptor	Gel structure		Ligand	Ligand concentration in the gel	Mode of elution of the receptor
	Matrix	Spacer arm			
μ-Opiate receptor from bovine striatum[1]	Sepharose® 4B	$-NH(CH_2)_5CO-$	NED[a]	20 — 40 mM	Naloxone 2.5 μM
μ-Opiate receptor from rat brain[2]	Affigel 102	$-CH_2CONH(CH_2)_2NHCO(CH_2)_2CO-$	Morphine	~10 mM	NaCl 0.35 M
μ-Opiate Receptor from rat brain[19]	Affigel 401	$-CH_2CONH(CH_2)_2NHCOCH(CH_2)_2S$ CH_3CONH	Hybromet[b]	~5 mM	Normorphine 1 μM
δ-Opiate receptor from NG108-15 cells[3]	Sepharose®	—	Anti-FIT[c]IgG	10 mM	FIT-Lysine 100 μM + propionic acid 1 M
Somatostatin receptor from rat brain[4]	Affigel 10	$-CH_2CONH(CH_2)_2NHCO(CH_2)_2CO-$	[D-Trp8]SRIF[d]	6 mM	Na acetate 50 mM pH5.5
Neurotensin receptor from mouse brain[5]	AcA22	$-(glutaraldehyde)_3-$	Neurotensin[2—13]	100 μM	NaCl 1M
Neurotensin receptor from bovine brain[6]	Affigel 10	$-CH_2CONH)CH_2)_2NHCO(CH_2)_2CO-$	Neurotensin	~5 mM	NaCl 0.25 M

a   NED, β-naltrexylethylenediamine,
b   Hybromet, 7α-[1(R)-hydroxy-1-methyl-3-*P*-[4-(3'-bromomercuri-2'-methoxy-propoxy)-phenyl]-prolyl]-6,14-endo-ethenotetrahydrothebaine,
c   FIT, fentanylisothiocyanate,
d   SRIF, somatostatin.

glutaraldehyde-activated AcA22 was found to be 50 to 70% for neurotensin$_{2-13}$ as compared to only 20 to 30% for the native peptide. Moreover, the affinity of the (2 — 13) sequence for the neurotensin receptor is slightly better than that of neurotensin itself.[5]

b. Properties of Affinity Columns

A common characteristic of gels that have been used successfully in ligand affinity chromatography of neuropeptide receptors is the very high concentration of ligand bound to the gel as compared to the affinity of the free ligand for the receptor. For example, the μ-opiate receptor was purified on affinity gels that contained about 10 μmol of ligand per milliliter of packed gel. This 10 mM concentration is about 10^6 times higher than the affinity of free opiate ligands for their receptor. Peptide concentrations in gels used to purify receptors of neurotensin and somatostatin were between 0.1 and 5 mM, whereas peptide-receptor dissociation constants are in the nanomolar range. This high ligand concentration of the affinity matrix seems to be essential, since dilution with unmodified gel progressively reduces the efficiency of the affinity column in retaining solubilized receptors.[22] In immunoaffinity chromatography, the concentration of antibody bound to the insoluble support is generally lower and of the same order of magnitude as the dissociation constant of the antigen-antibody complex.[3]

Most of the affinity gels are generally reusable after proper washing. The yield of μ-opiate receptor obtained from a 6-succinyl-morphine-Affigel-102 column decreased with repeated use.[23] By contrast, we found that the capacity of the neurotensin$_{2-13}$-Ultrogel® AcA22 matrix to bind soluble neurotensin receptors increased with time. This increase was directly correlated to a decrease of the leakage of neurotensin$_{2-13}$ bound to the gel, a parameter of primary importance in affinity chromatography.[5] These results show that the very low levels of ligand continuously released by the gel can compete efficiently for binding of the soluble receptor to the matrix and, thus, drastically limit the binding capacity of the affinity gel. Therefore, the amount of ligand released from the affinity column should be monitored continuously. This can be done easily if the ligand coupled to the gel is radio-labeled.[5,22] Contaminating IgG leaking from the antibody column was also observed during purification of the δ-opioid receptor.[3] Thus, ligand leakage seems to be a general phenomenon that can never be avoided completely.

c. Chromatography of Soluble Receptors

Retention of the solubilized material and elution from the affinity gel can be performed either in a batchwise manner[1] or by column chromatography.[2-6] Both techniques involve three consecutive steps: (1) adsorption of the soluble receptor to the affinity gel, (2) extensive washing of the gel, and (3) desorption of the purified receptor. In the last step, recovery of the receptor can be carried out by washing the gel with a buffer containing either a specific ligand of the receptor or a substance that decreases the affinity of the receptor for the affinity gel. As shown in Table 4, the μ-opiate receptor was specifically eluted from affinity gels by micromolar concentrations of naloxone or normorphine.[1,19] In a third study, the μ-opiate receptor could not be eluted from the affinity column with agonist or antagonist solutions. In this case, the receptor protein was recovered by washing the gel with high concentrations of NaCl, which inhibit receptor binding to the μ-agonist bound to the matrix.[2] The high- and low-affinity neurotensin receptors were also eluted with NaCl for the same reason; this salt specifically decreases the affinity of neurotensin for its membrane-bound or soluble receptor.[5,6] Elution with neurotensin solutions resulted in much lower yields of purified receptors. In the immunoaffinity chromatography step used to purify δ-opioid receptors, the covalent ligand-receptor complex was recovered from the antibody column by elution with a 100 μM solution of free ligand followed by a wash with 1 M propionic acid.[3] The variety of results just described shows that no general method can be prescribed *a priori* to elute neuropeptide receptors specifically and in high yields from their respective affinity gels.

TABLE 5
Purification of Neuropeptide Receptors from Mammalian Brain

Receptor	Purification step	Binding capacity (pmol/mg)	Purification factor	Purity (%)	Yield (%)
μ-Opiate receptor from bovine striatum[1]	Membranes	0.2	1	0.0013	100
	Detergent extract	0.19	0.95	0.0012	38.7
	Ligand affinity	914	4,570	5.94	9.6
	WGA[a]	13,000	65,000	84.5	5.8
	Theoretical purity	15,385	76,923	100	—
μ-Opiate receptor from rat brain[2]	Membranes	0.26	1	0.001	100
	Ligand affinity	110	423	0.6	70
	Gel filtration	2,200	8,460	13	39
	WGA[a]	9,200	35,380	53	17
	Isoelectric focusing	17,721	68,158	103	6
	Theoretical purity	17,241	66,313	100	—
δ-Opiate receptor from NG108-15 cells[3]	Detergent extract	0.7	1	0.0035	100
	WGA[a]	37	52.8	0.185	91
	Immunoaffinity	4,320	6,171	21.6	8
	SDS-PAGE	20,800	29,714	104	2
	Theoretical purity	20,000	28,571	100	—
Neurotensin receptor from mouse brain[5]	Membranes	0.25	1	0.0025	100
	Detergent extract	0.25	1	0.0025	100
	SP/HA[b]	0.50	2	0.005	95
	Ligand affinity	7,800	31,200	78	72
	SDS-PAGE	—	—	100	20
	Theoretical purity	10,000	40,000	100	—
Neurotensin receptor from bovine brain[6]	Membranes	0.35	1	0.0025	100
	Detergent extract	0.25	0.71	0.0018	21.2
	Ligand affinity	12,700	36,300	91.3	14.8
	Theoretical purity	13,900	53,460	100	—

[a]   Chromatography on wheat germ agglutinin-agarose,
[b]   Chromatography on two columns of sulfopropy-Sephadex® C-25 and hydroxylapatite connected in series.

3. Postaffinity Step

Although affinity chromatography is an extremely potent technique, it generally cannot provide homogeneous preparations of neuropeptide receptors. It is often necessary to terminate the purification by one or several additional steps (see Table 5). After affinity chromatography on a β-naltrexyl column, the μ-opiate receptor from bovine striatum was only 6% pure. Lectin chromatography of the affinity-purified fraction yielded a single protein with a binding capacity in the range of theoretical homogeneity.[1] To completely purify the μ-opiate receptor from rat brain, three further chromatographic steps were necessary after ligand affinity chromatography, namely, gel filtration, lectin adsorption, and isoelectric focusing.[2] Preparative gel electrophoresis is the technique of choice for the purification of membrane proteins. It has been used as a postaffinity step to obtain pure preparations of neurotensin and δ-opiate receptors (see Table 5).

TABLE 6
Structure of Purified Neuropeptide Receptors

Receptor	Subunit structure		Hydrodynamic structure	
	mol wt(kDa)	SDS-PAGE analysis	mol wt(kDa)	Method
μ-Opiate receptor from bovine striatum[1]	65	Silver staining, radioiodination and covalent labeling	300 — 350	Gel filtration
μ-Opiate receptor from rat brain[2]	58	Silver staining, radioiodination and covalent labeling	>350,110 and 58	Gel filtration
δ-Opiate receptor from NG108-15 cells[3]	60	Silver staining and covalent labeling	~200	Gel filtration and sucrose gradient centrifugation
Somatostatin receptor from rat brain[4]	60	Silver staining and covalent labeling	ND[a]	—
Neurotensin receptor from mouse brain[5]	100	Silver staining, radioiodination and covalent labeling	100	Gel filtration
Neurotensin receptor from bovine brain[6]	72	Silver staining	ND[a]	—

[a] ND, not determined.

4. Quantitative Results

Data concerning the purification of the neuropeptide receptors described earlier are summarized in Table 5. Affinity chromatography is by far the most potent purification process, since it can give a purification factor of several thousands in a single step. Although the efficacy of gel filtration or lectin chromatography appears comparatively limited, these additional methods are essential to achieve complete purification of neuropeptide receptors.

5. Structural and Functional Properties of Purified Neuropeptide Receptors
a. Binding Parameters and Specificity

Neuropeptide receptors purified in the free form bind to their specific radiolabeled ligands in a reversible and saturable manner and with high affinity.[1,2,5,6] The binding specificity does not seem to be affected after solubilization and purification of the receptor. For example, the order of affinity of various neurotensin analogues for the purified neurotensin receptor is the same as that observed with the membrane-bound or the crude soluble receptor.[5]

From the amount of protein estimated by silver staining or amino acid analysis, binding capacities of the purified receptors were calculated to be between 5 and 20 nmol/mg protein (Table 5). These values are close to the theoretical binding capacities of pure receptors calculated on the basis of one ligand binding site per molecule of receptor. Therefore, it seems that the μ- and δ-opioid receptors as well as the high- and low-affinity neurotensin receptors have been purified to apparent homogeneity.

b. Molecular Properties

Fractions of purified receptors corresponding to maximal binding capacities were further analyzed in order to verify that this material was homogeneous (see Table 6). After the last purification step described in Table 5, the μ-, δ,- and neurotensin receptors appeared as a single protein band on SDS-PAGE after Coomassie blue or silver staining. Radioautography of gels obtained from radiolabeled preparations of purified receptors also showed that the

TABLE 7

**Receptors of Peptide Hormones and Growth Factors Whose cDNA Have Been
Cloned after Complete Purification of the Protein Binding Site**

Peptide	Receptor source	Purification		cDNA cloning	
		Main step	Ref.	Precursor length (number of amino acid)	Ref.
Insulin	Human placenta	Immunoaffinity	24,25	1382	26
		Ligand affinity	27	1370	28
Insulin-like growth factor I	Human placenta	Immunoaffinity	29	1210	30
Insulin-like growth factor II	Rat liver cell line BRL-3A	Ligand affinity	31,32	2491	32
Epidermal growth factor	Human placenta or A431 cells	Immunoaffinity	33	1210	34
Platelet-derived growth factor	Mouse BALB/c 3T3 cells	Immunoaffinity	35	1098	36
Interleukin-2	Adult T-cell leukemia	Immunoaffinity or ligand affinity	37	272	37
Atrial natriuretic peptide	Bovine aortic smooth muscle cells	Ligand affinity	38	537	39

radioactivity was associated with a single band. Although these data indicate that the isolated receptors are nearly 100% pure, one could argue that the binding activity is actually associated with minor impurities not detectable by the techniques described earlier. To establish that the single protein band observed on gels is the material responsible for specific binding, the purified receptors were covalently labeled with a specific radioligand and analyzed by SDS-PAGE. The affinity-labeled material appeared in the same position as the silver stained protein band for the μ- and δ-opioid receptors as well as for the high-affinity neurotensin receptor and for the somatostatin receptor.[1-5] By this approach, the major protein constituent of each purified preparation was directly identified with the neuropeptide receptor.

Comparison of the molecular weights of receptor subunits observed on SDS-PAGE with those of the nondenatured receptors indicates that both the μ- and δ-opiate receptors are oligomeric structures. By contrast, the high-affinity neurotensin receptor seems to be constituted of a single protein chain of molecular weight about 100,000 (see Table 6).

E. CONCLUSION

The five neuropeptide receptors discussed earlier have been purified to near homogeneity. Four of them, namely, the μ-opiate receptor, the somatostatin receptor, and both the high- and low-affinity neurotensin binding sites, have been purified in a free and active form, whereas the δ-opiate receptor has been isolated as a covalent complex. These pure preparations are currently used to obtain partial sequences of the receptor proteins. Synthesis of the corresponding radiolabeled oligonucleotide probes as well as production of specific antibodies directed against these receptors will provide useful tools to undertake cloning of their cDNA genes. The methodology described previously has already been used successfully to purify receptors of other types of peptide ligands, such as hormones and growth factors, and to elucidate the complete nucleotide sequence of cDNAs encoding these receptors (see Table 7). In principle, it could be adapted to neuropeptide receptors of gastrointestinal origin, provided that sufficiently rich sources of gut receptors are available.

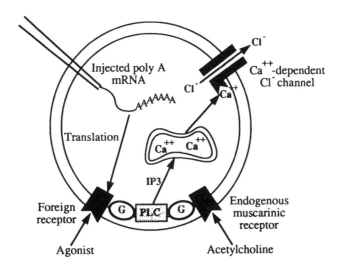

FIGURE 1. Mechanism of chloride channel opening by phospholipase
C-coupled receptors in the oocyte. Abbreviations: G, G proteins; PLC,
phospholipase C; IP_3, inositol trisphosphate.

III. RECEPTOR EXPRESSION IN THE *XENOPUS* OOCYTE: A PROMISING ALTERNATIVE APPROACH FOR RECEPTOR cDNA CLONING

From the preceding sections it is clear that peptide receptor purification is a most difficult task. The reasons for this have been discussed, one of them being the usually low abundancy of peptide receptors in tissues. Recently, an alternative strategy that does not require receptor purification has been used successfully for isolating cDNA encoding functional receptors, thus permitting establishment of the receptor primary structure from the nucleotide sequence of the cDNA. This strategy is based on the ability of *Xenopus* oocytes to functionally express foreign microinjected mRNA coding for a variety of proteins, including membrane receptors. The first receptor thus expressed was the nicotinic acetylcholine receptor following injection into the oocyte of mRNA from muscle tissue.[40] Since then, a number of neurotransmitter and neuropeptide receptors have been expressed upon injection of RNA from brain and other tissues.[7,41-50]

A class of receptors that can be readily expressed and detected at the level of a single oocyte are those receptors coupled in the cells normally expressing them to the phospholipase C pathway. The oocyte is endowed with such a receptor, i.e., the muscarinic acetylcholine receptor (see Figure 1). As in other systems muscarinic agonist binding leads to phospholipase C activation in the oocyte via a G protein. This results in polyphosphoinositide hydrolysis and formation of inositol trisphosphate, which acts as an intracellular second messenger to release calcium from the endoplasmic reticulum into the cytoplasm. Cytosolic calcium opens a calcium-operated chloride channel (see Figure 1), resulting in a strong depolarizing current that can be measured in a single oocyte by electrophysiological means.[44-47] Alternatively, calcium mobilization can be assessed by [$^{45}Ca^{2+}$] efflux measurement in small groups of oocytes.[48]

Injection into the oocyte of foreign mRNA coding for receptors that are coupled in their tissue of origin to phospholipase C usually leads to functional receptor expression, i.e., stimulation by the receptor agonist of calcium mobilization and activation of the calcium-dependent chloride channel. This is possible because the oocyte provides the necessary environment and all machinery required for receptor functioning. This includes the correct

TABLE 8

Cloning Strategy for Isolating cDNA Coding for Receptors in the Oocyte Expression System

1. Size fractionation of poly (A) RNA and assay in the oocyte
2. Construction of a cDNA expression library from active RNA fraction
3. *in vitro* RNA synthesis from pools of cDNA clones
4. Assay of synthetic RNA in the oocyte
5. Subdivision into smaller pools of a group of clones that responded to receptor agonist
6. Repetition of steps 2 - 5 until isolation of a single positive clone (sib selection)
7. Sequencing of cDNA insert

TABLE 9

Calcium Mobilizing Receptors that Have Been Expressed in the Oocyte

Peptide Receptor	mRNA source	Ref.
Angiotensin 2 receptor	AR42J cell line	48
Bombesin receptor	Rat liver	48
Cholecystokinin (CCK) peripheral receptor	AR42J cell line	48
Neurotensin receptor	Rat brain	44,45
Substance P NK-1 receptor	Rat brain	44,46
Neurokinin A (substance K) NK-2 receptor	Bovine stomach	46
TRH receptor	GH$_3$ cell line	47
Vasopressin receptor	Rat liver	48

trafficking and insertion of the foreign receptor into the oocyte plasma membrane, as well as a native phospholipase C system composed of the enzyme and the G protein(s) that couple that foreign receptor to the enzyme (see Figure 1).

So far, two such phospholipase C-coupled receptors, the neurokinin A (substance K) NK-2 receptor[7] and the serotonin 5HT1c receptor,[49,50] have been cloned using the oocyte as a bioassay for RNA purification and clone selection. The cloning procedure similar for the two receptors is summarized in Table 8. In both cases, mRNA was obtained from receptor-rich tissues, i.e., the bovine stomach for the NK-2 receptor and the rat brain choroid plexus for the 5HT1c receptor. In this way, large agonist-induced chloride currents were obtained following mRNA injection in the oocyte.[7,50]

As might be expected, the two receptors were related in primary structure to each other and to the larger family of G protein-coupled receptors, such as the α- and β-adrenergic receptors and the muscarinic acetylcholine M1 and M2 receptors.[51-54] All these receptors consist of a single polypeptide chain of 40,000 to 70,000 M^r with seven hydrophobic transmembrane domains of 20 to 25 amino acids in length and with an extracellular amino terminus and a cytoplasmic carboxy terminus. They are often referred to as being in the rhodopsin-like receptor family because of their similarity to the photoreceptor.[52]

Because this chapter deals with peptide receptors, a more specific description will be made of the tachykinin NK-2 receptor. Tachykinin receptors are divided into three subtypes: NK-1 (substance P preferring), NK-2 (neurokinin A preferring), and NK-3 (neurokinin B preferring).[55] The NK-2 receptor consists of a 384-amino-acid polypeptide chain with a molecular weight of 43,066 Da in its unglycosylated form. There are two potential *N*-glycosylation sites (Asn 11 and Asn 19) in its amino-terminal region. It shares 24%, 26%, and 22% of sequence identity with the β-adrenergic, the M1 muscarinic, and the M2 muscarinic receptors, respectively.[7]

Table 9 lists the neuropeptides that have been shown to stimulate calcium mobilization or chloride current following mRNA injection into the oocyte. The corresponding sources of mRNA are also indicated. Theoretically, the cloning procedure described earlier could be applicable to the receptors of these peptides. Because all these receptors share a common mechanism of transduction, i.e., activation of phospholipase C via G protein(s), it may be anticipated that they will also share some sequence homology with the NK-2 receptor and the other members of the super family of G protein-coupled receptors.

Although the NK-2 receptor represents the only neuropeptide receptor whose primary structure has been elucidated so far, it may be expected that the number will grow in the near future. Whether classical purification techniques or indirect cloning strategies such as the one described here will do the job remains to be seen.

ADDENDUM

While this paper was in press, three neuropeptide receptors have been cloned by Nakanishi and his group using the *Xenopus* oocyte expression system described here. These are NKI (Yokota et al., *J. Biol. Chem.*, 264, 17649, 1989), the NK2 (Shigemoto et al., *J. Biol. Chem.*, 265, 623, 1990) and the neurotensin receptors (Tanaka et al., *Neuron*, 4, 847, 1990).

ACKNOWLEDGMENTS

We wish to thank Valérie Dalmasso for expert secretarial assistance. This work was supported by the Centre National de la Recherche Scientifique, the Institut National de la Santé et de la Recherche Médicale (CRE 886017), and the Fondation pour la Recherche Médicale.

REFERENCES

1. **Gioannini, T. L., Howard, A. D., Hiller, J. M., and Simon, E. J.,** Purification of an active opioid-binding protein from bovine striatum, *J. Biol. Chem.*, 260, 15117, 1985.
2. **Cho, T. M., Hasegawa, J. I., Ge, B. L., and Loh, H. H.,** Purification to apparent homogeneity of a μ-type opioid receptor from rat brain, *Proc. Natl. Acad. Sci. U.S.A.*, 83, 4138, 1986.
3. **Simonds, W. F., Burke, T. R., Rice, K. C., Jacobson, A. E., and Klee, W. A.,** Purification of the opiate receptor of NG108-15 neuroblastoma-glioma hybrid cells, *Proc. Natl. Acad. Sci. U.S.A.*, 82, 4974, 1985.
4. **He, H. T., Johnson, K., Thermos, K., and Reisine, T.,** Purification of a putative brain somatostatin receptors, *Proc. Natl. Acad. Sci. U.S.A.*, 86, 1480, 1989.
5. **Mazella, J. ,Chabry, J., Zsürger, N., and Vincent, J. P.,** Purification of the neurotensin receptor from mouse brain by affinity chromatography, *J. Biol. Chem.*, 264, 5559, 1989.
6. **Mills, A., Demoliou-Mason, C. D., and Barnard, E. A.,** Purification of the neurotensin receptor from bovine brain, *J. Biol. Chem.*, 263, 13, 1988.
7. **Masu, Y., Nakayama, K., Tamaki, H., Harada, Y., Kuno, M., and Nakanishi, S.,** cDNA cloning of bovine substance-K receptor through oocyte expression system, *Nature (London)*, 329, 836, 1987.
8. **Martin, W. R., Eades, C. G., Thompson, J. A., Huppler, R. E., and Gilbert, P. E.,** The effects of morphine- and nalorphine-like drugs in the nondependent and morphine-dependent chronic spinal dog, *J. Pharmacol. Exp. Ther.*, 197, 517, 1976.
9. **Lord, J. A. H., Waterfield, A. A., Hughes, J., and Kosterlitz, H. W.,** Endogenous opioid peptides: multiple agonists and receptors, *Nature (London)*, 267, 495, 1977.
10. **Czernik, A. J. and Petrack, B.,** Somatostatin receptor binding in rat cerebral cortex. Characterization using a nonreducible somatostatin analog, *J. Biol. Chem.*, 258, 5525, 1983.
11. **Reubi, J. C., Perrin, M. H., Rivier, J., and Vale, W.,** High affinity binding sites for a somatostatin-28 analog in rat brain, *Life Sci.*, 28, 2191, 1981.
12. **Reubi, J. C., Perrin, M., Rivier, J., and Vale, W.,** High affinity binding sites for somatostatin to rat pituitary, *Biochem. Biophys. Res. Commun.*, 105, 1538, 1982.

13. **Srikant, C. B. and Patel, Y. C.,** Somatostatin receptors on rat pancreatic acinar cells. Pharmacological and structural characterization and demonstration of down-regulation in streptozotocin diabetes, *J. Biol. Chem.,* 261, 7690, 1986.

14. **Tran, V. T., Beal, M. F., and Martin, J. B.,** Two types of somatostatin receptors differentiated by cyclic somatostatin analogs, *Science,* 228, 492, 1985.

15. **Kitabgi, P., Rostène, W., Dusaillant, M., Schotte, A., Laduron, A., and Vincent, J. P.,** Two populations of neurotensin binding sites in murine brain: discrimination by the antihistamine levocabastine reveals markedly different radioautographic distribution, *Eur. J. Pharmacol.,* 140, 285, 1987.

16. **Kitabgi, P., Kwan, C. Y., Fox, J. E. T., and Vincent, J. P.,** Characterization of neurotensin binding to rat gastric smooth muscle receptor sites, *Peptides,* 5, 917, 1984.

17. **Kitabgi, P., Checler, F., Mazella, J., and Vincent, J. P.,** Pharmacology and biochemistry of neurotensin receptors, *Rev. Basic Clin. Pharmacol.,* 5, 397, 1985.

18. **Howells, R. D., Gioannini, T. L., Hiller, J. J., and Simon, E. J.,** Solubilization and characterization of active opiate binding sites from mammalian brain, *J. Pharmacol. Exp. Ther.,* 222, 629, 1982.

19. **Maneckjee, R., Zukin, R. S., Archer, S., Michael, J., and Osei-Gyimah, P.,** Purification and characterization of the μ-opiate receptor from rat brain using affinity chromatography, *Proc. Natl. Acad. Sci. U.S.A.,* 82, 594, 1985.

20. **Mazella, J., Chabry, J., Kitabgi, P., and Vincent J. P.,** Solubilization and characterization of active neurotensin receptors from mouse brain, *J. Biol. Chem.,* 263, 144, 1988.

21. **Checler, F., Vincent, J. P., and Kitabgi, P.,** Purification and characterization of a novel neurotensin-degrading peptidase from rat brain synaptic membranes, *J. Biol. Chem.,* 261, 11274, 1986.

22. **Gioannini, T. L., Howard, A. Hiller, J. M., and Simon, E. J.,** Affinity chromatography of solubilized opioid binding sites using CH-Sepharose modified with a new naltrexone derivative, *Biochem. Biophys. Res. Commun.,* 119, 624, 1984.

23. **Cho, T. M., Ge, B. L., and Loh, H. H.,** Isolation and purification of morphine receptor by affinity chromatography, *Life Sci.,* 36, 1075, 1985.

24. **Harrison, L. C. and Itin, A.,** Purification of the insulin receptor from human placenta by chromatography on immobilized wheat germ lectin and receptor antibody, *J. Biol. Chem.,* 255, 12066, 1980.

25. **Roth, R. A., Mesirow, M. L., and Cassell, D. J.,** Preferential degradation of the β subunit of purified insulin receptor. Effect of insulin binding and protein kinase activities of the receptor, *J. Biol. Chem.,* 258, 14456, 1983.

26. **Ebina, Y., Ellis, L., Jarnagin, K., Edery, M., Graf, L., Clauser, E., Ou, J. H., Masiarz, F., Kan, Y. W., Goldfine, I. D., Roth, R. A., and Rutter, W. J.,** The human insulin receptor cDNA: the structural basis for hormone-activated transmembrane signalling, *Cell,* 40, 747, 1985.

27. **Petruzzelli, L., Herrera, R., and Rosen, O. M.,** Insulin receptor is an insulin-dependent tyrosine protein kinase: copurification of insulin-binding activity and protein kinase activity to homogeneity from human placenta, *Proc. Natl. Acad. Sci. U.S.A.,* 81, 3327, 1984.

28. **Ullrich, A. Bell, J. R., Chen, E. Y., Herrera, R., Petruzzelli, L. M., Dull, T. J., Gray, A., Coussens, L., Liao, Y. C., Tsubokawa, M., Mason, A., Seeburg, P. H., Grunfeld, C., Rosen, O. M., and Ramachandran, J.,** Human insulin receptor and its relationship to the tyrosine kinase family of oncogenes, *Nature (London),* 313, 756, 1985.

29. **LeBon, T. R., Jacobs, S., Cuatrecasas, P., Kathuria, S., and Fujita-Yamaguchi, Y.,** Purification of insulin-like growth factor I receptor from human placental membranes, *J. Biol. Chem.,* 261, 7685, 1986.

30. **Ullrich, A., Gray, A., Tam, A. W., Yan-Feng, T., Tsubokawa, M., Collins, C., Henzel, W., Le Bon, T., Kathuria, S., Chen, E., Jacobs, S., Francke, U., Ramachandran, J., and Fujita-Yamaguchi, Y.,** Insulin-like growth factor I receptor primary structure: comparison with insulin receptor suggests structural determinants that define functional specificity, *EMBO J.,* 5, 2503, 1986.

31. **Hari, J., Pierce, S. B., Morgan, D. O., Sara, V., Smith, M. C., and Roth, R. A.,** The receptor for insulin-like growth factor II mediates an insulin-like response, *EMBO J.,* 6, 3367, 1987.

32. **Morgan, D. O., Edman, J. C., Standring, D. N., Fried, V. A., Smith, M. C., Roth, R. A., and Rutter, W. J.,** Insulin-like growth factor II receptor as a multifunctional binding protein, *Nature (London),* 329, 301, 1987.

33. **Downward, J. Yarden, Y., Mayes, E., Scrace, G., Totty, N., Stockwell, P., Ullrich, A., Schlessinger, J., and Waterfield, M. D.,** Close similarity of epidermal growth factor receptor and v-*erb*-B oncogene protein sequences, *Nature (London),* 307, 521, 1984.

34. **Ullrich, A., Coussens, L., Hayflick, J. S., Dull, T. J., Gray, A., Tam, A. W., Lee, J., Yarden, Y., Libermann, T. A., Schlessinger, J., Downward, J., Mayes, E. L. V., Whittle, N., Waterfield, M. D., and Seeburg, P. H.,** Human epidermal growth factor receptor cDNA sequence and aberrant expression of the amplified gene in A431 epidermoid carcinoma cells, *Nature (London),* 309, 418, 1984.

35. **Daniel, T. O., Tremble, P. M., Frackelton, A. R., and Williams, L.T.,** Purification of the platelet-derived growth factor receptor by using an anti-phosphotyrosine antibody, *Proc. Natl. Acad. Sci. U.S.A.,* 82, 2684, 1985.

36. **Yarden, Y. Escobedo, J. A., Kuang, W. J., Yang-Feng, T. L., Daniel, T. O., Tremble, P. M., Chen, E. Y., Ando, M. E., Harkins, R. N., Francke, U., Fried, V. A., Ullrich, A., and Williams, L. T.,** Structure of the receptor for platelet-derived growth factor helps define a family of closely related growth factor receptors, *Nature (London), 323,* 226, 1986.

37. **Nikaido, T., Shimizu, A., Ishida, N., Sabe, H., Teshigawara, K., Maeda, M., Uchiyama, T., Yodoi, J., and Honjo, T.,** Molecular cloning of cDNA encoding human interleukin-2 receptor, *Nature (London),* 311, 631, 1984.

38. **Schenk, D. B., Phelps, N. M., Porter, J. G., Fuller, F., Cordell, B., and Lewicki, J. A.,** Purification and subunit composition of atrial natriuretic peptide receptor, *Proc. Natl. Acad. Sci. U.S.A.,* 84, 1521, 1987.

39. **Fuller, F., Porter, J. G., Arfsten, A. E., Miller, J., Schilling, J. W., Scarborough, R. M., Lewicki, J. A., and Schenk, D. B.,** Atrial natriuretic peptide clearance receptor. Complete sequence and functional expression of cDNA clones, *J. Biol. Chem.,* 263, 9395, 1988.

40. **Barnard, E. A., Miledi, R., and Sumikawa, K.,** Translation of exogenous messenger RNA coding for nicotinic acetylcholine receptors produces functional receptors in *Xenopus* oocytes, *Proc. R. Soc. London.,* 215, 241, 1982.

41. **Gundersen, C. B., Miledi, R., and Parker, I.,** Serotonin receptors induced by exogenous messenger RNA in Xenopus oocytes, *Proc. R. Soc. London,* 219, 103, 1983.

42. **Gundersen, C. B., Miledi, R., and Parker, I.,** Properties of human brain glycine receptors expressed in *Xenopus* oocytes, *Proc. R. Soc. London,* 221, 1984.

43. **Sumikawa, K., Parker, I., and Miledi, R.,** Partial purification and functional expression of brain mRNAs coding for neurotransmitter receptors and voltage-operated channels, *Proc. Natl. Acad. Sci. U.S.A.,* 81, 7994, 1984.

44. **Parker, I., Sumikawa, K., and Miledi, R.,** Neurotensin and substance P receptors expressed Xenopus oocytes by messenger RNA from rat brain, *Proc. R. Soc. London,* 229, 151, 1986.

45. **Hirono, C., Ito, I., and Sugiyama, H.,** Neurotensin and acetylcholine evoke common responses in frog oocytes injected with rat brain messenger ribonucleic acid, *J. Physiol.,* 382, 523, 1987.

46. **Harada, Y., Takahashi, T., Kuno, M., Nakayama, K., Masu, Y., and Nakanishi, S.,** Expression of two different tachykinin receptors in *Xenopus* oocytes by exogenous mRNAs, *J. Neurosci.,* 7, 3265, 1987.

47. **Oron, Y., Straub, R. E., Traktman, P., and Gershengorn, M. C.,** Decreased TRH receptor mRNA activity precedes homologous downregulation: assay in oocytes, *Science,* 238, 1406, 1987.

48. **Williams, J. A., McChesney, D. J., Calayag, M. C., Lingappa, V. R., and Logsdon, C. D.,** Expression of receptors for cholecystokinin and other Ca^{2+} mobilizing hormones in *Xenopus* oocytes, *Proc. Natl. Acad. Sci. U.S.A.,* 85, 4939, 1988.

49. **Lubbert, H., Hoffman, B. J., Snutch, T. P., Van Dyke, T., Levine, A. J., Hartig, P. R., Lester, H. A., and Davidson, N.,** cDNA cloning of a serotonin 5-HT_{1C} receptor by electrophysiological assays of mRNA-injected *Xenopus* oocytes, *Proc. Natl. Acad. Sci. U.S.A.,* 84, 4332, 1987.

50. **Julius, D., McDermott, A. B., Axel, R., and Jessel, T. M.,** Molecular characterization of a functional cDNA encoding the serotonin 1c receptor, *Science,* 241, 558, 1988.

51. **Kobilka, B. K., Matsui, H., Kobilka, T. S., Yang-Feng, T. L., Francke, U., Caron, M. G., Lefkowtiz, R. J., and Regan, J. W.,** Cloning, sequencing, and expression of the gene encoding for the human platelet α_2-adrenergic receptor, *Science,* 238, 650, 1987.

52. **Dixon, R. A. F., Kobilka, B. K., Strader, D. J., Benovic, J. L., Dohlman, H. G., Frielle, T., Bolanowski, M. A., Bennett, C. D., Rands, E., Diehl, R. E., Mumford, R. A., Slater, E. E., Sigal, I. S., Caron, M. G., Lefkowitz, R. J., and Strader, C. D.,** Cloning of the gene and cDNA for mammalian β-adrenergic receptor and homology with rhodopsin, *Nature (London),* 321, 75, 1986.

53. **Kubo, T., Fukuda, K., Mikami, A., Maeda, A., Takahashi, H., Mishina, M., Haga, T., Haga, K., Ichiyama, A., Kangawa, K., Kojima, M., Matsuo, H., Hirose, T., and Numa, S.,** Cloning, sequencing and expression of complementary DNA encoding the muscarinic acetylcholine receptor, *Nature (London),* 323, 411, 1986.

54. **Kubo, T., Maeda, A., Sugimoto, K., Akiba, I., Mikami, A., Takahashi, H., Haga, T., Haga, K., Ichiyama, A., Kangawa, K., Matsuo, H., Hirose, T., and Numa, S.,** Primary structure of porcine cardiac muscarinic acetylcholine receptor deduced from the cDNA sequence, *FEBS Lett.,* 209, 367, 1986.

55. **Quirion, R.,** Multiple tachykinin receptors, *Trends Neurosci.,* 8, 183, 1985.

Chapter 9

GASTROINTESTINAL NEUROPEPTIDES AND SECOND MESSENGER SYSTEMS

Peter Kostka

TABLE OF CONTENTS

I.   Introduction 250

II.   General Considerations 250

III.   Second Messenger Systems Activated by Peptidergic Receptors 251
     A.   Vasoactive Intestinal Peptide 251
     B.   Somatostatin 253
     C.   Cholecystokinin 254
     D.   Opioid Peptides 256
     E.   Neurotensin 258
     F.   Substance P 259

IV.   Interactions between Second Messenger Systems 261

V.   Concluding Remarks 262

Appendix 263

Acknowledgment 263

References 263

I. INTRODUCTION

This chapter is intended to summarize the present knowledge about essential features of the mode of coupling of peptidergic receptors with second messenger systems in the gastrointestinal tissues. It is now well recognized that various tissues of the gastrointestinal tract contain diverse families of neuropeptides. Several of these peptides, besides their neuromodulatory role, also act as paracrine or endocrine agents.

Neuropeptides interact with a variety of target cells in the gastrointestinal tract and are involved in the control of multiple physiological functions. Such functions include the regulation of absorptive and secretory processes in epithelial cells, control of the contractile activity of gastrointestinal smooth muscles, modulation of neurotransmission in enteric ganglia, and the regulation of gastrointestinal blood flow. It is beyond the scope of this chapter to discuss the functional ramifications related to the activation of peptidergic receptors. Such information is presented elsewhere in this volume or can be found in several recent reviews.[1-5]

The literature survey for this chapter was concluded in March of 1989.

II. GENERAL CONSIDERATIONS

The majority of extracellular chemical signals (with the notable exception of steroid hormones) are recognized by specific receptors located on the plasma membrane of target cells. The propagation of these signals into the cellular interior is achieved by the interaction of activated receptors with discrete functional moieties commonly referred to as transducing systems. For most receptors, the transducing system involves a guanine nucleotide-binding protein (G-protein)[6-9] which couples the receptor recognition site to a particular effector, such as an enzyme or ion channel.

At present, three categories of signaling mechanisms have been clearly recognized:

1. Modulation of adenylate cyclase activity[10-12]
2. Activation of phospholipase C with subsequent enhancement of the turnover of phosphoinositides (PI) and Ca mobilization[13-15]
3. Alterations of ion fluxes across the plasma membrane[16,17]

The important feature of transmembrane signaling is that the activation of a particular transducing system in response to the extracellular mediator is primarily determined by the nature of the receptor rather than by the nature of the activating agent. Perhaps the best illustration of this point is offered by adrenergic receptors, where nonselective agents such as epinephrine or norepinephrine can either activate (through β-receptors),[18,19] or inhibit (through α_2-receptors)[12] the adenylate cyclase or cause Ca mobilization (through α_1-receptors).[20]

In spite of this potential heterogeneity of coupling mechanisms for a particular mediator, it appears that some families of neuropeptides activate similar transducing pathways in a variety of tissues. For example, cholecystokinin and gastrin have been shown to enhance the turnover of phosphoinositides in a number of experimental models. Similarly, vasoactive intestinal peptide (VIP) and structurally related peptides, such as secretin, peptide histidine methionine (PHM), and peptide histidine isoleucine (PHI), activate the adenylate cyclase in most systems studied so far. However, one can also argue that the clear-cut delineation of receptor subtypes for a number of neuropeptides is at the present time only tentative because selective antagonists for various peptidergic receptors are not readily available. Therefore, the extrapolation of findings about the mode of coupling from one experimental model to another may be unwarranted.

The studies of individual transducing systems and their components have different requirements in various experimental settings. The tissues of the gastrointestinal tract offer more or less suitable models for a particular type of study, resulting in an uneven progress in characterization of transducing mechanisms for various neuropeptides in different tissues. For example, dispersed pancreatic acinar cells represent a popular model for studies of agonist-induced activation of phospholipase C and Ca mobilization, and the signaling mechanism activated by cholecystokinin or acetylcholine in these cells is relatively well characterized. On the other hand, very little is known about the biochemical correlates underlying the neuromodulatory effects of neuropeptides in enteric ganglia.

In the following discussion of transducing pathways activated by a particular neuropeptide the emphasis will be on findings derived from studies on gastrointestinal tissues. The discussion also will include some of the findings derived from other experimental models, particularly for those peptides where the information about signaling mechanisms in gut is limited.

III. SECOND MESSENGER SYSTEMS ACTIVATED BY PEPTIDERGIC RECEPTORS

A. VASOACTIVE INTESTINAL PEPTIDE

The activation of receptors for VIP is linked to the stimulation of adenylate cyclase activity and subsequent elevation of intracellular levels of cAMP. Such a mechanism of transmembrane signaling has been identified in most systems studied so far, and in general there is a good correlation between the ability of VIP to activate adenylate cyclase and its ability to elicit the appropriate physiological response.

In the tissues of the gastrointestinal tract the presence of VIP-sensitive adenylate cyclase has been demonstrated in gastric and intestinal epithelia,[21-24] pancreas,[25,26] and liver.[27,28] The coupling of VIP receptors with the enzyme involves an intermediate G protein (G$_s$), as documented by the ability of GTP to modulate the receptor affinity for agonists[29,30] and to enhance the magnitude of stimulation of adenylate cyclase by VIP[31] and by the susceptibility of the G-protein to ADP-ribosylation by cholera toxin.[32]

The activation of VIP receptors in several gastrointestinal smooth muscles elicits a relaxing effect,[33-37] and this peptide is one of the candidates for nonadrenergic, noncholinergic inhibitory mediator.[38,39] At present, it is not clear whether the relaxing effect of VIP can be explained solely by the ability of this peptide to stimulate adenylate cyclase. The demonstration of elevated levels of cAMP in smooth muscles in response to VIP usually requires the inclusion of phosphodiesterase inhibitor.[40,41] In the absence of an inhibitor of phosphodiesterase, elevated levels of cAMP in response to exogenous VIP have been reported for the lower esophageal sphincter of the opossum.[42] However, in the same study, the relaxing response due to electrical field stimulation was accompanied by a rise in the level of cGMP, with no detectable increase in cAMP. Since the effects of exogenous VIP on the cyclic nucleotide content were studied in the absence of tetrodotoxin, the possibility remains that the elevated levels of cAMP could result from some indirect presynaptic effects as opposed to direct activation of VIP receptors on smooth muscle cells.

The difficulties with the reliable detection of VIP-stimulated adenylate cyclase in gastrointestinal smooth muscle may be related to a high activity of endogenous phosphodiesterase. In support of such a contention are recent observations that substance SK&F 94,120, a specific inhibitor of cAMP phosphodiesterase, significantly potentiates field-stimulated relaxation in the lower esophageal sphincter of the opossum.[43] Alternatively, the relaxing effect of VIP could be mediated by some additional transducing mechanism(s).

There is growing evidence that in several VIP-sensitive tissues the activation of VIP receptors is linked to the modulation of processes elicited by Ca-mobilizing agents, and vice versa. For example, in colonic epithelial cells, VIP-induced chloride secretion was found to be significantly potentiated in the presence of the calcium ionophore A23107.[44] Also, Laburthe et al.[45] have identified the presence of VIP receptors on mucus-secreting colonic cells. The activation of these receptors and subsequent rise in the level of cAMP did not have a significant effect on the basal mucin secretion; however, the secretory response elicited by carbachol was greatly potentiated in the presence of VIP. In pancreatic acinar cells, the secretory response elicited by VIP or dibutyryl cAMP was significantly enhanced by Ca-mobilizing secretagogues (the cholecystokinin [CCK] octapeptide, carbachol),[46] and VIP was reported to enhance the ability of CCK to activate the sodium pump.[47]

The biochemical correlates which underlie the synergism between VIP- and Ca-mediated events are at present only partially understood. In all likelihood, the potentiation mechanisms will involve multiple pathways which may vary from tissue to tissue. In some systems, the activation of VIP receptors was shown to elicit the transient rise in Ca_i concomitantly with the elevation of cAMP level. For example, in rat parotid cells, the VIP-induced secretory response was accompanied by an increase in Ca efflux, and the response was only partially inhibited by the removal of external Ca.[48] Since a substantially higher elevation of Ca efflux was observed in response to carbachol, it was proposed that the VIP-accessible Ca pool may be different from that mobilized by carbachol.

A similar VIP-induced increase in Ca_i was also demonstrated in rat pituitary cells.[49] In these cells, the Ca transient could be mimicked by forskolin or dibutyryl cAMP, indicating that the VIP-induced Ca release occurred secondarily to the activation of adenylate cyclase. On the other hand, the VIP-induced Ca release was absent in Swiss 3T3 cells, where VIP greatly potentiated the mitotic activity in the presence of insulin.[50]

In the pancreatic acinar cells, the synergistic effect of VIP and Ca-mobilizing agents apparently occurs at a site distal to the generation of second messengers because VIP did not elicit any rise in Ca_i and Ca-mobilizing agents exhibited no effect on the intracellular levels of cAMP.[46]

The interpretation of these findings in the integrated scheme of postreceptor molecular events is at present difficult. For several receptor types, the occupancy by an agonist will alter the equilibrium between free and associated subunits of a G protein. This event as such may have an effect on the response to another agonist. For example, VIP has been shown to increase the affinity of agonists for muscarinic receptors in the cat submandibular gland.[51] The receptor-induced generation of second messengers and subsequent alterations in the activities of protein kinases A and C will result in a modified pattern of phosphorylation of multiple intracellular proteins which may involve those participating in signal transduction (see Section IV).

The modulation of VIP-mediated responses also may involve endogenous mediators generated by the activation of phospholipase A_2. Such a mechanism has been demonstrated in the cerebral cortex, where the magnitude of stimulation of adenylate cyclase activity by VIP was enhanced by the activation of α_1-receptors.[52] This potentiating effect involved the release of arachidonate and could be mimicked by exogenous prostaglandins. Clearly, the elucidation of an overall pattern of molecular interactions elicited by activation of multiple receptors will require a more detailed characterization of individual steps involved in the propagation of a signal from the receptors to the physiological effects.

The VIP receptors in various tissues have been shown to be activated, to varying degrees, by structurally related peptides such as secretin, PHM, and PHI.[26,27,37] The relative potencies of VIP and secretin on pancreatic receptors provided a basis for the classification of VIP receptors in this tissue.[26,27] However, VIP- and secretin-preferring receptors seem to be similar in their ability to activate adenylate cyclase.

The studies of VIP receptors by ligand binding and/or covalent cross-linking in rat intestinal epithelium and liver also demonstrated the presence of VIP binding sites insensitive to GTP (or its analogues), and the activation of these receptors did not stimulate adenylate cyclase activity.[28,29] These receptors had a low affinity for VIP and a lower molecular weight as compared to high-affinity, adenylate cyclase-linked sites. These observations raised the possibility that the low-affinity sites may represent both structurally and functionally different types of VIP receptors. However, recent findings do not support such a contention. Rather, the low-affinity, low-molecular-weight VIP binding sites may represent receptors which are either internalized or uncoupled from a G protein.[32,53,54]

B. SOMATOSTATIN

Somatostatin (Som) is present in biological tissues in two molecular forms, Som 1—14 and Som 1—28. The relative ratio of these two forms varies from tissue to tissue. Both forms of Som share similar biological activities and seem to be acting on the same population of receptors.[55]

The initial studies on the mode of coupling of Som receptors with second messenger systems provided evidence that the activation of these receptors is linked to the inhibition of adenylate cyclase. Such a mechanism was documented for a variety of target tissues, including pancreatic acini,[56,57] pituitary secretory cells,[58-60] gastric and intestinal epithelia,[22,61,62] and S49 lymphoma cells.[63] The coupling of receptors with adenylate cyclase involves the intermediation by a G_i protein, as documented by the modulation of Som binding in the presence of GTP or its nonhydrolyzable analogues as well as by the sensitivity of the inhibitory mechanism to pertussis toxin.[63-65]

However, it soon became recognized that the somatostatin-mediated inhibition of adenylate cyclase cannot fully account for the ability of this peptide to modulate the intracellular responses. Such a contention was supported by observations that Som was capable of antagonizing the responses elicited by forskolin or dibutyryl cAMP or those triggered by agents linked to second messengers other than adenylate cyclase.[62,66,67] These findings suggested the presence of an additional signaling pathway acting independently of the interaction of Som with adenylate cyclase.

One such pathway was identified in pituitary and neuroblastoma × glioma hybrid cells NG108-15, where Som was shown to decrease the conductance of voltage-dependent Ca channels.[68,69] In pituitary cells, the inhibition of Ca channels occurred with no apparent involvement of cAMP and was mediated through a G protein which is a substrate for pertussis toxin.[70-72]

There is evidence to suggest that Som also can control the Ca homeostasis in other target tissues. For example, it has been demonstrated that Som can inhibit parietal cell proton secretion elicited by gastrin and carbachol.[62] The activating effects of these two mediators are known to be accompanied by an increased PI turnover and Ca mobilization. The blocking effect of Som on gastrin- and carbachol-induced proton secretion did not impair the ability of these secretagogues to activate the PI turnover, indicating that Som-induced inhibition of secretory response was exerted at a site distal to the receptor-linked activation of phospholipase C. Such a site may involve the hormone-sensitive Ca pool; however, such a hypothesis requires further evaluation.

Yet another putative signaling mechanism linked to Som receptors was elucidated along with the studies which examined the subcellular distribution of Som receptors. In fractionation studies of rat tissues such as pancreas and intestinal mucosa, a considerable proportion of Som receptors was found to be present in cytosol.[73,74] The cytosolic receptors displayed comparable binding affinity for Som as the membrane-bound receptors. The activation of cytosolic receptors was found to stimulate the phosphoprotein phosphatase activity, as documented by the ability of cytosol, in the presence of Som, to hydrolyze phosphorylated histone proteins. Interestingly, similar phosphoprotein phosphatase activity also was demonstrated for membrane-bound receptors in synaptosomal plasma membranes[75] and in the pancreatic cell line MIA PaCa-2.[76] In the latter experimental model, the activation of Som receptors was shown to dephosphorylate the tyrosine residue in the protein kinase domain of the receptor for epidermal growth factor. It was suggested that this mechanism may account for the ability of Som to antagonize the epidermal growth factor-mediated responses.

At present, it remains to be established whether the interaction of Som with multiple second messenger pathways is achieved by the activation of different receptor subtypes. In the pituitary cells this does not seem to be the case, since Som-mediated effects on adenylate cyclase and on Ca influx share similar pharmacological profiles and are equally sensitive to pertussis toxin.[66,67,72] However, in parietal cells the cAMP-independent effects were found to be insensitive to pertussis toxin.[62] Since these cells also have Som receptors coupled to the inhibition of adenylate cyclase, the cAMP-dependent and cAMP-independent effects of Som may be mediated by receptors with fundamentally different coupling to second messenger systems.

The pharmacological studies of Som receptors in various tissues have not as yet indicated substantial differences between receptors present in various tissues. In binding experiments, Som receptors from a variety of target tissues have similar affinities for Som (median K_d values are between 0.1 and 1 nM^{77}). Although displacement experiments with unlabeled Som often result in shallow displacement isotherms, the heterogeneity of binding sites seems to be primarily related to the agonist-induced desensitization and/or interconversion between high- and low-affinity states of the pharmacologically homologous receptor population.[78,79] This is supported by observation that the changes in the affinity of these receptors may also be elicited by GTP (or its stable analogues) or by treatment with pertussis toxin.[65,71]

It is well documented that Som is also involved in the control of gastrointestinal motility and intestinal transit.[80] The regulatory effects of Som seem to be primarily attributed to the neuromodulatory role of this peptide in the enteric nervous system.[81] In the myenteric plexus of guinea pig ileum, Som reduces the frequency of neuronal discharge and blocks the evoked release of acetylcholine.[82,83] The underlying biochemical correlates of these effects remain to be established.

C. CHOLECYSTOKININ

CCK belongs to the category of Ca-mobilizing mediators. The activation of CCK receptors in a variety of tissues is followed by stimulation of phospholipase C, increased turnover of phosphoinositides, and subsequent rise in Ca_i.

The signaling pathway activated by CCK has been studied extensively on dispersed pancreatic acinar cells. This experimental model is at present one of the best characterized systems activated by Ca-mobilizing agonists.

The CCK-mediated activation of phospholipase C in pancreatic acinar cells occurs at resting levels of Ca_i, indicating that this process does not require a prior activation of calcium influx.[84] In permeabilized acinar cells, the CCK-induced activation of phospholipase C was shown to be potentiated by stable analogues of GTP, indicating the involvement of a G protein in the coupling mechanism.[85] The nature of this protein is not clearly defined, and there are conflicting reports regarding its sensitivity to pertussis and cholera toxins. One group has reported that the potentiating effect of guanyl nucleotides was insensitive to the preincubation of cells with either cholera or pertussis toxin, and it was suggested that the transducing pathway may involve a G protein which is distinct from G_s or G_i.[85] Recently, another laboratory has reported that the G protein mediating the coupling of CCK receptors to phospholipase C in pancreatic acini is a substrate for cholera toxin. However, both molecular and chemical properties of the G protein were distinct from G_s.[86]

The CCK-mediated activation of PI turnover in pancreatic acini was demonstrated to be negatively regulated by arachidonic acid. Exogenous arachidonic acid was shown to decrease the steady-state labeling of [^{32}P]-phosphoinositol 4,5-bisphosphate concomitantly with reduction of incorporation of [^{3}H]myo-inositol into inositol phospholipids.[87] These effects also could also be reproduced by arachidonic acid liberated from endogenous phospholipids.[88]

Accelerated PI hydrolysis and elevated intracellular levels of inositol 1,4,5-trisphosphate (IP$_3$) result in the mobilization of Ca from intracellular stores, as documented by the insensitivity of Ca transient to the removal of external calcium.[89] The release of Ca by submaximal doses of agonist was shown to be further enhanced by the subsequent addition of a maximally effective agonist concentration,[90] suggesting that the amount of Ca released from the IP$_3$-sensitive pool is proportional to the amount of IP$_3$ generated by receptor activation. During the agonist stimulation, the intracellular Ca stores remain in a depleted state because the regeneration of the Ca pool (a process dependent on extracellular Ca) requires the termination of receptor activation.[90]

The propagation of signal to the secretory response seems to involve both IP$_3$- and protein kinase C-mediated processes. The CCK-induced Ca transient was shown to be accompanied by amylase release. However, the secretory response persisted even after the Ca$_i$ returned to the basal level.[89] The sustained phase of CCK-induced secretory response could be mimicked by the preincubation of acinar cells with phorbol esters. In phorbol ester-preincubated cells the initial phase of the secretory response, but not the sustained phase, could be further augmented by exogenous IP$_3$, thus indicating the synergistic effect of IP$_3$- and protein kinase C-sensitive pathways on the secretory response.[89] It appears that IP$_3$-induced Ca release is primarily involved in the mediation of short-term secretory activity, while diacylglycerol-activated protein kinase C may be involved in the regulation of long-term, sustained release.

In addition, the activation of protein kinase C was also implicated in the desensitization of CCK-mediated secretory responses. The prolonged exposure of rabbit pancreatic acini to phorbol esters was shown to attenuate the secretory response to CCK as compared to untreated controls.[91] Such an effect could not be attributed to the down-regulation or desensitization of CCK receptors, as suggested by the observation that the treatment did not impair the ability of CCK to promote the hydrolysis of phosphoinositides. However, treatment with phorbol ester compromised the ability of CCK to induce the increase in Ca$_i$.[91] It is conceivable that the treatment with phorbol esters could desensitize the intracellular receptor for IP$_3$. Such a possibility has yet to be examined.

CCK-mediated activation of protein kinase C was also implicated in the regulation of internal pH in acinar cells. Bastie et al.[92] have recently reported that the activation of CCK receptors activates the sodium:hydrogen exchange in dispersed acinar cells from guinea pig pancreas. Activation of the exchanger by CCK was independent of the elevation of Ca$_i$; however, it could be mimicked by a phorbol ester and diacylglycerol.

It is well recognized that a number of receptors linked to Ca mobilization also activate guanylate cyclase and elevate the intracellular levels of cGMP. The role of cGMP in the propagation of the signal to the secretory response in acinar cells is at present unclear. Rogers et al.[93] have demonstrated that permeable analogues of cGMP analogues or sodium nitroprusside inhibited the secretory response elicited by CCK. It has been suggested that cGMP may exert a negative control on the protein kinase C-mediated secretory pathway because the inhibitory effects were also present when the secretion was elicited by phorbol ester. However, these findings could not be reproduced in a subsequent study from another laboratory,[94] and it was suggested that the inhibitory effects of cGMP analogues may be related to their antagonistic properties for CCK receptors.

The CCK-mediated elevation in Ca$_i$ was also demonstrated in the isolated gastric glands.[95,96] This preparation consists mainly of parietal and chief cells. Both cell types respond to CCK by an increase in Ca$_i$; however, it appears that the cells utilize different sources of activating calcium. In isolated glands, the CCK-mediated Ca transient was shown to have a biphasic pattern. The initial rapid rise in Ca$_i$, which was independent of extracellular Ca, was followed by a secondary transient sensitive to blockade by nifedipine. The initial rise in Ca$_i$ was temporally correlated with the release of pepsin, but not with the uptake of [^{14}C]aminopyridine[96] (an indicator of secretory activity of parietal cells). It is likely that the activation of chief cells utilizes endogenous calcium stores while the activation of parietal cells requires Ca influx.

There is also evidence to suggest that secretory activities of parietal and chief cells are modulated differently by protein kinase C. In chief cells the secretory response could be duplicated by application of A23187 and phorbol ester,[96] indicating synergism between Ca mobilization and activation of protein kinase in the propagation of signal. On the other hand, the activation of protein kinase C in parietal cells has primarily been implicated in the down-regulation of the secretory response.[97-99]

The mobilization of intracellular Ca was also implicated in the CCK-mediated contractile responses of gastrointestinal smooth muscle. In isolated smooth muscle cells from human or guinea pig stomach, the initial component of the contractile response elicited by CCK was found to be independent of Ca influx, as documented by the insensitivity of contraction to methoxyverapamil and to the removal of extracellular calcium.[100] In the cells from circular muscle of guinea-pig stomach, the CCK-induced contractions were accompanied by a transient increase in Ca$_i$ which was independent of extracellular Ca, implicating the CCK-induced Ca release from intracellular stores.[101] The CCK-sensitive intracellular Ca stores in these cells seem to be identical to an IP$_3$-sensitive pool. In support of such a contention are observations that in saponin-permeabilized cells the contractile response elicited by CCK could be mimicked by exogenous IP$_3$.[102] However, the CCK-induced mobilization of Ca from the intracellular stores may not fully account for contractile responses in various smooth muscles. For example, the contractile studies on bovine and guinea pig gallbladder have shown that CCK-induced contractions are sensitive to Ca entry blockers.[103,104]

There are several lines of evidence to suggest that in some tissues the activation of CCK receptors results in altered intracellular levels of cAMP. In guinea pig gallbladder smooth muscle, CCK-induced contractions were accompanied by a decrease in the level of cAMP, and such an effect was due to the CCK-induced activation of phosphodiesterase.[105] Similar activating effects of CCK on phosphodiesterase also were identified in several rabbit tissues.[106] The increased rate of cAMP hydrolysis was shown to result from CCK-mediated conversion of a high-K$_m$ form of the enzyme to a low-K$_m$ form. In the pancreatic acinar cells, the activation of low-affinity CCK receptors were reported to elevate the levels of cAMP when studied in the presence of phosphodiesterase inhibitors.[4,107] It is likely that the CCK-induced modulation of cAMP level is a secondary transducing event which occurs subsequently to the generation of phospholipase C-derived second messengers.

In the gastrointestinal smooth muscles, CCK also regulates the contractile activity indirectly by modulating the pattern of enteric neurotransmission. CCK was demonstrated to stimulate the release of acetylcholine, substance P (SP), and τ-aminobutyric acid from the myenteric nerves.[108-109] The second messenger system(s) mediating the effects of CCK on myenteric nerves remain to be elucidated.

At present, it is not clear whether the differences in the dependency of CCK effects on external Ca in various tissues reflect tissue-specific differences in the pathways governing Ca homeostasis or can be related to the different coupling mechanisms of pharmacologically heterogeneous types of CCK receptors. Several studies have demonstrated that CCK receptors in various tissues may have different affinities for CCK analogues and gastrin,[110,111] and a tentative classification of CCK receptors into three categories has been suggested recently.[112]

D. OPIOID PEPTIDES

The molecular mechanisms of signal transduction activated by opioid receptors have been studied extensively in the tissues of the central nervous system. The second messenger systems coupled to opioid receptors include adenylate cyclase and ion channels. Their relative contribution to the signal transmission depends on the receptor subtype. According to the current classification scheme, the opioid receptors are divided into μ_1 δ_1 and κ categories.

The negative modulation of adenylate cyclase by opioid receptors is well documented in neuronal tissues and cell lines rich in δ-receptors.[113-119] The coupling mechanism involves the intermediate G_i protein, as documented by GTP and Na dependency of the inhibitory effects and their sensitivity to pertussis toxin.[120-123] Also, there is a good correlation between the potency of agonists to inhibit the adenylate cyclase and to stimulate the low-K_m GTPase.[124]

The inhibition of adenylate cyclase was also reported for μ-receptors.[122] However, the relative magnitude of μ-receptor-mediated inhibitory effects was lower than that elicited by δ-receptors. It was proposed that the inhibitory effect of μ-receptors may be mediated through the liberation of the $\beta\gamma$-subunit from an unspecified G-protein (G_i, G_o). Considering a substantial structural homogeneity of $\beta\gamma$-subunits from various G proteins, the $\beta\tau$-subunit liberated by the activation of μ-receptors would in turn associate with the α_s-subunit of G_s, thus inhibiting the α_s-mediated activation of adenylate cyclase. Alternatively, the differences in the magnitude of adenylate cyclase inhibition between δ-and μ-receptors may be attributed to the differences in receptor reserves. The comparison of binding affinities (K_d) of agonists with apparent inhibitory constants (K_i) for the inhibition of the enzyme revealed that the K_d/K_i ratio for δ-agonists was >4, thus suggesting the presence of a substantial receptor reserve.[125] For μ-agonists such a ratio was <2, indicating the presence of only a marginal fraction of spare receptors.

The κ-receptor-mediated inhibition of adenylate cyclase was demonstrated recently in cultured neurons from mouse spinal cord.[126] In these cells the κ-selective agonist U-50, 488H caused approximately a 30% decline in forskolin-stimulated adenylate cyclase activity in a pertussis toxin-sensitive manner.

The inhibition of adenylate cyclase by κ-selective agonists was also reported for guinea pig striatum.[127] In this study the activation of κ-receptors was shown to inhibit basal but not dopamine-stimulated adenylate cyclase activity, as opposed to μ- and δ-receptors, which inhibited both activities. It is difficult to assess whether the suppression of basal adenylate cyclase activity resulted from direct receptor-mediated inhibition of the enzyme or some another mechanism. In our experience with the assay of adenylate cyclase in neuronal tissues, the basal activity represents the sum of activating and inhibitory effects of endogenous mediators.[126] Thus, the effects of κ-agonist-mediated inhibition of basal activity eventually could be related to the suppression of the release of endogenous activators during the assay. Alternatively, the lack of κ-agonist-mediated inhibition of dopamine-stimulated activity may be due to differential cellular localization of κ-opioid and dopaminergic receptors.

As was indicated, the activation of opioid receptors also elicits several effects on ion conductances. In general, such effects are associated with the hyperpolarization of plasma membrane.[129] Both μ- and δ-receptors have been demonstrated to activate potassium channels in central neurons,[130-132] enteric ganglia,[133,134] and NG108-15 cells.[69,135] In the central neurons such effects were pertussis toxin-sensitive,[132] indicating the involvement of a G protein in the transducing mechanism. The modulation of potassium channels was insensitive to forskolin, dibutyryl cAMP, or phorbol esters, indicating the lack of involvement of other known second messengers.[134,136]

The κ-subtype of opioid receptors has been shown to attenuate the inward Ca current in myenteric plexus of guinea pig ileum[137] and in cultured spinal neurons.[138,139] In the cultured neurons from mouse dorsal root ganglion, the negative modulatory effects of κ-receptors were demonstrated for the N-type of Ca channels.[140] Recent evidence suggests that in these cells the κ-receptors also suppress the dihydropyridine-sensitive L channels.[141] These channels are positively modulated by the cAMP-dependent protein kinase,[142] and the inhibitory effects of κ-receptors apparently involve both cAMP-dependent and cAMP-independent mechanisms. The inhibition of L channels by κ-receptors was impaired by treatment of cells with pertussis toxin.

The activation of opioid receptors was also shown to elicit multiple changes in the pattern of Ca handling in neuronal tissues. The administration of opioid agonists was shown to result in altered Ca binding,[143,144] inhibition of Ca-transporting ATPase,[145] and decline in the amount of membrane-bound calmodulin.[146,147] Since these functions are regulated by several protein kinases, it is likely that the modulatory influences of opioids are exerted secondarily to the inhibition of adenylate cyclase and to the alterations in ion fluxes.

E. NEUROTENSIN

The coupling of neurotensin (NT) receptors with second messenger systems has been studied extensively in neuronal cells. Currently available evidence suggests that the activation of these receptors elicits multiple effects on intracellular mediators. Goedert et al.[148] have reported that the activation of NT receptors in rat brain slices stimulates phospholipase C and increases the turnover of inositol phospholipids. Such an effect of NT was present in several brain regions, and the magnitude of NT-induced stimulation of PI turnover showed good correlation with the receptor number identified by ligand binding. The acceleration of PI turnover was not accompanied by a detectable change in the basal or forskolin-stimulated adenylate cyclase activity.

These observations were further corroborated by Snider et al.[149] In neuroblastoma cell line N1E-115, the pharmacologically relevant doses of NT enhanced the incorporation of [^{32}P] into both PI and phosphatidic acid and increased, in the presence of lithium, the cytosolic levels of inositol phosphates.

The coupling of NT receptors with phospholipase C in N1E-115 cells involves an intermediate G protein, as documented by the sensitivity of binding isotherms to GTP and its stable analogues and by the ability of pertussis toxin to compromise the NT-induced stimulation of PI turnover.[150,151]

The activating effect of NT on PI turnover in N1E-115 cells was shown to be accompanied by elevated levels of Ca_i. However, the exact nature of the NT-sensitive Ca pool is not clearly defined. The NT-induced rise in Ca_i was reported to be dependent on extracellular Ca,[149] thus raising the possibility that such an effect was due to the increased Ca influx. Similarly, the NT-induced stimulation of secretory activity in rat pituitary was shown to be dependent on extracellular Ca and was accompanied by an increased ^{45}Ca influx.[152]

The activation of NT receptors was also shown to elevate the cytosolic levels of cGMP.[153,154] In general, the receptor-induced activation of guanyl cyclase is a secondary transducing event, occurring secondarily to the Ca mobilization. In N1E-115 cells, the activation of guanyl cyclase also seems to involve alterations in membrane phospholipids, in addition to elevated levels of Ca_i. In support of such a contention are observations that the receptor-induced activation of guanyl cyclase is susceptible to a blockade by lipoxygenase inhibitors.[155]

Amar et al.[151] have recently demonstrated that in permeabilized N1E-115 cells the NT-induced elevation in cytosolic levels of cGMP could be duplicated by the application of exogenous IP_3. Interestingly, the activating effect of IP_3 on guanyl cyclase was observed in the absence of extracellular calcium. Considering the obligatory role of Ca ions for the activation of guanyl cyclase, such an observation provides indirect evidence that N1E-115 cells contain an IP_3-sensitive intracellular Ca pool which is sufficient to activate the guanyl cyclase.

In N1E-115 cells, the activation of NT receptors has also been shown to inhibit PGE_1-stimulated adenylate cyclase activity.[150] The generation of phospholipase C-derived second messengers and the inhibition of adenylate cyclase seem to be mediated by an apparently homogeneous population of receptors. In all likelihood, the decreased sensitivity of adenylate cyclase to the stimulation of PGE_1 in the presence of NT may reflect a secondary transducing event occurring subsequently to the activation of phospholipase C.

There is evidence to suggest that NT receptors are also linked to the control of ion fluxes, independently of the interaction of these receptors with previously mentioned second messenger pathways. Nakagawa et al.[156] have reported the presence of NT receptors on neuroblastoma × glioma hybrid NG108-15 which, upon activation, depolarize the cells and increase their excitability. The effects of NT were not accompanied by a detectable change in the level of cAMP, PI turnover, or ^{45}Ca influx. The ionic basis of these effects remains to be determined.

The activation of NT receptors in gastrointestinal smooth muscle can elicit either stimulation or inhibition of the contractile response.[157] The underlying biochemical correlates of these effects are at present only partially understood. It is not clear whether the signaling mechanism involves the generation of intracellular metabolic messengers. For example, Kitabgi and Freychet[158] have reported that tetrodotoxin-insensitive relaxation of the longitudinal muscle of guinea pig ileum did not result in detectable changes in the levels of cAMP or cGMP.

It seems that the initial step in the transduction mechanisms in gastrointestinal smooth muscles may involve a transient increase in Ca permeability. Such a mechanism has been demonstrated in the guinea pig taenia coli[159] and rat gastric fundus[160] (muscles which contract in response to NT), as the NT-induced contractions were accompanied by increased Ca influx. The relaxing effect of NT observed in a number of smooth muscles seems to be attributed to the activation of Ca-dependent potassium channels subsequent to the transient rise in Ca_i. Such a concept is supported by observations that the blockade of potassium channels by apamin attenuates the relaxing effects of NT[161-164] or can reverse the inhibitory effect to the stimulation of contractile response.[165,166]

These observations suggest the possibility that both excitatory and inhibitory effects of NT in muscles of the gastrointestinal tract may be mediated by a single transduction mechanism — in particular, by the transient increase in Ca_i. Under conditions when the depolarizing effects of increased Ca_i are offset by increased K permeability, the overall response will result in membrane hyperpolarization and inhibition of contractile activity. On the other hand, in the absence of a substantial Ca-dependent K efflux (e.g., due to the blockage by apamin or by low concentration of Ca-dependent K channels), the Ca transient may be sufficient to activate the mechanisms which ultimately may result in the stimulation of contractile activity.

The NT-sensitive Ca pool in the smooth muscles of gastrointestinal tract remains to be identified. As mentioned earlier, NT activates the Ca conductance in guinea pig taenia coli, suggesting the possibility that the NT-mediated control of Ca influx may be a primary transducing event. Alternatively, considering the ability of NT to increase PI turnover in neuronal cells, the activation of Ca influx in smooth muscles may be preceded by the generation of phospholipase C-derived second messengers and by Ca release from IP_3-sensitive stores. In support of such mechanisms are recent observations that in the circular muscle of canine ileum the NT-induced, apamin-sensitive hyperpolarization is only partially sensitive to the removal of external calcium.[167]

NT also regulates the contractile activity of gastrointestinal smooth muscles by indirect presynaptic mechanisms. Such a regulation is achieved by the activation of NT receptors on enteric nerves and results primarily in the stimulation of contractile activity.[157] The presynaptic stimulatory effects of NT seem to be primarily attributed to the enhanced release of acetylcholine[168-170] and SP.[171]

The presynaptic inhibitory effects of NT have been demonstrated in canine small intestine and seem to result from enhanced release of norepinephrine.[172] The second messenger system(s) linked to NT receptors in enteric nerves remain to be elucidated.

F. SUBSTANCE P

SP belongs to a family of structurally related peptides collectively known as tachykinins.

These peptides can activate pharmacologically distinct types of receptors,[173] which are, according to currently accepted nomenclature, divided into NK-1, NK-2, and NK-3 categories. It appears that all these categories of tachykinin receptors activate a similar transducing pathway, namely, the phospholipase C-mediated enhancement of PI turnover.[174]

The ability of SP and related peptides to enhance PI turnover has been demonstrated in several gastrointestinal smooth muscles as well as in other tissues.[175-181] However, such observations have been made on isolated muscle strips or tissue slices, and the interpretation of these findings in relation to SP-linked signaling mechanisms is complicated. The difficulties with the interpretation arise for the following three reasons.

First, the experimental models contained heterogeneous types of cells, primarily nerves and muscle. Since both cellular types contain SP receptors, it is difficult to judge whether the effect of SP are due to the activation of presynaptic (neuronal) or postsynaptic (smooth muscle) receptors.

Second, the activation of presynaptic SP receptors was shown to enhance the release of acetylcholine from enteric neurons,[182] an agent which also can promote accelerated PI turnover in smooth muscles.[183] The participation of acetylcholine release in SP-induced effects on PI turnover can be evaluated by the inclusion of atropine or some other muscarinic antagonist. However, it is not clear whether the presynaptic effects of SP also may involve enhanced release of other noncholinergic mediators capable of activating phospholipase C.

Third, in order to ascertain that the activation of phospholipase C is a primary signaling event, SP should be able to enhance PI turnover at resting levels of Ca_i, irrespective of the absence or presence of extracellular calcium. If, for example, the activated receptor is linked to the opening of Ca channels, the enhanced PI turnover may result from nonspecific activation of phospholipase C by elevated levels of Ca_i.[184] In some studies on gastrointestinal smooth muscles, the Ca dependency of SP-induced PI turnover has not been indicated.

At present, no study is available which would address the previously mentioned considerations in a systematic manner. The ability of SP to activate PI hydrolysis in the absence of extracellular calcium has been demonstrated for intestinal muscles of guinea pig and rat.[175,176] Assuming that Ca-free media were sufficient to eliminate the SP-induced release of endogenous mediators, these studies would offer good evidence that SP-induced effects on PI hydrolysis are a direct consequence of the activation of SP receptors. However, these studies did not attempt to evaluate the cellular localization of such receptors.

It has been suggested that in the longitudinal smooth muscle/myenteric plexus of guinea pig ileum the SP-induced increase in PI turnover is mediated by both presynaptic and postsynaptic receptors. In this tissue the presynaptic receptors are of the NK-3 category, as opposed to postsynaptic receptors, which are of the NK-1 category.[185] Selective NK-3 and NK-1 agonists were shown to have an additive effect on the accumulation of inositol phosphates.[179] However, the effects have been demonstrated in the presence of extracellular calcium, thus not eliminating the possibility that such effects could be attributed in part to the stimulated Ca influx.

At present, it is safe to say that SP receptors in some gastrointestinal smooth muscles may activate phospholipase C. However, the implication of this process in signal transduction should be regarded as tentative until the relationship between the SP-mediated activation of phospholipase C and Ca mobilization is established in greater detail. Some indirect observations suggest that the activation of SP receptors can elicit the release of Ca from intracellular stores, thus implicating the phospholipase C-mediated generation of IP_3 in primary signaling mechanisms. For example, in the longitudinal smooth muscle/myenteric plexus of guinea pig ileum, SP was shown to elicit phasic contractions in the absence of extracellular Ca.[186] In the presence of external Ca, the activation of SP receptors in this preparation was shown to be accompanied by a transient decrease in ^{45}Ca influx,[187] indicating the superimposition of enhanced Ca efflux due to the mobilization of internal Ca. In isolated cells from toad stomach, the activation of Ca channels by SP was mimicked by an analogue of diacylglycerol, suggesting that the activation of protein kinase C may precede the stimulation of Ca influx.[188]

On the other hand, studies on gastrointestinal epithelia showed that SP-mediated effects on secretory response required extracellular Ca and were sensitive to Ca channel blockers.[5,189,190] These findings would suggest that epithelial SP receptors are primarily linked to the control of Ca influx.

IV. INTERACTIONS BETWEEN SECOND MESSENGER SYSTEMS

It is becoming increasingly evident that the primary transducing events triggered by receptor activation are subjected to a substantial integration at the intracellular level and that the individual transducing pathways can influence each other. This "cross-talk" between various signaling pathways likely enables a fine tuning of overall intracellular homeostasis, including the responsiveness of the cells to extracellular messages. For example, it is well documented that the activation of the phospholipase C-linked signaling pathway can either positively or negatively modulate the responses to agonists linked to adenylate cyclase. Such regulatory influences are primarily exerted by the protein kinase C-mediated phosphorylation of intracellular targets, and the direction of the regulatory effects depends on the cellular type.[191-195] Similarly, protein kinase A- or protein kinase C-mediated phosphorylation can modify the conductivity patterns of several ion channels.[142,196-201]

Alternatively, the activation of a single receptor can be directly linked to multiple transducing events. For example, such a mechanism of signal transduction has been documented for opioid receptors of μ- and δ-categories because these receptors are capable of simultaneous inhibition of adenylate cyclase and activation of potassium conductance.[119,134] These effects of opioids resemble the mode of coupling of cardiac muscarinic receptors, where both the inhibition of adenylate cyclase and the activation of potassium channels are well documented.[202,203]

In the context of the evaluation of transducing pathways activated by peptidergic receptors, the recognition of mutual interactions between various signaling systems emphasizes the importance of discriminating between primary and secondary transducing events. To illustrate this point, CCK has been shown to activate phosphodiesterase in smooth muscle,[106] elevate cAMP levels in pancreatic acinar cells,[4] and inhibit Ca channels in snail neurons.[204] Considering that phosphodiesterase is positively regulated by Ca/calmodulin[205] and the activities of adenylate cyclase and ion channels are modulated by protein kinase C-mediated phosphorylation, the previously mentioned effects of CCK conceivably may result from the well-documented ability of this agent to cause the activation of phospholipase C and Ca mobilization.

The experimental approaches which may distinguish the primary and secondary transducing events depend on the type of signaling pathway. The evidence that a particular receptor is directly linked to the control of adenylate cyclase can be substantiated by the direct measurement of agonist-induced modulation of enzyme activity in the homogeneous population of cells or in isolated membrane preparations of defined composition, as well as by examining the sensitivity of the effect to specific antagonists, GTP, and pertussis or cholera toxin.

The examination of receptor-induced modulation of ion conductances is studied primarily by electrophysiological techniques. The contention of a direct, receptor-mediated control of ion currents is mainly supported by observations that agonist-induced effects are independent of, or additive to, the effects exerted by metabolic second messengers. Yet the possibility still exists that in some systems the agonist-induced effects on ion channels may involve so far unidentified mediators. For example, recent developments in this area have demonstrated that product(s) of phospholipase A_2-mediated lipolysis can modify cardiac potassium channels.[206,207]

Perhaps the most important aspect of judging the relevance of enhanced turnover of phosphoinositides to signal transduction is the discrimination between the direct, receptor-linked activation of phospholipase C and secondary activating mechanisms such as the stimulation of the enzyme by elevated levels of Ca, due to agonist-induced stimulation of Ca influx. The experimental approaches which may address this area are not readily available. The examination of agonist-induced activation of phospholipase C and Ca mobilization requires the integrity of cellular components, since the signaling pathway employs both hydrophilic (IP_3) and lipophilic (diacylglycerol) mediators, Such an arrangement limits the studies of individual components of signaling pathways in isolation. Furthermore, the relationship between the receptor-linked activation of phospholipase C and the opening of Ca channels has yet to be establshed. In a number of systems, the receptor-induced acceleration of PI turnover is either partially or completely suppressed by the removal of external Ca. The different requirement for external Ca in various systems would suggest that Ca influx may serve as a positive feedback control regulating the amount of PI-derived second messengers and that the relative contribution of this mechanism to the activation of phospholipase C may be different for different receptors and different tissues. One possibility is that the stimulation of Ca influx is achieved by PI-derived mediators, inositol 1,3,4,5-tetrakisphosphate (IP_4) being the prime candidate for such a function.[208,209] However, the partial dependence of receptor-mediated activation of phospholipase C on external Ca also was observed in cells lacking IP_3 kinase (and, thus, incapable of generating IP_4).[210] Alternatively, the opening of Ca gates may result in some tissues from the protein kinase C-mediated phosphorylation of Ca channels. Currently available evidence suggests that activation of protein kinase C has tissue-specific effects on Ca channels, resulting either in stimulation,[188,211,212] inhibition,[204,213] or lack of change[199] in calcium conductance. In addition, the possibility remains that some mediators may simultaneously activate phospholipase C and Ca influx, the latter event being independent of the generation of phospholipase C-derived second messengers.

V. CONCLUDING REMARKS

The preceding discussion of signaling mechanisms activated by peptidergic receptors shows that various categories of neuropeptides interact with diverse transducing systems. The altered activities of such transducing systems are further propagated into the intracellular compartments primarily by alterations in phosphorylation-dephosphorylation equilibria of various targets, changes in intracellular ion concentrations, and modulation of electrical properties of plasma membrane.

The account of signaling pathways activated by neuropeptides was concentrated on several well-established categories of gastrointestinal neuropeptides. Recent progress in neuroanatomical studies has provided evidence for the presence of a much larger number of peptides with potential neurotransmitting/neuromodulatory roles.[214] The evidence about the interaction of relatively novel peptides with second messenger systems is either forthcoming[215,216] or awaits further studies.

It is also apparent that the interpretation of signaling mechanisms in relation to three major targets — adenylate cyclase, phosphoinositides, and ion channels — may be an oversimplification of biological reality. There are several indications to suggest that in the near future new insights into a potential role of phospholipids other than phosphoinositides in mechanisms of signal transduction will be gained. For example, some extracellular mediators can stimulate the hydrolysis of phosphatidylcholine,[217] and the activity of phospholipase A_2 was recently shown to be regulated by the βτ-subunit of G protein.[207,218]

APPENDIX

Abbreviations used — Ca_i, intracellular concentration of calcium ions; cAMP, cyclic adenosine 3′,5′-monophosphate; CCK, cholecystokinin; cGMP, cyclic guanosine 3′,5′-monophosphate; GTP, guanosine 5′-triphosphate; G protein, guanine nucleotide binding protein; IP_3, inositol 1,4,5-trisphosphate; IP_4, inositol 1,3,4,5-tetrakisphosphate; NT, neurotensin; PGE_1, prostaglandin E_i; PHI, peptide histidine isoleucine; PHM, peptide histidine methionine; PI, phosphoinositides; SOM, somatostatin; SP, substance P; and VIP, vasoactive intestinal peptide.

ACKNOWLEDGMENT

The author thanks Drs. E. E. Daniel, C. Y. Kwan, and S. Ahmad for their comments on the manuscript. This work was supported by the Medical Research Council of Canada.

REFERENCES

1. **Keast, J. R.,** Mucosal innervation and control of water and ion transport in the intestine, *Rev. Physiol. Biochem. Pharmacol.,* 109, 1, 1987.
2. **Kromer, W.,** Endogenous and exogenous opioids in the control of gastrointestinal motility and secretion, *Pharmacol. Rev.,* 40, 121, 1988.
3. **Rasmussen, H., Takuwa, Y., and Park, S.,** Protein kinase C in the regulation of smooth muscle contraction, *FASEB J.,* 1, 177, 1987.
4. **Gardner, J. D. and Jensen, R. T.,** Receptors and cell activation associated with pancreatic enzyme secretion, *Annu. Rev. Physiol.,* 48, 103, 1986.
5. **Brown, D. R.,** Intracellular mediators of peptide action in the intestine and airways: focus on ion transport function, *Am. Rev. Respir. Dis.,* 136, S43, 1987.
6. **Litosch, I. and Fain, J. N.,** Regulation of phosphoinositide breakdown by guanine nucleotides, *Life Sci.,* 39, 187, 1986.
7. **Gilman, A. G.,** G-proteins: transducers of receptor-generated signals, *Annu. Rev. Biochem.,* 56, 615, 1987.
8. **Weiss, E. R., Kelleher, D. J., Woon, C. W., Soparkar, S., Osawa, S., Heasley, L. E., and Johnson, G. L.,** Receptor activation of G proteins, *FASEB J.,* 2, 2841, 1988.
9. **Fain, J. N., Wallace, M. A., and Wojcikiewicz, R. J. H.,** Evidence for involvement of guanine nucleotide-binding proteins in the activation of phospholipases by hormones, *FASEB J.,* 2, 2569, 1988.
10. **Schramm, M. and Selinger, Z.,** Message transmission: receptor controlled adenylate cyclase system, *Science,* 225, 1350, 1984.
11. **Casperson, G. F. and Bourne, H. R.,** Biochemical and molecular genetic analysis of hormone-sensitive adenylyl cyclase, *Annu. Rev. Pharmacol. Toxicol.,* 27, 371, 1987.
12. **Limbird, L. E.,** Receptors linked to inhibition of adenylate cyclase: additional signalling mechanisms, *FASEB J.,* 2, 2686, 1988.
13. **Hokin, L. E.,** Receptors and phosphoinositide-generated second messengers, *Annu. Rev. Biochem.,* 54, 205, 1985.
14. **Fisher, S. K. and Agranoff, B. W.,** Receptor activation and inositol lipid hydrolysis in neural tissues, *J. Neurochem.,* 48, 999, 1987.
15. **Hirasawa, K. and Nishizuka, Y.,** Phosphatidylinositol turnover in receptor mechanism and signal transduction, *Annu. Rev. Pharmacol. Toxicol.,* 25, 147, 1985.
16. **Nicoll, R. A.,** The coupling of neurotransmitter receptors to ion channels in the brain, *Science,* 241, 545, 1988.
17. **Levitan, I. B.,** Modulation of ion channels in neurons and other cells, *Annu. Rev. Neurosci.,* 11, 119, 1988.
18. **Dohlman, H. G., Caron, M. G., and Lefkowitz, R. J.,** A family of receptors coupled to guanine nucleotide regulatory proteins, *Biochemistry,* 26, 2657, 1987.
19. **Levitzki, A.,** From epinephrine to cyclic AMP, *Science,* 241, 800, 1988.

20. **Legan, E., Chernow, B., Parrillo, J., and Roth, B. L.,** Activation of phosphatidylinositol turnover in rat aorta by α_1-adrenergic receptor stimulation, *Eur. J. Pharmacol.,* 110, 389, 1985.

21. **Simon, B. and Kather, H.,** Activation of human adenylate cyclase in the upper gastrointestinal tract by vasoactive intestinal polypeptide, *Gastroenterology,* 74, 722, 1978.

22. **Carter, R. F., Bitar, K. N., Zfass, A. M., and Makhlouf, G. M.,** Inhibition of VIP-stimulated intestinal secretion and cyclic AMP production by somatostatin in the rat, *Gastroenterology,* 74, 726, 1978.

23. **Amiranoff, B., Laburthe, M., Dupont, C., and Rosselin, G.,** Characterization of vasoactive intestinal peptide-sensitive adenylate cyclase in rat intestinal epithelial cell membranes, *Biochim. Biophys. Acta,* 544, 474, 1978.

24. **Prieto, J. C., Laburthe, M., Hui Bon Hoa, D., and Rosselin, G.,** Quantitative studies of vasoactive intestinal peptide (VIP) binding sites and VIP-induced adenosine 3':5'-monophosphate production in epithelial cells from duodenum, jejunum, ileum, coecum, colon and rectum in the rat, *Acta Endocrinol.,* 96, 100, 1981.

25. **Zhou, Z.-C., Gardner, J. D., and Jensen, R. T.,** Receptors for vasoactive intestinal peptide and secretin on guinea pig pancreatic acini, *Peptides,* 8, 633, 1987.

26. **Bissonnette, B. M., Collen, M. J., Adachi, H., Jensen, R. T., and Gardner, J. D.,** Receptors for vasoactive intestinal peptide and secretin on rat pancreatic acini, *Am. J. Physiol.,* 246, G710, 1984.

27. **Waelbroeck, M., Robberecht, P., De Neff, P., Chatelain, P., and Christophe, J.,** Binding of vasoactive intestinal peptide and its stimulation of adenylate cyclase through two classes of receptors in rat liver membranes, *Biochim. Biophys. Acta,* 678, 83, 1981.

28. **Couvineau, A. and Laburthe, M.,** The rat liver vasoactive intestinal peptide binding site: molecular characterization by covalent cross-linking and evidence for differences from the intestinal receptor, *Biochem. J.,* 225, 473, 1985.

29. **Laburthe, M., Breant, B., and Rouyer-Fessard, C.,** Molecular identification of receptors for vasoactive intestinal peptide in rat intestinal epithelium by covalent cross-linking, *Eur. J. Biochem.,* 139, 181, 1984.

30. **Amiranoff, B., Laburthe, M., and Rosselin, G.,** Characterization of specific binding for vasoactive intestinal peptide in rat intestinal epithelial cell membranes, *Biochim. Biophys. Acta,* 627, 215, 1980.

31. **Amiranoff, B., Laburthe, M., and Rosselin, G.,** Potentiation by guanine nucleotides of the VIP-induced adenylate cyclase stimulation in intestinal epithelial cell membranes, *Life Sci.,* 26, 1905, 1980.

32. **Couvineau, A., Amiranoff, B., and Laburthe, M.,** Solubilization of the liver vasoactive intestinal peptide receptor, *J. Biol. Chem.,* 261, 14482, 1986.

33. **Biancani, P., Walsh, J. H., and Behar, J.,** Vasoactive intestinal polypeptide. A transmitter for lower esophageal sphincter relaxation, *J. Clin. Invest.,* 73, 963, 1984.

34. **Grider, J. R., Cable, M. B., Said, S. I., and Makhlouf, G. M.,** Vasoactive intestinal peptide as a neural mediator of gastric relaxation, *Am. J. Physiol.,* 248, G73, 1985.

35. **Grider, J. R. and Makhlouf, G. M.,** Colonic peristaltic reflex: identification of vasoactive intestinal peptide as mediator of descending relaxation, *Am. J. Physiol.,* 251, G40, 1986.

36. **Biancani, P., Walsh, J. H., and Behar, J.,** Vasoactive intestinal peptide: a neurotransmitter for relaxation of the rabbit internal anal sphincter, *Gastroenterology,* 89, 867, 1985.

37. **Nurko, S., Dunn, B. M., and Rattan, S.,** Peptide histidine isoleucine and vasoactive intestinal polypeptide cause relaxation of opossum internal anal sphincter via two distinct receptors, *Gastroenterology,* 96, 403, 1989.

38. **Goyal, R. K., Rattan, S., and Said, S. I.,** VIP as a possible neurotransmitter of non-cholinergic, non-adrenergic inhibitory neurons, *Nature (London),* 288, 378, 1980.

39. **Daniel, E. E.,** Nonadrenergic, noncholinergic (NANC) neuronal inhibitory interactions with smooth muscle, in *Calcium and Contractility,* Grover, A. K. and Daniel, E. E., Eds., Humana Press, Clifton, NJ, 1985, 385.

40. **Ganz, P., Sandrock, A. W., Landis, S. C., Leopold, J., Gimbrone, M. A., Jr., and Alexander, R. W.,** Vasoactive intestinal peptide: vasodilatation and cyclic AMP generation, *Am. J. Physiol.,* 250, H755, 1986.

41. **Bitar, K. N. and Makhlouf, G. M.,** Relaxation of isolated gastric smooth muscle cell by vasoactive intestinal peptide, *Science,* 216, 531, 1982.

42. **Torphy, T. J., Fine, C. F., Burman, M., Barnette, M. S., and Ormsbee, H. S., III,** Lower esophageal sphincter relaxation is associated with increased cyclic nucleotide content, *Am. J. Physiol.,* 251, G786, 1986.

43. **Rattan, S. and Moummi, C.,** Influence of stimulators and inhibitors of cyclic nucleotides on lower esophageal sphincter, *J. Pharmacol. Exp. Ther.,* 248, 703, 1989.

44. **Cartwright, C. A., McRoberts, J. A., Mandel, K. G., and Dharmsathaphorn, K.,** Synergistic action of cyclic adenosine monophosphate- and calcium-mediated chloride secretion in a colonic epithelial cell line, *J. Clin. Invest.,* 76, 1837, 1985.

45. **Laburthe, M., Augeron, C., Rouyer-Fessard, C., Roumagnac, I., Maoret, J.-J., Grasset, E., and Laboisse, C.,** Functional VIP receptors in the human mucus-secreting colonic epithelial cell line CL.16E, *Am. J. Physiol.,* 256, G443, 1989.

46. **Collen, M. J., Sutliff, V. E., Pan, G.-Z., and Gardner, J. D.,** Postreceptor modulation of action of VIP and secretin on pancreatic enzyme secretion by secretagogues that mobilize cellular calcium, *Am. J. Physiol.,* 242, G423, 1982.

47. **Hootman, S. R., Ernst, S. A., and Williams, J. A.,** Secretagogue regulation of Na^+-K^+ pump activity in pancreatic acinar cells, *Am. J. Physiol.,* 245, G339, 1983.

48. **Scott, J. and Baum, B. J.,** Involvement of cyclic AMP and calcium in exocrine protein secretion induced by vasoactive intestinal polypeptide in rat parotid cells, *Biochim. Biophys. Acta,* 847, 255, 1985.

49. **Bjoro, T., Ostberg, B. C., Sand, O., Gordeladze, J., Iversen, J.-G., Torjesen, P. A., Gautvik, K. M., and Haug, E.,** Vasoactive intestinal peptide and peptide with N-terminal histidine and C-terminal isoleucine increase prolactin secretion in cultured rat pituitary cells (GH_4C_1) via a cAMP-dependent mechanism which involves transient elevation of intracellular Ca^{2+}, *Mol. Cell. Endocrinol.,* 49, 119, 1987.

50. **Zurier, R. B., Kozma, M., Sinnett-Smith, J., and Rozengurt, E.,** Vasoactive intestinal peptide synergistically stimulates DNA synthesis in mouse 3T3 cells: role of cAMP, Ca^{2+}, and protein kinase C, *Exp. Cell Res.,* 176, 155, 1988.

51. **Lundberg, J. M., Hedlund, B., and Bartfai, T.,** Vasoactive intestinal polypeptide enhances muscarinic ligand binding in cat submandibular salivary gland, *Nature (London),* 295, 147, 1982.

52. **Schaad, N. C., Schorderet, M., and Magistretti, P. J.,** Prostaglandins and the synergism between VIP and noradrenaline in the cerebral cortex, *Nature (London),* 328, 637, 1987.

53. **Turner, J. T., Bollinger, D. W., and Toews, M. L.,** Vasoactive intestinal peptide receptors/adenylate cyclase system: differences between agonist- and protein kinase C-mediated desensitization and further evidence for receptor internalization, *J. Pharmacol. Exp. Ther.,* 247, 417, 1989.

54. **Svoboda, M., De Neef, P., Tastenoy, M., and Christophe, J.,** Molecular characteristics and evidence for internalization of vasoactive-intestinal-peptide (VIP) receptors in the tumoral rat-pancreatic acinar cell line AR 4-2 J, *Eur. J. Biochem.,* 176, 707, 1988.

55. **Brazeau, P.,** Somatostatin: a peptide with unexpected physiological activities, *Am. J. Med.,* 81 (Suppl. 6B), 8, 1986.

56. **Taparel, D., Susini, C., Esteve, J. P., Diaz, J., Cazaubon, C., Vaysse, N., and Ribet, A.,** Somatostatin analogs: correlation of receptor affinity with inhibition of cyclic AMP formation in pancreatic acinar cells, *Peptides,* 6, 109, 1985.

57. **Viguerie, N., Tahiri-Jouti, N., Esteve, J.-P., Clerc, P., Logsdon, C., Svoboda, M., Susini, C., Vaysse, N., and Ribet, A.,** Functional somatostatin receptors on a rat pancreatic acinar cell line, *Am. J. Physiol.,* 255, G113, 1988.

58. **Dorflinger, L. J. and Schonbrunn, A.,** Somatostatin inhibits vasoactive intestinal peptide-stimulated cyclic adeonsine monophosphate accumulation in GH pituitary cells, *Endocrinology,* 113, 1541, 1983.

59. **Koch, B. D. and Schonbrunn, A.,** The somatostatin receptor is directly coupled to adenylate cyclase in GH_4C_1 pituitary cell membranes, *Endocrinology,* 114, 1784, 1984.

60. **Heisler, S., Reisine, T.D., Hook, V. Y. H., and Axelrod, J.,** Somatostatin inhibits multireceptor stimulation of cyclic AMP formation and corticotropin secretion in mouse pituitary tumor cells, *Proc. Natl. Acad.Sci. U.S.A.,* 79, 6502, 1982.

61. **Gespach, C., Hui Bon Hoa, D., and Rosselin, G.,** Regulation by vasoactive intestinal peptide, histamine, somatostatin-14 and -28 of cyclic adenosine monophosphate levels in gastric glands isolated from the guinea pig fundus or antrum, *Endocrinology,* 112, 1597, 1983.

62. **Park, J., Chiba, T., and Yamada, T.,** Mechanisms for direct inhibition of canine gastric parietal cells by somatostatin, *J. Biol. Chem.,* 262, 14190, 1987.

63. **Jakobs, K. H., Aktories, K., and Schultz, G.,** A nucleotide regulatory site for somatostatin inhibition of adenylate cyclase in S49 lymphoma cells, *Nature (London),* 303, 177, 1983.

64. **Mahy, N., Woolkalis, M., Thermos, K., Carlson, K., Manning, D., and Reisine, T.,** Pertussis toxin modifies the characteristics of both the inhibitory GTP binding proteins and the somatostatin receptor in anterior pituitary tumor cells, *J. Pharmacol. Exp. Ther.,* 246, 779, 1988.

65. **Enjalbert, A., Rasolonjanahary, R., Moyse, E., Kordon, C., and Epelbaum, J.,** Guanine nucleotide sensitivity of [^{125}I]-iodo-*N*-Tyr-somatostatin binding in rat adenohypophysis and cerebral cortex, *Endocrinology,* 113, 822, 1983.

66. **Yajima, Y., Akita, Y., and Saito, T.,** Pertussis toxin blocks the inhibitory effects of somatostatin on cAMP-dependent vasoactive intestinal peptide and cAMP-independent thyrotropin releasing hormone-stimulated prolactin secretion of GH_3 cells, *J. Biol. Chem.,* 261, 2684, 1986.

67. **Reisine, T., Wang, H. L., and Guild, S.,** Somatostatin inhibits cAMP-dependent and cAMP-independent calcium influx in the clonal pituitary tumor cell line AtT-20 through the same receptor population, *J. Pharmacol. Exp. Ther.,* 245, 225, 1988.

68. **Luini, A., Lewis, D., Guild, S., Schofield, G., and Weight, F.,** Somatostatin, an inhibitor of ACTH secretion, decreases cytosolic free calcium and voltage-dependent calcium current in a pituitary cell line, *J. Neurosci.,* 6, 3128, 1986.

69. **Tsunoo, A., Yoshii, M., and Narahashi, T.,** Block of calcium channels by enkephalin and somatostatin in neuroblastoma × glioma hybrid NG108-15 cells, *Proc. Natl. Acad. Sci. U.S.A.,* 83, 9832, 1986.

70. **Lewis, D. L., Weight, F. F., and Luini, A.,** A guanine nucleotide-binding protein mediates the inhibition of voltage-dependent calcium current by somatostatin in a pituitary cell line, *Proc. Natl. Acad. Sci. U.S.A.,* 83, 9035, 1986.

71. **Reisine, T. and Guild, S.,** Pertussis toxin blocks somatostatin inhibition of calcium mobilization and reduces the affinity of somatostatin receptors for agonists, *J. Pharmacol. Exp. Ther.,* 235, 551, 1985.

72. **Koch, B. D., Dorflinger, L. J., and Schonbrunn, A.,** Pertussis toxin blocks both cyclic AMP-mediated and cyclic AMP-independent actions of somatostatin, *J. Biol. Chem.,* 260, 13138, 1985.

73. **Reyl-Desmars, F. and Lewin, M. J. M.,** Evidence for an intracellular somatostatin receptor in pancreas: a comparative study with reference to gastric mucosa, *Biochim. Biophys. Res. Commun.,* 109, 1324, 1982.

74. **Arilla, E., Lopez-Ruiz, M. P., Guijarro, L. G., Prieto, J. C., Gomez-Pan, A., and Hirst, B.,** Characterization of somatostatin binding sites in cytosolic fraction of rat intestinal mucosa, *Biochim. Biophys. Acta,* 802, 203, 1984.

75. **Dokas, L. A., Zwiers, H., Coy, D. H., and Gispen, W. H.,** Somatostatin and analogs inhibit endogenous synaptic plasma membrane protein phosphorylation *in vitro, Eur. J. Pharmacol.,* 88, 185, 1983.

76. **Hierowski, M. T., Liebow, C., Du Sapin, K., and Schally, A. V.,** Stimulation by somatostatin of dephosphorylation of membrane proteins in pancreatic cancer MIA PaCa-2 cell line, *FEBS Lett.,* 179, 252, 1985.

77. **Lewin, M. J. M.,** Somatostatin receptors, *Scand. J. Gastroenterol.,* 21 (Suppl. 119), 42, 1986.

78. **Reisine, T.,** Somatostatin desensitization: loss of the ability of somatostatin to inhibit cyclic AMP accumulation and adrenocorticotropin hormone release, *J. Pharmacol. Exp. Ther.,* 229, 14, 1984.

79. **Mahy, N., Woolkalis, M., Manning, D., and Reisine, T.,** Characteristics of somatostatin desensitization in the pituitary tumor cell line AtT-20, *J. Pharmacol. Exp. Ther.,* 247, 390, 1988.

80. **Krejs, G. J.,** Physiological role of somatostatin in the digestive tract: gastric acid secretion, intestinal absorption, and motility, *Scand. J. Gastroenterol.,* 21 (Suppl. 119), 47, 1986.

81. **Furness, J. B. and Costa, M.,** Actions of somatostatin on excitatory and inhibitory nerves in the intestine, *Eur. J. Pharmacol.,* 56, 69, 1979.

82. **Williams, J. T. and North, R. A.,** Inhibition of firing of myenteric neurones by somatostatin, *Brain Res.,* 155, 165, 1978.

83. **Teitelbaum, D. H., O'Dorisio, T. M., Perkins, W. E., and Gaginella, T. S.,** Somatostatin modulation of peptide-induced acetylcholine release in guinea pig ileum, *Am. J. Physiol.,* 246, G509, 1984.

84. **Taylor, C. W., Merritt, J. E., Putney, J. W., Jr., and Rubin, R. P.,** Effect of Ca^{2+} on phosphoinositide breakdown in exocrine pancreas, *Biochem. J.,* 238, 765, 1986.

85. **Merritt, J. E., Taylor, C. W., Rubin, R. P., and Putney, J. W., Jr.,** Evidence suggesting that a novel guanine nucleotide regulatory protein couples receptors to phospholipase C in exocrine pancreas, *Biochem. J.,* 236, 337, 1986.

86. **Schnefel, S., Banfic, H., Eckhardt, L., Schultz, G., and Schulz, I.,** Acetylcholine and cholecystokinin receptors functionally couple by different G-proteins to phospholipase C in pancreatic acinar cells, *FEBS Lett.,* 230, 125, 1988.

87. **Chaudhry, A., Thompson, R. H., Rubin, R. P., and Laychock, S. G.,** Relationship between delta-9-tetrahydrocannabinol-induced arachidonic acid release and secretagogue-evoked phosphoinositide breakdown and Ca^{2+} mobilization of exocrine pancreas, *Mol. Pharmacol.,* 34, 543, 1988.

88. **Chaudhry, A., Laychock, S. G., and Rubin, R. P.,** The effects of fatty acids on phosphoinositide synthesis and myo-inositol accumulation in exocrine pancreas, *J. Biol. Chem.,* 262, 17426, 1987.

89. **Pandol, S. J., Schoeffield, M. S., Sachs, G., and Muallem, S.,** Role of free cytosolic calcium in secretagogue-stimulated amylase release from dispersed acini from guinea pig pancreas, *J. Biol. Chem.,* 260, 10081, 1985.

90. **Muallem, S., Pandol, S. J., and Beeker, T. G.,** Hormone-evoked calcium release from intracellular stores is a quantal process, *J. Biol. Chem.,* 264, 205, 1989.

91. **Willems, P. H. G. M., Van Nooij, I. G. P., Haenen, H. E. M. G., and De Pont, J. J. H. H. M.,** Phorbol ester inhibits cholecystokinin octapeptide-induced amylase secretion and calcium mobilization, but is without effect on secretagogue-induced hydrolysis of phosphatidylinositol 4,5-bisphosphate in rabbit pancreatic acini, *Biochim. Biophys. Acta,* 930, 230, 1987.

92. **Bastie, M. J., Delvaux, M., Dufresne, M., Saunier-Blache, J. S., Vaysse, N., and Ribet, A.,** Distinct activation of Na^+-H^+ exchange by gastrin and CCK peptide in acini from guinea pig, *Am. J. Physiol.,* 254, G25, 1988.

93. **Rogers, J., Hughes, R. G., and Matthews, E. K.,** Cyclic GMP inhibits protein kinase C-mediated secretion in rat pancreatic acini, *J. Biol. Chem.,* 263, 3713, 1988.

94. **Menozzi, D., Sato, S., Jensen, R. T., and Gardner, J. D.,** Cyclic GMP does not inhibit protein kinase C-mediated enzyme secretion in rat pancreatic acini, *J. Biol. Chem.,* 264, 995, 1989.

95. **Chew, C. S.,** Cholecystokinin, carbachol, gastrin, histamine, and forskolin increase [Ca^{2+}], in gastric glands, *Am. J. Physiol.,* 250, G814, 1986.
96. **Muallem, S., Fimmel, C. J., Pandol, S. J., and Sachs, G.,** Regulation of free cytosolic Ca^{2+} in the peptic and parietal cells of the rabbit gastric gland, *J. Biol. Chem.,* 261, 2660, 1986.
97. **Anderson, N. G. and Hanson, P. J.,** Involvement of calcium-sensitive phospholipid-dependent protein kinase in control of acid secretion by isolated rat parietal cells, *Biochem. J.,* 232, 609, 1985.
98. **Chiba, T., Fischer, S. K., Agranoff, B. W., and Yamada, T.,** Autoregulation of muscarinic and gastrin receptors on gastric parietal cells, *Am. J. Physiol.,* 256, G356, 1989.
99. **Hatt, J. F. and Hanson, P. J.,** Sites of action of protein kinase C on secretory activity in rat parietal cells, *Am. J. Physiol.,* 256, G129, 1989.
100. **Bitar, K. N., Burgess, G. M., Putney, J. W., Jr., and Makhlouf, G. M.,** Source of activator calcium in isolated guinea pig and human antral muscle cells, *Am. J. Physiol.,* 250, G280, 1986.
101. **Bitar, K. N., Bradford, P., Putney, J. W., Jr., and Makhlouf, G. M.,** Cytosolic calcium during contraction of isolated mammalian gastric smooth muscle cells, *Science,* 232, 1143, 1986.
102. **Bitar, K. N., Bradford, P. G., Putney, J. W., Jr., and Makhlouf, G. M.,** Stoichiometry of contraction and Ca^{2+} mobilization by inositol 1,4,5-trisphosphate in isolated gastric smooth muscle cells, *J. Biol. Chem.,* 261, 16591, 1986.
103. **Crochelt, R. F. and Peikin, S. R.,** Characterization of Ca channels mediating gallbladder contraction in the guinea pig, *Gastroenterology,* 90, 1787, 1986.
104. **Crochelt, R. F. and Peikin, S. R.,** Excitation-contraction coupling in the bovine gallbladder: measurement of ^{45}calcium (Ca) uptake and contraction stimulated by cholecystokinin-octapeptide (CCK) and potassium (K), *Gastroenterology,* 90, 1787, 1986.
105. **Andersson, K.-E., Andersson, R., and Hedner, P.,** Cholecystokinetic effect and concentration of cyclic AMP in gall-bladder *in vitro, Acta Physiol. Scand.,* 85, 511, 1972.
106. **Amer, M. S. and McKinney, G. R.,** Studies with cholecystokinin *in vitro.* IV. Effects of cholecystokinin and related peptides on phosphodiesterase, *J. Pharmacol. Exp. Ther.,* 183, 535, 1972.
107. **Gardner, J. D., Sutliff, V. E., Walker, M. D., and Jensen, R. T.,** Effects of inhibitors of cyclic nucleotide phosphodiesterase on actions of cholecystokinin, bombesin, and carbachol on pancreatic acini, *Am. J. Physiol.,* 245, G676, 1983.
108. **Hutchinson, J. B. and Dockray, G. J.,** Evidence that the action of cholecystokinin octapeptide on the guinea pig ileum longitudinal muscle is mediated in part by substance P release from the myenteric plexus, *Eur. J. Pharmacol.,* 69, 87, 1981.
109. **Sano, I., Taniyama, K., and Tanaka, G.,** Cholecystokinin, but not gastrin, induces gamma-aminobutyric acid release from myenteric neurons of guinea pig ileum, *J. Pharmacol. Exp. Ther.,* 248, 378, 1989.
110. **Innis, R. B. and Snyder, S. H.,** Distinct cholecystokinin receptors in brain and pancreas, *Proc. Natl. Acad. Sci. U.S.A.,* 77, 6917, 1980.
111. **Grider, J. R. and Makhlouf, G. M.,** Regional and cellular heterogeneity of cholecystokinin receptors mediating muscle contraction in the gut, *Gastroenterology,* 92, 175, 1987.
112. **Rattan, S. and Goyal, R. K.,** Structure-activity relationship of subtypes of cholecystokinin receptors in the cat lower esophageal sphincter, *Gastroenterology,* 90, 94, 1986.
113. **Goldstein, A., Cox, B. M., Klee, W. A., and Nirenberg, M.,** Endorphin from pituitary inhibits cyclic AMP formation in homogenates of neuroblastoma x glioma hybrid cells, *Nature (London),* 265, 362, 1977.
114. **Walczak, S. A., Wilkening, D., and Makman, M. H.,** Interaction of morphine etorphine, and enkephalins with dopamine-stimulated adenylate cyclase of monkey amygdala, *Brain Res.,* 160, 105, 1979.
115. **Law, P. Y., Wu, J., Koehler, J. E., and Loh, H. H.,** Demonstration and characterization of opiate inhibition of the striatal adenylate cyclase, *J. Neurochem.,* 36, 1834, 1981.
116. **Cooper, D. M. F., Londos, C., Gill, D. L., and Rodbell, M.,** Opiate receptor-mediated inhibition of adenylate cyclase in rat striatal plasma membranes, *J. Neurochem.,* 38, 1164, 1982.
117. **Klee, W. A., Koski, G., Tocque, B., and Simonds, W. F.,** On the mechanism of receptor-mediated inhibition of adenylate cyclase, *Adv. Cyclic Nucleotides Protein Phosphorylation Res.,* 17, 153, 1984.
118. **Bhoola, K. D. and Pay, S.,** Opioid inhibition of adenylate cyclase in the striatum and vas deferens of the rat, *Br. J. Pharmacol.,* 89, 109, 1986.
119. **Simonds, W. F.,** The molecular basis of opioid receptor function, *Endocr. Rev.,* 9, 200, 1988.
120. **Blume, A. J., Lichtshtein, D., and Boone, G.,** Coupling of opiate receptors to adenylate cyclase: requirement for Na^+ and GTP, *Proc. Natl. Acad. Sci. U.S.A.,* 76, 5626, 1979.
121. **Burns, D. L., Hewlett, E. L., Moss, J., and Vaughan, M.,** Pertussis toxin inhibits enkephalin stimulation of GTPase of NG108-15 cells, *J. Biol. Chem.,* 258, 1435, 1983.
122. **Milligan, G., Streaty, R. A., Gierschik, P., Spiegel, A. M., and Klee, W. A.,** Development of opiate receptors and GTP-binding regulatory proteins in neonatal rat brain, *J. Biol. Chem.,* 262, 8626, 1987.
123. **Kurose, H., Katada, T., Amano, T., and Ui, M.,** Specific uncoupling by islet-activating protein, pertussis toxin, of negative signal transduction via alpha-adrenergic, cholinergic, and opiate receptors in neuroblastoma × glioma hybrid cells, *J. Biol. Chem.,* 258, 4870, 1983.

124. **Koski, G., Streaty, R. A., and Klee, W. A.,** Modulation of sodium-sensitive GTPase by partial opiate agonists, *J. Biol. Chem.,* 257, 14035, 1982.
125. **Law, P. Y., Hom, D. S., and Loh, H. H.,** Opiate regulation of adenosine 3':5'-cyclic monophosphate level in neuroblastoma x glioma NG 108-15 hybrid cells, *Mol. Pharmacol.,* 23, 26, 1983.
126. **Makman, M. H., Dvorkin, B., and Crain, S. M.,** Modulation of adenylate cyclase activity of mouse spinal cord-ganglion explants by opioids, serotonin and pertussis toxin, *Brain Res.,* 445, 303, 1988.
127. **De Montis, G. M., Devoto, P., Preti, A., and Tagliamonte, A.,** Differential effect of mu, delta and kappa opioid agonists on adenylate cyclase activity, *J. Neurosci. Res.,* 17, 435, 1987.
128. **Kostka, P., Sipos, N. S., Kwan, C. Y., Niles, L. P., and Daniel, E. E.,** Identification and characterization of presynaptic and postsynaptic β-adrenoreceptors in the longitudinal smooth muscle/myenteric plexus of dog ileum, *J. Pharmacol. Exp. Ther.,* 251, 305, 1989.
129. **Pepper, C. M. and Henderson, G.,** Opiates and opioid peptides hyperpolarize locus coeruleus neurons *in vitro, Science,* 209, 394, 1980.
130. **North, R. A. and Williams, J. T.,** On the potassium conductance increased by opioids in rat locus coeruleus neurones, *J. Physiol. (London),* 364, 265, 1985.
131. **Williams, J. T., Egan, T. M., and North, R. A.,** Enkephalins open potassium channels on mammalian central neurones, *Nature (London),* 299, 74, 1982.
132. **Aghajanian, G. K. and Wang, Y.-Y.,** Pertussis toxin blocks the outward currents evoked by opiate and α_2-agonists in locus coeruleus neurons, *Brain Res.,* 371, 390, 1986.
133. **Mihara, S. and North, R. A.,** Opioids increase potassium conductance in submucous neurones of guinea-pig caecum by activating δ receptors, *Br. J. Phamacol.,* 88, 315, 1986.
134. **North, R. A., Williams, J. T., Suprenant, A., and Christie, M. J.,** μ and δ receptors belong to a family of receptors that are coupled to potassium channels, *Proc. Natl. Acad. Sci. U.S.A.,* 84, 5487, 1987.
135. **Hescheler, J., Rosenthal, W., Trautwein, W., and Schultz, G.,** The GTP-binding protein, G_o, regulates neuronal calcium channels, *Nature (London),* 325, 445, 1987.
136. **Karras, P. J. and North, R. A.,** Inhibition of neuronal firing by opiates: evidence against the involvement of cyclic nucleotides, *Br. J. Pharmacol.,* 65, 647, 1979.
137. **Cherubini, E. and North, R. A.,** Mu and kappa opioids inhibit transmitter release by different mechanisms, *Proc. Natl. Acad. Sci. U.S.A.,* 82, 1860, 1985.
138. **Werz, M. A. and McDonald, R. L.,** Dynorphin reduces voltage-dependent calcium conductance of mouse dorsal root ganglion neurons, *Neuropeptides,* 5, 253, 1984.
139. **McDonald, R. L. and Werz, M. A.,** Dynorphin A decreases voltage-dependent calcium conductance of mouse dorsal root ganglion neurones, *J. Physiol. (London),* 377, 237, 1986.
140. **Gross, R. A. and McDonald, R. L.,** Dynorphin A selectively reduces a large transient (N-type) calcium current of mouse dorsal root ganglion neurons in cell culture, *Proc. Natl. Acad. Sci. U.S.A.,* 84, 5469, 1987.
141. **Attali, B., Saya, D., Nah, S.-Y., and Vogel, Z.,** Kappa opiate agonists inhibit Ca^{2+} influx in rat spinal cord-dorsal root ganglion cocultures, *J. Biol. Chem.,* 264, 347, 1989.
142. **Rosenthal, W. and Schultz, G.,** Modulations of voltage-dependent ion channels by extracellular signals, *Trends Pharmacol. Sci.,* 8, 35, 1987.
143. **Ross, D. H.,** Calcium content and binding in synaptosomal subfractions during chronic morphine treatment, *Neurochem. Res.,* 2, 581, 1977.
144. **Yamamoto, H., Harris, R. A., Loh, H. H., and Way, E. L.,** Effects of acute and chronic morphine treatments on calcium localization and binding in brain, *J. Pharmacol. Exp.Ther.,* 205, 255, 1978.
145. **Pillai, N. P. and Ross, D. H.,** Effects of opiates on high-affinity Ca^{2+}, Mg^{2+}-ATPase in brain membrane subfractions, *J. Neurochem.,* 47, 1642, 1986.
146. **Nehmad, R., Nadler, H., and Simantov, R.,** Effects of acute and chronic morphine treatment on calmodulin activity of rat brain, *Mol. Pharmacol.,* 22, 389, 1982.
147. **Baram, D. and Simantov, R.,** Enkephalins and opiate antagonists control calmodulin distribution in neuroblastoma-glioma cells, *J. Neurochem.,* 40, 55, 1983.
148. **Goedert, M., Pinnock, R. D., Downes, C. P., Mantyh, P. W., and Emson, P. C.,** Neurotensin stimulates inositol phospholipids hydrolysis in rat brain slices, *Brain Res.,* 323, 193, 1984.
149. **Snider, R. M., Forray, C., Pfenning, M., and Richelson, E.,** Neurotensin stimulates inositol phospholipids metabolism and calcium mobilization in murine neuroblastoma clone N1E-115, *J. Neurochem.,* 47, 1214, 1986.
150. **Bozou, J.-C., Amar, S., Vincent, J.-P., and Kitabgi, P.,** Neurotensin-mediated inhibition of cyclic AMP formation in neuroblastoma N1E115 cells: involvement of the inhibitory GTP-binding component of adenylate cyclase *Mol. Pharmacol.,* 29, 489, 1986.
151. **Amar, S., Kitabgi, P., and Vincent, J.-P.,** Stimulation of inositol phosphate production by neurotensin in neuroblastoma N1E115 cells: implication of GTP-binding proteins and relationship with the cyclic GMP response, *J. Neurochem.,* 49, 999, 1987.

152. **Memo, M., Castelletti, L., Valerio, A., Missale, C., and Spano, P. F.,** Identification of neurotensin receptors associated with calcium channels and prolactine release in rat pituitary, *J. Neurochem.*, 47, 1682, 1986.

153. **Gilbert, J. A. and Richelson, E.,** Neurotensin stimulates formation of cyclic GMP in murine neuroblastoma clone N1E-115, *Eur. J. Pharmacol.*, 99, 245, 1984.

154. **Amar, S., Mazella, J., Checler, F., Kitabgi, P., and Vncent, J.-P.,** Regulation of cGMP levels by neurotensin in neuroblastoma clone N1E115, *Biochem. Biophys. Res. Commun.*, 129, 117, 1985.

155. **Snider, R. M., McKinney, M., Forray, C., and Richelson, E.,** Neurotransmitter receptors mediate cyclic GMP formation by involvement of arachidonic acid and lipoxygenase, *Proc. Natl. Acad. Sci. U.S.A.*, 81, 3905, 1984.

156. **Nakagawa, Y., Higashida, H., and Niki, M.,** A single class of neurotensin receptors with high affinity in neuroblastoma x glioma NG108-15 hybrid cells that mediate facilitation of synaptic transmission, *J. Neurosci.*, 4, 1653, 1984.

157. **Kitabgi, P., Checler, F., Mazella, J., and Vincent, J.-P.,** Pharmacology and biochemistry of neurotensin receptors, *Rev. Clin. Basic Pharmacol.*, 5, 397, 1985.

158. **Kitabgi, P. and Freychet, P.,** Neurotensin: contractile activity, specific binding, and lack of effect on cyclic nucleotides in intestinal smooth muscle, *Eur. J. Pharmacol.*, 55, 35, 1979.

159. **Kitabgi, P., Hamon, G., and Worcel, M.,** Electrophysiological study of the action of neurotensin on the smooth muscle of the guinea-pig taenia coli, *Eur. J. Pharmacol.*, 56, 87, 1979.

160. **Donoso, M. V., Huidobro-Toro, J. P., and Kullak, A.,** Involvement of calcium channels in the contractile activity of neurotensin but not acetylcholine: studies with calcium channel blockers and Bay K 8644 on the rat fundus, *Br. J. Pharmacol.*, 88, 837, 1986.

161. **Huidobro-Toro, J. P. and Yoshimura, K.,** Pharmacological characterization of the inhibitory effects of neurotensin on the rabbit ileum myenteric plexus preparation, *Br. J. Pharmacol.*, 80, 645, 1983.

162. **Fontaine, J. and Lebrun, P.,** Effects of neurotensin on the isolated mouse distal colon, *Eur. J. Pharmacol.*, 107, 141, 1985.

163. **Huidobro-Toro, J. P. and Zhu, Y. X.,** Neurotensin receptors on the ileum of the guinea-pig: evidence for the coexistence of inhibitory and excitatory receptors, *Eur. J. Pharmacol.*, 102, 237, 1984.

164. **Goedert, M., Hunter, J. C., and Ninkovic, M.,** Evidence for neurotensin as a non-adrenergic, non-cholinergic neurotransmitter in guinea-pig ileum, *Nature (London)*, 311, 59, 1984.

165. **Kitabgi, P. and Vincent, J.-P.,** Effects of apamin and calcium antagonists on the neurotensin-induced myogenic relaxation in guinea pig colon, *Regul. Peptides*, 5 (Suppl. 2), S85, 1983.

166. **Kitabgi, P. and Vincent, J.-P.,** Neurotensin is a potent inhibitor of guinea pig colon contractile activity, *Eur. J. Pharmacol.*, 74, 311, 1981.

167. **Christinck, F., Daniel, E. E., and Fox, J. E. T.,** Electrophysiological responses of canine ileal circular muscle to electrical stimulation and neurotensin (NT), *Gastroenterology*, 96, A680, 1989.

168. **Kitabgi, P. and Freychet, P.,** Effects of neurotensin on isolated intestinal smooth muscles, *Eur. J. Pharmacol.*, 50, 349, 1978.

169. **Kitabgi, P. and Freychet, P.,** Neurotensin contracts the guinea-pig longitudinal ileal smooth muscle by inducing acetylcholine release, *Eur. J. Pharmacol.*, 56, 403, 1979.

170. **Yau, W. M., Verdun, P. R., and Youther, M. L.,** Neurotensin: a modulator of enteric cholinergic neurons in the guinea pig small intestine, *Eur. J. Pharmacol.*, 95, 253, 1983.

171. **Monier, S. and Kitabgi, P.,** Substance P-induced autodensensitization inhibits atropine-resistant, neurotensin-stimulated contractions in guinea-pig ileum, *Eur. J. Pharmacol.*, 65, 461, 1980.

172. **Sakai, Y., Daniel, E. E., Jury, J., and Fox, J. E. T.,** Neurotensin inhibition of canine intestinal motility *in vivo* via α-adrenoceptors, *Can. J. Physiol. Pharmacol.*, 62, 403, 1984.

173. **Regoli, D., Drapeau, G., Dion, S., and D'Orleans-Juste, P.,** Pharmacological receptors for substance P and neurokinins, *Life Sci.*, 40, 109, 1987.

174. **Hunter, J. C., Goedert, M., and Pinnock, R. D.,** Mammalian tachykinin-induced hydrolysis of inositol phospholipids in rat brain slices, *Biochem. Biophys. Res. Commun.*, 127, 616, 1985.

175. **Watson, S. P.,** The action of substance P on contraction, inositol phospholipids and adenylate cyclase in rat small intestine, *Biochem. Pharmacol.*, 33, 3733, 1984.

176. **Watson, S. P. and Downes, C. P.,** Substance P induced hydrolysis of inositol phospholipids in guinea-pig ileum and rat hypothalamus, *Eur. J. Pharmacol.*, 93, 245, 1983.

177. **Bailey, S. J., Lippe, I. Th., and Holzer, P.,** Effect of the tachykinin antagonist, [D-Pro[4], D-Trp[7,9,10]] substance P-(4—11), on tachykinin- and histamine-induced inositol phosphate generation in intestinal smooth muscle, *Naunyn-Schmiedeberg's Arch. Pharmacol.*, 335, 296, 1987.

178. **Holzer, P. and Lippe, I. Th.,** Substance P action on phosphoinositides in guinea-pig intestinal muscle: a possible transduction mechanism?, *Naunyn-Schmiedeberg's Arch. Pharmacol.*, 329, 50, 1985.

179. **Guard, S., Watling, K. J., and Watson, S. P.,** Neurokinin$_3$-receptors are linked to inositol phospholipid hydrolysis in the guinea-pig ileum longitudinal muscle-myenteric plexus, *Br. J. Pharmacol.*, 94, 148, 1988.

180. **Best, L., Brooks, K. J., and Bolton, T. B.,** Relationship between stimulated inositol lipid hydrolysis and contractility in guinea-pig visceral longitudinal smooth muscle, *Biochem. Pharmacol.,* 34, 2297, 1985.

181. **Bristow, D. R., Curtis, N. R., Suman-Chauhan, N., Watling, K. J., and Williams, B. J.,** Effects of tachykinins on inositol phospholipid hydrolysis in slices of hamster urinary bladder, *Br. J. Pharmacol.,* 90, 211, 1987.

182. **Featherstone, R. L., Fosbraey, P., and Morton, I. K. M.,** A comparison of the effects of three substance P antagonists on tachykinin-stimulated [^{3}H]-acetylcholine release in guinea-pig ileum, *Br. J. Pharmacol.,* 87, 73, 1986.

183. **Salmon, D. M. W. and Bolton, T. B.,** Early events in inositol phosphate metabolism in longitudinal smooth muscle from guinea-pig intestine stimulated with carbachol, *Biochem. J.,* 254, 553, 1988.

184. **Jafferji, S. S. and Michell, R. H.,** Investigation of the relationship between cell-surface calcium-ion gating and phosphatidylinositol turnover by comparison of the effects of elevated extracellular potassium ion concentration on ileum smooth muscle and pancreas, *Biochem. J.,* 160, 397, 1976.

185. **Laufer, R., Wormser, U., Friedman, Z. Y., Gilon, C., Chorev, M., and Selinger, Z.,** Neurokinin B is a preferred agonist for a neuronal substance P receptor and its action is antagonized by enkephalin, *Proc. Natl. Acad. Sci. U.S.A.,* 82, 7444, 1985.

186. **Holzer, P. and Lippe, I. Th.,** Substance P can contract the longitudinal muscle of the guinea-pig small intestine by releasing intracellular calcium, *Br. J. Pharmacol.,* 82, 259, 1984.

187. **Holzer, P. and Petsche, U.,** On the mechanism of contraction and desensitization induced by substance P in the intestinal muscle of the guinea-pig, *J. Physiol. (London),* 342, 549, 1983.

188. **Clapp, L. H., Vivaudou, M. B., Singer, J. J., and Walsh, J. V., Jr.,** A diacylglycerol analogue mimics the action of acetylcholine and substance P on calcium currents in freshly dissociated smooth muscle cells, *J. Gen. Physiol.,* 90, 13a, 1987.

189. **Donowitz, M., Fogel, R., Battisti, L., and Asarkof, N.,** The neurohumoral secretagogues carbachol, substance P and neurotensin increase Ca^{++} influx and calcium content in rabbit ileum, *Life Sci.,* 31, 1929, 1982.

190. **Chang, E. B., Brown, D. R., Wang, N. S., and Field, M.,** Secretagogue-induced changes in membrane calcium permeability in chicken and chinchilla ileal mucosa, *J. Clin. Invest.,* 78, 281, 1986.

191. **Kikkawa, U. and Nishizuka, Y.,** The role of protein kinase C in transmembrane signalling, *Annu. Rev. Cell Biol.,* 2, 149, 1986.

192. **Yoshimasa, T., Sibley, D. R., Bouvier, M., Lefkowitz, R. J., and Caron, M. G.,** Cross-talk between cellular signalling pathways suggested by phorbol-ester-induced adenylate cyclase phosphorylation, *Nature (London),* 327, 67, 1987.

193. **Shuntoh, H., Taniyama, K., Fukuzaki, H., and Tanaka, C.,** Inhibition by cyclic AMP of phorbol ester-potentiated norepinephrine release from guinea pig brain cortical synaptosomes, *J. Neurochem.,* 51, 1565, 1988.

194. **Wiener, E. and Scarpa, A.,** Activation of protein kinase C modulates the adenylate cyclase effector system of B-lymphocytes, *J. Biol. Chem.,* 264, 4324, 1989.

195. **Tapia-Arancibia, L., Veriac, S., Pares-Herbute, N., and Astier, H.,** Activators of protein kinase C enhance cyclic AMP accumulation in cerebral cortical and diencephalic neurons in primary culture, *J. Neurosci. Res.,* 20, 195, 1988.

196. **Sadoshima, J.-I., Akaike, N., Kanaide, H., and Nakamura, M.,** Cyclic AMP modulates Ca-activated K channels in cultured smooth muscle cells of rat aortas, *Am. J. Physiol.,* 255, H754, 1988.

197. **Ewald, D. A., Williams, A., and Levitan, I. B.,** Modulation of single Ca^{2+}-dependent K$^+$-channel activity by protein phosphorylation, *Nature (London),* 315, 503, 1985.

198. **Cachelin, A. B., De Peyer, J. E., Kokubun, S., and Reuter, H.,** Ca^{2+} channel modulation by 8-bromocyclic AMP in cultured heart cells, *Nature (London),* 304, 462, 1983.

199. **Walsh, K. B. and Kass, R. S.,** Regulation of a heart potassium channel by protein kinase A and C, *Science,* 242, 67, 1988.

200. **McRoberts, J. A., Beuerlein, G., and Dharmsathaphorn, K.,** Cyclic AMP and Ca^{2+}-activated K$^+$ transport in a human colonic epithelial cell line, *J. Biol. Chem.,* 260, 14163, 1985.

201. **Kaczmarek, L. K.,** The role of protein kinase C in the regulation of ion channels and neurotransmitter release, *Trends Neurosci.,* 10, 30, 1987.

202. **Mattera, R., Yatani, A., Kirsch, G. E., Graf, R., Okabe, K., Olate, J., Codina, J., Brown, A. M., and Birnbaumer, L.,** Recombinant α$_i$-3 subunit of G protein activates G$_k$-gated K$^+$ channels, *J. Biol. Chem.,* 264, 465, 1989.

203. **Keen, M. and Nahorski, S. R.,** Muscarinic acetylcholine receptors linked to the inhibition of adenylate cyclase activity in membranes from the rat striatum and myocardium can be distinguished on the basis of agonist efficacy, *Mol. Pharmacol.,* 34, 769, 1988.

204. **Hammond, C., Paupardin-Tritsch, D., Nairn, A. C., Greengard, P., and Gershenfeld, H. M.,** Cholecystokinin induces a decrease in Ca^{2+} current in snail neurons that appears to be mediated by protein kinase C, *Nature (London),* 325, 809, 1987.

205. **Weishaar, R. E.**, Multiple molecular forms of phosphodiesterase: an overview, *J. Cyclic Nucleotide Protein Phosphorylation Res.*, 11, 463, 1987.
206. **Kurachi, Y., Ito, H., Sugimoto, T., Schimizu, T., Miki, I., and Ui, M.**, Arachidonic acid metabolites as intracellular modulators of the G protein-gated cardiac K^+ channel, *Nature (London)*, 337, 555, 1989.
207. **Kim, D., Lewis, D. L., Graziadei, L., Neer, E. J., Bar-Sagi, D., and Clapham, D. E.**, G-protein βγ-subunits activate the cardiac muscarinic K^+-channel via phospholipase A_2, *Nature (London)*, 337, 557, 1989.
208. **Irvine, R. F. and Moor, R. M.**, Micro-injection of inositol 1,3,4,5-tetrakisphosphate activates sea urchin eggs by a mechanism dependent on external Ca^{2+}, *Biochem. J.*, 240, 917, 1986.
209. **Taylor, C. W.**, Receptor regulation of calcium entry, *Trends Pharmacol. Sci.*, 8, 79, 1987.
210. **Horstman, D. A., Takemura, H., and Putney, J. W., Jr.**, Formation and metabolism of [^{3}H]inositol phosphates in AR42J pancreatoma cells, *J. Biol. Chem.*, 263, 15297, 1988.
211. **Farley, J. and Auerbach, S.**, Protein kinase C activation induces conductance changes in Hermissenda photoreceptors like those seen in associative learning, *Nature (London)*, 319, 220, 1986.
212. **Strong, J. A., Fox, A. P., Tsien, R. W., and Kaczmarek, L. K.**, Stimulation of protein kinase C recruits covert calcium channels in Aplysia bag cell neurons, *Nature (London)*, 325, 714, 1987.
213. **Rane, S. G. and Dunlap, K.**, Kinase C activator 1,2-oleoylacetylglycerol attenuates voltage-dependent calcium current in sensory neurons, *Proc. Natl. Acad. Sci. U.S.A.*, 83, 184, 1986.
214. **Llewellyn-Smith, I. J.**, Neuropeptides and the microcircuitry of the enteric nervous system, *Experientia*, 43, 813, 1987.
215. **Ryu, P. D., Gerber, G., Murase, K., and Randic, M.**, Calcitonin gene-related peptide enhances calcium current of rat dorsal root ganglion neurons and spinal excitatory synaptic transmission, *Neurosci. Lett.*, 89, 305, 1988.
216. **Reynolds, E. E. and Yokota, S.**, Neuropeptide Y receptor-effector coupling mechanisms in cultured vascular smooth muscle cells, *Biochem. Biophys. Res. Commun.*, 151, 919, 1988.
217. **Exton, J. H.**, Mechanisms of action of calcium-mobilizing agonists: some variations on a young theme, *FASEB J.*, 2, 2670, 1988.
218. **Bourne, H. R.**, Who carries the message?, *Nature (London)*, 337, 504, 1989.

Chapter 10

PEPTIDASES AND NEUROPEPTIDE-INACTIVATING MECHANISMS IN THE CIRCULATION AND IN THE GASTROINTESTINAL TRACT

Frédéric Checler

TABLE OF CONTENTS

I.   Introduction 274

II.   Technical Aspects 275
    A.   Identification and Characterization of Peptidases and
          Metabolites 275
          1.   Identification of Peptidases 275
          2.   Characterization of Peptides and Metabolites 276
               a.   Bioassay 276
               b.   Radioimmunoassay 276
               c.   Chromatographic Techniques 276
               d.   Choice of Substrate 277
    B.   Methods for Investigating Neuropeptide-Inactivating
          Mechanisms Occurring in the Circulation and in the
          Gastrointestinal Tract 277
          1.   Circulation 277
               a.   Degradation of Peptides in Serum, Plasma, or
                        Whole Blood *In Vitro* 277
               b.   Degradation of Peptides in the Circulation *In*
                        Vivo 277
          2.   Gastrointestinal Tract 277
               a.   Tissue Homogenates and Purified Membrane
                        Preparations 277
               b.   Tissue Slices 277
               c.   Isolated Organs 278
               d.   Perfused Organs *In Vitro* and *In Vivo* 278

III.   Neuropeptide Metabolism in the Circulation and in the Gastrointestinal
    Tract 278
    A.   Rates of Neuropeptide Disappearance in the Circulation 278
          1.   Degradation in Serum, Plasma, and Whole Blood *In*
               Vitro 278
          2.   Clearance of Neuropeptides in the Circulation *In Vivo* 280
    B.   Peptidases and Inactivating Processes Occurring in the
          Gastrointestinal Tract 283
          1.   Peptidases in the Gastrointestinal Tract 283
               a.   Organs Distribution 283
               b.   Subcellular Localization 285
               c.   Specificity 286
          2.   Mechanisms of Neuropeptide Inactivation 290
               a.   Cholecystokinin and Gastrins 290

        b.   Neurotensin 292

        c.   Opiates 294

        d.   Tachykinins 297

        e.   Other Peptides..298

IV.   Conclusions and Perspectives 299

Acknowledgment..300

References..302

I. INTRODUCTION

In recent years, the emergence of numerous newly discovered peptides has triggered many studies aimed at explaining their physiology. It rapidly became apparent that these peptides belonged to families of peptides that often exhibited a dual localization, in the brain and in the gastrointestinal tract. Besides their topographical distribution, studies have attempted to observe the processing of the precursors, the mechanisms by which peptides are released from neuron terminals, and their ability to bind to specific receptors. Cellular models were then used to correlate the receptor occupancy with a modification of subcellular events that constitute the message that is transmitted to the target cell. An important step in understanding the physiology of neuropeptides involved the elucidation of the mechanisms by which they are inactivated and the message therefore interrupted. Although this knowledge has been relatively slow to develop, there are now a growing number of reports that have contributed to a better understanding of the ways by which the regulation of peptide concentrations occurs. Unlike other classical neurotransmitters for which reuptake processes have been evidenced, there are no clues indicating that such a mechanism is responsible for the clearance of neuropeptides, and it is more generally admitted that peptides undergo proteolytic inactivation by various classes of ectoenzymes, including exo- and endopeptidases.

Immunohistochemical studies have established that peptides in the gastrointestinal tract are widely distributed within both muscular and nervous layers as well as in the mucosa of small intestine from which they can be released upon humoral or luminal stimuli. At this step, the concept of paracrine or endocrine function is directly related to the mechanisms of peptide inactivation. The fact that peptidases have been detected in smooth muscle tissues besides peptide receptors that are thought to be involved in the modulation of gastrointestinal smooth muscle contractility argues in favor of a paracrine role of some of these peptides that would be rapidly inactivated after their release and local interaction with specific receptors. By contrast, an endocrine function implies delayed mechanisms of degradation in order to allow the peptides to enter circulation and reach the target cells as native and biologically active entities. The present review will gather information concerning the presence, distribution, subcellular localization and specificities of the peptidases that have been detected in the circulation and in the gastrointestinal tract, and we will report on recent data related to the putative involvement of these enzymes in the metabolism of various brain-gut peptides.

TABLE 1
Fluorigenic Substrates and Specific Inhibitors to Identify Neuropeptide-hydrolyzing Peptidases

Peptidase	Substrate	Inhibitor	Ref.
Aminopeptidase			
A (AmA)	Glu–NA	Amastatin	1
B(AmB)	Arg–NA	Arphamenine	1
N(AmN)	Leu–NA	Bestatin, carbaphethiol	4, 2
		LY110947	116
Pyroglutamyl-peptide hydro-lase (pAm)	pyr–7AMC	Pyroglutamyl-ketone ester ox-oprolinal	5, 137
Postproline dipeptidyl amino-peptidase (DAP IV)	Gly–Pro–7AMC	Diprotin A	1
Postproline-cleaving enzyme (PPCE)	Z–Gly–Pro–7AMC	Z-pro-prolinal	6
Angiotensin-converting en-zyme (ACE)	Hipp–His–Leu	Captopril	7
Endopeptidase 24.11 (enke-phalinase)	Suc–Ala–Ala–Phe–7AMC	Thiorphan	8
		Acetorphan	9, 138
		Phosphoramidon	3
Endopeptidase 24.15	Bz–Gly–Ala–Ala–Phe–pAB	CPE–Ala–Ala–Phe–pAB	11
Carboxypeptidase A	Hipp–Phe	Arphamenine B	1
Dipeptidyl aminopeptidase (DAP)	Lys–Ala–7AMC	Kelatorphan	10

Note: NA, β-naphtylamide; pyr, pyroglutamic acid; 7AMC, 7 amido-4-methylcoumarin; Z, benzyloxycarbonyl; Hipp, hippuric acid; Suc, succinyl; pAB, *p*-amino-benzoate.

II. TECHNICAL ASPECTS

A. IDENTIFICATION AND CHARACTERIZATION OF PEPTIDASES AND METABOLITES
1. Identification of Peptidases

Recent studies concerned with the disappearance of natural peptides have led to the detection of numerous novel peptidases. It rapidly has become obvious that the development of highly specific substrates would make it more convenient to detect or monitor peptidases, particularly during the procedures of their purification. Table 1 summarizes a series of chromogenic peptides that are thought to behave as rather specific substrates of the indicated peptidases. The cleavage of these substrates liberates a chromophore (nitroanilide, *p*-aminobenzoate [pAB], or the 7-amido-4-methylcoumarinyl group [7AMC]) that allows sensitive detection and rapid quantification of peptidasic activities. Because of the possibility of these substrates being cleaved by unexpected peptidases, specific peptidase inhibitors were necessary to determine the inhibitor-sensitive fraction that could be ascribed to a given fluorigenic substrate-hydrolyzing activity. Several natural peptidase inhibitors have been purified from microbial culture filtrates.[1] These inhibitors, including amastatine,[1] arphamenine A and B,[1] bestatin,[2] diprotin A,[1] and phosphoramidon,[3] display both specificity and high affinities (IC$_{50}$ ranging between 0.01 and 5.5 μg/ml) for their respective peptidases.[1] Besides these natural agents, very potent synthetic molecules have been successfully developed that elicit half-maximal inhibitions of their corresponding enzyme (see Table 1) at the following concentrations: carbaphethiol, 5 nM;[4] oxoprolinal, 26 nM;[5] Z-pro-prolinal, 14 nM;[6] captopril, 20 nM;[7] thiorphan, 4.5 nM;[8] and acetorphan, 8.6 nM.[9] In addition to these monospecific inhibitors, an agent called kelatorphan has recently appeared that behaves as a mixed inhibitor with specificities directed toward endopeptidase 24.11 (K$_i$ = 1.4 nM), aminopeptidases

(K_i = 7 μM), and dipeptidylaminopeptidase (K_i = 1 nM).[10] In the absence of a specific inhibitor of the latter activity, kelatorphan was proved to be the only agent that allows the determination of the contribution of dipeptidylaminopeptidase by comparing the Lys–Ala–7AMC hydrolyzing activity recovered in the presence of either bestatin (to block the contribution of aminopeptidases) or kelatorphan.

It is interesting here to emphasize the fact that high concentrations of a specific inhibitor can also affect other peptidases belonging to the same class of proteases (acidic, serine, thiol, or metalloenzymes). For instance, captopril inhibits the angiotensin-converting enzyme with a K_i value of 20 nM, but can induce a 50% inhibition of endopeptidase 24.11 at a concentration of 10 μM.[8] In some cases, the difference between the relative potencies can be much more restricted. Thiorphan exhibits K_i values toward endopeptidase 24.11 and angiotensin-converting enzyme of 4.7 nM and 150 nM, respectively,[8] and CPE–Ala–Ala–Phe–pAB inhibits endopeptidase 24.15 and endopeptidase 24.16 with very close K_i values of 1.94 μM[11] and 10 μM,[12,13] respectively. Based on the above considerations, particular attention must be paid to the choice of the concentrations of inhibitors that are used. Although exploitation of data can be relatively simple when inhibition studies are carried out *in vitro*, in well-defined assay conditions, difficulties can appear when exploring *in vivo* models. Thus, it can become difficult to assess the identity of involved peptidases, since the initial concentration of inhibitor can be affected by unknown factors, such as metabolic stability and tissue accessibility, and therefore leads to erroneous interpretations.

It must be pointed out that for two recently isolated endopeptidases, endo-oligopeptidase A and endopeptidase 24.16, neither chromogenic substrate nor specific inhibitor has been developed yet. However, in both cases, monospecific polyclonal antibodies have been obtained[14,15] that allow immunological approaches that have proven to be convenient for the detection of these enzymes (as well as for other peptidases, see Section III.B.1.a).

2. Characterization of Peptides and Metabolites
a. Bioassay

This technique can be used to estimate the amount of peptide that remains intact after exposure to degradative conditions and, therefore, to derive the overall extent of peptidasic inactivation. However, this approach does not permit identification of the metabolites. Furthermore, the possibility of a conversion of the peptide into biologically active degradation products or its putative influence on the release of other endogenous peptides with biological properties interfering in the bioassay cannot be ruled out.

b. Radioimmunoassay

The sensitivity of this technique represents one of its main advantages and allows the detection of small quantities of endogenous substances. Careful attention must be paid to the characterization of antibody specificity in order to avoid misleading interpretations due to cross-reaction of parent peptides and metabolites. Antibodies displaying restricted specificity also can be coupled to chromatographic methods in order to characterize the peptide metabolites further.

c. Chromatographic Techniques

Separation of peptides and metabolites can be achieved readily by classical chromatographic techniques such as ion exchange and gel permeation. High pressure liquid chromatography (HPLC) allows rapid and efficient separation of various metabolites that can therefore be identified either by their retention times (by analogy with synthetic standards run under the same conditions), radioimmunoassay (RIA) or direct amino-acid analysis when sufficient amounts of peptide are recovered in eluted fractions.

d. Choice of Substrate

Unlabeled or radiolabeled substrates can be used to monitor peptide degradation. Tritiated, [^{14}C], or iodinated peptides have been synthesized successfully. These molecules can allow accurate quantitative determination of degradation rates provided that introduction of the label does not induce steric hindrance that prevents recognition and/or catalysis by the peptidases.

Labeled substrates can be detected after chromatographic separation on columns or thin layers. These molecules must be replaced by unlabeled molecules when the goal of the study is the characterization of the metabolites by direct amino-acid analysis.

B. METHODS FOR INVESTIGATING NEUROPEPTIDE-INACTIVATING MECHANISMS OCCURRING IN THE CIRCULATION AND IN THE GASTROINTESTINAL TRACT

1. Circulation

a. Degradation of Peptides in Serum, Plasma, or Whole Blood In Vitro

Experiments carried out with whole blood, plasma, or serum *in vitro* can give information about the stability of peptides in these biological compartments and, therefore, an estimation of their relative contribution to the catabolism of peptides that are injected *in vivo* in the circulation. Experiments are generally carried out with whole blood, heparinized plasma, or serum. Possibilities of peptidase inactivation after collection of the samples or, by contrast, enzyme activation during the process of clotting must not be underestimated.

b. Degradation of Peptides in the Circulation In Vivo

Peptides can be administered as a bolus injection or infused at a constant rate in the circulation. After bolus injection of unlabeled or radioactive peptides, venous effluents can be taken out after various time intervals, and the disappearance rate can be monitored by the previously described techniques according to the type of peptide that is used. Infusion of peptides at constant rates permits a steady-state corresponding to an equilibrium between peptide distribution and degradation. After cessation of infusion, it is possible to follow the decrease of the intact peptide. Whatever the administration mode, estimation of the half-life of peptides *in vivo* in the circulation becomes possible.

2. Gastrointestinal Tract

a. Tissue Homogenates and Purified Membrane Preparations

Homogenates of gastrointestinal tissues allow the determination of their complete content in neuropeptidases and their putative ability to catabolize neuropeptides. An important limitation of these preparations concerns the fact that no indication of the subcellular localization of the peptidases can be afforded. Furthermore, homogenization of the tissues can lead to the exposure of peptidases that are not accessible to the peptide *in vivo* or *in vitro* in intact tissues. Information dealing with the distribution of peptidases within muscular, nervous, and mucosal layers has become available with the development of procedures yielding highly purified and fully characterized membrane preparations.[16-19]

b. Tissue Slices

Tissue slices can be incubated or superfused with solutions containing physiological concentrations of labeled or unlabeled exogenous peptides. This technique allows the establishment of the contribution of cell-surface peptidases. Slices are also permitted to follow the metabolic processes of potassium-evoked released endogenous peptides. Although the overall inactivation mechanisms can reflect degradation occurring at locations besides the site of release, the contribution of peptidases present during the course of peptide diffusion or located at the surface of the slice can affect the initial pattern of peptide inactivation.

c. Isolated Organs

The isolated organ is one of the bioassays that is the most widely employed by pharmacologists. The main advantage of this technique in studies on peptide metabolism is the conservation of the integrity of the structure of the tissue. Furthermore, inactivation of the peptides can be directly appreciated by following the decrease in their biological activities. If experiments are performed with exogenous peptides, it is not possible to ascertain whether the loss of potency is due to peptidases encountered during peptide diffusion in the tissue. In some cases, experimental conditions, such as electrical-field stimulation, allow the investigation of the metabolism of endogenously released peptides that can be compared with those established for corresponding exogenous peptides.

d. Perfused Organs **In Vitro** and **In Vivo**

The ability of an organ to catabolize a peptide can be investigated by these techniques. The vascular arcade irrigating a restricted part of the gastrointestinal tract can be isolated and arteriovenous differences in the content of intact peptide can be monitored by the various detection methods described earlier. In both *in vitro* and *in vivo* approaches, organs can be irrigated by blood in order to mimic physiological conditions or flushed by isotonic buffers to prevent a putative participation of blood or plasmatic factors in the process of peptide degradation.

III. NEUROPEPTIDE METABOLISM IN THE CIRCULATION AND IN THE GASTROINTESTINAL TRACT

A. RATES OF NEUROPEPTIDE DISAPPEARANCE IN THE CIRCULATION
1. Degradation in Serum, Plasma, and Whole Blood *In Vitro*

Table 2 presents the estimates of the half-lives of various peptides *in vitro* after incubations with serum, plasma, or whole blood of various species. Human gastrin I and cholecystokinin (CCK) were shown by bioassay to be stable in cat plasma[20] and whole blood of dog.[21] Atrial peptides such as atrial natriuretic factor [ANF-(Ser^{99}–Tyr^{126})] and atriopeptin III were found to be relatively stable in rat plasma or whole blood whatever the method of detection of the intact peptide.[22,23] However, after prolonged incubations in plasma, ANF–(Ser^{99}–Tyr^{126}) was partially converted into ANF–(Ser^{103}–Tyr^{126}) by peptidasic mechanisms that have not been clarified. Thyrotropin-releasing hormone (TRH) was stable in dog and mouse plasma[24] and of shorter duration in rat and man.[24,25] It is interesting to note that most of these peptides exhibited either a pyroglutamyl group at their N-terminal side (TRH, gastrin I) or an amidated carboxy-terminal end (TRH, gastrin I, CCK) that could have conferred resistance to proteolytic attack by exopeptidases such as aminopeptidases, postproline dipeptidyl aminopeptidase (DAP IV), angiotensin-converting enzyme (ACE), and carboxypeptidases that correspond to the major activities detected in plasma and serum (see Table 3).

Studies on neurotensin inactivation indicate that this peptide undergoes a very slow proteolysis in rat whole blood[26] and human plasma.[27] HPLC analysis of tritiated neurotensin degradation led to the identification of neurotensin$_{1–8}$, neurotensin$_{1–11}$, and neurotensin$_{9–13}$.[26] A study carried out with unlabeled peptide established a half-life (226 min)[27] slightly higher than that obtained with the labeled substrate ($\cong$ 60 min).[26] Identification of the generated degradation products by combined HPLC and RIA also led to fragments chromatographically and immunologically identical to neurotensin$_{1–8}$, and neurotensin$_{1–11}$.[27] The loss of neurotensin appeared to be greatly reduced by ethylenediaminetetraacetic acid (EDTA), indicating that metallopeptidases were likely involved in the formation of these metabolites, although these activities were not definitely identified.

Studies on the inactivation of substance P revealed *in vitro* half-lives nearly identical in

TABLE 2
Half-life Estimates of Peptides Inactivation in Plasma or Whole Blood *In Vitro*

Peptides	Source	Species	Method of Assay	Half-Life (min)	Ref.
Angiotensin 1	Plasma	Dog	^{3}H peptide and	3	32
	Whole blood	Dog	paper electro-phoresis Bioassay	15	32
ANF	Whole blood plasma	Rat	[^{125}I] peptide and HPLC	>60	22
Atriopeptin III	Plasma	Rat	Bioassay	>30	23
Bradykinin	Plasma	Rat	Bioassay	5	29
	Plasma	Rat, rabbit and human	Bioassay	10 to 15	31
CCK	Whole blood	Dog	Bioassay	Stable after 30	21
Gastrin 1 (human)	Plasma	Cat	Bioassay	Stable	20
Leu-enkephalin	Plasma	Rat	Bioassay	2.5	33
Met-enkephalin	Plasma	Rat	Bioassay	2	33
Neurotensin	Whole blood plasma	Rat	[^{3}H] peptide and HPLC	≅60	26
		Human	HPLC + RIA	226	27
Substance P	Plasma	Human	RIA	15	28
		Rat	Bioassay	12	29
TRH	Plasma	Rat	[^{14}C] peptide and TLC	12.22	24, 25
		Dog		>1500	24
		Man		33	24
		Mouse		215	24
	Diluted ($^1/_{10}$) plasma	Rat	Bioassay	30	139

Note: ANF, atrial natriuretic factor ANF-(Ser^{99}-Tyr^{126}); atriopeptin III, ANF-(Ser^{103}-Tyr^{126}); CCK, cholecystokinin; TRH, thyrotropin-releasing hormone; TLC, thin layer chromatography; HPLC, high performance liquid chromatography.

TABLE 3
Peptidases Detected in Serum, Plasma, and Vascular Tissues

Source	Species	pAm	AmN	DAPIV	ACE	PPCE	Carboxypeptidase B	Ref.
Plasma	Human	+		+			+	28, 31, 140, 141
	Guinea pig				+			92
	Rat		+		+		+	30, 31, 34
	Rabbit						+	31
Serum	Human		+	+		+	+	35, 140, 142
Aorta and mes-enteric artery	Hog		+	+	+			143—145

Note: Abbreviations of peptidases are as in Table 1.

human or rat plasma,[28,29] but much shorter than those reported for the previously discussed peptides (see Table 2). Analysis of metabolites that were HPLC separated and partially sequenced established that substance P was rapidly converted into des[Arg^{1}Pro^2]-substance P and des[Arg^{1}Pro^{2}Lys^{3}Pro^4]-substance P in human plasma, indicating a sequential release of its two N-terminal dipeptides.[28] These data were consistent with the hypothesis of a postproline dipeptidyl aminopeptidase hydrolysis and were supported by the detection of a peptidase displaying such specificity in human plasma, i.e., DAP IV.[28] In rat plasma, SP inactivation was followed by evaluation of the disappearance of its biological activity on the guinea-pig isolated ileum.[29] Although the metabolites were not fully characterized, it was found that high concentrations of captopril induced a significant protection of intact SP[29] and, therefore, probably indicated an involvement of an angiotensin-converting enzyme-like activity. Interestingly, such activity had been detected previously in rat plasma.[30]

Captopril was used to protect bradykinin from degradation in the plasma of rat, rabbit, and human.[31] Unlike substance P, this specific inhibitor did not affect the rate of bradykinin disappearance.[31] This was in agreement with another study showing a very weak protection of this peptide in rat plasma.[29] By contrast, bradykinin appeared to be converted into des−Arg^9−bradykinin.[31] This could be indicative of the presence of an exopeptidase displaying a carboxypeptidase B-like activity. Here again, such a peptidase had been detected in the serum and plasma of various species including human, rat, and rabbit.[31] The fact that des−Arg^9−bradykinin was found to be completely stable after 2 h of incubation suggested that no other plasma peptidases could process this bradykinin fragment further.[31] Resistance to aminopeptidasic hydrolysis can be explained by a prolyl doublet located at the N-terminus of bradykinin that induces resistance to aminopeptidases and DAP IV.

The behavior of angiotensin I has been followed in plasma and whole blood of dog.[32] In both cases, the major metabolite that first appeared was identified as the C-terminal dipeptide histidyl-leucine[32] whose formation was followed by the emergence of free leucine that was shown to derive from the former dipeptide.[32] It is interesting to note that angiotensin I is biologically inactive and that the pressor peptide angiotensin II corresponds to the des−His−Leu angiotensin I molecule. Therefore, plasmatic degradation could be considered as a mechanism of activation. Little is known concerning the peptidases involved in the behavior of angiotensin II *in vitro* in plasma or whole blood.

Methionine and leucine enkephalins are two pentapeptides that exhibit free N- and C-termini and that undergo the most rapid disappearance *in vitro* (see Table 2). Half-lives in rat plasma were 2 and 2.5 min for Met- and Leu-enkephalins, respectively, as followed by their depressant effect on the electrically-induced contractions of the mouse vas deferens.[33] Thin layer chromatography analysis of the breakdown pattern of tritiated Met-enkephalin revealed the presence of the C-terminal tetrapeptide as the main labeled degradation product.[33] Such a product probably resulted from the release of tyrosine 1 by aminopeptidases that have been detected in rat plasma[34] and in human serum.[35]

2. Clearance of Neuropeptides in the Circulation *In Vivo*

Table 4 summarizes estimates of the *in vivo* half-lives of several neuropeptides in the circulatory system of various species. From the time course of the blood pressure changes it evoked, angiotensin II was shown to be catabolized rapidly ($t_{1/2} = 0.27$ min) in the general circulation of the rat.[36] This was in agreement with another study where the disappearance of iodinated angiotensin II was monitored by thin layer chromatography ($t_{1/2} = 0.21$ min[37]). By contrast, slightly higher durations were reported for the natural horse and hog angiotensin (Asp^1−Ile^5−angiotensin II) in the dog.[38]

Two studies established the nearly identical half-lives for angiotensin III[36,37] which were in the range of those obtained for angiotensin II (0.23 to 0.27 min, Table 4). Interestingly, Harding et al.[37] established that angiotensin III (des−Asp^1−angiotensin 2) was the obligatory product generated during the sequence of [^{125}I]angiotensin II conversion. This indicated a preferential processing of angiotensin II at its N-terminal side and was further corroborated by a study establishing an enhanced stability ($t_{1/2} = 6.4$ min) of an angiotensin II analogue in which the aspartyl in the N-terminal position 1 had been replaced by a saralasin residue.[36] By contrast, amidation of the C-terminal end of angiotensin II which is generally considered

TABLE 4
Half-life Estimates of Peptides Inactivation *In Vivo*

Peptide	Administration mode	Species	Method of assay	Half-life (min)	Ref.
Angiotensin II	Infusion	Rat	Bioassay	0.27	36
	Bolus	Rat	^{125}I peptide + HPLC	0.21	37
	Infusion	Dog	Bioassay	3	38
Angiotensin III	Infusion	Rat	Bioassay	0.23	36
	Bolus	Rat	^{125}I peptide + HPLC	0.27	37
ANF	Bolus	Rat	^{125}I peptide + HPLC	0.28	39
	Bolus	Human	RIA	2.5	146
Atriopeptin III	Infusion	Rat	RIA	0.5	147
	Bolus	Rat	^{125}I peptide + HPLC	0.37	39
	Bolus	Rat	^{125}I peptide + TLC	2.5	23
Bradykinin	Infusion	Cat	Bioassay	0.28	50
CCK	Infusion	Human	RIA	2.4	44
Gastrin 1 (human)	Bolus	Cat	Bioassay	2.65	20
	Bolus	Human	RIA	5.5—10.5	148
	Infusion	Dog	RIA	2.1	41
	Infusion	Dog	RIA	3.7	42
	Bolus	Dog	^{125}I peptide + chromatoelectrophoresis	3	40
Met-enkephalin	Bolus	Rat	^{3}H peptide + chromatography	0.08	49
Neurotensin	Bolus	Rat	RIA	0.55	26
	Bolus	Human	RIA	1.4	69
	Bolus	Human	Chromatography + RIA	1.5	46
Substance P	Infusion	Rat	RIA + HPLC	≤5	55
TRH	Bolus	Rat	^{14}C peptide + TLC	5[a]	25
	Bolus	Human	Bioassay	5	47
	Bolus	Human	RIA	5.3	48
Somatostatin	Infusion	Human	RIA	1.1—3	53
VIP	Infusion	Human	RIA	1	54

Note: VIP, vasoactive intestinal peptide; other abbreviations as in Table 2.

[a] Author's estimation.

to slow down or block carboxypeptidasic attacks, did not modify its half-life in the rat circulation (angiotensin II-amide $t_{1/2}$ = 15s).

Half-lives of atrial natriuretic factor (ANF) and atriopeptin III after bolus injection in the rat circulation were established to be 0.28 min and 0.37 min, respectively, after HPLC analysis of the iodinated peptides.[39] These values were slightly lower than those deriving from a quantification by either RIA (ANF: $t_{1/2}$ = 2.5 min) or thin layer chromatography(atriopeptin III: $t_{1/2}$ = 2.5 min). No data are available yet concerning the identity of the metabolites yielded during the rapid disappearance of atrial peptides *in vivo* in the circulation.

Several studies were aimed at investigating the fate of human gastrin I (G-17) and its related peptide, CCK, in the circulation of various species. A half-life of 2.65 min was determined by bioassay (acid secretion) after bolus injection of G-17 in the cat.[20] This method of peptide administration in human and dog led to values either in the same range of order, i.e., 3 min in the dog,[40] or noticeably higher (5.5, 8, and 10.5 min in three human subjects). When exogenous human gastrin I was infused intravenously for 2 h in the dog, it was shown that the mean gastrin level rose about sixfold above basal value and then returned to baseline almost 15 min after infusion.[41] These experiments resulted in an overall disappearance rate of 2.1 min,[41] a value in good agreement with another study ($t_{1/2}$ = 3.7 min) employing the same species, route of administration, and method of peptide detection.[42] Ion exchange chromatography of plasma samples taken during the plateau phase of G-17 after infusion in man revealed the main presence of G-17_{1-13}, and G-17_{1-16} accompanied by minor amounts of G-17_{5-17} and G-17_{1-11}.[43]

A sensitive RIA was developed to monitor the metabolic rate of CCK after i.v. infusion in man and dog and permitted derivation of a half-life of 2.5 min.[44] Until now, nothing is known about the identity of the degradation products of CCK.

All studies concerned with the metabolism of intravenously administered neurotensin were carried out by following the decrease of intact peptide by RIA alone or in combination with chromatographic characterization of the degradation products. Bolus injections of neurotensin in rat and human resulted in a rapid disappearance of the peptide, with half-lives varying between 0.55 and 1.5 min[26,45,46] when measured with C-terminal-directed antibodies. Interestingly, these values were significantly increased when clearance of neurotensin was followed with an N-terminal-directed antiserum, since half-lives corresponded to 5 min and 10 min in rat and human, respectively.[26,46] These discrepancies could be explained by the fact that the main neurotensin metabolites that were characterized were the N-terminal fragments neurotensin$_{1-8}$ and neurotensin$_{1-11}$. These peptides were shown to be rather resistant to degradation *in vivo* in the circulation ($t_{1/2}$ = 9 and 5 min, respectively), in contrast with the C-terminal counterpart neurotensin$_{9-13}$, which was catabolized extremely rapidly ($t_{1/2}$ = 0.5 min). Therefore, a cross-reaction of these fragments with the parent peptide in the N-terminal-directed RIA could have led to an overestimation of neurotensin half-life in the circulation.

The half-life of TRH after bolus injection in human was estimated to be 5 min and 5.3 min after bioassay[47] and RIA,[48] respectively. Administration of the [^{14}C] peptide in the rat and quantification of native peptide by thin layer chromatography led to a nearly identical half-life value.[25] Although the TRH metabolites were not identified, an interesting feature concerned the fact that the TRH analogue [^{14}C](3-MePro)-TRH displayed an increased stability *in vivo* (only 25% degradation after 15 min).[25] Therefore, it appears likely that the substitution of the proline amide group by the methylproline group has conferred the increased resistance to proteolysis, therefore indicating that the metabolic process occurs *in vivo* in the circulation, at the C-terminal side of TRH.

Degradation of tritiated Met-enkephalin is extremely rapid after its i.v. bolus administration to the rat.[49] Partition chromatography analysis of radioactivity recovered in the plasma 0.25 min after injection indicated that only 5% of the label comigrates with intact peptide.[49] These experiments indicated a half-life estimate of about 0.08 min (see Table 4).

Few studies were concerned with the disappearance of bradykinin, substance P (SP), somatostatin, and vasoactive intestinal polypeptide (VIP). Bradykinin was short-lived in the circulation of the cat, exhibiting a half-life of about 0.28 min.[50] This value was in good agreement with those determined in man[51] and dog.[52] Somatostatin disappeared linearly by 7 to 10 min after cessation of infusion in man,[53] then a much slower component was observed. The half-life of the first component varied between 1.1 and 3 min.[53] RIA measurements of VIP disappearance after cessation of infusion in man led to an average half-life degradation of about 1 min.[54] Finally, a study concerned with the inactivation of SP after infusion of various doses of this peptide in the rat circulation demonstrated that SP levels had returned to baseline values 5 min after the end of infusion, regardless of the dose of peptide employed.[55] No peptide metabolites had been characterized in the case of the four peptides mentioned above.

TABLE 5
Peptidases Detected or Purified from Organs of the Gastrointestinal Tract of Various Species

Peptidase	Esophagus	Stomach	Duodenum	Jejunum	Ileum	Cecum	Colon	Ref.
			Organ					
Aminopeptidase A					+			58
Aminopeptidase B	+	+	+	+	+	+	+	56,57
Aminopeptidase N	+	+	+	+	+	+	+	56—59,64
pGlu-aminopeptidase	−	−	−	−	−	−	−	56,64
DAP	+	+	+	+	+	+	+	56
DAPIV	+	+	+	+	+	+	+	56,58,64
PPCE	−	−/+	+	+	+	+	+	56,64,66
ACE	+	+	+	+	+	+	+	30,56,60
Endopeptidase 24.11	+	−/+	+	+	+	+	+	56,61—65
Endopeptidase 24.15		+			+			13,64
Endopeptidase 24.16		+	−	+	+	+	+	13,15
Carboxypeptidase					+			58

Note: Abbreviations of peptidases are as in Table 1.

A striking feature concerns the particularly slow rates of neuropeptide disappearance *in vitro* in plasma or serum. This probably can be explained by the fact that few peptidases have been detected in these biological fluids (see Table 3). A study concerned with the characterization of neuropeptide-metabolizing enzymes in homogenates prepared from vascular tissue has also revealed a poor ability to catabolize neuropeptides. Here again, this reflects a restricted enzymatic content in this tissue, as is shown in Table 3. By contrast, it was clear that the rates of neuropeptides disappearance in the circulation *in vivo* appeared considerably more rapid (see Table 4). Since it is improbable that the metabolism of neuropeptides *in vivo* is carried out by circulating enzymes or by membrane-associated blood vessel peptidases, the major contribution in the decrease in peptide concentrations can be attributed to tissue clearance, i.e., uptake and/or degradation mechanisms. In this context, recent attention has been paid to peptide inactivation in various organs, such as lung, liver, or kidney, that can fulfill such functions. Recent advances in the knowledge of the peptidases content and neuropeptides inactivating processes occurring in the gastrointestinal tract follow.

B. PEPTIDASES AND INACTIVATING PROCESSES OCCURRING IN THE GASTROINTESTINAL TRACT
1. Peptidases in the Gastrointestinal Tract
a. Organs Distribution

Table 5 illustrates the distribution of the peptidases that were either detected or purified from seven organs of the gastrointestinal tract. From a qualitative point of view, most of these organs exhibit a rich peptidasic content. Aminopeptidase B was detected fluorimetrically in all organs of the rat gastrointestinal tract[56] as well as in the porcine jejunum.[57] In the rat, this peptidase appeared to be abundant in the small intestine (duodenum, jejunum, and ileum), with a specific activity ($V_{max} \simeq 556$ nmol of 7AMC released per hour per milligram of protein) that was three- to fivefold higher than in the other organs.[56] Interestingly, the distribution of this enzyme paralleled that observed for aminopeptidase N, which was detected in high amounts in the rat small intestine,[56] while it was present in 20- to 30-fold lower amounts in the other organs.[56] These data corroborated previous studies showing that the highest level of specific activity of this enzyme was recovered in the distal part of the ileum.[58] Furthermore, immunohistochemical localization of aminopeptidase M had revealed that the peptidase was mainly present in the rat small gut, with stain particularly concentrated at the entire luminal surface.[59]

The richest content of DAP was detected in the jejunum and in the duodenum of the rat (487 and 234 nmol of 7AMC released per hour per milligram of protein, respectively).[56] A sixfold lower amount was also observed in rat ileum,[56] while the peptidase appeared poorly represented in other tissues, with specific activities varying within 11 and 28 nmol/h/mg protein.[56] Indeed, this activity was abundant in the rat small intestine,[56] particularly at the distal part of the ileum.[58] DAP was also observed in important quantities in the esophagus and in the colon, while the poorest areas of the gastrointestinal tract corresponded to the stomach and the cecum (specific activities of 88 and 115 nmol/h/mg of protein, respectively).[56]

The last enzymatic activity that was detected in all the gut organs was angiotensin-converting enzyme (ACE, see Table 5). In the rat, the fluorimetric detection of ACE led to nearly equal amounts of the peptidase, with specific activity varying within a fivefold factor.[56] This enzyme was also detected in the rat ileum by glass fiber paper separation and radiometric detection of its specific metabolite [^{14}C]His–Leu.[30] Finally, a monospecific antiserum directed toward human lung ACE was used to show the presence of the enzyme in human jejunum.[60]

A monoclonal antibody directed toward pig kidney endopeptidase 24.11 revealed the presence of this activity in pig jejunum[61] and duodenum.[62] These antibodies were used to develop an immunoradiometric assay that also detected this peptidase in pig ileum, but not in the whole stomach.[63] This activity was only observed in an enriched microsomal fraction prepared from the latter tissue,[63] indicating that endopeptidase 24.11 was probably present in very small amounts in the stomach. In agreement with this statement was the fact that this enzyme was not detected in purified smooth muscle plasma membranes from rat fundus.[64] In addition, fluorimetric assays established that the stomach was the rat gut organ that displayed the lowest endopeptidase 24.11 specific activity (13 nmol/h/mg of protein).[56] This activity was 40-, 50-, and 100-fold lower than those recovered in rat ileum, jejunum, and duodenum, respectively.[56] The presence of the enzyme in these peripheral tissues as well as in the esophagus and colon was confirmed by the radiometric assay of this peptidase.[65]

Postproline cleaving enzyme (PPCE) was detected fluorimetrically in high amounts in the homogenate of whole rabbit small intestine.[66] In the rat, a nearly identical amount of PPCE was detected in all gastrointestinal organs (specific activities varying within a twofold factor, except for the esophagus, which did not contain PPCE.[56]) Interestingly, although the enzyme was detected in the whole stomach, it was not detected in the smooth muscle plasma membranes from rat fundus.[64] These discrepancies could probably be explained by the fact that PPCE behaves as a cytosoluble enzyme that is easily released upon hypotonic treatment and, therefore, could be lost during fundic membrane purification.

Endopeptidase 24.16 was recently purified from rat ileum.[13] This activity was shown to cross-react with the IgG-purified fraction of a monospecific polyclonal antiserum that was raised against the brain endopeptidase 24.16.[13] Therefore, these antibodies, which corresponded to the only tool available to detect or monitor this neuropeptidase, were used to assess the presence of endopeptidase 24.16 in other rat peripheral organs. Concerning the gut, the enzyme appeared to be located in all the organs that were examined, with the noticeable exception of the duodenum.[15]

During the course of the purification of endopeptidase 24.16, it was possible to detect and chromatographically separate an activity that hydrolyzed neurotensin[13] with the same specificity as a previously described brain metallopeptidase, namely, endopeptidase 24.15,[67] and that was specifically blocked by the endopeptidase 24.15 inhibitor CPE–Ala–Ala–Phe–pAB (see Table 1). To date, this is the only indication of the presence of endopeptidase 24.15 in the intestine. Such activity was also observed in the fundus plasma membranes from rat stomach origin.[64] Finally, aminopeptidase A and a Co^{2+}-activated carboxypeptidase were recovered in the rabbit small intestine, with particularly high levels in the distal ileum,[58]

TABLE 6
Subcellular Localization of Neuropeptidases in the Gastrointestinal Tract

Membrane preparation	Specific activity (nmol/h/mg)[a]							Specific activity (fmol/h/mg)[b,c]			Ref.
	AmB	AmN	pAm	DAP	DAP IV	PPCE	ACE	Endo 24.11	Endo 24.15	Endo 24.16	
LSM	ND	107.8	0	ND	271.6	0	83.1	67.3	0	+ +	68
MP	8	57.4	0	24	747	0	15.6	325	+	+ + + +	69
CM	ND	107.8	0	ND	319.4	25.5	205.7	54.7	+ +	+ + +	68
DMP	22.7	114.7	0	42.7	1133	41.3	25.5	365.7	+	+ + + +	69
SMP	45.3	340	0	100	880	52	44.5	1339	+ + + +	+ +	69
LSM-PM	Yes	Yes		Yes				Yes			70, 71

Note: Abbreviations of peptidases are as in Table 1. LSM, longitudinal smooth muscle plasma membranes; MP, myenteric plexus; CM, circular smooth muscle plasma membranes; DMP, deep muscular plexus; SMP, submucous plexus; LSM-PM, myenteric plexus-containing longitudinal smooth muscle membranes; ND, not determined.

[a] Specific activities were measured by means of the specific fluorimetric substrates listed in Table 1 and correspond to the fraction of fluorescence that was found to be sensitive to saturating concentrations of the adequate specific inhibitor (see Table 1). Values are expressed as nanomoles of 7AMC or Hip (ACE) released per hour and mg of protein.
[b] Values correspond to femtomoles of neurotensin hydrolyzed per hour and mg of proteins.
[c] 0, not detectable; +, 0—300 fmol; + +, 300—600; + + +, 600—900; + + + +, > 1000.

while pyroglutamyl peptide hydrolase was not detected in the gut organ that was examined.[56,64]

b. Subcellular Localization

From Table 5 it appears that the gut axis is constituted by organs that are, from a qualitative point of view, particularly rich in peptidases. However, enzymatic specific activities have revealed noticeable quantitative discrepancies in the peptidasic contents of the various gut organs. Among these, the small intestine part of the gastrointestinal tract has appeared particularly rich in peptidase. Therefore, it was interesting to assess whether the wide variety of peptidases present in these tissues was distributed heterogeneously within the muscular, nervous, or mucosal layers. The study of the subcellular localization was made possible by the development of purification procedures of membranes originating from the circular (CM) and longitudinal smooth muscles (LSM)[16,19] and also by the isolation of highly enriched synaptosomal fractions from myenteric (MP), deep muscular (DMP), and submucous plexuses (SMP).[17-19] These membrane preparations have proved to be very useful in dissecting the topography of gut neuropeptidases. Table 6 illustrates the subcellular localization of enzymes in five purified membrane fractions from canine ileum.[68,69] Endopeptidase 24.11 appeared particularly abundant in the nervous layers, which exhibited 5- to 20-fold higher specific activities than those recovered in circular and longitudinal muscular layers (see Table 6). Such a feature was also observed for the distribution of DAP IV, which was detected in high amounts in the deep muscular plexus.[69] An example of the opposite of this preferential nervous distribution was ACE, whose most abundant quantities were observed in the circular and longitudinal smooth muscles.[68] Besides these three neuropeptidases exhibiting specific nervous or muscular localization, a group of enzymes including aminopeptidase N and endopeptidase 24.16 appeared more homogeneously distributed within ileal layers, without preferential association with one type of membrane preparation.[68,69]

Endopeptidase 24.15 was present in considerable amount in the submucous plexuses when compared to other muscular and nervous membranes (see Table 6). By contrast, this activity was nonexistent in the longitudinal smooth muscle as was the case for PPCE and pyroglutamyl peptide hydrolase (see Table 6), the latter enzyme being totally absent from the various canine membrane preparations.[69,68] Finally, aminopeptidase B and DAP were detected fluorimetrically in the dog ileum nervous membrane preparations[69,68] and in the homogenate of the myentheric plexus-containing longitudinal muscle layer of guinea-pig ileum[70] and bovine ileojejunum.[71]

In addition to the few studies that have examined the subcellular distribution of neuropeptide-hydrolyzing activities located in the muscular and nervous layers of the intestine, a number of reports dealing with the detection of peptidases in the intestinal brush border mucosa of various species are now available. Most of these studies were carried out with the whole small intestine whose mucosa was scraped off and homogenized (crude membranes) or with purified vesicles that consisted mainly of brush border as judged by phase-contrast microscopy.[72] Table 7 illustrates the qualitative variety of the peptidases that were detected or purified from the mucosa of various species. Three types of aminopeptidases were monitored by fluorimetry[73-76] or by direct immunocharacterization;[72,77] they displayed nearly identical specific activities in crude membrane preparations, while they were 15- to 20-fold enriched in the purified vesicles.[75,76] A highly active and electrophoretically homogeneous DAP was purified from the monkey small intestinal mucosa.[78] DAP IV was identified after fluorimetric measurements in the mucosa of rat[75] and hog.[72] In the latter species, the presence of DAP IV was confirmed at the level of dissociated enterocytes prepared from a section of the mid-jejunum by following the liberation of the N-terminal dipeptide (Arg–Pro) of substance P[79] and by immunoadsorption on Sepharose® conjugated to antibodies directed toward purified kidney DAP IV.[80]

Endopeptidase 24.11 was partially purified from pig intestinal microvilli[81] and later immonoadsorbed using antibodies raised against the kidney enzyme.[81] The presence of this peptidase was confirmed in two additional studies performed on rat purified vesicles[76] and hog isolated enterocytes.[79]

Purified vesicles prepared from hog small intestine mucosa were shown to display a 13-fold lower specific activity for ACE than those isolated from human.[82] The presence of ACE in the latter species was confirmed by immunohistochemistry carried out with antisera recognizing the human enzyme.[60]

To date, there are no data indicating the presence of PPCE, endopeptidase 24.15, and endopeptidase 24.16 in the small intestine mucosa.

Studies on nerve distribution have established that peptides are located in well-defined populations of neurons that project in restricted regions of the small intestine wall. This anatomical specificity could explain functional differences between the sites and the peptide mechanisms of action in the intestine. Thus, myogenic and neurogenic mechanisms indicate that peptides can exert their action through direct smooth muscle or nerve-mediated mechanisms. This indicates a heterogeneity of peptide receptor localization that is distributed within muscular or nervous layers. Before the diversity observed for the subcellular localization of peptides, receptors, and peptidases, putative colocalization is the first criterion of consideration prior to assigning a possible role for a peptidase in the physiological inactivation of a given peptide. In addition, since various activities have been detected in each intestinal layer, it is important to assess their specificity as well as the relative kinetic parameters (affinity, catalytic constant) they exhibit for various neuropeptides.

c. Specificity

Table 8 summarizes the susceptibility of various neuropeptides to *in vitro* hydrolysis by purified peptidases isolated from central or peripheral organs that were later detected in the gastrointestinal tract. Data clearly illustrate the fact that none of these enzymes displays an

TABLE 7

Peptidases Detected or Purified from the Intestinal Brush Border Mucosa of Various Species

Peptidases	Membrane preparation	Species	Specific activity, (nmol/min/mg)	Detection or purification methods	Ref.
AmA	Crude membranes	Rabbit	230	Fluorimetry	73
	Purified vesicles	Hog	1000	Fluorimetry	74
	Purified vesicles	Hog	—	Immunoelectrophoresis	72
AmN	Crude membranes	Rabbit	490	Fluorimetry	73
	Crude membranes	Rat	60	Fluorimetry	75
	Purified vesicles	Hog	2500	Fluorimetry	74
	Purified vesicles	Pig, rat, rabbit[a]	—	Immunoelectrophoresis	77
	Purified vesicles	Rat	920	Fluorimetry	75
	Purified vesicles	Rat	1139	Fluorimetry	76
AmB	Crude membranes	Rabbit	130	Fluorimetry	73
DAP	Purified enzyme	Monkey	4.5 to 9.3×10^6	Paper chromatography	78
DAP IV	Crude membranes	Pig	30	Fluorimetry	72
	Crude membranes	Rat	49	Fluorimetry	75
	Purified vesicles	Rat	745	Fluorimetry	75
	Isolated enterocytes	Hog[a]	—	Hydrolysis of substance P	79
	Purified enzyme	Hog	—	Immunoabsorption	80
Endo. 24.11	Purified enzyme	Hog	—	Hydrolysis of [125I]insulin B chain or immunoab-sorption	81, 149
					61
	Isolated enterocytes	Hog	—	Hydrolysis of substance P	79
	Purified vesicles	Rat	11, 73[b]	Hydrolysis of [125I]insulin B chain	76
ACE	Purified vesicles	Human	260	Fluorimetry	82
	Purified vesicles	Hog	19, 6	Fluorimetry	82
	Intact tissue	Human[a]	—	Immunohistochemistry	60

Note: Abbreviations of peptidases are as in Table 1.

[a] Activities were measured in preparations from the brush border mucosa of whole small intestine or only from the jejunum part of the gastrointestinal tractus.

[b] Specific activities were measured by means of their specific fluorigenic substrates or by means of a radiolabeled substrate.

TABLE 8
Specificity of the Neuropeptide-Hydrolyzing Peptidases Detected in the Gastrointestinal Tract

Peptide	Peptidases								
	AmN	AmB	DAP IV	PPCE	ACE	Endo A	Endo 24.11	Endo 24.15	Endo 24.16
Angiotensin I	+			+	+			+	+
Angiotensin II	+			+			+	+	+
ANF							+		
Atriopeptin III							+		+
Bradykinin	−		−	+	+		+	+	+
CCK-8	+/−					+	+		+
Dynorphin$_{1-8}$						+	+	+	−
Dynorphin$_{1-13}$							+		+
Gastrin 1 (human)				−			+		−
Leu-enkephalin	+	+			+	−	+	−	−
Met-enkephalin	+	+	+	+	+	−	+	−	−
Neurotensin	−			+	+	+	+		+
Substance P			+	+	+		+	+	+
Somatostatin	+						+	+	−
TRH	−			+					−
Ref.	83	84	85	66, 86—91	92—95	90, 102—104	43, 94, 96—100	67, 101	12, 13

Note: Abbreviations of peptidases and peptides are as in Tables 1 and 2.

exclusive specificity directed toward a single peptide. By contrast, peptidases can be separated into two major groups — namely, exo- and endopeptidases (according to the type of peptidasic attack they generate) — with a relatively large spectrum of cleavage. Aminopeptidases are exopeptidases that require free N-termini since they are blocked by pyroglutamyl groups. Aminopeptidase N displayed a high affinity for hydrophobic amino-acids and was blocked or drastically slowed down when the penultimate N-terminal residue was proline of sulfated tyrosine, respectively.[83] This explained the total resistance to proteolysis by aminopeptidase N of bradykinin, substance P, neurotensin, and TRH while the sulfated CCK octapeptide was hydrolyzed very weakly by this enzyme.[84] Another aminopeptidase (aminopeptidase B or MI) exhibited preferential affinity for basic N-terminal residues.[83] However, although this peptidase was shown to hydrolyze L-arginine-β-naphthylamide 17 times faster than the L-alanine-β-naphthylamide, this activity was also reported to release leucine and methionine residues from the two opiates pentapeptides,[84] Leu- and Met-enkephalins. DAP IV also acts as an exopeptidase that releases N-terminal dipeptides provided that the amino acid in the penultimate position corresponds to a prolyl residue. This feature that is shared by substance P makes it a good substrate for DAP IV.[85]

PPCE (also called proline endopeptidase, TRH-deamidating enzyme, or endooligopeptidase B) was shown to catalyze the cleavage of Pro–X peptide bonds located internally in a peptide sequence, with the exception of the Pro–Pro linkage.[86] This paper indicated that angiotensin 2 was cleaved by PPCE while human gastrin was not.[86] Later it was shown that all proline-containing peptides were hydrolyzed selectively at the carboxyl side of prolyl residues and that this enzyme could be considered as a general postproline cleaving enzyme.[66,87-91]

Angiotensin-converting enzyme (peptidyldipeptidase A, kininase II) was first reported to display an exopeptidasic activity that was responsible for the release of C-terminal dipeptides of natural substrates including angiotensin I and bradykinin.[92] Such specificity had been further documented by the ability of ACE to generate the C-terminal dipeptide of Met- and Leu-enkephalins[93] as well as neurotensin.[94] However, apart from its dipeptidylcarboxypeptidase activity, it had been reported that ACE was able to hydrolyze substance P, whose amidated C-terminus was expected to protect this peptide from exocarboxypeptidase attack.[94,95] In addition, ACE not only released the C-terminal amidated dipeptide of SP, but also generated its C-terminal tripeptide,[94,95] the latter fragment appearing to be generated predominantly by ACE, since the ratio of tripeptide to dipeptide released was 4:1.[94] This indicated that the initially described dipeptidylcarboxypeptidase activity of ACE is probably a restricted view of the genuine enzyme specificity that could also behave as an endopeptidase.

Endopeptidase 24.11 (neutral endopeptidase, enkephalinase) belongs to the endopeptidase family. A study with a series of synthetic substrates showed that the enzyme preferentially hydrolyzed peptidyl bonds in which the amino group was constituted by a hydrophobic residue.[96] In agreement with this statement, the hydrolysis of dynorphin$_{1-13}$,[96] Leu- and Met-enkephalins, angiotensins I and II, bradykinin,[96-98] neurotensin, substance P,[94,97] and somatostatin has been reported.[97] This wide range of natural peptides hydrolyzed by endopeptidase 24.11 was later completed by studies on the degradation of CCK8 and the heptadecapeptide gastrin I.[43,99,100]

Endopeptidase 24.15 (also called soluble metallopeptidase) appeared to cleave peptide bonds in which the carboxyl group was contributed by an aromatic amino acid residue.[67] Furthermore, the specificity studies had suggested that the active site of this enzyme could accommodate at least five amino-acids, with two and three residues located on the N-terminal and C-terminal sides of the scissile bond, respectively.[67] Sites of hydrolysis of several biologically active peptides[67,101] had been found to be consistent with this specificity, which was deduced from studies with synthetic model substrates (see Table 8).

Another neuropeptidase named endooligopeptidase A was particularly efficient in hydrolyzing various enkephalin-containing peptides, including dynorphin$_{1-8}$,[102-104] while Leu- and Met-enkephalins resisted degradation by this enzyme.[102] In addition, endooligopeptidase A was shown to hydrolyze bradykinin and neurotensin.[90] Although the specificity of endopeptidase A was not characterized extensively, it was striking that this activity strongly resembled endopeptidase 24.15, since the nature of the peptides together with the sites of hydrolysis generated by both enzymes appeared to be identical. However, a recent study indicated that endooligopeptidase A was not blocked by the endopeptidase 24.15 inhibitor CPE–Ala–Ala–Phe–pAB and that the latter activity could be physically separated from endopeptidase A by means of specific polyclonal anti-endo A antibodies.[103]

Finally, endopeptidase 24.16 was purified from rat brain[12] and ileum[13] by following the hydrolysis of neurotensin. From data obtained with neurotensin partial sequences, it appeared that one of the requirements of the enzyme for the recognition and/or catalysis of neurotensin was a C-terminal extension of at least three amino acids from the cleaved peptidyl bond.[12] Endopeptidase 24.16 could not be considered as an exclusive neurotensinase activity, since several other natural peptides were found to behave as excellent substrates (see Table 8).[12,13]

The terminology employed to name peptidasic activities was often related to the peptide whose degradation was examined (angiotensin-converting enzyme, angiotensinase, enkephalinase, atrial dipeptidyl carboxypeptidase, membrane-bound substance P-degrading enzyme, TRH deamidating enzyme); this implies an erroneous notion of exclusive specificity. By contrast, it is now generally admitted that a wide range of peptides can be (at least *in vitro*) hydrolyzed by a single peptidase. However, it is clear that the efficiency of substrate hydrolysis also must be taken into account. Particularly interesting was a recent review reporting that endopeptidase 24.11, which displayed the ability to cleave numerous natural peptides (see Table 8), could exhibit markedly different efficiencies, with k_{cat}/K_m values that could vary by two orders of magnitude.[105] Other examples of such peptidasic discrimination between peptide substrates are illustrated in Table 9. Endopeptidase 24.11 displayed a tenfold higher k_{cat}/K_m value for Leu-enkephalin than for neurotensin or substance P.[94,98] Concerning the latter peptide, it was interesting to note that endopeptidase 24.11 and ACE were found to be equipotent,[94] while these two enzymes could be distinguished according to their kinetic parameters for Leu-enkephalin and bradykinin. Thus, endopeptidase 24.11 was 100-fold more efficient than ACE in its ability to hydrolyze Leu-enkephalin, while it displayed a tenfold lower k_{cat}/K_m value than ACE for bradykinin.[93,98]

The physiological inactivation of a neuropeptide is the result of the combination of several factors: first, the presence of peptidases at the peptide afferences and their localization in the vicinity of its receptors; second, the ability for this peptide to be hydrolyzed by one or several peptidases that must always yield biologically inactive degradation products; and third, the competition between various colocalized peptidases displaying the same ability to hydrolyze the peptide, but present at different concentrations and exhibiting different kinetic parameters (i.e., affinity and catalytic constants).

As seen previously, several methods and tools have recently been developed to address the problem of peptide metabolism, and the following portion of this review will be concerned with the studies that have dealt with peptide inactivation in the gastrointestinal tract.

2. Mechanisms of Neuropeptide Inactivation

a. Cholecystokinin and Gastrins

The first report on the metabolism of the heptadecapeptide (G-17, human gastrin I) concerned the inactivation of this peptide in rat and dog small bowel.[106] Experiments consisted of infusion of exogenous G-17 into a cannulated branch of the superior mesenteric artery that supplied the first loop of the jejunum. The disappearance of human gastrin I was followed by monitoring acid secretion 40 min before and during the infusion. It was demonstrated

TABLE 9

Kinetic Parameters of Several Purified Peptidases Toward Peptides of the Gastrointestinal Tract

Peptidases	Bradykinin			Leu-enkephalin			Neurotensin			Substance P			Ref.
	kcat (s⁻¹)	Km (μM)	kcat/Km (M^{-1} × s⁻¹)	kcat (s⁻¹)	Km (μM)	kcat/Km (M^{-1} s⁻¹)	kcat (s⁻¹)	Km (μM)	kcat/Km (M^{-1} s⁻¹)	kcat (s⁻¹)	Km (μM)	kcat/Km (M^{-1} s⁻¹)	
Endo-oligo. A	9.8	8.78	1.12×10^6	—	—	—	3.55	12.49	2.8×10^5	—	—	—	102
Endo. 24.15	19.9	67	3×10^5	—	—	—	11.3	37	3.1×10^5	—	—	—	101
ACE	8.33	1	8.33×10^6	11.7	1000	1.2×10^4	0.57	14	4.1×10^4	3.75	25	1.5×10^5	93, 94
Endo. 24.11	79.5	120	6.6×10^5	78	74	1.1×10^6	6	39	1.5×10^5	33.9	190	1.78×10^5	94, 98

that acid secretion fell during infusion of G-17, indicating that the potency of the peptide was strongly reduced during transit through the small bowel vascular bed.[106] It was interesting to note that after infusion of G-17 in the portal vein, acid secretion did not decrease, suggesting that, unlike the gut, the liver did not participate in G-17 clearance. By contrast, portal vein infusions of the five-amino-acid C-terminal sequence pentagastrin resulted in an 87% inhibition of acid secretion.[106] These *in vivo* data were further supported by the fact that the rate of gastrin inactivation by homogenates of small bowel mucosa was over 20 times more rapid than in liver membranes.[106] By contrast, pentagastrin was efficiently destroyed by the latter preparation at a metabolic rate 500-fold higher than that observed for G-17.[106] This indicated a tissue-specific inactivating process, although both peptides belong to the same family.

Human gastrin I accounts for more than 95% of the biologically active forms of gastrins present in extracts of hog antral mucosa.[107] However, this percentage differed markedly in the porcine antral venous outflow (<60%) with a concomittant emergence of C-terminal fragments.[107] Recently, a study has approached the peptidasic conversion of G-17 after endogenous release from the antrum.[100] After ion exchange chromatography fractionation coupled to RIA using three region-specific gastrin antibodies, it was shown that four components sharing a C-terminal homology with the parent peptide were generated, representing about 43% of the total C-terminal immunoreactivity.[100] It is interesting to note that following prior infusion of the endopeptidase 24.11 inhibitor, phosphoramidon, the contribution of two C-terminal fragments was significantly lower, while immunoreactivity ascribed to intact G-17 was recovered in increased amounts.[100] The possible involvement of endopeptidase 24.11 was further supported by the fact that the purified enzyme rapidly hydrolyzed endogenous G-17 present in antral extracts by generating a post-DEAE cellulose pattern similar to that observed in antral venous plasma.[100] In addition, exogenous G-17 was shown to be hydrolyzed by purified kidney endopeptidase 24.11 at four peptide bonds according to the following scheme:

$$\downarrow \qquad\qquad \downarrow \qquad \downarrow \qquad\qquad \downarrow$$
$$pGlu\text{–}Gly\text{–}Pro\text{–}Trp - Leu\text{–}(Glu)_5\text{–}Ala - Tyr\text{–}Gly - Trp\text{–}Met\text{–}Asp - Phe\text{–}NH_2$$

with the predominant formation of $G\text{-}17_{1-13}$ due to rapid hydrolysis at the $Gly^{13}\text{–}Trp^{14}$ bond.[43] It is interesting to emphasize the fact that $G\text{-}17_{1-11}$, $G\text{-}17_{1-13}$, $G\text{-}17_{1-16}$, and $G\text{-}17_{5-16}$, representing the main products of G-17 hydrolysis by endopeptidase 24.11, appeared immunochemically and chromatographically related to the main fragments recovered after G-17 infusion in the human circulation.[43] These observations strongly suggest that endopeptidase 24.11 is involved in the gastric metabolism of gastrin I and that it leads to the formation of G-17 fragments that are probably released in the blood.

The gastrin-related peptides CCK-8 and CCK-7 share the same C-terminal pentapeptide than gastrin I. Both peptides were readily hydrolyzed by purified endopeptidase 24.11 at three peptide bonds[99,108] according to the following model:

$$\downarrow \quad \downarrow \qquad \downarrow$$
$$Asp\text{–}Tyr\text{–}Met\text{–}Gly - Trp - Met\text{–}Asp - Phe\text{–}NH_2$$

Whether the enzyme is involved in the process of CCK peptides in the gastrointestinal tract is still unknown.

b. Neurotensin

In 1982, Hammer and his collaborators[109] reported on the elevation of neurotensin-like immunoreactivity in peripheral plasma several hours after a meal. The authors demonstrated by combined Sephadex® G-25 and HPLC chromatography that most of the elevated plasma components corresponded to the biologically inactive N-terminal partial sequences neurotensin$_{1-8}$ and neurotensin$_{1-11}$, while neurotensin itself did not increase significantly.[109] These data could suggest that neurotensin had acted as a hormone after its increased secretion upon luminal stimuli. However, in light of the discrepancy observed between the neurotensin half-lives *in vitro* and *in vivo* in the circulation (see Tables 2 and 4), it is more probable that the peptide had been metabolized locally after having exerted a paracrine function. Supporting this view, the same group reported on an increased amount of neurotensin and neurotensin$_{1-8}$ recovered in the hepatic portal circulation after lipid infusion in the rat small intestine, while the levels of both peptides remained unchanged in the general circulation.[110] In addition, direct infusion of tritiated neurotensin in the superior mesenteric artery resulted in the formation of two radiolabeled degradation products whose retention times corresponded to neurotensin$_{1-8}$ and neurotensin$_{1-11}$. In such experimental conditions only 13% of radioactivity recovered behaved as intact neurotensin after HPLC.[110] The fact that neurotensin$_{1-8}$ was recovered in elevated amounts in portal, but not in the systemic circulation, possibly indicated that the small intestine might be the site of both release and metabolism of neurotensin.

In agreement with the previously mentioned study, we have reported that neurotensin$_{1-8}$ and neurotensin$_{1-11}$ were the two main N-terminal fragments that were recovered after bolus injection of tritiated neurotensin into one arcade of the superior mesenteric artery supplying an intestinal segment of canine ileum.[111] In addition, neurotensin$_{1-10}$ and free tritiated tyrosine were recovered, the latter product probably resulting from the secondary breakdown of the C-terminal fragments corresponding to the counterparts of neurotensin$_{1-8}$, neurotensin$_{1-10}$, and neurotensin$_{1-11}$; i.e., neurotensin$_{9-13}$, neurotensin$_{11-13}$, and neurotensin$_{12-13}$.[111] The apparent half-life of the peptide was between 2 and 6 min.[111] The same pattern was obtained when tritiated neurotensin was incubated with plasma membranes prepared from canine ileum longitudinal and circular smooth muscle[68] or with the synaptosomal fractions from dog ileum myenteric, deep muscular, and submucous plexuses.[69] An additional important amount of neurotensin$_{1-12}$ was observed particularly in the myenteric plexus.[69]

In addition to the examination of neurotensin metabolism by membranes from muscular and nervous origins, Shaw et al.[112] reported on the catabolism of this peptide in the epithelial layer of porcine small intestine. Metabolites produced by the incubation of the peptide with dispersed enterocytes were separated by HPLC and amino acid analyzed. The major fragments recovered were neurotensin$_{1-10}$ and neurotensin$_{1-11}$, which derived from sites of cleavages at the Pro^{10}–Tyr^{11} and Tyr^{11}–Ile^{12} peptide bonds; neurotensin$_{1-8}$ was not generated.[112] An additional minor pathway was the formation of neurotensin$_{4-13}$, which did not lead to biological inactivation of the parent peptide. Inhibition studies performed with porcine enterocytes revealed that formations of neurotensin$_{1-10}$ and neurotensin$_{1-11}$ were totally abolished by the endopeptidase 24.11 inhibitor, phosphoramidon.[112] This was in agreement with the observed effect of the other specific endopeptidase 24.11 inhibitor, thiorphan, which inhibited the formations of neurotensin$_{1-10}$ and neurotensin$_{1-11}$ *in vitro* in dog membrane preparations[68,69] or *in vivo* after canine ileum perfusion.[111] In both cases, this agent led to a significant increase in intact neurotensin recovery, although the protection of the peptide was far from complete.[69,111] The angiotensin-converting enzyme inhibitor captopril did not protect neurotensin from inactivation by canine ileum in either *in vitro* or *in vivo* models.[69,111] However, this inhibitor elicited a selective blockade of the conversion of neurotensin$_{1-10}$ into neurotensin$_{1-8}$.[69,111] Therefore, ACE could not be considered as a neurotensin-inactivating enzyme in the small intestine.

It is interesting to emphasize the fact that both purified preparations of endopeptidase 24.11 and angiotensin-converting enzyme have been shown to hydrolyze neurotensin at the

Tyr^{11}–Ile^{12} bond, leading to neurotensin$_{1-11}$.[94] The fact that only endopeptidase 24.11 appeared to be responsible for neurotensin$_{1-11}$ formation in dog ileum could illustrate the fact that the kinetic parameters for neurotensin (see Table 9), together with the concentration of endopeptidase 24.11 in this tissue, balanced the competition with ACE in favor of the former activity. This does not seem to be the case in the stomach. Indeed, Bunnett et al.[113] examined the catabolism of iodinated and tritiated neurotensin in the rat stomach wall of the conscious rat using an implanted catheter that allowed the delivery of the peptide to the tissues and collected putative metabolites by adjacent dialysis fiber. HPLC analysis indicated that neurotensin was mainly converted into neurotensin$_{1-8}$, neurotensin$_{1-11}$, neurotensin$_{9-13}$, and free tyrosine regardless of the type of the label, with an overall half-life that was estimated to be between 9 and 15 min.[114] However, unlike the canine ileum, the ACE inhibitors captopril and MK422 partially but significantly protected neurotensin from degradation, while phosphoramidon (endopeptidase 24.11 inhibitor) was totally ineffective.[114] It was then concluded that, in rat stomach, neurotensin was metabolized by an ACE-like activity, while endopeptidase 24.11 did not contribute to degradation mechanisms.[114] This was in complete agreement with the fact that little (whole stomach homogenate, see Table 5) if any (fundus plasma membranes)[64] endopeptidase 24.11 activity was detected in the rat stomach.

In conclusion, the above studies indicate that both ileum and stomach potently inactivate neurotensin by generating a nearly identical pattern of degradation products that appeared to be generated by different peptidases. However, in all cases, the partial inhibitory effect of specific peptidase inhibitors strongly suggests that the identified peptidases are not the only enzymes responsible for neurotensin metabolism in the gastrointestinal tract.

c. Opiates

Leu- and Met-enkephalin are probably the peptides whose inactivating mechanisms have been investigated the most thoroughly. Four distinct purified peptidases have been reported to inactivate Leu- and Met-enkephalins *in vitro* according to the following scheme:

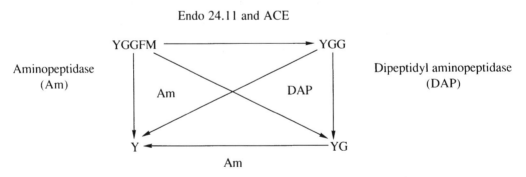

However, controversy exists over the precise route of inactivation of enkephalins that terminates their physiological central effects *in vivo*. In the gastrointestinal tract, several studies assessed the relative contribution of these enkephalin-degrading activities. Analysis of the degradation of tritiated Leu-enkephalin by guinea pig ileal broken cell preparations was reported.[115] Thin layer chromatography revealed that [³H]Leu enkephalin ([³H]YGGFL) was primarily converted into free [³H]-Y, although some [³H]YGG could be detected.[115] Interestingly, [³H]Y formation was reduced considerably by the aminopeptidase M inhibitor LY 110947.[116] In presence of this agent, the formation of [3H]YGG was increased by about tenfold and could be totally blocked by the endopeptidase 24.11 inhibitor thiorphan, while captopril (ACE inhibitor) did not affect the recovery of either [³H]Y or YGG.[115] These data agreed with those obtained with intact guinea pig ileum and myentheric plexus-containing longitudinal smooth muscle. A 10-min exposure of [³H]Leu-enkephalin with intact ileum led to 90% degradation of the peptide.[115] Of the metabolites, 70% corresponded to free [^{3}H]Y while less than 10% of the label was associated with [^{3}H]YGG.[115] The same pattern of Leu-enkephalin hydrolysis was observed with myenteric plexus longitudinal muscle, although the quantitative rate of peptide disappearance remained slower.[115] This report indicates that in the intact guinea pig ileum the major route of enkephalin inactivation occurs through an aminopeptidasic attack, while endopeptidase 24.11 only partially contributes to the degradation pattern. This confirmed a previous work showing that free tyrosine was the only metabolite that was recovered after TLC analysis of [^{3}H]Leu- and [^{3}H]Met-enkephalin destruction by intact ileum segments.[117] Further analysis of the nature of the aminopeptidases occurring in this tissue indicated that the selective inhibitor of aminopeptidase N, bestatin, increased [^{3}H]Leu-enkephalin content after incubation with intact guinea pig ileum and strongly reduced the formation of [^{3}H]Y. Such an effect was not elicited by pretreatment of the tissue with the more general aminopeptidase inhibitor, puromycin.[118] The explanation of the authors to clarify these discrepancies was that the puromycin-sensitive aminopeptidases are mainly located intracellularly, whereas bestatin-sensitive enzymes are probably cell-surface enzymes.[118] However, the effect of puromycin on the longitudinal smooth muscle appeared more complex, since other studies reported a considerable protection by puromycin of Leu-enkephalin in broken cells as well as in intact ileum.[119]

The guinea pig was also widely used in a more relevant physiological situation where metabolism could be correlated to enkephalin biological responses. It has been reported that Leu- and Met-enkephalins could inhibit the field-stimulated contractions of the guinea pig ileum and that this evoked response was rapid in onset, but of very short duration. This tissue that was amenable to prior treatment with an appropriate peptidase inhibitor was allowed to monitor its effect on complete dose-response curves elicited by exogenous enkephalins, but also on the protection of endogenously released enkephalins. Geary et al.[115] showed that thiorphan did not potentiate the effect of Met- and leu-enkephalins on the electrical stimulation of guinea pig ileum and myenteric plexus longitudinal smooth muscle[115] in agreement with another study.[120] In addition, they reported on the total lack of effect of captopril. This was further supported by a study showing that captopril was totally unable to modify the effect of exogenous enkephalins, and it also did not influence the action of endogenous enkephalins released after high-frequency electrical stimulation.[121] In apparent contradiction with the two above reports, Kuno et al.[122] demonstrated a very weak but significant enhanced potency of Met-enkephalin after pretreatment with phosphoramidon and captopril. Unlike endopeptidase 24.11 and angiotensin-converting enzyme, which showed little, if any, involvement in the inactivation of enkephalins, several reports indicated a major contribution of aminopeptidases. Bestatin enhanced the response to enkephalins in a concentration-dependent manner.[118] This inhibitor was also shown to be the most potent agent that prolonged the ability of Met-enkephalin to inhibit the electrical contractions of the myenteric plexus longitudinal smooth muscle,[120] while such maximal effect on intact ileum was achieved by amastatin.[122] Finally, puromycin was shown to enhance the depressant effect of this peptide, but did not shift the concentration-response curve to Met-enkephalin in either intact guinea pig ileum or longitudinal muscle.[118] It is interesting to note that whereas bestatin was the more potent agent, allowing a sixfold decrease of the IC_{50} values of Met-enkephalin in the myenteric plexus, a combination of thiorphan, captopril, and bestatin was provied to be more effective than bestatin alone, giving rise to a tenfold increase in the potency of Met-enkephalin in the field-stimulated myenteric plexus.[120] This observation was later supported by a study showing that a combination of amastatin and captopril was more efficient than amastatin alone, while addition of phosphoramidon to the former mixture did not induce additional protection of the peptide.[122] It can be concluded from the above observations that endopeptidase 24.11 and angiotensin-converting enzyme were detected only when a prior blockade of aminopeptidases was elicited. This showed that these two enzymes probably contributed in a minor way to the inactivation of Leu- and Met-enkephalin in the gastrointestinal tract while the major route of processing corresponded to an aminopeptidasic attack.

The relative importance of various neuropeptidases in the inactivating process of C-terminal extended enkephalins has been examined by means of the same field-stimulated guinea pig ileum preparations.[120,122] Interestingly, the nature of the enzymes that appeared to contribute to the metabolism of these opioid peptides differed markedly from that observed for their parent peptides. Met-enkephalin-Arg^6 inhibitory potency was drastically improved by pretreatment with either amastatin or captopril.[122] Furthermore, the combination of these two inhibitors led to an additive protection of the peptide.[122] By contrast, the endopeptidase 24.11 inhibitor phosphoramidon, either alone or in combination with amastatin and captopril, did not affect the efficiency of Met-enkephalin-Arg.[6,122] Based on the above reports, it seems clear that, besides aminopeptidases, angiotensin-converting enzyme can terminate the action of some enkephalin-containing peptides, while endopeptidase 24.11 does not. In agreement with this hypothesis, a previous work showed that captopril decreased the IC_{50} of the endogenous heptapeptide Met-enkephalin-Arg-Phe in the myenteric plexus from 152 to 73 nM.[120] Finally, the same authors reported that a combination of bestatin, captopril, and thiorphan could efficiently enhance the depressor effect on the electrically evoked contractions of the guinea pig myenteric plexus of other various opiates including Met-enkephalin-Lys (ME-Lys), ME-Arg-Arg, Me-Arg-Gly-Leu, Leu-enkephalin-Lys (LE-Lys), and LE-Arg.[120]

As is the case for enkephalins, β-endorphin, β-E_{1-31}, can be released from the pituitary and adrenal medulla into the systemic circulation. Several studies deal with the mechanisms of degradation/conversion of β-endorphin in the small intestine, which represents a putative target tissue for circulating β-endorphin. When exposed to a homogenate of myenteric plexus longitudinal smooth muscle prepared from guinea pig ileum, the time course of β-endorphin disappearance appeared biphasic, with half-lives of 11 and 24 min for the rapid and slow components, respectively.[123] The main degradation products that were identified by combined specific radioimmunoassay and HPLC analysis corresponded to γ-E (β-E_{1-17}), destyrosine^1 γ-E (DT-γ-E, β-E_{2-17}), α-E (β-E_{1-16}) and des-tyrosine^1 α-E (DT-α-E, β-E_{2-16}). Kinetics analysis revealed that the initial cleavage occurred in the central portion of the peptide, leading to γ-E and the complementary fragment of β-E.[123] Subsequently to this cleavage, γ-E was shown to undergo aminopeptidase and carboxypeptidase actions, resulting in the formation of DT-γ-E and α-E,[123] the latter fragment itself being converted into DT-α-E by aminopeptidases.[123] The effect of peptidase inhibitors on the intact tissue was examined with the electrically stimulated guinea pig myenteric plexus model.[120] Although the respective involvement of the neuropeptidases was not assessed by individual application of inhibitors, it was shown that a mixture of bestatin, captopril, thiorphan, and the dipeptide Leu–Leu caused a gradual potentiation of peptides inhibition provided that the length of the fragment was not longer than β-E_{1-17}. Indeed, potencies of β-E_{1-5} (Met-enkephalin), β-E_{1-9} and β-E_{1-17} were 19-, 4-, and 2-fold increased, respectively, while those of β-E_{1-19}, β-E_{1-27}, and β-E remained unchanged.[120] Interestingly, *in vivo* studies performed on dog isolated intestinal preparations[124,125] largely confirmed those obtained with guinea pig ileum homogenates.[123] After infusion of β-E through the mesenteric artery irrigating a portion of the midjejunum, HPLC analysis of the venous effluent indicated the processing of β-E into several related fragments, including γ-E, DT-γ-E, α-E, and DT-α-E.[124,125] In order to detect a putative subcellular specificity, mucosal and muscularis regions were dissected and examined for β-E metabolism.[124-126] It was shown that mucosal preparation displayed a higher efficiency of β-E hydrolysis and that this preparation generated a 3:1 ratio of γ-E vs. α-E, matching that recovered after intestinal perfusion.[124] Altogether, this could suggest that β-E processing initially takes place in the mucosal region of dog intestine.

d. Tachykinins

Proteolytic inactivation of substance P (SP_{1-11}) by isolated enterocytes and purified brush border membranes prepared from porcine mid-jejunum was reported recently.[79] The degradative HPLC patterns recovered with both preparations appeared qualitatively similar. Rapid cleavages between Gln^6–Phe^7, Phe^7–Phe^8 and Gly^9–Leu^{10} leading to the formation of the N-terminal degradation products SP_{1-6}, SP_{1-7}, and SP_{1-9} were observed with both membrane fractions.[79] Formation of SP_{3-11} that implies the release of the N-terminal di-peptide Arg–Pro was ascribed to postproline dipeptidyl aminopeptidase (DAP IV). Inhibition studies indicated that peptide disappearance was reduced by phosphoramidon in a dose-dependent manner while pretreatment with captopril was ineffective.[79] The participation of endopeptidase 24.11 as well as the apparent absence of ACE in the small intestine were further confirmed by other studies.[70,127,128] Iwamoto et al.[128] examined the degradation of tritiated substance P by homogenates of rat ileum and determined the effect of specific inhibitors on the specific binding of [^{3}H]SP on this preparation. Endopeptidase 24.11 in-hibitors thiorphan and phosphoramidon increased the recovery of intact peptide by 4.5-fold over control, while the specific binding of [^{3}H]SP in the presence of these agents represented 160% of that observed without inhibitors.[128] By contrast, captopril did not change either the recovery or the specific binding of substance P.[128] The metabolites of substance P recovered after incubation of the peptide with the myenteric plexus longitudinal muscle of guinea pig ileum were identical with those generated by the epithelial layer of pig small intestine.[70] Here again, phosphoramidon protected SP from degradation while captopril did not.[70] Finally, Watson[127] reported on the degradation of [^{3}H]SP after exposure to isolated bathed segments of whole guinea pig ileum. Thin layer chromatography analysis indicated a rapid disap-pearance of intact radiolabel, since after 15 s only 54% of the extracted radioactivity behaved as [^{3}H]SP. This extremely fast breakdown of SP was in agreement with the transient kinetics of SP pharmacological response that peaked after 15 s.[127] In this preparation, saturating concentrations of captopril displayed no apparent effect on the response of SP.[127] Altogether, it appears that, whereas both endopeptidase 24.11 and ACE are present in the small intestine (see Tables 5, 6, and 7), ACE cannot compete favorably with endopeptidase 24.11, which predominantly inactivates SP in the small intestine according to the following model:

$$\downarrow \qquad \downarrow \qquad \qquad \downarrow$$
$$Arg^1\text{–}Pro^2\text{–}Lys^3\text{–}Pro^4\text{–}Glu^5\text{–}Glu^6 - Phe^7 - Phe^8\text{–}Gly^9 - Leu^{10}\text{–}Met^{11}\text{–}NH_2$$

The metabolism of substance P in the stomach varies according to the nature of *in vitro* or *in vivo* approaches. Catabolism of SP by membrane preparations from porcine gastric corpus mimicked the pattern of degradation of this peptide by purified porcine endopepti-dase.[129] Phosphoramidon (20 μM) inhibited SP degradation by 85%, while such a concen-tration of captopril induced less than a 5% protection of SP.[129] The same degradation products were recovered with rat gastric corpus membranes.[130] *In vivo* studies were carried out by injecting tritiated SP through a catheter that was chronically implanted in the stomach wall of the rat.[114,130] Tritiated metabolites were recovered through an adjacent dialysis fiber and analyzed by HPLC. In opposition to *in vitro* experiments, the major identified catabolites were SP_{1-2} and SP_{3-4}, while smaller amounts of SP_{1-4}, SP_{1-6}, SP_{1-7}, and SP_{1-8} were also detected.[114,130] This indicated that the predominant site of enzymatic process occurred at the N-terminal side of SP, by a postproline dipeptidyl aminopeptidase attack that se-quentially released the two N-terminal dipeptides, Arg^1–Pro^2 and Lys^3–Pro^4. Such features led to degradation products that were not totally inactivated, but whose biological activity was markedly reduced (SP_{5-11} possesses about 30 to 40% of the SP activity measured by the isolated guinea pig ileum model). A partial but significant effect of MK 422 and captopril, but not phosphoramidon,[114] suggested that an additional peptidase resembling angiotensin-converting enzyme probably participated in the inactivation of SP at its C-terminal end.

Few studies have dealt with the gastrointestinal metabolism of other neuropeptides belonging to the tachykinin family. According to the fact that neurokinin A-like immuno-reactivity was detected mainly in the myenteric plexus longitudinal smooth muscle of the guinea pig small intestine,[131] Nau et al.[70] examined the metabolic processing of this peptide (NKA: His^1–Lys^2–Thr^3–Asp^4–Ser^5–Phe^6–Val^7–Gly^8–Leu^9–Met^{10}–NH_2) by membrane vesi-cles derived from this tissue. The rate of NKA disappearance was more rapid than that of substance P. The principal metabolites identified were NKA_{1-8}, NKA_{2-10} and NKA_{3-10}.[70] Captopril did not affect NKA metabolism in guinea pig small intestine. This agreed with a previous study showing that purified angiotensin-converting enzyme could hydrolyze sub-stance P, but not NKA.[132] The formations of NKA_{2-10} and NKA_{3-10} were completely blocked by the aminopeptidase inhibitor bestatin, indicating that NKA_{3-10} resulted from the sequential release of N-terminal residues instead of a single cleavage by an aminopeptidase.[70] It is interesting to note that the general aminopeptidase inhibitor puromycin did not protect substance P from degradation in guinea pig ileum.[127] The ability for aminopeptidases to distinguish between various tachykinins could be interpreted easily by the fact that unlike NKA, which includes a free N-terminus, substance P is protected from aminopeptidasic breakdown by a prolyl residue in the penultimate position. Nothing is yet known concerning the inactivation of NKA or other tachykinins *in vivo* in the gastrointestinal tract.

e. Other Peptides

A preparation of closed membrane vesicles prepared from longitudinal and circular muscles of porcine jejunum was employed to investigate the catabolism of the nonapeptide bradykinin (Arg^1–Pro^2–Pro^3–Gly^4–Phe^5–Ser^6–Pro^7–Phe^8–Arg^9).[133] The principal metabolite that was identified by amino acid analysis represented bradykinin$_{1-7}$. Phosphoramidon dras-tically (but incompletely) inhibited the formation of this product and its C-terminal dipeptide counterpart, bradykinin$_{8-9}$.[133] In agreement, endopeptidase 24.11 was previously shown to hydrolyze bradykinin not only at the Pro^7–Phe^8 bond, but also at Gly^4–Phe^5 after prolonged incubation. The product of the latter cleavage (bradykinin$_{5-9}$) was not recovered in this study, while low amounts of bradykinin$_{6-9}$ formation and des-Arg^1-bradykinin were also detected.[133] The fact that bradykinin$_{6-9}$ was shown to be partially inhibited by bestatin led the authors to suggest that at least a part of bradykinin$_{6-9}$ derived from the aminopeptidasic conversion of bradykinin$_{5-9}$, whose transitory formation was therefore not detected.[133] Des-Arg^1-bradykinin formation was not affected by a series of specific inhibitors including phosphoramidon, captopril, enalapril (an ACE inhibitor), and bestatin.[133] It is interesting to note that des-Arg^1-bradykinin derived from the release of the N-terminal residue remained totally bestatin insensitive. In fact, as had been previously documented for substance P, the N-terminal part of bradykinin is protected from aminopeptidasic attack by a penultimate prolyl residue. However, in the above membrane preparation this had not precluded the recovery of des-Arg^1-bradykinin probably generated by unknown peptidases[133] distinct from bestatin-sensitive aminopeptidases. Altogether, the above study indicates that bradykinin inactivation *in vitro* in the pig intestinal smooth muscle is primarily mediated through the action of endopeptidase 24.11. The participation of this activity in bradykinin metabolism *in vivo* remains to be demonstrated.

The gastrin-releasing peptide family includes GRP-27, the parent peptide, but also three molecular variants that were recently isolated from the canine small intestine.[134] Bunnett et al.[129] recently reported on the catabolism of the smallest form of these GRP congeners (GRP-10), representing the C-terminal decapeptide sequence (18—27) of GRP-27 (Gly^1–Asp^2–His^3–Trp^4–Ala^5–Val^6–Gly^7–His^8–Leu^9–Met^{10}–NH_2) which is common to all nat-urally occurring forms of GRP. The principal products of GRP-10 hydrolysis by membranes prepared from pig gastric corpus muscle were GRP-10$_{1-8}$ and GRP-10$_{9-10}$, consistent with a cleavage occurring at the His^8–Leu^9 peptidyl bond.[129] Prolonged incubations also yielded

GRP-10_{1-6} and its C-terminal counterpart GRP-10_{7-10}, while GRP-10_{1-8} remained relatively stable. The same degradation products of GRP-10 were generated after digestion of the peptide by membrane preparations originating from gastric mucosa.[129] Phosphoramidon strongly inhibited GRP-10 disappearance, with an IC_{50} of about 10 nM.[129] Both the IC_{50} value observed with phosphoramidon together with the pattern of GRP-10 hydrolysis matching that generated by purified endopeptidase 24.11 indicated that this activity was mainly involved in GRP-10 catabolism in gastric corpus muscle. However, inhibition was never complete even when high concentrations of phosphoramidon were employed, indicating that a phosphoramidon-insensitive peptidase distinct from endopeptidase 24.11 also contributed to GRP-10 catabolism in rat corpus.[129] Such a peptidase was not angiotensin-converting enzyme, as demonstrated by the lack of effect of captopril and MK 422.[129] To our knowledge no studies on the metabolism of GRP *in vivo* in intestinal tissues have been described. It is, however, interesting to report on a study that analyzed the catabolism of bombesin in the interstitial fluid of the rat stomach[135] because although this peptide does not occur in mammals, it represents the amphibian counterpart of GRP, with which it shares the same C-terminal amidated heptapeptide. The results show that four main products were detected after perfusion of [^{25}I]Tyr^4 bombesin into the submucosa of the gastric corpus.[135] One of them corresponded to free iodotyrosine, while the three others remained unidentified.[135] The fact that these catabolites did not bind to C-terminal-specific antibodies[135] suggested that they probably represented inactive fragments of bombesin, since biological activity of the peptide is borne by its C-terminal end.

In the present chapter, Tables 2 and 4 indicate that thyrotropin-releasing hormone (TRH: pGlu–His–Pro–NH$_2$) is one of the peptides that exhibit the longest half-lives *in vitro* and *in vivo* in the circulation. Two reports have provided evidence that this peptide is also extremely resistant to degradation by various organs of the gastrointestinal tract of several species.[24,136] Half-lives of TRH in the duodenum of rat, mouse, and man were 240, 145, and 162 min, respectively, while the peptide was found to be stable in the dog.[24] The gut appeared to be less potent that plasma, liver, and brain with regard to catabolism of TRH.[136] After 2 h of incubation with a homogenate from human gut, two ^{14}C products were recovered that behaved as free histidine and deamidated TRH after thin layer chromatography. However, the main product that was recovered comigrated with intact TRH, indicating a considerable stability[136] that is certainly due to the pyroglutamyl and amidated groups that conferred resistance to aminopeptidasic and carboxypeptidasic attacks.

IV. CONCLUSIONS AND PERSPECTIVES

The present chapter has reviewed the information available concerning peptidases present in the circulation and in the gastrointestinal tract and their relative contribution to the neuropeptide-inactivating mechanisms taking place in these biological compartments. It is clear that little degradation occurs in the circulation and that most of the peptide clearance occurs in various organs including those corresponding to the gastrointestinal tract.

The development of the knowledge of peptide metabolism has been slow to develop, and although an increasing number of papers now relate to this topic, one can consider that this field is still in its infancy. One of the challenges that has inspired some works on peptide metabolism is the hope that the elucidation of physiological inactivating processes could finally lead to diminishing the pharmacological doses of peptides that are often used and, therefore, uncovering additional physiological roles of neuropeptides that could remain unknown. Reaching this goal will be difficult due to technical limitations and is always subject to the development of specific tools. First of all, some peptidase inhibitors have been isolated or synthesized successfully. They must be potent and selective. However, most of the complexity derives from the fact that we have seen that peptidases display the ability to hydrolyze various neuropeptides. Therefore, *in vivo* studies necessitate the use of peptide antagonists that must reverse the protection induced by a peptidase inhibitor before the involvement of the corresponding peptidase in the physiological inactivation of a peptide can be established. Although this could clarify the mechanism of the degradation of one peptide, it will not preclude the possibility that other peptides are metabolized by this peptidase. To overcome this problem, another strategy could involve the development of peptides totally resistant to degradation. Table 10 lists a series of modified analogues that were recently developed and that display long lives *in vitro* and/or *in vivo*. The difficulty of this strategy is to confer increased metabolic stability of the peptide without loss of its ability to interact with its specific receptors. This is often difficult due to the fact that the region of the peptide that bears the biological activity obviously corresponds to the target of physiologically inactivating peptidases. Thus, any modification that renders the peptide resistant to degradation could alter the affinity for its specific receptor and, therefore, leads to the necessity to use high concentrations of the modified peptide.

The gate to the knowledge of peptide metabolism remains to be opened. The future emergence of numerous studies will highlight the critical role that is played by inactivating mechanisms and hopefully will lead to a better understanding of peptide physiology.

ACKNOWLEDGMENT

I wish to gratefully thank Valérie Dalmasso for expert secretarial assistance. This work was supported by the Centre National de la Recherche Scientifique, the Institut National de la Santé et de la Recherche Médicale (CRE 886017), and the Fondation pour la Recherche Médicale.

TABLE 10
Enzyme-Resistant Analogs of Peptides of the Gastrointestinal Tract

| Parent peptide | Modified analog | Metabolic stability | | Ref. |
		In vitro	In vivo	
Atrial natriuretic factor	[Mpr^7,Ala^{20},D-Arg^{27}]-rANF$_{7-27}$	Yes	No	150
CCK-8	[D-Trp^5]-CCK-8	Yes	ND	99
	[D-Met^6]-CCK-8	Yes	ND	99
	Boc–Tyr(SO$_3$H)–Nle–Gly–Trp– (N-Me)–Nle–Asp–Phe–NH$_2$	Yes	ND	151
	Boc–Tyr(SO$_3$H)–gNle–mGly–Trp– (N-Me)–Nle–Asp–Phe–NH$_2$	Yes	ND	151
Met-enkephalin	[D-Ala^2, MePhe^4, Met(o)-ol]-enkephalin	Yes	Yes	152
Neurotensin	[D-Tyr^{11}]-NT	Yes	Yes	153, 154
	[D-Trp^{11}]-NT	Yes	Yes	153, 154
	[Dopa^{11}]-NT	ND	Yes	154
Somatostatin	Acylated des-[Ala^1-Gly^2]-SRIF	Yes	Yes	155
Substance P	[pGlu^5, MePhe^8, Sar^9]-SP	Yes	Yes	156
	[pGlu^6-ψ-NH-CO)(RS)-Phe^7]-SP$_{6-11}$	Yes	ND	157
TRH	pGlu–His–Mep–NH$_2$	Yes	Yes	24, 25, 158

Note: Boc, *tert*-butyloxycarbonyl; Mep, methylproline; Mpr, 3-mercaptopropionic acid; Nle, norleucine; pyr, pyroglutamic acid; ND, not determined.

REFERENCES

1. **Umezawa, K. and Aoyagi, T.**, Elimination of protein degradation by use of protease inhibitors, in *Receptor Purification Procedures*, Alan R. Liss, New York, 1984, 139.
2. **Umezawa, H., Aoyagi, T., Suda, H., Hamada, M., and Takeuchi, T.**, Bestatin, an inhibitor of aminopeptidase B, produced by actinomycetes, *J. Antibiot.*, 29, 97, 1976.
3. **Suda, H., Aoyagi, T., Takeuchi, T., and Umezawa, H.**, A thermolysin inhibitor produced by actinomycetes: phosphoramidon, *J. Antibiot.*, 10, 621, 1973.
4. **Gros, C., Giros, B., Schwartz, J. C., Vlaiculescu, A., Costentin, J., and Lecomte, J. M.**, Potent inhibition of cerebral aminopeptidases by carbaphethiol, a parenterally active compound, *Neuropeptides*, 12, 111, 1988.
5. **Friedman, T. C., Kline, T. B., and Wilk, S.**, 5-oxoprolinal: transition-state aldehyde inhibitor of pyroglutamyl-peptide hydrolase, *Biochemistry*, 24, 3909, 1985.
6. **Friedman, T. C., Orlowski, M., and Wilk, S.**, Prolyl endopeptidase: inhibition in vivo by N-benzyloxycarbonyl-prolyl-prolinal, *J. Neurochem.*, 42, 237, 1984.
7. **Ondetti, M. A., Rubin, B., and Cushman, D. W.**, Design of specific inhibitors of angiotensin-converting enzyme: new class of orally antihypertensive agents, *Science*, 196, 441, 1977.
8. **Roques, B. P., Fournié-Zaluski, M. C., Soroca, E., Lecomte, J. H., Malfroy, B., Llorens, C., and Schwartz, J. C.**, The enkephalinase inhibitor thiorphan shows antinociceptive activity in mice, *Nature (London)*, 288, 286, 1980.
9. **Lecomte, A. M., Costentin, J., Vlaiculescu, A., Chaillet, P., Marcais-Collado, H., Llorens-Cortes, C., Leboyer, M., and Schwartz, J. C.**, Pharmacological properties of acetorphan, a parenterally active "enkephalinase" inhibitor, *J. Pharmacol. Exp. Ther.*, 237, 937, 1986.
10. **Fournié-Zaluski, M.,C., Chaillet, P., Bouboutou, R., Coulaud, A., Cherot, P., Waksman, G., Costentin, J., and Roques, B. P.**, Analgesic effects of kelatorphan, a new highly potent inhibitor of multiple enkephalin degrading enzymes, *Eur. J. Pharmacol.*, 102, 525, 1984.
11. **Chu, T. G. and Orlowski, M.**, Active site directed N-carboxymethyl peptide inhibitors of a soluble metalloendopeptidase from rat brain, *Biochemistry*, 23, 3598, 1984.
12. **Checler, F., Vincent, J. P., and Kitabgi, P.**, Purification and characterization of a novel neurotensin degrading peptidase from rat brain synaptic membranes, *J. Biol. Chem.*, 261, 11274, 1986.
13. **Barelli, H., Vincent, J. P., and Checler, F.**, Peripheral inactivation of neurotensin: isolation and characterization of a metallopeptidase from rat ileum, *Eur. J. Biochem.*, 175, 481, 1988.
14. **Carvalho, K. M. and Camargo, A. C. M.**, Purification of rabbit brain endooligopeptidases and preparation of anti-enzymes antibodies, *Biochemistry*, 20, 7082, 1981.
15. **Checler, F., Barelli, H., and Vincent, J. P.**, Tissue distribution of a novel neurotensin-degrading metallopeptidase. An immunological approach using monospecific polyclonal antibodies, *Biochem. J.*, 257, 549, 1989.
16. **Ahmad, S., Berezin, I., Vincent, J. P., and Daniel, E. E.**, Neurotensin receptors in canine intestinal smooth muscle: preparation of plasma membranes and characterization of ^{125}I-Tyr^3-neurotensin binding, *Biochem. Biophys. Acta*, 896, 224, 1987.
17. **Ahmad, S., Allescher, H. D., Manaka, H., Manaka, Y., and Daniel, E. E.**, [^{3}H]saxitoxin as a marker for canine deep muscular plexus neurons, *Am. J. Physiol.*, 255, G462, 1988.
18. **Ahmad, S., Allescher, H. D., Manaka, H., Manaka, Y., and Daniel, E. E.**, Biochemical studies on the opioid and β2 adrenergic receptors in canine submucosal neurons, *Am. J. Physiol.*, 256, 957, 1989.
19. **Kostka, P., Ahmad, S., Berezin, I., Kwan, C. Y., and Daniel, E. E.**, Subcellular fractionation of the longitudinal smooth muscle/myenteric plexus (LSM/MP) of dog ileum: dissociation of the distribution of two plasma membrane marker enzymes, *J. Neurochem.*, 49, 1124, 1987.
20. **Blair, E. L., Farra, Y., Richardson, D. D., and Steinbok, P.**, The half-life of exogenous gastrin in the circulation, *J. Physiol.*, 208, 299, 1970.
21. **Lehnert, P., Stahlheber, H., Forell, M. M., Fûllner, R., Fruhauf, S., Fritz, H., Hutzel, M., and Werle, E.**, Studies on the elimination of secretin and cholecystokinin with regard to the kinetics of exocrine pancreatic secretion, *Digestion*, 11, 51, 1974.
22. **Murthy, K. K., Thibault, G., Garcia, R., Gutkowska, J., Genest, J., and Cantin, M.**, Degradation of atrial natriuretic factor in the rat, *Biochem. J.*, 240, 461, 1986.
23. **Tang, J., Webber, R. J., Chang, D., Chang, J. K., Kiang, J., and Wei, E. T.**, Depressor and natriuretic activities of several atrial peptides, *Regul. Peptides*, 9, 53, 1984.
24. **Brewster, D., Humphrey, M. J., and Wareing, M. V.**, Metabolism and pharmacokinetics of THR and an analogue with enhanced neuropharmacological potency, *Neuropeptides*, 1, 153, 1981.
25. **Brewster, D. and Rance, M. J.**, An analogue of thyrotropin releasing hormone with improved biological stability both *in vitro* and *in vivo*, *Biochem. Pharmacol.*, 29, 2619, 1980.
26. **Aronin, N., Carraway, R. E., Ferris, C. F., Hammer, R. A., and Leeman, S. E.**, The stability and metabolism of intravenously administered neurotensin in the rat, *Peptides*, 3, 637, 1982.

303

27. **Lee, Y. C., Uttenthal, L. O., Smith, H. A., and Bloom, S. R.**, *In vitro* degradation of neurotensin in human plasma, *Peptides*, 7, 383, 1986.
28. **Conlon, J. M. and Sheehan, L.**, Conversion of substance P to C-terminal fragments in human plasma, *Regul. Peptides*, 7, 335, 1983.
29. **Couture, R. and Regoli, D.**, Inactivation of substance P and its C-terminal fragments in rat plasma and its inhibition by captopril, *Can. J. Physiol. Pharmacol.*, 59, 621, 1981.
30. **Huggins, C. G. and Thampi, N. S.**, A simple method for the determination of angiotensin I converting enzyme, *Life Sci.*, 7, 633, 1968.
31. **Marceau, F., Gendreau, M., Barabé, J., St-Pierre, S., and Regoli, D.**, The degradation of bradykinin (BK) and of des-Arg^9-BK in plasma, *Can. J. Physiol. Pharmacol.*, 59, 131, 1980.
32. **Oparil, S., Sanders, C. A., and Haber, E.**, *In vivo* and *in vitro* conversion of angiotensin I to angiotensin II in dog blood, *Circ. Res.*, 26, 591, 1970.
33. **Hambrook, J. M., Morgan, B. A., Rance, M. J., and Smith, C. F. C.**, Mode of deactivation of the enkephalins by rat and human plasma and rat brain homogenates, *Nature (London)*, 262, 782, 1976.
34. **Tamura, Y., Niinobe, M., Arima, T., Okuda, H., and Fujii, S.**, Studies on aminopeptidases in rat liver and plasma, *Biochim. Biophys. Acta*, 327, 437, 1973.
35. **Goldbarg, J. A. and Rutenburg, A. M.**, The colorimetric determination of leucine aminopeptidase in urine and serum of normal subjects and patients with cancer and other diseases, *Cancer*, 11, 283, 1958.
36. **Al-Merani, S. A. M. A., Brooks, D. P., Chapman, B. J., and Munday, K. A.**, The half-lives of angiotensin II, angiotensin II-amide, angiotensin III, sar^1-ala^8-angiotensin II and renin in the circulatory system of the rat, *J. Physiol.*, 278, 471, 1978.
37. **Harding, J. W., Yoshida, M. S., Dilts, R. P., Woods, T. M., and Wright, J. W.**, Cerebroventricular and intravascular metabolism of ^{125}I angiotensins in rat, *J. Neurochem.*, 46, 1292, 1986.
38. **Hodge, R. L., Ng, K. K. F., and Vane, J. R.**, Disappearance of angiotensin from the circulation of the dog, *Nature (London)*, 215, 138, 1967.
39. **Murthy, K. K., Thibault, G., Schiffrin, E. L., Garcia, R., Chartier, L., Gutkowska, J., Genest, J., and Cantin, M.**, Disappearance of atrial natriuretic factor from circulation in the rat, *Peptides*, 7, 241, 1986.
40. **Straus, E. and Yalow, R. S.**, Studies on the distribution and degradation of heptadecapeptide, big, and big big gastrin, *Gastroenterology*, 66, 936, 1974.
41. **Reeder, D. D., Jackson, B. M., Brandt, E. N., and Thompson, J. C.**, Rate and pattern of disappearance of exogenous gastrin in dogs, *Am. J. Physiol.*, 222, 1571, 1972.
42. **Schrumpf, E. and Semb, L. S.**, The metabolic clearance rate and half-life of synthetic human gastrin in dogs, *Scand. J. Gastroenterol.*, 8, 203, 1973.
43. **Deschodt-Lanckman, M., Pauwels, S., Najdovski, T., Dimaline, R., and Dockray, G. J.**, *In vitro* and *in vivo* degradation of human gastrin by endopeptidase 24.11, *Gastroenterology*, 94, 712, 1988.
44. **Thompson, J. C., Fender, H. R., Ramus, N. I., Villar, H. V., and Rayford, P. L.**, Cholecystokinin metabolism in man and dogs, *Ann. Surg.*, 182, 496, 1975.
45. **Lee, Y. C., Allen, J. M., Uttenthal, L. O., Walker, M. C., Shemilt, J., Gill, S. S., and Bloom, S. R.**, The metabolism of intravenously infused neurotensin in man and its chromatographic characterization in human plasma, *J. Clin. Endocrinol. Metab.*, 59, 45, 1984.
46. **Pedersen, J. H. and Fahrenkrug, J.**, Neurotensin-like immunoreactivities in human plasma: feeding responses and metabolism, *Peptides*, 7, 15, 1986.
47. **Leppäluoto, J., Virkkunen, P., and Lybeck, H.**, Elimination of TRH in man, *J. Clin. Endocrinol. Metab.*, 35, 477, 1972.
48. **Bassiri, R. M. and Utiger, R. D.**, Metabolism and excretion of exogenous thyrotropin-releasing hormone in humans, *J. Clin. Invest.*, 52, 1616, 1973.
49. **Dupont, A., Cusan, L., Garon, M., Alvarado-Urbina, G., and Labrie, F.**, Extremely rapid degradation of [^{3}H]methionine-enkephalin by various rat tissues *in vivo* and *in vitro*, *Life Sci.*, 21, 907, 1977.
50. **Ferreira, S. H. and Vane, J. R.**, The disappearance of bradykinin and eledoisin in the circulation and vascular beds of the cat, *Br. J. Pharmacol. Chemother.*, 30, 417, 1967.
51. **Saameli, K. and Eskes, K. A. B.**, Bradykinin and cardiovascular system: estimation of half-life, *Am. J. Physiol.*, 203, 261, 1962.
52. **McCarthy, D. A., Potter, D. E., and Nicolaides, E. D.**, An *in vivo* estimation of the potencies and half-lives of synthetic bradykinin and kallidin, *J. Pharmacol. Exp. Ther.*, 148, 117, 1965.
53. **Sheppard, M., Shapiro, B., Pimstone, B., Kronheim, S., Berelowitz, M., and Gregory, M.**, Metabolic clearance and plasma half-disappearance time of exogenous somatostatin in man, *J. Clin. Endocrinol. Metab.*, 48, 50, 1979.
54. **Domschke, S., Domschke, W., Bloom, S. R., Mitznegg, P., Mitchell, S., J., Lux, G., and Strunz, U.**, Vasoactive intestinal peptide in man: pharmacokinetics, metabolic and circulatory effects, *Gut*, 19, 1049, 1978.

55. **Holzer-Petsche, U., Saria, A., Rupitz, M., Schuligoi, R., and Lembeck, F.**, An International Symposium on Tachykinins, G., 20-23 July, Degradation of substance P, neurokinin A, and calcitonin gene-related peptide in rat plasma *in vivo, Regul. Peptide,* 22, 89, 1988.

56. **Barelli, H., Vincent, J. P., and Checler, F.**, Peptidase activities in various organs of the rat gastrointestinal tractus, in preparation, 1990.

57. **Maroux, S., Louvard, D., and Baratti, J.**, The aminopeptidase from hog intestinal brush border, *Biochim. Biophys. Acta,* 321, 282, 1973.

58. **Auricchio, S., Greco, L., De Vizia, B., and Buonocore, V.**, Dipeptidylaminopeptidase and carboxypeptidase activities of the brush border of rabbit small intestine, *Gastroenterology,* 75, 1073, 1978.

59. **Hersh, L. B., Aboukhair, N., and Watson, S.**, Immunohistochemical localization of aminopeptidase M in rat brain and periphery: relationship of enzyme localization and enkephalin metabolism, *Peptides,* 8, 523, 1987.

60. **Defendini, R., Zimmerman, E. A., Weare, J. A., Alhenc-Gelas, F., and Erdös, E. G.**, Angiotensin-converting enzyme in epithelial and neuroepithelial cells, *Neuroendocrinology,* 37, 32, 1983.

61. **Gee, N. S., Matsas, R., and Kenny, A. J.**, A monoclonal antibody to kieney endopeptidase-24.11, *Biochem. J.,* 214, 377, 1983.

62. **Matsas, R., Kenny, A. J. and Turner, A. J.**, An immunohistochemical study of endopeptidase-24.11 ("enkephalinase") in the pig nervous system, *Neuroscience,* 18, 991, 1986.

63. **Gee, N. S., Bowes, M. A., Buck, P., and Kenny, A. J.**, An immunoradiometric assay for endopeptidase-24.11 shows it to be a widely distributed enzyme in pig tissues, *Biochem. J.,* 228, 119, 1985.

64. **Checler, F., Barelli, H., Kwan, C. Y., and Vincent, J. P.**, Neurotensin-metabolizing peptidases in rat fundus plasma membranes, *J. Neurochem.,* 49, 507, 1987.

65. **Llorens, C. and Schwartz, J. C.**, Enkephalinase activity in rat peripheral organs, *Eur. J. Pharmacol.,* 69, 113, 1981.

66. **Orlowski, M., Wilk, E., Pearce, S., and Wilk, S.**, Purification and properties of a prolyl endopeptidase from rabbit brain, *J. Neurochem.,* 33, 461, 1979.

67. **Orlowski, M., Michaud, C., and Chu, T. G.**, A soluble metalloendopeptidase from rat brain. Purification of the enzyme and determination of specificity with synthetic and natural peptides, *Eur. J. Biochem.,* 135, 81, 1983.

68. **Checler, F., Ahmad, S., Kostka, P., Barelli, H., Kitabgi, P., Fox, J. E. T., Kwan, C. Y., Daniel, E. E., and Vincent, J. P.**, Peptidases in dog ileum circular and longitudinal smooth muscle plasma membranes: their relative contribution to the metabolism of neurotensin, *Eur. J. Biochem.,* 166, 461, 1986.

69. **Barelli, H., Ahmad, S., Kostka, P., Fox, J. E. T., Daniel, E. E., Vincent, J. P., and Checler, F.**, Neuropeptide-hydrolysing activities in synaptosomal fractions from dog ileum myenteric, deep muscular and submucous plexi. Their participation in neurotensin inactivation, *Peptides,* 10, 1055, 1989.

70. **Nau, R., Schäfer, G., Deacon, C. F., Cole, T., Agoston, D. V., and Conlon, J. M.**, Proteolytic inactivation of substance P and neurokinin A in the longitudinal muscle layer of guinea pig small intestine, *J. Neurochem.,* 47, 856, 1986.

71. **Hazato, T., Shimamura, M., Kase, R., Iijima, M., and Katayama, T.**, Separation of enkephalin-degrading enzymes from longitudinal muscle layer of bovine small intestine. Enzyme inhibition by arphamenine A, *Biochem. Pharmacol.,* 34, 3179, 1985.

72. **Danielsen, E. M., Sjöstrom, H, Norén, O., and Dabelsteen, E.**, Immunoelectrophoretic studies on pig intestinal brush border proteins, *Biochem. Biophys. Acta,* 494, 332, 1977.

73. **Andria, G., Marzi, A., and Auricchio, S.**, α-Glutamyl-β-naphthylamide hydrolase of rabbit small intestine. Localisation in the brush border and separation from other brush border peptidases, *Biochim. Biophys. Acta,* 419, 42, 1976.

74. **Benajiba, A. and Maroux, S.**, Purification and characterization of an aminopeptidase A from hog intestinal brush-border membrane, *Eur. J. Biochem.,* 107, 381, 1980.

75. **Morita, A., Chung, Y. C., Freeman, H. J., Erickson, R. H., Sleisenger, M. H., and Kim, Y. S.**, Intestinal assimilation of a proline-containing tetrapeptide. Role of a brush border membrane postproline dipeptidyl aminopeptidase IV, *J. Clin. Invest.,* 72, 610, 1983.

76. **Song, I. S., Yoshioka, M., Erickson, R. H., Miura, S., Guan, D., and Kim, Y. S.**, Identification and characterization of brush-border membrane-bound neutral metalloendopeptidases from rat small intestine, *Gastroenterology,* 91, 1234, 1986.

77. **Louvard, D., Maroux, S., Vannier, C., and Desnuelle, P.**, Topological studies on the hydrolases bound to the intestinal brush border membrane. I. Solubilization by papain and Triton X-100, *Biochim. Biophys. Acta,* 375, 236, 1975.

78. **Manjusri, D. and Radhakrishnan, A. N.**, Glycyl-l-leucine hydrolase, a versatile "master" dipeptidase from monkey small intestine, *Biochem. J.,* 135, 609, 1973.

79. **Nau, R., Schäfer, G., and Conlon, J. M.**, Proteolytic inactivation of substance P in the epithelial layer of the intestine, *Biochem. Pharmacol.,* 34, 4019, 1985.

80. **Svensson, B., Danielsen, M., Staun, M., Jeppesen, L., Noréen, O., and Sjöström, H.**, An amphiphilic form of dipeptidyl peptidase IV from pig small intestinal brush-border membrane, *Eur. J. Biochem.*, 90, 489, 1978.

81. **Fulcher, I. S., Chaplin, M. F., and Kenny, A. J.**, Endopeptidase-24.11 purified from pig intestine is differently glycoslylated from that in kidney, *Biochem. J.*, 215, 317, 1983.

82. **Ward, P. E., Sheridan, M. A., Hammon, K. J., and Erdös, E. G.**, Angiotensin I converting enzyme (kininase II) of the brush border of human and swine intestine, *Biochem. Pharmacol.*, 29, 1525, 1980.

83. **McDermott, J. R., Mantle, D., Lauffart, B., and Kidd, A. M.**, Purification and characterization of a neuropeptide-degrading aminopeptidase from human brain, *J. Neurochem.*, 45, 752, 1985.

84. **Hersh, L. B.**, Solubilization and characterization of two rat brain membrane-bound aminopeptidases active on met-enkephalin, *Biochemistry*, 20, 171, 1981.

85. **Kato, T., Nagatsu, T., Fukasawa, K., Harada, M., Nagatsu, I., and Sakakibara, S.**, Successive cleavage of a N-terminal Arg^1-Pro^2 and Lys^3-Pro^4 from substance P but no release of Arg^1-Pro^2 from bradykinin, by X-Pro-dipeptidyl-aminopeptidase, *Biochim. Biophys. Acta*, 525, 417, 1978.

86. **Koida, M. and Walter, R.**, Post-proline cleaving enzyme, *J. Biol. Chem.*, 251, 7593, 1976.

87. **Knisatchek, H. and Bauer, K.**, Characterization of "thyroliberin-deamidating enzyme" as a post-proline cleaving enzyme, *J. Biol. Chem.*, 254, 10936, 1979.

88. **Taylor, W. L. and Dixon, J. E.**, Catabolism of neuropeptides by a brain proline endopeptidase, *Biochem. Biophys. Res. Commun.*, 94, 9, 1980.

89. **Greene, L. J., Spadaro, A. C., Martins, A. R., Perussi de Jesus, W. D., and Camargo, A. C. M.**, Brain endo-oligopeptidase B: a post-proline cleaving enzyme that inactivates angiotensin I and II, *Hypertension*, 4, 178, 1982.

90. **Camargo, A. C. M., Caldo, H., and Emson, P. C.**, Degradation of neurotensin by rabbit brain endo-oligopeptidase A and endo-oligopeptidase B (proline endopeptidase), *Biochem. Biophys. Res. Commun.*, 116, 1151, 1983.

91. **Ward, P. E., Bausback, H. H., and Odya, C. E.**, Kinin and angiotensin metabolism by purified renal post-proline cleaving enzyme, *Biochem. Pharmacol.*, 19, 3187, 1987.

92. **Erdös, E. G.**, The angiotensin I converting enzyme, *Fed. Proc.*, 36, 1760, 1977.

93. **Stewart, T. A., Weare, J. A., and Erdös, E. G.**, Purification and characterization of human converting enzyme (kininase II), *Peptides*, 2, 145, 1981.

94. **Skidgel, R. A., Engelbrecht, S. Johnson, A. R., and Erdös, E. G.**, Hydrolysis of substance P and neurotensin by converting enzyme and neutral endopeptidase, *Peptides*, 5, 769, 1984.

95. **Hooper, N. M. and Turner, A. J.**, Isolation of two differentially glycosylated forms of peptidyl-dipeptidase A (angiotensin converting enzyme) from pig brain: a re-evaluation of their role in neuropeptide metabolism, *Biochem. J.*, 241, 625, 1987.

96. **Almenoff, J. and Orlowski, M.**, Membrane-bound kidney neutral metalloendopeptidase: interaction with synthetic substrates, natural peptides, and inhibitors, *Biochemistry*, 22, 590, 1983.

97. **Mumford, R. A., Pierzchala, P.A., Strauss, A. W., and Zimmerman, M.**, Purification of a membrane-bound metalloendopeptidase from procine kidney that degrades peptide hormones, *Proc. Natl. Acad. Sci. U.S.A.*, 78, 6623, 1981.

98. **Gafford, J. T., Skidgel, R. A., Erdös, E.G., and Hersh, L. B.**, Human kidney "enkephalinase", a neutral metalloendopeptidase that cleaves active peptides, *Biochemistry*, 22, 3265, 1983.

99. **Najdovski, T., De Pont, J. J. H. H. M., Tesser, G. I., Penke, B., Martinez, J., and Deschodt-Lanckman, M.**, Degradation of cholecystokinin octapeptide by the neutral endopeptidase EC 3.4.24.11 and design of proteolysis-resistant analogues of the peptide, *Neurochem. Int.*, 10, 459, 1987.

100. **Power, D. M., Bunnett, N., Turner, A. J., and Dimaline, R.**, Degradation of endogenous heptadeca-peptide gastrin by endopeptidase 24.11 in the pig, *Am. J. Physiol.*, 253, 33, 1987.

101. **Chu, T. G. and Orlowski, M.**, Soluble metalloendopeptidase from rat brain: action on enkephalin-containing peptides and other bioactive peptides, *Endocrinology*, 116, 1418, 1985.

102. **Camargo, A. C. M., Oliveira, E. B., Toffoletto, O., Metters, K. M., and Rossier, J.**, Brain endo-oligopeptidase A, a putative enkephalin converting enzyme, *J. Neurochem*, 48, 1258, 1987.

103. **Toffoletto, O., Metters, K. M., Oliveira, E. B., Camargo, A. C. M., and Rossier, J.**, Enkepahlin is liberated from metorphamide and dynorphin A_{1-8} by endo-oligopeptidase A, but not by metalloendopeptidase EC3.4.24.15, *Biochem. J.*, 252, 35, 1988.

104. **Cicilini, M. A., Ribeiro, M. J. F., Oliveira, E. B., Mortara, R. A., and Camargo, A. C. M.**, Endo-oligopeptidase A activity in rabbit heart: generation of enkephalin from enkephalin containing peptides, *Peptides*, 9, 945, 1988.

105. **Turner, A. J., Matsas, R., and Kenny, A. J.**, Are there neuropeptide-specific peptidases?, *Biochem. Pharmacol.*, 34, 1347, 1985.

106. **Temperley, J. M., Stagg, B. H., and Wyllie, J. H.**, Disappearance of gastrin and pentagastrin in the portal circulation, *Gut*, 13, 372, 1971.

107. **Gregory, R. A. and Tracy, H. J.,** Isolation of the two "big gastrins" from Zollinger-Ellison tumour tissue, *Lancet*, 2, 797, 1972.

108. **Durieux, C., Charpentier, B., Fellion, E., Gacel, G., Pelaprat, D., and Roques, B. P.,** Multiple cleavage sites of cholecystokinin heptapeptide by "enkephalinase", *Peptides*, 6, 495, 1985.

109. **Hammer, R. A., Carraway, R. E., and Leeman, S. E.,** Elevation of plasma neurotensin-like immunoreactivity after a meal. Characterization of the elevated components, *J. Clin. Invest.*, 70, 74, 1982.

110. **Ferris, C. F., Carraway, R. E., Hammer, R. A., and Leeman, S. E.,** Release and degradation of neurotensin during perfusion of rat small intestine with lipid, *Regul. Peptides*, 12, 101, 1985.

111. **Checler, F., Kostolanska, B., and Fox, J. E. T.,** *In vivo* inactivation of neurotensin in dog ileum: major involvement of endopeptidase 24-11, *J. Pharmacol. Exp. Ther.*, 244, 1040, 1988.

112. **Shaw, C., Göke, R., Bunnett, N. W., and Conlon, J. M.,** Catabolism of neurotensin in the epithelial layer of porcine small intestine, *Biochim. Biophys. Acta*, 924, 167, 1987.

113. **Bunnett, N. W., Mogard, M., Orloff, M. S., Corbett, H. J., Reeve, J. R., Jr., and Walsh, J. H.,** Catabolism of neurotensin in intestinal fluid of the rat stomach, *Am. J. Physiol.*, 246, 675, 1984.

114. **Orloff, M. S., Turner, A. J., and Bunnett, N. W.,** Catabolism of substance P and neurotensin in the rat stomach wall is susceptible to inhibitors of angiotensin converting enzyme, *Regul. Peptides*, 14, 21, 1986.

115. **Geary, L. E., Wiley, K. S., Scott, W. L., and Cohen, M. L.,** Degradation of exogenous enkephalin in the guinea-pig ileum: relative importance of aminopeptidase, enkephalinase and angiotensin converting enzyme activity, *J. Pharmacol. Exp. Ther.*, 221, 104, 1982.

116. **Miller, R. and Lacefield, W.,** Specific inhibitors of aminopeptidase M. Relationship to anti-inflammatory activity, *Biochem. Pharmacol.*, 28, 673, 1979.

117. **Craviso, G. L. and Musacchio, J. M.,** Inhibition of enkephalin degradation in the guinea pig ileum, *Life Sci.*, 23, 2019, 1978.

118. **Cohen, M. L., Geary, L. E., and Wiley, K. S.,** Enkephalin degradation in the guinea-pig ileum: effect of aminopeptidase inhibitors, puromycin and bestatin, *J. Pharmacol. Exp. Ther.*, 224, 379, 1983.

119. **Vogel, Z and Altsteim, M.,** The effect of puromycin on the biological activity of Leu-enkephalin, *FEBS Lett.*, 98, 44, 1979.

120. **McKnight, A. T., Corbett, A. D., and Kosterlitz, H. W.,** Increase in potencies of opioid peptides after peptidase inhibition, *Eur. J. Pharmacol.*, 86, 393, 1983.

121. **Buckett, W. R.,** The actions of enkephalins are not modified by the "kininase II" inhibitor captopril, *Eur. J. Pharmacol.*, 57, 267, 1979.

122. **Kuno, Y. and Oka, T.,** Estimation of relative importance of three enzymes in the inactivation of (Met^5)-enkephalin and (Met^5)-enkephalin-Arg^6 in three isolated preparations by employing the inhibitor specific for each enzyme, *Japan J. Pharmacol.*, 44, 241, 1987.

123. **Opmeer, F. A., Peter, J., Burbach, H., Wiegant, V. M., and Van Ree, J. M.,** β-endorphin proteolysis by guinea-pig ileum myentheric plexus membranes: increased γ-endorphin turnover after chronic exposure to morphine, *Life Sci.*, 31, 323, 1982.

124. **Hynes, M. R., Culling, A. J., Calligan, J. J., Burks, T. F., Schoemaker, H., Yamamura, H. I., and Davis, T. P.,** Processing of β-endorphin in the dog intestine: regional specificity, *Proc. West. Pharmacol. Soc.*, 26, 95, 1983.

125. **Davis, T. P., Culling, A. J., Schoemaker, H., and Galligan, J. J.,** β-Endorphin and its metabolites stimulate motility of the dog small intestine, *J. Pharmacol. Exp. Ther.*, 227, 499, 1983.

126. **Davis, T. P., Schoemaker, H., and Culling-Berglund, A. J.,** Characterization of *in vitro* proteolytic processing of β-endorphin by reversed-phase HPLC, *Peptides*, 5, 1037, 1984.

127. **Watson, S. P.,** Rapid degradation of [^{3}H]-substance P in guinea-pig ileum and rat vas deferens *in vitro*, *Br. J. Pharmacol.*, 79, 543, 1979.

128. **Iwamoto, I., Ueki, I., and Nadel, J. A.,** Effect of neutral endopeptidase inhibitors on ^{3}H-substance P binding in rat ileum, *Neuropeptides*, 11, 185, 1988.

129. **Bunnett, N. W., Kobayashi, R., Orloff, M. S., Reeve, J. R., Turner, A. J., and Walsh, J. H.,** Catabolism of gastrin releasing peptide and substance P by gastric membrane-bound peptidases, *Peptides*, 6, 277, 1985.

130. **Bunnett, N. W., Orloff, M. S., and Turner, A. J.,** Catabolism of substance P in the stomach wall of the rat, *Life Sci.*, 37, 599, 1985.

131. **Deacon, C. F., Agoston, D. V. Nau, R., and Conlon, J. M.,** Conversion of neuropeptide K to neurokinin A and vesicular colocalization of neurokinin A and substance P in neurons of the guinea pig small intestine, *J. Neurochem.*, 48, 141, 1987.

132. **Hooper, N. M., Kenny, A. J., and Turner, A. J.,** The metabolism of neuropeptides. Neurokinin A (substance K) is a substrate for endopeptidase 24.11 but not for dipeptidyl dipeptidase A (angiotensin-converting enzyme), *Biochem. J.*, 231, 357, 1985.

133. **Schäfer, G., Nau, R., Cole, T., and Conlon, J. M.,** Specific binding and proteolytic inactivation of bradykinin by membrane vesicles from pig intestinal smooth muscle, *Biochem. Pharmacol.*, 35, 3719, 1986.

134. **Reeve, J. R., Walsh, J. H., Chew, P., Clark, B., and Shively, J. E.**, Amino acid sequences of three bombesin like peptides from canine intestinal extracts, *J. Biol. Chem.*, 258, 5582, 1983.

135. **Bunnett, N. W., Reeve, J. R., and Walsh, J. H.**, Catabolism of bombesin in the interstitial fluid of the rat stomach, *Neuropeptides*, 4, 55, 1983.

136. **Brewster, D. and Waltham, K.**, TRH degradation rates vary widely between different animal species, *Biochem. Pharmacol.*, 30, 619, 1981.

137. **Wilk, S., Friedman, T. C., and Kline, T. B.**, Pyroglutamyl diazomethyl ketone: potent inhibitor of mammalian pyroglutamyl peptide hydrolase, *Biochem. Biophys. Res. Commun.*, 130, 662, 1985.

138. **Bado, A., Chicau-Chovet, M., Appia, F., Dubrasquet, M., Lecomte, J. M., and Rozé, C.**, Acetorphan, an enkephalinase inhibitor, decreases gastric secretion in cats, *Peptides*, 8, 89, 1987.

139. **Redding, T. W. and Schally, A. V.**, Studies on the inactivation of thyrotropin-releasing hormone (TRH), *Proc. Soc. Exp. Bio. Med.*, 131, 415, 1969.

140. **Erdös, E. G., Sloane, E. M., and Wohler, I. M.**, Carboxypeptidase in blood and other fluids. I. Properties, distribution, and partial purification of the enzyme, *Biochem. Pharmacol.*, 13, 893, 1964.

141. **Mori, M., Mallik, T., Prasad, C., and Wilber, J. F.**, Histidyl-proline diketopiperazine cyclo (his-pro): measurement by radioimmunoassay in human blood in normal subjects and in patients with hyper- and hypothyroidism, *Biochem. Biophys. Res. Commun.*, 109, 541, 1982.

142. **Hino, M., Nagatsu, T., Kakumu, S., Okuyama, S., Yoshii, Y., and Nagatsu, I.**, Glycylprolyl β-naphthylamidase activity in human serum, *Clin. Chim. Acta*, 62, 5, 1975.

143. **Palmieri, F. E. and Ward, P. E.**, Mesentery vascular metabolism of substance P, *Biochim. Biophys. Acta*, 755, 522, 1983.

144. **Palmieri, F. E., Petrelli, J. J., and Ward, P. E.**, Vascular, plasma membrane aminopeptidase M, metabolism of vasoactive peptides, *Biochem. Pharmacol.*, 34, 2309, 1985.

145. **Ward, P. E.**, Immunoelectrophoretic analysis of vascular, membrane-bound angiotensin I converting enzyme, aminopeptidase M, and dipeptidylaminopeptidase IV, *Biochem. Pharmacol.*, 33, 3183, 1984.

146. **Yandle, T. G., Richards, A. M., Nicholls, M. G., Cuneo, R., Espiner, E. A., and Livesey, J. H.**, Metabolic clearance rate and plasma half-life of alpha-human atrial natriuretic peptide in man, *Life Sci.*, 38, 1827, 1986.

147. **Katsube, N., Schwartz, D., and Needlemen, P.**, Atriopeptin turnover: quantitative relationship between *in vivo* changes in plasma levels and atrial content, *J. Pharmacol. Exp. Ther.*, 239, 174, 1986.

148. **Ganguli, P. C., Elder, J. B., Smith, I. S., and Hunter, W. M.**, The half-life ($T_{1/2}$) of synthetic human gastrin I in man, *Br. J. Surg.*, 57, 750, 1970.

149. **Danielsen, E. M., Vyas, J. P. and Kenny, A. J.**, A neutral endopeptidase in the microvillar membrane of pig intestine, *Biochem. J.*, 191, 645, 1980.

150. **Berman, J. M., Chen, T. M., Sargent, R., Buck, S. H., Shea, P., Heminger, E. F., and Broersma, R. J.**, Design and synthesis of metabolically stable atrial natriuretic factor analogs. Amino-and carboxy-terminal stabilization, *FEBS Lett.*, 237, 76, 1988.

151. **Charpentier, B., Durieux, C., Pelaprat, D., Dor, A., Reibaud, M., Blanchard, J. C., and Roques, B. P.**, Enzyme-resistant CCK analogs with high affinities for central receptors, *Peptides*, 9, 835, 1988.

152. **Roemer, D., Buescher, H. H., Hill, R. C., Pless, J., Bauer, W., Cardinaux, F., Closse, A., Hauser, D., and Huguenin, R.**, A synthetic enkephalin analogue with prolonged parenteral and oral analgesic activity, *Nature (London)*, 268, 547, 1977.

153. **Checler, F., Vincent, J. P., and Kitabgi, P.**, Neurotensin analogs D-Tyr[11] and D-Phe[11] neurotensin resist degradation by brain peptidases *in vitro* and *in vivo*, *J. Pharmacol. Exp. Ther.*, 227, 743, 1983.

154. **Jolicoeur, F. B., Saint-Pierre, S., Aubé, C., Rivest, R., and Gagué, M. A.**, Relationships between structure and duration of neurotensin's central action: emergence of long acting analogs, *Neuropeptides*, 4, 467, 1984.

155. **Brazeau, P., Vale, W., Rivier, J., and Guillemin, R.**, Acylated des-(Ala[1]-Gly[2])-somatostatin analogs: prolonged inhibition of growth hormone secretion, *Biochem. Biophys. Res. Commun.*, 60, 1202, 1974.

156. **Eison, A. S., Iversen, S. D., Sandberg, B. E. B., Watson, S. P., Hanly, M. R., and Iversen, L. L.**, Substance P analog, DiMe-C7: evidence for stability in rat brain and prolonged central actions, *Science*, 215, 188, 1982.

157. **Chorev, M., Rubini, E., Hart, Y., Gilon, C., Wormser, U., and Selinger, Z.**, Metabolically stable analogues of substance P: persistent action of partially modified retro-inverso analogues of substance P on rat parotid and hypothalamic slices, *Eur. J. Pharmacol.*, 127, 187, 1986.

158. **Brewster, D., Dettmar, P. W., Lynn, A. G., Metcalf, G., Morgan, B. A., and Rance, M. J.**, Modification of the proline residue of TRH enhances biological activity and inhibits degradation, *Eur. J. Pharmacol.*, 66, 65, 1980.

Chapter 11

POSTULATED PHYSIOLOGICAL AND PATHOPHYSIOLOGICAL ROLES ON MOTILITY

Hans-Dieter Allescher and Sultan Ahmad

TABLE OF CONTENTS

I.   Introduction 311

II.   Neurotransmitter-Paracrine Substance-Hormone 311
        A.   Physiological Action — Pathophysiological Action 313
                1.   Molecular Differences 313
                2.   Receptor Location and Receptor Subtypes 314
                3.   Mode of Action 314
        B.   Conclusion 314

III.   Gastrin 315
        A.   Molecular Forms of Gastrin 315
        B.   Gastrin Receptors and Receptor Antagonists 315
        C.   Gastrin Release 316
        D.   Physiological Role 316
        E.   Actions along the Gastrointestinal Tract 316
                1.   Esophagus and LES 316
                2.   Stomach 317
                3.   Pylorus 318
                4.   Small Intestine 318
                5.   Effects on the MMC 318
                6.   Colon and Sphincter Ani Internus 318
                7.   Biliary System and Sphincter of Oddi 318
        F.   Conclusion 318

IV.   Cholecystokinin 319
        A.   Background 319
        B.   Presence and Molecular Form of Endogenous CCK 319
        C.   Release of CCK 320
        D.   CCK Receptors 320
        E.   CCK Receptor Antagonists 320
        F.   General Physiological and Pathophysiological Roles of CCK 321
        G.   Actions along the Gastrointestinal Tract 322
                1.   Esophagus 322
                2.   Stomach 323
                3.   Pylorus 323
                4.   Gastric Emptying 324
                5.   Small Intestine 325
                6.   Colon 327
                7.   Biliary Tract and Sphincter of Oddi 328
        H.   Conclusion 328

V.   Substance P and Related Neurokinins..328
       A.   Precursor Forms...330
       B.   Tissue Localization...331
       C.   Substance P Receptors 333
       D.   Physiological Action of Substance P and Neurokinins in the
             Gastrointestinal Tract..335
       E.   Actions along the Gastrointestinal Tract................................336
             1.   Esophagus and LES...336
             2.   Stomach 336
             3.   Small Intestine 338
             4.   Colon and Sphincter Ani Internus..................................339
             5.   Biliary Tract and Sphincter of Oddi...............................339
       F.   Pathophysiological Role 339
       G.   Conclusion 340

VI.   Opioid Peptides..340
       A.   Sources of Opioids 340
             1.   Intrinsic to Gastrointestinal Tract 340
             2.   Exogenous Opioids — Exorphines....................................342
             3.   Presence of Opioid-Like Peptides in the Gastrointestinal
                   Tract...342
       B.   Opioid Receptors..342
             1.   Multiple Opioid Receptors 342
             2.   Opioid Receptor Location..344
       C.   Mechanisms of Action 344
       D.   General Physiological and Pathophysiological Role 345
       E.   Opioid Actions along the Proximal Gastrointestinal Tract 347
             1.   Esophagus 347
             2.   Stomach 347
             3.   Gastric Emptying 348
             4.   Pylorus...348
             5.   Small Intestine 348
       F.   Opioid Actions on Integrated Gastrointestinal Activities................353
             1.   Role of Endogenous Opioids for Peristaltic Activity 353
             2.   Role of Opioids for the MMC 354
             3.   Gastrointestinal Transit 354
       G.   Opioid Actions on Distal Gastrointestinal Tract and Biliary
             System 354
             1.   Ileocecal Sphincter...354
             2.   Colon and Sphincter Ani Internus..................................355
             3.   Colonic Peristalsis 355
             4.   Action of Opioids on Extraintestinal Sites to Affect
                   Motility 356
             5.   Biliary Tract and Sphincter of Oddi...............................356
       H.   Pathophysiological Action of Opioids on Gastrointestinal Motor
             Function..356
       I.   Conclusion 356

VII.   Motilin..357
       A.   Erythromycin and Motilides 357
       B.   Motilin Receptors 358

C.   Release of Motilin and Its Relation to the MMC......................358
D.   Motilin Actions along the Gastrointestinal Tract.......................361
         1.   Esophagus and LES...361
         2.   Gastric Emptying 361
         3.   Stomach and Small Intestine..................................361
         4.   Colon and Sphincter Ani Internus.............................362
         5.   Gallbladder, Bile Duct, and Sphincter of Oddi 362
E.   Pathophysiological Role...362
F.   Conclusion 362

VIII.   Neurotensin...363
A.   Molecular Forms and Degradation of Neurotensin......................363
B.   Related Peptides..365
C.   Neurotensin Receptors...365
D.   Release of Neurotensin..365
E.   Actions of Neurotensin in the Gastrointestinal Tract 366
         1.   Esophagus and LES...366
         2.   Stomach 366
         3.   Small Intestine...368
         4.   Colon 370
         5.   Biliary Tract and Sphincter of Oddi...........................370
F.   Conclusion 370

References..371

I. INTRODUCTION

In this chapter and the next, we will attempt to estimate the possible role of various neuropeptides and peptide hormones for the physiological and pathophysiological function of gut motility. Before considering evidence about single peptides, some general considerations and definitions are necessary.

II. NEUROTRANSMITTER-PARACRINE SUBSTANCE-HORMONE

The occurrence of several peptides in endocrine cells and in enteric neurons changed the rather strict concept of hormones and neurotransmitters because it could be speculated that the same substance might act as a hormone and, when released from another site, might also act as a neurotransmitter. This makes it difficult to determine a physiological role because higher concentrations are more likely to be persistent in synaptic clefts than in the arterial blood. Thus, the physiological concentrations of an agent acting as a hormone are irrelevant to neural activation on nearby receptors. However, the action of a hormonally released peptide may be at a receptor inaccessible to neurally released peptides, and these receptors might mediate different or even opposite actions.

Classically defined, neurotransmitters are substances which are released from neural structures mediating their effect across a synapse or a nerve muscle junction. Whereas synapses between neural structures can be identified in the myenteric plexus and the submucous plexus, there are no specialized synaptic structures present between motor neurons and smooth muscles. Peptides and neurotransmitters are released from nerve varicosities close to smooth muscle which will then diffuse to receptors on the smooth muscle surface. There is no definite synaptic cleft between nerves and smooth muscle such as the motor end plate present in striated muscle. Similar nerve varicosities also exist within the neural plexus of the intrinsic nervous system without a defined synapse. The functional significance of these nerve varicosities without synaptic relationship to either somas, dendrites, or other synapses remains unclear.

Several requirements have to be fulfilled to demonstrate that a substance is acting as a neurotransmitter:

1. The substance should be present in neurons, and a mechanism for biosynthesis should exist.
2. The substance should be released from neurons in response to an adequate stimulus, and exogenous application of the putative transmitter should mimic the cellular action of the endogenous transmitter when applied in close vicinity to the receptor. (Note that responses such as contraction or relaxation can each be mediated by multiple cellular mechanisms.)
3. Specific blockade of the action of the exogenous agent should block the action of the endogenous transmitter and vice versa.
4. There should be mechanisms to limit the action of the substance by either degradation, reuptake, or molecular modification which will result in an inactive or less active compound. Their manipulation should similarly affect the actions of the endogenous and the exogenous putative mediators.

For some endogenous peptides, such as endogenous opioids and substance P, the neurotransmitter role has been established. For others, such as cholecystokinin (CCK), somatostatin, CGRP, galanin, gastrin-releasing peptide (GRP), vasoactive intestinal peptide (VIP), and many others, definite proof has not yet been achieved.

In contrast to the very localized action of neurotransmitters a more remote action or systemic action is the criterion for paracrine substances or hormones. Paracrine substances are peptides or transmitters released into the interstitium or lumen of an organ from endocrine cells with a local action in the neighborhood. Hormonal substances, in contrast, are substances released into the circulation with a remote and/or systemic action. The criteria for identification of hormone actions as physiological are well known and similar to those for identification of neurotransmitters, with the exception that the relevant concentrations of hormone are those achieved in the arterial blood after a physiological stimulus. The criteria for a physiological role of a paracrine substance are similar to those for hormones except that the relevant concentrations are those achieved in the lumen after physiological stimuli in the interstitium. These are usually much more difficult to measure than blood levels.

The enteric nervous system contains a variety of neurons which are characterized by a specific pattern of neuropeptides present within the neuron,[1] often together with the classical transmitters. This so-called "chemical coding" of neurons also has some functional importance. Are these colocalized peptides released alone or together with other transmitters? Are they costored within the same vesicle and released in a defined relationship, or are they stored in different vesicles? The best current hypothesis suggests that peptides are usually stored in large granular vesicles, while acetylcholine (ACh) is stored in small clear vesicles, and serotonin and norepinephrine may be present in small or large granular vesicles.

There is clear evidence that costored peptides can modify and influence the action of the single peptides.

1. Costored peptides might compete for the same receptor with different affinity, therefore modulating the effect by action as partial and full agonist (e.g., VIP and PHI).

2. Costored peptides might influence the degradation process of each other, either prolonging or shortening their half-lives (e.g., CGRP and substance P).
3. Costored peptides might act synergistically or antagonistically. This could occur at the postsynaptic receptor level by activation of different or identical second messenger systems or by changing the receptor affinities (e.g., ACh and substance P). Coreleased peptides might also act on different postsynaptic structures, e.g., smooth muscle, blood vessels, mast cells, or epithelia.
4. Costored peptides may function as presynaptic modulators of the release of one another or classical mediators.

The main factor which was hampering the research on peptide physiology was (and is) the lack of highly selective, potent, stable receptor antagonists. There are some peptide systems in which specific receptor antagonists are available. Naloxone, which is a potent opioid antagonist, will block nearly all opioid-mediated responses, blocking the three main receptor subtypes with only slightly different affinity. However, if activation of each of two receptor subtypes leads to opposite actions, the blockage of all receptor subtypes would obscure their individual role. Even activation of a single receptor subtype could have an opposite physiological action on receptors at different loci.

Recently two new classes of potent and selective CCK antagonist using either proglumide derivatives or nonpeptide substances have been developed.[2] There are also reputed antagonists for substance P, bombesin, and VIP available, but their affinity and selectivity for their receptors is still rather poor.

A. PHYSIOLOGICAL ACTION — PATHOPHYSIOLOGICAL ACTION

A physiological action on gastrointestinal motility must involve a process which can occur under normal conditions in the control of gut motor function. Peptides or hormones which mediate or modulate this function under normal conditions can be regarded as physiological transmitters or messengers.

Several methods have been used to demonstrate a physiological action of a peptide:

1. Infusion studies mimicking physiological levels of the respective hormone in the plasma or the levels achieved in the plasma in response to a given stimulus
2. Comparisons of the endogenously occurring effect with the effect of exogenously applied mediator (e.g., electrophysiological studies, effect on intracellular second messengers)
3. Removal of the endogenous peptide or hormone by immunoneutralization using antibodies
4. Use of selective and specific receptor antagonists

Due to the specific and complex arrangement of the enteric nervous system and the neuropeptides present within it, several general problems have to be considered in the interpretation of the results obtained.

1. Molecular Differences

Differences of the peptides sequences might be present either due to species differences or due to various molecular forms produced by different molecular processing (enzymatic cleavage of precursors). Different molecular forms present within one species also can be caused by other posttranslational processing mechanisms (e.g., sulfated or nonsulfated forms of CCK). These heterogeneous peptides might be released differently (neurotransmitter-hormone), could have different pharmacokinetics (different metabolizing rate, half-life, accessibility), could have different affinities for various receptor types, or could have different interaction with antibodies.

TABLE 1
Summary of Peptide Effects *In Vivo*, *In Vitro*, and in Isolated Cells

Species	*In vivo*	*In vitro*	Ref.
Gastrin	Neural excitation	No neural effect	3
	No smooth muscle effect	Smooth muscle excitation	4
Bombesin (GRP)	Neural excitation	No neural excitation	3
	No smooth muscle effect	Smooth muscle excitation	
CCK	Neural excitation	No neural effect	5
	Smooth muscle effect	Smooth muscle effect	6
Motilin	Neural excitation	No neural effect	3
	No smooth muscle effect	No smooth muscle effect	
Substance P	Neural excitation	No neural effect	3
	Smooth muscle effect	Smooth muscle effect	7
Opioids	Neural excitation	No neural excitation	8, 9
	Neural inhibition	Neural inhibition	8
	Smooth muscle effect?	No smooth muscle effect	10

2. Receptor Location and Receptor Subtypes

Differences in the receptor location within the body (e.g., the brain and the gut or myenteric ganglia and gut) could cause multiple additive and (including antagonistic) adverse effects. Several subtypes of a peptide receptor might be present (e.g., opioids, tachykinins) which might all be recognized by the natural ligand, but with different affinities and efficacies. Moreover, various receptor subtypes can be located differently and mediate various effects.

3. Mode of Action

The peptide investigated can have different modes of action, e.g., as neurotransmitter, as paracrine transmitter, or as hormone. Even more striking differences can be observed when peptide effects are investigated *in vivo* and *in vitro* or under varying experimental conditions. Some of these discrepancies are listed in Table 1, and their explanations are discussed in Chapter 5.

For example, motilin or gastrin each exerts a contractile effect *in vivo* via a neural response releasing acetylcholine and other excitatory mediators such as opioids. When tested in isolated strips *in vitro*, motilin shows no excitatory effect at all.[3] Similar responses were observed for opioids.[8] When gastrin was tested *in vitro*, it only showed a direct muscle response. However, both motilin and gastrin contract isolated cells.[4]

From these results it is clear that the effect of a given peptide is highly dependent on the study conditions. Whether this difference is due to a different receptor expression *in vivo* or *in vitro* or to the absence *in vitro* of certain modulatory mechanisms (e.g., a tonic inhibitory innervation in the form of a tonic VIP release *in vivo*) remains to be investigated. Also, the discrepancies observed in isolated cells compared to *in vitro* strips and the high sensitivity of isolated cells to many stimuli (often three orders of magnitude) remain unclear. Changes in the microenvironment affecting agonist access, in receptor affinity, in receptor expression, and in methodology used (isometric vs. isotonic unloaded) could account for this. All of these points lead to the question whether a true physiological action can ever be determined *in vitro*. Perhaps *in vitro* studies should focus on cellular and molecular actions of peptides while those *in vivo* may be able to define physiological actions.

B. CONCLUSION

This short discussion elucidates the difficulties for determination of the physiological effect of a peptide on smooth muscle activity. If global functions such as gut transit are considered, matters are more complex. Transit results from integrated activities of the entire gut — not only motor but also secretory activities. Attempts to determine physiological effects of peptides on transit are unlikely to be able to cope with their diverse actions at various sites; also, excitation does not necessarily mean improved propulsion, and contraction inhibition does not always mean decreased function or stasis. The coordinated activity of adjacent segments and of longitudinal and circular muscle layers working against downstream resistances will determine the net effect of the contractile activity. In trying to estimate the physiological role of peptides in the gastrointestinal tract, it also has to be considered that the gastrointestinal system is a complex system with redundant mechanisms which might compensate for the removal of one. Thus, even lack of any clear effect by an antagonist does not provide absolute proof of the absence of a physiological role.

III. GASTRIN

Gastrin is structurally related to CCK and shares a common C-terminal sequence with CCK. In contrast to CCK, which is also present in the intrinsic nervous system and can act as a hormone and/or as a possible neurotransmitter, gastrin is confined to endocrine cells, the G cells. Therefore, gastrin acts by either a hormonal or paracrine mechanism.[15] There is, however, evidence that gastrin might be present in vagal nerve fibers[16] and in the central nervous system (CNS).[17]

In dogs and other mammalian species the majority of G cells are localized in the antrum,[18] whereas in man there are also many G cells in the proximal duodenum.[19,20] Gastrin is produced from a precursor peptide by enzymatic cleavage and, like CCK, several molecular forms can be extracted and measured in the plasma. The antral cells contain mainly the 17-amino-acid form, G-17, whereas the duodenum contains mainly the N-terminally extended form, G-34 (big-gastrin).[21] G-17 and G-34 are of similar potency, but G-17 is degraded more rapidly than the more stable G-34. Consequently, in the fasted state the majority of serum gastrin is G-34 (two thirds of total gastrin), whereas after gastrin release by a meal the amounts of G-17 and G-34 are almost equal.[22-25]

A. MOLECULAR FORMS OF GASTRIN

Gastrin is present in several molecular forms, with G-17 and G-34 being the major forms. Both molecules exist in sulfated and unsulfated forms which show no significant differences in biological activity, in contrast to CCK (see Section IV.B). Recently, the cDNA sequences of porcine and human gastrin precursors were determined, consisting of 104 and 101 residues, respectively.[26-28] This precursor consists of a signal peptide, N-terminal and C-terminal flanking peptides (cryptic peptide A and B, intervening peptides), and the G-34 sequence. Cleavage of the G-34 leads to the N-terminal fragment of G-34 (NT-G-34) and to G-17. This posttranslational processing was demonstrated in ultrastructural studies using different specific antibodies, demonstrating that the precursor peptide occurs in the rough endoplasmic reticulum, Golgi apparatus, and dense-cored secretory granules. In contrast, antibodies to G-17 and G-34, the major active products, predominantly labeled secretory granules, suggesting that the final processing of progastrin occurred after the secretory products had left the Golgi apparatus.[29] Smaller forms of G-17 have been isolated, such as "minigastrin" G-14[30] and the C-terminal hexapeptide.[31] The commonly used pentapeptide pentagastrin and the C-terminal tetrapeptide "tetrin"[17,32,33] still exhibit full biological activity, however, with less potency in some systems.

In additon to G-34 and G-17, peptides with glycine-extended C-terminal forms have been identified,[34] and there is some evidence that these glycine-extended forms of gastrin (gastrin G) are cosecreted with gastrin.[35]

B. GASTRIN RECEPTORS AND RECEPTOR ANTAGONISTS

Gastrin receptors have mainly been studied in the gastric mucosa.[36] Only a few receptor binding studies in smooth muscle have been conducted. Using partially purified plasma membranes of antral muscle, specific [^{125}I]gastrin binding could be demonstrated.[37] Due to the similarity of gastrin to CCK, the relationship of these receptors with those for CCK is unclear. Recent developments of selective CCK antagonists might improve knowledge about the gastrin receptor (see Section IV.E). There is some recent evidence that in addition to proglumide, which acts as a rather weak and not very selective gastrin antagonist, new gastrin antagonists such as L-365,260 might be available.[38] L-365,260, which had no agonistic activity, showed a potent inhibition of gastrin-induced contractions of gastric smooth muscle cells, with a half-maximal inhibition of 28 nM.[38]

C. GASTRIN RELEASE

Gastrin release can be modified by three different mechanisms: (1) via neural factors involving extrinsic and intrinsic neural mechanisms; (2) via hormonal or paracrine mechanisms, e.g., GRP, CCK, GIP, glucagon, secretin, and somatostatin; and (3) via luminal stimuli, e.g., nutrients, pH (see Reference 19). Gastrin release can be stimulated and inhibited by cholinergic mechanisms.[39-41] Other peptidergic neurotransmitters, such as gastrin-releasing peptide, VIP, or galanin, were shown to influence gastrin release. The study of mechanisms controlling release and regulation of synthesis has been facilitated by the isolation and culture of G cells.[42]

Gastrin is potently released by digested small peptides[43] or amino acids (acting from the lumen), by neural pathways (e.g., bombesin), or by circulating factors (e.g., catecholamines). Fat or glucose, on the other hand, is a weak stimulus for gastrin release.[19]

D. PHYSIOLOGICAL ROLE

Besides its potent stimulatory effect on gastric acid secretion, gastrin has been shown to exert a variety of motility effects in the gastrointestinal tract. Gastrin is released in response to a meal and to various other stimuli. Its main physiological effect is to stimulate gastric acid secretion via a direct action on parietal cells. It also has a variety of other gastrointestinal effects including trophic effects on parietal cells. Influences on GI motility have also been speculated to be physiological effects of gastrin. The evidence is usually based on showing a correspondence between levels of gastrin during infusion and those found during a meal-rendered response. Whether these effects described below are physiological is still controversial. While most effects are elicited directly in the gut, gastrin also might have effects on neural sites outside the gut or on the CNS which could modify gastrointestinal motility.[44] A potent, highly selective gastrin receptor antagonist, lacking any action on CCK receptors, would help resolve these controversies.

E. ACTIONS ALONG THE GASTROINTESTINAL TRACT
1. Esophagus and LES

Gastrin G-17, G-34, and the smaller synthetic pentapeptide, pentagastrin, cause an increase of the lower esophageal sphincter pressure (LESP) *in vitro*[45-49] and *in vivo*.[50-56] There has long been controversy about whether the effects of gastrin on the lower esophageal sphincter (LES) are physiological (for review, see References 57 to 59). In early studies, a physiological role of gastrin in the generation and modulation of the LESP was postulated because gastrin increased LESP at concentrations which occur under physiological conditions after meal stimulation[54,55,60] and because immunoneutralization of gastrin using rabbit antiserum decreased the LESP in the opossum *in vivo*.[47] It was noted that the overall response to gastrin was smaller than the contractile response after a meal, suggesting that the effect was not mediated exclusively by gastrin.[60] However, the physiological significance of gastrin for the LESP has been questioned.[59,61] Goyal and McGuigan[62] could not demonstrate any effect of gastrin antiserum on LESP in a careful double-blind study. In another study in dogs, Jennewein et al.[63] observed an increase of LESP only at supraphysiological gastrin G-17 levels, while physiological concentrations had no effect. These results were confirmed by human studies using pentagastrin[64-66] or gastrin G-17 and G-34[67] infusion. There was also no correlation between serum gastrin levels and LESP in patients with hypergastrinemia.[68-70] Therefore, a contributing role of gastrin to the resting LESP seems unlikely even though postprandial gastrin concentrations might be able to influence LESP under certain conditions.

Patients with hiatus hernia showed decreased responses of the LES to pentagastrin,[71] whereas in patients suffering from achalasia and diffuse esophageal spasms the results are contradictory (see Reference 59).

2. Stomach

The effect of gastrin on the stomach is dependent on the region studied (fundus-corpus-antrum-pylorus), on the species used, and on whether fasted or postprandial conditions are studied. The overall effect observed *in vivo* and *in vitro* is an excitatory one in the corpus or antrum. *In vitro* it is mediated on circular muscle by a direct muscle effect,[4,12,72] but *in vivo* there is also a response at lower doses mediated by the release of acetylcholine.[3,4,73,74] *In vitro*, longitudinal muscle also shows a neural as well as a direct response.[75]

In vitro an excitatory effect of gastrin or pentagastrin in the stomach was found in rats,[76] guinea pigs,[77,78] cats,[79] dogs,[3,75,80-83] and humans.[45,86] This excitatory effect *in vitro* was mediated mainly by a direct smooth muscle action[3,4,46,48,76,79] in circular muscle, but there was also evidence for a nerve-mediated response of longitudinal muscle *in vitro*.[75,81] In the guinea pig fundus, gastrin induced an increase in gastric pressure which could not be blocked by either atropine or tetrodotoxin (TTX), but was antagonized by the H_1 receptor antagonist mepyramine,[77,78] suggesting the release of histamine from gastric mast cells.

In vivo, gastrin increased the tone and/or frequency of gastric circular muscle contractions of dogs[73,85,86] and humans.[87] The excitatory effect *in vivo* was reduced or blocked by atropine.[4,85] The discrepancies of the gastrin effects in *in vivo, in vitro,* and isolated cell systems have been mentioned.[3,4] Antral muscle *in vitro* and isolated smooth muscle cells contracted in response to pentagastrin,[4] and gastrin receptors have been demonstrated on smooth muscle membranes of the canine antrum,[37] even though intraarterial gastrin had no effect on antral contractions after administration of TTX or atropine.[4] The reason why gastrin acts mainly via neural mechanisms *in vivo* and via direct smooth muscle *in vitro* is unknown, but this is common in actions of several peptides in the gastrointestinal tract.[3,4]

In vivo inhibitory effects of gastrin on phasic activity in the proximal stomach were demonstrated in the canine[88,89] and porcine[90] stomach.

Despite its usual stimulatory function on antral motor activity, gastrin, like CCK, delayed gastric emptying of liquid and solids.[91-93] Intragastric pressure, which is one major determinant of liquid gastric emptying, is regulated by the ability of the fundus to accommodate. In the dog, pentagastrin inhibited phasic contractions in the proximal stomach,[88] leading to an increased gastric accommodation and, consequently, to a decrease of intragastric pressure. While inhibiting the proximal stomach, gastrin increases the frequency of antral slow waves.[88,94] It is unclear what action of gastrin on motor activity of the stomach accounts for the delay in emptying; the best hypothesis is that this is due to the effect on the fundus. In contrast to its actions in healthy volunteers, gastrin was reported to increase gastric emptying in gastric and duodenal ulcer patients.[95,96]

The physiological significance of the stimulatory effect of gastrin on the distal stomach was suggested by the fact that the ED_{50} concentrations of gastrin G-17 which caused antral excitation and gastric acid secretion (a known physiological action) were almost identical in the dog.[83,85] Furthermore, the antral region seems to be the most sensitive region when compared with other parts of the gastrointestinal tract.[59]

3. Pylorus

The effects of gastrin on pyloric motor function are still unclear. Whereas some studies reported no effect or an inhibitory effect of exogenous and endogenous gastrin,[46,97,98] others reported an excitatory effect.[99-102] These controversial results probably reflect varied and inadequate techniques for recording pyloric activity.

4. Small Intestine

In the small intestine, as in the stomach, gastrin was found in most studies to be mainly excitatory either by direct muscle action (e.g., in the cat) or via release of acetylcholine from intrinsic nerves.[103,104] Vizi et al.[104] reported that gastrin caused neurally dependent, reciprocal responses of excitation and inhibition occurring sequentially in the longitudinal and circular muscle. It was speculated that gastrin and CCK might release substance P in addition to acetylcholine in the guinea pig ileum because the excitatory responses were reduced by substance P desensitization or TTX as well as by atropine.[105]

5. Effects on the MMC

When given in the fasted state, gastrin interrupted the MMC and replaced it by a digestive-type pattern of intermittent spiking and contraction. The disruption lasted as long as the infusion was continued,[66,88,106,107] and the effect could be antagonized by proglumide.[108] Gastrin released endogenously by irrigation of the antrum with acetylcholine inhibited the interdigestive motility pattern both in intact and in an extrinsically denervated autotransplanted fundic pouch.[109] However, the gastrin levels accompanying these responses were well above normal physiological postprandial levels.

In dogs the MMC was inhibited by ingestion of meals such as arachis oil and medium-chain triglycerides, which did not cause an increase of gastrin levels. Furthermore, using other meals such as milk or sucrose, the first MMC complexes reappeared before the gastrin levels had returned to basal levels.[110] These observations suggest that elevation of gastrin levels may be sufficient but not necessary for the postprandial interruption of the MMC.

6. Colon and Sphincter Ani Internus

Gastrin[111,112] and pentagastrin[111,113,114] have a stimulatory effect on the colonic motility in humans. Even though a possible direct effect on human colonic smooth muscle has been proposed,[115] the participation of a neural pathway is more important.[116] This is supported by the fact that, in patients suffering from megacolon caused by Chagas' disease, gastrin has no stimulatory effect on the dilated segment.[117] In the cat, pentagastrin caused a dose-dependent contraction which was completely blocked by TTX and hexamethonium.[118] Atropine abolished the contractile response in the distal colon, but caused only a partial blockade in the proximal colon. This remaining excitatory effect could be abolished by the addition of the opiate antagonist naloxone, suggesting a possible involvement of endogenous opiates.[118]

7. Biliary System and Sphincter of Oddi

Intravenous infusion of gastrin (0.01 to 0.1 μg/kg) in humans increased phasic and tonic activity in the sphincter of Oddi.[119,120] A similar response was demonstrated in the opossum together with a decreased bile flow.[121]

F. CONCLUSION

Gastrin is a potent stimulator of gastric acid secretion. Its role in the regulation of gastrointestinal motility is far less characterized. Excessive secretion of gastrin can be observed due to a gastrinoma, a gastrin-producing tumor, or as a consequence of reduced acid secretion caused by Billroth II surgery, atrophic gastritis, or drugs (H_2 blockers or omeprazol). No specific motility changes in these hypergastrinemia conditions have been reported, and the diarrhea observed in gastrinoma patients probably is primarily due to the changes in the secretory responses and, to a minor extent, to motility changes. The possible physiological role in the regulation of antral motility and in the maintenance of the basal lower esophageal sphincter pressure awaits further confirmation with new selective gastrin receptor antagonists.

IV. CHOLECYSTOKININ

A. BACKGROUND

Cholecystokinin was named by Ivy and Oldberg after a proposed hormonal factor which caused gallbladder contraction,[122] but it took almost 50 years until the peptide was purified from the mucosa of the porcine duodenum.[123-125] Independently, Erspamer and Melchiorri[126] reported the isolation of a peptide named caerulein, closely resembling CCK, from the skin of the tree frog *Hyla caerulea*. CCK and gastrin and their amphibian analogue, caerulein, are related peptides which share a common C-terminal pentapeptide (R–Gly–Trp–Met–Asp–Phe–NH_2) which also constitutes the biologically active sequence in these peptides.

B. PRESENCE AND MOLECULAR FORM OF ENDOGENOUS CCK

Like many other peptides, CCK is derived from a precursor peptide from which the biologically active forms are produced by enzymatic cleavage and posttranslational processing. The cDNA of preprocholecystokinin was first isolated from a rat medullary thyroid carcinoma.[127] Molecular cloning of the human and porcine preprocholecystokinin revealed a molecule of 115 and 114 amino acid residues, respectively.[128,129] Porcine preprocholecystokinin is composed of a C-terminal signal peptide and pro-CCK, consisting of a spacer peptide (20—43), CCK-58(44—102), and a C-terminal fragment (102—114).[130] Southern blot hyridization suggested that CCK is encoded by a single copy gene and, therefore, that the same gene is expressed in the brain and the gut.[129] Molecular cloning of the cDNA from porcine brain and gut did reveal identical precursor forms.[128] This suggests that the differential distribution and presence of various molecular CCK-forms must be due to differential posttranslational processing.[128]

Several molecular forms have been demonstrated to occur in the intestine *in vivo* or *in vitro*, including CCK-83,[131] CCK-58,[132] CCK-39,[133] CCK-33,[123] CCK-22,[134,135] CCK-12,[17,136,137] and CCK-8.[136] These larger forms can be processed further to CCK-7, CCK-5,[138] and CCK-4.[17,137] There is some evidence that C-terminal extensions of the CCK (amino acid residues 102 to 105, glycine-extended CCK) might exist and might be present in the gut. The full biological information seems to be contained in the C-terminal tetrapeptide-amide which is identical for gastrin and CCK, whereas the N-terminal portion determines receptor affinity and degradation kinetics of the molecular forms of CCK and gastrin peptides.[137]

In mongrel dogs, the two main molecular forms released in response to a peptone meal were characterized as CCK-33 and CCK-8.[139]

Gastrin and cholecystokinin are present in the CNS[137,140,141] and have been shown to occur in the extrinsic vagal nerve fibers.[16,142] In contrast to gastrin, which is only present in endocrine cells in the gut, CCK is also present within intrinsic myenteric neurons.[143] Therefore, endogenous CCK may have dual roles, one as a hormone which is released systemically and one as a possible neurotransmitter within the myenteric nervous system. Endocrine cells containing CCK are located mainly in the mucosa of the duodenum and upper jejunum.[133,144-146] Since most CCK antibodies cross-react with the different molecular forms of CCK, a clear distinction of the distribution of the various molecular forms of CCK in the myenteric neurons was not possible so far.

TABLE 2
Suggested CCK Receptor Subclassification

Receptor-type	Rank order	Antagonist
"A"	sCCK-8 >>> gastrin, CCK-8	L-364,718, CR-1409
"B"	sCCK-8 (10 ×) > gastrin, CCK-8	L-365,260
Gastrin	sCCK-8 = gastrin, CCK-8	L-365,260

C. RELEASE OF CCK

CCK is released postprandially from endocrine cells of the upper gut, with fat and protein being the most potent stimuli.[147-149] However, CCK release was also demonstrated in response to phenylalanine,[150-152] glucose,[153] and acid.[154] Isolation and short-term culture of CCK-containing endocrine cells revealed that these cells contain and release mainly CCK-8 and CCK-33/39 in response to stimulation with KCl, forskolin, dibutyl-cAMP, and L-tryptophan.[155] Neural transmitters, such as GRP or neuromedin B and C, released CCK in an isolated perfused rat duodenum, suggesting that the release of CCK is also regulated by neural mechanisms.[156]

D. CCK RECEPTORS

In analogy with the structural similarity of gastrin and CCK, both peptides show affinity for the same receptor, but with different affinities in various tissues and locations. These differences in affinity depend on structural properties of the N-terminal extensions of the common pentapeptide. According to their relative affinities to various ligands, these receptors are termed gastrin receptors (e.g., on parietal cells) or CCK receptors (e.g., on pancreas acinar cells) (for review see References 157 and 158). There is considerable evidence that two classes of CCK receptors are present.[159] The first type, mainly present in pancreatic acini and peripheral tissues, was named CCK_A (A = alimentary); the other, mainly present in the central nervous system, was named the CCK_B (B = brain) receptor.[160] This definition of CCK receptors was primarily based on the fact that sulfated CCK (sCCK) is about three to four orders of magnitude more potent on CCK_A receptors than nonsulfated CCK-8 and CCK-4, but all three peptides have affinities almost equal to the CCK_B receptor. Newly developed CCK antagonists such as CR-1409 and L-364,718 (see Section IV.E) show high affinity for CCK receptors, but are rather poor ligands for CCK_B receptors.

Recent studies suggest that the CCK_B receptor also occurs in peripheral tissue and that the CCK_A receptor also occurs in the brain.[161] The two receptors and the gastrin receptors are defined by a characteristic rank order of potencies (see Table 2).

E. CCK RECEPTOR ANTAGONISTS

In the last 20 years four different groups of CCK receptor antagonists have been described:[162-164]

1. Cyclic nucleotides
2. Amino acid derivatives
3. Peptide derivatives of CCK
4. Nonpeptide compounds

The "first generation" of these antagonists consisted of derivatives of the cyclic nucleotides, such as dibutyryl-cGMP,[165] and derivatives of amino acids, such as proglumide (as a derivative of glutamic acid) and benzotript (as a derivative of tryptophan).[166] The usefulness of these substances was limited because of their low affinity and/or selectivity for CCK-receptors in the periphery or CNS and their interaction with the gastrin receptor.[164] As a

TABLE 3
CCK Antagonists — Comparison of Inhibition of CCK-Induced Smooth Muscle
Effects and [^{125}I]CCK Binding in Guinea Pig Brain[163,164]

	Smooth muscle contraction pA$_2$	[^{125}I]CCK binding	
		Pancreas IC$_{50}$ (nM)	Brain IC$_{50}$
Proglumide	3.3[a]	250,000	800,000
Aspercilin	4.6[a]	184	>100,000
CR-1409	7.2[a]	18	2,230
L-364,718	9.9[b]	0.08	245

[a] Guinea pig gallbladder.
[b] Guinea pig ileum and colon.

Data taken from Rovati, L. C. and Makovec, F., in *Cholecystokinin Antagonists,* Alan R. Liss, New York, 1988; and Chang, R. S. L. and Lotti, V. J., in *Cholecystokinin Antagonists,* Alan R. Liss, New York, 1988. With permission.

further group of CCK antagonists, peptide derivatives of the C-terminal or N-terminal region of CCK, including the C-terminal heptapeptide CCK-27-32-NH$_2$, were described.[167-169]

Systemic modification of glutaramic and aspartic derivatives led to the discovery of a whole group of new potent and selective CCK antagonists. Two of these substances, CR-1409 (lorglumide) — D,L-4-(3,4-dichlorobenzoylamino)-5-(dipentylamino)-5-oxo-pentaoic acid — and CR-1505 (loxiglumide) — D,L-4-(3,4-dichlorobenzoylamino)-5-(N-3-methoxypropylpentilamino)-5-oxo-pentanoic acid — were characterized by rather high affinity and selectivity for the peripheral CCK receptor[170,171] (see Table 3).

The isolation of the microbiological substance aspercillin,[172] which showed antagonistic activity at the CCK receptor and the systematic modification of the benzodiazepine structure within this molecule,[173] led to the synthesis of a new nonpeptidal CCK antagonist, L-364,718, [3S(−)-N-(2,3-dihydro-1-methyl-2-ox-5-phenyl-1H-1,4-benzodiazepine-3-yl)-1H-indole-2-carboxamide]. This substance showed a high potency to block peripheral CCK receptors *in vitro* and *in vivo*. In addition to L-364,718, which has a 55-fold higher affinity for CCK$_A$ receptors than for CCK$_B$ receptors, a related compound (L-365,260) was shown to exhibit a 90-fold higher affinity for gastrin and CCK$_B$ receptors than for CCK$_A$ receptors.[174]

These new CCK antagonists blocked CCK-induced effects *in vitro*, such as contraction of gallbladder,[170] ileum, and colon[175] and pancreatic secretion,[170,176] and *in vivo*, such as delay of gastric emptying and bile secretion, contraction of gallbladder, and contraction of antrum, pylorus, and duodenum (see Table 4). When tested on gastrointestinal function without application of exogenous CCK, L-364,718 accelerated gastric emptying of a solid meal in cats,[177] but had no effect on oleate-induced inhibition of gastric emptying in dogs.[178] CR-1505 (loxiglumide) increased gastric emptying in healthy volunteers[179,180] and decreased colonic transit time, but had no influence on small intestinal transit.[179]

L-364,718 or CR-1409 and analogues had no influence on gastric emptying of saline in rats, guinea pigs, and dogs; they failed to antagonize the delay of gastric emptying in response to amino acids,[176] hyperosmolar solutions, and acid[181] in rats, to oleic acid in dogs,[178] and the pyloric motor response to intraduodenal acid.[5]

F. GENERAL PHYSIOLOGICAL AND PATHOPHYSIOLOGICAL ROLES OF CCK

The postprandial release of CCK suggested a variety of physiological functions concomitant with responses to food, e.g., changes of secretion (see Chapter 13) and motility, such as gallbladder contraction,[182,183] inhibition of the sphincter of Oddi,[119] contraction of

TABLE 4
Effects of New Potent CCK Antagonists on Gastrointestinal Motility *In Vivo*

Species	CCK antagonist	Effect	Ref.
		Stomach, gastric emptying (GE), small intestine, colon	
Rat	Proglumide	Blocked CCK-induced, motility in stomach and duodenum; increased L-GE	188
Rat	L-364,718	Blocked CCK and peptone effect on L-GE; no effect on L-GE of saline, acid, osmotic	181
Rat	L-364,718, asperci-lin	Reduced CCK effect on L-GE	176
Rat	L-364,718	Reduced CCK effect on L-GE and food intake; no effect on GE of amino acids	176
Cat	L-364,718	Increased S-GE of a meal	177
Dog	L-364,718	Reduced CCK effect on L-GE; no effect on GE of amino acids	176
Dog	L-364,718	No effect of L-GE on sodium oleate; decrease of bile outflow	178
Dog	CR-1392	Reduced CCK-effect on motility; no effect on acid-induced pyloric motility	5
Dog	L-364,718	No effect on MMC; reduced postprandial spiking activity; blocked CCK-induced "fed" motility pattern	189
Dog	CR-1505	Reduced phasic and tonic activity of colon; phase II activity in small intestine and motor response to feeding	190 191
Rabbit	L-364,718	Reduced CCK-induced contractions of colon	176
Human	CR-1505	Increased L-GE and S-GE	180
Human	CR-1505	Increased S-GE and transit time in colon; no effect on small intestinal transit	179
		Gallbladder and bile flow	
Guinea pig	CR-1409, CR-1392	Reduced CCK-induced contractions of the gallbladder	170
Cat	L-364,718	Reduced CCK contractions of the gallbladder	176
Human	L-364,718	Reduced CCK and meal-induced gallbladder contractions	192
Human	CR-1505	Increased fasting gallbladder volume	193

Note: GE, gastric emptying; S, solids; L, liquids.

the pylorus,[184] changes of gastric motility and relaxation of the gastric fundus,[185] retardation of gastric emptying in response to various nutrients,[186] and satiety.[187] These functions were speculated because of evidence that infusion studies mimicking postprandial CCK-levels or CCK levels producing submaximal to maximal stimulation of pancreatic secretion produced these responses. This is a necessary but insufficient criterion for a hormonal action, but may be irrelevant to determination of neuronal actions. However, the recent development of potent and specific CCK antagonists makes it possible to block endogenous hormonal or neural CCK specifically and to characterize the role of endogenous CCK on gastrointestinal motility (see Table 4).

Further studies using these selective antagonists are needed to determine the physiological roles of endogenous CCK as a neural substance and as a circulating hormone for gastrointestinal function.

G. ACTIONS ALONG THE GASTROINTESTINAL TRACT
1. Esophagus

In most studies *in vivo,* CCK applied as a bolus was shown to decrease the lower esophageal sphincter pressure in the cat[194,195] and in humans.[196-198] In the cat LES, CCK caused a relaxation by stimulating noncholinergic, nonadrenergic inhibitory neurons; i.e., the inhibitory response was blocked by TTX, but not by cholinergic or adrenergic blockers.[194] After neural blockade, CCK induced an excitatory response by a direct action on the smooth muscle. Rattan and Goyal[195] confirmed these results and demonstrated that sulfated and desulfated CCK-8, G-17, and CCK-4 (i.v. or intra-arterially [i.a.]) showed different potencies at the neuronal receptor and at the smooth muscle receptor, suggesting the presence of a CCK_A receptor on nerves and a CCK_B receptor on the smooth muscle.[195] In contrast, in the opossum, CCK-8 (400 ng/kg i.v.) not only contracted the longitudinal and circular smooth muscle of the esophageal body, but also increased the LESP (measured by a Dent sleeve) and the intragastric pressure.[199] In this study the limitations of side hole recording, which was mainly used in other studies, for studying sphincter responses were demonstrated. The opossum LES also showed an excitatory effect to caerulein; this effect was enhanced after neural blockade using TTX, suggesting that an additional action of caerulein on inhibitory neurons had been prevented.[58] In the opossum LES, the contractile effect of CCK was locally partially antagonized by TTX, but not by atropine, phentolamine, or pyrilamine.[199] A contractile response to CCK was also found in LES strips *in vitro*.[198] However, this response was blocked by atropine and TTX. Using a Dent sleeve sensor in humans, the depressant effect of CCK-8 on LESP reported in earlier studies[196-198] was confirmed.[200,201] It has been speculated that CCK might act only as a partial agonist compared with gastrin at the LES, thus potentially antagonizing the gastrin effect.[198,202] This was not supported by comparison of the actions of various molecular forms of CCK and gastrin in the cat LES.[195]

2. Stomach

In vitro, CCK showed an excitatory response on the corpus and the antrum of stomach of the guinea pig,[77,78,203] rat,[204] cat,[205] dog,[81,206] and human.[207] A similar effect was found for caerulein.[208]

These excitatory effects were mediated by a direct action on smooth muscle[205] and/or were reduced for atropine and TTX.[77,78] The presence of smooth muscle receptors for CCK was supported by the fact that CCK-8 contracted isolated smooth muscle cells from the guinea pig stomach.[209] In the guinea pig stomach, CCK-8 contracted muscle strips from the lesser curvature by a partly cholinergic mechanism, but inhibited strips from the greater curvature by a TTX-sensitive, nonadrenergic, noncholinergic mechanism, suggesting regional differences in responses to CCK within the stomach.[210]

When tested *in vivo*, CCK had a potent stimulatory effect on the antrum[5,94,211-214] and pylorus (see Section IV.G.3), whereas it apparently inhibited the proximal stomach.[215]

There has been some controversy about whether the excitatory effect of CCK is mediated by the release of a neural transmitter such as acetylcholine[214,216] or via a direct action on the smooth muscle as shown in *in vitro* experiments.[6,204,206] In recent studies in the dog, CCK given close intra-arterially to the antrum, duodenum, and pylorus caused excitation via a neural release of acetylcholine as well as via a TTX- and atropine-insensitive, presumably direct muscular action.[5] Atropine shifted the dose response curve to the right, and higher amounts of CCK were needed after administration of atropine and TTX to cause excitation, suggesting that the neural receptors might be more sensitive or more accessible to CCK and, therefore, could constitute the physiological site of action *in vivo*.

3. Pylorus

CCK has been shown in several studies to cause a contraction of the pyloric muscle *in vivo*[5,97,102,184,216,217] and *in vitro*.[6,97,204] Only the feline pylorus was reported to relax in response to CCK and to contract in response to gastrin.[218] The CCK response in the rat pylorus was partly reduced by α-adrenergic blockers, suggesting an activation of adrenergic mechanisms by CCK.[204]

In the canine pylorus the excitatory response was mediated by the release of acetylcholine

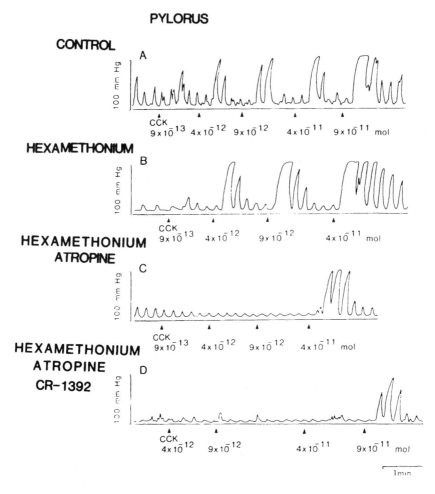

FIGURE 1.   Effect of CCK-OP on canine pyloric motility and the effect of hexamethonium, atropine, and the CCK antagonist CR-1392 on the CCK-induced contractions. Hexamethonium had no influence on the dose response to CCK, whereas atropine and CR-1392 cause a rightward shift of the dose-response curve of CCK. (From Allescher, H.-D., Daniel, E. E., Rovati, L. A., Kostolanska, F., and Fox, J. E. T., *J. Pharmacol. Exp. Ther.*, 25, 1134, 1989. With permission).

and by a direct TTX-insensitive effect, and both responses could be antagonized by the specific receptor antagonist CR-1392[5] (see Figure 1). However, the CCK-receptor antagonist did not reduce the motor response of the pylorus to intraduodenal acid infusion,[5] suggesting that endogenous CCK is not involved in the neural reflex response of the pylorus to small amounts of intraluminal acid in the duodenum.[5]

4. Gastric Emptying

CCK-8 is a potent inhibitor of liquid and solid gastric emptying in man,[187,219-222] dog,[186,223] and rat.[224,225] Yamagishi and Debas[223] reported that pyloroplasty and antrectomy prevented the inhibitory action of physiological doses of CCK-OP on gastric emptying. The inhibition of gastric emptying was also abolished by complete vagotomy, suggesting that CCK inhibits gastric emptying through a vagally dependent mechanism acting on the pylorus and stomach.[223] However, vagotomy also could influence motility of the duodenum. In elegant studies in conscious dogs, the delay of gastric emptying was attributed to a decrease of the contractile force of antral contractions, a reduced opening of the pylorus, and decreased contraction force and frequency in the duodenum. CCK further decreased the duodenal lumen and changed the contractile pattern from propulsive to segmenting activity.[226] Besides the effect on pylorus and antrum, CCK-OP or its analogues, such as caerulein, increase duodenal pressure and decrease intragastric pressure.[227] In rats this inhibitory effect on gastric pressure is also dependent on integrity of vagal innervation[228] and is reduced by guanethidine, phentolamine, hexamethonium, or vagotomy.

Interestingly, the action of CCK on gastric emptying and on pyloric CCK receptors was suggested as a possible mechanism whereby CCK could exert its effect on satiety.[187] Correlation of gastric emptying and satiety effects of CCK supported the view that lower doses of CCK might act through this mechanism.[229] Autoradiographic studies have demonstrated a high density of CCK-binding sites in the circular muscle of the pyloric sphincter. However, the method used could not distinguish a neural from a muscular location.[230] There is evidence that CCK might have a direct excitatory effect on gastric afferent vagal fibers arising from gastric mechanoreceptors.[231] CCK receptors are present in vagal fibers and are transported from the nodose ganglion to the periphery, suggesting their presence on vagal afferent fibers.[232-234] Therefore, a direct action of hormonally released CCK on CCK receptors on afferent fibers from gastric mechanoreceptors running in the vagus could provide a pathway for the peripheral action of CCK to induce satiety. This would be consistent with the dependency of the response on vagal integrity, with the role of CCK_A-type receptors in this response, and with the inhibitory effect of pylorectomy.

5. Small Intestine

In vitro CCK had an excitatory effect on the guinea pig ileum, and this effect was dependent on the presence of nervous structures since it was abolished by atropine and tetrodotoxin.[235,236] Also, longitudinal muscle without myenteric plexus did not respond to CCK.[237] This excitatory effect of CCK is probably due to a release of acetylcholine from the myenteric plexus.[103,238,239] The excitatory CCK effect was not completely atropine-sensitive, and the remaining effect could be blocked with the substance P antagonist D-Pro[2]-D-Trp[7,9]-substance P, suggesting the release of substance P by CCK. Opioids, peptides, and morphine inhibited the CCK-induced effects, probably by inhibiting the release of the neural mediators.[240] In electrophysiological studies in the guinea pig myenteric plexus,[241] CCK-8 caused depolarization in a subgroup (both AH/type 2 and S/type 1 neurons) of the recorded neurons leading to an increased excitability. However, there were some neurons (mainly AH/type 2) which were hyperpolarized by CCK-8, indicating that CCK-8 also might have an inhibitory action in the myenteric plexus.[241] CCK-8 simulated the slow postsynaptic excitatory potentials (slow EPSP) in the myenteric plexus of the guinea-pig small intestine.[242,243]

The participation of CCK in the peristaltic activity was hypothesized from the fact that CCK-IR is released into the vascular system of the isolated perfused guinea pig small intestine in measurable amounts and that the release increased during peristaltic activity evoked by increased intraluminal pressure.[244] In earlier studies in the guinea pig, the peristaltic reflex was reduced by dibutyryl cyclic GMP at concentrations antagonizing CCK effects at its receptor.[245] On the other hand, caerulein and CCK initiated peristaltic activity and increased the efficiency of the peristaltic reflex.[246-248]

There is also some evidence that CCK might have an inhibitory effect on small intestinal motor activity *in vivo*, especially of the proximal duodenum. This inhibition of motor activity was shown in the cat,[249] in dogs,[225] and in humans.[250-252]

In the distal small intestine, CCK and caerulein were stimulatory and decreased the small bowel transit time.[202,235,252] Figure 2 shows an example of its excitatory effects in the jejunum. A possible therapeutic action of CCK for treatment of paralytic ileus and other motor disturbances was speculated.[202]

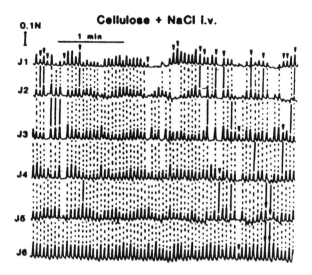

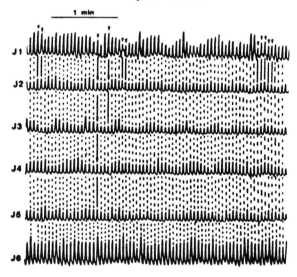

FIGURE 2.   Influence of CCK-OP i.v. on canine small intestinal motility and propagation of a cellulose meal. CCK-OP increases the number of propagated waves indicated by the dashed lines. (From Schemann, M., Siegle, M. L., Sahyoun, H., and Ehrlein, H. J., *Z. Gastroenterol.*, 24, 262, 1986. With permission.)

Besides the action via the circulation or as a neurotransmitter, CCK also might act via luminal release, since intraluminal CCK was shown to stimulate the electrical activity in the rabbit ileum.[253] To date, there has been no follow-up on this interesting observation.

In the canine small intestine *in vivo* neither of the selective antagonists CR-1409 or L364,718 had any significant influence on the interdigestive motor pattern, suggesting that this motility phenomenon is independent of endogenous CCK.[189] Both antagonists blocked the "fed-type" motility pattern induced by exogenous CCK, gastrin, or bombesin infusion and restored the interdigestive motility pattern (see Figure 3). When tested after ingestion of a meal, both antagonists reduced postprandial spiking activity, but neither restored the

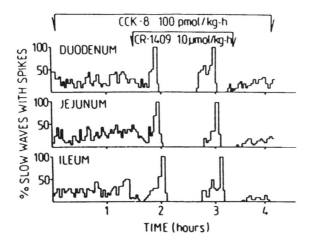

FIGURE 3. Effect of CCK and the CCK antagonist on the interdigestive motility in the canine small intestine. The CCK antagonist CR-1409 (10 μmol/kg/h) blocked the CCK-induced fed-motility pattern and restored the interdigestive pattern. (From Thor, P., Laskiewicz, J., Konturek, P., and Konturek, S. J., *Am. J. Physiol.*, 255, G498, 1988. With permission.)

fasted pattern.[189] From these *in vivo* data several conclusions can be drawn. First, endogenous CCK and gastrin are not major factors in the interruption of the interdigestive motility. Exogenous CCK probably acts on peripheral CCK_A-type receptors to induce the fed-type motility pattern.[189,254] Endogenous CCK could be acting directly or by stimulating the release of other peptides or transmitters to regulate the postprandial motor activity. Similar effects on canine small intestine motor function in dogs using CR-1505 (loxiglumide) were reported in preliminary form by another group.[191]

In contrast to these findings on electrical activity of small bowel in the dog, no significant effect of CR-1505 (loxiglumide) was found on small bowel transit in healthy volunteers.[179] The results in humans suggest that endogenous CCK does not play a major role in small bowel function in man. However, the test used (hydrogen breath test) might be too insensitive to detect changes in the motility pattern. Furthermore, the importance of CCK could be rather variable in response to the composition of the applied meal, as suggested by release studies. Further studies with specific CCK-antagonists using different test meals, especially with high fat content and more elaborate techniques for monitoring gut function, are needed in humans and in laboratory animals to establish the precise role of endogenous CCK in the postprandial state.

6. Colon

In vitro, human colonic muscle responded to caerulein and CCK with an excitatory response which was mainly neurally mediated. A small direct smooth muscle action seemed to be present.[208,255]

Infusion of CCK increased the spike activity in the human rectosigmoid colon.[113,115] A similar effect was found for caerulein in the guinea pig colon, and the excitatory response was almost abolished by atropine, suggesting a mainly nerve-dependent action.[255] The intestinal phase of the gastrocolonic response seemed to be mediated via chemoreceptors and among various stimuli fat seemed to be the major stimulant.[256,257] Intraduodenal application of amino acids and sodium oleate was shown to increase the motor activity of the sigmoid colon of normal subjects.[258] Because fat is also a potent releaser of CCK,[153] it was suggested that CCK may play a role in the modulation of postprandial colonic motility and in the gastrocolonic reflex.[202] This requires evaluation with selective CCK antagonists.

7. Biliary Tract and Sphincter of Oddi

CCK was named for its potent contractile effect on the gallbladder,[123,259] and the Ivy dog unit (equivalent to about $^1/_3$ μg of CCK-33[123]), formerly used to describe the potency of CCK extract, was defined as the dose of CCK producing a maximal contraction of the canine gallbladder. CCK-8, CCK-33, and CCK-58 seem to have similar affinities in receptor binding studies and similar excitatory effects on gallbladder smooth muscle *in vitro*.[260,261] In functional studies, the effect of CCK on gallbladder contraction in the guinea pig was mediated by release of acetylcholine[262] and by a direct smooth muscle effect which might involve two different CCK-receptor subtypes.[263] The CCK receptor present in the guinea pig gallbladder was reported to be of the CCK$_A$ type because it resembles the pancreatic CCK receptors and shows similar binding characteristics.[264,265] Techniques used to prepare nerve and muscle membranes and study CCK receptor binding have not been applied in the gallbladder; thus, revision of these findings may have to be made.

Besides eliciting a gallbladder contraction and subsequently producing increased bile flow, CCK also relaxes the sphincter of Oddi *in vivo*.[119,259] The inhibitory effect on the sphincter of Oddi (see Figure 4) is mediated through an inhibitory neuronal pathway, and CCK produces a contraction of the sphincter of Oddi after neuronal blockade.[259] This is analogous to the dual inhibitory and excitatory effect described for the LES.[195] In humans a paradoxical excitatory effect of CCK on sphincter of Oddi pressure *in vivo* has been suggested as a possible pathogenetic mechanism for symptoms associated with sphincter of Oddi dyskinesia.[266] The CCK antagonist L-364,718 or CR-1505 reduced the postprandial bile outflow in dogs[178] and humans[267] and the postprandial gallbladder contraction in humans,[191,192] suggesting a participation of endogenous CCK in the postprandial gallbladder contraction and regulation of bile output. On the other hand, bile salts have been shown to inhibit CCK release and might serve as a feedback regulator for CCK release.[268] Interestingly, the application of the antagonist loxiglumide increased the volume of the gallbladder in human volunteers even in the fasted state.[191] Kellow et al.[269] reported that the gallbladder was more sensitive to CCK in patients with irritable bowel.

A possible participation of decreased responsiveness of the gallbladder to CCK in the pathogenesis of gallstone formation was speculated, but no definite proof has been obtained that these changes are not secondary.[270]

H. CONCLUSION

CCK and related peptides are present in both endocrine cells and in nerves. CCK receptors are of two types; CCK$_A$ receptors are predominant in the gastrointestnal tract, and CCK$_B$ receptors are predominant in the central nervous system. Receptor antagonists of high potency and selectivity are available for CCK$_A$ receptors. These agents block nearly all peripheral hormonal/neural actions of CCK and its analogues. Their actions against effects of endogenous CCK are not fully clarified as yet, but the current data suggest CCK involvement in gastric emptying and in gallbladder contraction and sphincter of Oddi relaxation. Availability of these antagonists should quickly clarify a number of proposed physiological roles of CCK. In considering roles of neurally released CCK, it should be kept in mind that evidence that an antagonist blocks systemic CCK effects, but not an effect due to putative release of CCK from nerves, is not conclusive. This is because high concentrations of CCK may be released into the cleft between a nerve and its target organ, compared to those obtained when the substance diffuses from the bloodstream.

V. SUBSTANCE P AND RELATED NEUROKININS

In 1930 von Euler and Gaddum[271] described a compound isolated from the equine intestine and brain which contracted the atropinized rabbit jejunum. This compound, which

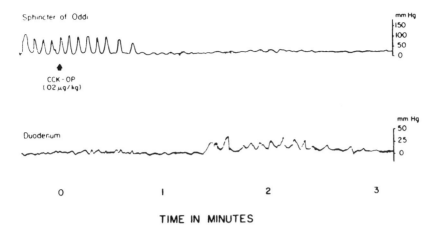

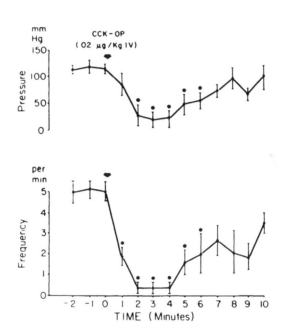

FIGURE 4. (A) Effect of cholecystokinin-octapeptide (CCK-OP) on sphincter of Oddi (SO) pressure and duodenal pressure. (B) Inhibitory effect of CCK-OP on the amplitude and the frequency of phasic pressure waves recorded from human sphincter of Oddi. (From Geenen, J. E., Hogan, W. J., Dodds, W. J., Steward, E. T., and Arndorfer, R. C., *Gastroenterology*, 78, 317, 1980. With permission.)

was suggested to be of peptide nature and named substance P (SP), was isolated and sequenced 50 years later as an undecapeptide from bovine hypothalami.[272-274] Substance P-like material was subsequently isolated from the intestines of horse, guinea pig, cat, human, rabbit, and rat.[275-279] Similar to substance P, nonmammalian tachykinins, such as eledoisin, physaleamin, and kassinin were isolated and sequenced from various mollusca and amphibia.[280]

Recently, two novel mammalian tachykinins, neurokinin A (NKA) and neurokinin B

TABLE 5
Amino Acid Sequences of Substance P, Neurokinin A, Neurokinin B, Neuropeptide K, and Various Nonmammalian Tachykinins

Mammals

Substance P	Arg–Pro–Lys–Pro–Gln–Gln–Phe–Phe–Gly–Leu–Met–NH$_2$
Neurokinin B	Arg–Met–His–Asp–Phe–Phe–Val–Gly–Leu–Met–NH$_2$
Neurokinin A	His–Lys–Thr–Asp–Ser–Phe–Val–Gly–Leu–Met–NH$_2$
Neuropeptide K	R–His–Lys–Thr–Asp–Ser–Phe–Val–Gly–Leu–Met–NH$_2$

Mollusca

Eledoisin	pGlu–Pro–Ser–Lys–Asp–Ala–Phe–Ile–Gly–Leu–Met–NH$_2$

Fishes

Scyliorhinin I	Ala–Lys–Phe–Asp–Lys–Phe–Tyr–Gly–Leu–Met–NH$_2$

Amphibia

Physalaemin	pGlu–Ala–Asp–Pro–Asn–Lys–Phe–Tyr–Gly–Leu–Met–NH$_2$
Uperolein	pGlu–Pro–Asp–Pro–Asn–Ala–Phe–Tyr–Gly–Leu–Met–NH$_2$
Phyllomedusin	pGlu–Asn–Pro–Asn–Arg–Phe–Tyr–Gly–Leu–Met–NH$_2$
Kassinin	Asp–Val–Pro–Lys–Ser–Asp–Gln–Phe–Val–Gly–Leu–Met–NH$_2$
Hylambatin	Asp–Pro–Pro–Asp–Pro–Asp–Arg–Phe–Tyr–Gly–Met–Met–NH$_2$

Note: R = Asp–Ala–Asp–Ser–Ser–Ile–Glu–Lys–Gln–Val–Ala–Leu–Leu–Lys–His–Ser–Ile–Gln–Gly–His–Gly–Tyr–Leu–Ala–Lys–Arg.

(NKB), which are structurally related to kassinin, have been identified and isolated from the mammalian spinal cord.[281-285] In addition, an N-terminally extended form of neurokinin A with 36 amino acids was isolated from the porcine brain and named neuropeptide K (NP-K).[286] There is some evidence that additional, so far unidentified tachykinin-like peptides might be present.[287-291] Substance P, neuropeptide K, neurokinin A, and neurokinin B have a common C-terminal amino acid sequence.

$$R–Phe–X–Gly–Leu–Met–NH_2$$

which they share with other nonmammalian tachykinins such as eledoisin, physaleamin, kassinin, scyliorhinin, and phyllomedusin.[280,292]

There has been a confusing nomenclature for the possible receptor subtypes involved in the physiological responses to tachykinins. In this chapter the nomenclature for tachykinins and tachykinin receptors as proposed by Henry[293] is used, and an overview is given in Table 6.

A. PRECURSOR FORMS

Neurokinin A and B, substance P, and neuropeptide K are derived from larger precursor peptides by posttranslational processing. The cDNA and DNA sequences of two precursor peptides, preprotachykinin A and preprotachykinin B, have been described in the central nervous system.[296] These two preprotachykinins, which share structural similarity, are located on two genes. Preprotachykinin A (PPTA) encoded the sequence of SP, neuropeptide K, and NKA, while preprotachykinin B (PPTB) encoded NKB. The expression of the two preprotachykinins varies in different tissues. (See Figure 5.) Due to different RNA-processing (splicing), three different mRNAs are generated from the PPTA. α-PPTA encodes only the sequence of substance P; β-PPTA and γ-PPTA encode both SP and NKA. The expression of the preprotachykinins and the mRNA-splicing are tissue specific. From the occurrence of β- and γ-PPTA, it can be concluded that substance P and neurokinin A coexist in a subpopulation of neurons. Deacon et al.[297] have reported that neuropeptide K is converted into neurokinin A during packaging into vesicles for axonal transport.[297]

TABLE 6
Nomenclature of Substance P and Related
Neurokinins Acting on the Proposed
Receptor Subtypes

Peptide		Receptor type
Substance P		NK-1 = NK-P
		= SP-P[a]
Neurokinin A	= Neurokinin-α	NK-2 = NK-A
	= Neuromedin L	= SK[b]
	= Substance K	
Neurokinin B	= Neuromedin-β	NK-3 = NK-B
	= Neuromedin K	= SP-N
		SP-E[a]
Neuropeptide K		

Note: The terms used in this review are underlined. The term "tachykinin" is mainly used to describe non-mammalian peptides (e.g., kassinin, eledoisin, and physaleamin).

[a] Lee et al.[294] suggested two receptor subtypes which they labeled according to their preferential ligands, SP-P and SP-E receptors. The SP-E receptor is not (!) identical with the NK-3 receptor, but shows some characteristics which have also been described for NK-3 receptor.
[b] Buck et al.[295] With permission.

B. TISSUE LOCALIZATION

Numerous studies have been carried out to locate SP-like immunoreactivity (SP-LI) in various tissues including the gastrointestinal tract. SP-LI has been demonstrated in nerve fibers, nerve cell bodies, and endocrine cells of the intestinal mucosa[298-301] and in the vagal nerve.[302] It has to be mentioned that most immunohistochemical studies were carried out using antibodies which would not distinguish substance P-, neuropeptide K-, neurokinin A-, and neurokinin B-like immunoreactivity. Therefore, the results on the distribution of substance P and neurokinins have to be interpreted cautiously in view of the probable cross-reactivity of antibodies with them. Furthermore, in contrast to Dale's principle, neurons contain and release a variety of neuropeptides and neurotransmitters (for review see References 303 and 304), and substance P has been shown to coexist in various subtypes of enteric neurons which can be identified by their neurochemical coding. In the guinea pig ileum, substance P is present in the submucous plexus, coexisting with acetylcholine (about 10% of all neurons). In the myenteric plexus, three main types of substance P-containing nerves can be identified. In the first type, accounting for about 30% of all neurons, substance P is present either alone or coexists with enkephalins and/or dynorphin. In the second type, substance P coexists with somatostatin and cholecystokinin (about 10% of all neurons), and in the third type, with dynorphin, VIP, gastrin-releasing peptide, and cholecystokinin (<2% of all neurons). Even though the majority of the substance P-containing neurons and nerve fibers seem to be of intrinsic origin, there are some fibers which mainly supply the submucous plexus ganglia and the arteries, which originate from the dorsal root ganglia.[303,305]

There is clear evidence that NKA-like immunoreactive material is also present in the myenteric, deep muscular, and submucous plexuses of the gut, as has been shown by radioimmunoassay of extracted synaptosomes from the dog ileum.[306] In the synaptosomes of the myenteric plexus and of the deep muscular plexus of the canine ileum almost equal

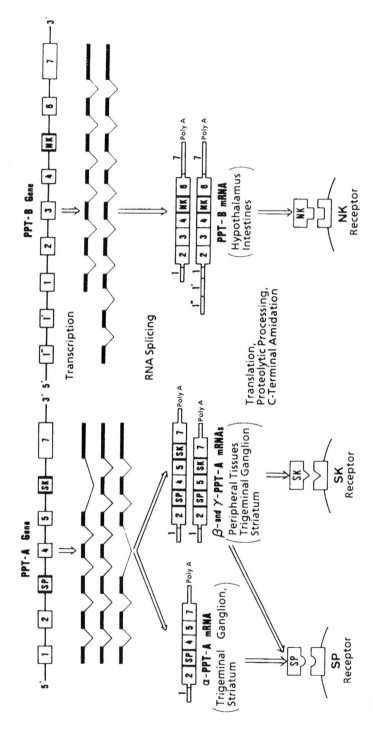

FIGURE 5.   Expression of preprotachykinin genes. PPT-A, preprotachykinin A; PPT-B, preprotachykinin B; SP, substance P; SK, substance K; NK, neuromedin K. (SK = NKA, and NK = NKB.)

amounts of substance P and neurokinin A were present, whereas in the submucous plexus the concentration of substance P was almost twice the concentration of NKA. However, to date no convincing evidence for the presence of NKB-like material in synaptosomes of canine intestine could be obtained. In addition, in the rat,[307] as in the guinea pig[291,297] and the canine small intestine,[306] little or no neurokinin B-like material could be detected. However, the reported expression of the preprotachykinin B (which encodes NKB) in the intestine, suggests that NKB-like material could be present within the enteric nervous system.[285,308,309]

C. SUBSTANCE P RECEPTORS

It is well established that substance P and neurokinins act on distinct receptors in the gastrointestinal tract.[310] Using the potency differences of SP, NKA, KNB, nonmammalian tachykinins, and peptide fragments in pharmacological studies[294,311,313-315] and receptor-binding studies,[295] three different tachykinin receptor subtypes were suggested. Similar to the bioassays used for opioid receptor subtypes, Regoli and co-workers[311-313,315] identified various *in vitro* smooth muscle preparations which demonstrated high selectivity in content of each of the proposed receptor subtypes. Using these bioassays they designed analogues of substance P, neurokinin A, and neurokinin B which showed marked potency differences in the various assay systems, suggesting a selective interaction with the various receptor subtypes.[315-317] Selective ligands for SP-P and NK-3 (NKB) receptors have also been developed and reported by other groups[318-322] (see Table 7). Using selective agonists or antagonists, the probable existence of tachykinin receptors selective for various tachykinins has been demonstrated by various studies.[7,323-325]

Receptor binding studies using [³H]SP have demonstrated binding sites in the myenteric plexus-longitudinal muscle of the guinea pig.[328,329] The saturation data suggested the interaction of SP with either two binding sites[328] or a single high-affinity binding site. However, displacement studies revealed only a poor correlation of some analogues and tachykinins in their ability to displace SP binding and induce contraction or inositol phosphate hydrolysis, suggesting that they might interact with different receptor types.[329] When highly purified membranes of the canine small intestinal muscle were used for [¹²⁵I]substance P and [¹²⁵I]eledoisin binding, only [¹²⁵I]Tyr⁸-SP showed specific binding to a high-affinity receptor. However, comparison of the binding data with functional data suggested the existence of an additional, probably NK-2 type receptor[330] to which the SP ligand had no affinity.

Binding sites for substance P in the guinea pig[331] and for substance P and NKA in the rat[332] have been visualized autoradiographically in the longitudinal and circular muscle layer of the small intestine. In a subsequent study using iodinated substance P, neurokinin A, and eledoisin for receptor binding studies to membranes and autoradiography, three distinct binding sites in the guinea pig intestine were postulated.[332] In the rat intestine only two receptor types (NK-1 and NK-2) seemed to be present, and eledoisin was suggested to interact with these receptors.[332] However, substance P as well as NKA and eledoisin will interact with all receptor subtypes as suggested in functional studies.[315] Using the selective analogue 3H-Senktide for receptor binding studies, evidence for NKB receptors was reported in the guinea pig ileum.[333] More detailed studies using selective [³H]NKB ligands in the gastrointestinal tract need to be conducted.

Using (β-Ala⁴, Sar⁹, Met[O₂]¹¹)-SP$_{4-11}$, [Nle¹⁰]NKA$_{4-10}$, and [β-Asp⁴, MePhe⁷]-NKB$_{4-10}$ a selective ligands for NK-1, NK-2, and NK-3 receptors, respectively, in the canine antro-duodenal region *in vivo* suggested the presence of all three receptor subtypes in the canine gastroduodenal region *in vivo*. Whereas NK-2 (NKA) receptors seemed to be mainly located postjunctionally, NK-3 (NKB) receptors were suggested to be located preferentially on neurons.[7] This study, together with several others supported the concept that NK-3 receptors are mainly located prejunctionally and NK-2 receptors are mainly located postjunctionally, whereas NK-1 receptors occur at both locations.

TABLE 7
Neurokinin Receptor Subtypes, Proposed Selective *In Vitro* Bioassays Models,
Selective Ligands, and Characteristic Rank Order of Potencies for Tachykinins

Receptor subtype	Bioassay	Ligand	Rank order of potencies
NK-1	Dog carotid artery	$[Sar^9, Met(O_2)^{11}]SP$	P > E> K
		$[\beta\text{-}Ala^4,Sar^9,Met(O_2)^{11}]\text{-}SP_{4\text{-}11}$	SP > NKA > NKB
		$Ac\text{-}[Arg^6,Sar^9,Met(O_2)^{11}]\text{-}SP_{6\text{-}11}$	
		$[pGlu^6,Pro^9]SP_{6\text{-}11}$ = Septide[a]	
		SP-OMe	
		Antagonist:[c]$[D\text{-}Pro^4, D\text{-}Trp^{7,9},Phe^{11}]SP_{4\text{-}11}$	
NK-2	Rat pulmonary artery	$NKA_{4\text{-}10}$	E ≥ K > P
		$[Nle^{10}]NKA_{4\text{-}10}$	NKA > NKB > SP
		Antagonist:[b,c] cyclo(Gln–Trp–Phe–Gly–Leu–Met) = L-659,877 (pA$_2$ 8.0) Ac–Leu–Met–Gln–Trp–Phe–NH$_2$ = L-659,874 (pA2 6.8)	
NK-3	Rat portal vein	$[\beta Asp^4,MePhe^7]\text{-}NKB_{4\text{-}10}$	K > E > P
		$[MePhe^7]NKB$	NKB > NKA > SP
		$Succ\text{-}(Asp^{5,6},MePhe^8)\text{-}SP_{6\text{-}11}$ = Senktide[6]	
		$[Cys^{2,5}]NKB$	
		Antagonist:[c] $[D\text{-}Pro^2, D\text{-}Trp^{6,8},Nle^{10}]NKB$	

Note: P, physaleamin; K, kassinin; E, eledoisin; SP, substance P, NKA, neurokinin A; NKB, neurokinin B.

[a]   From Reference 326.
[b]   From References 319 and 320.
[c]   For review of antagonists see Reference 327.

The proof for distinct neurokinin receptors was established by Masu and colleagues.[334] They used the *Xenopus* oocyte in which they injected exogenous mRNA isolated from bovine stomach. This "foreign" mRNA was then expressed by the oocyte. By studying various fractions of RNA it was shown that neurokinin A receptor (NK-2) and the substance P receptor (NK-1) were encoded by different mRNAs, and these mRNAs were synthesized differentially. By measuring electrophysiological changes in response to neurokinin after the expression of the NK-2 receptors, the mRNA encoding the NK-2 receptor could be identified. Subsequently, the cDNA and a peptide sequence of 384 amino acids which constituted the NK-2 receptor was determined.[334] Interestingly, this peptide receptor showed some homology and structural similarity to other rhodopsin-type receptors such as M1 and M2 muscarinic and β-1 and β-2 adrenoceptors, as well as to the MAS oncogene, suggesting that all membrane receptors might share this common structural principal (see Figure 6). A characteristic feature is seven hydrophobic α-helical membrane-spanning domains, with the N-terminus being extracellular and the C-terminus being intracellular. The second messenger mechanisms of NK-2 receptors involve a G protein-dependent phospholipase activating the phosphoinositol turnover.[334]

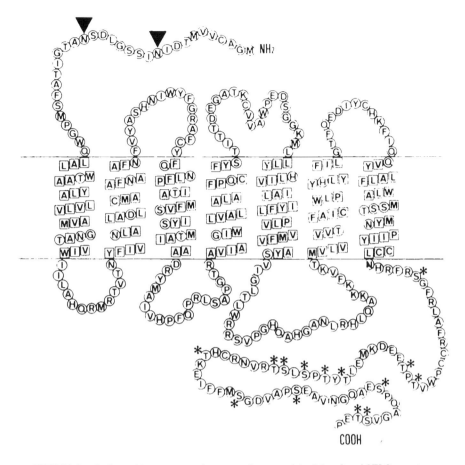

FIGURE 6. Amino acid sequence and transmembrane model of the cloned NK-2 receptor. Triangles, potential glycosylation sites; asterisks, possible phospholilation site; square amino acids, hydrophobic amino acids in the putative transmembrane domains. (From Masu, Y., Nakayama, K., Tamaki, H., Harada, Y., Kuno, M., and Nakanishi, S., *Nature (London)*, 329, 836, 1987. With permission.)

D. PHYSIOLOGICAL ACTION OF SUBSTANCE P AND NEUROKININS IN THE GASTROINTESTINAL TRACT

There are afferent substance P-containing nerve fibers which have their cell bodies located in the dorsal root spinal ganglia or in the nodose ganglia and their fibers running with the splanchnic nerves or the vagal nerve to the gut. These afferent nerve fibers are sensitive to the depletion of SP by capsaicin.[335] However, it must be kept in mind that capsaicin is not specific for SP-containing neurons and will also release other transmitters from afferent myelinated neurons, such as CGRP or somatostatin, etc.[336] Also, intrinsic SP-LI containing intrinsic enteric nerves are not depleted by capsaicin. In the guinea pig small intestine more than 90% of the SP-LI was not depleted by capsaicin, suggesting its location mainly in intrinsic nerves.[279] These intrinsic nerves containing SP can be subclassified according to their (1) neurochemical coding; (2) projections to mucosa, blood vessels, muscularis mucosae, and circular or longitudinal muscle layer; (3) oral or anal projections; (4) structural cell types; (5) electrophysiological characteristics; and (6) relation to calcium binding proteins (see Chapter 4). There are several lines of evidence that substance P acts as a neurotransmitter in the gastrointestinal intrinsic nervous system (for review see Reference 337). Besides its possible role as an afferent neurotransmitter in extrinsic nerves and its possible vasomotor-, secretomotor-, and immunomodulating functions, three major effects have been attributed to substance P.

First, there is evidence that SP might be involved in the nervous control of peristalsis[248,338-342] and in the peristaltic reflex.[343] Substance P is released during peristaltic activity,[339] and after cholinergic blockade the atropine resistant component of the ascending reflex could be blocked by a substance P antagonist.[343]

Second, SP might be involved in the modulation of neural activity by releasing other neurotransmitters (e.g., acetylcholine, 5-HT) or by changing the excitability of other neurons.[344,345] There is accumulating evidence that SP might be one mediator for the slow synaptic responses (slow EPSP) which cause a long-lasting change in the membrane potential.[346,347] This mechanism could contribute to the gating mechanisms present in the myenteric plexus.

Daniel et al.[348] observed in canine intestine that substance P-induced (given intra-arterially) contractions were antagonized by atropine or tetrodotoxin at low doses, while higher doses caused atropine- and TTX-independent contractions. Later it was found[349] that the lowest doses of substance P given intra-arterially during field stimulation caused inhibition. Clearly this peptide has multiple neural-modulating effects as well as a direct action on intestinal muscle.

Third, SP (and/or neurokinin A) might act as one nonadrenergic, noncholinergic (NANC) excitatory neurotransmitter in motor neurons exciting the smooth muscle of longitudinal and circular muscle and muscularis mucosae.[248,350-352] Substance P desensitization blocked the atropine-resistant contraction to electrical field stimulation in the guinea pig ileum[105] and partially blocked naloxone-induced contractions of the opiate-dependent guinea pig.[353] This indicates a possible participation of substance P as a (NANC)-excitatory transmitter in various conditions. This function of SP may be the same as in the peristaltic reflex (see above), but expressed in a less complete system, i.e., in the strips rather than in segments.

E. ACTIONS ALONG THE GASTROINTESTINAL TRACT
1. Esophagus and LES
In the opossum esophagus the highest density of SP-like immunoreactivity was demonstrated in the muscularis mucosa even though the longitudinal muscle contained some immunoreactivity. The circular muscle and the LES showed a relatively sparse SP-like innervation. The varying SP content was paralleled by a decreased sensitivity of the longitudinal and circular muscle as compared to the muscularis mucosae.[354] Substance P was suggested as a possible NANC mediator which was released from a capsaicin-sensitive afferent nerve[354,355] in response to field stimulation, eliciting a tonic contraction of the muscularis mucosae. The receptor subtype eliciting the contraction of the muscularis mucosae, which was suggested by potency studies as an SP (NK-1)[356] was identified as an NK-1-type using selective analogues for the various receptor types[357] (see Figure 7).

In contrast to the opossum, a dense SP-like innervation of the LES was revealed in cat, pig, and human.[358] Intravenous administration of SP increased the LESP in opossum[359] or cat[360,361] *in vivo*. In the cat the participation of substance P in a reflex of the LES in response to esophageal irritation with luminal acid was postulated.[360] In a subsequent study it was demonstrated that bombesin also might participate in this reflex, since bombesin tachyphylaxis reduced this response. The contractile effect of bombesin in the cat LES could be antagonized by substance P antagonists and substance P tachyphylaxis, suggesting that substance P could be the final mediator.[361] The role of substance P in human LES is unknown.

2. Stomach
The excitatory effect of substance P in the stomach of various species was shown to be mediated either by a direct smooth muscle effect or by the release of acetylcholine because the effect could be partly blocked by atropine or TTX.[3,7,362,363] However, higher doses caused atropine-insensitive contractions. Delbro and co-workers[364-366] suggested that afferent SP-

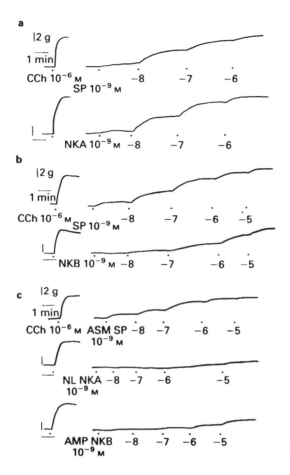

FIGURE 7. Typical cumulative responses to substance P, NKA, and NKB showing that each naturally occurring tachykinin causes full contractile responses (compared to a supramaximal dose of carbachol [CCh]). However, only the analogue selective for NK-1 receptors was effective when agonists highly selective for NK-1, NK-2, and NK-3 receptors were used: [β-Ala^4, Sar^9, MetO$_1$^{11}]SP$_{4-11}$ or AMS-SP receptors for NK-Y; [Nle^{10}] NKA$_{4-10}$ or NL-NKA for NK-2 receptors; [β-Asp^4, MePhe^7]-NKB$_{4-10}$ for NK-3 receptors. (From Daniel, E. E., Cipris, S., Manaka, Y., Bowker, P., and Regoli, D., *Br. J. Pharmacol.*, 97, 1013, 1989. With permission.)

containing nerve fibers could influence gastric motility by an NANC mechanism in which afferent fibers release substance P by an axon reflex. This mechanism was further speculated to take part in the gastric nociceptive reflex which is mediated by the splanchnic nerve.[366]

The different functional divisions of the canine stomach (fundus, corpus, and antrum) show differential responses of their smooth muscle to substance P *in vitro*.[368] Whereas the fundus contracted only tonically to substance P, the antrum contracted only phasically and the corpus had a mixed contractile response. In the conscious dog, SP stimulated gastric motility, increasing intragastric pressure, spike discharge, and propagation velocity of slow waves.[368]

Either substance P or NKA given intravenously in the rat delayed gastric emptying in the initial phase (3 min), but accelerated gastric emptying and gastrointestinal transit in a later phase (15 min).[341] This time-dependent effect of substance P on gastric emptying[341] could explain why either acceleration,[369] no effect, or even inhibition of gastric emptying was reported in other studies.[370] The initial inhibitory effect of substance P on gastric emptying was attributed to the contractile effect of substance P on the pylorus, as demonstrated in the cat by measuring the transpyloric flow *in vivo*.[362] Subsequent studies suggested that substance P also might be involved in the physiological regulation of the feline pylorus *in vivo*, since contractions induced by vagal stimulation or by substance P were blocked by the substance P antagonist (D-Pro^2–D-Trp^{7,9})–SP.[371-374] The delaying effect of substance P on gastric emptying was time dependent and could be partly reversed by atropine.[375] The SP antagonist (D-Pro^2-D-Trp^{7,9})– SP delayed the gastric emptying which occurred after 15 min, further suggesting that endogenous SP might be involved under physiological conditions to regulate gastric emptying.[376]

Using selective analogues for substance P (β-Ala^4, Sar^9, Met[O$_2$]^{11})-SP$_{4-11}$, for neurokinin A (Nle^{10})-NKA$_{4-10}$, and for neurokinin B (NKB) (β-Asp^4, MePhe^7)-NKB$_{4-10}$, the presence of three different subtypes of tachykinin receptors in the gastroduodenal region of the dog were demonstrated. The analogues contracted antrum, pylorus, and duodenum with different potencies, and their responses were affected differently by cholinergic blockade with atropine or neural blockade with TTX. It was suggested that these analogues were acting on different receptor sites.[7] The antrum was characterized by its high-affinity NKB (NK-3) and SP-P (NK-1) receptors, which are located on cholinergic neurons and lead to release of acetylcholine (responses blocked by atropine or TTX). The pylorus was excited by substance P- or neurokinin A-selective analogues acting postjunctionally. A small neural component of the response to the substance P analogue was found. The duodenum was characterized by its response to the substance P, NKA, and neurokinin B analogues which could be partially blocked by atropine or TTX, suggesting that pre- and postjunctional receptors of all classes existed. A limitation of this and similar studies is the difficulty in identifying prejunctional receptors which are responsive to higher doses of tachykinins than are postjunctional receptors. Such receptors will be missed if neural receptors are identified by the inhibitory effects of TTX or atropine.

These results are in agreement with earlier studies which showed the existence of neural responses to natural substance P[3,363,377] and neurokinins A and B[363] and postsynaptic responses to substance P in the canine stomach.[377] A similar distribution of neurokinin A and substance P receptors was also found in the rat stomach.[324] Consistent with these studies, a high density of binding sites for NKA (substance K) and for substance P (with lesser density) were observed in the circular muscle of the rat stomach, pylorus, and proximal duodenum using autoradiography.[332] Such light-microscopic studies cannot, however, distinguish binding to receptors on nerves within the muscle.

3. Small Intestine

The tachykinin-induced contraction of the guinea pig ileum *in vitro* is mediated in part by a direct action on smooth muscle and partly by stimulation of acetylcholine release from enteric neurons[345,378] and also might involve release of endogenous substances such as prostaglandins.[312] Laufer et al.[318] demonstrated that neuronal receptors releasing acetylcholine were activated preferentially by neurokinin B (50 times more potent than SP and NKA), suggesting that the prejunctional receptor could be of the NKB type. Interestingly, the excitatory effect of NKB could be antagonized using a Met-enkephalin analogue, suggesting that the same cholinergic neuron was regulated by an inhibitory opioid and an excitatory NKB receptor. Similar results have been obtained using the more selective ligand (MePhe^7)-NKB.[379] On the other hand, there is also an inhibitory action of opioid-containing neurons on substance P release.[339,380] In rat duodenum, neurokinin A has also been reported to be a potent agonist *in vivo* and *in vitro*.[381] Receptor binding studies and functional studies in isolated cells of the guinea pig intestine suggested a NK-1 receptor on the muscle, mediating the contractile response[382,383] on the basis of a characteristic pattern of potency of agonists.

The effect of substance P given intra-arterially in the dog to the small intestine was dependent on the activity state of the gut segment.[377] In the ileum, substance P induced contractions at lower concentrations by a neural postganglionic pathway, presumably by release of acetylcholine since the response was blocked by atropine.[384] At higher concentrations a TTX- and atropine-insensitive excitatory effect was present.[377,384] In contrast, when contractile activity was induced by electrical field stimulation of intrinsic nerves, substance P inhibited it at concentrations of substance P (threshold 10^{-13} mol) even lower than those causing excitation in the same muscle. This inhibitory effect of substance P could be reduced by atropine or pirenzepine and mimicked by acetylcholine and McNeil A343, an M1-muscarinic agonist. These observations suggest that the inhibitory effect is due in part to the release of acetylcholine acting on M1 receptors and causing an autoinhibition.[377] Similar excitatory and inhibitory effects of substance P on basal and stimulated [³H]ACh release were demonstrated in the guinea pig ileum.[344,385] This inhibitory effect also was observed in the antroduodenal region of dogs using an NK-3-selective analogue suggesting that this response is mediated via a prejunctional NKB receptor.[7]

The organization of the SP-containing nerves in the human small intestine resembles that of other mammalian species, such as the guinea pig[303,386] and the dog,[348] at the light and the electron microscopic levels.[387] Perhaps the human small intestine has responses and circuities similar to those found in animal models, but further study is required.

4. Colon and Sphincter Ani Internus

The effects of substance P in the large intestine resemble the responses reported in the small intestine. The mainly excitatory responses are either directly on smooth muscle or mediated by the release of acetylcholine or another excitatory mediator and consist of phasic or tonic contractions varying from species to species.[388-391] Substance P applied to the cat colon *in vivo* elicited strong contractions; the distal colon was more sensitive than the proximal colon.[392] These substance P-induced contractions in the cat colon could be blocked by (D-Arg¹–D–Pro²–D–Trp⁷,⁹–Leu¹¹)–SP, but were unaffected by various other blocking agents including TTX, atropine, hexamethonium, and naloxone.[389] In the dog colon,[390,391] atropine or TTX antagonized contractions to substance P *in vivo* and *in vitro*, but high amounts contract by an atropine-insensitive mechanism. In the guinea pig taenia coli, (D-Pro²–D–Trp⁷,⁹)–SP antagonized the NANC contractile response *in vitro* induced by electrical field stimulation of nerves.[393] In the longitudinal muscle of the guinea pig colon, contractions induced by CCK-8 and serotonin were partially antagonized by a substance P antagonist, suggesting that these responses might be mediated in part by the release of substance P.[388]

5. Biliary Tract and Sphincter of Oddi

Substance P which is present in the ganglionated plexus of the gallbladder[394,395] has an excitatory effect on the gallbladder smooth muscle of the dog, cat, rabbit, and guinea pig.[396-399] There seem to be some species differences as to whether this contractile effect is directly on the smooth muscle[398] or indirect through neural mechanisms.[396] Recently it has been demonstrated that isolated smooth muscle cells from the canine gallbladder would contract in response to substance P and that this response could be antagonized by a substance P antagonist.[400] A contractile response of substance P was also found in the sphincter of Oddi.[397]

F. PATHOPHYSIOLOGICAL ROLE

As have other peptides, substance P has been identified in immunohistochemical and radioimmunological studies to be present in carcinoid tumors[401] and in the plasma of patients with the carcinoid syndrome.[402,403] Elevation of plasma SP levels induced alterations of gastrointestinal motility similar to those observed in patients with the carcinoid syndrome.[403]

Recently neurokinin A and neurokinin A fragments also have been isolated from carcinoid tumors,[404,405] whereas neurokinin B-like immunoreactive material has been isolated from human pheochromocytoma.[406] Whether these two peptides, in addition to substance P, are released and produce symptoms in these patients has not been established.

There is also some evidence that substance P and neurokinins present in afferent neurons might be involved in the pathogenesis of irritable bowel syndrome. Autoradiographic studies demonstrated increased receptor binding sites for substance P and for neurokinin A in the colon and small intestine taken from diseased parts of the colon from patients with ulcerative colitis when compared to "control" tissues from patients who underwent surgery for resection of colon carcinoma.[407,408] Especially the arterioles and venules of the colon and the lymph nodules showed a significantly increased number of tachykinin receptors in inflamed tissue.[407,408] Therefore, it was speculated that the extrinsic neurons, which are probably afferent, could release substance P or neurokinin A by an axon reflex or by antidromic excitation, thus inducing changes of the immune status, vascular permeability, and motility changes. In this context it is noteworthy that substance P also was speculated to participate in the pathogenesis of arthritis, which is a common symptom of inflammatory bowel disease patients.[409] Further experimental studies are needed to confirm this exciting hypothesis.

The participation of substance P in a viscerocirculatory reflex which causes changes in blood pressure and heart rate in response to topical warming of the mucosa demonstrates the possible influence of afferent information from the gut on systemic body function.[410]

At present, there is almost no information on the involvement of substance P in the pathophysiology of motility disorders. Definitive study of the role of tachykinins in normal motility or its disorders awaits development of high-affinity ligands selective for each receptor since, as this chapter demonstrates, there are multiple receptor subtypes in multiple locations and the naturally occurring tachykinins are not highly selective for the various receptor subtypes. Physiological selectivity is most likely a geographic phenomenon (release of a particular tachykinin at a particular site) and not a pharmacological one. Thus, there is a need for potent antagonists highly selective for each receptor subtype to interfere with actions of each tachykinin.

G. CONCLUSION

There are three neuropeptides related to substance P in the gastrointestinal tract: substance P, neurokinin A, and (probably) neurokinin B. Neurokinin B remains to be definitively located there. Each has a separate receptor: NK-1 for substance P, NK-2 for neurokinin A, and NK-3 for neurokinin B. Each of these neuropeptides can activate all three receptor types, but has some selectivity for its own receptor. Peptides with higher selectivities have been designed, but highly selective and potent antagonists do not exist to date. Substance P is probably a transmitter of slow excitatory events within the enteric nerve plexus. It and neurokinin A are also probably involved in excitatory neurotransmission from enteric nerves to smooth muscles. Extrinsic and afferent nerves containing substance P are apparently involved in some excitatory events at gastrointestinal and vascular smooth muscle, releasing substance P by axon reflexes. Future progress will depend upon the development of highly selective antagonists to each receptor subtype and on exact localization of each of these neuropeptides and its receptors.

VI. OPIOID PEPTIDES

A. SOURCES OF OPIOIDS
1. Intrinsic to Gastrointestinal Tract

Opium derivatives have been used almost since the third century B.C. as an antidiarrheal remedy because of their well-known constipating effect. The mode of action of exogenous

Pro-opiomelanocortin

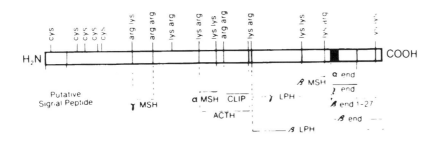

Pre-Pro-enkephalin A

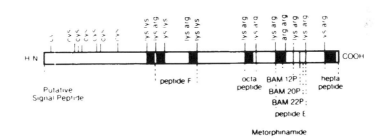

Pre-Pro-enkephalin B

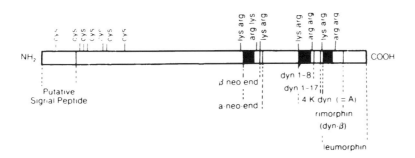

FIGURE 8. Structure of the three opioid precursor molecules. (Top) bovine prepro-opiomelanocortin (265 amino acids); (middle) preproenkephalin A; (bottom) preproenkephalin B opioid sequences are shown in black (see text for details).

opiates from various plants was clarified by the discovery of endogenous substances with opiate-like action in the pig brain by Hughes et al.[411] and the isolation of "endorphins".[412] Endogenous opioid peptides are characterized by the typical N-terminal structure of enkephalins:

$$\text{R–Tyr–Gly–Gly–Phe–Met or R–Tyr–Gly–Gly–Phe–Leu}$$

and until now more than 20 opioid peptides have been described;[413,414] see Figure 8.

Like other neuropeptides and peptide hormones, endogenous opioids are derived from larger precursor forms by posttranslational processing. Currently three precursor forms of opioid peptides have been described and cloned: pro-opiomelanocortin,[415] proenkephalin A,[416,417,418] and proenkephalin B.[419]

Pro-opiomelanocortin contains β-endorphin (31 amino acids, corresponding to the C-terminal extension of Met-enkephalin) besides other nonopioid peptide hormones. Pro-opiomelanocortin is mainly expressed in the pituitary gland and the CNS. Proenkephalin A is mainly expressed in the CNS, in the digestive tract, and in the adrenal medulla. It contains four copies of Met-enkephalin, one copy of Leu-enkephalin, and several C-terminal extended forms of Met-enkephalin (Met–enkephalin–Arg^6–Phe^7, Met-enkephalin–Arg^6–Gly^7-Leu^8, peptide F, peptide E, BAM-22P, BAM-20P, BAM-12P, amidorphin, and metorphamid) which are derived from various modes of processing.

Proenkephalin B, which is also termed prodynorphin and is expressed in the CNS and in the digestive tract, contains a variety of opioid peptides (α-neoendorphin, β-neoendorphin, dynorphin A [1—17, 1—13, 1—9, 1—8], dynorphin B [1—29, 1—12], and rimorphin).

Some of these opioid peptides act as classical neurotransmitters whereas others (e.g., β-endorphin) might act via release in the circulation and are putative hormones.[420]

2. Exogenous Opioids — Exorphines

Besides these endogenous sources, there are exogenous opioids which might be of physiological importance as constituents of food such as digested gluten[421,422] and milk.[423-425] β-casomorphin, a peptide consisting of seven amino acids, has been demonstrated as a cleavage product of the casein protein of milk and might be a physiological source of exogenous opioid-like material,[425] These exorphins have been shown to cause changes in peptide and hormone release and exocrine responses of the pancreas probably via a systemic action.[426,427]

3. Presence of Opioid-Like Peptides in the Gastrointestinal Tract

Several studies have demonstrated the presence of opioid-like substances using radioimmunoassay and immunocytochemistry studies[428] (see Table 8 for summary and Reference 1 for review). Opioid-containing nerve cells are present in the myenteric plexus and the submucous plexus; however, there seem to be significant distinct species differences. The opioid-containing nerve fibers are mainly projected to the circular muscle layer; fewer are projected to the longitudinal muscle layer, and they are almost absent in the mucosa.

There is also evidence for a colocalization of opioid-like peptides with other peptide and nonpeptide neurotransmitters such as substance P,[429,430] VIP,[1,431] PHI,[430] NPY, GRP, and CCK.[1] There even seems to be colocalization of dynorphin as well as Met-enkephalin and its C-terminally extended forms.[1,431] For details see Chapter 4.

B. OPIOID RECEPTORS
1. Multiple Opioid Receptors

On the basis of different activity profiles of various opioids *in vivo* in the chronic spinal dog[455] and *in vitro* on the mouse vas deferens,[456] the concept of multiple opioid receptor types was suggested,[455-456] and meanwhile four different main types of opioid receptors have been described and named μ-, δ-, κ-, and ε-receptors.[413,458] The σ-receptor as well as several other suggested receptor subtypes have somewhat different pharmacological characteristics, are not readily blocked by naloxone, and might actually be related to other neurotransmitter receptor systems. Some authors have presented evidence for a further subclassification of opioid receptors such as $κ_1$- and $κ_2$-receptors[459] or $μ_1$- and $μ_2$-receptors.[460]

Receptor-binding studies in the gastrointestinal tract,[461-464] autoradiographic studies,[465-467] and functional studies[468-472] have demonstrated the presence of opioid receptors in

TABLE 8
Demonstration of Opioid-Like Peptides in the Gastrointestinal Tract with Radioimmunoassay (RIA) or Immunocytochemistry (IMC)

	Method	Species	Region	Ref.
Met^5-enk[a]	IMC	Human	GI tract	432
	IMC		GI tract	433
	RIA	Human	Stomach	434
	RIA	Human	Stomach	435
	IMC		GI tract	436
	IMC	Guinea pig	Ileum	437
	IMC	Guinea pig	Small Int.	438
	IMC	Guinea pig	GI tract	439
	IMC	Cat	Ileum	429
	IMC	Rat	GI tract	386
	IMC	Rat	GI tract	439
	IMC	Dog	GI tract	440
	IMC	Dog	GI tract	348
	IMC	Dog	Ileum	441
	IMC	Cat	Sphincter of ani	442
Leu^5-enk[a]	IMC	Rat	GI tract	443
	IMC		GI tract	444
	IMC		GI tract	433
	RIA	Human	Stomach	435
	IMC	Cat	Esophagus	445
	IMC	Guinea pig	Esophagus	445
	IMC	Opossum	Esophagus	445
	IMC	Pig	Esophagus	445
	IMC	Monkey	Esophagus	445
Met^5-enk-Arg^6-Gly^7-Leu^8	RIA	Human	Stomach	435
	RIA	Human	GI tract	446
	IMC	Guinea pig	Duodenum	447
	IMC	Guinea pig	Colon	448
	IMC	Rat	GI tract	449
	IMC	Rat	GI tract	450
	IMC	Pig	GI tract	450
Met^5-enk-Arg^6-Phe^7	RIA	Human	Stomach	435
	IMC		GI tract	436
β-Endorphin	IMC	Guinea pig	GI tract	451
	IMC	Rat	Duodenum	452
Dynorphin	RIA	Pig	Duodenum	453
	IMC	Rat	GI tract	439
	IMC	Guinea pig	GI tract	439
	IMC	Rat	Duodenum	454

[a] In most studies, the antibodies used were not absolutely specific for either Leu- or Met-enkephalin, so there might be some cross-reactivity.

the GI-tract and particularly on myenteric neurons.[461,473,474] Intracellular microelectrode recordings of myenteric neurons in the guinea pig ileum have demonstrated that two different opioid receptors (μ and κ) can even exist on a single neuron.[475-477]

Unlike some synthetic peptides, endogenous opioid peptides do not show a clear receptor selectivity and have only small affinity and efficacy differences at the various receptor subtypes.[413] However, the known difference in the affinity might suggest that some peptides could act as endogenous ligands for the various receptors (see Table 9). Dynorphin was suggested to be the physiological ligand for κ-receptors.[486-488] There is also a slightly higher

TABLE 9
Opioid Receptor Subtypes, Endogenous Ligands, Selective Agonists and Antagonists, and Bioassay Systems

Receptor	Endogenous agonist	Selective agonist	Selective antagonist	Bioassay system
μ	Met-enkephalin β-Endorphin	Morphiceptin PL017 Morphine DAGO Sufetanyl	CTP Funaltrexamine Naloxone	Guinea pig ileum
δ	Leu-enkephalin	DPDPE DPLPE	ICI 174,864 ICI 139,462 ICI 154,129	Mouse vas deferens
κ	Dynorphin A$_{1-13}$	U-50488 Tifluadome Ethylketocy- cloazosine (EKC)	TENA MR 2266	Rabbit vas deferens
ε	β-Endorphin			

Note: CTP, cyclic-D-Pen–Cys–Tyr–D-Trp–LYS–Thr–Pen–Thr–NH$_2$;[478] TENA, 6β,6′β(ethylene-*bis*[oxy-ethyleneimino])*bis*(17-[cyclopropylmethyl]-4,5a-epoxymorphinan-3,14-diol);[479,480] ICI 174864;[481] morphiceptin;[482] DPDPE (D-Pen^{2,5}-enkephalin);[483] ICI 154129;[484] U-50488-H.[485,479]

affinity of Leu-enkephalin for δ-receptors and of Met-enkephalin and β-endorphin for μ- and ε-receptors, respectively.[413] However, the observation that Leu- and Met-enkephalins are often present together in, and released together from, nerves raises questions about how selective activation of μ- and δ-receptors is achieved *in vivo*.

2. Opioid Receptor Location

Opioid peptides administered exogenously *in vivo* have both excitatory and inhibitory effects on gastrointestinal motor function.[8,384] Besides their local action in the gastrointestinal tract, opioids also might act centrally, at the level of the spinal cord, or at the level of prevertebral ganglia to modify gastrointestinal motor function.[489,490] The distribution of opioid receptors presents broad variability depending on the species studied and the intestinal segment considered.[491]

It is still controversial whether opioid receptors are located exclusively on neuronal tissue or also occur postjunctionally on the smooth muscle cells of the gut. Contractility studies using isolated smooth muscle cells of the guinea pig stomach[13,14] and human intestine[14] and *in vivo* studies,[8,9,384,492] suggested that opioid receptors might be present on smooth muscle. However, there is also evidence against a postsynaptic localization:

1. Excitatory TTX-insensitive *in vivo* effects which were ascribed to direct muscle action are absent *in vitro* before or after neural blockade with TTX.
2. Receptor binding studies could not detect opioid receptors in either highly purified smooth muscle plasma membranes[464] or on longitudinal smooth muscle when the myenteric plexus was removed.[461]
3. Alternative actions of opiates to explain TTX-resistant excitation *in vivo* have been demonstrated.[493]
4. Electrophysiological studies could not demonstrate any effect of opioid peptides on the membrane potential of smooth muscle cells.[494,495]

C. MECHANISMS OF ACTION

Opiates and opioid active compounds inhibited electrically induced contractions in the guinea pig ileum longitudinal muscle/myenteric plexus preparation. This effect was due to a presynaptic inhibition of acetylcholine release.[496-498] Releases of a variety of other neurotransmitters have been demonstrated to be inhibited by exogenous or endogenous opioid peptides; these include substance P,[470] serotonin,[499] VIP,[493,500] and opioids themselves.[501] Furthermore, there is clear evidence for an interaction with neural excitatory and inhibitory effects of other neuropeptides such as CCK and caerulein,[502] neurotensin,[503] and substance P.[353,503,504]

The action of opioids on μ-receptors to inhibit acetylcholine release was demonstrated to be caused by hyperpolarization secondary to a potassium conductance increase. This was a calcium-dependent (using internal calcium stores) potassium channel which could hyperpolarize S or AH-neurons.[505,506] The inhibitory action of κ-receptors resulted from direct inhibition of a calcium inward current which was responsible for the action potential in the soma.[576] The mechanism of this effect could be either a decrease in the absolute number of open calcium channels or a decrease in their mean opening time as suggested by studies in dorsal root ganglia cells. Only activation of the κ-receptors decreased the TTX-insensitive, calcium-mediated action potential; it was unaffected by the activation of μ-receptors.[576]

So far only inhibitory effects of opioid peptides on myenteric neurons have been described, and there is no evidence that opioid receptors can elicit action potential by a direct mechanism. This leaves unexplained the frequent observation of opioid excitation of intestine *in vivo* which is reduced by atropine. Part of the excitatory effects of opiates on intestine *in vivo* may result from suppression of a tonic inhibitory tone in the myenteric nervous system. Such a tone was suggested to explain how neural blockade with either TTX, lidocaine, or ω-conotoxin leads to an activation or increase of motor activity;[507,811] *In vivo* studies using isolated perfused ileal segments have shown that there is a tonic release of VIP in high amounts which is abolished by TTX,[493,508] suggesting that VIP might be a mediator of this tonic inhibitory response. It is likely that part or all of the excitatory effect of opioid described *in vivo* is due to the inhibition of this inhibitory tone, and it has been demonstrated that opioid agonists which induce motor activity *in vivo* will decrease the endogenous tonic VIP release[508] (see Figure 9). An inhibitory influence on the release of VIP from isolated gut segments, associated with activation of contraction, has also been demonstrated in the rat colon.[500] Such an inhibitory effect of opioids on endogenous tonic inhibition may be lost in isolated muscle strips: this could explain TTX-insensitive excitation present *in vivo* but absent *in vitro*.

Excitatory effects of morphine and opioid peptides were also found in an isolated dog ileum *ex vivo* preparation[492] and in studies of the small intestine of dogs[509,510] and humans.[509,511]

D. GENERAL PHYSIOLOGICAL AND PATHOPHYSIOLOGICAL ROLE

Several specific problems have to be considered in evaluating opioid functions. First, most studies investigating opioid function were carried out in the isolated guinea pig ileum, which had great historical importance as the bioassay system for the characterization of opioid actions. However, the inhibitory effects of opioids in guinea pig may be peculiar and not representative of the excitatory opioid actions on the intestine of other species and in humans. Moreover, these excitatory actions were all demonstrated *in vivo*, and no *in vivo* studies on the effect of opioids in the guinea pig are known to us. Second, early studies were carried out using what were assumed to be "specific" ligands (e.g., Met-enk-μ, DADLE-8, EKC-κ), which have since been shown to have interactions with other opioid receptor subtypes. Therefore, some of the conclusions drawn with these ligands have to be reevaluated for their validity using current knowledge of receptor selectivity of ligands. Third, the effects of exogenously applied opioids depend on the receptor selectivity of the compound, the species studied, and the study environment. It has not always been clear whether so-called species differences in the effects of opioids which have been demonstrated are in fact related to differences between species or between study conditions (*in vivo* vs. *in vitro*). Fourth, many opiates (e.g., morphine or etorphine) or opioid antagonists (naloxone)

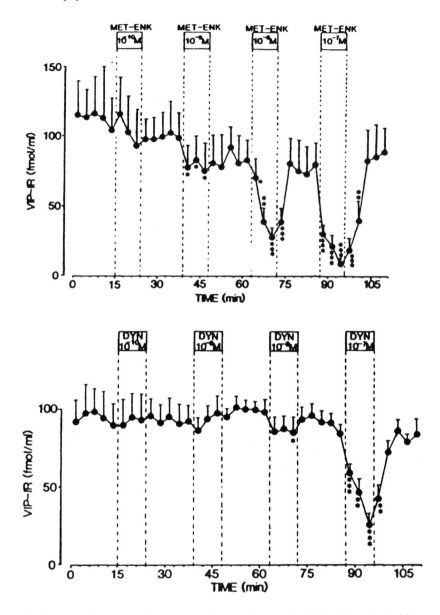

FIGURE 9.   Effect of opioid peptides on the basal VIP-IR release into the perfusate. Met-enkephalin and dynorphin were infused at the final concentration of 10^{-10}, 10^{-9}, 10^{-8}, or 10^{-7} *M*. Met-enkephalin and lower potency dynorphin caused a dramatic decrease of the basal VIP release which recovered after the opioid infusion was stopped. (From Manaka, H., Manaka, Y., Kostolanska, F., Fox, J. E. T., and Daniel, E. E., *Gastroenterology*, 92, 128, 1988. With permission.)

can cross the blood-brain barrier and can act at a central, spinal, or prevertebral level. Consequences of such actions might be different from local peripheral actions of the same agent.[512] Therefore, a clear distinction between local effects in the gastrointestinal tract and changes of extrinsic nervous or hormonal influences has to be made, and adequate models to study this problem have to be developed (for review see Reference 513).

When opiate receptor blockade is carried out in normal volunteers or in animals, no substantial change in gastrointestinal motility can be observed. There seems to be no change in gastric emptying,[514-517] fed gastric or duodenal motility,[511] gastrointestinal transit, or gastrointestinal propulsion.[518-521] This might suggest that endogenous opioids do not play a major role in the overall regulation of gastrointestinal motor function. There are reports that naloxone will block the gastrocolonic reflex observed after ingestion of food, but opioid receptors outside the GI tract appear to be involved in this reflex response.[522] A further limitation to the conclusion that endogenous opioids play no physiological role in GI motility is that no highly selective κ-antagonists are available, and naloxone is only partially selective for μ-over δ-receptors.

E. OPIOID ACTIONS ALONG THE PROXIMAL GASTROINTESTINAL TRACT
1. Esophagus
The data on the effect of exogenous and endogenous opioids on esophageal motility are conflicting. In the esophagus of the guinea pig, κ-opioid receptors were demonstrated to decrease acetylcholine release to the muscle of the muscularis mucosae.[523] However, in the cat esophagus distal body circular muscle (1 cm above LES), the inhibitory effect of exogenous opioids on electrically induced contractions was attributed to a prejunctional inhibition of adrenergic excitatory fibers.[445] Using a stable opioid analogue Stacher et al.[524] found an increase in amplitude and duration of esophageal peristaltic contractions in humans. In monkeys and man, morphine sulfate was found to decrease lower esophageal sphincter pressure (LESP).[525,256] Consistent with the existence of an inhibitory opioid control, naloxone (10 μg/kg/h) caused a significant increase in basal LES pressure in humans without a significant effect on the motility of the esophageal body.[527,528] However, both substances, morphine as well as naloxone, can cross the blood-brain barrier, and a central action of opioids cannot be excluded. Endogenous opioids did not seem to be involved in the NANC-mediated relaxation of the LES, since naloxone did not block swallow-induced LES relaxation in vivo[528] or NANC-induced inhibition of LES tone in vitro.[529] There are, however, reports that exogenous opioids might be able to reduce swallow-induced LES relaxation.[530,531] When given to human volunteers, the enkephalin analogue FK-33824 inhibited the relaxation of the lower esophageal sphincter, and this effect could be antagonized by naloxone.[531] Based on this experiment, the authors speculated that endogenous opioids might be involved in the pathogenesis of achalasia, but no conclusive data demonstrating any involvement of endogenous opioids has been obtained so far.

In the opossum, morphine caused a relaxation of the LES which was reversed into contraction after neural blockade with TTX.[532] A detailed study in the anesthetized opossum of opioid receptors was reported by Rattan and Goyal.[533] Based on the effect of several opioid agonists injected in the esophageal branch of the left gastric artery in anesthetized opossum, they described five distinct opioid receptors of μ-, δ-, κ-, ε-, and σ-type with locations on nerves and muscle. However, the opioid ligands used in that study are not as specific as the authors assumed, and the existence of various opioid receptors and their location will have to await further confirmation with more highly selective ligands.

In conclusion, endogenous opioids might contribute to the basal tone of the lower esophageal sphincter, and systemically released opioids might modulate LES tone and LES relaxation. However, there is no evidence for an involvement of opioids in the motility of the esophageal body or as NANC mediators in the LES.

2. Stomach
In the cat excitatory effects of opioids administered into the celiac arterial supply were demonstrated in the stomach[534] and the pylorus.[534-536] However, only inhibitory actions could be found in the canine stomach[8] and canine pylorus[10] when opioids were given close intra-arterially. When given intravenously, Met-enkephalin,[537] morphine,[538] and a κ-selective agonist, U-50488,[539] were shown to increase gastric contractions and motor activity. How-

ever, in these studies a central action or an action outside the gastrointestinal tract could not be excluded or was probable.

3. Gastric Emptying

As mentioned earlier, opioid blockade with naloxone did not influence gastric emptying[514,515,517,526]or postprandial upper gut motility.[511] Naloxone also did not improve chronic idiopathic gastric stasis or gastric motility disorders in combination with diabetic gastroparesis.[515] However, naloxone improved symptoms and motility patterns in patients with duodenal dyskinesia.[515] Therefore, under normal circumstances opioid peptides seem not to be of major physiological importance for gastric motility.

On the other hand, exogenous opioids cause a delay of gastric emptying.[517,526,540] In a recent study in dogs, Gue et al.[541] showed that orally administered κ-agonists (tifluadom or U-50,488) significantly accelerated solid emptying, but slowed the emptying of liquids. However, this effect was not observed when the κ-agonists were given intravenously.[541] Morphine and DAGO had no effect orally, and only higher doses of morphine given i.v. decreased gastric emptying.[541] In mice, naloxone reversed the delay of liquid gastric emptying induced by a fat meal and by exogenous CCK-8, suggesting an involvement of endogenous opioids in the delaying effect of a high-fat meal.[542] These recent studies suggest that opioid peptides might have differential effects on solid or liquid gastric emptying and that the opioid effects depend on the receptor subtype. Therefore, further studies using receptor-selective opioid agonists and antagonists and studies monitoring solid and liquid emptying of various nutrient meals, especially with high fat content, are necessary to determine the role of endogenous opioids on gastric emptying.

4. Pylorus

The action of opioid peptides on the pyloric motor activity appears to exhibit some species variations, but might depend on the route by which the opioids are applied and whether they reach the duodenum and/or antrum. Studies in the cat suggested that endogenous opioids might be involved in the excitatory response of intraduodenal stimuli on pyloric motor activity[535,536] and in the excitatory action of vagal stimulation.[534] This was put forward as one explanation for the delaying effect of exogenous opioids on gastric emptying.

In the dog, when pyloric motor activity was monitored with a Dent sleeve and exogenous opioids were given close intra-arterially, they were not excitatory for the pylorus, but inhibited neurally stimulated pyloric motility.[10] Endogenous opioids also seemed not to be involved in the excitation ascending from the duodenum to the pylorus because naloxone did not block neural stimulation of the pylorus[529] or pyloric excitation in response to intraduodenal acid.[474] Receptor binding studies demonstrated the presence of opioid receptors on the neural elements of the pyloric muscle. When exogenous opioids were given to the proximal duodenum (see Figure 10), they activated duodenal and pyloric motor activity via a cholinergic reflex pathway.[10] Intravenous administration of β-endorphin in doses which produced blood levels observed during stress situations caused increased pyloric motor activity in humans;[511] see Figure 11. The apparent increase in pyloric motor activity induced by i.v. opioid administration may be explained by a strong excitatory effect on the proximal duodenum[10] in combination with a potent antral inhibition.[8,511] These serve as stimuli to activate nerve pathways to the pylorus from the duodenum, causing contraction, and to inactivate a nerve pathway from the antrum, causing relaxation.[10] In man, pyloric motor responses induced by intraduodenal fat were not blocked by naloxone, suggesting that endogenous opioids are not involved.[542]

5. Small Intestine

The guinea pig ileum longitudinal muscle/myenteric preparation has been used classically for the opiate receptor bioassay. This preparation has been studied extensively with phar-

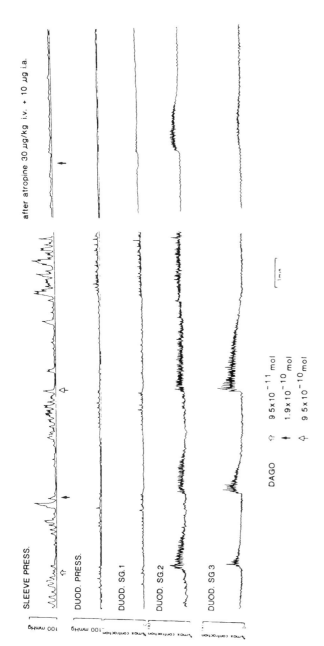

FIGURE 10. Excitatory effect of opioid peptides in the proximal canine duodenum when given close intra-arterially (i.a.) to the duodenal bulb. The perfusion area did not reach the second (SG2) and third (SG3) strain gauge, which were at a distance of 5 and 9 cm from the pylorus, respectively. Pyloric activity was recorded with a Dent-sleeve sensor. Note that the μ-agonist DAGO causes a delayed excitation in the pylorus after an initial inhibition. This excitatory response is abolished after cholinergic blockade with atropine (30 μg/kg i.v. and 10 μg i.a.). (From Allescher, H.-D., Ahmad, S., Daniel, E. E., Dent, J., Fox, J. E. T., and Kostolanska, F., *Am. J. Physiol.*, 255, G352, 1988. With permission.)

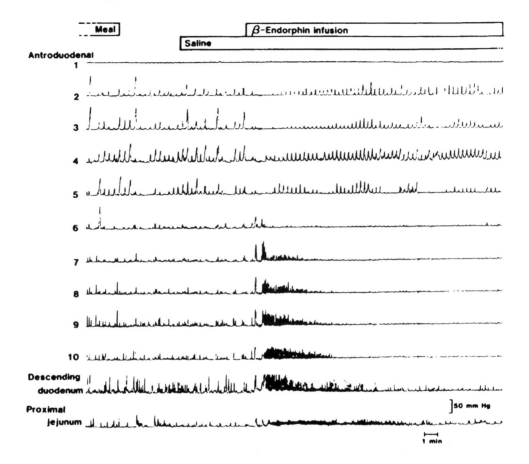

FIGURE 11.   Effect of β-endorphin infusion on human upper gastrointestinal pressure activity. β-Endorphin reduces the contractile activity in the antrum and enhances the pressure activity recorded from the proximal duodenum and ileum. (From Camillieri, M., Malagelada, J.-R., Stanghellini, V., Zinsmeister, A. R., Kao, P. C., and Li, C. H., *Am. J. Physiol.*, 251, G147, 1986. With permission.)

macological, immunocytochemical, and electrophysiological techniques. The presynaptic inhibition of neural acetylcholine release by opioids in the preparation became the model response of the gut.[496] In contrast to the huge number of *in vitro* studies with this tissue, almost no data are available on the effect of either exogenous or endogenous opioids in the guinea pig small intestine *in vivo*. In contrast to other species such as cat, dog, rabbit, rat, and man studied *in vivo*, the guinea pig studied *in vitro* shows no excitatory response to opioids.[544] This demonstrates that although the guinea pig ileum preparation is a very valuable tool to study opioid effects and opioid pharmacology, it might not be a representative model to determine the physiological role of opioids for gastrointestinal motility. Whether the guinea pig intestine would have excitatory responses to opioids given i.v. is not known.

Intra-arterial or i.v. application of morphine or other opioids excites the canine[8-10,384,492,509,510,545,546] and human small intestine.[509,511] It has been demonstrated that cholinergic[384,510] as well as serotonergic[546] mechanisms are involved in this excitatory response. Met-enkephalin given intra-arterially caused excitation of the canine small intestine *in vivo*, but excitation did not occur *in vitro*.[8] This excitation to intra-arterially injected opioids was partially abolished by atropine, suggesting that opioids might cause the release of acetylcholine. When opioids were given intra-arterially during excitation using field stimulation, the same opioids were inhibitory; see Figure 12. Dynorphin, which has little or no excitatory effect, was the most potent inhibitor.[8,9] Interestingly, the excitatory effect

351

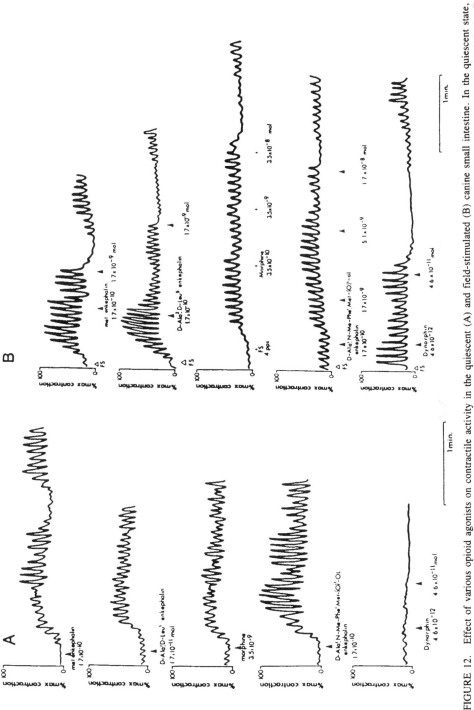

FIGURE 12. Effect of various opioid agonists on contractile activity in the quiescent (A) and field-stimulated (B) canine small intestine. In the quiescent state, opioid agonists cause an excitatory response (except dynorphin, which showed no effect of inhibition). In the activated state (during electrical field stimulation), all opioid peptides caused an inhibition of the contractile activity. (From Fox, J. E. T. and Daniel, E. E., *Am. J. Physiol.*, 253, G179, 1987. With permission.)

of motilin involved an NANC component which was blocked by naloxone, suggesting the release of endogenous opioids by motilin.[9]

In anesthetized dogs, contractions of the circular muscle of the duodenum were increased by μ-agonists (normorphine) and Met-enkephalin, but not by putative κ-agonists such as dynorphin$_{1-13}$ bremazocine, and U-50488.[547] Thus, in the dog intestine, κ-agonists appeared to inhibit acetylcholine release while μ- or δ-selective agonists appeared to enhance it. Why these agonists inhibited contractions when cholinergic nerves were stimulated[8,9] is unclear.

In contrast to the exciting action of opioids on unstimulated intestine *in vivo*, no such excitatory effect of μ- or δ-agonists was observed *in vitro*. This raises the question how opioids can cause excitation *in vivo*, especially since the response was only reduced and not abolished by TTX. It has been reported that isolated cells from circular muscle of human and guinea pig small intestine were contracted by opioid agonists.[13,14] κ-Agonists such as dynorphin were very potent agents causing contraction of the isolated cells. These results were interpreted as strong evidence for the presence of opioid receptors on the smooth muscle.[13,14] Further confirmation of such receptors would be desirable. In contrast to these findings, opioid receptor binding studies in the canine small intestine failed to demonstrate any postsynaptically located opioid receptors in the longitudinal or the circular muscle layer. When compared with the neuronal marker [^{3}H]saxitoxin, the opioid receptors seemed to be located exclusively on neuronal membranes. The opioid receptor subtypes identified by displacement studies using selective ligands revealed about 40 to 45% μ- and δ-receptors and about 10 to 15% κ-receptors.[474] Interestingly, the overall distribution of opioid receptors was not different in the synaptosomes from myenteric and from the deep muscular plexus of the canine small intestine. In submucous plexus a different distribution of opioid receptors was found, with an increased proportion of μ- and a decreased proportion of δ-receptors.[548]

The absence of direct excitatory effects of opioids *in vitro* was supported by electro-physiological studies in the canine duodenum which demonstrated that dynorphin$_{1-13}$ had no effect on the spontaneous mechanical and intracellular electrical activity.[494] Dynorphin$_{1-13}$, but not the κ-agonist U-50,488, significantly decreased the amplitude of the inhibitory junction potential (IJP), and this effect could be blocked by the δ-selective antagonist ICI-174,864 (10^{-5} M). At this concentration, ICI-174, 864 also could show some interaction with other opioid receptors. It was speculated that opioid-containing nerve fibers might function *in vivo* to decrease the release of one or more inhibitory transmitters.[494]

How, then, can opioids cause excitation of the small intestine *in vivo*, but fail to do so *in vitro*? Recently Manaka et al.[493] reported that in the *ex vivo* perfused canine small intestine there is a high, continuous basal release of VIP, probably driven by tonic cholinergic neurons acting through nicotinic receptors. Application of TTX caused a drastic decrease of basal VIP release and induced excitation of muscle which began to return to baseline levels[493,508] before field stimulation could again activate cholinergic nerves. The presence of a tonic inhibitory innervation of the small intestine has already been suggested by Wood and Marsh.[507] Application of opioids or other agents, such as motilin or α-adrenoreceptor agonists, caused a marked decrease of this high basal VIP release concurrent with an excitation of motor activity. Met-enkephalin was a more potent agonist for suppression of VIP release than dynorphin, which was ten times less potent. The effects of both opioids were blocked by naloxone. Subsequent studies[812] showed that either highly selective μ- (morphiceptin) or δ-(D-pen^{2,5}-enkephalin) agonists could suppress VIP release and cause motor excitation. There-fore, opioid receptors, of δ- or of μ-type, present on VIP neurons can decrease a tonic (VIP-mediated) inhibition of the circular muscle.[493,494] This could explain why VIP has little inhibitory effect *in vivo*, and the absence of this basal release of VIP *in vitro* also could explain the absence of the excitatory effect under these conditions. Moreover, since VIP output recovers before field stimulation of cholinergic nerves after TTX, the TTX-resistant response to opioids may be a result of opioid inhibition of VIP output in the partially blocked enteric nervous system.

F. OPIOID ACTIONS ON INTEGRATED GASTROINTESTINAL ACTIVITIES
1. Role of Endogenous Opioids for Peristaltic Activity

Endogenous opioid peptides were suggested to participate in the peristaltic activity and the peristaltic reflex.[339,549,550-555] During peristaltic activity of the isolated guinea pig intestine, the ongoing release of dynorphin immunoreactivity[549] and Met-enkephalin[553,556] is reduced. Furthermore, when opioid receptors were blocked with naloxone, the frequency of peristaltic contractions increased[550,551] and the induction of the peristaltic reflexes by luminal distention was facilitated.[552] This could be due to a disinhibition of cholinergic neurons because opioid receptor blockade with naloxone increased the acetylcholine release from the guinea pig ileum.[497]

Some peptides, such as CRF, neurotensin, bombesin, or CCK_{26-33}, caused excitation in the guinea pig ileum myenteric plexus longitudinal muscle preparation. These peptides seemed to release opioid-like material which then suppressed the release of acetylcholine or other noncholinergic excitatory mediators.[556,557] This was suggested because naloxone caused an excitatory response when given after these excitatory peptides, whereas the opioid agonist FK-33-824 suppressed their excitatory responses.[558] These excitatory responses to naloxone and various peptides were observed before and, to a smaller extent, also after atropine, suggesting that outputs of noncholinergic as well as cholinergic mediators such as substance P were inhibited by opioids.[470,504]

Opioid-like material can be released by electrical stimulation of the myenteric plexus as shown in the guinea pig ileum.[501,519] This release was inversely proportional to the frequency of stimulation and could be reduced by pretreatment with morphine. Naloxone, on the other hand, caused an increase of the basal release of Met-enkephalin-like material, suggesting that neurons which transmit via Met-enkephalin are under opioid regulation.[501]

By studying the nerve projections in the canine small intestine using the spread of an excitatory response to electrical field stimulation of myenteric nerves it could be demonstrated that endogenous opioids contribute to the cholinergic transmission both orally and anally, facilitating the spread of excitation[555] It was speculated that stimulation at low frequencies caused opioid release onto VIP nerves and/or acetylcholine release, causing excitation, whereas stimulation at higher frequencies caused release of dynorphin, which inhibits proximal excitation. There also seemed to be inhibition of distal relaxation and facilitation of excitation at higher frequencies of nerve stimulation, possibly by release of Met-enkephalin to inhibit VIP or other nonadrenergic nerves.

From these results obtained in the canine small intestine, the following model can be hypothesized for the action of endogenous opioids. κ-Opioid receptors are present presynaptically on cholinergic neurons, inhibiting the cholinergic nerves and the motor activities they mediate such as responses to electrical field stimulation. μ- and/or δ-receptors are present on VIP neurons and presumably on other nonadrenergic, noncholinergic neurons to inhibit the tonic release of the inhibitory mediator(s). Orad contraction during the peristaltic reflex is enhanced via orad-projecting opioid neurons which turn off the VIP-release via μ- and δ-receptors (possible mediator, Met-enkephalin), facilitating and increasing the cholinergic influence. Anal relaxation during the peristaltic activity is facilitated by decreasing cholinergic tone via κ-opioid receptors (possible mediator, dynorphin) on cholinergic neurons and by increase of VIP release due to a decreased release of opioids (possible mediator, Met-enkephalin) on μ- and δ-receptors.

Similar but not identical effects of endogenous opioids in the descending relaxation and ascending contraction of the peristaltic reflex were reported in the rat colon.[500] More information about enteric nerves, their projections, the locus of receptors to their mediators, and the responses they initiate is required. In addition, it appears that the system has to be studied *in vivo* to elicit its full range of effects.

2. Role of Opioids for the MMC

In humans and omnivore animals such as dogs, a typical cycling electrical pattern, migrating myoelectric complex (MMC), is present in the fasted state which is disrupted by feeding and replaced by regularly occurring action potentials in the stomach and irregular spiking activity throughout the small intestine.[559] Small bolus doses of morphine given i.v. in the fasted state initiate premature MMC complexes in dog[560-561] and man.[563] When given in the fed state, morphine is able to interrupt the fed pattern and induce MMC activity.[562,564] Telford[562] demonstrated that an opioid receptor antagonist which can not cross the blood-brain barrier would block this effect, suggesting that the action of morphine is caused by peripheral opioid receptors. Systemic application of the Met-enkephalin analogue Hoe-825 in humans prematurely induced phase III of the MMC starting ectopically in the duodenum.[565] In dogs, i.v. morphine increased the frequency of the interdigestive cycle, whereas Met-enkephalin decreased its occurrence.[560] Systemic application of naloxone (1 to 2 mg/kg bolus plus 0.2 to 1 mg/kg/h) or Win-44,441 ($-$) in conscious dogs significantly prolonged the cycle time of MMCs in the stomach and duodenum from 103 ± 7 to 219 ± 29 min.[565] Naloxone did not completely suppress the occurrence of MMC activity, suggesting that additional mechanisms for the initiation of the MMC might be present.[566,567] Interestingly, naloxone did not block the initiation of MMC activity by motilin, suggesting that the effect of motilin on MMC is not mediated primarily via opioid receptors.[567,568] The dose of naloxone used in these studies was rather high, and nonspecific effects of this naloxone dose[8,9] cannot be ruled out. In another study, naloxone (40 µg/kg/h) abolished the interdigestive motor activity in one half of a cohort of human volunteers and also reduced antroduodenal motility.[569]

In conclusion, endogenous opioids and peripheral opioid receptors might be involved in the initiation of MMCs in the duodenum. The action of motilin on the MMC is apparently mediated by a nonopioid mechanism even though some motility phenomena observed after motilin administration are due to the release of endogenous opioids and to the suppression of tonic VIP release.[9,493] In the proximal small intestine, the same study[9] found motilin to act more prominently through cholinergic than through opioid mechanisns, thus possibly reconciling these several observations.

3. Gastrointestinal Transit

Opioids delay gastrointestinal transit, one mechanism whereby opioids act as antidiarrheal agents.[570] Thus, delay in gastrointestinal transit is mainly by peripheral mechanisms,[571] but centrally administered opioids also can inhibit gastrointestinal transit (for review see Reference 513). Antidiarrheal drugs such as loperamide (which does not readily cross the blood-brain barrier, but interacts with peripheral opioid receptors) are in clinical use for treatment of diarrhea.[572] Other compounds which are used for regulation of gastrointestinal motility, such as trimebutine, have been shown to stimulate gastrointestinal motility via opiate receptors.[573] The differential effects of selective opiate agonists on various gastrointestinal functions might provide the basis for the development of more specific therapeutic agents. Rational development in this area will require the insights about opioid nerves and their receptors and functions noted previously.

G. OPIOID ACTIONS ON DISTAL GASTROINTESTINAL TRACT AND BILIARY SYSTEM

1. Ileocecal Sphincter

Studies carried out in the cat ileocecal sphincter (ICS) showed that the feline ICS contracted to exogenous opiates via µ-receptors located on cholinergic nerves and via a TTX-insensitive pathway involving κ-receptors which was speculated to be located on the smooth muscle of the sphincter.[574] It was also speculated that opioid peptides might participate in the excitatory response of the ICS in response to ileal contractions and during vagal stimulation.[575] If no opioid receptors are really present on gastrointestinal muscle (see Section IV.B.2), it will be necessary to consider the possibility that opioids act in this sphincter to inhibit the output of inhibitory mediator(s) under circumstances in which cholinergic nerves appear to be blocked by TTX (see Section IV.E.5). In humans no effect of morphine on the tone of the ileocecal sphincter was found.[527]

2. Colon and Sphincter Ani Internus

The *in vitro* effects of opioids are rather conflicting. Whereas Met- and Leu-enkephalin are inhibitory in isolated segments of the guinea pig colon[576] and the canine colon,[577] both peptides cause excitation in the isolated distal rat colon.[578-580] The excitatory effect of opioids in the rat colon has been attributed to the release of serotonin or adrenergic transmitters[578,581] or to the involvement of prostaglandins or thromboxanes.[582] More recently Grider and Makhlouf[500] showed that opioids could cause excitation of the isolated (rat) colon by inhibiting VIP output as in the perfused segments of canine small intestine (see Section IV.E.5).

In vivo overall excitatory effects on tonic or phasic activity were observed in rat,[583,584] cat,[585,586] rabbit,[586-588] and dog[589,590] when opioids were administered intravenously. Other studies failed to demonstrate an excitatory effect when opioids were administered intraarterially in the canine colon *in vivo*.[591] The possibility that excitatory actions of opioids in the colon involve sites of action outside the colon wall or that opioids excite by inhibiting release of an inhibitory mediator *in vivo* has not been explored adequately.

In man, morphine decreased or abolished colonic propulsion and peristalsis and increased the tone and intraluminal pressure *in vivo*.[545,592] However, *in vitro* studies failed to demonstrate any excitatory effect of morphine.[593,594] The slowing of the transit was associated with the increase of short spike bursts[595] which are regarded as mixing contractions acting as a brake for propulsion.[596] It was reported that in some conditions, such as in patients with diverticulosis[592] and colitis ulcerosa,[597] that the increased pressure produced by exogenous opioids could be dangerous as a factor contributing to perforation.[598]

3. Colonic Peristalsis

Naloxone caused a concentration-dependent increase of the descending relaxation in the isolated guinea pig or rat colon, whereas the ascending contraction induced by abroad stretching was decreased.[500] The opioid agonist Met-enkephalin and, with less potency, dynorphin and morphiceptin decreased descending relaxation and increased the ascending contraction. Additionally, the release of endogenous dynorphin, the putative mediator, was decreased during the descending relaxation and increased during the ascending contraction.[500] This experiment suggested that endogenous opioids exert a continuous restraint on intrinsic VIP neurons and that the descending relaxation could be mediated in part by the elimination of this opioid restraint.[500] The authors argue that increased opioid activity during the ascending reflex would increase contraction by direct muscle action,[500] but alternative explanations for this excitatory effect seem possible, as discussed earlier.

Morphine significantly delayed transit in the cecum and ascending colon and decreased the number of bowel movements.[599] Naloxone, on the other hand, accelerated transit in the transverse colon and rectosigmoid colon, but had no influence on the number of bowel movements.[599] This suggested that endogenous opioids may play a tonic inhibitory role in the regulation of colonic transit in normal humans.[599]

Leu-enkephalin, Met-enkephalin, and morphine decreased the spontaneous electrical activity of the internal anal sphincter of the cat as well as its excitatory responses to stimulation of the hypogastric nerve *in vivo*.[442] Naloxone also blocked the inhibitory influence of visceral afferents on postganglionic cholinergic fibers running in the hypogastric nerve exciting the internal anal sphincter.[442] It was hypothesized that enkephalins act on intramural sympathetic nerve endings by reducing the release of noradrenaline.[442] In contrast, in the opossum, loperamide caused a naloxone-sensitive rise in the internal anal sphincter pressure and a decrease in the rectal distention induced internal sphincter relaxation.[600]

In humans, morphine caused excitation of the rectum which was blocked by naloxone, but not by atropine or by spinal injury.[601] The excitatory effect of morphine was absent in the aganglionic segment of patients with Hirschsprung's disease, indicating that it has no direct effect on muscle.[601]

4. Action of Opioids on Extraintestinal Sites to Affect Motility

Endogenous opioids not only act on intrinsic neurons affecting gut motility, but also influence efferent and afferent nerve impulses from the prevertebral ganglia,[489] the spinal cord,[602,603] and the central nervous system.[604-606] Neurotransmitter release of afferent neurons induced by colonic distention (releasing probably ACH and SP) can be inhibited by activation of enkephalinergic spinal neurons.[489,490] A detailed analysis of these actions is beyond the scope of this chapter.

5. Biliary Tract and Sphincter of Oddi

Morphine exerts a profound effect on the biliary tract, resulting in spasm and increased biliary pressure;[607] this effect can partly be prevented by anticholinergic drugs. A detailed study on the effect of endogenous and exogenous opioids in the neural control was carried out in the sphincter of Oddi (SO) of the cat. Leu- and Met-enkephalin caused a biphasic response with an initial sphincter of Oddi contraction followed by a prolonged inhibition. Pharmacological analysis suggested that two receptor sites are present, one on serotonergic nerves releasing 5-hydroxytryptamine and the other on noncholinergic, nonadrenergic inhibitory neurons.[608] Naloxone caused a dose-dependent reduction of motor activity in the sphincter of Oddi, suggesting that opioids might participate in the neural control of the basal SO motor activity.[608]

H. PATHOPHYSIOLOGICAL ACTION OF OPIOIDS ON GASTROINTESTINAL MOTOR FUNCTION

So far only a little is known about a possible pathophysiological role of endogenous opioid for gut motility disorders. Kreek et al.[609] reported that in two patients with idiopathic constipation, opioid receptor blockage with naloxone relieved the clinical symptoms even when naloxone was given orally. A beneficial effect of naloxone has been confirmed in eight elderly constipated patients.[610] Since naloxone has weak systemic actions after enteral application, an action at peripheral opioid receptors is likely. Therefore, an increased activity of endogenous opioids might be responsible for the clinical symptoms in these patients.

The opioid antagonist naloxone was able to block motor effects induced by stress and labyrinth stimulation,[611] and a hormonal action of β-endorphin was suggested as a physiological mediator of this response. In addition, β-endorphin is released during stress,[611,612] and i.v. infusion of β-endorphin, in doses mimicking blood levels observed after stress, can induce similar changes of gastrointestinal motility.[511]

I. CONCLUSION

That opioid peptides act as neurotransmitters in the intrinsic nervous system is generally accepted. The effect of opioid peptides is highly dependent on the region of the gastrointestinal tract, the species, the study conditions, and the compounds used. There are several endogenous opioids in the gastrointestinal tract and several receptor subtypes. The various opioid agonists and their receptors have different loci, and this, together with the lack of highly selective antagonists for opioid receptor subtypes, complicates analysis. There is some evidence that endogenous opioids might be involved in severe forms of constipation. This ability to constipate is beneficial when opioid agonists are used as antidiarrheal agents. Regarding the physiological involvement in motility, the data available using naloxone as an antagonist suggest some physiological role in the gastrocolonic response[522] and in inter-digestive motility. It is, however, necessary to consider that naloxone can block all opiate receptor subtypes, although with higher affinity for μ-receptors. The effects of a truly selective antagonist at only one receptor type are unknown. There is evidence to suggest that different opioid receptors will exert a different, sometimes even opposite action on gastrointestinal motility. Selective blockade or activation of each of these receptor types might help to understand its action. In this context, further development of selective antag-onists such as ICI-174,864, CTP, or TENA and application of experiments using these antagonists under physiological conditions are needed to characterize the physiological role of endogenous opioid peptides for gastrointestinal motility. Concerning the mechanisms underlying opioid effects, recent research strongly indicates the need to evaluate the locus of their receptors carefully and to discover whether all their actions can be explained by inhibition of release of either excitatory or inhibitory neurotransmitters.

VII. MOTILIN

Motilin is a 22-amino-acid peptide which was first isolated[613] and sequenced from the porcine duodenum.[614] In the gut, motilin is mainly (if not exclusively) localized in endocrine cells.[615,616] Motilin is present in specialized motilin cells (Mo cells). These are relatively small cells located mainly in the crypts with small osmiophilic granules. Motilin also occurs in the heterogenous group of EC-cells; these cells are characterized by their ultrastructural features and their content of 5-HT.[617] The highest concentration of motilin immunoreactivity is in the mucosa of duodenum and proximal jejunum. There is some indirect evidence that motilin-like immunoreactive material might be present in neural elements of the intrinsic nervous system[618] as well as in the CNS.[619-621] There is also some evidence that precursor forms or C-terminally extended forms might be present as a higher molecular weight form in human gut and plasma.[622] Recently, the sequence of the human cDNA encoding motilin and its 115-amino-acid precursor, prepromotilin, has been reported.[623] The motilin peptide is flanked by an N-terminal 25-amino-acid signal peptide and by a 68-amino-acid C-terminal peptide of unknown function.

Human and porcine motilin have an identical amino acid sequence, whereas canine motilin is characterized by differences in five amino acids at residues 7, 8, 12, 13, and 14.[623,624] Studies on the structure-activity relationship suggest that the biological action of motilin is dependent on the whole peptide sequence[625,626] even though replacements of certain amino acids (e.g., 13-Nle-motilin) did not change biological activity of motilin.[625]

A. ERYTHROMYCIN AND MOTILIDES

Erythromycin (EM), a widely used macrolide antibiotic, has been shown to induce gastrointestinal activity similar to the IMC when given i.v. or orally. Since this effect was associated with an increase of plasma motilin, it was first speculated that erythromycin might act via release of endogenous motilin.[627]

Recent studies, however, demonstrated that EM and more potent synthesized analogues, which were called motilides, act directly on the motilin receptor, as demonstrated by receptor binding studies.[628,629] Changes of the chemical structure produced potent nonpeptide agonists for the motilin receptor, some more than 2000 times more potent than EM; e.g., the compound EM-536 (8,9-anhydroerythromycin A 6,9-hemiketal-propagyl-bromide) showed an EC_{50} of 5.3 nM in exciting the isolated rabbit duodenum and displaced [^{125}I]motilin binding with an IC_{50} of 3 to 40 nM.[628] Erythromycin was shown previously to be a potent stimulant of gastric motility[630,631] which would induce phase III activity in normal subjects and in patients suffering from diabetic gastroparesis.[630] In these patients, erythromycin (200 mg i.v.) accelerated gastric emptying of liquids and solids to normal levels, suggesting that erythromycin and especially its more potent analogues might have therapeutic value in patients with impaired gastric emptying.[632]

B. MOTILIN RECEPTORS

Using receptor binding studies, motilin receptors have been demonstrated in the corpus, antrum, and upper duodenum, but not in the fundus and ileum of the rabbit[633] and in human intestinal smooth muscle.[634,635] Using porcine motilin, which is structurally different from canine motilin,[636] canine duodenal or ileal circular muscle was excited *in vivo* by a TTX-sensitive mechanism probably involving interaction with both cholinergic and opioid neurons.[9] Circular muscle did not respond *in vitro*. Poitras et al.[636] reported that canine motilin had a direct excitatory effect *in vitro* on longitudinal smooth muscle of canine intestine. The concentrations required to contract this muscle were high (10^{-6} to $10^{-2} M$). Receptor binding studies using iodinated porcine and canine motilin demonstrated motilin receptors in the human gut, but failed to demonstrate them in the canine antroduodenal region.[635] Such receptors are expected to occur only on nerves in the dog.[3,637] and the techiques used may miss them. Comparison of the biological activity of canine and porcine motilin in the rabbit duodenum revealed no significant difference in potency.[638] Studies with isolated cells from the guinea pig stomach demonstrated an excitatory effect of motilin, suggesting the presence of muscular receptors.[639] There seem to be species differences as well as possible differences related to study environment regarding the presence of functional motilin receptors in the muscle of the antrum and intestine.

C. RELEASE OF MOTILIN AND ITS RELATION TO THE MMC

There also seem to be marked species differences in the luminal stimulus for the motilin release. In dogs, duodenal alkalization was shown to release motilin and to increase antral and duodenal motor activity, whereas acid was either ineffective or produced only a small increase in motilin levels when high amounts were used.[640-642] In contrast, in man, duodenal alkalization was ineffective, but duodenal acidification caused release of motilin.[643-649] Strunz et al.[650] reported that motilin is released from duodenal mucosa *in vivo* and *in vitro* by bile salts and luminal acid. Intraduodenal fat proved to be an even more potent stimulus for motilin release[643,649] than acid in humans, whereas carbohydrates inhibited motilin release.[651] There is also some evidence that gastric distention might cause motilin release.[652] Interestingly, mixed meals seem to have little or no effect on plasma motilin concentrations.[653]

In dogs, release of immunoreactive-motilin from the canine duodenum was stimulated by vagal stimulation, intra-arterial injection of carbachol, or field stimulation[654] and was reduced by atropine. The increase of IR-motilin at low parameters of stimulation was only detectable in the portal venous blood, not in the systemic circulation. Motilin release from the duodenum was also increased by stimulation of the N. laterjet and excitation of the antrum by injection of carbachol.

In subsequent studies in dogs, Hall et al.[655] demonstrated using two different radioimmunoassays for canine and porcine motilin that exogenous porcine motilin will promote endogenous motilin release by a vagally-independent, mainly (but not exclusively) cholinergic mechanism.[655] This suggests that motilin release may be characterized by positive feedback; consideration of such a mechanism for its release might contribute to the problem in sorting out whether motilin release causes or is caused by the MMC (see Section VII.C).

Plasma motilin levels normally show a cycling pattern in all mammal studies. High motilin levels parallel the occurrence of the interdigestive motor pattern[626,656-659] (see Figures 13 and 14). Peak levels occur in the dog when the phase III activity of the migrating motor complex (MMC) is traveling through the proximal duodenum.[641,660] In detailed studies in

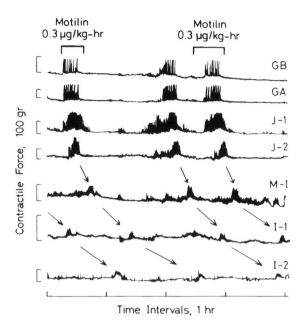

FIGURE 13. Effect of i.v. infusion of motilin in a dose of 0.3 µg/kg/h during the interdigestive state. Motilin induced a premature MMC cycle which resembled the physiologically occurring MMC. (From Itoh, Z., Honda, K., Hiwatshi, K., Takeuchi, S., Aizawa, I., Takayanagi, R., and Couch, E. F., *Scand. J. Gastroenterol.,* 11 (Suppl. 39), 93, 1976. With permission.)

humans, the motilin peaks seemed to precede or coincide with the duodenal MMC activity.[661] The mechanism by which motilin release occurs cyclically and whether it is the cause or the consequence of the MMC are still controversial. The periodic release of motilin is not due to a periodic extrusion of acid into the duodenum during the MMC activity, since it is not affected by blocking gastric acid secretion with cimetidine.[662] It has been shown that cyclic motilin release is mediated by a non-vagal mechanism. Cyclic occurrence of plasma peaks of motilin persisted in dogs after vagotomy near the diaphragm and during vagal cooling at the cervical level.[663,655]

When the MMC activity in the upper gut was interrupted by i.v. infusion of gastrin or CCK and a digestive motility pattern of irregular activity was produced, the cycling changes of motilin were not abolished.[641] A similar result was obtained when the MMC was interrupted by intracerebroventricular injection of CRF.[664] This indicates that cyclic changes of motilin are not sufficient to induce MMC activity and that the cyclic changes of motilin are unlikely to be only the consequence of the cyclic motility changes during the MMC, as postulated.[665,660]

There is good evidence that the activity fronts originating in the stomach and in the duodenum coincide with elevated plasma motilin levels and are at least partly dependent on circulating motilin,[657,666] but ileal MMC activity and propagation are apparently independent of motilin blood levels.[667] When motilin levels were suppressed using pancreatic polypeptide, MMC activity started from lower regions of the small bowel, and an additional infusion of motilin along with PP restored the MMC activity in the stomach.[661] Similar results were obtained in humans using somatostatin to suppress motilin levels.[668] The strongest evidence for a physiological role of endogenous motilin in the regulation of the MMC was obtained by immunoneutralization using rabbit antimotilin serum in dogs. Immunoneutralization abolished the regular occurrence of the MMC in the antrum and duodenum.[669] It also caused a

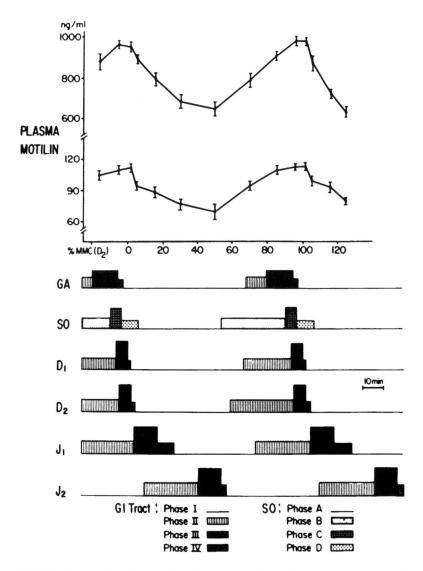

FIGURE 14. Periodical changes of motilin plasma levels synchronous to the MMC in the proximal gastrointestinal tract of the opossum. GA, gastric antrum; SO, sphincter of Oddi; D1 and D2, proximal and distal duodenum; J1 and J2, proximal and distal jejunum. The motilin plasma levels measured with two different radioimmunoassays are given in the upper panel. (From Takahashi, I., Honda, R., Dodds, W. J., Sarna, S., Toouli, J., Itoh, Z., Chey, W. Y., Hogan, W. J., Greiff, D., and Baker, K., *Am. J. Physiol.*, 245, G476, 1983. With permission.)

decrease of plasma motilin levels and an abolition of the cyclic plasma changes of motilin and produced a motility pattern similar to the digestive pattern. Antral and duodenal MMC reappeared when the motilin levels reached their usual peak concentrations again. In contrast to the proximal gut, the distal jejunum and ileum developed an irregular MMC-like activity. This activity usually did not propagate aborally, suggesting that motilin might be of importance for coordination of the MMC propagation.[669] An earlier study using antimotilin serum given at the onset of the MMC suggested that the motor pattern was not influenced, but this could have been due to the inadequate amount of antisera used in this study.[670]

The effect of exogenous motilin infusion *in vivo* is influenced by the fasting and fed state of the animal. In the fed state, i.v. administration of motilin has little effect. In the fasted state, motilin induces premature activity fronts of the MMC of the upper gastrointestinal tract in dogs[626,671] and humans[657] which are similar to spontaneously-occurring motor complexes. They could be suppressed by feeding,[626,641] by atropine,[672] or by somatostatin.[673]

D. MOTILIN ACTIONS ALONG THE GASTROINTESTINAL TRACT
1. Esophagus and LES
In vitro the canine LES did not respond to motilin,[674] whereas the opossum LES showed an atropine-sensitive excitatory response to motilin.[675]

Intravenous infusion of motilin *in vivo* caused an increase in the canine LES pressure together with a marked increase in intragastric pressure.[674,676] This effect involved muscarinic and nicotinic neurotransmission in the dog[677] and in the opossum.[675] A similar increase in LESP was demonstrated in the opossum[678] and in humans.[679]

Initially, a good correlation between the rise of plasma motilin levels induced by duodenal acidification and lower esophageal sphincter pressure was reported,[680] but this could not be verified in a subsequent study.[645] The excitation of the LES in response to i.v. motilin infusion is related to the induction of the MMC; therefore, a functional role of motilin for the control of interdigestive motor activity of the LES has been suggested.[675-677,681]

2. Gastric Emptying
Intravenous infusion of 13-Nle-motilin in concentrations mimicking physiological plasma levels[682] decreased liquid gastric emptying in three healthy volunteers.[683] Subsequent studies, however, demonstrated an acceleration of solid emptying in man[653] and of liquid emptying in dogs.[684] The increase in gastric emptying can, at least in part, be attributed to the rise of intragastric pressure observed after motilin infusion.[227] Besides increasing gastric emptying, motilin also seems to accelerate gastrointestinal transit.[685] As mentioned before, erythromycin, which has been shown to act on motilin receptors, normalized gastric emptying in patients with diabetic gastroparesis.[632] How motilin affects the pylorus is unknown, and its site(s) of action in accelerating gastric emptying are not entirely clear.

3. Stomach and Small Intestine
The *in vitro* effect of motilin was highly dependent on species and on study environment. No effect of porcine motilin was found *in vitro* on the stomach or small intestine of the dog,[3,674,686] guinea pig, or rat.[687,688] Porcine motilin increased the myoelectrical activity in the extrinsically denervated stomach, but not in the extrinsically denervated jejunum in dogs.[689,690] In transected and reanastomosed duodenal, jejunal, and ileal segments, where the MMC cycles are uncoupled (probably due to uncoupled intrinsic oscillators), a distally decreasing sensitivity for motilin to induce premature MMCs was demonstrated.[691] Motilin given intra-arterially in canines was shown to act entirely by tetrodotoxin-dependent mechanisms.[637] There was decreasing sensitivity to its excitant effects distally, and these effects were more susceptible to naloxone distally and to atropine proximally. Both antagonists were required for blocking excitation in both regions. In another study of the local effects of intra-arterially injected peptide, motilin exhibited a unique ability to stimulate contractions which spread orally from the perfusion site. These propagated contractions could be blocked by hexamethonium, TTX, or mucosal application of lidocaine.[692] It was concluded that motilin might be involved in the intrinsic mucosal reflex. Motilin has also been shown to increase luminal 5-HT release in isolated canine jejunum by a cholinergic pathway, and the 5-HT release seemed to be responsible for the motility effects induced by motilin.[693] Since luminal 5-HT has been suggested to participate in myenteric reflex activity,[694] this could add additional support to the involvement of motilin in this activity. This interesting hypothesis awaits further evaluation and confirmation.

In the canine ileum *in vivo,* motilin was shown to excite motor activity by both a cholinergic and an opiate-mediated pathway releasing endogenous opiates.[9] Subsequently, it was demonstrated in *in situ* perfused ileal segments that motilin decreases the tonic release of VIP immunoreactivity from the canine ileum, probaly via an opiate-mediated pathway. This could explain the noncholinergic, TTX-sensitive excitatory effect of motilin observed in the canine ileum *in vivo*.[695] This also could be relevant to its action in initiating MMC activity and in promoting the myenteric reflex.

4. Colon and Sphincter Ani Internus

The effect of motilin is mainly confined to the proximal gut, and little information is available that it influences colonic muscle and motility.[696]

5. Gallbladder, Bile Duct, and Sphincter of Oddi

Intravenous infusion of motilin-induced gallbladder and sphincter of Oddi contractions in association with the onset of the MMC in dogs and the opossum;[658,697,698] see Figure 14. Human gallbladder strips did not respond to motilin *in vitro*.[699] In the cat, i.v. motilin increased sphincter of Oddi tonic pressure and phasic activity *in vivo* in concentrations which were ineffective on gallbladder contraction.[700] This caused a decrease of flow through the sphincter of Oddi. The effect was independent of extrinsic, but dependent on intrinsic, innervation and was mediated by multiple neural pathways involving cholinergic, opiate, and serotinergic mechanisms.[700] The physiological role of motilin on biliary tract motility, if any, remains uncertain.Since the MMC cycle is preceded by an increase in bile flow, motilin might play a role in the coordination of the exocrine changes occurring during the MMC.

E. PATHOPHYSIOLOGICAL ROLE

It has been speculated that motilin, which was found to be increased in patients suffering from diarrhea, could contribute to the motility changes observed in these patients.[701,702] More than 50% of patients with carcinoid syndrome had increased plasma levels of motilin which correlated with increased incidence of diarrhea in these patients. Since the carcinoid tumor itself did not contain motilin-IR, it was speculated that motilin is released by another hormonal mechanism.[703] Elevated plasma levels for motilin have also been reported in patients with small cell carcinoma of the lung[704] and in patients with renal failure.[705] Interestingly, elderly patients were shown to have increased plasma motilin levels with an almost complete absence of gastric phase III activity.[706] It was speculated that this might either be due to receptor down-regulation similar to that found in type 2 diabetes or to the presence of autoantibodies in the plasma of these patients.

Patients with diabetic gastroparesis were shown to lack antral phase III activity in spite of elevated and periodically cycling motilin levels. Metoclopramide induced phase III activity in these patients, but apparently caused a fall of plasma motilin levels.[707] In recent studies, infusion of erythromycin proved beneficial for diabetic gastroparesis. It increased gastric emptying[630] and was able to reverse the delayed emptying of liquids and solids in these patients.[632]

In contrast to diabetic patients, those with an idiopathic delay in gastric emptying also had reduced gastric antral activity, but this was accompanied by a reduction of motilin peaks and decreased integrated motilin output.[708] Whether these disease states are caused primarily by impaired motilin release or by an impaired motor activity leading secondarily to impaired motilin release needs further study. However, stimulation of motilin receptors (e.g., using erythromycin) seems to be a possible way to improve these motor abnormalities.[632]

F. CONCLUSION

Motilin seems to be derived primarily from endocrine cells in the upper intestine. Very little is known about the physiological role of motilin for the LES, biliary tract, and colonic motility, but there are several lines of evidence supporting a physiological role of motilin in the occurrence of the MMC.

Initiation of the MMC in the upper GI tract is dependent on intrinsic neural activity,[665] vagal innervation[709] and activity,[710] and motilin.[669] Motor activity initiated by vagal activity[710] or intrinsic neural activity[665] will release motilin which cannot be detected in the peripheral plasma due to the high background release of the jejunum, the clearance function of the liver, and the relatively high clearance rate ($t_{1/2}$ = 4.8 min) in the peripheral circulation.[682] The released motilin could exert a positive feedback mechanism on the motor activity of the stomach and the proximal duodenum by (1) releasing ACH from intrinsic or extrinsic cholinergic neurons which could again stimulate motilin release, (2) stimulating the smooth muscle either to excite the smooth muscle or to potentiate the action of acetylcholine,[711] (3) further promoting the release of motilin (positive feedback), and (4) the release of endogenous opioids[9] which will turn off tonic release of VIP.[695] These factors, which would all lead to an enhancement of contractile activity, provide a possible interacting mechanism to explain how the initiation of MMC activity originating in the upper gut could be consolidated and amplified.

VIII. NEUROTENSIN

The tridecapeptide neurotensin (NT) was first isolated from bovine hypothalamic extracts[712] and has subsequently been localized and isolated from bovine[713,714] and human intestine.[715] It is mainly localized in open-type endocrine N-cells throughout the gut, with the highest density in the distal ileum.[716-718] Besides its localization in endocrine cells, there is some evidence for the presence of NT-immunoreactive fibers in the intrinsic nervous system of the rat,[386] guinea pig,[451,719] and dog.[440,720] The highest density of nerve fibers occurred in the stomach and in the duodenum. Using a monoclonal antibody against the C-terminal end of neurotensin, NT-containing nerve cells were mainly found in the submucous plexus in immunocytochemical studies, and neurotensin-like immunoreactive material could be released from submucosal nerves in short-term culture.[721] However, other studies using several antibodies against NT found NT only in endocrine cells and failed to detect neurotensin-like immunoreactivity (NTLI) in myenteric neurons.[348,716,717,722] Since in only one study was a chemical characterization of NT-like material in nerves attempted,[277] the presence *in vivo* of true neurotensin, rather than a related compound recognized by some antibodies against neurotensin in nerves, needs further chemical confirmation.

A. MOLECULAR FORMS AND DEGRADATION OF NEUROTENSIN

The majority of NTLI present in the plasma in the basal state or after meal ingestion is due to N-terminal fragments of neurotensin NT_{1-13}, especially NT_{1-8} and NT_{1-11}; see Table 10.[723-727] Determination of the percentage of NT_{1-13} of total NTLI depends on the use of both N-terminally or C-terminally directed antibodies and varies from 20 to 50%[723] down to 5%.[727] The relationship varies in the basal and stimulated state; in some studies NT_{1-13} levels in the basal state were below detection levels whereas NT_{1-8} was present.[725] These N-terminal fragments are probably the consequence of rapid degradation of NT_{1-13}, since the total content of NT_{1-8} and NT_{1-11} in ileal extracts is <1% of total NTLI.[715] A rapid degradation of NT_{1-13} even within the gut wall has been demonstrated,[728-731] generating as the major breakdown products the N-terminal fragments NT_{1-11} and NT_{1-8}.[723,731]

The main degrading enzymes involved in the canine small intestine were endopeptidase-24.11 and angiotensin-converting enzyme (as a secondary enzyme degrading NT_{1-11}).[731,732] Actions of these enzymes generate N-terminal fragments which are not biologically active; the C-terminal hexapeptide of NT, NT_{8-13}, contains all structural requirements for binding to NT-receptors and expression of biological activity.[733-735] This C-terminal region is highly

TABLE 10

Neurotensin (NT), Neurotensin Fragments, and Related Peptides

NT	pGlu-Leu-Tyr-Glu-Asn-Lys-Pro-Arg-Arg-Pro-Tyr-Ile-Leu-OH
(Gln⁴)-NT	pGlu-Leu-Tyr-Gln-Asn-Lys-Pro-Arg-Arg-Pro-Tyr-Ile-Leu-OH
Fragments	
NT_{1-8}	pGlu-Leu-Tyr-Glu-Asn-Lys-Pro-Arg-OH
NT_{1-11}	pGlu-Leu-Tyr-Glu-Asn-Lys-Pro-Arg-Arg-Pro-Tyr-OH
Neuromedin N	Lys-Ile-Pro-Tyr-Ile-Leu-OH
Avian peptides	
[Ser⁷]NTᶜ	Glu-Leu-Tyr-Glu-Asn-Lys-Ser-Arg-Arg-Pro-Tyr-Ile-Leu-OH
LANT-6 = Lys⁸,Asn⁹, NT_{8-13}	Lys-Asn-Pro-Tyr-Ile-Leu-OH
Xenopsin	pGlu-Gly-Lys-Arg-Pro-Trp-Ile-Leu-OH
Neurotensin-related peptide	
Kinetensin = NRP	Ile-Ala-Arg-Arg-His-Pro-Tyr-Phe-Leu

preserved in the different species. The half-life of exogenously infused NT was 1.4 min, whereas the NT_{1-8}-fragment was much more stable in the circulation ($t_{1/2}$ = 30 min).[724] The rapid degradation of NT_{1-13} raises the question as to how neurotensin exerts a physiological role as a hormone. Perhaps neurotensin also acts as a paracrine mediator or as a neurotransmitter.

B. RELATED PEPTIDES*

A hexapeptide with an identical C-terminal tetrapeptide sequence has been isolated from the porcine spinal cord and named neuromedin N.[736] Neuromedin N was shown in receptor binding studies to have almost the same affinity to the neurotensin receptor as did neurotensin itself.[737] Additionally, Folkers and co-workers[738] suggested that (Gln^4)-neurotensin also might be a naturally occurring form of neurotensin. The substitution of glutamate for glutamine in position 4 caused no change in the biological activity of the peptide.

From the chicken intestine a slightly different neurotensin[739] as well as a modified smaller peptide (LANT-6) have been isolated and sequenced.[740] This hexapeptide showed a weak agonist activity at the mammalian neurotensin receptor.[735] As for most other neuropeptides and hormones, no potent antagonist is available for neurotensin. A compound substituted at the isoleucine^M with D-tryptophan retained its binding activity, but showed some inhibitory effect.[741,742]

C. NEUROTENSIN RECEPTORS

Distinct neurotensin receptors have been identified with receptor binding studies in the rat stomach[735] and in the canine small intestine.[743] There is clear evidence in both tissues for high- and low-affinity binding site.[735,743] Purification of the rat stomach receptor with covalent cross-linking[744] or determination of the molecular weight by radiation inactivation[745] revealed a single receptor with a molecular weight of 110,000 Da. In the canine small intestine, neurotensin receptors could be identified on purified circular smooth muscle membranes as well as on purified synaptosomal membranes from the deep muscular plexus and the submucous plexus, suggesting a dual location of neurotensin receptors on nerves and smooth muscle membranes (see Chapters 7 and 8). No binding to neurotensin receptors was found in longitudinal muscle membranes or synaptosomes of the myenteric plexus. Besides the presence of neurotensin receptors on some smooth muscles and neural tissues, there is also strong evidence of neurotensin receptors on mast cells,[746,747] and it has been suggested that systemic and gut effects of NT are mediated partly by release of mediators from mast cells.[746-748]

D. RELEASE OF NEUROTENSIN

Neurotensin-like immunoreactivity (NTLI) is released continuously under basal conditions from the distal ileum, and this basal release can be stimulated by ingestion of a meal.[724,725,749-751] The most potent stimulus for the neurotensin release is intraluminal fat[752] while glucose and amino acids are less or ineffective.[723,752,753] Ileal resection abolished the release of NT.[754] Neurotensin is not only released into the circulation, but also into the lumen of the gut.[749]

As mentioned earlier, NT is rapidly broken down in the circulation into inactive N-terminal fragments, and the amount of neurotensin detected by radioimmunoassay is highly dependent on whether a C-terminally directed or an N-terminally directed antibody is used. The total amount of NT measured with an N-terminal directed antibody can be more than 20 times the amount of the biological active form.[727]

The plasma levels of total NTLI rise relatively rapidly after ingestion of a meal (10 to

---

* See Table 10.

15 min),[725,755] with concentrations reaching a plateau after 20 to 30 min.[725,752,756] It is improbable that the NT release is due to direct contact of food with neurotensin-containing N-cells because these are mainly situated in the lower jejunum and ileum and neurotensin levels rise before food enters that part of the intestine. This is supported by the fact that jejunal perfusion with fat causes a neurotensin release, while perfusion of the ileum or colon has no[757,758] or only a diminished effect.[751] There is good evidence that both a neural mechanism and a hormonal mechanism are responsible for the release of neurotensin. Neurotensin release can be blocked with atropine in humans[759] and in dogs,[760] suggesting the involvement of cholinergic, neural mechanisms for its release. In the rat the release of NT was not blocked by atropine; however, it was inhibited by ganglionic blockade with hexamethonium and by administration of opioids.[761] The mechanisms regulating NT release were studied in isolated endocrine N cells in short-term culture.[762,763] NT released from N cells was stimulated by β-adrenergic stimulation and elevation of cellular cAMP levels as well as by bombesin.[763] A similar stimulating effect of bombesin was found in humans[764] and in the rat after bombesin infusion either into the mesenteric artery[765] or into the isolated, vascularly perfused ileum.[766]

Recently, it has been shown that dietary fatty acids may regulate the genetic expression of neurotensin. Rats with a low or a high fatty acid diet showed significant differences in NT tissue content and the relative amount of NT mRNA, whereas the NT cell density was unchanged.[767] Neurotensin, on the other hand, increased the translocation of fatty acids from the lumen of the intestine into the lymph.[768]

E. ACTIONS OF NEUROTENSIN IN THE GASTROINTESTINAL TRACT

Neurotensin causes different effects on gastrointestinal motility in different parts of the gut. There are also species differences and differences related to the mode of application of neurotensin and the experimental conditions, e.g., *in vivo* vs. *in vitro*.[3] Besides its hormonal effects, paracrine effects and actions as neurotransmitter seem possible.[769] A more detailed discussion of these issues follows.

1. Esophagus and LES

(Gln[4])-neurotensin infusion in man reduced the lower esophageal sphincter pressure significantly.[770,771] After cholinergic blockade with atropine, the inhibitory effect of NT was blocked.[772] Because atropine itself lowers the basal LES pressure in humans, the loss of the inhibitory response to NT could be due to the inability of neurotensin to reduce the basal pressure below that achieved after administration of atropine. Alternatively, neurotensin reduces the LES pressure generated by tonic activity of the cholinergic innervation in man. There is evidence that the LESP decreases after ingestion of fat;[773] however, there is no clear evidence that the release and subsequent action of neurotensin are involved in the physiological mediation of this response.

2. Stomach

In strips from the rat stomach studied *in vitro*, neurotensin caused a contraction by a direct smooth muscle action,[774] and NT-receptors with similar affinity were demonstrated on smooth muscle membranes from the rat stomach.[775]

In vivo vagally innervated antral pouches of conscious dogs were more sensitive to the inhibition of motor activity induced by i.v. neurotensin than were vagally denervated fundic pouches, suggesting that an intact extrinsic innervation participates in this inhibition.[776]

Circular muscle strips of the canine gastric corpus responded *in vitro* with a relaxation at low concentrations and a biphasic response with initial excitation, followed by a relaxation at higher concentrations. The contractile effect in higher concentrations was inhibited by histamine antagonists and was probably due to the degranulation of mast cells, whereas the inhibition was due to a direct (nerve-independent) action on smooth muscle.[777]

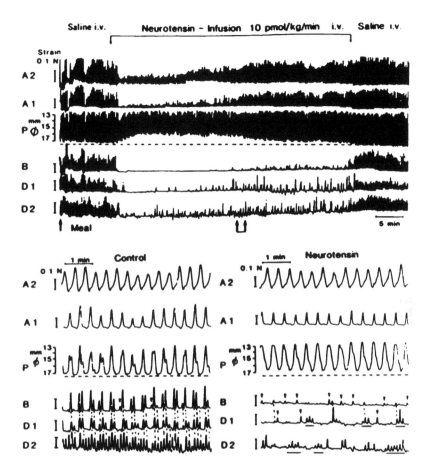

FIGURE 15. Effect of neurotensin i.v. on the canine antroduodenal region and pyloric diameter with low (A) and with high (B) paper speed. The pyloric diameter during saline infusion is marked with a horizontal dashed line. Sequential duodenal contractions at adjacent recording sites representing "propagated waves" are marked with dotted lines. Isolated duodenal contractions at one recording site without contractions of adjacent sites ("stationary individual contractions") are marked with arrowheads. Groups of 3 to 5 repetitive contractions (clustered contractions) are marked with horizontal lines. (From Keinke, D., Wulschke, S., and Ehrlein, H. J., *Digestion*, 34, 281, 1986. With permission.)

In humans, neurotensin infusion increased contraction frequency of the duodenum, decreased contraction frequency of the antrum, and reduced the number of antral contractions migrating into the duodenum.[772] The mean pressure gradient between antrum and duodenum increased in spite of an unchanged baseline pressure.[778] This change of the pressure gradient together with the reduced antroduodenal coordination could explain the delayed gastric emptying after neurotensin infusion.[750,779,780] In electrophysiological studies of the human antral muscle using intracellular recording, neurotensin increased tonic contractile activity through both cholinergic and capsaicin-sensitive nerves, but increased phasic activity by a direct smooth muscle action.[781] There was no explanation for the difference between *in vitro* and *in vivo* effects of neurotensin.

The effects of i.v. neurotensin in the canine gastroduodenal region were analyzed in an elegant study with combined measurement of gastric emptying, gastric and duodenal motility, and pyloric diameter together with fluoroscopic studies;[780] see Figure 15. Neurotensin reduced antral contractions without altering antral frequency, it decreased the diameter of the pylorus and duodenal lumen size, it inhibited duodenal contractile activity, and it reduced both coordinated propagated contraction in the duodenum and gastroduodenal coordinated activity.[780] These factors contributed to an increased outflow resistance and were held responsible for the observed delay in gastric emptying.[780] A similar delay of gastric emptying was observed in humans[750] and in rats.[779] It was also speculated that neurotensin could be involved in the delay of gastric emptying due to fat because fat is a very potent releaser of neurotensin.[782,783] On the other hand, patients suffering from dumping syndrome have an excessive rise of NTLI, presumably occurring as a result of the rapid passage of nutients[784] through the intestine.

Neurotensin effects on the pyloric sphincter have been studied only in dogs. When pyloric activity was studied *in vivo* using a combined sleeve side hole manometry in anesthetized dogs, NT given by close intra-arterial injections caused excitation of the basal motor activity and inhibited neurally activated pyloric motor activity.[813] Whether effects of neurotensin on pyloric tone contribute to causing delay in gastric emptying is unknown.

3. Small Intestine

When given locally in the canine small intestine by close intra-arterial injection, NT inhibited field-stimulated, atropine-sensitive contractile responses in the duodenum and in the ileum. α-Adrenergic blockade with phentolamine (nonselective for α-adrenoceptor subtypes) or yohimbine (α_2-selective), but not prazosin (α_1-selective), reduced the inhibitory effect of NT.[785] Neural blockade with TTX also shifted the dose-response curve of NT to the right. However, full inhibition could still be obtained, suggesting an additional, less sensitive smooth muscle receptor.[785]

The neurotensin response in the canine small intestine *in vitro* consisted of three concentration-dependent effects. At low concentrations (10^{-12} to 10^{-9} M), circular muscle responded with an increase in frequency and amplitude of spontaneous phasic contractions at the frequency of slow waves. In concentrations ranging from 10^{-8} to 10^{-7} M, neurotensin inhibited spontaneous phasic activity; at even higher concentrations, a tonic contraction occurred.[748] None of these responses were blocked by tetrodotoxin, and a participation of arachidonic acid metabolites in the excitatory response was suggested using inhibitors of its transformation.

In the single sucrose gap, neurotensin caused hyperpolarization in concentrations which had a direct smooth muscle effect. This was apamine sensitive and not totally dependent on external calcium[786] and was attributable to release of intracellular Ca^{2+} together with entrance of Ca^{2+}, which opened K^+ channels. Before the hyperpolarization ceased, there was the initiation of contraction. This was blocked by inhibitors of L-type Ca channel activity and was attributed to opening of such a channel by neurotensin. These findings accounted for the inhibition of phasic contractions and the production of tonic contractions by neurotensin. Production of increased phasic contractions by low concentrations of neurotensin probably required the presence of slow waves, which were usually absent in isolated circular muscle strips in the sucrose gap. Neuromedin N produced identical responses with complete cross-tachyphylaxis with neurotensin.

The inhibitory effect of NT on circular muscle might be responsible for the retardation of the transit of chyme through the canine small intestine and the delayed emptying through the ileocecal valve.[779,787] Using a computerized analysis of motility data, neurotensin infusion was shown to decrease the transit rate of a cellulose meal by increasing stationary intestinal contractions in dogs;[788] see Figure 16. Recent experiments suggest that slowing of gastric emptying and gastrointestinal transit in dog also occurs in rats and that an intact vagal innervation is of importance for this effect.[789]

As a result of these motility changes, the chyme is kept longer in the stomach and the small intestine. This delay might facilitate digestion and absorption of nutrients. This function of neurotensin in the postprandial state has been speculated as a possible physiological role of the peptide.[752,783,790]

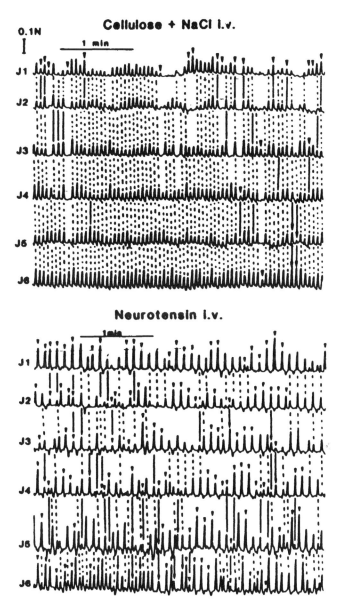

FIGURE 16. Influence of neurotensin (i.v.) on canine small intestinal motility and propagation of a cellulose meal. Neurotensin (lower panel) decreased the number of propagated waves, indicated by the dashed lines when compared to the saline infusion (upper panel). (From Schemann, M., Siegle, M. L., Sahyoun, H., and Ehrlein, H. J., Z. *Gastrolenterol.*, 24, 262, 1986. With permission.)

In the guinea pig ileum *in vitro*, neurotensin causes a biphasic response consisting of a relaxation followed by a contraction. The contractile effect was neurogenic, since it was blocked completely by tetrodotoxin and partially by atropine.[791,792] The atropine-resistant component of contraction could be antagonized by substance P desensitization.[793] The relaxant action of neurotensin was apparently directly on myogenic receptors,[733,792] involving activation of the Ca^{2+}-dependent K^+ channel as in the canine intestine.[794-796]

Intravenous neurotensin infusion in man changed the normal interdigestive motility in the duodenum[797] and the antroduodenal region[772] into an irregular fed-type motility pattern. In rats a similar effect was observed, and the inhibition of the interdigestive migrating myoelectric complex (MMC) was blocked by atropine or hexamethonium. Whereas the inhibition of the MMC seemed to involve cholinergic nerves, neurotensin induced irregular spiking activity by a direct myogenic action.[798] The effect of i.v. neurotensin is in sharp contrast to its effect given intracerebroventricularly (ICV) in the fasted and fed state. When given ICV, this peptide restored the fasting motor pattern, and this effect was mediated vagally. This suggests that the digestive influences responsible for the fed pattern are mediated through a central structure on which neurotensin can either act directly or indirectly.[799]

4. Colon

In vivo i.v. infusion of neurotensin into cats increased muscular tone and antiperistaltic activity in the colon.[392] These neurotensin-induced contractions could be abolished by TTX, hexamethonium, or atropine.[389] In dogs, i.v. (but not ICV) administration of the peptide stimulated colonic motility.[799] The excitatory effect, which was more pronounced in the proximal colon, probably involved prostaglandins because pretreatment with cyclooxygenase inhibitors reduced the neurotensin-induced hypermotility.[800] The prostaglandin synthetase inhibitors also reduced meal-induced motility responses, e.g., the gastrocolonic reflex.[800] In patients and human volunteers, Thor and Rosell[801] reported increased colonic motor activity, with the proximal colon being more sensitive to neurotensin infused i.v. than the distal part. It was also shown that peptide infusion, which produced supraphysiological plasma levels as compared to those observed postprandially, induced defecation in healthy volunteers.[802] Similar results were obtained in humans in other studies.[753,770,797,801] An increased discharge of fecal volume was also observed after neurotensin infusion in patients with colostomy[801] or ileostomy.[803] From these data, a possible participation of this peptide in gastrocolonic reflex was speculated.

Depending on the species, the excitatory effects observed *in vitro* were either direct on smooth muscle, as in the rabbit,[804] or indirect on neural structures, as in the guinea pig.[714] Besides this excitatory effect, there was also evidence for potent inhibitory action of neurotensin in the guinea pig colon.[805] Studies of neurotensin effects on human colon *in vitro* have not been reported.

5. Biliary Tract and Sphincter of Oddi

Neurotensin induced a dose-dependent contraction in the isolated gallbladder, bile duct, and ampulla of guinea pigs. This response was reduced by atropine or TTX.[806] Neurotensin also contracted the gallbladder of dogs *in vivo*.[807] In contrast, Walker et al.[808] demonstrated a dilatation of the gallbladder during NT infusion in humans. Gallstone patients showed a significantly greater release of neurotensin than normal volunteers, and it was speculated that this effect together with a decreased CCK release in these patients could be involved in the pathogenesis of gallstone formation.[808]

F. CONCLUSION

Ingestion of fat decreases lower esophageal sphincter pressure, inhibits the MMC, delays gastric emptying, and produces a strong increase in colonic motility in humans.[113,773,809,810] Similar responses were obtained after i.v. infusion of (Gln⁴)-neurotensin at doses resulting in plasma concentrations similar to those observed after fat ingestion.[392,750,761,770,771,797,801] Therefore, it has been speculated that neurotensin or related analogues might be physiologically involved in these responses[801] and that neurotensin might act as a hormone to coordinate postprandial gastrointestinal activity[778,783] and to act as an enterogastrone, inhibiting gastric acid secretion. Due to the lack of a specific antagonist of neurotensin and the rapid breakdown of neurotensin, definite proof that neurotensin fulfills this role under physiological conditions has not been achieved. Another problem is the uncertainty about whether neurotensin functions only as a hormone released from endocrine cells or might also be released from enteric nerves or (in a paraverine fashion) from endocrine cells. If neurotensin is released from nerves or locally, plasma levels of undegraded peptide may be irrelevant to determination of its physiological role. However, neurotensin is a good hormone and/or neuromediator candidate. In general, the motor effects of neurotensin released by fat in the intestine delay the passage of chyme. Inhibition of gastric emptying and ileal motility and transit could lead to an increased absorption. Because resection of the ileum results in an abolition of its postprandial release, neurotensin could be involved in the motility phenomena observed in short bowel syndrome.

REFERENCES

1. **Furness, J. B. and Costa, M.**, *The Enteric Nervous System*, Churchill Livingstone, Edinburgh, 1987.
2. **Wang, R. Y. and Schoenfeld, R.**, *Cholecystokinin Antagonists*, Vol. 47, Neurology and Neurobiology, Alan R. Liss, New York, 1988.
3. **Fox, J. E. T., Daniel, E. E., Jury, J., Fox, A. E., and Collins, S.**, Sites and mechanisms of action of neuropeptides on canine gastric motility differ *in vivo* and *in vitro*, *Life Sci.*, 33, 817, 1983.
4. **Fox, J. E. T., Collins, S. M., and Daniel, E. E.**, Expression of peptide responses depends upon study environment, in *Regulatory Peptides in Digestive, Nervous and Endocrine Systems*, Lewin, J. and Bonfils, S., Eds., Elsevier, Amsterdam, 1985, 265.
5. **Allescher, H. D., Daniel, E. E., Fox, J. E. T., Kostolanska, F., and Rovati, L. A.**, Effect of the novel CCK-receptor antagonist CR-1392 on CCK induced antro-duodenal and pyloric motor activity *in vivo*, *J. Pharmacol. Exp. Ther.*, 251, 1134, 1989.
6. **Murphy, R. B., Smith, G. P., and Gibbs, J.**, Pharmacological examination of cholecystokinin (CCK-8)-induced contractile activity in the rat isolated pylorus, *Peptides*, 8, 127, 1987.
7. **Allescher, H. D., Kostolanska, F., Tougas, G., Fox, J. E. T., Regoli, D., Drapeau, G., and Daniel, E. E.**, The actions of neurokinins and substance P receptors in canine pylorus, antrum and duodenum, *Peptides*, 10, 671, 1989.
8. **Fox, J. E. T. and Daniel, E. E.**, Exogenous opiates: their local mechanisms of action in the canine small intestine and stomach, *Am. J. Physiol.*, 253, G179, 1987.
9. **Fox, J. E. T. and Daniel, E. E.**, Activation of endogenous excitatory opiate pathways in the canine small intestine by field stimulation and motilin, *Am. J. Physiol.*, 253, G189, 1987.
10. **Allescher, H. D., Ahmad, S., Daniel, E. E., Dent, J., Fox, J. E. T., and Kostolanska, F.**, Inhibitory opioid receptors in canine pylorus *in vivo* and *in vitro*, *Am. J. Physiol.*, 255, G352, 1988.
11. **Collins, S. M.**, Calcium utilization by dispersed canine gastric smooth muscle cells, *Am. J. Physiol.*, 251, G195, 1986.
12. **Collins, S. M. and Gardner, J. D.**, Cholecystokinin-induced contraction of dispersed smooth muscle cells, *Am. J. Physiol.*, 243, G497, 1982.
13. **Bitar, K. N. and Makhlouf, G. M.**, Specific opiate receptors on isolated mammalian gastric smooth muscle cells, *Nature (London)*, 297, 72, 1982.
14. **Bitar, K. N. and Makhlouf, G. M.**, Selective presence of opiate receptors on intestinal circular muscle cells, *Life Sci.*, 37, 1545, 1985.
15. **Yalow, R. and Berson, S.**, Further studies on the nature of immunoreactive gastrin in human plasma, *Gastroenterology*, 60, 203, 1971.
16. **Uvnäs-Wallensten, K., Rehfeld, I. F., Larsson, L. I., and Uvnäs, B.**, Heptadecapeptide gastrin in the vagal nerve, *Proc. Natl. Acad. Sci. U.S.A*, 74, 5707, 1977.
17. **Rehfeld, J. F. and Larsson, L. I.**, The predominating molecular form of gastrin and cholecystokinin in the gut is a small peptide corresponding to their COOH-terminal tetrapeptide amide, *Acta Physiol. Scand.*, 105, 177, 1979.
18. **McGuigan, J.**, Gastrin mucosal intracellular localization of gastrin by immunofluorescence, *Gastroenterology*, 55, 315, 1968.
19. **Walsh, J. H.**, Gastrointestinal hormones — gastrin, in *Physiology of the Gastrointestinal Tract*, Johnson, L. R., Ed., Raven Press, New York, 1987, 181.
20. **Lamers, C., Walsh, J., Janssen, J., Harrison, A., Ippoliti, A., and van Tongeren, J.**, Evidence that gastrin 34 is preferentially released from the human duodenum, *Gastroenterology*, 83, 233, 1982.

21. **Dockray, G., Vaillant, C., and Hopkins, C.,** Biosynthetic relationships of big and little gastrins, *Nature (London),* 273, 770, 1978.
22. **Straus, E. and Yalow, R.,** Studies on the distribution and degradation of heptadecapeptide, big and big-big gastrin, *Gastroenterology,* 66, 936, 1974.
23. **Rehfeld, J., Stadil, F., and Vikelsoe, J.,** Immunoreactive gastrin components in human serum, *Gut,* 15, 102, 1974.
24. **Taylor, I., Dockray, G., Calam, J., and Walker, R.,** Big and little gastrin responses to food in normal and ulcer subjects, *Gut,* 20, 957, 1979.
25. **Walsh, J., Debas, H., and Grossman, M.,** Pure human big gastrin. Immunochemical properties, disappearance half time, and acid stimulating action in dogs, *J. Clin. Invest.,* 54, 477, 1974.
26. **Yoo, O. J., Powell, C. T., and Agrawal, K. L.,** Molecular cloning and nucleotide sequence of full-length cDNA coding for porcine gastrin, *Proc. Natl. Acad. Sci. U.S.A.,* 79, 1049, 1982.
27. **Boel, E., Vuust, J., Norris, F., Wind, A., Rehfeld, J. F., and Marcker, K. A.,** Molecular cloning of human gastrin cDNA: evidence for evolution of gastrin by gene duplication, *Proc. Natl. Acad. Sci. U.S.A.,* 80, 2866, 1983.
28. **Wiborg, O., Berglund, L., Boel, E., Norris, F., Norris, K., Rehfeld, J., Marcker, K., and Vuust, J.,** Structure of the human gastrin gene, *Proc. Natl. Acad. Sci. U.S.A.,* 81, 1067, 1984.
29. **Rahier, J., Pauwels, S., and Dockray, G. J.,** Biosynthesis of gastrin-localization of the precursor and peptide products using electron microscopic-immunogold methods, *Gastroenterology,* 92, 1146, 1987.
30. **Gregory, R. A., Tracy, H. J., Harris, J. L., and Runswick, M. J.,** Minigastrin: corrected structure and synthesis, *Hoppe-Seyler's Z. Physiol. Chem.,* 369, 73, 1979.
31. **Gregory, R. A., Dockray, G., Reeve, J., Shively, J., and Miller, C.,** Isolation from porcine antral mucosa of a hexapeptide corresponding to the C-terminal sequence of gastrin, *Peptides,* 4, 319, 1983.
32. **Tracy, H. J. and Gregory, R. A.,** Physiological properties of a series of synthetic peptides structurally related to gastrin I, *Nature (London),* 204, 935, 1964.
33. **Rehfeld, J. F.,** Tetrin, in *Gut Hormones,* Bloom, S. R. and Polak, J. M., Eds., Churchill Livingstone, Edinburgh, 1981, 240.
34. **Sugano, K., Aponte, G., and Yamada, T.,** Identification and characterization of glycine-extended post-translational processing intermediates of progastrin in porcine stomach, *J. Biol. Chem.,* 260, 11724, 1985.
35. **Azuma, T., Taggart, T., and Walsh, J. H.,** Effects of bombesin on the release of glycine-extended progastrin (gastrin G) in rat antral tissue culture, *Gastroenterology,* 93, 322, 1987.
36. **Takeuchi, K., Speir, G., and Johnson, L.,** Mucosal gastrin receptor. II. Physical characteristics of binding, *Am. J. Physiol.,* 237, E295, 1979.
37. **Baur, S. and Bacon, V.,** A specific gastrin receptor on plasma membranes of antral smooth muscle, *Biochem. Biophys. Res. Commun.,* 73, 928, 1976.
38. **Huang, S. C., Zhang, L., Chiang, V., Wank, S. A., Maton, P. N., Gardner, J. D., and Jensen, R. T.,** Benzodiazepine analogues L-365,260 and L-364,718 as gastrin and pancreatic CCK receptor antagonists, *Am. J. Physiol.,* 257, G169, 1989.
39. **Feldman, M., Richardson, C., Taylor, I., and Walsh, J. H.,** Effect of atropine on vagal release of gastrin and pancreatic polypeptide, *J. Clin. Invest.,* 71, 715, 1979.
40. **Fox, J. E. T., Daniel, E. E., Jury, J., Track, N. S., and Chiu, S.,** Cholinergic control of canine antral immunoreactive gastrin release and motility, *Can. J. Physiol. Pharmacol.,* 60, 893, 1982.
41. **Walsh, J. H., Yalow, R., and Berson, S.,** The effect of atropine on plasma gastrin response to feeding, *Gastroenterology,* 60, 16, 1971.
42. **Yamada, T., Soll, A. H., Sugano, K., Chiba, T., and Todisco, A.,** Isolated canine gastric mucosal cells as models for study of neurohumoral regulation of gastric acid secretion, in *Regulatory Peptides: Mode of Action on Digestive, Nervous and Endocrine Systems,* Lewin, M. J. M. and Bonfits, J., Eds., Elsevier, Amsterdam, 1985, 19.
43. **Elwin, C. E.,** Gastric acid responses to antral application of some amino acids, peptides, and isolated fractions of a protein hydrolysate, *Scand. J. Gastroenterol.,* 9, 239, 1974.
44. **Grovum, W. and Chapman, H.,** Pentagastrin in the circulation acts directly on the brain to depress motility of the stomach in the sheep, *Requl. Peptides,* 5, 35, 1982.
45. **Bennett, A., Misiewicz, J. J., and Waller, S. L.,** Analysis of the motor effects of gastrin and pentagastrin on the human alimentary tract *in vitro, Gut,* 8, 470, 1967.
46. **Lipshutz, W. and Cohen, S.,** Interaction of gastrin I and secretin on gastrointestinal circular muscle, *Am. J. Physiol.,* 222, 775, 1971.
47. **Lipshultz, W., Hughes, W., and Cohen, S.,** The genesis of the lower esophageal sphincter pressure: its identification through the use of gastrin antiserum, *J. Clin. Invest.,* 51, 522, 1972.
48. **Lipshutz, W., Tuch, A. F., and Cohen, S.,** A comparison of the site of action of gastrin I on lower esophageal sphincter and antral circular smooth muscle, *Gastroenterology,* 61, 545, 1971.
49. **Burleigh, D. E.,** The effects of drugs and electrical field stimulation on the human lower esophageal sphincter, *Arch. Int. Pharmacodyn. Ther.,* 240, 169, 1979.

50. **Castell, D. O. and Harris, L. D.**, Hormonal control of the gastro-esophageal sphincter strength, *N. Engl. J. Med.*, 282, 886, 1970.

51. **Cohen, S. and Green, F. E.**, The mechanics of esophageal muscle contraction. Evidence of an inotropic effect of gastrin, *J. Clin. Invest.*, 52, 2029, 1973.

52. **Rattan, S., Coln, D., and Goyal, R. K.**, The mechanisms of action of gastrin on the lower esophageal sphincter, *Gastroenterology*, 70, 828, 1976.

53. **Zwick, R., Bowes, K. L., Daniel, E. E., and Sarna, S. K.**, Mechanism of action of pentagastrin on the lower esophageal sphincter, *J. Clin. Invest.*, 57, 1644, 1976.

54. **Heil, T., Mattes, P., and Raptis, S.**, Effects of somatostatin and human gastrin I on the lower esophageal sphincter in man, *Digestion*, 15, 461, 1977.

55. **Domschke, W., Lux, G., Domschke, S., Strunz, U., Bloom, S. R., and Wünsch, E.**, Effect of vasoactive intestinal peptide on resting and pentagastrin-stimulated lower esophageal sphincter pressure, *Gastroenterology*, 75, 9, 1978.

56. **Corazziari, E., Pozzessere, C., Dani, S., Anzini, F., and Torsoli, A.**, Lower oesophageal sphincter response to intravenous infusions of pentagastrin in normal subjects antrectomized and achalasic patients, *Gut*, 19, 1121, 1977.

57. **Christensen, J.**, Pharmacology of the esophageal motor function, *Annu. Rev. Pharmacol. Toxicol.*, 15, 243, 1975.

58. **Goyal, R. K. and Rattan, S.**, Neurohumoral, hormonal, and drug receptors for the lower esophageal sphincter, *Gastroenterology*, 74, 598, 1978.

59. **Bertaccini, G.**, Peptides: gastrointestinal hormones — gastrin, in *Mediators and Drugs in Gastrointestinal Motility, Vol. 2*, Bertaccini, G., Ed., Springer-Verlag, Berlin, 1982, 11.

60. **Freeland, G. R., Higgs, R. H., Castell, D. O., and McGuigan, J. E.**, Lower esophageal sphincter (LES) and gastric acid (GA) response to intravenous infusion of synthetic human gastrin heptadecapeptide I (HGH), *Gastroenterology*, 68, A894, 1975.

61. **Sturdenvant, R.**, Is gastrin a major regulator of lower esophageal sphincter pressure?, *Gastroenterology*, 67, 551, 1974.

62. **Goyal, R. K. and McGuigan, J. E.**, Is gastrin a major determinant of basal lower esophageal sphincter pressure? A double blind controlled study using high titer gastrin antiserum, *J. Clin. Invest.*, 57, 291, 1976.

63. **Jennewein, H. M., Hummelt, H., Siewert, R., and Waldeck, F.**, The effect of intravenous infusion of synthetic human gastrin-I on lower esophageal sphincter (LES) pressure in the dog and its relation to gastrin level, *Digestion*, 14, 376, 1976.

64. **Calvert, C. H., Parks, T. G., and Buchanan, K. D.**, The relationship of lower esophageal sphincter pressure to plasma gastrin concentration, *Gut*, 16, A403, 1975.

65. **Walker, C. O., Frank, S. A., Manton, J., and Fordtran, J. S.**, Effect of continuous infusion of pentagastrin on lower esophageal sphincter pressure and gastric acid secretion in normal subjects, *J. Clin. Invest.*, 56, 218, 1975.

66. **Itoh, Z., Takayanagi, R., Takeuchi, S., and Issigiki, S.**, Interdigestive motor activity of Heidenhain pouches in relation to main stomach in conscious dogs, *Am. J. Physiol.*, 234, E333, 1978.

67. **Jensen, D. M., McCallum, R., and Walsh, J. H.**, Failure of atropine to inhibit gastrin-17 stimulation of the lower esophageal sphincter in man, *Gastroenterology*, 75, 825, 1978.

68. **Cohen, S. and Harris, L. D.**, The lower esophageal sphincter, *Gastroenterology*, 63, 1066, 1972.

69. **Siewert R., Jennewein, H. M., Arnold, R., and Creutzfeldt, W.**, The lower oesophageal sphincter in the Zollinger-Ellison-Syndrom, *Dtsch. Med. Wochenschr.*, 3, 101, 1973.

70. **McCallum, R. W. and Walsh, J. H.**, Relationship between lower esophageal sphincter pressure and serum gastrin concentration in Zollinger Ellison syndrome and other clinical settings, *Gastroenterology*, 76, 65, 1979.

71. **Siewert, R., Weiser, F., Jennewein, H. M., and Waldeck, F.**, Clinical and manometric investigations of the lower esophageal sphincter and its reactivity to pentagastrin in patients with hiatus hernia. LES-pentagastrin-test, *Digestion*, 10, 287, 1974.

72. **Collins, S. M., Jung, C. Y., and Grover, A. K.**, Calcium utilization by dispersed canine gastric smooth muscle cells, *Am. J. Physiol.*, 251, G195, 1986.

73. **Gregory, R. A. and Tracy, H. J.**, The constitution and properties of two gastrins extracted from hog antral mucosa, *Gut*, 5, 103, 1964.

74. **Jacoby, H. I. and Marshall, C. H.**, Gastric motor-stimulating activity of gastrin-tetrapeptide in dogs, *Gastroenterology*, 56, 80, 1969.

75. **Szurszewski, J. H.**, Mechanism of action of pentagastrin and acetylcholine on longitudinal muscle of the canine antrum, *J. Physiol. (London)*, 252, 335, 1975.

76. **Bennett, A.**, Effect of gastrin on isolated smooth muscle preparations, *Nature (London)*, 208, 170, 1965.

77. **Gerner, T. and Haffner, J. F. W.**, The role of local cholinergic pathways in the motor response to cholecystokinin and gastrin in isolated guinea pig fundus and antrum, *Scand. J. Gastroenterol.*, 12, 751, 1977.

78. **Gerner, T., Haffner, J. F. W., and Norstein, J.**, The effects of mepyramine and cimetidine on the motor responses to histamine, cholecystokinin and gastrin in the fundus and antrum of isolated guinea pig stomachs, *Scand. J. Gastroenterol.*, 14, 65, 1979.

79. **Ohkawa, H. and Watanabe, M.**, Effects of gastrointestinal hormones on the electrical and mechanical activity of the cat stomach, *Tohoku J. Exp. Med.*, 122, 287, 1977.

80. **Fara, J. W. and Berkowitz, J. M.**, Effects of histamine and gastrointestinal hormones on dog antral smooth muscle, *in vitro, Scand. J. Gastroenterol.*, 13 (Suppl. 49), 60, 1978.

81. **Fara, J. W., Praissman, M., and Berkowitz, J. M.**, Interaction between gastrin, CCK, and secretin on canine antral smooth muscle *in vitro, Am. J. Physiol.*, 236, E39, 1979.

82. **El-Sharkawy, T. Y. and Szurszewski, J. H.**, Modulation of canine antral circular smooth muscle by acetylcholine, noradrenaline, and pentagastrin, *J. Physiol. (London)*, 279, 309, 1978.

83. **Morgan, K., Schmalz, P., Go, V., and Szurszewski, J.**, Effect of pentagastrin, G17 and G34 on the electrical and mechanical activities of the canine antral smooth muscle, *Gastroenterology*, 75, 405, 1978.

84. **Cameron, A. J., Phillips, S. F., and Summerskill, W. H. J.**, Comparison of effects of gastrin, cholecystokinin-pancreozymin, secretin, and glucagon on human stomach, *in vitro, Gastroenterology*, 59, 539, 1970.

85. **Strunz, U. T. and Grossman, M. I.**, Antral motility stimulated by gastrin: a physiological action affected by cholinergic activity, in *Endocrinology of the Gut*, Chey, W. Y. and Brooks, F. P., Eds., Slack, Thorofara, 1977, 233.

86. **Strunz, U., Code, C., and Grossman, M.**, Effect of gastrin on electrical activity of antrum and duodenum, *Proc. Soc. Exp. Biol. Med.*, 161, 25, 1979.

87. **Misiewicz, J. J., Holdstock, D. J., and Waller, S. L.**, Motor responses of the human alimentary tract to near maximal infusion of pentagastrin, *Gut*, 8, 463, 1967.

88. **Kelly, A.**, The effect of pentagastrin on canine myoelectric and motor activity, in *Gastrointestinal Hormones*, Thompson, G., Ed., University of Texas Press, Austin, 1975, 381.

89. **Kowalewski, K., Zajac, S., and Kolodej, A.**, Effect of release of endogenous gastrin on myoelectrical and mechanical activity of isolated canine stomach, *Pharmacology*, 13, 56, 1975.

90. **Kowalewski, K., Zajac, S., and Kolodej, A.**, The effect of drugs on the electrical and mechanical activity of the isolated porcine stomach, *Pharmacology*, 13, 86, 1975.

91. **Hunt, J. N. and Ramsbottom, N.**, Effect of gastrin II on gastric emptying and secretion during a test meal, *Br. Med. J.*, 4, 386, 1967.

92. **Hamilton, S. G., Sheiner, H. J., and Quinlan, M. F.**, Continuous monitoring of the effect of pentagastrin on gastric emptying of solid food in man, *Gut*, 17, 273, 1976.

93. **Dorzois, R. R. and Kelly, K. A.**, Gastrin pentapeptide and delayed gastric emptying, *Am. J. Physiol.*, 221, 113, 1971.

94. **Schuurkes, J. A. J. and Charbon, G. A.**, Motility and hemodynamics of the canine gastrointestinal tract. Stimulation by pentagastrin, cholecystokinin and vasopressin, *Arch. Int. Pharmacodyn. Ther.*, 236, 214, 1978.

95. **Gamblin, G. T., Dubois, A., and Castell, D. O.**, Contrasting effect of pentagastrin on gastric emptying in normals and patients with gastric ulcer, *Clin. Res.*, 25, A17, 1976.

96. **Dubois, A. and Castell, D. O.**, Abnormal gastric emptying response to pentagastrin in duodenal ulcer, *Scand. J. Gastroenterol.*, 13, (Suppl. 49), 50, 1978.

97. **Fisher, R. S., Lipshutz, W., and Cohen, S.**, The hormonal regulation of pyloric sphincter function, *J. Clin. Invest.*, 52, 1289, 1973.

98. **Fisher, R. S., and Boden, G.**, Gastrin inhibition of the pyloric sphincter, *Am. J. Dig. Dis.*, 21, 468, 1976.

99. **White, C. M. and Keighley, M. R. B.**, An explanation of the paradoxial effect of pentagastrin on gastric motility, *Gut*, 19, A343, 1978.

100. **Munk, J. F., Hoare, M., and Johnson, A. G.**, Hormonal influence of pyloric diameter and antral motility in man, *Gut*, 19, A435, 1978.

101. **Mantovani, P. and Bertraccini, G.**, Action of caerulein and related substances on gastrointestinal tract motility of anaesthetized dog, *Arch. Int. Pharmacodyn. Ther.*, 193, 362, 1971.

102. **Bertaccini, G., Impicciatore, M., and DeCaro, G.**, Action of caerulein and related substances on the pyloric sphincter of the anaesthetized rat, *Eur. J. Pharmacol.*, 22, 320, 1973.

103. **Vizi, E., Bertraccini, G., Impicciatore, M., and Knoll, J.**, Evidence that acetylcholine released by gastrin and related polypeptides contributes to their effect on gastrointestinal motility, *Gastroenterology*, 64, 268, 1973.

104. **Vizi, E., Bertraccini, G., Impicciatore, M., Mantovani, P., Zséli, J., and Knoll, J.**, Structure-activity relationship of some analogues of gastrin and cholecystokinin on intestinal smooth muscle of the guinea pig, *Naunyn-Schmiedebergs Arch. Pharmacol.*, 284, 233, 1974.

105. **Hutchinson, J. B. and Dockray, G. J.**, Evidence that the action of cholecystokinin octapeptide on the guinea pig ileum longitudinal muscle is mediated in part by substance P release from myenteric plexus, *Eur. J. Pharmacol.*, 69, 87, 1981.

106. **Marik, F. and Code, C. F.**, Control of the interdigestive myoelectric activity in dogs by the vagus nerves and pentagastrin, *Gastroenterology*, 69, 387, 1975.

107. **Weisbrodt, N. W., Copeland, E. M., Kearly, R. W., Moore, E. P., and Johnson, L. R.**, Effect of pentagastrin on electrical activity of small intestine of the dog, *Am. J. Physiol.*, 227, 425, 1974.

108. **Erckenbrecht, J. F., Caspari, I., and Wienbeck, M.**, Pentagastrin induced motility pattern in the human upper gastrointestinal tract is reversed by proglumide, *Gut*, 25, 953, 1984.

109. **Thomas, P. A., Schang, J. C., Kelly, K. A., and Go, V. L. W.**, Can endogenous gastrin inhibit canine interdigestive gastric motility?, *Gastroenterology*, 78, 716, 1980.

110. **Eeckhout, C., DeWever, I., Peeters, T., Hellemans, J., and Vantrappen, G.**, Role of gastrin and insulin in postprandial disruption of migrating complex in dogs, *Am. J. Physiol.*, 235, E666, 1978.

111. **Connel, A. M. and Logan, C. J. H.**, The role of gastrin in the ileocolonic response, *Am. J. Dig. Dis.*, 12, 277, 1967.

112. **Snape, W. J., Matarazzo, S. Z., and Cohen, S.**, Effect of eating and gastrointestinal hormones on human colonic myoelectrical and motor activity, *Gastroenterology*, 75, 373, 1978.

113. **Taylor, I., Duthie, H. L., Smallwood, R., Brown, B. H., and Linkens, D.**, The effect of stimulation on the myoelectrical activity of the rectosigmoid in man, *Gut*, 15, 599, 1974.

114. **Snape, W. J., Carlson, G. M., and Cohen, S.**, Human colonic myoelectric activity in response to prostigmin and gastrointestinal hormones, *Am. J. Dig. Dis.*, 22, 881, 1977.

115. **Bennett, A. and Whitney, B.**, A pharmacological study of the motility of the human gastrointestinal tract, *Gut*, 7, 307, 1966.

116. **Smith, A. N. and Hogg, D.**, Effect of gastrin II on the motility of the gastrointestinal tract, *Lancet*, 1, 403, 1966.

117. **Meneghelli, U. G., Godoy, R. A., Oliveira, R. B., Santos, J. C. M., Dantas, R. O., and Troncon, L. E. A.**, Effect of pentagastrin on the motor activity of the dilated and nondilated sigmoid and rectum in Chagas' disease, *Digestion*, 27, 152, 1983.

118. **Hellström, P. M.**, Atropine and naloxone block the colonic contraction elicited by cholecystokinin and pantagastrin, *Acta Physiol. Scand.*, 124, 25, 1985.

119. **Geenen, J. E., Hogan, W. J., Dodds, W. J., Steward, E. T., and Arndorfer, R. C.**, Intraluminal pressure recording of the human sphincter of Oddi, *Gastroenterology*, 78, 317, 1980.

120. **Nebel, O. T.**, Manometric evaluation of the papilla of vater, *Gastrointest. Endosc.*, 21, 126, 1975.

121. **Becker, J. M., Moody, F. G., and Zinsmeister, A. R.**, Effect of gastrointestinal hormones on the biliary sphincter of the opossum, *Gastroenterology*, 81, 1300, 1982.

122. **Ivy, A. C. and Oldberg, E. A.**, A hormone mechanism for gallbladder contraction and evacuation, *Am. J. Physiol.*, 86, 599, 1928.

123. **Mutt, V. and Jorpes, J.**, Structure of porcine cholecystokinin-pancreozymin, *Eur. J. Biochem.*, 6, 156, 1968.

124. **Jorpes, J. E.**, The isolation and chemistry of secretin and cholecystokinin, *Gastroenterology*, 55, 157, 1968.

125. **Jorpes, J. E. and Mutt, V.**, Secretin and cholecystokinin (CCK), in *Secretin, Cholecystokinin, Pancreozymin and Gastrin, Handbook of Experimental Pharmacology 34*, Jorpes, J. E. and Mutt, V., Eds., Springer-Verlag, Berlin, 1973, 1.

126. **Erspamer, V. and Melchiorri, P.**, Active polypeptides of the amphibian skin and their synthetic analogues, *Pure Appl. Chem.*, 35, 463, 1973.

127. **Deschenes, R. J., Lorenz, L. J., Haun, R. S., Roos, B. A., Collier, K. J., and Dixon, J. E.**, Cloning and analysis of a cDNA encoding rat preprocholecystokinin, *Proc. Natl. Acad. Sci. U.S.A.*, 81, 726, 1984.

128. **Gubler, U., Chua, A. O., Hoffman, B. J., Collier, K. J., and Eng, J.**, Cloned cDNA to cholecystokinin mRNA predicts an identical proprocholecystokinin in pig brain and gut, *Proc. Natl. Acad. Sci. U.S.A.*, 81, 4307, 1984.

129. **Takahashi, Y., Kato, K., Hayashizaki, Y., Wakabayashi, T., Ohtsuka, E., Matsuki, S., Ikehara, M., and Matsubara, K.**, Molecular cloning of the human cholecystokinin gene by use of a synthetic probe containing deoxyinosine, *Proc. Natl. Acad. Sci. U.S.A.*, 82, 1931, 1985.

130. **Rehfeld, J. F. and Hansen, F.**, Characterization of preprocholecystokinin products in the porcine cerebral cortex — evidence of different processing pathways, *J. Biol. Chem.*, 261, 5832, 1986.

131. **Eberlein, G. A., Eysselein, V. E., Lee, T. D., Shively, J. E., Davis, M., Schaeffer, M., Niebel, W., Zeeh, J., Moessner, A., Meyer, H. E., Grandt, D., Goebell, H., and Reeve, J.**, Processing of human preprocholecystokinin by signal peptidase: formation of cholecystokinin-83, *Gastroenterology*, 96, A134, 1989.

132. **Eysselein, V. E., Reeve, J. R., Shively, J. E., Hawke, D., and Walsh, J. H.**, Partial structure of a large canine cholecystokinin (CCK-58): amino acid sequence, *Peptides*, 3, 687, 1982.

133. **Rehfeld, J. F.**, Immunochemical studies on cholecystokinin. I. Development of sequence-specific radioimmunoassay for porcine triacontatriapeptide cholecystokinin, *J. Biol. Chem.*, 253, 4016, 1978.

134. **Eng, J., Du, B. H., Pan, Y. C. E., Chang, M., Hulmes, J. D., and Yalow, R. S.,** Purification and sequencing of a rat intestinal 22 amino acid C-terminal CCK fragment, *Peptides*, 5, 1203, 1984.
135. **Zhou, Z. Z., Eng, J., Pan, Y. C. E., Chang, M., Hulmes, J. D., Raufman, J. P., and Yalow, R. S.,** Unique cholecystokinin peptides isolated from guinea pig intestine, *Peptides*, 6, 337, 1985.
136. **Dockray, G. J.,** Immunoreactive component resembling cholecystokinin octapeptide in intestine, *Nature (London)*, 270, 359, 1977.
137. **Rehfeld, J. F.,** Four basic characteristics of the gastrin-cholecystokinin system, *Am. J. Physiol.*, 240, G255, 1981.
138. **Shively, J., Reeve, J. R., Eysselein, V. E., Ben-Avram, C., Vigna, S. R., and Walsh, J. H.,** CCK-5: sequence analysis of a small cholecystokinin from canine brain and intestine, *Am. J. Physiol.*, 252, G272, 1987.
139. **Wolfe, M. and McGuigan, J. E.,** Immunochemical characterization of gastrin and cholecystokininlike peptides released in dogs in response to a peptone meal, *Gastroenterology*, 87, 323, 1984.
140. **Dockray, G. J.,** Immunochemical evidence of cholecystokininlike peptides in brain, *Nature (London)*, 264, 586, 1976.
141. **Eysselein, V. E., Reeve, J., Shively, J., Miller, C., and Walsh, J.,** Isolation of a large cholecystokinin precursor from canine brain, *Proc. Natl. Acad. Sci. U.S.A.*, 81, 6565, 1984.
142. **Dockray, G. J., Gregory, R. A., and Tracy, H.,** Cholecystokinin octapeptide in the dog vagus nerve: identification and accumulation on the cranial side of ligatures, *J. Physiol. (London)*, 301, 50P, 1980.
143. **Larsson, L. I. and Rehfeld, J. F.,** Localization and molecular heterogeneity of cholecystokinin in the central and peripheral nervous system, *Brain Res.*, 165, 201, 1979.
144. **Buchan, A., Polak, J., Solcia, E., Capella, C., Hudson, D., and Pearse, A.,** Electron immunohistochemical evidence for the human I cell as the source of CCK, *Gut*, 19, 403, 1978.
145. **Buffa, R., Solcia, E., and Go, V.,** Immunohistochemical identification of the cholecystokinin cell in the intestinal mucosa, *Gastroenterology*, 70, 528, 1976.
146. **Polak, J., Pearse, A., Bloom, S., Buchan, A., Rayford, P., and Thompson, J.,** Identification of cholecystokinin-secreting cells, *Lancet*, 2, 1016, 1975.
147. **Meyer, J. and Jones, R.,** Canine pancreatic responses to intestinally perfused fat and products of fat digestion, *Am. J. Physiol.*, 226, 1178, 1974.
148. **Konturek, S., Tasler, J., and Obtulowicz, W.,** Localization of cholecystokinin release in intestine of the dog, *Am. J. Physiol.*, 222, 16, 1972.
149. **Konturek, S. J., Tasler, J., Bilski, J., DeJong, A. J., Jansen, J. B. M. J., and Lamers, C. B.,** Physiological role and localization of cholecystokinin release in dogs, *Am. J. Physiol.*, 250, G391, 1986.
150. **Konturek, S., Radecki, T., Thor, P., and Dembinski, A.,** Release of cholecystokinin by amino-acids, *Proc. Soc. Exp. Biol. Med.*, 143, 305, 1973.
151. **Go, V., Hoffmann, A., and Summerskill, W.,** Pancreozymin bioassay in man based on pancreatic enzyme secretion: potency of specific amino acids and other digestive products, *J. Clin. Invest.*, 49, 1558, 1970.
152. **Meyer, J., Kelly, G., Spingola, L., and Jones, R.,** Canine gut receptors mediating pancreatic responses to luminal L-amino acids, *Am. J. Physiol.*, 231, 669, 1976.
153. **Liddle, R., Goldfine, I., Rosen, M., Taplitz, R., and Williams, J.,** Cholecystokinin bioactivity in human plasma: molecular forms, responses to feeding, and relationship to gallbladder contraction, *J. Clin. Invest.*, 75, 1144, 1985.
154. **Chen, Y. F., Chey, W. Y., Chang, T. M., and Lee, K. Y.,** Duodenal acidification releases cholecystokinin, *Am. J. Physiol.*, 249, G29, 1985.
155. **Barber, D. L., Walsh, J. H., and Soll, A. H.,** Release and characterization of cholecystokinin from isolated canine jejunal cells, *Gastroenterology*, 91, 627, 1986.
156. **Nakano, I., Miyazaki, K., Funakoshi, A., Tateishi, K., Mamaoka, T., and Yajima, H.,** Gastrin-releasing peptide stimulates cholecystokinin secretion in perfused rat duodenum, *Regul. Peptides*, 23, 153, 1988.
157. **Daniel, E. E., Collins, S. M., Fox, J. E. T., and Huizinga, J.,** Pharmacology of neuroendocrine peptides — gastrin and cholecystokinin, in *Handbook of Physiology*, in press, 1989.
158. **Walsh, J. H.,** Gastrointestinal hormones — cholecystokinin, in *Physiology of the Gastrointestinal Tract*, Johnson, L. R., Ed., Raven Press, New York, 1987, 195.
159. **Innis, R. B. and Snyder, S. H.,** Distinct cholecystokinin receptors in brain and pancreas, *Proc. Natl. Acad. Sci. U.S.A.*, 77, 6917, 1980.
160. **Moran, T. H. and Robinson, P. H., Goldrich, M. S., and McHugh, P. R.,** Two brain CCK receptors: implications for behavioral action, *Brain Res.*, 362, 175, 1986.
161. **Hill, D. R. and Shaw, T. M.,** Autoradiographic localization of CCK-A receptors in mammalian brain using the selective CCK-antagonist L-364,718, in *Cholecystokinin Antagonists*, Wang, R. Y. and Schoenfield, R., Eds., Alan R. Liss, New York, 1988, 133.
162. **Jensen, R. T., Zhou, Z.-C., Murphy, R. B., Jones, S. W., Setnikar, I., Rovati, L. A., and Gardner, J. D.,** Structural features of various proglumide-related cholecystokinin receptor antagonists, *Am. J. Physiol.*, 44, G839, 1986.

377

163. **Rovati, L. C. and Makovec, F.**, New pentanoic acid derivatives with potent CCK antagonistic properties: different activity on the periphery vs. central nervous system, in *Cholecystokinin Antagonists,* Wang, R. Y. and Schoenfield, R., Eds., Alan R. Liss, New York, 1988, 1.

164. **Chang, R. S. L. and Lotti, V. J.**, L-364,718: A review of biochemical and pharmacological characterization as a highly potent peripherally selective CCK-antagonist, in *Cholecystokinin Antagonists,* Wang, R. Y. and Schoenfield, R., Eds., Alan R. Liss, New York, 1988, 13.

165. **Peiken, S. R., Costenbader, C. L., and Garner, J. D.**, Actions of derivative cyclic nucleotides on dispersed acini from guinea pig pancreas, *J. Biol. Chem.,* 254, 5321, 1979.

166. **Hahne, W. F., Jensen, R. T., Lemp, G. F., and Gardner, J. D.**, Proglumide and benzotript: members of a different class of cholecystokinin receptor antagonists, *Proc. Natl. Acad. Sci. U.S.A.,* 78, 6304, 1981.

167. **Jensen, R. T., Jones, S. W., and Gardner, J. D.**, COOH-terminal fragments of cholecystokinin: a new class of cholecystokinin receptor antagonists, *Biochim. Biophys. Acta,* 761, 269, 1983.

168. **Gardner, J. D. and Jensen, R. T.**, Cholecystokinin receptor antagonists, *Am. J. Physiol.,* 246, G471, 1984.

169. **Sparnarkel, M., Martinez, J., Briet, C., Jensen, R. T., and Gardner, J. D.**, Cholecystokinin-27-32-amide. A member of a new class of cholecystokinin receptor antagonists, *J. Biol. Chem.,* 258, 6746, 1983.

170. **Makovec, F., Bani, M., Chisté, R., Revel, L., Rovati, L. C., and Rovati, L. A.**, Differentiation of central and peripheral cholecystokinin receptors by new glutaramic acid derivatives with cholecystokinin-antagonistic activity, *Drug Res.,* 36, 98, 1986.

171. **Makovec, F., Chisté, R., Bani, M., Pacini, M. A., Setnikar, I., and Rovati, L. A.**, New glutaramic acid derivatives with potent competitive and specific cholecystokinin-antagonistic activity, *Drug Res.,* 35, 1048, 1985.

172. **Chang, R. S. L., Lotti, V. J., Monaghan, R. L., Birnbaum, J., Stapley, E. O., Goetz, M. A., Albers-Schonberg, G., Patchett, A. A., Liesch, J. M., Hensens, O. D., and Springer, J. P.**, A potent nonpeptide cholecystokinin antagonist selective for peripheral tissues isolated from *Aspergillus Alliaceus, Science,* 230, 177, 1985.

173. **Evans, B. E., Bock, M. G., Rittle, K. E., DiPardo, R. M., Whitter, W. L., Veber, D. F., Anderson, P. S., and Friedinger, R. M.**, Design of potent, orally effective, non-peptidal antagonists of the peptide hormone cholecystokinin, *Proc. Natl. Acad. Sci. U.S.A.,* 83, 4918, 1986.

174. **Huang, S. C., Wank, S. A., Gardner, J. D., and Jensen, R. T.**, Comparison of the ability of nonpeptide antagonists L-365,260 and L-364718 to selectively interact with gastrin and cholecystokinin (CCK) receptors, *Gastroenterology,* 96, A220, 1989.

175. **Chang, R. S. L. and Lotti, V. J.**, Biochemical and pharmacological characterization of an extremely potent and selective nonpeptide cholecystokinin antagonist, *Proc. Natl. Acad. Sci. U.S.A.,* 83, 4923, 1986.

176. **Lotti, V. J., Pendelton, R. G., Gould, R. J., Hanson, H. M., Chang, R. S., and Clineschmidt, B. V.**, *In vivo* pharmacology of L-364,718, a new potent nonpeptide peripheral cholecystokinin antagonist, *J. Pharmacol. Exp. Ther.,* 241, 103, 1987.

177. **Gould, R. J., Cook, P. G., Fioravanti, C., and Solomon, H. F.**, L-364,718, a cholecystokinin antagonist, promotes gastric emptying in cats, *Clin. Res.,* 35, 590A, 1987.

178. **Pendleton, R. G., Bedesky, R. J., Schaffer, L., Nolan, T. E., Gould, R. J., and Clineschmidt, B. V.**, Roles of endogenous cholecystokinin in biliary, pancreatic and gastric function: studies with L-364,718, a specific cholecystokinin receptor antagonist, *J. Pharmacol. Exp. Ther.,* 241, 110, 1987.

179. **Werth, B., Meyer, B., Beglinger, C., Hildebramd, P., and Stadler, G. A.**, Loxiglumide (LOX), a new cholecystokinin (CCK) receptor antagonist, affects regional gastrointestinal transit times in healthy subjects, *Gastroenterology,* 96, A542, 1989.

180. **Beglinger, C., Meyer, B., Zach, D., Rovati, L., Setnikar, I., and Stalder,** Cholecystokinin (CCK) is a physiological regulator of gastric emptying in man, *Biomed. Res.,* 9, 55, 1988.

181. **Green, T., Dimaline, R., Peikin, S., and Dockray, G. J.**, Action of the cholecystokinin antagonist L364,718 on gastric emptying in the rat, *Am. J. Physiol.,* 255, G685, 1988.

182. **Wiener, I., Inoue, K., Fagan, C., Lilja, P., Watson, L., and Thompson, J.**, Release of cholecystokinin in man: correlation of blood levels with gallbladder contraction, *Ann. Surg.,* 194, 321, 1981.

183. **Byrnes, D., Borody, T., Daskalopoulos, G., Boyle, M., and Benn, I.**, Cholecystokinin and gallbladder contraction: effect of CCK-infusion, *Peptides,* 2, 259, 1981.

184. **Isenberg, J. I. and Csendes, A.**, Effect of octapeptide of cholecystokinin on canine pyloric pressure, *Am. J. Physiol.,* 222, 428, 1972.

185. **Schang, J. C. and Kelly, K.**, Inhibition of canine interdigestive proximal gastric motility by cholecystokinin octapeptide, *Am. J. Physiol.,* 240, G217, 1981.

186. **Debas, H. T., Farooq, O., and Grossman, M. I.**, Inhibition of gastric emptying is a physiological action of cholecystokinin, *Gastroenterology,* 68, 1211, 1975.

187. **Moran, T. H. and McHugh, P. R.**, Cholecystokinin suppresses food intake by inhibiting gastric emptying, *Am. J. Physiol.,* 242, R491, 1982.

188. **Shillabeer, G. and Davison, J. S.,** Proglumide, a cholecystokinin antagonist, increases gastric emptying in rats, *Am. J. Physiol.,* 252, R353, 1987.

189. **Thor, P., Laskiewicz, J., Konturek, P., and Konturek, S. J.,** Cholecystokinin in the regulation of intestinal motility and pancreatic secretion in dogs, *Am. J. Physiol.,* 255, G498, 1988.

190. **Karaus, M. and Niederau, C.,** Effects of CCK-receptor antagonist on colonic motility in the dog, *Gastroenterology,* 96, A542, 1989.

191. **Niederau, C. and Karaus, M.,** Effects of CCK-receptor blockade on fed and fasted small intestinal motor activity, *Gastroenterology,* 96, A366, 1989.

192. **Liddle, R. A., Kanayama, S., and Beccaria, L.,** The ability of a new cholecystokinin receptor antagonist, L-364,718, to inhibit CCK- and meal-stimulated gallbladder contraction in humans, *Biomed. Res.,* 9, 59, 1988.

193. **Niederau, C., Heindges, T., and Rovati, L.,** Blockade of the CCK-receptor does not only abolish meal induced gallbladder emptying but increases fasting gallbladder volume in healthy humans, *Gastroenterology,* 96, A365, 1989.

194. **Behar, J. and Biancani, P.,** Effect of cholecystokinin octapeptide on lower esophageal sphincter, *Gastroenterology,* 73, 57, 1977.

195. **Rattan, S. and Goyal, R. K.,** Structure-activity relationship of subtypes of cholecystokinin receptors in the cat lower esophageal sphincter, *Gastroenterology,* 90, 94, 1986.

196. **Resin, H., Stren, D. H., Sturdevant, R. A. L., and Isenberg, J. I.,** Effect of the C-terminal octapeptide of cholecystokinin on lower esophageal sphincter pressure in man, *Gastroenterology,* 64, 946, 1973.

197. **Sturdevant, R. A. L. and Kun, T.,** Interaction of pentagastrin and octapeptide of cholecystokinin on the human lower esophageal sphincter, *Gut,* 15, 700, 1974.

198. **Fisher, R. S., DiMagno, A. J., and Cohen, S.,** Mechanism of cholecystokinin inhibition of the lower esophageal sphincter pressure, *Am. J. Physiol.,* 228, 1469, 1975.

199. **Dent, J., Dodds, W. J., Hogan, W. J., Arndorfer, R. C., and Teeter, B. C.,** Effect of cholecystokinin-octapeptide on opossum lower esophageal sphincter, *Am. J. Physiol.,* 239, G230, 1980.

200. **Dent, J., Dodds, W. J., Hogan, W. J., and Arndorfer, R. C.,** CCK-OP: a useful agent for evaluating lower esophageal sphincter of the opossum, *Gastroenterology,* 78, 1978, 1979.

201. **Dent, J., Dodds, W. J., Hogan, W. J., and Arndorfer, R. C.,** CCK-OP: a useful agent for evaluating lower esophageal sphincter (LES) denervation in human, *Gastroenterology,* 74, A1025, 1978.

202. **Bertaccini, G.,** Peptides: gastrointestinal hormones-cholecystokinin, in *Mediators and Drugs in Gastrointestinal Motility,* Vol. 2, Bertraccini, G., Ed., Springer-Verlag, Berlin, 1982, 40.

203. **Gerner, T. M.,** Pressure responses to OP-CCK compared to CCK-PZ in the antrum and fundus of isolated guinea pig stomachs, *Scand. J. Gastroenterol.,* 14, 73, 1979.

204. **Scheurer, U. L., Varga, L., Drack, E., Burki, H. R., and Halter, F.,** Mechanisms of action of CCK-octapeptide on rat antrum, pylorus and duodenum, *Am. J. Physiol.,* 244, G266, 1983.

205. **Ohkawa, H. and Watanabe, M.,** Effects of gastrointestinal hormones on the electrical and mechanical activities of the cat small intestine, *Jpn. J. Physiol.,* 27, 71, 1977.

206. **Morgan, K. G., Schmalz, P. F., Go, V. L. M., and Szurszewski, J. H.,** Electrical and mechanical effects of molecular variants of CCK on antral smooth muscle, *Am. J. Physiol.,* 235, E324, 1978.

207. **Cameron, A. J., Phillips, S. F., and Summerskill, W. H. J.,** Effect of cholecystokinin on motility of human stomach and gallbladder muscle "in vitro", *Clin. Res.,* 15, 416, 1967.

208. **Bertaccini, G., Agosti, A., and Impicciatore, M.,** Caerulein and gastrointestinal motility in man, *Rend. Gastroenterol.,* 3, 23, 1971.

209. **Bitar, K. N., Saffouri, B., and Makhlouf, G. M.,** Cholinergic and peptidergic receptors on isolated human antral smooth muscle cells, *Gastroenterology,* 82, 832, 1982.

210. **Kantoh, M., Takahashi, T., Kusunoki, M., Yamamura, T., and Utsunomiya, J.,** Dual action of cholecystokinin-octapeptide on the guinea-pig antrum, *Gastroenterology,* 92, 376, 1987.

211. **Chey, W. Y., Yoshimori, M., Hendricks, J., and Kimani, S.,** Effects of C-terminal octapeptide of cholecystokinin (CCK) on the motor activities of the antrum and pyloric sphincter in dogs, *Gastroenterology,* 62, 733, 1972.

212. **Lee, K. Y., Hendricks, J., and Chey, W. Y.,** Effects of gut hormones on motility of the antrum and duodenum in dogs, *Gastroenterology,* 66, A729, 1974.

213. **Castresana, M., Lee, K. Y., Chey, W. Y., and Yajiam, H.,** Effects of motilin and octapeptide of cholecystokinin on antral and duodenal myoelectric extivity in the interdigestive state and during inhibition by secretin and gastric inhibitory polypeptide, *Digestion,* 17, 300, 1987.

214. **Kuwahara, A., Ozawa, K., and Yanaihara, N.,** Effects of cholecystokinin-octapeptide on gastric motility of anesthetized dogs, *Am. J. Physiol.,* 251, G678, 1986.

215. **Valenzuela, J. E. and Grossman, M. I.,** Effect of pentagastrin and caerulein on intragastric pressure in the dog, *Gastroenterology,* 69, 1383, 1975.

216. **Telford, G. L., Mir, S. S., Mason, G. R., and Ormsbee, H. S., III,** Neural control of the canine pylorus, *Am. J. Surg.,* 137, 92, 1979.

217. **Phaosawasdi, K. and Fisher, R. S.**, Hormonal effects on the pylorus, *Am. J. Physiol.*, 243, G330, 1982.

218. **Behar, J., Biancani, P., and Zabinski, P.**, Characterization of feline gastroduodenal junction by neural and hormonal stimulation, *Am. J. Physiol.*, 236, E45, 1979.

219. **Chey, W. Y., Hitanant, S., Hendricks, J., and Lorber, S. H.**, Effect of secretin and cholecystokinin on gastric emptying and gastric secretion in man, *Gastroenterology*, 58, 820, 1970.

220. **Sterz, P., Guth, P., and Sturdevant, R.**, Gastric emptying in man: delay by octapeptide of cholecystokinin and L-tryptophan, *Clin. Res.*, 22, A174, 1974.

221. **Valenzuela, J. E. and Defilippi, C.**, Inhibition of gastric emptying in humans by secretin, octapeptide of cholecystokinin, and intraduodenal fat, *Gastroenterology*, 81, 898, 1981.

222. **Liddle, R. A., Morita, E. T., Conrad, C. K., and Williams, J. A.**, Regulation of the gastric emptying in humans by cholecystokinin, *J. Clin. Invest.*, 77, 992, 1986.

223. **Yamagishi, T. and Debas, H. T.**, Cholecystokinin inhibits gastric emptying by acting on both proximal stomach and pylorus, *Am. J. Physiol.*, 234, E375, 1978.

224. **Anika, M. S.**, Effects of cholecystokinin and caerulein on gastric emptying, *Eur. J. Pharmacol.*, 85, 195, 1982.

225. **Mangel, A. W. and Koegel, A.**, Effects of peptides on gastric emptying, *Am. J. Physiol.*, 246, G342, 1984.

226. **Keinke, O., Ehrlein, H. J., and Wulschke, S.**, Mechanical factors regulating gastric emptying examined by the effects of exogenous cholecystokinin and secretin on canine gastroduodenal motility, *Can. J. Physiol. Pharmacol.*, 65, 287, 1987.

227. **Valenzuela, J. E.**, Effect of intestinal hormones and peptides on intragastric pressure in dogs, *Gastroenterology*, 71, 766, 1976.

228. **Raybould, H. E., Roberts, M. E., and Dockray, G. J.**, Reflex decreases in intragastric pressure in response to cholecystokinin in rats, *Am. J. Physiol.*, 253, G165, 1987.

229. **Moran, T. H. and McHugh, P. R.**, Gastric and nongastric mechanisms for satiety action of cholecystokinin, *Am. J. Physiol.*, 254, R628, 1988.

230. **Smith, G. T., Moran, T. H., Coyle, J. T., Kuhar, M. J., O'Donahue, T. L., and McHugh, P. R.**, Anatomic localization of cholecystokinin receptors to the pyloric sphincter, *Am. J. Physiol.*, 246, R127, 1984.

231. **Raybould, H. E., Gayton, R. J., and Dockray, G. J.**, CNS effects of circulating CCK8: involvement of brainstem neurons responding to gastric distention, *Brain Res.*, 342, 187, 1985.

232. **Zarbin, M. A., Warmsley, J. T., Innis, R. B., and Kuhar, M. J.**, Cholecystokinin receptors: presence and axonal flow in the rat vagus nerve, *Life Sci.*, 29, 697, 1981.

233. **Moran, T. H., Smith, G. P., Hostetler, A. M., and McHugh, P. R.**, Transport of cholecystokinin binding sites in subdiaphragmatic vagal branches, *Brain Res.*, 415, 149, 1987.

234. **Moran, T. H. and McHugh, P. R.**, Anatomical and pharmacological differentiation of pyloric, vagal, and brain stem cholecystokinin receptors, in *Cholecystokinin Antagonists*, Wang, R. Y. and Schoenfield, R., Alan R. Liss, New York, 1988, 117.

235. **Hedner, P.**, Effect of the C-terminal octapeptide of cholecystokinin on guinea pig ileum and gall-bladder *in vitro*, *Acta Physiol. Scand.*, 78, 232, 1970.

236. **Hedner, P. and Rorsman, G.**, Structure essential for the effect of cholecystokinin on the guinea pig small intestine, *Acta Physiol. Scand.*, 74, 58, 1968.

237. **Hutchinson, J. B., Dimaline, R., and Dockray, G. J.**, Neuropeptides in the gut: quantification and characterization of cholecystokinin octapeptide-, bombesin- and vasoactive intestinal polypeptide immunoreactivities in the myenteric plexus of the guinea pig small intestine, *Peptides*, 2, 23, 1981.

238. **Yau, W. M., Makhlouf, G. M., Edwards, L. E., and Farrar, J. T.**, The action of cholecystokinin and related peptides on guinea pig small intestine, *Can. J. Physiol.*, 52, 1974, 1974.

239. **Vizi, E. S., Bertaccini, Impicciatore, M., and Knoll, J.**, Acetylcholine releasing effect of gastrin and related polypeptides, *Eur. J. Pharmacol.*, 17, 175, 1972.

240. **Zetler, G.**, Antagonism of cholecystokinin-like peptides by opioid peptides, morphine or tetrodotoxin, *Eur. J. Pharmacol.*, 60, 67, 1979.

241. **Nemeth, P. R., Zafirov, D. H., and Wood, J. D.**, Effects of cholecystokinin, caerulein and pentagastrin on electrical behaviour of myenteric neurons, *Eur. J. Pharmacol.*, 116, 263, 1985.

242. **Wood, J. D. and Mayer, C. J.**, Intracellular study of electrical activity of Auerbach's plexus in guinea pig small intestine, *Pflügers Arch.*, 374, 265, 1978.

243. **Wood, J. D. and Mayer, C. J.**, Intracellular study of tonic-type enteric neurons in guinea pig small intestine, *J. Neurophysiol.*, 42, 582, 1979.

244. **Donnerer, J., Meyer, D. K., Holzer, P., and Lembeck, F.**, Release of cholecystokinin-immunoreactivity into the vascular bed of the guinea pig small intestine during peristalsis, *Naunyn-Schmiedeberg's Arch. Pharmacol.*, 328, 324, 1985.

245. **Hutchinson, J. B. and Dockray, G. J.**, Inhibition of the action of cholecystokinin octapeptide on the guinea pig ileum myenteric plexus by dibutyryl cyclic guanosine monophosphate, *Brain Res.*, 202, 501, 1980.

246. **Holzer, P. and Lembeck, F.,** Effect of neuropeptides on the efficiency of the peristaltic reflex, *Naunyn-Schmiedeberg's Arch. Pharmacol.,* 307, 257, 1979.

247. **Chijikawa, J. B. and Davison, J. S.,** The action of gastrin-like polypeptides in the peristaltic reflex in the guinea pig intestine, *J. Physiol. (London),* 238, 68P, 1974.

248. **Bartho, L., Holzer, P., Donnener, J., and Lembeck, F.,** Effects of substance P, cholecystokinin octapeptide, bombesin and neurotensin on the peristaltic reflex of the guinea-pig ileum in the absence and in the presence of atropine, *Naunyn-Schmiedeberg's Arch. Pharmacol.,* 321, 321, 1982.

249. **Persson, G. G. A. and Ekman, M.,** Effect of morphine, cholecystokinin, and sympathomimetics on the sphincter of Oddi and intraluminal pressure in cat duodenum, *Scand. J. Gastroenterol.,* 7, 345, 1972.

250. **Osnes, M.,** The effect of secretin and cholecystokinin on the duodenal motility in man, *Scand. J. Gastroenterol.,* 10 (Suppl. 35), 22, 1975.

251. **Labo, G. and Bortolotti, M.,** Effect of gut hormones on myoelectrical and manometric activity of the duodenum in man, *Rend. Gastroenterol.,* 8, 64, 1976.

252. **Bertaccini, G. and Agosti, A.,** Action of caerulein on intestinal motility in man, *Gastroenterology,* 60, 55, 1971.

253. **Sninsky, C. A., Wolfe, M. M., McGuigan, J. E., and Mathias, J. R.,** Alterations in motor function of the small intestine from intravenous and intraluminal cholecystokinin, *Am. J. Physiol.,* 247, G724, 1984.

254. **Schang, J. C. and Kelly, K. A.,** Inhibition of canine interdigestive proximal gastric motility by cholecystokinin-octapeptide (CCK-OP), *Gastroenterology,* 78, 1253, 1980.

255. **Del Tacca, M., Soldani, G., and Crema, A.,** Experiments on the mechanism of action of caerulein at the level of the guinea pig ileum and colon, *Agents Actions,* 1, 176, 1970.

256. **Renny, A., Snape, W. J., Jr., Sun, E. A., London, R., and Cohen, S.,** Role of cholecystokinin in the gastrocolonic response to a fat meal, *Gastroenterology,* 85, 17, 1983.

257. **Wiley, J., Tatum, D., Keinath, R., and Owyang, C.,** Participation of gastric mechanoreceptors and intestinal chemoreceptors in the gastrocolonic response, *Gastroenterology,* 94, 1144, 1988.

258. **Meshkinpour, H., Dinoso, V. P., and Lorber, S. H.,** Effect of intraduodenal administration of essential amino acids and sodium oleate on motor activity of the sigmoid colon, *Gastroenterology,* 66, 373, 1974.

259. **Behar, J. and Biancani, P.,** Effect of cholecystokinin and the octapeptide of cholecystokinin on the feline sphincter of Oddi and gallbladder, *J. Clin. Invest.,* 66, 1231, 1980.

260. **Solomon, T. E., Yamada, T., Elashoff, J., Wood, J., and Beglinger, C.,** Bioactivity of cholecystokinin analogues: CCK-8 is not more potent than CCK-33, *Am. J. Physiol.,* 247, G105, 1984.

261. **Eysselein, V. E., Deveney, C., Sankaran, H., Reeve, J., Jr., and Walsh, J.,** Biological activity of canine intestinal cholecystokinin-58, *Am. J. Physiol.,* 245, G313, 1983.

262. **Yamamura, T., Takahashi, T., Kusunoki, M., Kantoh, M., Ishikawa, Y., and Utsunomiya, J.,** Cholecystokinin octapeptide-evoked (^{3}H)acetylcholine release from guinea pig gallbladder, *Neurosci. Lett.,* 65, 167, 1986.

263. **Takahashi, T., Yamamura, T., Kusonoki, M., Kantoh, M., Ishikawa, Y., and Utsunomiya, J.,** Differences between muscular and neural receptors for cholecystokinin octapeptide in the guinea-pig gallbladder, *Eur. J. Pharmacol.,* 136, 255, 1987.

264. **Von Schrenck, T., Moran, T. H., Heinz-Erian, P., Gardner, J. D., and Jensen, R. T.,** Cholecystokinin receptors on gallbladder muscle and pancreatic acinar cells a comparative study, *Am. J. Physiol.,* 255, G512, 1988.

265. **Steigerwalt, R. W., Goldfine, I. D., and Williams, J. A.,** Characterization of cholecystokinin on bovine gallbladder membranes, *Am. J. Physiol.,* 247, G709, 1984.

266. **Hogan, W. J., Geenen, J. E., Dodds, W. J., Touuli, J., Venu, R., and Helm, J. F.,** Paradoxial motor response to cholecystokinin octapeptide (CCK-OP) in patients with suspected sphincter of Oddi dysfunction, *Gastroenterology,* 82, 1085, 1982.

267. **Cicala, M., Scopinaro, F., and Corazziari, E.,** Postprandial hepatic biliary secretion is inhibited by a specific CCK receptor antagonist in man, *Gastroenterology,* 96, A585, 1989.

268. **Gomez, G., Upp, J. P., Lluis, F., Alexander, R. W., Poston, G. J., Greeley, G. H., and Thompson, J. C.,** Regulation of the release of cholecystokinin by bile salts in dogs and humans, *Gastroenterology,* 94, 1036, 1988.

269. **Kellow, J. E., Miller, L. J., Phillips, S. F., Zinmeister, A. R., and Charboneau, J. W.,** Altered sensitivity of the gallbladder to cholecystokinin octapeptide in irritable bowel syndrome, *Am. J. Physiol.,* 253, G650, 1987.

270. **Upp, J. R., Nealon, W. H., Singh, P., Fagan, C. F., Jonas, A. S., Greeley, G. H., and Thompson, J. C.,** Correlation of cholecystokinin receptors with gallbladder contractility in patients with gallstones, *Ann. Surg.,* 205, 641, 1987.

271. **von Euler, U. S. and Gaddum, J. H.,** An unidentified depressor substance in certain tissue extracts, *J. Physiol. (London),* 72, 74, 1931.

272. **Chang, M. M., and Leeman, S.,** Isolation of a sialogogic peptide from borine hypothalamic tissue and its characterization as substance P, *J. Biol. Chem.,* 245, 4784, 1970.

381

273. **Chang, M. M., Leeman, S. and Niall, H. D.**, Amino acid sequence of substance P, *Nature (London)*, 232, 86, 1971.
274. **Leeman, S. E.**, Substance P and neurotensin: discovery, isolation, chemical characterization and physiological studies, *J. Exp. Biol.*, 89, 193, 1980.
275. **Brodin, E., Linfors, N., Dalsgaard, C.-J., Theodorsson-Norheim, E., and Rosell, S.**, Tachykinin multiplicity in rat central nervous system as studied using antisera raised against substance P and neurokinin A, *Regul. Peptides*, 13, 253, 1986.
276. **Franco, R., Costa, M., and Furness, J. B.**, Evidence for the release of endogenous substance P from intestinal nerves, *Naunyn-Schmiedeberg's Arch. Pharmacol.*, 306, 195, 1979.
277. **Holzer, P., Bucsics, A., Saria, A., and Lembeck, F.**, A study on the concentrations of substance P and neurotensin in the gastrointestinal tract of various mammals, *Neuroscience*, 7, 2919, 1982.
278. **Studer, R. O., Trzeciak, A., and Lergier, W.**, Isolierung und Aminosäuresequenz von Substanz P aus Pferdedarm, *Helv. Chim. Acta*, 56, 860, 1973.
279. **Murphy, R., Furness, J. B., Beardsley, A. M., and Costa, M.**, Characterization of substance P-like immunoreactivity in peripheral sensory nerves and enteric nerves by high pressure liquid chromatography and radioimmunoassay, *Regul. Peptides*, 4, 203, 1982.
280. **Erspamer, V.**, The tachykinin peptide family, *Trends Neurosci.*, 4, 267, 1981.
281. **Kangawa, K., Minamino, N., Fukuda, A., and Matsuo, H.**, Neuromedin K: a novel mammalian tachykinin identified in porcine spinal chord, *Biochem. Biophys. Res. Commun.*, 114, 533, 1983.
282. **Kimura, S., Okada, M., Sugita, Y., Kanazawa, I., and Munekata, E.**, Novel neuropeptides neurokinin α- and β-isolated from porcine spinal chord, *Proc. Jpn. Acad. Sci.*, 59, 101, 1983.
283. **Hunter, J. C. and Maggio, J. E.**, Pharmacological characterization of a novel tachykinin isolated from mammalian spinal chord, *Eur. J. Pharmacol.*, 97, 159, 1984.
284. **Nawa, H., Doteuchi, M., Igano, K., Inoue, K., and Nakanishi, S.**, Substance K: a novel mammalian tachykinin that differs from substance P in its pharmacological profile, *Life Sci.*, 34, 1153, 1984.
285. **Nawa, H., Kotani, H., and Nakanishi, S.**, Tissue specific generation of two preprotachykinin in RNA's from one gene by alternative RNA splicing, *Nature (London)*, 312, 729, 1984.
286. **Tatemoto, K., Lundberg, J. M., Jörnvall, H., and Mutt, V.**, Neuropeptide K: isolation, structure and biological activities of a novel brain tachykinin, *Biochem. Biophys. Res. Commun.*, 128, 947, 1985.
287. **Toresson, G., Brodin, E., Wahlstrom, A., and Bertilsson, L.**, Detection of N-terminally extended substance P but not of substance P in human cerebrospinal fluid: quantitation with HPLC-radioimmunoassay, *J. Neurochem.*, 50, 1701, 1988.
288. **Hua, Y.-Y., Theodorsson-Norheim, E., Brodin, E., Lundberg, J. M., and Hökfelt, T.**, Multiple tachykinins (neurokinin A, neuropeptide K and substance P) in capsaicin sensitive sensory neurons in the guinea pig, *Regul. Peptides*, 13, 1, 1985.
289. **Bowers, C. W., Jan, L. Y., and Jan, Y. N.**, A substance P-like peptide in bullfrog autonomic nerve terminals: anatomy biochemistry and physiology, *Neuroscience*, 19, 343, 1986.
290. **Lazarus, L. H., Linnoila, R. I., Hernandez, O., and DiAugustine, R.**, A neuropeptide in mammalian tissues with physalaemin-like immunoreactivity, *Nature (London)*, 287, 555, 1980.
291. **McDonald, T. J., Christofi, F. L., Brooks, B. D., Barnett, W., and Cook, M. A.**, 'Characterization of content and chromatographic forms of neuropeptides in purified nerve varicosities prepared from guinea pig myenteric plexus, *Regul. Peptides*, 21, 69, 1988.
292. **Negri, L. and Melchiori, P.**, Nonmammalian tachykinins their contribution to the discovery and biological characterization of the mammalian neurokinin system, *Regul. Peptides*, 22, 13, 1988.
293. **Henry, J. L.**, Discussion and nomenclature for tachykinins and tachykinin receptors, in *Substance P and Neurokinins*, Proc. Symp. Substance P and Neurokinins, Henry, J. L., Conture, A. K., Cuello, A. K., Pelletier, Quirion, R., and Regoli, D., Eds., Springer-Verlag, New York, 1987, 13.
294. **Lee, C. M., Iversen, L. L., Hanley, M. R., and Sandberg, B. E. B.**, The possible existence of multiple receptors for substance P, *Naunyn-Schmiedeberg's Arch. Pharmacol.*, 318, 281, 1982.
295. **Buck, S. H., Burcher, E., Shults, C. W., Lovenberg, W., and O'Donohue, T. L.**, Novel pharmacology of substance K-binding sites: a third type of tachykinin receptor, *Science*, 226, 987, 1984.
296. **Nawa, H., Hiose, T., Takashima, H., Inayama, S., and Nakanishi, S.**, Nucleotide sequences of cloned cDNAs for two types of brain substance P precursors, *Nature (London)*, 306, 32, 1983.
297. **Deacon, C. F., Agoston, D. V., Nau, R., and Conlon, J. M.**, Conversion of neuropeptide K to neurokinin A and vesicular colocalization of neurokinin A and substance P in neurons of the guinea pig small intestine, *J. Neurochem.*, 48, 141, 1987.
298. **Pearse, A. G. E. and Polak, J.**, Immunocytochemical localization of substance P in mammalian intestine, *Histochemistry*, 41, 373, 1975.
299. **Nilsson, G., Larsson, L. I., Hakanson, R., Brodin, E., Pernow, B., and Sundler, F.**, Localization of substance P-like immunoreactivity in mouse gut, *Histochemistry*, 43, 97, 1975.
300. **Oki, M. and Daniel, E. E.**, Distribution of substance P-like immunoreactivity in both intrinsic and extrinsic nerves of the gastrointestinal tract, *Nippon Heikatsukin Gakkai Zasshi*, 16, 75, 1980.

301. **Leander, S., Brodin, E., Håkanson, R., Sundler, F., and Uddman, R.,** Neuronal substance P in the esophagus. Distribution and effects on motor activity, *Acta. Physiol. Scand.*, 115, 427, 1982.

302. **Lundberg, J. M., Hökfelt, T., Kewenter, T., Pettersson, G., Ahlman, H., Edin, R., Dahlström, A., Terenius, L., Uvnäs-Wallensten, K., and Said, S.,** Substance P, VIP and enkephalin like immunoreactivity in the human vagus nerve, *Gastroenterology*, 77, 468, 1979.

303. **Furness, J. B. and Costa, M.,** *The Enteric Nervous System*, Churchill Livingstone, Edinburgh, 1987.

304. **Furness, J. B., Morris, L., Gibbins, I. L., and Costa, M.,** Chemical coding of neurons and plurichemical transmission, *Annu. Rev. Pharmacol. Toxicol.*, 29, 289, 1989.

305. **Costa, M., Furness, J. B., Llewellyn-Smith, I. J., and Cuello, A. C.,** Projections of substance P containing nerves within the guinea-pig small intestine, *Neuroscience*, 6, 411, 1981.

306. **McDonald, T. J., Ahmad, S., Allescher, H. D., Kostka, P., Daniel, E. E., Barnett, W., and Brodin, E.,** Canine myenteric, deep muscular, and submucosal plexus preparations of purified nerve varicosities: content and chromatographic forms of certain neuropeptides, *J. Neurochem.*, submitted, 1989.

307. **Maggio, J. E.,** Tachykinins, *Ann. Rev. Neurosci.*, 11, 13, 1988.

308. **Nakanishi, S.,** Substance P precursor and kininogen: their structures, *Physiol. Rev.*, 67, 1117, 1987.

309. **Masu, Y., Tamaki, H., Yokota, Y., and Nakanishi, S.,** Tachykinin precursors and receptors: molecular genetic studies, *Regul. Peptides*, 22, 9, 1988.

310. **Buck, S. H. and Burcher, E.,** The tachykinins: a family of peptides with a brood of 'receptors', *Trends Pharmacol. Sci.* 7, 65, 1986.

311. **Regoli, D., D'Orleans-Juste, P., Escher, E., and Mizrahi, I.,** Receptors for Substance P, I. Pharmacological preparations, *Eur. J. Pharmacol.*, 97, 161, 1984.

312. **Regoli, D., Mizrahi, I., D'Orléans-Juste, P., and Escher, E.,** Receptors for Substance P. II. Classification by agonist fragments and homologues, *Eur. J. Pharmacol.*, 97, 171, 1984.

313. **Regoli, D., Escher, E., Drapeau, G., D'Orléans-Juste, P., and Mizrahi, I.,** Receptors for Substance P. III. Classification by competitive antagonists, *Eur. J. Pharmacol.*, 97, 179, 1984.

314. **Quirion, R.,** Multiple tachykinin receptors, *Trends Neurosci.*, 8, 183, 1985.

315. **Regoli, D., Drapeau, G., Dion, S., and Couture, R.,** New selective agonists for neurokinin receptors: pharmacological tools for receptor characterization, *Trends Neurol. Sci.*, 9, 290, 1988.

316. **Drapeau, G., D'Orleans-Juste, P., Dion, S., Rhaleb, N. E., Rouissi, N. E., and Regoli, D.,** Selective agonists for substance P and neurokinins, *Neuropeptides*, 10, 43, 1987.

317. **Drapeau, G., D'Orléans-Juste, P., Rhaleb, N. E., Dion, S., and Regoli, D.,** Specific agonists for neurokinins B receptors, *Eur. J. Pharmacol.*, 136, 401, 1987.

318. **Laufer, R., Wormser, U., Friedman, Z. Y., Gilon, C., and Chorev, M.,** Neurokinin B is a preferred agonist for a neural substance P receptor and its action is antagonized by enkephalin, *Proc. Natl. Acad. Sci. U.S.A.*, 82, 7444, 1985.

319. **Williams, B. J., Curtis, N. R., McKnight, A. T., Maguire, J., Foster, A., and Tridgett, R.,** Development of NK-2 selective antagonists, *Regul. Peptides*, 22, 189, 1988.

320. **McKnight, A. T., Maguire, J. J., Williams, B. J., Foster, A. C., Tridgett, R., and Iversen, L. L.,** Pharmacological specificity of synthetic peptides as antagonists at tachykinin receptors, *Regul. Peptides*, 22, 127, 1988.

321. **Rovero, P., Pestellini, V., Patacchini, R., Giuliani, S., Santicioli, P., Maggi, C. A., Meli, A., and Giachetti, A.,** A potent and selective agonist for NK-2 tachykinin receptor, *Peptides*, 10, 593, 1989.

322. **Watson, S. P., Sandberg, B. E. B., Hanley, M. R., and Iversen, L. L.,** Tissue selectivity of substance P alkyl esters: suggesting multiple receptors, *Eur. J. Pharmacol.*, 87, 77, 1983.

323. **Jacoby, H. I., Lopez, I., Wright, D., and Vaught, J. L.,** Differentiation of multiple neurokinin receptors in the guinea pig ileum, *Life Sci.*, 39, 1995, 1986.

324. **Holzer-Petsche, U., Lembeck, F., and Seitz, H.,** Contractile effects of substance P and neurokinin A on the rat stomach *in vivo* and *in vitro*, *Br. J. Pharmacol.*, 90, 273, 1987.

325. **Laufer, R., Gilon, C., Chorev, M., and Selinger, Z.,** Desensitization with a selective agonist discriminates between multiple tachykinin receptors, *J. Pharmacol. Exp. Ther.*, 245, 639, 1988.

326. **Laufer, R., Gilon, C., Chorev, M., and Selinger, Z.,** (pGlu^6,Pro^8)-Sp$_{6-11}$Jis a selective agonist for the substance P P-receptors subtype, *J. Med. Chem.*, 29, 1284, 1986.

327. **Regoli, D., D'Orléans-Juste, P., Drapeau, G., Dion, S., and Escher, E.,** Pharmacological characterization of substance P antagonists, in *Tachykinin Antagonists*, Håkanson, R. and Sundler, F., Eds., Elsevier, Amsterdam, 1985, 277.

328. **Buck, S. H., Maurin, Y., Burks, T. F., and Yamamura, H. I.,** High affinity ^{3}H-substance P binding to longitudinal muscle membranes of the guinea pig small intestine, *Life Sci.*, 34, 497, 1984.

329. **Watson, S. P. and Iversen, L. L.,** ^{3}H-substance P binding to guinea pig ileum longitudinal smooth muscle membranes, *Regul. Peptides*, 8, 273, 1984.

330. **Muller, M. J., Sato, H., Bowker, P., and Daniel, E. E.,** Receptors for tachykinins in canine intestine circular muscle, *J. Pharmacol. Exp. Ther.*, 246, 739, 1988.

331. **Mantyh, P. W., Goedert, M., and Hunt, S. P.,** Autoradiographic visualization of receptor binding sites for substance P in the gastrointestinal tract of the guinea pig, *Eur. J. Pharmacol.,* 100, 133, 1984.

332. **Burcher, E., Buck, S. H., Lovenberg, W., and O'Donohue, T. L.,** Characterization and autoradiographic localization of multiple tachykinin binding sites in gastrointestinal tract and bladder, *J. Pharmacol. Exp. Ther.,* 236, 819, 1986.

333. **Guard, S., Watling, K. J., and Watson, S. P.,** ³H-Senktide binding in the guinea pig ileum: a comparison with ¹²⁵I-Bolton-Hunter substance P and ¹²⁵I-Bolton-Hunter eledoisin, *Regul. Peptides,* 22, 75, 1988.

334. **Masu, Y., Nakayama, K., Tamaki, H., Harada, Y., Kuno, M., and Nakanishi, S.,** cDNA cloning of bovine substance-K receptor through oocyte expression system, *Nature (London),* 329, 836, 1987.

335. **Buck, S. H. and Burks, T. F.,** The neuropharmacology of capsaicin: review of some recent observations, *Pharmacol. Rev.,* 38, 179, 1986.

336. **Holzer, P.,** Local effector functions of capsaicin-sensitive nerve endings: involvement of tachykinins, calcitonin gene-related peptide and other neuropeptides, *Neuroscience,* 24, 739, 1988.

337. **Bartho, L. and Holzer, P.,** Search for a physiological role of substance P in gastrointestinal motility, *Neuroscience,* 16, 1, 1985.

338. **Yokohama, S. and North, R. A.,** Electrical activity of longitudinal and circular muscle during peristalsis, *Am. J. Physiol.,* 244, G83, 1983.

339. **Donnerer, J., Holzer, P., and Lembeck, F.,** Release of dynorphin, somatostatin and substance P from the vascularly perfused small intestine of the guinea-pig during peristalsis, *Br. J. Pharmacol.,* 83, 919, 1984.

340. **Costa, M., Furness, J. B., Pullin, C. O., and Bornstein, J.,** Substance P enteric neurons mediate non-cholinergic transmission to the circular muscle of the guinea pig intestine, *Naunyn-Schmiedeberg's Arch. Pharmacol.,* 328, 446, 1985.

341. **Holzer, P.,** Stimulation and inhibition of gastrointestinal propulsion induced by substance P and substance K in the rat, *Br. J. Pharmacol.,* 86, 305, 1985.

342. **Holzer, P. and Lembeck, F.,** Effect of neuropeptides on the efficiency of the peristaltic reflex, *Naunyn-Schmiedeberg's Arch. Pharmacol.,* 307, 257, 1979.

343. **Holzer, P.,** Ascending enteric reflex: multiple neurotransmitter systems and interactions, *Am. J. Physiol.,* 256, G540, 1989.

344. **Kilbinger, H., Sharp, S., Erlhof, I., and Holzer, P.,** Antagonist discrimination between subtypes of tachykinin receptors in the guinea-pig ileum, *Naunyn-Schmiedeberg's Arch. Pharmacol.,* 334, 181, 1986.

345. **Fosbraey, P., Featherstone, R. L. and Morton, I. K. M.,** Comparison of substance P and related peptides on ³H-acetylcholine release, and contractile actions, in the guinea pig ileum, *Naunyn-Schmiedeberg's Arch. Pharmacol.,* 326, 111, 1984.

346. **Katayama, Y. and North, R. A.,** Does substance P mediate slow synaptic excitation within the myenteric plexus?, *Nature (London),* 274, 387, 1978.

347. **Katayama, Y., North, R. A., and Williams, J. T.,** The action of substance P on neurones of the myenteric plexus of the guinea-pig intestine, *Proc. R. Soc. London Ser. B,* 206, 191, 1979.

348. **Daniel, E. E., Costa, M., Furness, J. B., and Keast, J. R.,** Peptide neurons in the canine small intestine, *J. Comp. Neurology,* 237, 227, 1985.

349. **Fox, J. E. T., McDonald, T. J., Alford, L., and Kostlanska, F.,** Tachykinin activation of muscarinic inhibition in canine small intestine is SPP in nature, *Life Sci.,* 39, 1123, 1986.

350. **Holzer, P.,** Different contractile effects of substance P on the intestine of mammals, *Naunyn-Schmiedeberg's Arch. Pharmacol.,* 320, 217, 1982.

351. **Holzer, P.,** An enquiry into the mechanism by which substance P facilitates the phasic longitudinal contractions of the rabbit ileum, *J. Physiol. (London),* 325, 377, 1982.

352. **Bartho, L., Holzer, P., Lembeck, F., and Szolcsanyi, J.,** Evidence that the contractile response of the guinea-pig ileum to capsaicin is due to release of substance P, *J. Physiol. (London),* 332, 157, 1982.

353. **Gintzler, A. R.,** Substance P involvement in the expression of gut dependence on opiates, *Brain Res.,* 182, 224, 1980.

354. **Domoto, T., Jury, J., Berezin, I., Fox, J. E. T., and Daniel, E. E.,** Does substance P comediate with acetylcholine in nerves of opossum esophageal muscularis mucosae, *Am. J., Physiol.,* 245, G19, 1983.

355. **Robotham, H., Jury, J., and Daniel, E. E.,** Capsaicin effects on muscularis mucosa of opossum esophagus: substance P release from afferent nerves?, *Am. J. Physiol.,* 248, G655, 1985.

356. **Daniel, E. E., Jury, J., and Robotham, K. H.,** Receptors for neurotransmitters in opossum oesophagus muscularis mucosa, *Br. J. Pharmacol.,* 88, 707, 1986.

357. **Daniel, E. E., Cipris, S., Bowker, P. and Regoli, D.,** Classification of tachykinin receptors in muscularis mucosae of opossum oesophagus, *Br. J. Pharmacol.,* 97, 1013, 1989.

358. **Aggestrup, S., Uddman, R., Jensen, S. L., Hakanson, R., Sundler, F., Schafalitzky de Muckadell, O., and Emson, P.,** Regulatory peptides in lower esophageal sphincter of pig and man, *Dig. Dis. Sci.,* 31, 1370, 1986.

359. **Mukhopadhyay, A. K.,** Effect of substance P on the lower esophageal sphincter of the opossum, *Gastroenterology,* 75, 278. 1978.

360. **Reynolds, J. C., Dukehart, M. R., Ouyang, A., and Cohen, S.,** Interactions of bombesin and substance P at the feline lower esophageal sphincter, *J. Clin. Invest.,* 77, 436, 1986.

361. **Reynolds, J. C., Ouyang, A., and Cohen, S.,** A lower esophageal sphincter reflex involving substance P, *Am. J. Physiol.,* 246, G346, 1984.

362. **Edin, R., Lundberg, J. M., Lidberg, P., Dählstrom, A., and Ahlman, H.,** Atropine sensitive contractile motor effects of substance P on the feline pylorus and stomach *in vivo, Acta Physiol. Scand.,* 110, 207, 1980.

363. **Kuwahara, A. and Yanaihara, N.,** Action of the newly discovered mammalian tachykinins, substance K and neuromedin K, on gastroduodenal motility of anesthetized dogs, *Regul. Peptides,* 17, 221, 1987.

364. **Delbro, D., Fändriks, L., Lisander, B., and Andersson, S. A.,** Gastric atropine-sensitive excitation by peripheral vagal stimulation after hexamethonium. Antidromic activation of afferents?, *Acta Physiol. Scand.,* 114, 433, 1982.

365. **Delbro, D., Fändriks, L., Rosell, S., and Folkers, K.,** Inhibition of antidromically induced stimulation of gastric motility by substance P receptor blockade, *Acta Physiol. Scand.,* 118, 309, 1983.

366. **Delbro, D., Lisander, B., and Andersson, S. A.,** Atropine-sensitive gastric excitation by local heating — the possibility of a visceral axon reflex arrangement, *Acta Physiol. Scand.,* 114, 319, 1982.

367. **Milenov, K. and Golenhofen, K.,** Differentiated contractile responses of gastric smooth muscle to substance P, *Pflügers Arch.,* 397, 29, 1983.

368. **Milenov, K., Oehme, P., Bienert, M., and Bergmann, J.,** Effect of substance P on mechanical and myoelectrical activities of stomach and small intestines in conscious dog, *Arch. Int. Pharmacodyn.,* 233, 251, 1978.

369. **Mangel, A. W.,** Potentiation of colonic contractility to cholecystokinin and other peptides, *Eur. J. Pharmacol.,* 100, 285, 1984.

370. **Bertaccini, G., De Castiglione, R., and Scarpignato, C.,** Effects of substance P and its natural analogues on gastric emptying in the conscious rat, *Br. J. Pharmacol.,* 72, 221, 1981.

371. **Lidberg, P.,** On the role of substance P and serotonin in the pyloric motor control. An experimental study in cat and rat, *Acta Physiol. Scand.,* 538, (Suppl.), 1, 1985.

372. **Lidberg, P., Dahlström, A., and Ahlman, H.,** On the nature of the contractile motor responses of the rat stomach elicited by serotonin or substance P, *J. Neural Transm.,* 63, 73, 1985.

373. **Lidberg, P., Edin, R., Lundberg, J. M., Dahlström, A., Rosell, S., Folkers, K., and Ahlman, H.,** The involvement of substance P in the vagal control of the feline pylorus, *Acta Physiol. Scand.,* 114, 307, 1982.

374. **Lidberg, P., Dahlström, A., Lundberg, J. M., and Ahlman, H.,** Different modes of action of substance P in the motor control of the feline stomach and pylorus, *Regul. Peptides,* 7, 41, 1983.

375. **Holzer, P. and Lippe, I. T.,** Substance P actions on phosphoinositides in guinea-pig intestinal muscle: a possible transduction mechanism?, *Naunyn-Schmiedeberg's Arch. Pharmacol.,* 329, 50, 1985.

376. **Holzer, P., Holzer-Petsche, U., and Leander, S.,** A tachykinin antagonist inhibits gastric emptying and gastroduodenal transit in the rat, *Br. J. Pharmacol.,* 89, 453, 1986.

377. **Fox, J. E. T. and Daniel, E. E.,** Substance P: a potent inhibitor of the canine small intestine *in vivo, Am. J. Physiol.,* 250, G217, 1986.

378. **Holzer, P. and Lembeck, F.,** Neurally mediated contraction of ileal longitudinal muscle by substance P, *Neurosci. Lett.,* 17, 101, 1980.

379. **Dion, S., D'Orléans-Juste, P., Drapeau, G., Rhaleb, N. E., Rouissi, N., Tousignant, C., and Regoli, D.,** Characterization of neurokinin receptors in various isolated organs by the use of selective agonists, *Lige Sci.,* 41, 2269, 1987.

380. **Bartho, L., Sebok, B., and Szolcsanyi, J.,** Indirect evidence for the inhibition of enteric substance P neurones by opiate agonists but not by capsaicin, *Eur. J. Pharmacol.,* 77, 273, 1982.

381. **Maggi, C. A., Giuliani, S., Manzini, S., Santicioli, P., and Meli, A.,** Motor effects of neurokinins on the rat duodenum: evidence for the involvement of substance K and substance P receptors, *J. Pharmacol. Exp. Ther.,* 238, 341, 1986.

382. **Souquet, J. C., Bitar, K. N., Grider, J. R., and Maklouf, G. M.,** Receptors for substance P on isolated intestinal smooth muscle cells of the guinea pig, *Am. J. Physiol.,* 253, G666, 1987.

383. **Souquet, J. C., Bitar, K. N., Grider, J. R., and Maklouf, G. M.,** Receptors for mammalian tachykinins on the isolated intestinal smooth muscle cells, *Am. J. Physiol.,* 249, 533, 1985.

384. **Daniel, E. E., Gonda, T., Domoto, T., Oki, M., and Yanaihara, N.,** The effects of substance P and met⁵-enkephalin in dog ileum, *Can. J. Physiol. Pharmacol.,* 60, 830, 1982.

385. **Yau, W. M. and Youther, M. L.,** Direct evidence for a release of acetylcholine from the myenteric plexus of guinea-pig small intestine by substance P, *Eur. J. Pharmacol.,* 81, 665, 1982.

386. **Schultzberg, M., Höklfelt, T., Nilsson, G., Terenius, L., Rehlfeld, J. F., Brown, M., Elde, R., Goldstein, M., and Said, S.,** Distribution of peptide and catecholamine-containing neurones in the gastrointestinal tract of rat and guinea pig: immunohistochemical studies with antisera to substance P vasoactive intestinal polypeptide, enkephalins, somatostatin, gastrin-cholecystokinin, neurotensin and dopamine beta-hydroxylase, *Neuroscience,* 5, 689, 1980.

387. **Llewellyn-Smith, I. J., Furness, J. B., Murphy, R., O'Brien, P. E., and Costa, M.,** Substance P-containing nerves in the human small intestine, *Gastroenterology,* 86, 421, 1984.

388. **Wiley, J. and Owyang, C.,** Participation of serotonin and substance P in the action of cholecystokinin on colonic motility, *Am. J. Physiol.,* 252, G431, 1987.

389. **Hellström, P. M.,** Pharmacological analysis of the mechanism of action for colonic contraction induced by neurotensin, substance P and methionine-enkephalin, *Acta Physiol. Scand.,* 125, 13, 1985.

390. **Gonda, T., Daniel, E. E., Kostolanska, F., Oki, M., and Fox, J. E. T.,** Neural control of canine colon motor function: studies *in vivo, Can. J. Physiol. Pharmacol.,* 66, 350, 1988.

391. **Gonda, T., Daniel, E. E., Kostolansks, F., Oki, M., and Fox, J. E. T.,** Neural control of canine colon motor function: studies *in vitro, Can. J. Physiol. Pharmacol.,* 66, 359, 1988.

392. **Hellström, P. M. and Rosell, S.,** Effects of neurotensin, substance P and methionine-enkephaline on colonic motility, *Acta Physiol. Scand.,* 113, 147, 1981.

393. **Leander, S., Håkanson, R., Rosell, S., Folkers, K., Sundler, F., and Tornquist, K. A.,** A specific substance P antagonist blocks smooth muscle contractions induced by non-cholinergic non-adrenergic stimulation, *Nature (London),* 294, 467, 1981.

394. **Cai, W. Q., Gu, J., Huang, W., McGregor, G. P., Ghatei, M. A., Bloom, S. R., and Polak, J. M.,** Peptide immunoreactive nerves and cells of the guinea pig gallbladder and biliary pathways, *Gut,* 24, 1186, 1983.

395. **Keast, J. R., Furness, J. B., and Costa, M.,** Distribution of certain peptide-containing nerve fibers and endocrine cells in the gastrointestinal mucosa in five mammalian species, *J. Comp. Neurol.,* 236, 403, 1985.

396. **Bjorck, S., Eriman, S., Dahlström, A., and Svanvik, J.,** Immunocytochemical localization and transport and motility effects of substance P and serotonin in the feline gallbladder, *Gastroenterology,* 88, 1651, 1985.

397. **Guo, Y. S., Singh, P., Lluis, F., Gomez, G., and Thompson, J. C.,** Contractile response of gallbladder and sphincter of Oddi to substance P and related peptides *in vitro* compared to CCK-8, *Fed. Proc.,* 45, 291, 1986.

398. **Lembeck, F. and Juan, H.,** Comparative action of peptides on the gallbladder and the sphincter of Oddi, *Adv. Exp. Med. Biol.,* 21, 337, 1972.

399. **Mate, L., Sakamoto, T., Greeley, G. H., Jr., and Thompson, J. C.,** Effect of substance P on contractions of the gallbladder, *Surg. Gynecol. Obstet.,* 163, 163, 1986.

400. **Severi, C., Grider, J. R., and Maklouf, G. M.,** Identification of separate bombesin and substance P receptors on isolated muscle cells from canine gallbladder, *J. Pharmacol. Exp. Ther.,* 245, 195, 1988.

401. **Yeo, C. J., Jaffe, B. M., and Zinner, M. J.,** The effect of intravenous substance P infusion on hemodynamics and regional blood flow in conscious dogs, *Surgery,* 95, 175, 1984.

402. **Kaneto, A., Kaneto, T., and Kajimuna, H.,** Effects of substance P and neurotensin infused intrapancreatically on glucagon and insulin secretion, *Endocrinology,* 102, 393, 1978.

403. **Zinner, M. J., Yeo, C. J., and Jaffe, B. M.,** The effect of carcinoid levels of serotonin and substance P on hemodynamics, *Ann. Surg.,* 199, 197, 1984.

404. **Theodorsson-Norheim, E., Jornvall, H., Andersson, M., Norheim, I., Oberg, K., and Jacobsson, G.,** Isolation and characterization of neurokinin A, neurokinin A(3—10) and neurokinin A (4—10) from a neutral water extract of a ileal carcinoid tumour, *Eur. J. Biochem.,* 166, 693, 1987.

405. **Conlon, J. M., Deacon, C. F., Richter, G., Schmidt, W. E., Stockamnn, F., and Creutzfeldt, W.,** Measurement and partial characterization of the multiple forms of neurokinin A-like immunoreactivity in carcinoid tumors, *Regul. Peptides,* 13, 183, 1986.

406. **Kage, R. and Conlon, J. M.,** Neurokinin B in a human pheochromocytoma measured with a specific radioimmunoassay, *Peptides,* 10, 713, 1988.

407. **Mantyh, P. W., Mantyh, C., Gates, T., Vigna, S. R., and Maggio, J. E.,** Receptor binding sites for substance P and substance K in the canine gastrointestinal tract and their possible role in inflammatory bowel disease, *Neuroscience,* 23, 817, 1988.

408. **Mantyh, P. W., Catton, M. D., Boehmer, C. G., Welton, M. L., Passaro, E. P., Maggio, J. E., and Vigna, S. R.,** Receptors for sensory neuropeptides in human inflammatory diseases: implications for the effector role of sensory neurons, *Peptides,* 10, 627, 1989.

409. **Levine, J. D., Dardick, S. J., Roezen, M. F., Helms, C., and Basbaum, A. I.,** Contribution of sensory afferents and sympathetic efferents to joint injury in experimental arthritis, *J. Neurosci.,* 6, 3423, 1986.

410. **Rosza, Z., Mattila, J., and Jacobson, E. D.,** Substance P mediates a gastrointestinal thermoreflex in rats, *Gastroenterology,* 95, 265, 1988.

411. **Hughes, J., Smith, T. W., Kosterlitz, H. W., Fothergill, L. A., Morgan, B. A., and Morris, H. R.**, Identification of two related pentapeptides from the brain with potent opiate agonist activity, *Nature (London)*, 258, 577, 1975.

412. **Li, C. H. and Chung, D.**, Isolation and structure of an untriakonta peptide with opiate activity from camel pituitary glands, *Proc. Natl. Acad. Sci. U.S.A.*, 73, 1145, 1976.

413. **Chang, K. J.**, Opioid receptors: multiplicity and sequelae of ligand-receptor interactions, in *The Receptors*, Conn, P. M., Ed., Academic Press, Orlando, FL, 1984, 1.

414. **Höllt, V.**, Opioid peptide processing and receptor selectivity, *Annu. Rev. Pharmacol. Toxicol.*, 26, 59, 1986.

415. **Nakanishi, S., Inove, A., Kita, T., Nakamura, M., Chang, A. C. Y., Cohen, S. H., and Numa, S.**, Nucleotide sequence of cloned cDNA for bovine corticotropin-lipotropin precursor, *Nature (London)*, 278, 423, 1979.

416. **Comb, M., Seeburg, P., Adelman, J., Eiden, L., and Herbert, E.**, Primary structure of the human Met- and Leu-enkephaline precursor and its mRNA, *Nature (London)*, 295, 663, 1982.

417. **Gubler, V., Seeburg, P., Hoffman, B. J., Gage, L. P., and Undenfriend, S.**, Molecular cloning establishes proenkephalin as precursor of enkephalin-containing peptides, *Nature (London)*, 295, 206, 1982.

418. **Noda, M., Furuntani, Y., Takahashi, H., Toyosato, M., Hirose, T., Inayama, S., Nakanishi, S., and Numa, S.**, Cloning and sequences analysis of cDNA for bovine adrenal preproenkephaline, *Nature (London)*, 295, 202, 1982.

419. **Kakidani, H., Furutani, Y., Takahashi, H., Noda, M., Nakanishi, S., Morimoto, Y., Hirose, T., Asai, M., Inayama, S., and Numa, S.**, Cloning and sequence analysis of cDNA for porcine β-neo endorphine/dynorphin precursor, *Nature (London)*, 298, 245, 1982.

420. **Guillemin, R., Vargo, T., Rossier, J., Minick, S., Ling, N., Rivier, C., Vale, V., and Bloom, F.**, Beta-endorphin and adrenocorticotropin are secreted concomitantly by the pituitary gland, *Science*, 197, 1367, 1977.

421. **Zioudrou, C., Streaty, R. A., and Klee, W. A.**, Opioid peptides derived from food proteins: the exorphins, *J. Biol. Chem.*, 254, 2446, 1979.

422. **Morley, J. E., Levine, A. S., Yamada, T., Gebhard, R. L., Prigge, W. F., Shafer, R. B., Goetz, F. C., and Silvis, S. E.**, Effect of exorphins on gastrointestinal function, hormonal release, and appetite, *Gastroenterology*, 84, 1517, 1983.

423. **Hazum, E., Sabatka, J. J., Chang, K., Brent, D. A., Findlay, J. W. A., and Cuatrecasas, P.**, Morphine in cow and human milk: could dietary morphine constitute a ligand for specific morphine (u) receptors, *Science*, 213, 1010, 1981.

424. **Brantl, V. and Teschemacher, H.**, A material with opioid activity in bovine milk and milk products, *Naunyn-Schmiedeberg's Arch. Pharmacol.*, 306, 301, 1979.

425. **Brantl, V., Teschemacher, H., Bläsig, J., Henschen, A., and Lottspeich, F.**, Opioid activities of beta-casomorphins, *Life Sci.*, 28, 1903, 1981.

426. **Schusdziarra, V., Henrichs, I., Holland, A., Klier, M., and Pfeiffer, E. F.**, Evidence for an effect of exorphins on plasma insulin and glucagon levels in dogs, *Diabetes*, 30, 362, 1981.

427. **Schusdziarra, V., Schick, R., de la Fuente, A., Holland, A., Brantl, V., and Pfeiffer, E. F.**, Effect of beta-casomorphines on somatostatin release in dogs, *Endocrinology*, 112, 1948, 1983.

428. **Hughes, J., Kosterlitz, H. W., and Smith, T. W.**, The distribution of methionine-enkephalin and leucine-enkephalin in the brain and peripheral tissues, *Br. J. Pharmacol.*, 61, 639, 1977.

429. **Domoto, T., Gonda, T., Oki, M., and Yanaihara, N.**, Coexistence of substance P- and methionine[5]-enkephalin like immunoreactivity in nerve cells of the myenteric ganglia in the cat ileum, *Neurosci. Lett.*, 47, 9, 1984.

430. **Uchida, T., Kobayashi, S., and Yanaihara, N.**, Occurrence and projections for three subclasses of met-enkaphaline-Arg[6]-Gly[7]-Leu[8] neurons in the guinea pig duodenum: immunoelectron microscopic study on the co-storage of met-enkephalin-Arg[6]-Gly[7]-Leu[8] with substance P or PHI (1—15), *Biomed. Res.*, 6, 415, 1985.

431. **Costa, M., Furness, J. B., and Cuello, A. C.**, Separate populations of opioid containing neurons in the guinea-pig intestine, *Neuropeptides*, 5, 445, 1985.

432. **Polak, J. M., Sullivan, S. N., Bloom, S. R., Facer, P., and Pearse, A. G. E.**, Enkephaline like immunoreactivity in the human gastrointestinal tract, *Lancet*, 1, 972, 1977.

433. **Larsson, L.-I., Childers, S., and Synder, S. H.**, Met- and Leu-enkephalin immunoreactivity in separate neurons, *Nature (London)*, 282, 407, 1979.

434. **Feuerle, G. E., Helmstaedter, V., and Weber, U.**, Met- and Leu-enkephalin immuno- and bioreactivity in human stomach and pancreas, *Life Sci.*, 31, 2961, 1982.

435. **Sakamoto, M., Nakao, K., Yoshimasa, T., Ikeda, Y., Suda, M., Takasu, K., Shimbo, S., Yanaihara, N., and Imura, H.**, Occurrence of methionine-enkephalin-Arg[6]-Gly[7]-Leu[8] with methionine-enkephalin, leucine-enkephalin and methionine-enkephalin-Arg[6]-Phe[7] in human gastric antrum, *J. Clin. Endocrinol. Metab.*, 56, 202, 1983.

436. **Bu'Lock, A. J., Vaillant, C., and Dockray, G. J.,** Immunohistochemical studies on the gastrointestinal tract using antisera to Met-enkephalin and Met-enkephalin-Arg[6]-Phe[7], *J. Histochem. Cytochem.*, 31, 1356, 1983.

437. **Furness, J. B., Costa, M., and Miller, R. J.,** Distribution and projections of nerves with enkephalin-like immunoreactivity in the guinea pig small intestine, *Neuroscience*, 8, 653, 1983.

438. **Bornstein, J. C., Costa, M., Furness, J. B., and Lees, G. M.,** Electrophysiology and enkephalin immunoreactivity of identified myenteric plexus neurons of guinea pig small intestine, *J. Physiol.*, 351, 313, 1984.

439. **Vincent, S. R., Dalsgaard, C. L., Schultzberg, M., Hökfelt, T., Christensson, L., and Terenius, L.,** Dynorphin-immunoreactive neurons in the autonomic nervous system, *Neurosciences*, 11, 973, 1984.

440. **Tange, A.,** Distribution of peptide containing endocrine cells and neurons in the gastrointestinal tract of the dog: immunohistochemical studies using antisera to somatostatin, substance P, vasoactive intestinal polypeptide, Met-enkephalin, and neurotensin, *Biomed. Res.*, 4, 9, 1983.

441. **Daniel, E. E., Furness, J. B., Costa, M., and Belbeck, L.,** The projections of chemically identified nerve fibres in canine ileum, *Cell Tissue Res.*, 247, 377, 1987.

442. **Bouvier, M., Kirschner, G., and Gonella, J.,** Actions of morphine and enkephalins on the internal anal sphincter of the cat: relevance for the physiological role of opiates, *J. Auton. Nerv. Syst.*, 16, 219, 1986.

443. **Elde, R., Hökfelt, T., Johansson, O., and Terenius, L.,** Immunohistochemical studies using antibodies against leucine enkephalin: initial observation on the nervous system of the rat, *Neurosciences*, 1, 349, 1976.

444. **Alumets, J., Hakanson, R., Sundler, F., and Chang, K. J.,** Leu-enkephalin-like material in nerves and enterochromaffin cells in the gut, *Histochemistry*, 56, 187, 1978.

445. **Uddman, R., Aluments, J., Hakanson, R., Sundler, F., and Walles, B.,** Peptidergic (enkephalin) innervation of the mammalian esophagus, *Gastroenterology*, 78, 732, 1980.

446. **Ferri, G. L., Morreale, R. A., and Dockray, G. J.,** Met[5]-enkephalin-Arg[6]-Gly[7]-Leu[8] immunoreactivity in the human gut, *Peptides*, 7, 737, 1986.

447. **Kobayashi, S., Suzuki, M., Uchida, T., and Yanaihara, N.,** Enkephalin neurons in the guinea pig duodenum: a light and electron microscopic immunocytochemical study using an antiserum to methionine-enkephalin-Arg[6]-Gly[7]-Leu[8], *Biomed. Res.*, 5, 498, 1984.

448. **Kobayashi, S., Suzuki, M., and Yanaihara, N.,** Enkephalin neurons in the guinea pig proximal colon: an immunocytochemical study using an antiserum to methionine-enkephalin-Arg[6]-Gly[7]-Leu[8], *Arch. Histol. Jpn.*, 48, 27, 1985.

449. **Wang, Y. N. and Lindberg, I.,** Distribution and characterization of opioid octapeptide met[5]-enkephalin-arg[6]-gly[7]-leu[8] in the gastrointestinal tract of the rat, *Cell Tissue Res.*, 244, 77, 1986.

450. **Nihei, K. and Iwanaga, T.,** Localisation of Met-enkephalin-Arg[6]-Gly[7]-Leu[8]-like immunoreactivity in the gastrointestinal tract of rat and pig, *J. Histochem. Cytochem.*, 33, 1001, 1985.

451. **Leander, S., Ekman, R., Uddman, R., Sundler, R., and Hakanson, R.,** Neuronal cholecystokinin, gastrin-releasing peptide, neurotensin and beta-endorphine in the intestine of the guinea pig. Distribution and possible motor functions, *Cell Tissue Res.*, 235, 521, 1984.

452. **Wolter, H. J.,** Alpha-melanotropin and beta-endorphine immunoreactivities are contained within neurons and nerve fibers of the rat duodenum, *Brain Res.*, 295, 378, 1984.

453. **Tachibana, S., Araki, K., Ohya, S., and Yoshida, S.,** Isolation and structure of dynorphin, an opioid peptide, from porcine duodenum, *Nature (London)*, 295, 339, 1982.

454. **Wolter, H. J.,** Dynorphin-A (1—8) is contained within perikarya, nerve fibers and nerve reminals of rat duodenum, *Biochem. Res. Commun.*, 127, 610, 1985.

455. **Martin, W. R., Eedes, C. G., Thompson, J. A., Huppler, R. E., and Gilbert, G. E.,** The effects of morphine- and neomorphine-like drugs in the non-dependent and morphine dependent chronic spinal dog, *J. Pharmacol. Exp. Ther.*, 197, 517, 1976.

456. **Lord, J. A. H., Waterfield, A. A., Hughes, J., and Kosterlitz, H. W.,** Endogenous opioid peptides: multiple agonists and receptors, *Nature (London)*, 267, 495, 1977.

457. **Chang, K. J. and Cuatrecasas, P.,** Multiple opiate receptors: enkephalins and morphine bind to receptors of different specificity, *J. Biol. Chem.*, 254, 2610, 1979.

458. **Martin, W. R.,** Pharmacology of opioids, *Pharmacol. Rev.*, 35, 283, 1984.

459. **Zukin, R. S., Eghbali, M., Olive, Unterwald, E. M., and Tempel, A.,** Characterization and visualization of rat and guinea pig brain kappa opioid receptors: evidence for kappa[1] and kappa[2] opioid receptors, *Proc. Natl. Acad. Sci. U.S.A.*, 85, 4061, 1988.

460. **Lutz, R. A., Cruciani, R. A., Munson, P. J., and Rodbard, D.,** MU[1]: a very high affinity subtype of enkephalin binding sites in the rat brain, *Life Sci.*, 36, 2233, 1985.

461. **Pert, C. B. and Synder, S. H.,** Opiate receptor; demonstration in nervous tissue, *Science*, 179, 1011, 1973.

462. **Leslie, F. M., Chavkin, C., and Cox, B. M.,** Opioid binding properties of brain and peripheral tissues: evidence for heterogeneity in opioid ligand binding sites, *J. Pharmacol. Exp. Ther.*, 214, 395, 1980.

463. **Glasel, J. A., Bradbury, W. M., and Venn, R. F.,** Opiate binding to subcellular fractions from guinea pig ileum, *Life Sci.,* 34, 345, 1984.

464. **Allescher, H. D., Ahmad, S., and Daniel, E. E.,** The distribution of opioid receptors in canine small intestine: implication for function, *Am. J. Physiol.,* 256, G966, 1989.

465. **Dashwood, M. R., Degman, E. S., Bagnall, J., and Thompson, C. S.,** Autoradiographic localization of opiate receptors in rat small intestine, *Eur. J. Pharmacol.,* 107, 267, 1985.

466. **Nishimura, E., Buchan, A. M. J., and McIntosh, C. H.,** Autoradiographic localization of opiate receptors in the rat stomach, *Neurosci. Lett.,* 50, 73, 1984.

467. **Nishimura, E., Buchan, A. M. J., and McIntosh, C. H. S.,** Autoradiographic localization of mu- and delta-type opioid receptors in the gastrointestinal tract of the rat and guinea pig, *Gastroenterology,* 91, 1084, 1986.

468. **Gintzler, A. R. and Hyde, D.,** Multiple opiate receptors in the guinea-pig enteric nervous system: unmasking the copresence of receptor subtypes, *Proc. Natl. Acad. Sci. U.S.A.,* 81, 2252, 1984.

469. **Gintzler, A. R. and Scalisi, J. A.,** Physiological correlates of multiple subtypes of enteric opiate receptor; functional analysis of myenteric & receptors, *Brain Res.,* 238, 254, 1982.

470. **Gintzler, A. R. and Scalisi, J. A.,** Effects of opioids on non-cholinergic excitatory responses of the guinea pig isolated ileum: inhibition of the release of enteric substance P, *Br. J. Pharmacol.,* 75, 199, 1982.

471. **Ward, S. J. and Takemori, A. E.,** Relative involvement of receptor subtypes in opioid-induced inhibition of gastrointestinal transit in mice, *J. Pharmacol. Exp. Ther.,* 224, 359, 1983.

472. **Takemori, A. E. and Portoghese, P. S.,** Receptors for opioid peptides in the guinea pig ileum, *J. Pharmacol. Exp. Ther.,* 235, 389, 1985.

473. **Creese, I. and Snyder, S. H.,** Receptor binding and pharmacological activity of opiates in the guinea pig intestine, *J. Pharmacol. Exp. Ther.,* 194, 205, 1975.

474. **Allescher, H. D., Daniel, E. E., Dent, J., Kostolanska, F., and Fox, J. E. T.,** Neural reflex of the canine pylorus to intraduodenal acid infusion, *Gastroenerology,* 96, 18, 1989.

475. **Cherubini, E., Morita, K., and North, R. A.,** Opioid inhibition of synaptic transmission in the guinea-pig myenteric plexus, *Br. J. Pharmacol.,* 85, 805, 1985.

476. **Cherubini, E. and North, R. A.,** μ and κ opioids inhibit transmitter release by different mechanisms, *Proc. Natl. Acad. Sci. U.S.A.,* 82, 1860, 1985.

477. **North, R. A.,** Opioid receptor types and membrane ion channels, *Trends Neuro Sci.,* 7, 114, 1986.

478. **Pelton, J. T., Gulya, K., Hruby, V. J., Duckles, S. P., and Yamamura, H. I.,** Conformationally restricted analogues of somatostatin with high mu-opiate receptor specificity, *Proc. Natl. Acad. Sci. U.S.A.,* 82, 236, 1985.

479. **Takemori, A. E., Ikeda, M., and Portoghese, P. S.,** The mu, kappa and delta properties of various opioid agonists, *Eur. J. Pharmacol.,* 123, 357, 1986.

480. **Portoghese, P. S. and Takemori, A. E.,** TENA, a selective kappa opioid receptor antagonist, *Life Sci.,* 36, 801, 1985.

481. **Cotton, R., Giles, M., Miller, L., Shaw, J. S., and Timms, D.,** ICI 174,864: a highly selective antagonist at the opioid delta receptor, *Eur. J. Pharmacol.,* 97, 331, 1984.

482. **Chang, K. J., Wei, E. T., Killian, A., and Chang, J.-K.,** Potent morphiceptin analogues: structure activity relationship and morphine-like activities, *J. Pharmacol. Exp. Ther.,* 227, 403, 1983.

483. **Mosberg, H. I., Hurst, R., Hruby, V. J., Gee, K., Yamamura, H. I., Galligan, J. J., and Burks, T. F.,** Bis-penicillamine enkephalins possess highly improved specificity toward δ opioid receptors, *Proc. Natl. Acad. Sci. U.S.A.,* 80, 5871, 1983.

484. **Cowan, A. and Gmerek, D. E.,** *In vivo* studies with ICI 154,129, a putative delta receptor antagonist, *Life Sci.,* 31, 2213, 1982.

485. **Lathi, R. A., VonVoigtlander, P. F., and Barsuhn, C.,** Properties of a selective kappa agonist, U-50, 488, *Life Sci.,* 31, 2257, 1982.

486. **Chavkin, C., James, I. F., and Goldstein, A.,** Dynorphin is a specific endogenous ligand of the κ opioid receptor, *Science,* 215, 413, 1982.

487. **James, I. F., Fischli, W., and Goldstein, A.,** Opioid receptor selectivity of dynorphin gene products, *J. Pharmacol. Exp. Ther.,* 228, 88, 1984.

488. **Yoshimura, K., Huidobro-Toro, J. P., Lee, N. M., Loh, H. H., and Way, E. L.,** Kappa opioid properties of dynorphin and its peptide fragments on the guinea-pig ileum, *J. Pharmacol. Exp. Ther.,* 222, 71, 1982.

489. **Kreulen, D. L. and Peters, S.,** Non-cholinergic transmission in a sympathetic ganglion of the guinea pig elicited by colon distension, *J. Physiol. (London),* 374, 314, 1986.

490. **Kennedy, C. and Krier, J.,** (Met5)-enkephalin acts via delta-opioid receptors to inhibit pelvic nerve evoked contractions of cat distal colon, *Br. J. Pharmacol.,* 92, 291, 1987.

491. **Oka, T.,** Enkephalin (opiate) receptors in the intestine, *Trends Pharmacol. Sci.,* 2, 328, 1981.

492. **Burks, T. F., Hirning, L. D., Galligan, J. J., and Davis, T. P.,** Motility effects of opioid peptides in dog intestine, *Life Sci.,* 31, 2237, 1982.

493. **Manaka, H., Manaka, Y., Kostolanska, F., Fox, J. E. T., and Daniel, E. E.,** Release of VIP and substance P from isolated perfused canine ileum, *Am. J. Physiol.,* 257, G182, 1989.

494. **Bauer, A. J. and Szursweski, J. H.,** Dynorphin presynaptically inhibits neuromuscular transmission via delta opioid receptors in circular muscle of canine duodenum, *Gastgroenterology,* 96, A33, 1989.

495. **Huizinga, J. D. and Den Hertog, A.,** The effect of enkephalins on the intramural inhibitory non-adrenergic nerve responses of smooth muscle, *Eur J. Pharmacol.,* 54, 389, 1979.

496. **Paton, W. D. M.,** The action of morphine and related substances on contraction and on acetylcholine output of coaxially stimulated guinea-pig ileum, *Br. J. Pharmacol.,* 12, 119, 1957.

497. **Waterfield, A. A. and Kosterlitz, H. W.,** Stereospecific increase by narcotic antagonists of evoked acetylcholine output in guinea pig ileum, *Life Sci.,* 16, 1787, 1975.

498. **Vizi, E. S., Ono, K., Adam-Vizi, V., Duncalf, D., and Foldes, F. F.,** Presynaptic inhibitory effect of met-enkephalin on [^{14}C] acetylcholine release from the myenteric plexus and its interaction with muscarinic negative feedback inhibition, *J. Pharmacol. Exp. Ther.,* 230, 493, 1984.

499. **Gintzler, A. R.,** Serotonin participation in gut withdrawal from opiates, *J. Pharmacol. Exp. Ther.,* 211, 7, 1979.

500. **Grider, J. R. and Makhlouf, G. M.,** Role of opioid neurons in the regulation of intestinal peristalsis, *Am. J. Physiol.,* 253, G226, 1987.

501. **Glass, J., Chan, W. C., and Gintzler, A. R.,** Direct analysis of the release of methionine-enkephalin from guinea pig myenteric plexus: modulation by endogenous opioids and exogenous morphine, *J. Pharmacol. Exp. Ther.,* 239, 742, 1986.

502. **Yau, W. M., Lingle, P. F., and Youther, M. L.,** Interaction of enkephalin and caerulein on guinea pig intestine, *Am. J. Physiol.,* 244, G65, 1983.

503. **Yau, W. M., Dorsett, J. A., and Youther, M. L.,** Inhibitory peptidergic neurons: functional difference between somatostatin and enkephalin in myenteric plexus, *Am. J. Physiol.,* 250, G60, 1986.

504. **Holzer, P., Lippe, I., Bartho, L., and Lembeck, F.,** (D-met^2, pro^5)-Enkephalinamide and dynorphine (1—13) inhibit the cholinergic contraction induced in the guinea pig ileum by substance P, *Eur. J. Pharmacol.,* 91, 83, 1983.

505. **North, R. A., Katayama, Y., and Williams, J. T.,** On the mechanism and site of action of enkephalin on single myenteric neurons, *Brain Res.,* 165, 67, 1979.

506. **Morita, K. and North, R. A.,** Opiates and enkephalin reduce the excitability of neuronal processes, *Neuroscience,* 6, 1943, 1981.

507. **Wood, J. D. and Marsh, D. R.,** Effects of atropine, tetrodotoxin, and lidocain on rebound excitation of guinea pig small intestine, *J. Pharmacol. Exp. Ther.,* 184, 590, 1970.

508. **Manaka, H., Manaka, Y., Kostolanska, F., Fox, J. E. T., and Daniel, E. E.,** Release of VIP from isolated perfused canine ileum, *Gastroenterology,* 92, 128, 1988.

509. **Daniel, E. E., Sutherland, W. H., and Bogoch, A.,** Effects of morphine and other drugs in motility of terminal ileum, *Gastroenterology,* 36, 510, 1959.

510. **Daniel, E. E.,** Further studies of the pharmacology of the pyloric region. Analysis of the effects of intra-arterial histamine, serotonin, phenylbiguanide, morphine, and other drugs on the antrum and duodenal bulb, *Can. J. Physiol. Pharmacol.,* 44, 981, 1966.

511. **Camilleri, M., Malagelada, J.-R., Stanghellini, V., Zinsmeister, A. R., Kao, P. C., and Li, C. H.,** Dose-related effects of synthetic human β-endorphin and naloxone on fed gastrointestinal motility, *Am. J. Physiol.,* 251, G147, 1986.

512. **Porreca, F., Mosberg, H. I., Hurst, R., Hruby, V. J., and Burks, T. F.,** Roles of mu, delta and kappa opioid receptors in spinal and supraspinal mediation of gastrointestinal transit effects and hot-plate analgesia in the mouse, *J. Pharmacol. Exp. Ther.,* 230, 341, 1984.

513. **Manara, L. and Biachetti, A.,** The central and peripheral influences of opioids on gastrointestinal propulsion, *Annu. Rev. Pharmacol. Toxicol.,* 25, 249, 1985.

514. **Frank, E. B., Lange, R., Plankey, M., and McCallum, R. W.,** Effect of morphine and naloxone on lower esophageal sphincter pressure and gastric emptying in man, *Gastroenterology,* 82, 1060, 1982.

515. **Narducci, F., Bassotti, G., and Granata, M. T.,** Functional dyspepsia and chronic idiopathic stasis — role of endogenous opiates, *Arch. Int. Med.,* 146, 716, 1986.

516. **Feldman, M., Walsh, J. H., and Taylor, I. L.,** Effect of naloxone and morphine on gastric acid secretion and on serum gastrin and pancreatic polypeptide concentrations in humans, *Gastroenterology,* 79, 492, 1980.

517. **Shea-Donohue, P. T., Adams, N., Arnold, J., and Dubois, A.,** Effects of met-enkephalin and naloxone on gastric emptying and secretion in rhesus monkeys, *Am. J. Physiol.,* 245, G196, 1983.

518. **Parolaro, D., Sala, M., and Gori, E.,** Effects of intracerebroventricular administration of morphine upon intestinal motility in rats and its antagonism with naloxone, *Eur. J. Pharmacol.,* 46, 329, 1977.

519. **Schultz, R., Wüster, M., and Herz, A.,** Centrally and peripherally mediated inhibition of intestinal motility by opioids, *Naunyn-Schmiedeberg's Arch. Pharmacol.,* 308, 255, 1979.

520. **Tavani, A., Bianchi, G., and Manara, L.,** Morphine no longer blocks gastrointestinal transit but retains antinociceptive action in diallylnormorphine-pretreated rats, *Eur. J. Pharmacol.,* 59, 151, 1979.

521. **Bianchi, G., Fiocchi, R., Tavani, A., and Manara, L.,** Quaternary narcotic antagonists' relative ability to prevent antinociception and gastrointestinal transit inhibition in morphine-treated rats as an index of peripheral selectivity, *Life Sci.,* 30, 1875, 1982.

522. **Sun, E. A., Snape, E. J., Cohen, S., and Renny, A.,** The role of opiate receptors and cholinergic neurons in the gastrocolonic response, *Gastroenterology,* 82, 689, 1982.

523. **Kamikawa, Y. and Shimo, Y.,** Pharmacological characterization of the opioid receptor in the submucous plexus of the guinea-pig oesophagus, *Br. J. Pharmacol.,* 78, 693, 1983.

524. **Stacher, G., Bauer, P., Steinringer, H., Schmierer, G., Langer, B., and Winklehner, S.,** Dose-related effects of the synthetic met-enkephalin analogue FK 33-824 on esophageal motor activity in healthy humans, *Gastroenterology,* 83, 1057, 1982.

525. **Hall, A. W., Moossa, A. R., Clark, J., Cooley, G. R., and Skinner, D. B.,** The effect of premedication drugs on the lower oesophageal high pressure zone and reflux status of rhesus monkeys and man, *Gut,* 16, 347, 1975.

526. **Mittal, R. K., Frank, E. B., Lange, R. C., and McCallum, R. W.,** Effects of morphine and naloxone on esophageal motility and gastric emptying in man, *Dig. Dis. Sci.,* 31, 936, 1986.

527. **Borody, T. J., Quigley, E. M. M., Phillips, S. F., Wienbeck, M., Tucker, R. L., Haddad, A., and Zinsmeister, A. R.,** Effects of morphine and atropine on motility and transit in the human ileum, *Gastroenterology,* 89, 562, 1985.

528. **Wienbeck, M.,** Involvement of enkaphalins and other endogenous opioids in the regulation of esophageal motility, *Gastroenterol. Clin. Biol.,* 11, 1987, 52B.

529. **Allescher, H.-D., Berezin, I., Jury, J., and Daniel, E. E.,** Characteristics of the canine lower esophageal sphincter: a new electrophysiological tool, *Am. J. Physiol.,* 255, G453, 1988.

530. **Dowlathahi, K., Evander, A., Walther, B., and Skinner, D. B.,** Influence of morphine on the distal oesophagus and the lower esophageal sphincter — a manometric study, *Gut,* 26, 802, 1985.

531. **Howard, J. M., Belsheim, M. R., and Sullivan, S. N.,** Enkephalin inhibits relaxation of the lower esophageal sphincter, *Br. Med. J.,* 285, 1605, 1982.

532. **McCallum, R. W., Dodds, J., Osborne, H. P. and Bianciani, P.,** Effect of enkephalin and other opiates on opossum lower esophageal sphincter (LES), in *Gastrointestinal Motility,* Christensen, J., Ed., Raven Press, New York, 1980, 37.

533. **Rattan, S. and Goyal, R. K.,** Identification and localization of opioid receptors in the opossum lower esophageal sphincter, *J. Pharmacol. Exp. Ther.,* 224, 391, 1983.

534. **Edin, R., Lundberg, J., Terenius, L., Dahlström, A., Hökfelt, T., Kewenter, J., and Ahlman, H.,** Evidence for vagal enkephalinergic control of the feline pylorus and stomach, *Gastroenterology,* 78, 492, 1980.

535. **Reynolds, J. C., Ouyang, A., and Cohen, S.,** Evidence for an opiate-mediated pyloric sphincter reflex, *Am. J. Physiol.,* 246, G130, 1984.

536. **Reynolds, J. C., Ouyang, A., and Cohen, S.,** Opiate nerves mediate feline pyloric response to intraduodenal amino acids, *Am. J. Physiol.,* 248, G307, 1985.

537. **Konturek, S., Pawlik, W., Walus, K., Coy, D. H., and Schally, A. V.,** Methionine-enkephalin stimulates gastric secretion and gastric mucosal blood flow, *Proc. Soc. Exp. Biol. Med.,* 158, 156, 1978.

538. **Fioramonti, J., Fargeas, M. J., and Bueno, L.,** Comparative effects of morphine and cyclazocine on gastrointestinal motility in conscious dogs, *Arch. Int. Pharmacodyn. Ther.,* 270, 141, 1984.

539. **Duranton, A. and Bueno, L.,** Central opiate mechanism involved in gastrointestinal motor disturbances induced by *E. coli* endotoxin in sheep, *Life Sci.,* 34, 1795, 1984.

540. **Sullivan, S. N., Lamki, L., and Corcoran, P.,** Inhibition of gastric emptying by enkephalin analogue, *Lancet,* 2, 86, 1981.

541. **Gue, M., Fioramonti, J., Honde, C., Pascaud, X., Junien, J. L., and Bueno, L.,** Opposite effects of kappa-opioid agonists on gastric emptying of liquids and solids in dogs, *Gastroenterology,* 95, 927, 1988.

542. **Fioramonti, J., Fargeas, M. J., and Bueno, L.,** Involvement of endogenous opiates in regulation of gastric emptying of fat test meals in mice, *Am. J. Physiol.,* 255, G158, 1988.

543. **Bovell, K. T., Tougas, G., Dent, J., Collins, S. M., and Hunt, R. H.,** Effect of naloxone on lipid induced pyloric motor responses in humans, *Gastroenterology,* 92, 12325, 1987.

544. **Furness, J. B. and Costa, M.,** Identification of gastrointestinal neurotransmitters, in *Mediators and Drugs in Gastrointestinal Motility I. Morphological Basis and Neurophysiological Control,* Bertaccini, G., Ed., Springer-Verlag, Berlin, 1982, 384.

545. **Plant, O. H. and Miller, G. H.,** Effects of morphine and some other opium alkaloids on the muscular activity of the alimentary canal. I. Action on the small intestine of the unanesthetized dog and man, *J. Pharmacol. Exp. Ther.,* 27, 361, 1926.

546. **Burks, T. F.,** Mediation by 5'-hydroxytryptamine of morphine stimulate actions in dog intestine, *J. Pharmacol. Exp. Ther.,* 185, 530, 1973.

547. **Vaught, J. L., Cowan, A., and Gmerek, D. E.,** A species difference in the slowing effect of intrathecal morphine on gastrointestinal transit, *Eur. J. Pharmacol.,* 94, 81, 1983.
548. **Ahmad, S., Allescher, H. D., Manaka, H., Manaka, Y., and Daniel, E. E.,** Biochemical studies on opioid and alpha₂-adrenoceptors in canine submucosal neurons, *Am. J. Physiol.,* 256, G957, 1989.
549. **Kromer, W., Höllt, V., Schmitt, H., and Herz, A.,** Release of immunoreactive dynorphin from the isolated guinea pig ileum is reduced during peristaltic activity, *Neurosci. Lett.,* 25, 53, 1981.
550. **Kromer, W. and Petzlaff, W.,** In vitro evidence for the participation of intestinal opioids in the control of peristalsis in the intestine of the guinea pig small intestine, *Naunyn-Schmiedeberg's Arch. Pharmacol.,* 309, 153, 1979.
551. **Kromer, W., Petzlaff, W., and Woinoff, R.,** Opioids modulate periodicity rather than efficacy of peristaltic waves in the guinea pig ileum *in vitro, Life Sci.,* 26, 1857, 1980.
552. **Clark, S. J. and Smith, T. W.,** Modulation of the peristaltic reflex *in vivo* by endogenous opioids, *Br. J. Pharmacol.,* 74, 953P, 1981.
553. **Clark, S. J. and Smith, T. W.,** Peristalsis abolishes the release of methionine-enkephalin from guinea pig ileum *in vitro, Eur. J. Pharmacol.,* 70, 421, 1981.
554. **Schultz, R., Wuester, M., Simantov, R., Snyder, S., and Herz, A.,** Electrically stimulated release of opiate like material from the myenteric plexus of the guinea pig ileum, *Eur. J. Pharmacol.,* 41, 347, 1977.
555. **Daniel, E. E. and Kostolanska, F.,** Functional studies of nerve projections in the canine intestine, *Can. J. Physiol. Pharmacol.,* in press, 1990.
556. **Clark, S. J. and Smith, T. W.,** The release of met-enkephalin from the guinea pig ileum at rest and during peristaltic activity, *Life Sci.,* 33 (Suppl.), 465, 1983.
557. **Höllt, V., Garzon, J., Schulz, R., and Herz, A.,** Corticotropin-releasing factor is excitatory in the guinea pig ileum and activates an opioid mechanism in this tissue, *Eur. J. Pharmacol.,* 101, 165, 1984.
558. **Garzon, J., Höllt, V., Sanchez-Blaquez, P., and Herz, A.,** Neural activation of opioid mechanisms in guinea pig ileum by excitatory peptides, *J. Pharmacol. Exp. Ther.,* 240, 642, 1987.
559. **Szurszeswski, J. H.,** A migrating electric complex of the canine small intestine, *Am. J. Physiol.,* 217, 1757, 1969.
560. **Konturek, S. J., Thor, P., Krol, R., Dembinski, A., and Schally, A. V.,** Influence of methionine-enkephalin and morphine activity on myoelectric of small bowel, *Am. J. Physiol.,* 238, G384, 1980.
561. **Sarna, S., Northcott, P., and Belbeck, L.,** Mechanism of cycling and migrating myoelectric complexes: effect of morphine, *Am. J. Physiol.,* 242, G588, 1982.
562. **Telford, G. L., Hoshmonai, M., Moses, A. J., and Szurszewski, J. H.,** Morphine initiates migrating myoelectric complexes by acting on peripheral opioid receptors, *Am. J. Physiol.,* 249, G557, 1985.
563. **Waterfall, W. E.,** Electrical patterns in the human jejunum with and without vagotomy: migrating myoelectric complexes and the influence of morphine, *Surgery,* 94, 186, 1983.
564. **Sarna, S. K. and Condon, R. E.,** Morphine-initiated migrating myoelectrical complexes in the fed state in dogs, *Gastroenterology,* 86, 662, 1984.
565. **Jian, R., Janssens, J., Vantrappen, G., and Ceccatelli, P.,** Influence of metenkephalin analogue on motor activity of the gastrointestinal tract, *Gastroenterology,* 93, 114, 1987.
566. **Telford, G. L., Condon, R. E., and Szurszewski, J. H.,** Opioid receptors and the initiation of migrating myoelectric complexes in dog, *Am. J. Physiol.,* 256, G72, 1989.
567. **Telford, G. L. and Szurszewski, J. H.,** Blockade of migrating myoelectric complexes by naloxone, *Gastroenterology,* 86, 1278, 1984.
568. **Sarna, S. K., Condon, R. E., and Cowles, V.,** Morphine versus motilin in the initiation of migrating myoelectric complexes, *Am. J. Physiol.,* 245, G217, 1983.
569. **Rees, W. D., Sharpe, G. R., Christofides, N. D., Bloom, S. R., and Turnberg, L. A.,** The effects of an opiate agonist and antagonist on the human upper gastrointestinal tract, *Eur. J. Clin. Invest.,* 13, 221, 1983.
570. **Powell, D. W.,** Muscle or mucosa: the site of action of antidiarrheal opiates?, *Gastroenterology,* 80, 406, 1981.
571. **Manara, L., Bianchi, G., Ferretti, P., and Tavani, A.,** Inhibition of gastrointestinal transit by morphine in rats results primarily from direct drug action on gut opioid sites, *J. Pharmacol. Exp. Ther.,* 237, 945, 1986.
572. **Piercey, M. F. and Ruwart, M. J.,** Naloxone inhibits the antidiarrhoeal activity of loperamide, *Br. J. Pharmacol.,* 66, 373, 1979.
573. **Daniel, E. E., Kostolanska, F., Allescher, H. D., Ahmad, S., and Fox, J. E. T.,** Local actions of trimebutine maleate in canine small intestine, *J. Pharmacol. Exp. Ther.,* 245, 1002, 1988.
574. **Ouyang, A., Vos, P., and Cohen, S.,** Sites of action of mu-, kappa- and sigma-opiate receptor agonists at the feline ileocecal sphincter, *Am. J. Physiol.,* 254, G224, 1988.
575. **Ouyang, A., Clain, C. J., Snape, W. J., and Cohen, S.,** Characterization of opiate mediated responses of the feline ileum and ileocecal sphincter, *J. Clin. Invest.,* 69, 507, 1982.

576. **Leander, S., Hakanson, R., and Sundler, F.**, Nerves containing substance P, vasoactive intestinal polypeptide, enkephalin and somatostatin in the guinea pig teania coli, *Cell Tissue Res.*, 215, 21, 1981.
577. **North, R. A.**, Hyperpolarization of myenteric neurons by enkephalins, *Br. J. Pharmacol.*, 59, 504P, 1977.
578. **Nijkamp, F. P. and Van Ree, J.**, Effects of endorphin on different parts of the gastrointestinal tract of rat and guinea pig *in vitro*, *Br. J. Pharmacol.*, 68, 599, 1980.
579. **Gillian, M. G. C. and Pollock, D.**, Acute effects of morphine and opioid peptides on the motility on the responses of rat colon to electrical stimulation, *Br. J. Pharmacol.*, 68, 381, 1980.
580. **Scheurer, U., Drack, E., Varga, L., and Halter, F.**, Morphine-like action of enkephalin analog FK 33-824 on motility of the isolated rat colon, *J. Pharmacol. Exp. Ther.*, 219, 534, 1981.
581. **Huidobro-Toro, J. P. and Way, E. L.**, Contractile effects of morphine and related opioid alkaloids, β-endorphin and methionine enkephalin on the isolated colon from Long Evans rats, *Br. J. Pharmacol.*, 74, 681, 1981.
582. **Scheurer, U., Drack, E., and Halter, F.**, Cyclooxygenase inhibitors affect met-enkephalin- and acetycholine-stimulated motility of the isolated rat colon, *J. Pharmacol. Exp. Ther.*, 234, 742, 1985.
583. **Pascaud, X. B., Genton, M. G., Remond, G., and Vincent, M.**, Antral to colonic motility responses to intracerebroventricular administration of D-Ala²-enkephalinamide, beta-endorphin, methionine enkephalin and fentanyl in anesthetized rats, in *Gastrointestinal Motility*, Christensen, J., Ed., Raven Press, New York, 1980, 459.
584. **Ruckebusch, Y., Ferre, J. P., and Du, C.**, *In vivo* modulation of intestinal motility and sites of opioid effects in the rat, *Regul. Peptides*, 9, 109, 1984.
585. **Hellström, P. M. and Rosell, S.**, Effects of neurotensin, substance P and methionine-enkephalin on colonic motility, *Acta Physiol. Scand.*, 113, 147, 1981.
586. **Blanquet, F., Bouvier, M., and Gonella, J.**, Effects of enkephalins and morphine on spontaneous electrical activity and on junction potentials elicited by parasympathetic nerve stimulation in cat and rabbit colon, *Br. J. Pharmacol.*, 77, 419, 1982.
587. **Ruckebusch, Y., Bardon, T., and Pairet, M.**, Opioid control of the ruminant stomach motility, functional importance of μ, κ and δ receptors, *Life Sci.*, 35, 1731, 1984.
588. **Blanquet, F., Bouvier, M., and Gonella, J.**, Action of trimebutine in cat and rabbit colon: evidence of an opioid-like effect, *J. Pharmacol. Exp. Ther.*, 234, 708, 1985.
589. **Bickel, M.**, Stimulation of colonic motility in dogs and rats by an enkephalin analogue pentapeptide, *Life Sci.*, 33, 469, 1983.
590. **Bueno, L., Fioramonti, J., and Ruckebusch, Y.**, Comparative effects of morphine and nalorphine on colonic motility in the conscious dog, *Eur. J. Pharmacol.*, 75, 239, 1981.
591. **Daniel, E. E., Fox, J. E. T., Allescher, H. D., Ahmad, S., and Kostolanska, F.**, Peripheral actions of opiates in canine gastrointestinal tract: actions on nerves and muscles, *Gastroenterol. Clin. Biol.*, 11, 35B, 1987.
592. **Painter, N. S. and Truelove, S. C. L.**, The intraluminal pressure pattern in diverticulosis of the colon. II. The effects of morphine, *Gut*, 5, 207, 1964.
593. **Burleigh, D. E. and D'Mello, A.**, Neural and pharmacologic factors affecting motility of the internal anal sphincter, *Gastroenterology*, 84, 409, 1983.
594. **Burleigh, D. E. and Trout, S. J.**, Morphine attenuates cholinergic nerve activity in human isolated colonic muscle, *Br. J. Pharmacol.*, 88, 307, 1986.
595. **Schang, J. C., Hemond, M., Hebert, M., and Pilote, M.**, How does morphine work on colonic motility? An electromyographic study in the human left and sigmoid colon, *Life Sci.*, 38, 671, 1986.
596. **Bueno, L. and Fioramonti, J.**, Enkephalins, other endogenous opioids and colonic motility in dog and man, *Gastroenterol. Clin. Biol.*, 11, 69B, 1987.
597. **Garret, J. M., Sauer, W. G., and Moertel, C. G.**, Colonic motility in ulcerative colitis after opiate administration, *Gastroenterology*, 53, 93, 1967.
598. **Konturek, S. J.**, Opiates and the gastrointestinal tract, *Am. J. Gastroenterol.*, 74, 285, 1980.
599. **Kaufman, P. N., Krevsky, B., Malmud, L. S., Maurer, A. H., Somers, M. B., Siegel, J. A., and Fisher, R. S.**, Role of opiate receptors in the regulation of colonic transit, *Gastroenterology*, 94, 1351, 1988.
600. **Rattan, S. and Culver, P. J.**, Influence of loperamide on the internal anal sphincter in the opossum, *Gastroenterology*, 93, 121, 1987.
601. **Bouvier, M., Grimaud, J. C., Naudy, B., and Salducci, J.**, Effects of morphine on electrical activity of the rectum in man, *J. Physiol. (London)*, 388, 153, 1987.
602. **Porreca, F. and Burks, T. F.**, The spinal cord as a site of opioid effects on gastrointestinal transit in the mouse, *J. Pharmacol. Exp. Ther.*, 227, 22, 1983.
603. **Porreca, F., Cowan, A., Raffa, R. J., and Tallarida, R. J.**, Ketoazocines and morphine: effects on gastrointestinal transit after central and peripheral administration, *Life Sci.*, 32, 1785, 1983.
604. **Bueno, L., Fioramonti, J., Honde, C., Fargeas, M. J., and Primi, M. P.**, Central and peripheral control of gastrointestinal and colonic motility by endogenous opiates in conscious dogs, *Gastroenterology*, 88, 549, 1985.

605. **Steward, J. J., Weisbrodt, N. W., and Burks, T. F.,** Centrally mediated intestinal stimulation by morphine, *J. Pharmacol. Exp. Ther.,* 202, 174, 1977.
606. **Steward, J. J., Weisbrodt, N. W., and Burks, T. F.,** Central and peripheral actions of morphine on intestinal transit, *J. Pharmacol. Exp. Ther.,* 205, 547, 1978.
607. **Economou, G. and Ward-McQuaid, J. N.,** A cross over comparison of the effect of morphine, pethidine, pentazocine and phenazocine on biliary pressure, *Gut,* 12, 218, 1971.
608. **Behar, J. and Biancani, P.,** Neural control of the sphincter of Oddi-physiologic role of enkephalins on the regulation of basal sphincter of Oddi motor activity in the cat, *Gastroenterology,* 86, 134, 1984.
609. **Kreek, M. J., Hahn, E. F., Schaefer, R. A., and Fishman, J.,** Naloxone, a specific opioid antagonist, reverses chronic idiopathic constipation, *Lancet,* 1, 261, 1983.
610. **Kreek, M. J., Paris, P., Bartol, M. A., and Müller, D.,** Effect of short term oral administration of the specific opioid antagonist naloxone on fecal evacuation in geriatric patients, *Gastroenterology,* 86, 1144, 1984.
611. **Stanghellini, V., Malagelada, J. R., Zinsmeister, A. R., Go, V. L. W., and Kao, P. C.,** Effect of opiate and adreneric blockers on the gut motor response to central acting stimuli, *Gastroenterology,* 87, 1104, 1984.
612. **Camilleri, M., Malagelada, J.-R., Kao, P. C., and Zinsmeister, A. R.,** Effects of somatovisceral reflexes and selective dermatomal stimulation on postcibal antral pressure activity, *Am. J. Physiol.,* 247, G703, 1984.
613. **Brown, J. C., Mutt, V., and Dryburgh, J. R.,** The further purification of motilin, a gastric motor activity stimulating polypeptide from the mucosa of the small intestine of hogs, *Can. J. Physiol. Pharmacol.,* 49, 399, 1971.
614. **Brown, J. C., Cook, M. A., and Dryburgh, J. R.,** Motilin a gastric motor activity stimulating polypeptide: the complete amino acid sequence, *Can. J. Biochem.,* 51, 533, 1973.
615. **Helmstaedter, V., Kreppein, W., Domschke, W., Mitznegg, P., Yanaihara, N., Wuensch, E., and Forssmann, W. G.,** Immunohistochemical localization of motilin in endocrine non-enterochromaffin cells of the small intestine of humans and monkey, *Gastroenterology,* 76, 897, 1979.
616. **Pearse, A. G. E.,** The cellular origin of motilin in the gastrointestinal tract, *Scand. J. Gastroenterol.,* 11, 35, 1976.
617. **Solcia, E., Polak, J. M., Larsson, L. I., Buchan, A. M. J., and Capella, C.,** Update on Lausanne classification of endocrine cells, in *Gut Hormones,* Bloom, S. R. and Polak, J. M., Eds., Churchill Livingstone, Edinburgh, 1981, 96.
618. **Fox, J. E. T., Track, N., and Daniel, E. E.,** Motilin: its presence and function in muscle layers of the gastrointestinal tract, in *Gastrointestinal Motility,* Christensen, J., Ed., Raven Press, New York, 1980, 59.
619. **Yanaihara, N., Yanaihara, C., Nagai, H., Sato, H., Shimizu, F., Yamaguchi, K., and Abe, K.,** Motilin-like immunoreactivity in porcine, canine and human and rat tissue, *Biomed. Res.,* 1, 76, 1980.
620. **Fox, J. E. T., Track, N., Daniel, E. E., and Yanaihara, N.,** Immunoreactive motilin is not exclusive to the gastrointestinal mucosa, *Biomed. Res.,* 2, 321, 1981.
621. **Beinfeld, M. C. and Bailey, G. J.,** The distribution of motilin-like peptides in Rhesus monkey brain as determined by radioimmunoassay, *Neurosci. Lett.,* 54, 345, 1985.
622. **Christofides, N. D., Bryant, M. G., Ghatei, M. A., Kishimoto, S., Buchan, A. M. J., Polak, J. M., and Bloom, S. R.,** Molecular forms of motilin in the mammalian and human gut and human plasma, *Gastroenterology,* 80, 292, 1981.
623. **Seino, Y., Tanaka, K., Takeda, J., Takahashi, H., Mitani, T., Kurono, M., Kayano, T., Koh, G., Fukumoto, H., Yano, H., Fujita, Inagaki, N., Yamada, Y., and Imura, H.,** Sequence of an intestinal cDNA encoding human motilin precursor, *FEBS Lett.,* 223, 74, 1987.
624. **Poitras, P., Reeve, J. R., Hunkapiller, M. W., Hood, L. E., and Walsh, J. H.,** Purification and characterization of canine intestinal motilin, *Regul. Peptides,* 5, 197, 1983.
625. **Wuensch, E.,** Synthesis of motilin analogues, *Scand. J. Gastroenterol.,* 11 (Suppl. 39), 19, 1976.
626. **Itoh, Z., Takeuchi, S., Aizawa, I., Mori, K., Taminato, T., Seino, Y., Imura, H., and Yanaihara, N.,** Changes in plasma motilin concentrations and gastrointestinal contractile activity in conscious dogs, *Am. J. Dig. Dis.,* 23, 929, 1978.
627. **Itoh, Z., Nakaya, M., Suzuki, T., Arai, H., and Wakabayashi, K.,** Erythromycin mimics exogenous motilin in gastrointestinal contractile activity in the dog, *Am. J. Physiol.,* 247, G688, 1984.
628. **Itoh, Z., Kondo, Y., and Omura, S.,** Motilide: a new family of macrolide compounds mimicking motilin, *Biomed. Res.,* (Suppl. 1), 8, 1988.
629. **Peeters, T. L., Matthijs, G., Depoortere, I., Cachet, T., Hoogmartens, J., and Vantrappen, G.,** Erythromycin and its derivates are motilin receptor agonists, *Biomed. Res.,* (Suppl. 1), 95, 1988.
630. **Vantrappen, G., Janssens, J., Tack, J., Muls, E., Bouillon, R., and Peeters, T.,** Erythromycin is a potent gastrokinetic in diabetic gastroparesis, *Gastroenterology,* 96, A525, 1989.
631. **Sarna, S. K., Soergel, K. H., Koch, T. R., Stone, J. E., Wood, C. M., Ryan, J. H., Cavanaugh, J. H., Nellans, H. N., and Lee, M. B.,** Effects of erythromycin on human gastrointestinal motor activity in the fasted and fed state, *Gastroenterology,* 96, A440, 1989.

632. **Janssens, J., Vantrappen, G., Urbain, J. L., De Roo, M., Bouillon, R., Muls, E., and Peeters, T.,** The motilin agonist erythromycin normalises impaired gastric emptying in diabetic gastroparesis, *Gastroenterology,* 96, A237, 1989.

633. **Bormans, V., Peeters, T. L., and Vantrappen, G.,** Motilin receptors in rabbit stomach and small intestine, *Regul. Peptides,* 15, 143, 1986.

634. **Peeters, T. L., Bormans, V., and Vantrappen, G.,** Regional and temporal variations of motilin receptor density in the human and rabbit gastrointestinal tract, *Dig. Dis. Sci.,* 30, 658, 1985.

635. **Peeters, T. L., Bormans, V., and Vantrappen, G.,** Comparison of motilin binding to crude homogenates of human and canine gastrointestinal smooth muscle tissue, *Regul. Peptides,* 23, 171, 1988.

636. **Poitras, P., Lahaic, R. G., St.-Pierre, S., and Trudel, L.,** Comparative stimulation of motilin duodenal receptor by porcine or canine motilin, *Gastroenterology,* 92, 658, 1987.

637. **Fox, J. E. T., Daniel, E. E., Jury, J., and Robotham, H.,** The mechanisms of motilin excitation of the canine small intestine, *Life Sci.,* 34, 1001, 1984.

638. **Peeters, T. L., Bormans, V., Matthijs, G., and Vantrappen, G.,** Comparison of the biological activity of canine and porcine motilin in rabbit, *Regul. Peptides,* 15, 333, 1986.

639. **Louie, D. S. and Owyang, C.,** Motilin receptors on isolated gastric smooth muscle cells, *Am. J. Physiol.,* 254, G210, 1988.

640. **Dryburgh, J. R. and Brown, J. C.,** Radioimmunoassay for motilin, *Gastroenterology,* 68, 1169, 1975.

641. **Lee, K. Y., Kim, M. S., and Chey, W. Y.,** Effects of a meal and gut hormones on plasma motilin and duodenal motility in dog, *Am. J. Physiol.,* 238, G280, 1980.

642. **Fox, J. E. T., Track, N. S., and Daniel, E. E.,** Relationship of plasma motilin concentration to fat ingestion, duodenal acidification and alkalization, and migrating motor complexes in dogs, *Can. J. Physiol. Pharmacol.,* 59, 180, 1981.

643. **Mitznegg, P., Bloom, S. R., Christofides, N., Besterman, H., Domschke, W., Domschke, S., Wuensch, E., and Demling, L.,** Release of motilin in man, *Scand. J. Gastroenterol.,* 11, (Suppl. 39), 53, 1976.

644. **Mitznegg, P., Bloom, S. R., Domschke, W., Domschke, S., Wuensch, E., and Demling, L.,** Release of motilin after duodenal acidification, *Lancet,* 1, 888, 1976.

645. **Hellemans, J., Vantrappen, G., and Bloom, S. R.,** Endogenous motilin and LES pressure, *Scand. J. Gastroenterol.,* 11 (Suppl. 39), 67, 1976.

646. **Strunz, U., Mitznegg, P., Domschke, W., Subramanian, N., Domschke, S., and Wunsch, E.,** Acid releases motilin from human duodenum *in vitro, Acta Hepato-Gastroenterol.,* 24, 456, 1977.

647. **Lewis, T. D., Collins, S. M., Fox, J. E. T., and Daniel, E. E.,** Initiation of duodenal acid-induced motor complexes, *Gastroenterology,* 77, 1217, 1979.

648. **Daniel, E. F., Fox, J. E. T., Collins, S. M., Lewis, T. D., Meghji, M., and Track, N. S.,** Initiation of migrating myoelectric complexes in human subjects: Role of duodenal acidification and plasma motilin, *Can. J. Physiol. Pharmacol.,* 59, 173, 1981.

649. **Collins, S. M., Lewis, T. D., Fox, J. E. T., Track, N. S., Meghji, M. M., and Daniel, E. E.,** Changes in plasma motilin concentration in response to manipulation of intragastric and intraduodenal contents in man, *Can. J. Physiol. Pharmacol.,* 59, 188, 1981.

650. **Strunz, U., Neeb, S., and Mitznegg, P.,** Somatostatin but not atropine inhibits motilin secretion *in vitro, Gastroenterology,* 76, 1256, 1979.

651. **Christofides, N. D., Bloom, S. R., Besterman, H. S., Adrian, T. E., and Ghatei, M. A.,** Release of motilin by oral and intravenous nutrients in man, *Gut,* 20, 102, 1979.

652. **Christofides, N. D., Larson, D. L., Alburquerque, R. H., Ghatei, M. A., Modlin, I. M., and Bloom, S. R.,** Release of gastrointestinal hormones following on oral water load, *Experientia,* 35, 1521, 1979.

653. **Christofides, N. D., Modlin, I. M., Fitzpatrick, M. L., and Bloom, S. R.,** Effect of motilin on the rate of gastric emptying and gut hormone release during breakfast, *Gastroenterology,* 76, 908, 1979.

654. **Fox, J. E. T., Daniel, E. E., Jury, J., Track, N. S., and Chiu, S.,** Cholinergic control mechanisms for immunoreactive motilin release and motility in the canine duodenum, *Can. J. Physiol. Pharmacol.,* 61, 1042, 1983.

655. **Hall, K. E., Greenberg, G. R., El Sharkawy, T. Y., and Diamant, N. E.,** Relationship between porcine motilin-induced migrating motor complex like activity, vagal integrity and endogenous motilin release in dogs, *Gastroenterology,* 87, 76, 1984.

656. **Itoh, Z., Honda, K., Hiwatshi, K., Takeuchi, S., Aizawa, I., Takayanagi, R., and Couch, E. F.,** Motilin-induced mechanical activity in the canine alimentary tract, *Scand. J. Gastroenterol.,* 11 (Suppl. 39), 93, 1976.

657. **Vantrappen, G., Janssens, J., Peeters, T. L., Bloom, S. R., Christofides, N. D., and Hellemans, D.,** Motilin and the interdigestive migrating motor complex in man, *Dig. Dis. Sci.,* 24, 497, 1979.

658. **Takahashi, I., Honda, R., Dodds, W. J., Sarna, S., Toouli, J., Itoh, Z., Chey, W. Y., Hogan, W. J., Greiff, D., and Baker, K.,** Effect of motilin on the opossum upper gastrointestinal tract and sphincter of Oddi, *Am. J. Physiol.,* 245, G476, 1983.

659. **You, C. H., Chey, W. Y., and Lee, K. Y.**, Studies on plasma motilin concentrations and interdigestive motility of the duodenum in humans, *Gastroenterology*, 79, 62, 1980.

660. **Sarna, S., Chey, W. Y., Condon, R. E., Dodds, W. J., Myers, T., and Chang, T. M.**, Cause-effect relationship between motilin and migrating myoelectric complexes, *Am. J. Physiol.*, 245, G277, 1983.

661. **Janssens, J., Vantrappen, G., and Peeters, T. L.**, The activity front of the migrating motor complex of the human stomach but not of the small intestine is motilin dependent, *Regul. Peptides*, 6, 363, 1983.

662. **Rees, W. D. W., Malagelada, J. R., Miller, L. J., and Go, V. L. W.**, Human interdigestive and postprandial motor and gastrointestinal hormone patterns, *Dig. Dis. Sci.*, 27, 321, 1982.

663. **Hostein, J., Janssens, J., Vantrappen, G., Peeters, T. L., Vandeweerd, M., and Leman, G.**, Somatostatin induces ectopic activity fronts of the migrating motor complex via local intestinal mechanism, *Gastroenterology*, 87, 1004, 1984.

664. **Bueno, L., Fargeas, M. J., Gue, M., Peeters, T. L., Bormans, V., and Fioramonti, J.**, Effects of corticotropin-releasing factor on plasma motilin and somatostatin levels and gastrointestinal motility in dogs, *Gastroenterology*, 91, 884, 1986.

665. **Sarna, S., Condon, R. E., and Cowles, V.**, The enteric mechanisms of initiation of migrating myoelectric complexes (MMC's) in dogs, *Gastroenterology*, 84, 814, 1983.

666. **Lee, K. Y., Chey, W. Y., Tai, H. H., and Yajima, H.**, Radioimmunoassay of motilin, validation and studies on the relationship between plasma motilin levels and interdigestive myoelectric activity of the duodenum of dog, *Am. J. Dig. Dis.*, 23, 789, 1978.

667. **Poitras, P., Steinbach, J. H., VanDeventer, G., Code, C. F., and Walsh, J. H.**, Motilin-independent ectopic fronts of the interdigestive myoelectric complex in dogs, *Am. J. Physiol.*, 239, G215, 1980.

668. **Peeters, T. L., Janssens, J., and Vantrappen, G.**, Somatostatin and the interdigestive migrating motor complex in man, *Regul. Peptides*, 5, 209, 1983.

669. **Lee, K. Y., Chang, T. M., and Chey, W. Y.**, Effect of rabbit antimotilin serum on myoelectric activity and plasma motilin concentration in fasting dog, *Am. J. Physiol.*, 245, G547, 1983.

670. **Borody, T., Byrnes, D., Slowiaczek, J., and Titchen, D.**, Immunoneutralization of motilin, *Horm. Metab. Res.*, 13, 470, 1981.

671. **Wingate, D. L., Ruppin, H., Green, W. E. R., Thompson, H. A., Domschke, W., Wuensch, E., Demling, L., and Ritchie, H. D.**, Motilin-induced electrical activity in the canine gastrointestinal tract, *Scand. J. Gastroenterol.*, 11 (Suppl. 39), 111, 1976.

672. **Ormsbee, H. S., Telford, G. L., and Mason, D. T.**, Required neural involvement in control of canine migrating motor complex, *Am. J. Physiol.*, 237, E451, 1979.

673. **Ormsbee, H. S., Hoehler, S. L., and Telford, G. L.**, Somatostatin inhibits motilin induced interdigestive contractile activity in the dog, *Dig. Dis. Sci.*, 23, 781, 1978.

674. **Jennewein, H. M., Hummelt, H., Siewert, R., and Waldeck, F.**, The motor stimulant effect of natural motilin and the lower esophageal sphincter, fundus, antrum, and duodenum in dogs, *Digestion*, 13, 246, 1975.

675. **Holloway, R. H., Blank, E., Takahashi, I., Dodds, J., and Layman, R. D.**, Motilin: a mechanism incorporating the opossum lower esophageal sphincter into the migrating motor complex, *Gastroenterology*, 89, 507, 1985.

676. **Jennewein, H. M., Bauer, R., Hummelt, H., Lepsin, G., Siewert, R., and Waldeck, F.**, Motilin effects on gastrointestinal motility and esophageal sphincter (LES) pressure in dogs, *Scand. J. Gastroenterol.*, 11 (Suppl. 39), 63, 1976.

677. **Meissner, A. J., Bowes, K. L., Zwick, R., and Daniel, E. E.**, Effect of motilin on the lower esophageal sphincter, *Gut*, 17, 925, 1976.

678. **Gutierrez, J. G., Thanik, K. D., Chey, W. Y., and Yajima, H.**, Effect of motilin on the lower esophageal sphincter of the opossum, *Dig. Dis. Sci.*, 22, 402, 1977.

679. **Lux, G., Roesch, W., Domschke, S., Domschke, W., Wünsch, E., Jaeger, E., and Demling, L.**, Intravenous 13-Nle-motilin increases the human lower esophageal sphincter pressure, *Scand. J. Gastroenterol.*, 11 (Suppl. 39), 75, 1976.

680. **Domschke, W., Lux, G., Mitznegg, P., Rösch, W., Domschke, S., Bloom, S. R., Wünsch, E., and Demling, L.**, Relationship of plasma motilin response to lower esophageal sphincter pressure in man, *Scand. J. Gastroenterol.*, 11 (Suppl. 39), 81, 1976.

681. **Dent, J., Dodds, W. J., Sekiguchi, T., Hogan, W. J., and Arndorfer, R. C.**, Interdigestive phasic contractions of the human lower esophageal sphincter, *Gastroenterology*, 84, 453, 1983.

682. **Mitznegg, P., Bloom, S. R., Domschke, W., Domschke, S., Wuensch, E., and Demling, L.**, Pharmacokinetics of motilin in man, *Gastroenterology*, 72, 413, 1977.

683. **Ruppin, H., Domschke, S., Domschke, W., Wuensch, E., Jaeger, E., and Demling, L.**, Effects of 13-Nle-motilin in man-inhibition of gastric evacuation and stimulation of pepsin secretion, *Scand. J. Gastroenterol.*, 10, 199, 1975.

684. **Debas, H. T., Yamagishi, T., and Dryburgh, J. R.**, Motilin enhances gastric emptying of liquids in dogs, *Gastroenterology*, 73, 777, 1977.

685. **Ruppin, H., Sturm, G., Westhoff, D., Domschke, S., Domschke, W., Wuensch, E., and Demling, L.,** Effect of 13-Nle-motilin on small intestinal transit time in healthy subjects, *Scand. J. Gastroenterol.,* 11 (Suppl. 39), 85, 1976.

686. **Domschke, W.,** Motilin: spectrum and mode of gastrointestinal actions, *Dig. Dis.,* 22, 454, 1977.

687. **Strunz, U., Domschke, W., Mitznegg, P., Domschke, S., Schubert, E., Wünsch, E., Jaeger, E., and Demling, L.,** Analysis of the motor effects of 13-nor-leucine motilin on the rabbit, guinea pig, rat, and human alimentary tract *in vitro, Gastroenterology,* 68, 1485, 1975.

688. **Strunz, U., Domschke, W., Mitznegg, P., Wuensch, E., Jaeger, E., and Demling, L.,** Gastroduodenal motor response to natural motilin and synthetic position 13-substituted motilin analogues: a comparative *in vitro* study, *Scand. J. Gastroenterol.,* 11 (Suppl. 39), 199, 1976.

689. **Thomas, P., Kelly, K., and Go, V. L. W.,** Does motilin regulate canine interdigestive gastric motility?, *Dig. Dis. Sci.,* 24, 577, 1979.

690. **Sarr, M. G., Kelly, K. A., and Go, V. L. W.,** Motilin regulation of canine interdigestive intestinal motility, *Dig. Dis. Sci.,* 28, 249, 1983.

691. **Matsumoto, T., Sarna, S. K., Condon, R. E., Cowles, V. E., and Frantzides, C.,** Differential sensitivities of morphine and motilin to initiate migrating motor complex in isolated intestinal segments. Regeneration of intrinsic nerves, *Gastroenterology,* 90, 61, 1986.

692. **Lang, I. M., Sarna, S. K., and Condon, R. E.,** Myoelectric and contractile effects of motilin on dog small intestine *in vivo, Dig. Dis. Sci.,* 31, 1062, 1986.

693. **Kellum, J. M., Maxwell, R. J., Potter, J., and Kummerle, J. F.,** Motilin's induction of phasic contractile activity in canine jejunum is mediated by the luminal release of serotonin, *Surgery,* 100, 445, 1986.

694. **Bülbring, E. and Lin, R. C. Y.,** The effect of intraluminal application of 5-hydroxytryptamine and 5-hydroxytryptophan on peristalsis, the local production of 5-hydroxytryptamine and its release in relation to intraluminal pressure and propulsive activity, *J. Physiol. (London),* 140, 381, 1958.

695. **Fox, J. E. T., Manaka, H., Cypris, S., and Daniel, E. E.,** Nonmuscarinic motilin-induced motility results from naloxone sensitive inhibition of VIP release, *Gastroenterology,* 96, A156, 1989.

696. **Rennie, J. A., Christofides, N. D., Bloom, S. R., and Johnson, A. G.,** Stimulation of human colonic activity by motilin, *Gut,* 20, A912, 1979.

697. **Takahashi, I., Suzuki, T., Aizawa, I., and Itoh, Z.,** Comparison of gallbladder contractions induced by motilin and cholecystokinin in dogs, *Gastroenterology,* 82, 419, 1982.

698. **Muller, E. L., Grace, P. A., Conter, R., Roslyn, J. J., and Pitt, H. A.,** Influence of motilin and cholecystokinin on sphincter of Oddi and duodenal motility, *Am. J. Physiol.,* 253, G679, 1987.

699. **Pomeranz, I. S., Davison, J. S., and Shaffer, E. A.,** *In vitro* effects of pancreatic polypeptide and motilin on contractility of human gallbladder, *Dig. Dis. Sci.,* 28, 539, 1983.

700. **Behar, J. and Biancani, P.,** Effects and mechanisms of action of motilin on the cat sphincter of Oddi, *Gastroenterology,* 95, 1099, 1988.

701. **Bloom, S. R., Christofides, N. D., and Besterman, H. S.,** Raised motilin in diarrhoea, *Gut,* 19, A959, 1978.

702. **Besterman, H. S., Christofides, N. D., Welsby, P. D., Adrian, T. E., Larson, D. L., and Bloom, S. R.,** Gut hormones in acute diarrhoea, *Gut,* 24, 665, 1983.

703. **Oberg, K., Theordorsson-Norheim, E., and Norheim, I.,** Motilin in plasma and tumour tissues from patients with carcinoid syndrome. Possible involvement in the increased frequency of bowel movements, *Scand. J. Gastroenterol.,* 22, 1041, 1987.

704. **Noseda, A., Peeters, T. G., Delhaye, M., Bormans, V., Couvreur, Y., Vandermoten, G., DeFranquen, P., Rocmans, P., and Yernault, J. C.,** Increased plasma motilin concentrations in small cell carcinoma of the lung, *Thorax,* 42, 784, 1987.

705. **McLeod, R. S., Track, N. S., and Reynold, L. E.,** Plasma motilin concentrations in chronic renal disease, *Can. Med. Assoc. J.,* 121, 268, 1979.

706. **Bortolotti, M., Frada, G., Vezzadini, P., Bonora, G., Barbagallo-Sangiorgi, G., and Labo, G.,** Influence of gastric acid secretion on interdigestive gastric motor activity and serum motilin in elderly, *Digestion,* 38, 226, 1987.

707. **Achem-Karam, S. R., Funakoshi, A., Vinik, A. I., and Owyang, C.,** Plasma motilin concentration and interdigestive migrating motor complex in diabetic gastroparesis: effect of metoclopramide, *Gastroenterology,* 88, 492, 1985.

708. **Labo, G., Bortolotti, M., Vezzadini, P., Bonora, G., and Bersani, G.,** Interdigestive gastroduodenal motility and serum motilin levels in patients with idiopathic delay in gastric emptying, *Gastroenterology,* 90, 20, 1986.

709. **Hall, K. E., El-Sharkawy, T. Y., and Diamant, N. E.,** Vagal control of the migrating motor complex in the dog, *Am. J. Physiol.,* 243, G276, 1982.

710. **Miolan, J. P. and Roman, C.,** Discharge of efferent vagal fibers supplying gastric antrum: indirect study by nerve suture technique, *Am. J. Physiol.,* 235, E366, 1978.

711. **Strunz, U., Domschke, W., Mitznegg, P., Wuensch, E., Jaeger, E., and Demling, L.,** Potentiation between 13-Nle-motilin and acetylcholine on rabbit pyloric muscle *in vitro, Scand. J. Gastroenterol.,* 11 (Suppl. 39), 29, 1976.

712. **Carraway, R. and Leeman, S. E.,** The isolation of a new hypotensive peptide, neurotensin, from bovine hypothalami, *J. Biol. Chem.,* 248, 6854, 1973.

713. **Carraway, R. E., Kitabgi, P., and Leeman, S.,** The amino acid sequence of radioimmunoassayable neurotensin from bovine intestine, *J. Biol. Chem.,* 253, 7996, 1978.

714. **Kitabgi, P., Carraway, R. E., and Leeman, S. E.,** Isolation of a tridecapetide from bovine intestine tissue and its partial characterization as a neurotensin, *J. Biol. Chem.,* 251, 7053, 1976.

715. **Hammer, R. A., Leeman, S. E., Carraway, R. E., and Williams, R. H.,** Isolation of human intestinal neurotensin, *J. Biol. Chem.,* 255, 2476, 1980.

716. **Orci, L., Baetens, O., Rufener, C., Brown, M., Vale, W., and Guillemin, R.,** Evidence for immunoreactive neurotensin in dog intestinal mucosa, *Life Sci.,* 19, 559, 1976.

717. **Sundler, F., Alumets, J., Hakanson, R., Carraway, R. E., and Leeman, S. E.,** Ultrastructure of the gut neurotensin cell, *Histochemistry,* 53, 25, 1977.

718. **Polak, J. M., Sullivan, S. N., Bloom, S. R., Buchan, A. M. J., Facer, P., Brown, M. R., and Pearse, A. G. E.,** Specific localization of neurotensin to the N cell in human intestine by radioimmunoassay and immunocytochemistry, *Nature (London),* 270, 183, 1977.

719. **Reinecke, M., Forssmann, W. G., Thierkötter, G., and Triepel, J.,** Localization of neurotensin immunoreactivity in the spinal cord and the peripheral nervous system of the guinea pig, *Neurosci. Lett.,* 37, 37, 1983.

720. **Buchan, A. J. M. and Barber, D. L.,** Neurotensin containing nerves in the canine enteric innervation, *Neurosci. Lett.,* 76, 13, 1987.

721. **Barber, D. L., Buchan, A. M. J., and Soll, A. H.,** Neurotensin containing canine submucosal neurons: characterization and response to exogenous stimuli, *Can. J. Physiol. Pharmacol.,* 64 (Suppl.), 130, 1986.

722. **Helmstaedter, V., Tuagner, C., Feuerle, G., and Forssmann, W. G.,** Localization of neurotensin immunoreactive cells in the small intestine of man and various mammals, *Histochemistry,* 53, 35, 1977.

723. **Hammer, R. A., Carraway, R. E., and Leeman, S. E.,** Evaluation of plasma neurotensin-like immunoreactivity after a meal. Characterization of the elevated components, *J. Clin. Invest.,* 70, 74, 1982.

724. **Lee, Y. C., Allen, J. M., Uttenthal, L. O., Walker, M. C., Shemilt, J., Gill, S. S., and Bloom, S. R.,** The metabolism of intravenous infused neurotensin in man and its chromatographic characterization in human plasma, *J. Clin. Endocrinol. Metab.,* 59, 45, 1984.

725. **Lee, Y. C., Allen, J. M., Uttenthal, L. O., Roberts, P. M., Gill, S. S., and Bloom, S. R.,** Quantitation and characterization of human plasma neurotensin-like immunoreactivity in response to a meal, *Dig. Dis. Sci.,* 30, 129, 1985.

726. **Ferris, C. F., Carraway, R. E., Hammer, R. G., and Leeman, S. E.,** Release and degradation of neurotensin during perfusion of rat small intestine with lipid, *Peptides,* 12, 101, 1985.

727. **Theodorsson-Norheim, E. and Rosell, S.,** Characterization of human plasma neurotensin-like immunoreactivity after fat ingestion, *Regul. Peptides,* 6, 207, 1983.

728. **Bunnett, N. W., Mogard, M., Orloff, M. S., Corbett, H. J., Reeve, J. R., Jr., and Walsh, J. H.,** Catabolism of neurotensin in interstitial fluid of the rat stomach, *Am. J. Physiol.,* 246, G675, 1984.

729. **Checler, F., Ahmad, S., Kostka, P., Barelli, H., Kitabgi, P., Fox, J. E. T., Kwan, C.-Y., Daniel, E. E., and Vincent, J.-P.,** Peptidases in dog ileum circular and longitudinal smooth muscle plasma membranes: their relative contribution to the metabolism of neurotensin, *Eur. J. Biochem.,* 166, 461, 1987.

730. **Checler, F., Barelli, H., Kwan, C. Y., and Vincent, J. P.,** Neurotensin-metabolizing peptidases in rat fundus plasma membranes, *J. Neurochem.,* 49, 507, 1987.

731. **Checler, F., Kostolanska, F., and Fox, J. E. T.,** *In vivo* inactivation of neurotensin in dog ileum: major involvement of endopeptidase 24-11, *J. Pharmacol. Exp. Ther.,* 244, 1040, 1988.

732. **Barelli, H., Ahmad, S., Kostka, P., Fox, J. E. T., Daniel, E. E., Vincent, J. P., and Checler, F.,** Neuropeptide-hydrolysing activities in synaptosomal fractions from dog ileum myenteric, deep muscular and submucous plexi. Their participation in neurotensin inactivation, *Peptides,* 10, 1055, 1989.

733. **Kitabgi, P.,** Effects of neurotensin on intestinal smooth muscle: application to the study of structure activity relationships, *Ann. N.Y. Acad. Sci.,* 400, 37, 1982.

734. **Kitabgi, P., Poustis, C., Granier, C., Van Rietschoten, J., Rivier, J., Morgat, J. L., and Freychet, P.,** Neurotensin binding to extraneural and neural receptors: comparison with biological activity and structure-activity relationships, *Mol. Pharmacol.,* 18, 11, 1980.

735. **Kitabgi, P., Checler, F., and Vincent, J. P.,** Comparison of some biological properties of neurotensin and its natural analogue LANT-6, *Eur. J. Pharmacol.,* 99, 357, 1984.

736. **Minamino, N., Kangawa, K., and Matsuo, H.,** Neuromedin N: a novel neurotensin like peptide identified in porcine spinal cord, *Biophys. Biochem. Res. Commun.,* 122, 542, 1984.

737. **Checler, F., Magella, J., Kitabgi, P., and Vincent, J. P.,** High affinity receptor sites and rapid proteolytic inactivation of neurotensin in primary cultured neurons, *J. Neurochem.,* 47, 1742, 1986.

738. **Folkers, K., Chang, D., Humphries, J., Carraway, R., Leeman, S. E., and Bowers, C. Y.**, Synthesis and activities of neurotensin, and its acid and amide analogs: possible natural occurrence of (Gln4)-neurotensin, *Proc. Natl. Acad. Sci. U.S.A.*, 73, 3833, 1976.

739. **Carraway, R. E.**, A critical analysis of three approaches of radioimmunoassay of peptides: applications to the study of the neurotensin family, *Ann. N.Y. Acad. Sci.*, 400, 17, 1982.

740. **Carraway, R. E. and Ferris, C. F.**, Isolation, biological and chemical characterization, and synthesis of a neurotensin-related hexapeptide from chicken intestine, *J. Biol. Chem.*, 258, 2475, 1983.

741. **Rioux, F., Kerouac, R., and St.-Pierre, S.**, Characterization of the inhibitory effect of (D-Trp11)-NT toward some biological actions of neurotensin in rats, *Neuropeptides*, 3, 345, 1983.

742. **Ukai, M., Itatsu, T., Shibata, A., Rioux, F., and St. Pierre, S.**, Inhibition of neurotensin (NT)-induced glucagon release by (D-Trp11)-NT, *Experientia*, 38, 122, 1982.

743. **Ahmad, S., Berezin, I., Vincent, J. P., and Daniel, E. E.**, Neurotensin receptors in canine intestinal smooth muscle: preparation of plasma membranes and characterization of ^{125}I-Tyr^3-neurotensin binding, *Biochim. Biophys. Acta*, 896, 224, 1987.

744. **Mazella, J., Kitabgi, P., and Vincent, J.**, Molecular properties of neurotensin receptors in rat brain, *J. Biol. Chem.*, 260, 508, 1985.

745. **Ahmad, S., Kwan, C. Y., Vincent, J. P., Grover, A. K., Jung, C. Y., and Daniel, E. E.**, Target size analysis of neurotensin receptors, *Peptides*, 8, 195, 1987.

746. **Lazarus, L. H., Perrin, M. H., and Brown, M. R.**, Mast cell binding of neurotensin-iodination of neurotensin and characterization of the interaction of neurotensin with mast cell receptor sites, *J. Biol. Chem.*, 20, 7174, 1977.

747. **Lazarus, L. H., Perrin, M. H., and Brown, M. R.**, Mast cell binding of neurotensin-molecular conformation of neurotensin involved in the stereospecific binding to mast cell receptor sites, *J. Biol. Chem.*, 20, 7180, 1977.

748. **Fox, J. E. T., Kostolanska, F., Daniel, E. E., Allescher, H. D., and Hanke, T.**, Mechanism of excitatory actions of neurotensin on canine small intestinal circular muscle *in vivo* and *in vitro*, *Can. J. Physiol. Pharmacol.*, 65, 2254, 1987.

749. **Mashford, M. L., Nilsson, G., Rökaeus, Å., and Rosell, S.**, Release of neurotensin-like immunoreactivity (NTLI) from the gut in anesthetized dogs, *Acta Physiol. Scand.*, 104, 375, 1978.

750. **Blackburn, A. M., Bloom, S. R., Long, R. G., Fletcher, D. R., Christofides, N. D., Fitzpatrick, M. L., and Baron, J. M.**, Effect of neurotensin on gastric function in man, *Lancet*, 1, 987, 1980.

751. **Go, V. L. W. and Demol, P.**, Role of nutrients in the gastrointestinal release of immunoreactive neurotensin, *Peptides*, 2 (Suppl.), 267, 1981.

752. **Rosell, S. and Rökaeus, Å.**, The effect of ingestion of amino acids, glucose and fat on circulating neurotensin-like immunoreactivity (NTLI) in man, *Acta Physiol. Scand.*, 107, 263, 1979.

753. **Flaten, O. and Hanssen, L. E.**, Concentration of neurotensin in human plasma after glucose, meals and lipids, *Acta Physiol. Scand.*, 114, 311, 1982.

754. **Walker, J. P., Fujimura, M., Sakamoto, T., Greeley, G. H., Townsend, C. M., and Thompson, J. C.**, Importance of the ileum in neurotensin released by fat, *Surgery*, 98, 224, 1985.

755. **Shaw, C. and Buchanan, K. D.**, Intact neurotensin in human plasma: response to oral feeding, *Regul. Peptides*, 7, 145, 1983.

756. **Ferris, C. F., Hammer, R. A., and Leeman, S. E.**, Elevation of plasma neurotensin during lipid perfusion of rat small intestine, *Peptides*, 2 (Suppl.), 263, 1981.

757. **Read, N. W., McFarlane, A., Kinsman, R. I., Bates, T. E., Blackhall, N. W., Farrar, G. B. J., Hall, J. C., Moss, G., Morris, A. P., O'Neill, B., Welch, I., Lee, Y., and Bloom, S. R.**, Effect of infusion of nutrient solutions into the ileum on gastrointestinal transit and plasma levels of neurotensin and enteroglucagon, *Gastroenterology*, 86, 274, 1984.

758. **Fujimura, M., Khalil, T., Greeley, G. H., Townsend, C. M., and Thompson, J. C.**, Release of neurotensin: effect of selective jejunal or ileal perfusion with sodium oleate in the dog, *Gastroenterology*, 86, 1083, 1984.

759. **Feuerle, G. E., Baca, I., and Knauf, W.**, Atropine depresses release of neurotensin and its effect on the exocrine pancreas, *Regul. Peptides*, 4, 75, 1982.

760. **Fletcher, D. R., Shulkes, A., Bladin, P. H. D., Booth, D., and Hardy, K. J.**, Cholinergic inhibition of meal stimulated plasma neurotensin like immunoreactivity in man, *Life Sci.*, 33, 863, 1983.

761. **Al-Saffar, A., Theodorsson-Norheim, E., and Rosell, S.**, Nervous control of the release of neurotensin like immunoreactivity from the small intestine of the rat, *Acta Physiol. Scand.*, 112, 1, 1984.

762. **Barber, D. L., Buchan, A. M. J., Walsh, J. H., and Soll, A.**, Isolated canine ileal mucosal cells in short-term culture: a model for study of neurotensin release, *Am. J. Physiol.*, 250, G374, 1986.

763. **Barber, D. L., Buchan, A. M. J., Walsh, J. H., and Soll, A.**, Regulation of neurotensin release from canine enteric primary cell cultures, *Am. J. Physiol.*, 250, G385, 1986.

764. Ghatei, M. A., Jung, R. T., Stevenson, J. C., Hillyard, C. J., Adrian, T. E., Lee, Y. C., Chrsitofides, M. D., Sarson, D. L., Mashiter, K., MacIntyre, I., and Bloom, S. R., Bombesin: action on gut hormones and calcium in man, *J. Clin. Endocrinol. Metab.*, 54, 980, 1982.
765. Rökaeus, Å., Yanaihara, N., and McDonald, T. J., Increased concentration of neurotensin-like immunoreactivity (NTLI) in rat plasma after administration of bombesin and bombesin related peptides (porcine and chicken gastrin releasing peptides), *Acta Physiol. Scand.*, 114, 605, 1982.
766. Cuber, J. C., Herrman, C., Kitabgi, P., Bernard, C., Velecela, E., and Chayvialle, J. A., Lack of corelease of ileal neuromedin N with neurotensin upon luminal nutrients and vascular bombesin infusion in rats, *Gastroenterology*, 96, A103, 1989.
767. Barber, D. L., McGuire, M. E., and Buchan, A. M. J., Dietary fatty acids regulate neurotensin expression *in vivo*, *Gastroenterology*, 96, A25, 1989.
768. Armstrong, M. J., Parker, M. C., Ferris, C. F., and Leeman, S., Neurotensin stimulates (^{3}H) oleic acid translocation across rat small intestine, *Am. J. Physiol.*, 251, G823, 1986.
769. Reinecke, M., Neurotensin-immunohistochemical localization in central and peripheral nervous system and in endocrine cells and its functional role as neurotransmitter and endocrine hormone, *Prog. Histochem. Cytochem.*, 16, 1, 1985.
770. Rosell, S., Thor, K., Rökaeus, Å., Nyquist, O., Lewenhaupt, A., Kager, L., and Folkers, K., Plasma concentration of neurotensin-like immunoreactivity (NTLI) and lower esophageal sphincter (LES) pressure in man following infusion of (Gln^4)-neurotensin, *Acta Physiol. Scand.*, 109, 369, 1980.
771. Theodorsson-Norheim, Thor, K., E., and Rosell, S., Relation between the lower esophageal sphincter (LES) pressure and the plasma concentration of neurotensin during intravenous infusion of neurotensin (1—13) and (Gln^4)-neurotensin in man, *Acta Physiol. Scand. Suppl.*, 515, 29, 1983.
772. Thor, K. and Rökaeus, Å., Antro-duodenal motor response induced by (Gln4)-neurotensin in man, *Acta Physiol. Scand.*, 118, 269, 1983.
773. Nebel, O. T. and Castell, D. O., Lower esophageal sphincter pressure changes after food ingestion, *Gastroenterology*, 63, 778, 1972.
774. Quirion, R., Regoli, D., Rioux, F., and St. Pierre, S., The stimulatory effects of neurotensin and related peptides on rat stomach strips and guinea pig atria, *Br. J. Pharmacol.*, 68, 83, 1980.
775. Kitabgi, P., Kwan, C. Y., Fox, J. E. T., and Vincent, J. P., Characterization of neurotensin binding to rat gastric smooth muscle receptor sites, *Peptides*, 5, 917, 1984.
776. Andersson, S., Rosell, S., Hjelmquist, U., Chang, D., and Folkers, K., Inhibition of gastric and intestinal motor activity in dogs by [Gln^4]neurotensin, *Acta Physiol. Scand.*, 100, 231, 1977.
777. McLean, J. and Fox, J. E. T., Mechanisms of action of neurotensin on motility of canine gastric corpus *in vitro*, *Can. J. Physiol. Pharmacol.*, 61, 29, 1983.
778. Rosell, S., Al-Saffar, A., and Thor, K., The role of neurotensin in gut motility, *Scand. J. Gastroenterol.*, 19 (Suppl. 96), 69, 1984.
779. Hellström, P. M., Nylander, G., and Rosell, S., Effect of neurotensin on the transit of gastrointestinal contents in the rat, *Acta Physiol. Scand.*, 115, 239, 1982.
780. Keinke, O., Wulschke, S., and Ehrlein, H. J., Neurotensin slows gastric emptying by a transient inhibition of gastric and a prolonged inhibition of duodenal motility, *Digestion*, 34, 281, 1986.
781. Yamanaka, K. and Szurszewski, J. H., Electrophysiological effects of neurotensin on human gastric antrum, *Gastroenterology*, 96, A555, 1989.
782. Rökaeus, Å., The effect of fat instillation in the proximal gastrointestinal tract upon circulating neurotensin-like immunoreactivity (NTLI) in rat, *Regul. Peptides*, (Suppl. 1), 93, 1980.
783. Rosell, S. and Rökaeus, Å., Actions and possible hormonal functions of circulating neurotensin, *Clin. Physiol.*, 1, 3, 1981.
784. Blackburn, A. M., Lawaetz, O., and Bloom, S. R., Plasma neurotensin release and gastric emptying in the dumping syndrome, *Life Sci.*, 32, 833, 1982.
785. Sakai, Y., Daniel, E. E., Jury, J., and Fox, J. E. T., Neurotensin inhibition of canine intestinal motility *in vivo* via α_2-adrenoreceptors, *Can. J. Physiol. Pharmacol.*, 62, 403, 1984.
786. Christinck, F., Daniel, E. E., and Fox, J. E. T., Electrophysiological responses of the canine ileal circular muscle to electrical stimulation and neurotensin (NT), *Gastroenterology*, 96, A680, 1989.
787. Wilen, T., Gustavsson, S., and Jung, B., Effects of neurotensin on small bowel propulsion in intact and vagotomized rats, *Regul. Peptides*, 4, 191, 1982.
788. Schemann, M., Siegle, M. L., Sahyoun, H., and Ehrlein, H. J., Computer analysis of intestinal motility: effects of cholecystokinin and neurotensin on jejunal contraction patterns, *Z. Gastroenterol.*, 24, 262, 1986.
789. Hellström, P. M., Vagotomy inhibits the effect of neurotensin on gastrointestinal transit in the rat, *Acta Physiol. Scand.*, 128, 47, 1986.
790. Rökaeus, Å., Studies on neurotensin as a hormone, *Acta Physiol. Scand. Suppl.*, 501, 1, 1981.
791. Kitabgi, P. and Freychet, P., Effects of neurotensin on isolated intestinal smooth muscles, *Eur. J. Pharmacol.*, 50, 349, 1978.

792. **Kitabgi, P. and Freychet, P.,** Neurotensin: contractile activity, specific binding, and lack of effect on cyclic nucleotides in intestinal smooth muscle, *Eur. J. Pharmacol.,* 55, 35, 1979.

793. **Monier, S. and Kitagbi, P. C.,** Substance P-induced auto desensitization inhibits atropine resistant, neurotensin-stimulated contractions in guinea-pig ileum, *Eur. J. Pharmacol.,* 65, 461, 1980.

794. **Goedert, N., Hunter, J. C., and Ninkovic, M.,** Evidence for neurotensin as a non-adrenergic, non-cholinergic neurotransmitter in guinea pig ileum, *Nature (London),* 311, 59, 1984.

795. **Huidobro-Toro, J. P. and Zhu, Y. X.,** Neurotensin receptors on the ileum of the guinea pig: evidence for the coexistence of inhibitory and excitatory receptors, *Eur. J. Pharmacol.,* 102, 237, 1984.

796. **Huidobro-Toro, J. P. and Yoshimura, K.,** Pharmacological characterization of the inhibitory effects of neurotensin on the rabbit ileum myenteric plexus preparation, *Br. J. Pharmacol.,* 80, 645, 1983.

797. **Thor, K., Rosell, S., Rökaeus, A., and Lager, L.,** (Gln^4)-neurotensin changes the motility pattern of the duodenum and the proximal jejunum from a fasting-type to a fed-type motility in man, *Gastroenterology,* 83, 569, 1982.

798. **Al-Saffar, A.,** Analysis of the control of intestinal motility in fasted rats, with special reference to neurotensin, *Scand. J. Gastroenterol.,* 19, 422, 1984.

799. **Bueno, L., Fioramonti, J., Fargeas, M. J., and Primi, M. P.,** Neurotensin: a central neuromodulator of gastrointestinal motility in the dog, *Am. J. Physiol.,* 248, G15, 1985.

800. **Bardon, T. and Ruckebusch, Y.,** Neurotensin-induced colonic motor responses: a mediation by prostaglandins, *Regul. Peptides.,* 10, 107, 1985.

801. **Thor, K. and Rosell, S.,** Neurotensin increases colonic motility, *Gastroenterology,* 90, 27, 1986.

802. **Calam, J., Unwin, R., and Peart, W. S.,** Neurotensin stimulates defecation, *Lancet,* 2, 737, 1983.

803. **Wiklund, B., Liljeqvist, L., and Rökaeus, A.,** Neurotensin increases net fluid secretion and transit rate in the small intestine in man, *Regul. Peptides,* 8, 33, 1984.

804. **Snape, W. J., Hyman, P. E., Mayer, E. A., Sevy, N., Kao, H. W., and Root, D.,** Calcium dependence of neurotensin stimulation of circular colonic muscle of the rabbit, *Gastroenterology,* 93, 823, 1987.

805. **Kitabgi, P. C. and Vincent, J. P.,** Neurotensin is a potent inhibitor of guinea pig colon contractile activity, *Eur. J. Pharmacol.,* 74, 311, 1981.

806. **Yamasato, T. and Nakayama, S.,** Effects of neurotensin on the motility of the isolated gallbladder, bile duct and ampulla in guinea pigs, *Eur. J. Pharmacol.,* 148, 101, 1988.

807. **Fujimura, M., Sakamoto, T., and Khalil, T.,** Physiological role of neurotensin in gallbladder contraction in the dog, *Surg. Forum,* 35, 192, 1984.

808. **Walker, J. P., Khalil, T., Wiener, I., Fagan, C. J., Townsend, C. M., Greeley, G. H., and Thompson, J. C.,** The role of neurotensin in human gallbladder motility, *Ann. Surg.,* 201, 687, 1985.

809. **Schang, J. C., Dauchel, J., Sava, P., Angel, R., Bouchet, P., Lambert, A., and Grenier, J. F.,** Specific effects of different food components on intestinal motility, *Eur. Surg. Res.,* 10, 425, 1978.

810. **Wright, S. H., Snape, W. J., Battle, W. M., and Cohen, S.,** Effect of dietary components on the gastrocolonic response, *Am. J. Physiol.,* 238, G228, 1980.

811. **Allescher, H. D., Willis, S., Schusdtiarra, V., and Classen, M.,** Omega coustoxin GVIA specifically blocks neuro mechanisms in rat ileum, *Neuropeptides,* 13, 253, 1989.

812. **Daniel, E. E. and Fox, J. E. T.,** unpublished results.

813. **Allescher, H.-D., Daniel, E. E., Dent, J., and Fox, J. E. T.,** Modulation of canine pyloric motor function by regulatory peptides, *Biomed. Res.,* Suppl. 1, 120, 1988.

Chapter 12

PHYSIOLOGICAL AND PATHOPHYSIOLOGICAL ROLES OF VIP, SOMATOSTATIN, OPIOIDS, GALANIN, GRP, AND SECRETIN

Jan D. Huizinga and Julio Pintin-Quezada

TABLE OF CONTENTS

I. Vasoactive Intestinal Polypeptide 402
  A. Introduction 402
  B. Intracellular Locus...402
  C. Receptors..402
  D. Antagonists..403
  E. Role of VIP in Neural Control of Motility 404
  F. Evidence for VIP Involvement in Intestinal Reflexes..................404
  G. Properties of VIP Neurons 404
  H. VIP Action on Nerves...405
  I. Mechanism of VIP Effects on Smooth Muscle 405
  J. VIP as Mediator of Nonadrenergic, Noncholinergic Neural Inhibition..406
  K. Clinical Aspects 407

II. Somatostatin 408
  A. Introduction 408
  B. Receptors..408
  C. Somatostatin and Gut Function..408
  D. Localization 408
  E. Role of Somatostatin in the Peristaltic Reflex 409
  F. Somatostatin and Interdigestive Motor Activity 409
  G. Somatostatin as an Inhibitor of Acetylcholine Release..................409
  H. Mechanism of Somatostatin Action on Nerves 409
  I. Intracellular Mechanisms of Somatostatin 410
  J. Clinical Studies...410

III. Opioids 411
  A. Introduction 411
  B. Effects of Receptor Stimulation on Ionic Conductance 411
  C. Inhibition of Acetylcholine Release 411
  D. Mechanism of Reduction of Neural Excitability 412
  E. Actions and Interactions of the Opioids 412
  F. Opioids and Intestinal Peristalsis 413
  G. Organ Specific Actions of the Opioids 413

IV. Galanin 415
  A. Excitatory Activity on Smooth Muscle 416
  B. Inhibitory Activity on Smooth Muscle.....................................416
  C. Action on Nerves 416
  D. Clinical Studies...417

V.   Gastrin-Releasing Polypeptide (GRP)/Bombesin Family 417
        A.   Introduction 417
        B.   Receptors, Antagonists, and Degradation.............................417
        C.   GRP/Bombesin Affect Neural Activity 417
        D.   Bombesin Effects on Smooth Muscle 418

VI.   Calcitonin Gene-Related Peptide..419
        A.   Histological Characteristics..419
        B.   Inhibitory Activity of CGRP 419
        C.   Excitatory Action of CGRP 420
        D.   Action on Myenteric Nerves...420

VII.   Secretin...420
        A.   Secretin and Gastric Emptying 420

Acknowledgments..421

References..421

I. VASOACTIVE INTESTINAL POLYPEPTIDE

A. INTRODUCTION

Vasoactive intestinal polypeptide (VIP) was isolated from extracts of porcine small intestine in the early 1970s. It has been shown to be similar in structure in all mammalian species studied so far — porcine, canine, bovine, human, and rat — but differs in the chicken.[1] A number of other peptides have been identified which have some homology with VIP. They include the mammalian peptides secretin and glucagon, rat and human growth hormone-releasing factor (rGRF and hGRF), peptide histidine-isoleucine (PHI), peptide histidine-methionine (PHM), gastric (glucose) inhibitory peptide (GIP), oxyntomodulin, and the Gila monster venom peptide helodermin.[2] These peptides constitute a family of VIP/secretin peptides.[3] Some aspects of the effects of VIP on intestinal motility were reviewed recently.[1]

B. INTRACELLULAR LOCUS

Using immunohistochemical methods the intracellular locus of VIP in the gastrointestinal tract has been shown to be the large granular vesicles of nerve varicosities by the protein A gold method in the dog[4] and the rat.[5] When the peroxidase antiperoxidase method was used in the guinea pig colon, both large and small vesicles were stained, probably by the diffusible reaction product. However, it is possible that there is a low concentration of VIP in these small vesicles which is not detected by the immunogold method.[5]

C. RECEPTORS

Radioautographic studies reveal low levels of radioactive VIP binding in gastrointestinal muscle, but insufficient evidence for receptors on smooth muscle has been obtained so far.[6] Most studies on the receptors/binding sites for the VIP family of peptides have been conducted on other tissues such as liver membranes, enterocyte cells, or membranes from brain tissue. As reviewed by Rosselin,[3] each member of the VIP family cited above possesses a specific recognition site; the expression of the recognition site is cell specific according to the function of the cell, and there are three such groups in the family of receptors: (1) those which recognize glucagon and oxyntomodulin; (2) those which recognize VIP, secretin, rat and human GRF and PHI/PHM; and (3) those which recognize GIP. Peptides acting on receptors in one group do not interact with those in another group. Within a given receptor type, the same structure/function relationship has been found to be retained across the different mammalian species and within tissues. For the VIP receptor, the order is VIP > (PHI, PHM) > secretin. All parts of the molecule are required both for binding[3] and function,[7] however, within the molecule certain amino acids can be substituted without changing the activity, while others are critical for the function of the molecule either in enhancing or suppressing the action.[8] Rosselin[3] suggests that two sites of the molecule interact to promote receptor binding, one which involves a few sequences in the molecule interacting with the receptor directly and another constraining the molecule in a specific position when the primary site is bound.

The characterization of the VIP receptor has been carried out in a variety of nonmuscle systems. There appear to be both low- and high-affinity receptors in some types of cells and a single class of receptors in other cells.[9] These differences may be methodological rather than biological. VIP receptors have been shown to be internalized rapidly.[10] However, occupation of VIP receptors appears in all cases to stimulate adenylate cyclase.[11] The ability to activate adenylate cyclase appears to depend upon the N-terminal histidine residue. Its modification/deletion results in a decrease in potency and efficacy of cyclase stimulation of both rat enterocytes and membrane from these cells, but has little effect on the binding properties of VIP.[12] In human colonic epithelial cells, deleting the N terminal histidine resulted in a drastic loss of affinity for VIP receptors. In these cells, a low and high affinity binding site for VIP were characterized, and it was shown that high affinity low capacity binding sites are playing the major role in activation of adenylate cyclase.[13]

D. ANTAGONISTS

Manipulation of the structural requirements of the VIP receptors for various analogues has led to the development of analogues of rGRF as competitive antagonists of VIP. These have not been tested on motility functions to date. AC-Tyr[1]hGRF was shown by Laburthe et al.[13] to antagonize VIP stimulation of adenylate cyclase activity competitively in the rat enterocyte membranes, with a pA_2 of 6.5. However, it had partial VIP and rGRF agonist activity. Another antagonist, [4Cl-D-Phe[6]-Leu[7]]VIP, was able to decrease VIP-stimulated amylase release and cyclic AMP production in dispersed guinea pig pancreatic acini and cyclic AMP production in dispersed rat hepatocytes. It also inhibited VIP-induced short-circuit current increases in monolayers of a colonic tumor cell line. There was, however, a partial agonist effect on amylase release. The values of pA_2 were not calculated, but the authors suggested that since the shift of the dose response curve to VIP was parallel the antagonism was competitive.[14] None of these agonists have a high enough affinity or are sufficiently free of agonist activity to serve as good antagonists for mechanistic studies.

Prior to the development of antagonists the only compounds available as putative antagonists were antisera to VIP. These have been used extensively to attempt to prove that VIP is a neurotransmitter at various sites. However great the structural selectivity of certain antisera, there are a number of limitations to the use of antisera as antagonists. These include difficulties in getting antisera to the cleft between nerve and muscle and the fact that antisera given *in vivo* could bind the VIP and eliminate it from the circulation, but not occupy the receptor site on the muscle itself. In addition, one cannot distinguish whether the released VIP is acting at a proximal site or directly on smooth muscle. Only electrophysiological and biochemical studies as well as use of selective antagonists to block the muscle receptor are able to distinguish between proximal or direct action of VIP.

E. ROLE OF VIP IN NEURAL CONTROL OF MOTILITY

VIP is one of the most abundant peptides found in the nerves of the gastrointestinal system. Since in many systems VIP produces inhibition of motor activity (although excitatory actions have been described), VIP has been proposed as a nonadrenergic, noncholinergic (NANC) inhibitory neurotransmitter. VIP-immunoreactive nerves have been shown to be present in most organs, including the opossum esophagus,[15] and VIP is present in nerves of the stomach of most species.[8] It can be shown to be released by vagal stimulation in pig[16] and field stimulation in the guinea pig.[17] VIP may determine, in part, the contractile state of intestinal motility. There is much evidence that VIP is released continuously, thereby determining to a large extent the contractile state of the organs. In studying the peristaltic reflex in the guinea pig and rat colon, Grider and Makhlouf[21] observed that VIP antiserum augmented ascending contraction, suggesting that spontaneously released VIP suppressed activity. Continuous activity of inhibitory nerves is observed in many experimental conditions,[18,19] and VIP neurons appear to be involved in long-lasting inhibition.[20]

F. EVIDENCE FOR VIP INVOLVEMENT IN INTESTINAL REFLEXES

The possible participation of VIP in descending relaxation as part of the peristaltic reflex was studied in the guinea pig and rat colon.[21] Graded radial stretch of the extreme orad end of isolated whole segments of colon caused descending relaxation accompanied by significant release of VIP. Radial stretch of the caudad end caused contraction without VIP release. VIP antiserum inhibited descending relaxation. Hexamethonium and tetrodotoxin (TTX) abolished both relaxation and contraction, indicating the participation of cholinergic nerves in both phenomena. In the guinea pig small intestine the presence of VIP has been shown in intramural neurons, and fibers from these neurons are caudad oriented and innervate other neurons and smooth muscle cells.[22] The conclusion was that cholinergic nerves coupled to VIP motor neurons regulate descending relaxation.

Participation of VIP in the relaxation of the internal anal sphincter (IAS) in response to rectal distention was studied using the VIP antagonists [4Cl-D-Phe^6-Leu^{17}]-VIP and (N-Ac-Tyr,D-Phe)-GRF$_{1-29}$ NH$_2$. VIP was shown to inhibit sphincter muscle directly, and the VIP antagonists reduced the VIP response as well as IAS relaxation caused by the recto-anal reflex.[23] The reflex inhibition was not inhibited at low levels of stimulation, suggesting to the authors the participation of neuromediators other than VIP in the reflex. These VIP antagonists also caused inhibition of relaxation due to local electrical stimulation of inhibitory nerves. VIP immunoreactivity has been shown in neurons in the IAS.[24]

G. PROPERTIES OF VIP NEURONS

Some characteristics of VIP neurons were studied in submucous neurons of the guinea pig small intestine.[25] Intracellular recordings were made from neurons showing immuno-histochemical evidence for VIP content. Inhibitory synaptic potentials were recorded from these neurons. The origin of this input was either extrinsic noradrenergic neurons (with cell bodies in the celiac ganglion) or myenteric neurons. The VIP nerves also received slow excitatory input from neurons intrinsic to the intestine. The picture emerges that VIP nerves are interneurons connecting different neuronal systems. It is therefore likely that VIP nerves will interact with many other neural systems. For example, inhibitory sympathetic input to VIP nerves is suspected in rats, since chemical destruction of peripheral sympathetic neurons elevated VIP levels in small intestine and colon.[26] In the sphincter of Oddi, evidence has been presented that relaxation by cholecystokinin (CCK) is mediated by VIP nervous activity.[27]

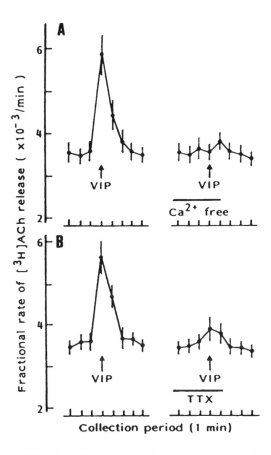

FIGURE 1. VIP evokes acetylcholine release in myenteric
plexus of small intestine. The acetylcholine release is blocked
by TTX and Ca-Free medium. (From Kusunoki, M., Tsai,
L. H., Taniyama, K., and Tanaka, C., *Am. J. Physiol.*,
251, G51, 1986. With permission.

H. VIP ACTION ON NERVES

Although VIP causes relaxation of most intestinal muscle, a notable exception is the
longitudinal muscle of the guinea pig small intestine.[28] This action is in part TTX sensitive
and may be mediated in part through the release of acetylcholine (see Figure 1). VIP induced
a concentration-dependent increase in the release of [³H]acetylcholine. This VIP-induced
release was inhibited by TTX and removal of Ca^{2+} from the superfusion medium. These
experiments suggest direct action of VIP on postganglionic cholinergic neurons. An excitatory presynaptic action of VIP on cholinergic nerves was also proposed by Yau et al.[29]
Those authors observed that VIP increased acetylcholine release from isolated myenteric
varicosities of the same preparation.

Consistent with the above excitatory activity of VIP in the guinea pig small intestine is
the observation that VIP depolarizes myenteric neurons in this tissue. When 129 neurons
were studied, VIP increased the firing rate in 45%. Some cells were excited by concentrations
as low as 1 p*M*.[30]

I. MECHANISM OF VIP EFFECTS ON SMOOTH MUSCLE

In the internal anal sphincter it was demonstrated that *N*-ethylmaleimide strongly inhibited relaxation by forskolin (increasing intracellular cAMP), VIP- and nerve-mediated re-

laxation. This suggested that both VIP-induced relaxation and nerve-mediated relaxation were, at least in part, mediated by an increase in intracellular cAMP.[31] In support of this, cAMP levels were found to be increased in the sphincter of Oddi upon addition of VIP in a dose-dependent manner.[27] Studies in opossum LES showed that VIP-induced relaxation was accompanied by an increase in cAMP. Interestingly, NANC nerve-induced relaxation was accompanied by an increase in cGMP (and not cAMP).[32]

J. VIP AS MEDIATOR OF NONADRENERGIC, NONCHOLINERGIC NEURAL INHIBITION

Electrical stimulation of intrinsic nerves (in the presence of atropine and adrenergic blocking agents) causes relaxation of most intestinal smooth muscle associated with membrane hyperpolarization. The hyperpolarization caused by one or a small train of 0.5-ms-duration pulses is fast and 10 to 30 mV in amplitude. It is referred to as the inhibitory junction potential (IJP). The mediator of this neurally mediated response is in most cases unknown. In some tissues there is evidence accumulating that VIP is the (main) neurotransmitter involved. In other tissues the available evidence suggests that VIP may not be the (main) neurotransmitter.

The positive evidence is similar in many tissues studied. In the cat gastric fundus, relaxation induced by VIP as well as intramural nerve stimulation were reduced selectively by VIP antiserum.[33] A similar response with VIP antiserum was observed in the opossum lower esophageal sphincter (LES),[34] cat LES,[35] dog gastric muscularis mucosa,[36] and guinea pig stomach.[37]

An interesting study was performed in the muscularis mucosa of the canine stomach, which responds with relaxation to field stimulations. These responses were blocked by α-chymotrypsin or by preincubation for 20 min with VIP antisera. VIP was shown to be present in the muscle and to be released by field stimulation in a tetrodotoxin- and calcium-sensitive fashion. VIP relaxed the muscle, and this response was also blocked by VIP antisera exposure for 20 min, but not by exposure to tetrodotoxin.[24] Furthermore, electrophysiological studies demonstrated that VIP hyperpolarized the tissue, as did the IJPs induced by field stimulation.[38] This is the strongest evidence to date that VIP is a direct neurotransmitter to muscle.

VIP injected intra-arterially caused relaxation of the lower esophagus sphincter in the anesthetized opossum. This relaxation was not blocked by administration of TTX. Since vagal stimulation also produced relaxation and intra-arterial infusions of VIP antisera reduced the vagal-stimulated, field-stimulated, and intra-arterial nicotine-induced inhibitory response, VIP was proposed as the mediator of the NANC, relaxation of the opossum esophagus.[39,40]

VIP is present in nerves of the stomach of most species.[8] It can be shown to be released by vagal stimulation in pig[16] and field stimulation in the guinea pig.[19] TTX blocked the field-stimulated release of immunoreactive VIP in the guinea pig. VIP relaxed the stomach of the ferret[41] and the cat[42] *in vivo* in a manner similar to vagal stimulation. In the guinea pig stomach, isolated strips relaxed to VIP and VIP antisera reduced the response to low parameters of field stimulation.[19,43] Isolated cells of the guinea pig stomach relaxed to VIP, and the threshold for relaxation was augmented by isobutyl methylxanthine. The relaxation was accompanied by an increase in intracellular cAMP.[44] These results partially fulfill a requirement for identity of mechanical action of real and putative transmitter, but do not characterize the cellular actions of exogenous VIP and endogenous neurotransmitter.

In some tissues there is enough evidence to suggest that VIP may not be the main neurotransmitter mediating the IJP; VIP may not be the direct mediator of nerve-stimulated relaxation in the opossum esophagus. *In vitro*[45,46] in the circular muscle of the body of the esophagus, VIP causes modest hyperpolarization or no change in membrane potential followed by repetitive membrane depolarizations accompanied by spikes and contraction as measured in the single sucrose gap. In contrast, field stimulation elicited large inhibitory junction potentials. VIP-induced effects, but not IJPs, were eliminated in a chloride-free medium. Catecholamines (dopamine, dopa, adrenaline, and noradrenaline) caused electrophysiological responses of the esophagus similar to those induced by VIP. These effects were haloperidol or bulbocapnine sensitive and were abolished after scorpion venom had blocked nerve responses. Elevation of the Mg^{2+}/Ca^{2+} ratio also blocked these responses. This suggested release of a VIP-like material by catecholamines through a mechanism involving dopamine (D_1-like) receptors.[46] Moreover, responses to VIP and the endogenous substance released by catecholamines were abolished by removal of chloride accompanied by increased membrane resistance. The IJPs were accompanied by an increase in K^+ conductance.[47] Thus, VIP has electrophysiological effects inconsistent with a direct effect on the muscle like the inhibitory mediator. In the LES, VIP and the inhibitory mediator also differed in their actions; one striking finding was that VIP was nearly ineffective in relaxing contractions by carbachol, while NANC nerve stimulation was effective.[48]

The elucidation of the role of VIP in inhibitory innervation is clearly clouded by the problem of how to establish identity of cellular action of mediator and VIP. A recent study on esophageal circular smooth muscle of the cat[49] showed that in this tissue VIP acts directly on muscle. Highly specific VIP antiserum antagonized both VIP- and electrically induced TTX-sensitive relaxation. This kind of evidence has been obtained in several tissues and in several species. To date no one has obtained evidence that the action of VIP on smooth muscle is similar to that of the mediator. Since relaxation can be initiated by a variety of mechanisms, it does not provide sufficient proof of identity of cellular action. Even the use of a highly potent and selective antagonist (although not yet available) cannot establish such an identity. What it can do is provide evidence for the involvement of the antagonized substance in the response. Identity of cellular action requires evidence of electrophysiological and/or biochemical nature of the cellular response to both real and putative mediators. In the case of VIP, evidence has accumulated that VIP nerve endings are frequently located near a special network of cells, interstitial cells of Cajal, rather than smooth muscle.[50,51] These cells are interconnected to one another and to smooth muscle cells by gap junctions. It has been suggested[52] that these cells are the main sites of VIP action in the neurally mediated inhibitory response. This possibility has not yet been evaluated, but it could explain a difference in the cellular action of VIP and endogenous neurotransmitters.

K. CLINICAL ASPECTS

In evaluating a possible role for VIP in intestinal motor disturbances, it is important to realize the powerful effects of VIP on intestinal secretion. There is evidence that VIP may be involved in the watery diarrhea syndrome.[53] Infusion of VIP into ambulant pigs caused watery diarrhea and hypokalemia at plasma VIP levels comparable to that observed in humans (Verner-Morrison syndrome). When the VIP infusion was discontinued, the diarrhea ceased abruptly. Possible effects of VIP on intestinal motor effects were not studied in these experiments. In separate studies, VIP was seen to decrease gastric emptying.[54] VIP has been implicated in the local pathophysiology of human small bowel obstruction;[55] as obstruction progressed in a subacute canine small bowel obstruction model, portal VIP levels markedly increased to levels which caused hypersecretion and splanchnic vasodilation. In some patient populations, VIP serum levels have been found to be markedly altered; e.g., increased in cystic fibrosis[56] and reduced in smooth muscle in hypertrophic pyloric stenosis.[57] In human inflammatory bowel disease, the VIP content was shown to be markedly reduced.[58] A comparison of VIP binding in normal tissue with that of the colon in Crohn's disease revealed no difference.[59] However, colonic tissue from ulcerative colitis patients saw a marked reduction in binding sites in both the luminal and basal portions of the mucosa. No difference in muscular binding sites was noted.

In summary, evidence is increasing in support of an important role for VIP in neurally mediated inhibitions of intestinal smooth muscle. It seems likely that VIP is not the only inhibitory transmitter and may not be the neurotransmitter mediating IJPs in all tissues. In fact, definite proof of the latter awaits further electrophysiological and biochemical studies. Recent evidence points to the interesting possibility that VIP action on smooth muscle is mediated by interstitial cells of Cajal.

II. SOMATOSTATIN

A. INTRODUCTION

Somatostatin or growth hormone release-inhibiting hormone, a 14-amino-acid peptide isolated in 1973 from rat hypothalamus at the Salk Institute,[60] is a member of a peptide family which has members widely distributed in the central and peripheral nervous systems as well as in endocrine cells in the gastrointestinal tract.[1] Besides the mammalian somatostatin 14, there are amino terminal-extended somatostatin forms (somatostatin 28), several species-specific variants, and their prohormones. With the isolation of the somatostatin gene a number of posttranslational products have been described, and the biosynthetic pathways are being determined.[61] Somatostatins have been shown to be present in every vertebrate class and in primitive invertebrates, although not all structures have been evaluated.[62,63] Somatostatin is present throughout the gastrointestinal tract in nerve cells of the myenteric plexus and submucous plexus[52,64] in the dog (see Reference 8 for studies in rodents and Reference 65 for a review on the guinea pig). Some nerves with somatostatin immunoreactivity are present in muscle layers and mucosa.

B. RECEPTORS

There are no studies on binding of somatostatin to gastrointestinal muscle or nerves. However, binding studies have been carried out on brain synaptosomes, pituitary membranes, cell lines, tumor membranes, adrenal capsules, pancreatic membranes, isolated pancreatic cells and acini, and isolated mucosal cells of the stomach.[1] Only high-affinity binding sites (K_ds 0.1 to 1.0 nM) have been identified in the brain sites while both high- and low- (K_ds 5 to 12 nM) affinity sites have been identified in the pancreas and stomach.[66]

C. SOMATOSTATIN AND GUT FUNCTION

Somatostatin exerts an extraordinary range of physiological effects in the gut, including the suppression of secretion of all known gut peptides from both nerves and endocrine cells, including VIP, gastrin, cholecystokinin, glucagon, and secretin. Somatostatin also suppresses exocrine secretion of both gut and pancreas, blood flow, absorptive activity, and motor activity.[67] Somatostatin secretion is stimulated by glucagon, secretin, gastrin, and bombesin and is inhibited by substance P, the endorphins, and acetylcholine.

D. LOCALIZATION

Numerous nerve fibers with somatostatin-like immunoreactivity are found in the enteric plexuses of the small intestine, and most of these are of intrinsic origin.[8,65,68] Costa and Furness[69] proposed that somatostatin also functions as a cotransmitter for noradrenaline in a subpopulation of noradrenergic neurons. Somatostatin may "code" this subpopulation of noradrenergic neurons to allow them to project to specific target tissue. Costa and Furness found somatostatin and dopamine β-hydroxylase to be colocalized in the celiac-mesenteric ganglia, in mesenteric nerves, and in the wall of the guinea pig small intestine. They used nerve lesions to determine the source of the nerves. Most of these nerve endings were found in the submucous ganglia, but these nerves also supplied the myenteric plexus and the circular muscle.

E. ROLE OF SOMATOSTATIN IN THE PERISTALTIC REFLEX

Somatostatin neurons may act as interneurons to facilitate descending inhibition in intestinal peristalsis. Evidence for this has been provided recently.[70] Stretch-induced descending relaxation was studied in the rat colon. Release of somatostatin and VIP increased significantly during this procedure. Preincubation with somatostatin antiserum decreased VIP release and descending relaxation. Somatostatin increased VIP release and relaxation in a dose-dependent manner. Somatostatin did not have a direct effect on smooth muscle contractility. From this study it has been proposed that somatostatin acts on VIP nerves to regulate their activity.

F. SOMATOSTATIN AND INTERDIGESTIVE MOTOR ACTIVITY

The interdigestive migrating motor complex (MMC) is one phase of a recurring motility pattern of stomach and small intestine in the fasting state characterized by a quiescent period (phase I), intermittent activity (phase II), and a period of rhythmic contractions occurring at maximal frequency (MMC or phase III). The activity in phase III normally begins in the LES and stomach and frequently migrates to the terminal ileum. Peeters et al.[71] proposed a physiological role for somatostatin in the regulation of this activity in man. Somatostatin inhibited all gastric activity and accelerated the progression of phase III in the small intestine. Hormonal, neural, and myogenic factors are involved in the regulation of the MMC. Somatostatin probably is involved, but related to many other factors. Motilin may be involved in initiation of phase III in the stomach. Somatostatin inhibits release of motilin and the effect of motilin on the proximal gut.[72] In some studies, somatostatin levels are increased during phase III.[72,73] Bueno et al.[73] studied the role of somatostatin in the pig. Although somatostatin inhibited gastrointestinal activity in the pig, no role was obvious in the regulation of the MMC. However, the dietary habits of the pig (snacking without real fasting periods) suggest that regulation of the MMC must differ from that in other species.

G. SOMATOSTATIN AS AN INHIBITOR OF ACETYLCHOLINE RELEASE

Somatostatin relaxes guinea pig ileum and distal colon, a response sensitive to nerve conduction blockade.[74] Since no direct action on smooth muscle was observed and since somatostatin inhibited cholinergic twitch responses, it was proposed that somatostatin-induced inhibition was mediated by inhibition of release of acetylcholine from cholinergic nerves and activation of nonadrenergic inhibiting neurons. Inhibition of acetylcholine release by somatostatin was demonstrated directly by Teitelbaum et al.[75] and Yau et al.[76]

Somatostatin also inhibited the release of acetylcholine induced by superfusion of caerulein and neurotensin, but not that released by substance P in the guinea pig small intestine.[74,75] No direct effect of somatostatin on small intestinal smooth muscle has been found.

In the dog antrum, somatostatin inhibited acetylcholine release induced by electrical and substance P stimulation.[77] In this tissue, evidence was obtained that part of the inhibitory effect of somatostatin was mediated by prostaglandins. The complexity of the intrinsic neuronal circuitry is further illustrated by reports that inhibition by somatostatin of electrical-induced twitch contractions and release of [^{3}H] acetylcholine was prevented by Baclofen, a $GABA_B$ antagonist. In addition, somatostatin induced a TTX-sensitive release of ^{3}H-GABA in the guinea pig ileum.[78]

H. MECHANISM OF SOMATOSTATIN ACTION ON NERVES

Somatostatin depolarized a subset of neurons in the myenteric plexus of the guinea pig ileum[79] (see Figure 2). The depolarization was associated with an increase in membrane resistance due to inactivation of the resting K conductance. Although the neurons depolarized by somatostatin were not identified in this study, the data are compatible with somatostatin-

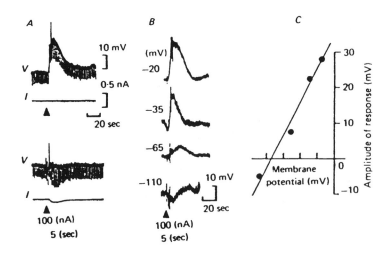

FIGURE 2. Somatostatin can depolarize myenteric neurons, which is associated with an increase in cell input resistance, suggesting inhibition of K conductance. (A) Change in membrane resistance during the depolarizing response to ionophoresis of somatostation. V, membrane potential; I, transmembrane current. The depolarizing response to somatostatin (applied at triangle 100 nA, 5 s) was associated with an increase in membrane resistance and action potential firing (upper panel). When the response was clamped manually at the initial membrane potential by passing current, a clear increase in the membrane resistance was observed (lower panel). (B) Depolarizing response to somatostatin at various potential levels. Somatostatin (100 nA, 5 s) was applied by ionophoresis (filled triangle, indicated for lower trace only). The response was enhanced by membrane depolarization, reduced by membrane hyperpolarization, and reversed by strong hyperpolarization. During the depolarizing responses at −20 and −35 mV there were action potentials which were not recorded faithfully by the pen recorder. (C) The amplitude of the response by B was plotted against the membrane potential at which the response was evoked. In this neuron the reversal potential for the somatostatin depolarization was −95 mV. (See Reference 79.)

activating intrinsic inhibitory neurons as proposed by Furness and Costa.[65] A smaller subset of neurons was hyperpolarized with somatostatin. Hyperpolarization was associated with a decrease in membrane resistance involving K^+ and Cl^- ions. Inhibition of acetylcholine release may be mediated by the latter mechanism.

I. INTRACELLULAR MECHANISMS OF SOMATOSTATIN

There are no studies on the intracellular mechanisms whereby somatostatin affects motility; however, in other sites somatostatin appears to activate the GTP-dependent inhibitory subunit of adenylate cyclase, leading to inhibition of intracellular cAMP production. Somatostatin also activates cytosolic phosphoprotein phosphatases.[66]

J. CLINICAL STUDIES

The action of somatostatin on stomach motility appears to be complex. The effect of somatostatin was studied using a long-acting somatostatin analogue (SMS 201-995). In eight volunteers, gastric emptying was accelerated as measured by 99mTc isotope disappearance after subcutaneous (s.c.) injection of the analogue; however, the mouth-to-cecum transit time was markedly prolonged. Using the same isotopic method, somatostatin was shown to block stomach emptying of solid food completely and to inhibit emptying of liquid food.[81] This study also supported suppression by somatostatin of insulin, glucagon, PP, motilin, and glucose levels. Reduction of gastric emptying in man was also observed by Johansson et al.[82] This study also showed marked prolongation of transit in the small intestine. Despite delayed transit, glucose absorption was markedly decreased by somatostatin, suggesting there may be a direct action of this peptide on mucosal transport.

An increasing number of studies link mucosal inflammation with altered intestinal motility. There may be peptidergic pathways involved in the communication between muscle and mucosa, and somatostatin is one of the possible mediators.[83]

There are other interesting clinical findings which require further study. Motility disturbances in premature infants have been suggested to be mediated by increased levels of somatostatin.[84] An increase in somatostatin immunoreactive fibers was found in the aganglionic segment of patients with Hirschsprung's disease.[85]

In summary, the most prominent role of somatostatin in motility seems to be that of neurotransmitter in interneurons of the intestinal tract. Activity of somatostatin nerves seems to inhibit acetylcholine release and to enhance VIP release.

III. OPIOIDS

A. INTRODUCTION

As many as 18 bioactive peptides have been found to belong to the opioid peptide family, which share the COOH-terminal part of enkephalin.[86] The receptors involved in control of gastrointestinal function are μ-, μ_1-, δ-, and κ-receptors.[87,88] μ-Receptors have the chemicals morphine and morphoceptin as agonists. Enkephalins appear to be recognized by δ-receptors as well as μ-receptors, but D-pen^2,D-pen^5-enkephalin is highly selective for δ-receptors. κ-Receptors selectively recognize the endogenous prodynorphin products.[89-91] See Chapter 11 for more details about selective opioid agonists and other general aspects of the opioids.

B. EFFECTS OF RECEPTOR STIMULATION ON IONIC CONDUCTANCE

Activation of μ-receptors located presynaptically on cholinergic neurons decreased release of acetylcholine by increasing a Ca^{2+}-dependent K^+ conductance that was blocked by application of external Ba^{2+} or quinine or by intracellular Cs.[92,93] When κ-receptors are activated (with μ-receptors blocked), they depress the excitatory postsynaptical potential by an action on the presynaptic neuron. Its proposed mechanism is a direct reduction of the inward Ca^{2+} conductance rather than an increase in Ca^{2+}-dependent K^+ conductance. Thus, reduction of the inward Ca^{2+} is either by decreasing the absolute number of Ca^{2+} channels or by decreasing their mean open time.[94] Two types of receptors may occur in the same nerve, utilizing different ion channels, but both subserving the same ultimate action.[95]

An unusual characteristic of the μ-receptor is that glucose, in concentrations often used in the Krebs solution,[96] reduces the potency of the μ-agonists to depress the release of acetylcholine. This effect of glucose appears to be related to metabolic effects and not to the osmotic effect of glucose or its ability to be transported. The precise mechanism is not well understood; it might occur because of alterations in the conformation of the opiate receptor, a decrease in the number of opiate receptors, or an alteration in the postreceptor intracellular mechanism,[96,97] such as ATP-induced inhibition of membrane conductance.

C. INHIBITION OF ACETYLCHOLINE RELEASE

Most opioid receptor agonists and endogenous opioids modulate the action of acetylcholine during induction of peristaltic activity.[98] Inhibition of acetylcholine release by opioid alkaloids has been regarded as the mechanism of opioid inhibition of peristalsis in the guinea pig ileum.[99,100] This conclusion was based on studies by Paton,[99] who demonstrated that morphine in low concentrations has a depressant action on the postganglionic cholinergic nerve fibers of guinea pig ileum, reducing both the resting output of acetylcholine and the output from the nerve endings when they are stimulated. Schaumann[100] suggested that the reduced release of acetylcholine is neither by an inhibition of synthesis nor by stabilization of the bound form of acetylcholine in the tissues. Bornstein and Fields[101] showed that opioids act presynaptically to inhibit transmitter release from nerves by reducing influx of Ca^{2+}. Kromer et al.[102] demonstrated that increasing the extracellular concentration of calcium attenuated the excitatory influence of the opiate antagonist naloxone.

D. MECHANISM OF REDUCTION OF NEURAL EXCITABILITY

The electrophysiological effects of opioids on vertebrate neurons appear to be exclusively "indirect".[95] In these cases, the transmitter interacts with its cell surface receptor and sets in train a series of intracellular transduction steps which eventually lead to the opening or closing of ion channels and, therefore, changes in cell excitability. It has been shown that potassium channels affected by opioids can also be opened by norepinephrine (acting on α_2-adrenoceptors), implying that the different cell surface receptors share the same intracellular transduction process. Furthermore, when opioiods open potassium channels and thereby hyperpolarize neurons and reduce their rate of firing, they appear to do so by acting on highly specific cell surface receptors, but events subsequent to receptor activation may very well be shared with other transmitters.

Four mechanisms of action have been proposed for the inhibitory effects of enkephalins.[93] First, enkephalins may interact directly with the excitatory conductance change produced by a neurotransmitter. This does not seem to occur in myenteric neurons because the effect of enkephalin occurs in conditions in which no synaptic transmission persists. Second, enkephalins cause postsynaptic inhibition off the neurons whose activity is being recorded. North el al.[93] give evidence for the occurrence of this mechanism, but the functional significance of enkephalin remains unclear in this regard because hyperpolarizing inhibiting synaptic potentials are rarely observed.[93] The third action of opiates is presynaptic inhibition of transmitter release by hyperpolarization of the same membrane. Fourth, opiates and enkephalin may block action potential propagation by local hyperpolarization or conductance increase.[103] Morita and North[103] evoked action potentials in the soma either by stimulating a cell process at a distance of up to 100 μm from the soma and allowing the spike to propagate along the process or by directly depolarizing the soma membrane by passing current through the recording electrode. Enkephalin and morphine prevented the appearance in the soma of the action potential following distant stimulation without changing the action potential caused by direct depolarization of the neuron soma. These effects were prevented by naloxone. The opiates and enkephalin sometimes caused the action potential to fractionate, suggesting that an important site of action in many neurons may be the proximal part of the process. The blockage of action potential propagation which occurred at the proximal process or more distal parts of the nerve process was sometimes, but usually not, associated with the hyperpolarization of the membrane of the soma.

E. ACTIONS AND INTERACTIONS OF THE OPIOIDS

Studies of the interactions of other endogenous control systems with opioid actions suggest that the opioids take part in a complex network of interactions within the enteric nervous system ultimately resulting in ordered intestinal motility. The excitatory effect of naloxone on distention-induced peristalsis is blocked by desensitization of the intestine to serotonin. This suggests that serotonin is a possible mediator of opioid-induced changes in intestinal motility.[104] This hypothesis is consistent with earlier studies of Burks[105] using *ex vivo* perfused canine intestine showing that opiates release serotonin. Indirect pharmacological evidence suggests that biosynthesis and release of myenteric opioids may be under the inhibitory control of receptors activated by dopamine.[106,107] Enkephalins have been shown to inhibit neuromuscular transmission in the human colon *in vitro*,[108] which may involve ATP and VIP being putative inhibitory neurotransmitters.[109-112] Endogenous opioids and VIP coexist within a subpopulation of guinea pig myenteric neurons.[105] In the guinea pig and rat colon *in vitro*, opioids were suggested to inhibit distal relaxation and augment proximal contraction in response to radial stretch by inhibition of VIP release.[114]

Substance P is an endogenous excitatory compound whose release, similar to that of acetylcholine, is inhibited by endogenous opioids.[115] Since substance P and [Met^5]-enkephalin coexist within a subpopulation of myenteric neuron,[115,117] the opioid may serve as a cotransmitter substance mediating negative feedback of substance P release.

Motilin, an endocrine excitatory substance, stimulates the release of neurotransmitters such as acetylcholine within the myenteric plexus and may, at least in the dog, be responsible for initiation of phase III activity fronts of the migrating motility complex.[118] Since the release of motilin is impaired by opioids in man,[119] it has been suggested that this might be the mode of action behind the constipating effect of opioids. On the other hand, in the canine intestine *in vivo*, motilin effects, especially in the ileum, are inhibited by naloxone.[120] Recent studies[20] suggest that motilin releases enkephalins to cause excitation by turning off VIP release.

Acetylcholine release is inhibited by activation of both opioid and α-receptors on nerve terminals in the guinea pig myenteric plexus.[100,121,-123] This may explain why the α-receptor antagonist yohimbine functionally antagonized the antitransit effect of morphine in the mouse *in vivo*.[124]

F. OPIOIDS AND INTESTINAL PERISTALSIS

Naloxone enhances distention-induced peristalsis in the guinea pig ileunm.[98] This excitatory effect of naloxone is blocked by TTX, hexamethonium, and atropine, suggesting that endogenous opioids may control peristalsis at the neuronal level.[93] When the intestinal wall is distended by intestinal contents, neural reflexes are initiated causing contraction oral to the distention and relaxation distal to it. Relaxation may occur through deactivation of excitatory impulses or activation of inhibitory ones.

Opioids depress the peristaltic reflex in two ways, presynaptic inhibition of acetylcholine release and postsynaptic inhibition of the action of acetylcholine.[98-100,125] The postsynaptic inhibition was demonstrated by Kromer and Schmide,[98] who demonstrated that acetylcholine and naloxone reinitiated peristaltic activity in the isolated guinea pig ileum after the reflex had been abolished by hexamethonium. Acetylcholine effects were dose dependently inhibited by preapplication of normorphine, but naloxone augmented the action of it upon peristalsis. Naloxone per se enhances peristaltic activity, but fails to initiate circular muscle contractions in the presence of TTX or absence of distension stimulus. In the latter case, the reflex arc is not activated. These data suggest that opioid receptors can be involved in the regulation of the peristaltic reflex.

G. ORGAN SPECIFIC ACTIONS OF THE OPIOIDS

The porcine and human LES are supplied with VIP-, substance P-, enkephalin-, and NPY-immunoreactive nerve fibers, suggesting that neuronal transmitters from these fibers may influence LES functions in both species.[126] Peptidergic innervation of the porcine and human LES are similar. Furthermore, in both pig and man, smooth muscle predominates in the LES. Enkephalin-containing nerve fibers are relatively scarce in both the porcine and human LES.[126] They are not found in the striated muscle. Generally, immunoreactive nerves are found in the muscularis mucosa and in the external muscle coat. In the submucosa, single immunoreactive nerves occurred close to small blood vessels. Reserpine was found to deplete the adrenergic nerve terminals of norepinephrine without visibly affecting the enkephalin nerve supply.[127,128] Howard et al.[129] reported that enkephalin reduced LES relaxation in healthy volunteers upon swallowing, without any effect on LES pressure. Uddman el al.[130] concluded that enkephalin is a potent presynaptic blocker of adrenergic-mediated

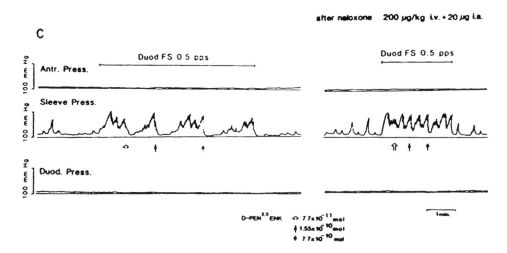

FIGURE 3. D-Pen²·⁵-enkephalin is a δ-agonist and inhibits activity induced by electrical stimulation of the intrinsic nerves in the canine pylorus *in vivo*. (See Reference 135.)

contraction of esophagus smooth muscle through inhibition of the release of the adrenergic transmitter.

Opioids have been shown to delay gastric emptying and to inhibit gastric contractions in the dog and man.[131,132] The pyloric sphincter deserves special mentioning because it contains the highest density of enkephalinergic innervation of all sphincters of the human gastrointestinal tract.[133,189] It was hypothesized that opioids mediated in excitatory reflex pathway to the pylorus from data on the cat pylorus where opiates are excitatory.[134] However, opioids do not excite the canine pyloric muscle, and endogenous opioids are not involved in the excitatory pyloric motor response elicited by duodenal field stimulation in the anesthetized dog. In fact, an inhibitory function of opioids has been shown in the canine pylorus[135] (See Figure 3).

The circular muscle of the small intestine of many species, including man, is excited by opiates *in vivo*.[136] Daniel et al. reported that strips of terminal ileum from the dog, guinea pig, human, rabbit, and rat are unresponsive to morphine. Methionine, leucine, and enkephalin depress the spontaneous contractions of the rat ileum in a dose-related way. This inhibitory effect is fast in onset and short in duration.[137] Morphine stimulates propagating "migrating motor complex (MMC)-like" activity in the upper intestine of dogs and humans.[138-140] Morphine contracts the rat ileum, possibly by releasing serotonin,[141] but it does not have such effect on serotonin release in the guinea pig ileum.[142] Burks[143] studied the effect of morphine using vascular perfusion of the dog small intestine. Morphine induced hypermotility that appeared to result from release of local intestinal 5-HT. The 5-HT liberated by morphine exerted both direct and indirect excitatory effects on the smooth muscle. The indirect effect was mediated by cholinergic nerves.

Morphine decreases the transit through the colon[144,145] while increasing tone and intraluminal pressure.[146,147] Schang et al.[148] showed that morphine actually alters the colonic smooth muscle activity in two ways: it promotes stationary spiking activity, stimulating the contractile activity in doing so; it suppresses the production of propagating spike bursts, in this way inhibiting colonic propulsion (See Figure 4). The failure of naloxone alone to alter colonic motility patterns may be related to a counter balance between a blockade of inhibitory and excitatory opioid mechanisms.[148] These might be active in an alternating fashion in order to modulate the periodicity of peristalsis.[149] The gastrocolonic response, a vagally mediated increase in distal colonic spiking activity after eating, is abolished by naloxone in healthy volunteers. This suggests a stimulatory role of endogenous opioids on colon activity.[150,151]

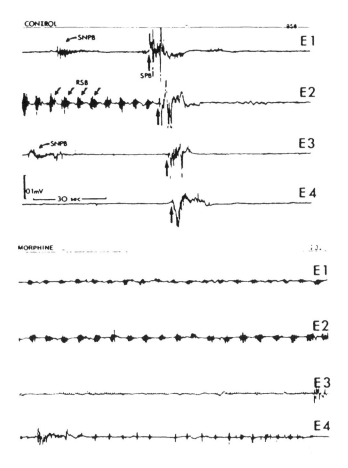

FIGURE 4. Morphine altered human colonic smooth muscle electrical activity in two ways: enhancement of stationary spiking activity and suppression of propagating spike bursts. (See Reference 148.)

Morphine produces anal sphincter contractions in the cat and dog, and loperamide caused a rise in internal anal sphincter pressure and a decrease in internal anal sphincter relaxation upon rectal distension in the opossum. These effects are antogonized by naloxone.[152-154]

In summary, opioids play an important role in inhibition of neural activity in the intrinsic nervous system of the gastrointestinal tract. However, opioids also directly and indirectly excite smooth muscle. Only precise characterization both *in vivo* and *in vitro* of indirect and direct effects for each organ will reveal the complex picture of opioid control of motility.

IV. GALANIN

Galanin, a 29 amino-acid peptide, was isolated and characterized by Tatemoto et al.[155] from extracts of porcine small intestine at the Karolinska Institute. Galanin immunoreactivity has been detected in the mouse, rat, guinea pig, cow, dog, monkey, human, fish, and frog.[1] Although only the porcine form has been sequenced, deviations in the standard radioimmunoassay curves for rat, cow, and human galanin and comparison of cDNA probes from porcine and d bovine sources suggest that the N-terminal portion may be preserved between species while the mid- and COOH-terminal portions may have some amino acid substitutions.[156] Immunohistochemical studies have shown it to be found exclusively in nerves, both in the central and enteric nervous systems,[157,158] including those of man.[159] In the rat gas-

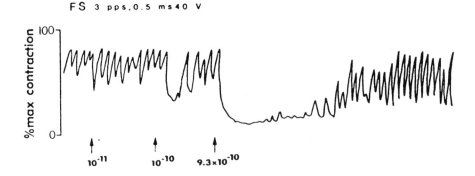

FIGURE 5.   Galanin injected intra-arterially during phasic activity of the canine small intestine *in vivo* produced inhibition of contractile activity induced by field stimulation of intrinsic nerves. Pharmacological studies suggested direct action on smooth muscle. (From Fox, J. E. T., McDonald, T. J., Kostolanska, F., and Tatemoto, K., *Life Sci.*, 390, 103, 1986. With permission.)

trointestinal tract, immunohistochemistry revealed a dense network of galanin-immunoreactive nerve fibers in the submucosa, smooth muscle layers, and intramural ganglia.[160] In the smooth muscle, the density of innervation was lower in the colon compared to the small intestine. Enteric galanin-immunoreactive fibers seem to be intrinsic to the gut, since their distribution and frequency were unaffected by extrinsic denervation. Myectomy at the mid-jejunal level resulted in total loss of galanin-immunoreactive nerve fibers 5 mm anally to the lesion, with a gradual return of such fibers up to 15 to 20 mm farther anally; there was no overt loss of fibers of the lesion orally. These findings support a role of galanin in control of motor reflexes in the intestinal smooth muscle.[160] Galanin-containing nerves were seen in close association with smooth muscle cells (<40 nm) in rat small intestine.[161,162]

A. EXCITATORY ACTIVITY ON SMOOTH MUSCLE

Galanin acts directly on longitudinal smooth muscle of rat jejunum. It induces tonic and superimposed phasic contractile activity.[163] Similar activity is produced in porcine small intestine.[155] Galanin at nanomolar concentrations was also found to have direct excitatory effects on rat ileal longitudinal muscle strips.[164]

B. INHIBITORY ACTIVITY ON SMOOTH MUSCLE

Galanin inhibits the canine small intestine, both *in vitro* and *in vivo* (see Figure 5). This occurs in the presence of TTX, suggesting a direct action on the smooth muscle cell membrane.[165] Galanin at a concentration of $10^{-9}M$ abolished phasic contractile activity of circular muscle strips and reduced responses to intra-arterial acetylcholine. More recently, galanin was shown to inhibit canine corpus and antrum circular muscle, but the site of action was not fully defined.[166] These data are consistent with galanin functioning as an inhibitory neurotransmitter in the small intestine.

C. ACTION ON NERVES

Galanin did not directly affect contractile activity of the guinea pig taenia coli, nor did it affect responses evoked by exogenously applied acetylcholine on substance P.[163] However, galanin could eliminate responses evoked by stimulation of intrinsic nerves, mediated by acetylcholine or substance P. Thus, the site of action is possibly prejunctional, suppressing release of acetylcholine and substance P. Gonda et al.[166] also found inhibition of galanin of nerve-mediated responses in canine stomach and intestine, but it was not established if the effect was pre - or postjunctional. Allescher et al.[167] found that galanin inhibited neural excitation of the canine pyloris.

In the guinea pig small intestine myenteric neurons, galanin simulates slow synaptic inhibition as measured intracellularly.[168] The hyperpolarization (5 to 20 mV) was accompanied by suppression of excitability and a reduction in input resistance. The neurons studies were of the AH/type 2 variety. These data support the hypothesis that galanin functions as an inhibitory neurotransmitter within the enteric nervous system.

Two mechanisms of action of galanin were proposed.[169] One is to open K^+ channels, decrease input resistance, and hyperpolarize the membrane toward the equilibrium potential. The second is to block voltage gated Ca^{2+} channels and suppression of the after-hyperpolarization by indirect prevention of opening of Ca-dependent K channels. The action of galanin on Ca^{2+} channels was proposed because galanin inhibits Ca^{2+} spikes evoked in the presence of blockage of K^+ channels by Cs^+.

D. CLINICAL STUDIES

In large bowel samples from children with Hirschsprung's disease a marked reduction of peptide-containing nerves is found, including galanin-containing nerves.[170]

In summary, the physiological role of galanin still needs to be defined. Galanin can act directly on smooth muscle of some tissues. Its major role, however, may be that of suppression of excitability in intrinsic neuronal circuits.

V. GASTRIN-RELEASING POLYPEPTIDE (GRP)/BOMBESIN FAMILY

A. INTRODUCTION

A number of bioactive peptides with similar structures have been isolated from amphibian skin (bombesin, alytesin, litorin, and ranatensin[171]), mammalian gastrointestinal tract (GRP),[172] and porcine spinal cord (neuromedin B, ranatensin/litorin-like,[173] neuromedin C, GRP_{18-27}[174]). The C-terminal seven amino acids and bombesin/GRP/alytesin were shown to be necessary for bioactivity.[171] Immunoreactive bombesin or GRP has been found mainly in nerves of the gastrointestinal tract.[52,64,175] There is good evidence to suggest that bombesin/GRP acts at interneurons in a variety of sites, a role which is consistent with the locus and long distal projections of GRP/bombesin-immunoreactive nerves in the myenteric plexus, with a less dense distribution in the muscle layers.[52,64,175,176]

B. RECEPTORS, ANTAGONISTS, AND DEGRADATION

There are no studies on binding of these peptides to gastrointestinal muscle tissues, although binding studies have been carried out on pancreatic acinar membranes.[177,178]

Since the active C-terminal of bombesin/GRP ($-Gly-His-Leu-Met-NH_2$) bears some similarity to the C-terminal of the tachykinins ($-Phy/Tyr-Gly-Leu-Met-NH_2$), antagonist to substance P have been studies for their antagonism of bombesin activity, particularly on pancreatic acinar cells and the urinary bladder.[1] Spantide was a weaker competitive antagonist against bombesin than substance P,[179] [D-Pro^{4}Lys^6,D-Trp^{7,9,10}] and [D-Pro^4,D-Trp^{7,9,10},Phe^{11}]substance P$_{4-11}$ were weak antagonist of both substance P and bombesin, with similar activity against both peptides acting on the urinary bladder to cause contraction. [D-Pro^2,D-Trp^{7,9,10}]substance P was more active against bombesin than substance P, but was noncompetitive against bombesin-induced contractions of the urinary bladder.[180] All pA$_2$s were <6.5.[1]

C. GRP/BOMBESIN AFFECT NEURAL ACTIVITY

Intracellular recording methods were used to study the actions of bombesin and gastrin-releasing peptide on electrical behavior of AH/type 2 myenteric neurons in guinea pig ileum *in vitro*[182] (see Figure 6). Both peptides evoked membrane depolarization associated with

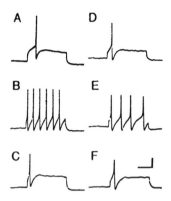

FIGURE 6. Enhanced excitability of myenteric neurons by VIP and bombesin. Single depolarizing current pulses are shown while recording intracellular electrical activity from a nerve. A,D and C,F are control and recovery, respectively. B in the presence of 1 μ*M* VIP; E in the presence of 0.5 μ*M* bombesin; vertical calibration, 20 mV; horizontal calibration, 50 ms. (From Zafirov, D. H., Palmer, J. M., et al., *Eur. J. Pharmocol.*, 115, 103, 1985. With permission.)

FIGURE 7. Bombesin increases contractile activity of circular antral smooth muscle strips. The excitation was not associated with activation of cholinergic, adrenergic, substance P on CCK receptors. The response was not sensitive to TTX. B, addition of 5 × 10⁻⁸ *M* bombesin. (From Mayer, E. A., Elashoff, J., and Walsh, J. H., *Am. J. Physiol.*, 243, G141, 1982. With permission.)

increased input resistance, enhanced excitability, and suppression of hyperpolarizing afterpotentials. The effects of the peptides simulated slow synaptic excitation in the myenteric plexus and are consistent with a neurotransmitter or neuromodulatory function.

Close intra-arerial injection of GRP or bombesin induced contractile activity in the canine stomach.[183] This effect was mediated by neural activity and probably involved release of acetylcholine. Excitatory effects were also observed in the guinea pig ileum. Studied for their possible involvement in the peristaltic reflex, bombesin effects were shown to be mediated by release of acetylcholine and substance P.

Both in the opossum and the cat, bombesin contracted the lower esophageal sphincter by releasing noradrenaline[184] and substance P.[185]

D. BOMBESIN EFFECTS ON SMOOTH MUSCLE

Bombesin is not seen to act directly on many muscle preparations. The canine stomach *in vitro* is a notable exception. Bombesin increased the frequency and amplitude of spontaneously occurring contractions of muscle strips from antrum and corpus of the canine stomach *in vitro*[186] (see Figure 7). The effect of bombesin on frequency of contractions in this system was myogenic, was unrelated to receptors for acetylcholine, norepinephrine, substance P, or cholecystokinin, and was independent of extracellular calcium. The effect of bombesin on the amplitude in the circular muscle was unaffected by TTX and atropine, but was blocked by verapamil. The contrasting effects of GRP/bombesin *in vivo* on canine antrum (neural action[1]) and *in vitro* (myogenic action[186]) have been discussed further in Chapter 5. In the circular antral muscle, bombesin was approximately 10,000 times more potent than acetylcholine at threshold concentrations and was equipotent to cholecystokinin. The high potency of bombesin and its known presence in gastric nerve fibers make it a candidate for a neurotransmitter function in regulation of gastric motility.

VI. CALCITONIN GENE-RELATED PEPTIDE (CGRP)

A. HISTOLOGICAL CHARACTERISTICS

CGRP (a 37-amino-acid peptide)-immunoreactive nerve fibers are widespread in the central and peripheral nervous systems.[187] In the gastrointestinal tract, it has been localized in nerve cell bodies and nerve fibers in the myenteric and submucous plexuses as well as in the nerve fibers in smooth muscle. CGRP-containing nerve fibers are extrinsic in origin, but some were seen to be of intrinsic origin.[187] In the rat, treatment with the sensory neurotoxin capsaicin reduced CGRP immune reactivity by 95% in the esophagus and stomach and by 40% in the lower gut.[188] CGRP was seen to be colocalized with substance P in some varicose nerve fibers in the circular muscle of the esophagus, LES, and fundus of the cat.[189] These are likely to be sensory nerves of extrinsic origin. CGRP-containing nerve processes in the guinea pig small intestine were seen very close (<40 nm) to smooth muscle cells.[190] Myenteric CGRP neurons were found to issue ascending and descending projections to myenteric ganglia and smooth muscle.[191] Submucous CGRP neurons project orally to other submucous ganglia, to the mucosa and submucosa, and also to circular smooth muscle.[191]

B. INHIBITORY ACTIVITY OF CGRP

Intestinal peristalsis, i.e., phasic changes in intraluminal pressure of isolated segments of guinea pig small intestine, induced by saline infusion, was inhibited by CGRP. In studying balloon-induced ascending contraction, it was observed that CGRP inhibited the atropine-resistant component of this reflex.[192]

CGRP (0.8 to 8 nM) produces concentration-dependent relaxation of the longitudinal muscle of the guinea pig small intestine.[192] The effect was studied on histamine- or bethanechol-induced contractions and appeared to involve a direct action on smooth muscle.[193]

Duodenal longitudinal muscle strips from rats relaxed by CGRP (1 to 100 nM). This relaxation was inhibited (40%) by TTX, but not by atropine. Therefore, in addition to a neurally mediated effect, CGRP may have direct action on this smooth muscle.

Inhibition was also observed on the LES pressure in cats where CGRP was given intra-arterially.[189] This inhibition was in part a direct effect on muscle and, in part, mediated by nerves, but these nerves were not cholinergic, adrenergic, or opioid.[191]

Rat and human α and β-calcitonin gene-related peptides, in the concentration range 1 to 100 nM, produced sustained relaxations of longitudinal muscle from the rat fundus and guinea pig gastric corpus.[194] The peptides were equipotent and equally effective. TTX, adrenoceptor and purine receptor antagonists, somatostatin, apamin, and Tyr-rat α-CGRP$_{28-37}$ peptide did not modify the action of the CGRP peptides. The CGRP-induced responses were inhibited by verapamil and potentiated by Bay K-8644. Incubation of the tissues with indomethacin markedly reduced the magnitude of the CGRP- and adrenaline-induced relaxations, but their responsiveness was restored by addition of prostaglandins E_1, E_2, and $F_{2\alpha}$ in concentrations that alone did not affect the motility of the indomethacin-treated strips. It is suggested that an inhibitory receptor for CGRP on gastric smooth muscle cells is linked to calcium channels and may be activated or sensitized by endogenous prostaglandins. Studies

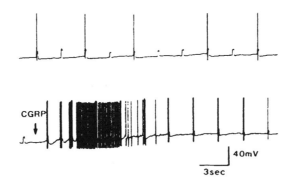

FIGURE 8.   Excitatory effects of CGRP on electrical behavior
of an AH/type 2 myenteric neuron. Intracellular recording of
electrical activity of the neuron with depolarizing current pulses.
Application of CGRP evoked membrane depolarization, repetitive spike discharge to depolarizing pulses, and the occurrence
of spontaneous spikes. (From Palmer, J. M., Schemann, M.,
Tamjra, K., and Wood, J. D., *Eur. J. Pharmacol.*, 132, 163,
1986. With permission.)

comparing the sites of action of CGRP in *in vivo* and *in vitro* in the same tissue are missing,
and it is therefore unclear whether major differences occur (see Chapter 5).

C. EXCITATORY ACTION OF CGRP

CGRP (0.3 to 30 nM) induced phasic contractile activity of circular muscle preparations
from the guinea pig small intestine. These contractions were inhibited by TTX and atropine,
suggesting mediation by cholinergic nerves.[192]

D. ACTION ON MYENTERIC NERVES

CGRP (1 nM to 1 μM) was applied by addition to solutions superfusing longitudinal
muscle-myenteric plexus preparation, and intracellular methods were used to record electrical
behavior of the myenteric neurons in guinea pig ileum.[195] CGRP evoked a long-lasting
depolarization of the cell membranes that was dose dependent. (ED[50] = 50 nM) and was
associated with an increase in the input resistance, a suppression of post spike hyperpolarizing
potential, and an enhanced excitability in all neurons that were tested (see Figure 8). These
findings are consistent with closure of K^+ channels by CGRP. Enhanced excitability was
apparent as a train of spikes that appeared at the crests of the CGRP-induced depolarization.
The excitatory action of CGRP simulated slow synaptic excitation. The results are consistent
with a neruotransmitter or neuromodulator role for CGRP in the enteric nervous system and
suggest that it may participate in local neurohumoral regulation of gastrointestinal effector
systems.

In summary, CGRP actions in gastrointestinal muscle are so far incompletely described,
but appear to involve mostly neural and possible some non-neural actions.

VII. SECRETIN

Secretin is a 27-amino-acid peptide. Secretin belongs to the same "peptide family" as
VIP. Secretin is released from endocrine cells in the small intestine. Secretin stimulates
pancreatic secretion and bile flow, but it inhibits gastric acid secretion.

A. SECRETIN AND GASTRIC EMPTYING

Intravenous administration of secretin decreased the rate of gastric emptying of 500 ml
saline by 80%.[196] Secretin increased the half-emptying time of a 2069-kcal pancake meal

in man by 133% when the plasma concentration was raised from 0.8 pM (control) to 9.8 pM.[197] Normally, postprandial plasma concentrations are ~7 pM or less. This study showed a rather poor correlation between plasma secretin concentration and gastric emptying in the individuals studied. In the rat, i.p injection of secretin inhibited gastric emptying and small intestinal transit of both solids and liquids.[198] In this study, VIP injected i.p. did not affect transit. No clear evidence that secretin is a neurotransmitter in the enteric nervous system has been obtained.

ACKNOWLEDGMENTS

This work was supported by the Medical Research Council (MRC) of Canada, and the secretarial assistance of Mrs. Mary Lambert is gratefully acknowledged.

REFERENCES

1. **Daniel E. E., Collins, S. M., Fox, J. E. T., and Huizinga, J. D.,** Pharmacology of neuroendocrine peptides, in *Handbook of Physiology*, Schultz, S. G., Wood, J. D., and Rauner, B. B., Eds., Oxford University Press, New York, 1989, 759.
2. **Robberech, P., Waelbroek, M., Dehaye, J. P., Winand, J., Vandermeers, A., Vandermeers-Piret, M. C., and Christophe, J.,** Evidence that helodermin, a newly extracted peptide from Gila monster venom, is a member of the secretin/VIP/PHI family of peptides with an original pattern of biological properties, *FEBS Lett.*, 166, 277, 1984.
3. **Rosselin, G.,** The receptors of the VIP family peptides (VIP, secretin, GRF, PHI, PHM, GIP, glucagon, and oxyntomodulin). Specificities and identity, *Peptides*, (Suppl. 1), 89, 1986.
4. **Berezin, I., Sheppard, S., Daniel, E. E., and Yanaihara, N.,** Ultrastructural immunocytochemical distribution of VIP-like immunoreactivity in dog ileum *Regul. Peptides*, 11, 287, 1985.
5. **Loesch, A. and Burnstock, G.,** Ultrastructural identification of VIP-containing nerve fibers in the myenteric plexus of rat ileum, *J. Neurocytol.*, 14, 327, 1985.
6. **Daniel, E. E.,** Pharmacology of adrenergic, cholinergic and drugs acting on other receptors, in *Mediators and Drugs in Gastrointestinal Motility*, Vol. 2, Bertaccini, G., Ed., Springer-Verlag, Berlin, 1982, 249.
7. **Fournier, A., Saunders, J. K., and St. Pierre, S.,** Synthesis conformational studies and biological activities of VIP and related fragments, *Peptides*, 5, 169, 1984.
8. **Schultzberg, M., Hokfelt, T., Nilsson, G., Terenius, L., Rehfeld, J. F., Brown, M., Elde, R., Goldstein, M., and Said, S.,** Distribution of peptide- and catecholamine-containing neurons in the gastrointestinal tract of rat and guinea-pig: immunohistochemical studies with antisera to substance P, vasoactive intestinal polypeptide, enkephalins, somatostatin, gastrin/cholecystokinin, neurotensin and dopamine-β-hydroxylase, *Neuroscience*, 5, 689, 1980.
9. **Marvaldi, J., Luis, J., Muller, J. M., El-Battari, A., Fantini, J., Martin, B., Martin, J. M., Abadie, B., Tirard, A., and Pichon, J.,** Characterization of the vasoactive intestinal peptide (VIP) binding sites: a biochemical and an immunological approach, *Peptides*, 7 (Suppl. 1), 1986, 137.
10. **Muller, J. M., El-Battari, E., Ah-Kye, J., Luis, J., Ducret, F., Pichon, J., and Marvaldi, J.,** Internalization of the vasoactive intestinal peptide (VIP) in a human adenocarcinoma cell line (HT29), *Eur. J. Biochem.*, 152, 107, 1985.
11. **Christophe, J., Svoboda, M., Lambert, M., Wallbrock, M., Winand, J., DeHaze, J. P., Wandermeers-Piret, M. C., Wandermeers, A., and Robbrecht, P.,** Effector mechanism of peptides of the VIP family, *Peptides*, 7 (Suppl. 1), 101, 1986.
12. **Prieto, J. C., Laburthe, M., and Rosselin, G.,** Interaction of vasoactive intestinal peptide with isolated intestinal cells from rat. I. Characterization, quantitative aspects and structural requirements of binding sites, *Eur. J. Biochem.*, 96, 229, 1979.
13. **Broyart, J.-P., Dupont, C., Laburthe, M., and Rosselin, G.,** Characterization of vasoactive intestinal peptide receptors in human colonic epithelial cells, *J. Clin. Endocrinol. Metab.*, 52, 715, 1981.
14. **Pandol, J., Dharmsathaphorn, K., Schoeffield, M.S., Vale, W., and Rivier, J.,** Vasoactive intestinal peptide receptor antagonist [4C1-D-Phe^{6}Leu^{17}] VIP, *Am. J. Physiol.*, 250, G553, 1986.
15. **Rattan, S., Walsh, J. H., and Goyal, R. K.,** Distribution of vasoactive intestinal polypeptide (VIP) in the opossum esophagus, *Dig. Dis. Sci.*, 25, 729, 1980.

16. **Fahrenkrug, J., Galbo, H., Holst, J. J., and Schaffalitzky de Muckadell, O., B.,** Influence of the autonomic nervous system of the release of vasoactive intestinal polypeptide from the porcine gastrointestinal tract, *J. Physiol. (London),* 280, 422, 1978.

17. **Bortoff, A., and Muller, R.,** Stimulation of intestinal smooth muscle by atropine, procaine, and tetrodotoxin, *Am. J. Physiol.,* 229, 1609, 1975.

18. **Wood, J. D.,** Excitation of intestinal muscle by atrophine, tetrodotoxin and xylocaine, *Am. J. Physiol.,* 222, 118, 1972.

19. **Grider, J. R., Cable, M. B., Said, S. I., and Makhlouf, G. M.,** Vasoactive intestinal peptide as a neural mediator of gastric relaxation, *Am. J. Physiol.,* 248, G73, 1985.

20. **Manaka, H., Manaka, Y., Kostolanska, F., Fox, J. E. T., and Daniel, E. E.,** Release of VIP and substance P from isolated perfused canine ileum, *Am. J. Physiol.,* 257, G182, 1989.

21. **Grider, J. R. and Makhlouf, G. M.,** Colonic peristaltic reflex: identification of vasoactive intestinal peptide as mediator of descending relaxation, *Am. J. Physiol.,* 251, G40, 1986.

22. **Costa, M. and Furness, J. B.,** The origins, pathways and terminations of neurons with VIP-like immunoreactivity in the guinea pig small intestine, *Neuroscience,* 8, 665, 1983.

23. **Nurko, S. and Rattan, S.,** Role of vasoactive intestinal polypeptide in the internal anal sphincter relaxation of the opossum, *J. Clin. Invest.,* 81, 1146, 1988.

24. **Alumets, J., Fahrenkrug, J., Hakanson, R., Schaffalitzky de Muckadell, O., Sundler, F., and Vodman, R.,** A rich VIP nerve supply is characteristic of sphincters, *Nature (London),* 280, 155, 1979.

25. **Bornstein, C., Costa a uess, J. B.,** Intrinsic and extrinsic inhibitory synaptic inputs to submucous neurones of the guinea-pig small intestine, *J. Physiol. (London),* 398, 371, 1988.

26. **Nelson, D. K., Service, J. E., Studelska, D. R., Brimijoin, S. and Go, V. L.,** Gastrointestinal neuropeptide concentrations following quanethidine, *J. Auton. Nerv. Syst.,* 22, 203, 1988.

27. **Willey, J. W., O'Dorisio, T. M., and Owyang, C.,** Vasoactive intestinal polypeptide mediates cholecystokinin-induced relaxation of the sphincter of Oddi, *J. Clin. Invest.,* 81, 1920, 1988.

28. **Kusunoki, M., Tsai, L. H., Taniyama, K., and Tanaka, G.,** Vasoactive intestinal polypeptide provokes acetylcholine release from the myenteric plexus, *Am. J. Physiol.,* 251, G51, 1986.

29. **Yau, W. M., Youthier, M. L., and Verdun, P. R.,** A presynaptic site of action of substance P and vasoactive intestinal polypeptide on myenteric neurons, *Brain Res.,* 330, 382, 1983.

30. **Williams, J. T. and North, R. A.,** Vasoactive intestinal polypeptide excites neurons of the myenteric plexus, *Brain Res.,* 175, 174, 1979.

31. **Moummi, C. and Rattan, S.,** Effect of methylene blue and N-ethylmaleimide on internal anal sphincter relaxation, *Am. J. Physiol.,* 255, G571, 1988.

32. **Torphy, T. J., Fine, C. F., Burman, M., Barnette, M. S., and Ormsbee, M. S. III,** Lower esophageal sphincter relaxation is associated with increased cyclic nucleotide content, *Am. J. Physiol.,* 251, G786, 1986.

33. **D'Amato, M., DeBeurme, F. A., and Lefebvre, A.,** Comparison of the effect of vasoactive intestinal polypeptide and non-adrenergic non-cholinergic neurons stimulation in the cat gastric fundus, *Eur. J. Pharmacol.,* 152, 41, 1988.

34. **Goyal, R. K., Rattan, S., and Said S. I.,** VIP as a possible neurotransmitter of non-cholinergic nonadrenergic inhibitory neurons, *Nature (London),* 288, 378, 1980.

35. **Biancani, P., Walsh, J. H., and Behar, J.,** Vasoactive intestinal polypeptide, a neurotransmitter for lower esophageal sphincter relaxation, *J. Clin. Invest.,* 73, 963, 1984.

36. **Angel, F., Go, V. L. W., Schmalz, P. F., and Szurszewski, J. H.,** Vasoactive intestinal polypeptide: a putative transmitter in the canine gastric muscularis mucosa, *J. Physiol. (London),* 341, 641, 1983.

37. **Grider, J. R., Cable, M. B., Bitar, K. N., Said, S. N., and Makhlouf, G. M.,** Vasoactive intestinal peptide relaxant neurotransmitter in taenuia coli of the guinea pig, *Gastroenterology,,* 89, 36, 1985.

38. **Morgan, K. G., Angel, F., Schmalz, P. F., and Szurszewski, J. H.,** Intracellular electrical activity of muscularis mucosa of the dog stomach, *Am. J. Physiol.,* 249, G256, 1985.

39. **Holmgren, S., Grove, D. J. and Nilsson, S.,** Substance P acts by releasing 5-hydroxy-tryptamine from enteric neurons in the stomach of the rainbow trout, *Salmo gairdneiri, Neuroscience,* 14(2), 683, 1985.

40. **Rattan, S., Said, S. I., and Goyal, R. K.,** Effect of vasoactive intestinal polypeptide on lower esophageal sphincter pressure, *Proc. Soc. Exp. Biol. Med.* 155, 40, 1977.

41. **Andrews, P. L. R. and Lawes, I. N. C.,** Characteristics of the vagally driven non-adrenergic, noncholinergic inhibitory innervation of the ferret gastric corpus, *J. Physiol. (London),* 363, 1, 1985.

42. **Eklund, J., Jodal, M., Lundgren, O., and Sjoquist, A.,** Effects of vasoactive intestinal polypeptide on blood flow, motility and fluid transport in the gastrointestinal tract of the cat, *Acta. Physiol. Scand.,* 105, 461, 1979.

43. **Makhlouf, G. M., Said, S., and Grider, J. R.,** Vasoactive intestinal peptide as a neural transmitter of gastric relaxation, *Am. J. Physiol.,* 248, G73, 1985.

44. **Bitar, K. N. and Mikhlouf, G. M.,** Relaxation of isolated gastric smooth muscle cells by basoactive intestinal peptide, *Science,* 216, 531, 1982.

45. **Daniel, E. E., Helmy-Elkholy, A., Jager, L. P., and Kannan, M. S.,** Neither a purine nor VIP is the mediator of inhibitory nerves of opossum oesophageal smooth muscle, *J. Physiol. (London),* 336, 243, 1983.

46. **Daniel, E. E., Jager, L. P., and Jury, J.,** Catecholamines release mediators in the opossum oesophageal circular smooth muscle, *J. Physiol. (London),* 382, 489, 1987.

47. **Jury, J., Jager, L. P., and Daniel, E. E.,** Unusual potassium channels mediate non-adrenergic non-cholinergic nerve mediated inhibition in opossum esophagus, *Can. J. Physiol. Pharmacol.,* 63, 107, 1985.

48. **Daniel, E. E., Jager, L. P., and Jury, J.,** Vasoactive intestine polypeptide and non-adrenergic, non-cholinergic inhibition in lower esophageal sphincter of opossum, *Br. J. Pharmacol.,* 96, 746, 1989.

49. **Behar, J., Guenard, V., Walsh, J. H., and Biancani, P.,** VIP and acetylcholine: neurotransmitters in esophageal circular smooth muscle, *Am. J. Physiol.,* 257, G380, 1989.

50. **Berezin, I., Huizinga, J. D., and Daniel, E. E.,** Interstitial cells of cajal in canine colon, a special communication network at the inner border of the circular muscle, *J. Comp. Neurol.,* 273, 42, 1988.

51. **Berezin I, Huizinga J. D., Farraway, L., and Daniel, E. E.,** Innervation of interstitial cells of Cajal in canine colon by VIP nerves, *Can. J. Physiol.,* 68, 922, 1990.

52. **Daniel, E. E., Costa, M., Furness, J. B., Keast, J. R.,** Peptide neurons in the canine small intestine, *J. Comp. Neurol.,* 237, 227, 1985.

53. **Modlin, I. M., Bloom, S. R., and Mitchell, S. J.,** Experimental evidence for vasoactive intestinal peptide as the cause of the watery diarrhea syndrome, *Gastroenterology,* 75, 1051, 1978.

54. **Ebert, R.,** Control of gastric emptying by regulatory peptides, *Gastroenterology,* 23, 165, 1988.

55. **Basson, M. D., Fielding, L. P., Bilchik, A. J., Zucker, K. A., Ballantyne, G. H., Sussman, S., Adrian, T. E., and Modlin, I. M.,** Does vasoactive intestinal polypeptide mediate the pathophysiology of bowel obstruction? *Am. J. Surg.,* 109, 15, 1989.

56. **Teufel, M., Luik, G., and Niessen, K. H.,** Gastrin secretin, VIP and motilin in children with mucoviscidosis and Crohn's disease, *Monatsschr. Kinderheilkd.,* 134, 132, 1986.

57. **Malmfors, G. and Sundler, F.,** Peptidergic innervation in infantile hypertrophic pyloric stenosis, *J. Pediatr. Surg.,* 21, 303, 1986.

58. **Koch, T. R., Carney, J. A., and Go, V. L. W.,** Distribution and quantitation of gut neuropeptides in normal intestine and inflammatory bowel diseases, *Dig. Dis. Sci.,* 32, 369, 1987.

59. **Mantyk, P. W., Catton, M. D., Boehmer, C. G., Walton, M. L., Passaro, E. P., Maggio, J. E., and Vigna, S. R.,** Receptors for sensory neuropeptides in human inflammatory diseases; implications for the effector role of sensory neurons, *Peptides,* 10, 627, 1989.

60. **Brazeau, P., Vale, W. L., Burgus, R., Ling, N., Butcher, M., Rivier, J., and Guillemin, R.,** Hypothalamic polypeptide that inhibits the secretion of immunoreactive pituitary growth hormone, *Science,* 179, 77, 1973.

61. **Patel, Y. C. and Tannenbaum, G. S.,** *Somatostatin,* Plenum, Press, New York 1985.

62. **Reichlin, S.,** Somatostatin, I, *N. Engl. J. Med.,* 309, 1495, 1983.

63. **Reichlin, S.,** Somatostatin, II, *N. Engl. J. Med.,* 309, 1556, 1983.

64. **Daniel, E. E., Furness, J. B., Costa, M., and Belbeck, L.,** The projections of chemically identified nerve fibers in canine ileum, *Cell Tissue Res.,* 247, 377, 1987.

65. **Furness, J. B. and Costa, M.,** Action of somatostatin on excitatory and inhibitory nerves in the intestine, *Eur. J. Pharmacol.,* 56, 69, 1979.

66. **Lewin, M. J. M.,** Somatostatin receptors, *Scand. J. Gastroenterol.,* 21, 42, 1986.

67. **Reichlin, S.,** Somatostatin, historical aspects, *Scand. J. Gastroenterol.,* 119, (Suppl.) 1, 1986.

68. **Costa M., Furness, J. B. Llewellyn Smith, I. J., Davies, B., and Oliver, J.,** An immunohistochemical study of the projections of somatostatin-containing neurons in the guinea pig intestine, *Neuroscience,* 5, 841, 1980.

69. **Costa, M. and Furness, J. B.,** Somatostatin is present in a subpopulation of noradrenergic nerve fibres supplying the intestine, *Neuroscience,* 13, (3) 911, 1984.

70. **Grider, J. R. and Makhlouf, G. M.,** Role of opioid neurons in the regulation of intestinal peristalsis, *Am. J. Physiol.,* 253, G226, 1987.

71. **Peeters, T. L., Janssen, J., and Vantrappen, G. R.,** Somatostatin and the interdigestive migrating motor complex in man, *Regul. Peptides,* 5, 209, 1983.

72. **Poitras, P., Lemoyne, M., Tasse, E., and Trudel, L.,** Variations in the plasma motilin, somatostatin, and pancreatic polypeptide concentrations and the interdigestive myoelectric complex in dog, *Can. J. Physiol. Pharmacol.,* 63, 1495, 1985.

73. **Bueno, L., Fioramonti, J., Rayner, V., and Ruckebusch, Y.,** Effects of motilin somatostatin and pancreatic polypeptide on the migrating myoelectric complex in pig and dog, *Gastroenterology,* 82, 1395, 1982.

74. **Furness, J. B. and Costa, M.,** Action of somatostatin on excitatory and inhibitory nerves in the intestine, *Eur. J. Pharmacol.,* 56, 69, 1979.

75. **Teitelbaum, D. H., O'Dorisio, T. M., Perkins, W. E., and Gaginella, T. S.,** Somatostatin modulation of peptide-induced acetylene release in guinea pig ileum, *Am. J. Physiol.,* 246, G509, 1984.
76. **Yau, W. M., Lingle, P. F., and Youther, M. L.,** Modulation of cholinergic neurotransmitter release from myenteric plexus by somatostatin, *Peptides,* 4, 49, 1983.
77. **Koelbel, C. B., van-Deventer, G., Khawaja, S., Mogard, M. Walsh, J. H., and Mayer, E. A.,** Somatostatin modulates cholinergic neurotransmission in canine antral muscle, *Am. J. Physiol.,* 254, G201, 1988.
78. **Takeda, T., Taniyama, K., Baba, S., and Tanaha, C.,** Putative mechanisms involved in excitatory and inhibitory effects of somatostatin on intestinal motility, *Am. J. Physiol.,* 257, G532, 1989.
79. **Katayama, Y. and North, R. A.,** The action of somatostatin on neurons of myenteric plexus of the guinea pig ileum, *J. Physiol. (London),* 303, 315, 1980.
80. **Fuessl, H. S., Carolan, G., Williams, G., and Bloom, S. R.,** Effect of long acting somatostatin analog (SMS201-995) on postprandial gastrin emptying of 99mTc-tin colloid and mouth to calcium transit time in man, *Digestion,* 36, 101, 1987.
81. **Peeters, J. M., Saltzman, M., Sherwin, R. S., Lange, R., and McCallum, R. W.,** Somatostatin inhibits gastric emptying of solids and liquids in man, *Dig. Dis. Sci.,* 29(8), 65S, 1984.
82. **Johansson, C., Wisen, O., Efendic, S., and Uunas-Wallenstein, K.,** Effects of somatostatin on gastrointestinal propagation and absorption of oral glucose in man, *Digestion,* 22, 126, 1981.
83. **Mayer, E. A., Raybould, H., and Koelbel, C.,** Neuropeptides, inflammation, and motility, *Dig. Dis. Sci.,* 33 (Suppl. 3), 71S, 1988.
84. **Marchini, G., Lagercratz, H., Milerad, J., Winberg, J., and Uvnas-Moberg, K.,** Plasma levels of somatostatin and gastrin in sick infants and small for gestational age infants, *J. Pediatr. Gastroenterol. Nutr.,* 7(5), 641, 1988.
85. **Lolova, I., Davidoff, M., Itzev, D., Apostolov, A., and Ivanchev, I.,** Distribution of substance P-, methionine-enkephalin-, somatostatin- and *Zentralbl. Allg. Pathol.,* 132, 25, 1986.
86. **Fox, J. -A.,** Control of gastrointestinal motility by peptides; old peptides, new tricks — new peptides old tricks, *Gastroenterol. Clin. North Am.,* 18(2), 163, 1989.
87. **Garzon, J., Schulz, R., and Herz, A.,** Evidence for the type of ε-opioid receptor in the rat vas deferens, *Mol. Pharm.,* 28, 1, 1985.
88. **Tam, S. W.,** Naloxone-inaccessible α-receptor in rat central nervous system, *Proc. Natl. Acad. Sci. U.S.A.* 80, 6703, 1983.
89. **Miller, L., Shaw, J. S., and Whiting, E. M.,** The contribution of intrinsic activity to the action of opioids *in vitro, Br. J. Pharmacol.,* 87, 595, 1986.
90. **Takemori, A. E., Ikeda, M., and Portoghese, P. S.,** The μ- κ- and δ-properties of various opioid agonist, *Eur. J. Pharmacol.,* 123, 357, 1986.
91. **Yoshimura, K., Huidobro-Toro, J. P., Lee, N. M., Loh, H. H., and Way, E. L.,** Kappa-opioid properties of dynorphin and its peptide fragments on the guinea-pig ileum, *J. Pharmacol. Exp. Ther.,* 222, 71, 1982.
92. **Hirst, G. D. S., Holman, M. E., and Spence, I.,** Two types of neurones in the myentereic plexus of the duodenum of the guinea pig, *J. Physiol. (London),* 236, 303, 1974.
93. **North, R. A., Katayama, Y., and Williams, J. T.,** On the mechanism and site of action of enkephalin on single myenteric neurons, *Brain Res.,* 165, 67, 1979.
94. **MacDonald, R. L. and Werz, M. A.,** Dynorphin A decreases voltage-dependent calcium conductance of mouse doral root ganglion neurones, *J. Physiol. (London),* 377, 237, 1986.
95. **North, R. A.,** Opioid receptor types and membrane ion channels, *Trends Neurosci.,* 7, 114, 1986.
96. **Shook, J. E., Kachur, J. F., Brase, D. A., and Dewey, W. A.,** Morphine dependence and diabetes. II. Alterations of normorphine potency in the guinea-pig ileum and mouse vas deferens and of ileal morphine dependence by changes in glucose concentration, *J. Pharamacol. Exp. Ther.,* 237, 848, 1986.
97. **Shook, J. E. and Dewey, W. L.,** Morphine dependence and diabetes. I. The development of morphine dependence in streptozotocin-diabetic rats and spontaneously diabetic C57BL/KsJ mice, *J. Pharmacol. Exp. Ther.,* 237, 841, 1986.
98. **Kromer, W. and Schmidt, H.,** Opioids modulate intestinal peristalsis at a site of action additional to that modulating acetylcholine release, *J. Pharmacol. Exp. Ther.,* 223, 271, 1982.
99. **Paton, W. D. M.,** The action of morphine and related substances on contraction and on acetylcholine output of coaxially stimulated guinea-pig ileum, *Br. J. Pharmacol.,* 11, 119, 1957.
100. **Schaumann, W.,** Inhibition by morphine of the release of acetylcholine from the intestine of the guinea-pig, *Br. J. Pharmacol.,* 12, 115, 1957.
101. **Bornstein, J. C. and Fields, J. L.,** Morphine presynaptically inhibits a ganglionic cholinergic synapse, *Nerosci. Lett.,* 15, 77, 1979.
102. **Kromer, W., Holt, V., Schmidt, H., and Herz, A.,** Release of immunoreactive-dynorphin from isolated guinea pig small intestine is reduced during peristaltic activity, *Neurosci. Lett.,* 25, 53, 1981.

103. **Morita, K. and North, R. A.,** Opiates and enkephalins reduce the excitability of neuronal processes, *Neuroscience,* 1981a, 6, 1943.

104. **Kromer, W.,** Endogenous and exogenous opioids in the control of gastrointestinal motility and secretion, *Pharmacol. Rev.,* 40(2), 121, 1988.

105. **Burks, T. F., Hirning, L. D., Galligan, J. J., and Davis, T. P.,** Motility effects of opioid peptides in dog intestine, *Life Sci.,* 31, 2237, 1982.

106. **Vargas, M. L., Martinez, J. A., and Milanes, M. V.,** Pharmacological evidence for the existence of interactions between dopaminergic and opioid peptidergic systems in guinea-pig ileum myenteric plexus, *Eur. J. Pharmacol.,* 128, 259, 1986.

107. **Vargas, M. L., Martinez, J. A., and Milanes, M. V.,** Effects of droperidol on the biosynthesis and release of endogenous opioid peptides in guinea-pig ileum, *Gen. Pharmacol.,* 18, 283, 1987.

108. **Hoyle, C. H., Burnstock, G., Jass, J., and Lennard-Jones, J. E.,** Enkelphalins inhibit non-adrenergic, non-cholinergic neuromuscular transmission in the human colon, *Eur. J. Pharmacol.,* 131, 159, 1986.

109. **Gershon, M. D. and Erde, S. M.,** The nervous system of the gut, *Gastroenterology,* 80, 1571, 1981.

110. **Kosterlitz, H. W. and Lees, G. M.,** Pharmacological analysis of intrinsic intestinal reflexes, *Pharmacol. Rev.,* 16, 301, 1964.

111. **Daniel, E. E.,** Pharmacology of the gastrointestinal tract, in *Handbook of Physiology,* Vol. 4, Vissher, M. B., Hastings, A., and Pappenheimer, J. R., Eds., Williams & Wilkins, Baltimore, 1969, 2267.

112. **Kosterlitz, H. W. and Watt, A. J.,** The peristaltic reflex, in *Methods in Pharmacology,* Vol. 3, Daniel, E. E. and Paton, D. M., Eds., Plenum Press, New York, 1975.

113. **Costa, M., Furness, J. B., and Cuello, A. C.,** Separate populations of opioid containing neurons in the guinea-pig intestine, *Neuropeptides,* 5, 445, 1985.

114. **Grider, J. R., Arimura, A., and Makhlauf, G. M.,** Role of somatostatin neurons in intestinal peristalsis:facilitatory interneurons in descending pathways, *Am. J. Physiol.,* 253, G434, 1987.

115. **Bartho, L., Sebok, B., and Szolcsanyi, J.,** Indirect evidence for the inhibition of enteric substance P neurones be opiate agonist but not by capsaicin, *Eur. J. Pharmacol.,* 77, 273, 1982.

116. **Domoto, T., Gonda, T., Oki, M., and Yanaihara, N.,** Coexistence of substance P and methionine[5] enkephalin-like immunoreactivity in nerve cells of the myenteric ganglia in the cat ileum, *Neurosci. Lett.,* 47, 9, 1984.

117. **Uchida, T., Kobayashi, S., and Yanaihara, N.,** Occurrence of met-enkephalin-Arg[6]-Gly[7]-Lev[8] neurons in the guinea pig duodenum: immuno-electron microscopic study on the costorage of met-enkephalin - Arg[5]-Gly[7]-Lev[8] with substance P or PHI (1 — 15), *Biomed. Res.,* 6, 415, 1985.

118. **Fox, J. E. T.,** Motilin — an update, *Life Sci.,* 35, 695, 1984.

119. **Sekiya, K., Funakoshi, A., Nakano, I., Nawata, H., Kato, K., and Ibayashi, H.,** Effect of methionine-enkephalin analog (FK33-824) on plasma motilin, *Gastroenterology,* 21, 344, 1986.

120. **Fox, J. E. T., and Daniel, E. E.,** Activation of endogenous excitatory opiate pathways in the canine small intestine and stomach, *Am. J. Physiol.,* 253, 189, 1987.

121. **Paton, W. D. M. and Visi, E. S.,** The inhibitory action of noradrenaline and adrenaline on acetylcholine output by guinea-pig ileum longitudinal muscle strip, *Br. J. Pharmacol.,* 35, 10, 1969.

122. **Kosterlitz, H. W., Lydon, R. J., and Watt, A. J.,** The effects of adrenaline, noradrenaline and isoprenaline or inhibitory alpha- and β-adrenoceptors in the longitudinal muscle of the guinea-pig ileum, *Br. J. Pharmacol.,* 39, 398, 1969.

123. **Stark, K.,** Regulation off noradrenaline release by presynaptic receptor systems, *Rev. Physiol. Biochem. Pharmacol.,* 77, 1, 1977.

124. **Wong, C. L.,** The possible involvement of adrenoceptors in the intestinal effect of morphine in mice, *Clin. Exp. Pharmacol. Physiol.,* 11, 605, 1984.

125. **Vizi, E., Bertaccini, G., Impicciatore, M., and Knoll, J.,** Evidence that acetylcholine released by gastrin and related polypeptides contributes to their effect on gastrointestinal motility, *Gastroenterology,* 64, 268, 1973.

126. **Aggestrup, S., Vodman, R., Jensen, S. T., Hakanson, R., Sundler, F., Schaffalitzkey de Muckadell, O., and Emson, P.,** Regulatory peptides in lower esophageal sphincter of pig and man, *Dig. Dis. Sci.,* 31, 1370, 1986.

127. **Carlsson, A., Rosenglyen, E., and Bertler, A.,** Effect of reselpine on the the catechol amines, in *Psychomopic Drugs,* S. Garattini, S. and Ghetti, V., Eds., Elsevier, Amsterdam, 1957, 363.

128. **Swedin, G.,** Studies on neurotransmission mechanisms in the rat and guinea-pig vas deferens, *Acta Physiol. Scand.,* Suppl. 369, 1, 1971.

129. **Howard, J. M., Belsheim, M. R., and Sullivan, S. N.,** Enkephalin inhibits relaxation of the lower oesophageal sphincter, *Br. Med. J.,* 285, 1605, 1982.

130. **Uddman, R., Alumets, J., Hakanson, R., Sundler, F., and Walles, B.,** Peptidergic (enkephalin) innervation of the mammalian esophagus, *Gastroenterology,* 78, 732, 1980.

131. **Fioramonti, J., Fargeas, M. J., and Bueno, L.,** Comparative effects of centrally-administered morphine on colonic motility in dogs by the benzodiazepine receptor antagonist RO 15-1788, *Life Sci.,* 41, 1449, 1987.

132. **Mittal, R. K., Frank, E. G., Lang, R. C., and McCallum, R. W.,** Effects of morphine and naloxone on esophageal motility and gastric emptying in man, *Dig. Dis. Sci.,* 31, 936, 1986.

133. **Ferri. G. L., Morreale, R. L., and Soimero, L.,** Intramural distribution of met 5-enkephalin-AR56-gly[7]-level in sphincter regions of the human gut, *Neurosci. Lett.,* 74, 304, 1989.

134. **Reynolds, J. C., Ouyang, A., and Cohen, S.,** Interactions of bombesin and substance P at the feline lower oesophageal sphincter, *J. Clin. Invest.,* 77, 436, 1986.

135. **Allescher, H. -D., Ahmad, S., Daniel, E. E., Dent, J., Kostolanska, F., and Fox, J. E. T.,** Inhibitory opioid receptors in canine pylorus, *Am. J. Physiol.,* 255, G352, 1988.

136. **Bailey, S. J., Featherstone, R. L., Jordan, C. C., and Morton, I. K. M.,** An examination of the phararmacology of two substance P antagonists and the evidence for tachykinin receptor subtypes, *Br. J. Pharmacol.,* 87, 79. 1986.

137. **Nakatus, K., Goldenberg, E., Penning, D., and Jhmandas, K.,** Enkephalin-induced inhibition of the isolated rat ileum is not blocked by naloxone, *Can. J. Physiol. Pharmacol.,* 59, 901, 1981.

138. **Konturek, S. J., Thor, P., Krol, R., Dembinsky, A., and Schally, A. V.,** Influence of methionine-enkephalin and morphin on myoelectric activity of small bowel, *Am J. Physiol.,* 238, G384, 1980.

139. **Imgram, D. M. and Catchpole, B. N.,** Effect of opiates on gastroduodenal motility following surgical operation, *Dig. Dis. Sci.,* 26, 989, 1981.

140. **Waterfall, W. E.,** Electric patterns in the human jejunum with and without vagotomy; migrating myoelectrical complexes and the influence of morphine, *Surgery,* 94, 186, 1983.

141. **Burks, T. F.,** Acute effects of morphine on rat intestinal motility, *Eur. J. Pharmacol.,* 40, 279, 1976.

142. **Schultz, R. and Cartwright, C.,** Effect of morphine on serotonin release from myenteric plexus of the guinea pig, *J. Pharmacol. Exp. Ther.,* 190, 420, 1974.

143. **Burks, R. G.,** Mediation by 5-hydroxytryptamine of morphine stimulant actions in doe intestine, *J. Pharmacol. Exp. Ther.,* 185, 530, 1973.

144. **Jaffe, J. J. and Martin, W. R.,** Opioid analgesics and antagonists, in *Pharmacological Basis of Therapeutics,* 7th ed. Gilman, A. G., Goodman, L. S., Rall, T. W., and Murad, F., Eds., Macmillan, New York, 1985, 491.

145. **Stewart, J. J.,** Temporal effects of morphine on rat intestinal transit, *Pharmacology,* 29, 47, 1984.

146. **Plant, O. H. and Miller, G. H.,** Effect of morphine and some other opium alkaloids on the muscular activity of the alimentary canal, *J. Pharmacol. Exp. Ther.,* 28, 245, 1926.

147. **Blanquet, F., Bouvier, M., and Gonella, J.,** Effects of enkephalins an morphine on spontaneous electrical activity and on junction potentials elicted by parasympathetic nerve stimulation in cat and rabbit colon, *Br. J. Pharmacol.* 77, 419, 1982.

148. **Schang, J. C., Hemond, M., Herbert, M., and Pilote, M.,** How does morphine work on colonic motility? An electromyographic study in the human left and sigmoid colon, *Life Sci.,* 38, 671.a, 1986.

149. **Kromer, W., Pretzlaff, W., and Woinoff, R.,** Opioids modulate periodicity rather than efficacy of peristaltic waves in the guinea-pig ileum *in vitro, Life Sci.,* 26, 1857, 1980.

150. **Sun, E. A., Snape, W. J. Jr., Cohen, S., and Renny, A.,** The role of opiate receptors and cholinergic neurons in the gastrocolonic response, *Gastroenterology,* 82, 689, 1982.

151. **Lundberg, I. M., Hokfelt, T., Kewenter, I., Pettersson, G., Ahlman, H., Edin, R., Dahlstrom, A., Nilsson, G., Terenius, L., Uvnas-Wallensten, K., and Said, S.,** Substance P-, VIP-, and enkephalin-like immunoreactivity in the human vagus nerve, *Gastroenterology,* 77, 468, 1979.

152. **Kopponyi, T. and Murphy, W. S.,** The effect of morphine on the anal sphincters, *Science,* 78, 14, 1933.

153. **Rattan, S. and Culver, P. J.,** Influence of loperamide on the internal anal sphincter in the opossum, *Gastroenterology,* 93, 121, 1987.

154. **Wuster, M. and Herz, A.,** Opiate agonist action of antidiarrheal agents *in vitro* and *in vivo* findings in support for selective action, *Naunyn-Schmiedelberg's Arch. Pharmacol.,* 301, 187, 1978.

155. **Tatemoto, K., Rokaeus, A., Jornvall, H., McDonald, T. J., and Mutt, V.,** Galanin — a novel biologically active peptide from porcine intestine, *FEBS Lett.,* 164, 124, 1983.

156. **Rökaeus, Å.,** Galanin — a newly isolated biologically active neuropeptide, *Trends Neurosci.,* 10, 158, 1987.

157. **Rökaeus, Å., Melander, T., Hokfelt, T., Rundbert, J. M., Tatemoto, K., Carlquist, M., and Mutt, V.,** A galanin-like peptide in the central nervous system of the rat, *Neurosci. Lett.,* 47, 161, 1984.

158. **Melander, T., Hokfelt, T., Rökaeus, Å., Fahrenkrug, J., Tatemoto, K., and Mutt, V.,** Distribution of galanin-like immunoreactivity in the gastrointestinal tract of several species, *Cell Tissue Res.,* 239, 253, 1985.

159. **Bishop, A. E., Polak, J. M., Bauer, F. E., Christofides, N. D., Carlei, F., and Bloom, S. R.,** Occurrence and distribution of a newly discovered peptide, galanin, in the mammalian enteric nervous system, *Gut,* 27(7), 849, 1986.

160. **Ekblad, E., Rökaeus, A., Hakanson, R., and Sundler, F.,** Galanin nerve fibers in the rat gut: distribution, origin and projections, *Neuroscience,* 16(2), 355, 1985.

161. **Feher, E. and Burnstock, G.,** Galanin like immunoreactive nerve elements in the small intestine of the rat, *Neurosci. Lett.,* 92, 137, 1988.
162. **Daniel, E. E., Furness, J. B., Costa, M., and Belbeck, L.,** The projections of chemically identified nerve fibres in canine ileum, *Cell Tissue Res.,* 247, 379, 1987.
163. **Ekblad, E., Hakanson, R., Sundler, F., and Wahlestedt, C.,** Galanin: neuromodulatory and direct contractile effects on smooth muscle preparations, *Br. J. Pharmacol.,* 86, 241, 1985.
164. **Muramatsu, I. and Yanalhara, N.,** Contribution of galanin to non-cholinergic non-adrenergic transmission in rat ileum, *Br. J. Pharmacol.,* 94, 1241, 1988.
165. **Fox, J. E. T., McDonald, T. J., Kostolanska, F., and Tatemoto, K.,** Galanin: an inhibitory neural peptide in the canine small intestine, *Life Sci.,* 390, 103, 1986.
166. **Gonda, T., Daniel, E. E., McDonald, T. J., Fox, J. E. T., Brooks, B. D., and Oki, M.,** Distribution and function of enteric GAL-IR nerves in dogs: comparison with VIP, *Am. J. Physiol.,* 256, G884, 1989.
167. **Allescher, H. D., Daniel, E. E., Dent, T., Fox, J. E. T., and Manaka, H.,** Inhibitory functions of VIP-PHI and galanin in canine pyloris, *Am. J. Physiol.,* 256, G789, 1989.
168. **Palmer, J. M., Schemann, M., Tamura, K., and Wood, J. D.,** Galanin mimics slow synaptic inhibition of myenteric neurons, *Eur. J. Pharmacol.,* 124, 379, 1986.
169. **Tauramura, K., Palmer, J. T., Winkelmann, C. K., and Wood, J. D.,** Mechanism of action of galanin on myenteric neurons, *J. Neurophysiol.,* 60, 966, 1988.
170. **Hamada, Y., Bishop, A. E., Federici, G., Zivosecchi, M., Talbot, I. C., and Polak, J. M.,** Increased neuropeptide Y-immunoreactive innervation of aganglionic bowel in Hirschsprung's disease, *Virchows Arch.,* 411(4), 369, 1987.
171. **Erspamer, V.,** Peptides of the amphibian skin active in the gut. II Bombesin-like peptides: isolation, structure and basic function, *Gastrointestinal Hormones,* Jerzy Glass, G. B., Ed., Raven Press, New York, 1980, 343.
172. **McDonald, T. J., Nilsson, G., Vagne, M., Ghatei, M. A., Bloom, S. R., and Mutt, V.,** A gastrin releasing peptide from the porcine nonantral gastric tissue, *Gut,* 19, 767, 1978.
173. **Minamino, N., Kangawa, K., and Matsuo, H.,** Neuromedin B: another bombesin-like peptide identified in porcine spinal cord, *Biochem. Biophys, Res. Commun,.* 114, 541, 1983.
174. **Minamino, N., Kangawa, K., and Matsuo, H.,** Neuromedin C: bombesin-like peptide identified in porcine spinal cord, *Biochem. Biophys. Res. Commun.,* 119, 14, 1984.
175. **Costa, M., Furness, J. B., Yanahara, N., Yanaihara, C., and Moody, T. W.,** Distribution and projections of neurons with immunoreactivity for both gastrin-releasing peptide and bombesin in the guinea-pig small intestine, *Cell Tissue Res.,* 235, 285, 1984.
176. **Ekblad, E., Ekman, R., Hakanson, R., and Sundler, F.,** GRP neurons in the rat small intestine issue long anal projections, *Regul. Peptides,* 9, 279, 1984.
177. **Gardner, J. D., and Jensen, R. T.,** Receptor for secretagogues on pancreatic acinar cells, *Am. J. Physiol.,* 238, G63, 1980.
178. **Jensen, R. T., Moody, T., Pert, C., Rivier, J. E., and Gardner, J. D.,** Interactions of bombesin and litorin with specific membrane receptors on pancreatic acinar cells, *Proc. Natl. Acad. Sci. U.S.A.,* 75, G139, 1978.
179. **Jensen, R. T., Jones, S. W., Folkers, K., and Gardner, J. D.,** A synthetic peptide that is a bombesin antagonist, *Nature (London),* 309, 61, 1984.
180. **Mizrahi, J., Dion, S., D'Orleans-Juste, P., and Regoli, D.,** Activities and antagonism of bombesin on urinary smooth muscles, *Eur. J. Pharmacol.* 111, 339, 1985.
181. **Bunnett, N. W., Kobayashi, R., Orloff, M. S., Reeeve, J. R., Turner, A. J., and Walsh, J. H.,** Catabolism of gastrin releasing peptide and substance P by gastric membrane bound peptidases, *Peptides,* 6, 277, 1985.
182. **Zafirov, D. H., Palmer, J. M., Nemeth, P. R., and Wood, J. D.,** Bombesin, gastrin releasing peptide and vasoactive intestinal peptide excite myenteric neurons, *Eur. J. Pharmacol.,* 115, 103, 1985.
183. **Fox, J. E. T. and McDonald T. J.,** Motor effects of gastrin releasing peptide (GRP) and bombesin in the canine stomach and small intestine, *Life Sci.,* 35, 1667, 1984.
184. **Mukhopadhyay, A. K. and Kunnemann, M.,** Mechanism of lower esophageal sphincter stimulation by bombesin in the opossum, *Gastroenterology,* 76, 1409, 1979.
185. **Reynolds, J. C., Dukehart, M. R., Ouyang, A., and Cohen, S.,** Interactions of bombesin and substance P at the feline lower esophageal sphincter, *J. Clin. Invest.,* 77, 436, 1986.
186. **Mayer, E. A., Elashoff, J., and Walsh, J. H.,** Characterization of bombesin effects on canine gastric muscle, *Am. J. Physiol.,* 243, G141, 1982.
187. **Belai, A. and Burnstock, G.,** Selective damage of intrinsic calcitonin gene-related peptide-like immunoreactive enteric nerve fibers in streptozotocin-induced diabetic rate, *Gastroenterology,* 92, 730, 1987.
188. **Sternini, C., Reeve, J. R., and Brecha, N.,** Distribution and characterization of calcitonin gene-related peptide immunoreactivity in the digestive system of normal and capsaicin-treated rats, *Gastroenterology,* 93, 852, 1987.

189. **Parkman, H. P., Reynolds, J. C., Elfman, K. S., and Ogorek, C. P.,** Calcitonin gene-related peptide: a sensory and motor neurotransmitter in the feline lower esophageal sphincter, *Regul. Peptides,* 25, 131, 1989.

190. **Feher, E., Burnstock, I., Varndell, M., and Polak, J. M.,** Calcitonin gene-related peptide-immunoreactive nerve fibres in the small intestine of the guinea-pig; electron-microscopic immunocytochemistry, *Cell Tissue Res.,* 245, 353, 1986.

191. **Ekblad, E., Winther, R. Hakanson, R., and Sunder, F.,** Projections of peptide-containing neurons in rat small intestine, *Neuroscience,* 20, 169, 1987.

192. **Holzer, P., Bartho, L., Matusak, O., and Bauer, V.,** Calcitonin gene-related peptide action on intestinal circular muscle, *Am. J. Physiol.,* 256, G546, 1989.

193. **Bartho, L., Lambeck, F., and Holzer, P.,** Calcitonin gene-related peptide is a potent relaxant of intestinal smooth muscle, *Eur. Pharmacol.,* 135, 449, 1987.

194. **Katsoulis, S. and Conlon, J. M.,** Calcitonin-gene-related peptides relax guinea pig and rat gastric smooth muscle, *Eur. J. Pharmacol.,* 161, 129, 1989.

195. **Palmer, J. M., Schemann, M., Tamjra, K., and Wood, J. D.,** Calcitonin gene-related peptide excites myenteric neurons, *Eur. J. Pharmacol.,* 132, 163, 1986.

196. **Chey, W. Y., Hitanant, S., Hendricks, J., and Lorber, S. H.,** Effect of secretin and cholecystokinin on gastric emptying and gastric secretion in man, *Gastroenterology,* 58, 820, 1970.

197. **Kleibeuker, J. H., Beekhvis, H., Piers, D. A., and Schaffalitzey de Muckaddell, O. B.,** Retardation of gastric emptying of solid food by secretin, *Gastroenterology,* 94, 122, 1988.

198. **Murthy, S. N. S. and Ganiban, G.,** Effect of the secretin family of peptides on gastric emptying and small intestinal transit in rats, *Peptides,* 9, 583, 1988.

Chapter 13

EFFECTS OF NEUROPEPTIDES ON INTESTINAL ION TRANSPORT

P. K. Rangachari

TABLE OF CONTENTS

I.   Introduction 430

II.   Overview of Intestinal Ion Transport 430

III.   Questions Considered Individually 433
       A.   Which Neuropeptides Should Be Considered? 433
       B.   Where Are They Found (Location)? 433
       C.   What Do Neuropeptides Do to Mucosal Transport Processes
             (Effect) and How Do They Do It (Mechanism)? 434
       D.   Are the Effects of Physiological/Pathophysiological
             Significance? 436

IV.   Neuropeptide Y And Peptide YY: Effects on Mucosal Transport
       Process 439

V.   Coda 442

Acknowledgments 442

References 443

I. INTRODUCTION

This chapter has been written from the perspective of a transport physiologist-*cum*-pharmacologist whose major interest in neuropeptides lies in their effects on intestinal ion transport. In preparing this chapter, this author is acutely aware of several excellent reviews that have recently appeared in a variety of books and journals dealing with intestinal ion transport as well as the nervous control of the gut.[1-14] Neuropeptides have received significant attention in several of the above-mentioned reviews, and no attempt will be made to duplicate material; however, some overlap is inevitable. References are made to these reviews, where appropriate, regarding details and critical comments for issues barely addressed. In addition, a number of chapters in the current volume have direct relevance. The organization of this chapter is as follows: It begins with a brief overview of gastrointestinal mucosal transport to provide a framework for the discussion that follows. Then a series of questions are listed which are believed to be relevant to a discussion of neuropeptide effects in the gut. Each of these is elaborated on. Then, neuropeptide Y and peptide YY are discussed in particular. Finally the chapter is concluded with a personal comment.

II. OVERVIEW OF INTESTINAL ION TRANSPORT

Symmetrical cells, such as smooth muscle cells, transport ions into and out of the same cell, but are not organized to produce net transport of ions across the cell. Epithelial cells are specially designed for such purposes by nature of their asymmetrical membranes and their geometric organization into sheets. Thus, the transport characteristics of the apical and basolateral membranes differ, but in order to understand epithelial function it is necessary to consider the properties of the tight junctions as well. In this context, Diamond's simple but picturesque model of the epithelial sheet as a six-pack has considerable heuristic value.[15] (See Figure 1)

Transport of water across epithelial sheets is, in general, secondary to the transport of ions. Across the intestinal epithelium, bidirectional transport of ions and water occurs, i.e., from blood to lumen and lumen to blood. Code[16] argued that a specific terminology must be developed to describe such processes, a terminology that "defines the direction of movement without connoting the mechanism involved". He proposed that the body as represented by the blood compartment be chosen as the term of reference. Movement into the blood was termed "insorption" and that out of the blood as "exsorption". Thus, where insorption exceeds absorption, net absorption could occur, whereas the converse could be termed "enterosorption". Such terminology could describe direction of movement rather than define a mechanism. Unfortunately, this logical approach has never found favor, and conventionally the terms absorption and secretion are used, the latter referring to what Code would term enterosorption. The term secretion has an implicit mechanistic connotation, a pitfall that Code wished to avoid. Bridges and Rummel[6] have noted that "it is useful to think of the function of the mucosa as varying along a continuum from a maximal absorption to a maximal secretory state." Clearly, then, the apparent effect of any modulator will depend on the position of the epithelium along the continuum. Thus, the final amount of water and electrolytes excreted is the result of all absorptive and secretory processes occurring along the entire length of the gut.

In general, the intestine receives 9 l of fluid per day, 1.5 to 2 l from ingested food and liquids and the rest from endogenous secretions (salivary, gastric, pancreatic, biliary, and intestinal). The small bowel absorbs 84% of the fluid, with the colon absorbing 15%, leading to a net fecal output of 150 to 200 ml. This pattern closely resembles the organization of the renal tubule, where the proximal segments reabsorb the major fraction, leaving the final modification to the distal segments.[6-10]

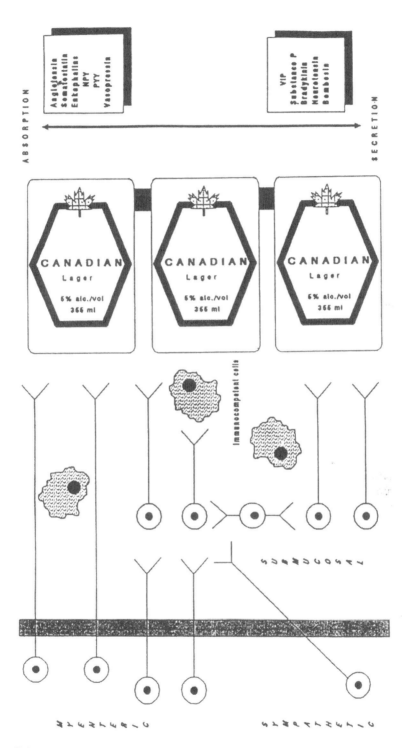

FIGURE 1. A schematic diagram of the intestinal epithelium and its innervation. The epithelium is modeled as a six-pack of beer cans (see Reference 15). The cans represent the columnar epithelial cells and the rim of plastic the tight junctions. To emphasize the author's institutional affiliations, Canadian content has been introduced. The projections of the myenteric and submucosal neurons are shown. The neurotransmitters shown are those defined for the guinea pig small intestine (see References 3, 4, 8, and 13,). Species variations do exist (see References 18 and 19). Absorption and secretion form a continuum, and peptides that promote one or the other are shown (see Reference 6 for details).

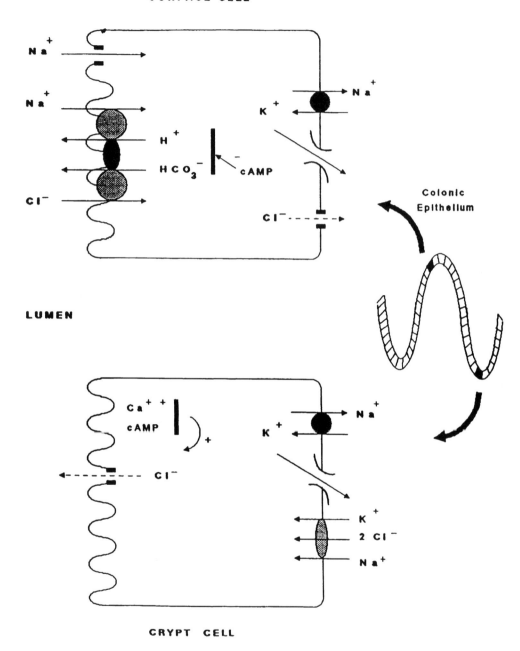

SURFACE CELL

LUMEN

CRYPT CELL

FIGURE 2.   The crypt-villus model of ion transport. The specific transporting elements are shown (refer to text and References 2, 6, 7, and 11 for details).

A widely accepted model of small intestinal transport links absorptive processes to the villus cell and secretory processes to the crypt cells.[2,6-8,10,11] Other models have been proposed. In both villus and crypt cells (see Figure 2), the driving forces are provided by a serosally located Na$^+$/K$^+$ ATPase. In the villus cell, this created an inward directed Na$^+$ gradient, allowing Na$^+$ to enter by a variety of carrier-mediated processes (Na$^+$ glucose symport, Na$^+$ amino acid symport, Na$^+$/H$^+$ antiport etc). The Na$^+$/H$^+$ antiport serves to

TABLE 1
Framework for the Analysis of Peptide Effects

Which neuropeptide is being considered?
Where is the peptide found in the gut(location)?
What does the peptide do (effect) and how does it do it(mechanism)?
  Effect
    Does the peptide alter absorption?
    Does it alter secretion?
  Mechanism
    Does it act directly on the transporting epithelial cell?
    Which transport process does it affect?
    Does it act indirectly:
    By altering blood flow?
    By releasing other neurotransmitters?
    By releasing mediators from other cells?
    By modulating the actions of other transmitters?
Are the effects of physiological/pathophysiological significance?

extrude H^+ produced within the cell, allowing Na^+ to be absorbed. The operation of this antiport is closely linked to the operation of a Cl^-/HCO_3^- symport, the result being net NaCl absorption. That these transporters are distinct is suggested by genetic defects. Cl^- secretion by the secretory cell is also driven by the serosally located Na^+/K^+ ATPase which, in conjunction with an $Na^+/K^+/2\ Cl^-$ symport, allows Cl^- to enter the cell.[17] The Cl^- accumulates above its electrochemical equilibrium and leaves the apical membrane by Cl^- channels. Intracellular mediators such as cAMP and Ca^{2+} modulate the transport processes. In addition, transport processes for K^+ exist, as shown by Figure 2.

Neuropeptides can modulate absorptive and secretory processes by modulating the transport processes mentioned above.[6,11] Given the large number of neuropeptides present in the gut and the complexity of these effects, it is appropriate to develop a framework for discussing their effect. A series of questions that can provide such a framework is shown in Table 1. Each of these questions is discussed below in greater detail.

III. QUESTIONS CONSIDERED INDIVIDUALLY

A. WHICH NEUROPEPTIDES SHOULD BE CONSIDERED?

In recent years, a large number of neuropeptides have been identified in various parts of the gastrointestinal tract. Regional and species variations exist. Among the peptides that can be tested as transmitter candidates are calcitonin gene-related peptide (CGRP), cholecystokinin (CCK), dynorphin (DYN), enkephalin (ENK), galanin (GAL), gastric-releasing peptide (GRP), neurokinin A (NKA), neuropeptide Y (NPY), neurotensin, peptide histidine isoleucine (PHI), somatostatin (SOM), substance P (SP), and vasoactive intestinal peptide (VIP). In addition, peptides such as angiotensin and vasopressin have effects on mucosal transport as well. Such lists are partial at best, since novel peptides are being identified.

B. WHERE ARE THEY FOUND (LOCATION)?

That regional and species variations exist in the distribution of neuropeptides has been mentioned already. Elegant mapping studies[5,9,18,19] have described in detail the location of different neuropeptides in a variety of species. The gastrointestinal tract contains nerve fibers from both extrinsic and intrinsic nerves, with neuropeptides being present in fibers from both types. Those neurons that are intrinsic to the gut will have their cell bodies located in either submucosal or myenteric ganglia, whereas extrinsic ganglia will be the location for cell bodies of extrinsic fibers.

With particular reference to control of mucosal transport processes, it is relevant to question whether a given peptide is present in nerves in submucosal ganglia, since these project to the epithelium, blood vessels, endocrine cells, muscularis mucosa, and other submucosal, and myenteric ganglia.[4,13,14] Myenteric neurons, on the other hand, project to external muscle layers and are primarily concerned with altering motility, although interconnections do occur and processes are sent to the submucosal ganglia.

A number of neuropeptides have been identified in submucosal neurons, including VIP, PHI, dynorphin, galanin, substance P, CCK-octapeptide, somatostatin, NPY, ACTH/β-endorphin, etc. Myenteric neurons contain substance P, gastric-releasing peptide/bombesin, and enkephalins. Neuropeptides are also present in other cell types, variously labeled as enteroendocrine or paracrine cells.[8] Such cells are scattered throughout the intestinal crypts and villus as well as in the lamina propria. The neuropeptides released from such cells can act locally in a paracrine manner or diffuse through the capillaries and act in an endocrine fashion. Neuropeptides present in such cells include secretin, CCK-33, somatostatin, neurotensin, substance P, motilin, PYY, and glucagon/enteroglucagon. These localizations are not unique, and neuropeptides such as VIP, substance P, and somatostatin are present both in neurons and enteroendocrine cells. Interestingly, in intrinsic neurons, CCK is present predominantly in the form of the octapeptides (CCK-8), whereas the whole molecule (CCK-33) is the primary form in endocrine cells.[2,8]

The coexistence of neurotransmitters is intriguing, and the possible physiological/pathophysiological significance often is a matter for speculation. Simultaneous release of neurotransmitters from sections of neurons[8] would provide for important interactions.

C. WHAT DO NEUROPEPTIDES DO TO MUCOSAL TRANSPORT PROCESSES (EFFECT) AND HOW DO THEY DO IT (MECHANISM)?

Neuropeptides can alter mucosal transport processing by altering absorption and/or secretion. These effects can, in turn, be produced by either direct effects on the transporting epithelial cells or indirect effects mediated by release of neurotransmitters, mediators, alterations in blood flow, etc. The transport processes affected would include any of those mentioned earlier.

The effects of neuropeptides on ion transport have been studied using both *in vivo* and *in vitro* techniques. Each technique used has obvious advantages and disadvantages, and the information obtained from them should be regarded as complementary.

The perfusion of intestinal segments *in situ* is usually performed on anesthetized animals.[20] The segment to be studied is isolated and cannulated for intraluminal perfusion. The segments are returned to the abdominal cavity and perfused with appropriate solutions.[20] The perfusates are collected and samples are analyzed for electrolytes. The perfusing solution also contains a nonabsorbable dilution marker of water absorption, usually [14C]-polyethylene glycol. The outputs from intestinal segments can be calculated relatively easily using standard formulas. These procedures, used with care, mimic the natural environment and as such yield valuable information However, hemodynamic parameters are difficult to control, and the effects of anesthesia have to be monitored. An alternative procedure employs tied-off loops of bowel, the net movements of fluids being estimated gravimetrically.

An isolated, perfused, *ex vivo* preparation from the rabbit ileum has been used to study the effects of serotonin and substance P. This preparation permits stricter control of total intestinal blood flow, keeping intact the enteric neural plexus, but dissociating the preparation from central controls.[21]

In vitro procedures, although artificial, do permit better definition of mechanisms. The procedure most commonly used to study the effects of neuropeptides on intestinal ion transport is the Ussing chamber technique[2-4,7,8] Using this procedure, preparations from varying regions of the gut can be mounted between two halves of a Lucite® chamber. This permits the maintenance of the normal vectorial properties of the epithelial sheet. The responsiveness of the epithelial cells to added agonists is monitored by changes in electrical parameters, such as transmucosal potential differences (P. D.), transmucosal resistances, and more often the short circuit current (variously abbreviated as SCC or Isc). When tissues are bathed by solutions of identical composition, the current passed to clamp the spontaneously developed P. D. to zero (short-circuit current) is a measure of active ion transport processes. In general, for many intestinal epithelia, the Isc is given by the relation

$$Isc = J_{net}^{Na} - J_{net}^{Cl} + J_{net}^{Res} \tag{1}$$

where J_{net}^{Na} is the net flux of Na^+, J_{net}^{Cl} is the net of Cl^-, and J_{net}^{Res} is the residual net flux of other ions such as HCO_3^-. Equation 1 can be further expanded to.

$$Isc = (J_{ms}^{Na} - J_{sm}^{Na}) - (J_{ms}^{Cl} - J_{sm}^{Cl}) + J_{net}^{Res} \tag{2}$$

where J_{ms} refers to the flux of an ion from the mucosal to the serosal solution and J_{sm} refers to fluxes in the opposite direction. It is imperative that in each preparation the ions contributing to the Isc be determined.

The great advantage of the Ussing chamber technique is the ease with which the composition of bathing solution can be changed and the facility for adding agonists and antagonists to either serosal or mucosal solutions. Ion fluxes can be monitored, and the preservation of vectorial properties is a major advantage. Edge damage is a problem, however, and should be considered. Since tissues are isolated, transport processes are relatively unaffected by changes in blood flow, but neural elements clearly persist. Certain modifications to the conventional technique have permitted further exploration of intestinal function. Hubel[22] developed a technique to study the effects of intrinsic nerves on transport parameters using electrical field stimulation, a technique that had been used by smooth muscle physiologists since the 1950s. This procedure has been used to study the role of enteric nerves in a number of species, including mice, guinea pigs, rabbits, humans, ground squirrels, dogs, etc.[8,23-36] Electrical stimulation leads to release of neurotransmitters which in turn lead to alterations in Isc. Several different modifications of the above technique have been developed (for details see Reference 8). This procedure can be used in conjunction with desensitization to identify putative neurotransmitters. The logic underlying the desensitization procedure is as follows: the isolated preparations are exposed repeatedly to the peptide in question until desensitization occurs, i.e., no further responses can be obtained. If a response is then obtained following electrical field stimulation, it is unlikely that the peptide used was a neurotransmitter. This procedure is particularly useful where specific selective antagonists are not available.[23].

Another modification has been the development of aganglionated preparations. Several different preparations from the gut can be set up in Ussing chambers. Whole thickness preparations contain both myenteric and submucosal ganglia. Removal of circular and longitudinal muscles leads to a preparation that contains the epithelial layer and the muscularis mucosa with attendant submucous plexuses. This is the preparation that has been used most often for *in vitro* studies. In such preparations, the effects of added agonists could be due to direct effects on transporting cells as well as indirect effects due to release of neurotransmitters through modulation of enteric mucosal activity. To obtain aganglionated preparations, the dissection procedure is extended to remove the muscularis mucosae and attendant submucosal plexus as well. Such procedures have been used to obtain aganglionated preparations from guinea pigs,[28] rats,[29,30] and dogs.[31-33] These preparations have been variously termed aganglionated,[28] mucosal,[29] or epithelial.[31] The use of such preparations in conjunction with more conventional mucosal or mucosal-submucosal preparations permits analysis of neuronally mediated effects. Other added tools in the pharmacologist's armamentarium include neurotoxins such as TTX, scorpion venom, veratridine, etc.[2,8,22] None of these approaches are free of pitfalls, and interpretations must be made with caution. Furthermore, as Brown et al.[11] note, use of denervated procedures "does not rule out the possibility that a given transmitter acts secondary to the release of other bioactive substances from endocrine or paracrine-type cells in the gut epithelium-subepithelium."

This reductionist approach can be carried further with the use of isolated cells, isolated membranes, etc. Such approaches clearly provide the great advantages of better definition at the molecular level, but need to be tempered with approaches at more systemic levels to place the information in appropriate context (see Reference 11 for further comments).

Exploration of indirect effects of neuropeptides could include analysis of putative neurotransmitters. Since the indirect effects of neuropeptides could involve the release of an intermediary substance, it is important to define the substance(s) responsible. For this, we can apply the procedures customarily used to define the involvement of a neurotransmitter. Werman[37] has explained that there are essentially only two criteria — identity of action and collectibility. The Hubel procedure of desensitization can be profitably employed in this context. Identification of the mediators involved can thus be achieved by a combination of physiological and pharmacological techniques.

Stimulation of enteric nerves *in vitro* leads to increases in Isc in several mammalian species.[28,29,31,34,36] Addition of TTX leads to a damping of oscillations observed in several mammalian preparations.[31,36] Such preparations appear to be under tonic neural inhibition, since denervation by stripping or addition of TTX leads to an increase in absorption. Thus, based on the criterion of identity of action, neuropeptides that do not have either anti-absorptive or secretory effects are unlikely to be the physiologically relevant neurotransmitters at neuroepithelial junctions. This excludes somatostatin, dynorphin, or NPY, but others such as VIP or substance P are likely candidates. Further identification can use a combination of electrical field stimulation and desensitization, as mentioned earlier.[8] In addition to neural elements, the lamina propria of the gut contains a variety of immuno-competent cells.[38] Neuropeptide effects induced by interaction with receptors in such cells could in turn modulate transport functions of the gut. Recent studies have shown the existence of receptors for substance P, somatostatin, and VIP on lymphocyte populations in the gut.[38-40] Furthermore, substance P is known to release histamine from mucosal mast cells[39] and induce the release of interleukin 1, tumor necrosis factor, and interleukin 6 from human blood monocytes.[41] Since these have potent biological actions, the possibility for complex interactions exists.

Neuropeptides could affect absorption and secretion by alterations in mucosal blood flow. Powell[2] has noted that it is not easy to investigate the relationship between blood flow and transport. Most agents that stimulate absorption and secretion also increase total gut blood flow. The neuropeptides that do so include VIP, CCK, pancreatic polypeptides, and gastric inhibiting peptide. (See Reference 2, Table 4, and Chapter 4.)

D. ARE THE EFFECTS OF PHYSIOLOGICAL/PATHOPHYSIOLOGICAL SIGNIFICANCE?

It has been mentioned earlier that the transport functions of the gut can be thought of as a continuum ranging from maximal secretion to maximal absorption. Peptides can promote secretion as well as absorption.[6] (See Figure 2.)

Although it is relatively easy to demonstrate that a given neuropeptide produces an effect on intestinal transport and even to dissect out the underlying mechanism, the real problem is to place the information obtained within a physiological or pathophysiological context. Given the variety of effects that neuropeptides can exert at different levels, a synthesis is often speculative at best, requiring a willing suspension of disbelief.

In a recent study, Hubel et al.[42] studied the effects of norepinephrine on the responses of the rabbit ileum to electrical field stimulation and secretagogues. Since norepinephrine, along with somatostatin and neuropeptide Y, is one of the major neurotransmitters enhancing absorption of ions and water, the studies had obvious relevance. Their discussion of the physiological relevance of this observation is worth quoting directly:

> In the absence of knowledge of the concentrations of norepinephrine at the adrenergic receptors of the neurons or enterocytes, the relevance of *in vitro* studies must remain open to question. The *in vitro* studies discussed here tell us only the possible mechanisms by which norepinephrine might increase absorption by the enterocytes but they provide little measure of the probability of the physiologic importance of such mechanisms. It is reasonable to assume that the physiological relevance of a response to a neurotransmitter depends on the concentration required to cause a response, i.e., the lower the required concentration, the greater the probability of physiological evidence. By that criterion, the relatively low EC_{50} of 13 nM for the effect of norepinephrine on basal Isc suggests a physiological role.[42]

Another major problem in defining the physiological significance of neuropeptides stems from the coexistence of neurotransmitters within the same neuron. Such coexistence has been demonstrated both in central and peripheral nervous systems.[43] Peptides have been shown to colocalize not only with classical neurotransmitters, such as acetylcholine, norepinephrine, or dopamine, but also with other peptides. Thus, VIP and NPY coexist in enteric neurons. NPY coexists with a variety of transmitters in different neuronal populations. In cat and mouse gut, NPY is found with VIP, in guinea pig submucous ganglia, with acetylcholine, and in extrinsic sympathetic fibers, with norepinephrine.[2,8,13,43]

Coexistence raises the possibility that neural stimulation could lead to corelease of the transmitters. The pattern of transmitter release may, however, be dependent on the frequency of stimulation. In the parasympathetic vasodilator innervation of the cat salivary glands there is evidence that low-frequency stimulation leads to release of acetylcholine, whereas stimulation with higher frequencies leads to VIP release.[43-47]

However, if stimulation leads to release of both transmitters, potential for interactions occur. Campbell[43] argues cogently that cotransmission involves the clear demonstration "that two or more coreleased transmitters act on the same target cell, so that the net result of transmission incorporates interactive effects of the transmitters". His analysis thus holds both coexistence and corelease to be necessary but insufficient conditions for demonstrating cotransmission. Such interactions could be synergistic (additive) or antagonistic. Corelease may not necessarily imply release from the same site. In the feline jejunum, 5-HT and substance P were released into the lumen following vagal stimulation, but pharmacological antagonists were used to demonstrate that the release of the two transmitters was independent and from two different sources (the EC cell for 5-HT and peptidergic neurons for substance P).[48]

In both the central and peripheral nervous systems, studies have shown that coexistence occurs. Potential interactions between neurotransmitters have been defined largely by considering the effects of two or more different neurotransmitters on the target cell. Cooke et al.[49] have studied the interaction of VIP and acetylcholine in the guinea pig ileum, noting that prior treatment with VIP enhanced secretory responses to cholinergic stimulation. In the guinea pig distal colon, NPY acts as a potent inhibitor of acetylcholine-mediated increases in short circuit current as well as reponses to exogenous VIP.[50]

The term neuromodulation[51,52] has been applied to effects that are different from synergistic or antagonistic effects where the modulator itself has no direct effect on the target cell or an effect that is independent of the modulation. This raises numerous possibilities, and these have been reviewed recently by Kow and Pfaff.[52] Peptides can exert both neurotransmitter and neuromodulatory actions, and it may be possible to distinguish these by different means such as time course, lack of causal relationship, or different order of sensitivity.

Neuromodulation can be demonstrated at sites where coexistence does not occur. Thus, in the central nervous system, the neurons of the ventral tegmental area contain both CCK and dopamine and project to the caudal, medial nucleus accumbens, amygdala, and olfactory tubercle. In these regions, CCK appears to attenuate K^+-evoked dopamine release, to increases the number of binding sites, potentiates inhibitory actions of dopamine and apomorphine, and attenuates adenylate cyclase activation by dopamine. Such interactions are also seen in sites where the two neurotransmitters do not coexist.[53-57]

This welter of complexity may be exciting to the investigator, since numerous fruitful areas for investigation exist. However, it is clearly difficult to assign physiological roles for given neuropeptides, a problem that is clearly enhanced where pathophysiological mechanisms are concerned.[58-71]

Increases in the levels of secretory peptides would be expected to produce diarrhea. In watery diarrhrea syndrome associated with non-B-islet cell tumors, VIP is a prime candidate.[2,7,58] Other peptides that could be included are calcitonin, GIP, secretin glucagon, and enteroglucagon.[58] Although diarrhea would not be anticipated with somatostatinomas since this peptide promotes absorption, diarrhea has been reported.[59,64-69] It is possible in this case that other peptides, such as calcitonin, could be responsible. The involvement of different peptides in tumors of endocrine cells has been well documented and reviewed. The reported incidence of gastroenteropancreatic (GEP) tumors is around 1.5 cases for 100,000 of the general population.[59] Carcinoids account for 55%, insulinomas 17%, tumors of unknown types 15%, gastrimonas 9%, and VIPomas 2%.[70] It is believed that these tumors arise from a primitive stem cell.

Neuropeptides have also been implicated in inflammatory bowel diseases. Bishop et al.[60] reported an increase in VIP nerves in the gut wall and an increase in total content of VIP in Crohn's disease, but not in ulcerative colitis. In left-sided UC, substance concentration of mucosal-submucosal layers was increased. However, the concentration abnormalities could not clearly differentiate between the two conditions. More recently, Mantyth et al.[62,63] used quantitative receptor autoradiography to examine abnormalitites in tachykinin binding sites in colonic samples from patients with ulcerative colitis and Crohn's disease using[125]I-labeled Bolton-Hunter conjugates of NK, SK, and substance P. In colonic tissue from patients with inflammatory bowel disease (IBD), high concentrations of substance P receptor binding sites were found in arterioles and venules in the submucosa as well as in the muscularis mucosa, and external circular and longitudinal muscles. More interestingly, very high concentrations were found within the germinal center of lymph nodules that bordered the muscularis mucosa. It would be interesting to determine whether increased receptors were found on mucosal mast cells. These observations suggest the possible involvement of different neuropeptides in inflammatory bowel disease.

For the purpose of pharmacological intervention, the preabsorptive or antisecretory peptides are of greater interest. Particular attention has been focused on somatostatin and the enkephalins.[64-80] Somatostatin has inconsistent effects on basal transport, but does appear to inhibit induced secretion.[64-67] Both somatostatin and a long-acting analogue, SMS-201-995, have been used clinically to treat patients with watery diarrhea and carcinoid syndrome. Variable results have been reported. Krol et al.[70] noted improvements in 76% of their patients, whereas Vinik and Moattari[59] noted long-term improvements only in 44%. However, acute administration of the somatostatin analogue changed the overall pattern from a secretory to an absorptive state. Cook et al.[71] noted dramatic therapeutic effects of SMS-201-995 in an acquired immunodeficiency syndrome (AIDS) patient who had secretory diarrhea resulting from cryptosporidial infection. Fedorak and Allen[72] have noted that the analogue had marked effects on rat intestines *in vitro*, but systemic administration for five consecutive days porduced a paradoxical decrease in basal colonic fluid absorption. Thus, the *in vitro* effects did not translate into similar effects *in vivo*. Another analogue, BIM 23014, has been studied for its effects on the secretion of growth hormone, thyrotropin, and gastropancreatic hormones in normals.

Opiate agonists such as morphine or enkephalins alter both gut motility and mucosal transport.[73-80] Different receptor subtypes may be involved, μ-receptors mediating effects on motility and δ-receptors mediating transport effects. Recently, the effects of a novel enkephalin-like pentapeptide, BW 942C, have been studied on the rabbit ileum.[73] The opiate has also been used in a clinical trial against traveler's diarrhea. The agonist has effects on both serosal and luminal addition and may produce its effects by multiple mechanisms. The receptor subtype involved was not clear, since the agonist is more selective against μ-receptors and larger doses of naloxone were needed to antagonize the effects. The authors note that although the opiates have significant effects on electrolyte transport in the intestine, the major antidiarrheal mechanism of action of currently available opiates may be less on transport than on alterations in transit time. More knowledge is needed on the specific opiate receptor subtype mediating transport effects. It must be noted that the effects of opiates are complex, and they could exert their effects on intestinal ion transport through a central mechanism, since intracerebroventricular administration of enkephalins inhibits responses to cholera toxin as well as prostaglandins.[79,80]

IV. NEUROPEPTIDE Y AND PEPTIDE YY: EFFECTS ON MUCOSAL TRANSPORT PROCESS

This section of the chapter will focus on the effects of neuropeptide Y (NPY) and peptide YY on intestinal ion transport. These two peptides will be used as worked examples in an attempt to discuss them within the framework proposed.[81-96]

The choice of NPY and PYY as examples was guided by the following considerations. Both peptides are fairly recent entrants, having been discovered only in the 1980s. Both are present in the gut in different locations. A significant proportion of nerve fibers in submucous ganglia of different species contain NPY, suggesting a role for this peptide in modulation intestinal ion transport. PYY, on the other hand, is localized predominantly to endocrine cells in distal and small intestine.[8,81-84]

The effects of NPY/PYY on intestinal ion transport have been studied both *in vitro* and *in vivo*. Saria and Beubler[85] studied the effects of close intra-arterial infusion of NPY on PGE_2-induced secretion in tied-off loops of rat jejunum *in vivo*. Net fluid transport rates were determined gravimetrically. PGE_2 converted net absorption to secretion in a dose-dependent fashion. NPY infused intra-arterially had no significant intra-absorptive effects of its own, but markedly inhibited secretion of fluid produced by PGE_2. PYY had similar effects, but appeared to be less potent. MacFadyen et al.[86] found that NPY inhibited the net secreting effects of simultaneously infused VIP in proximal diverticulum and jejunum in an *in vivo* preparation.

Recently, Buell and Harding[87] reported that PYY infusion led to a pronounced redistribution of blood flow away from the muscularis toward the mucosal-submucosal compartment. As noted earlier, increases in total gut blood flow have been noted with agents that either stimulate absorption or promote secretions.[2]

In vitro studies have been used to explore in greater detail the underlying mechanisms. Rabbit, guinea pig, rat, and pig intestines have been studied. In all four species, NPY/PYY caused decreases in short circuit current. In rabbit ileum, NPY produced dose-dependent decreases in Isc, with EC_{50}s between 3 to 5×10^{-8} M. PYY appeared to be more potent, with an EC_{50} of 2×10^{-9} M. Hubel and Renquist[92] noted that TTX did not inhibit effects of NPY, and Friel et al.[90] showed that yohimbine had no significant inhibitory effects either. The logic underlying the use of the latter antagonist was the observation that α_2-adrenoceptor effects on rabbit intestines were similar to those seen with NPY. Densensitization of responses were noted with NPY.[92] However, interestingly, a desensitized tissue still responded to transmural field stimulation. These studies suggested that the effects of NPY were largely direct in keeping with the observed topographical location of NPY-immunoreactive nerves close to the epithelium. On the porcine jejunum,[91] effects of NPY were inhibited by both TTX and the opiate antagonist naloxone, suggesting the possible involvement of opioid interneurons.

To determine the ionic bases for the alterations in short circuit currents, ion flux studies were done under conventional short-circuiting conditions. Friel et al.[90] found that decreases in Isc were accompanied by significant increases in the Na^+ and Cl^- fluxes. This resulted from significant increases in mucosal to serosal fluxes with a reduction in the serosal to mucosal fluxes of both ions. Thus, the peptides appeared to increase absorption and reduce secretion. No significant increases were noted in residual ion fluxes. Hubel and Renquist's[92] studies showed a major increase in net Cl^-, once again resulting from a combination of an increase in J_{ms}^{Cl} and a decrease in J_{sm}^{Cl}. However, there was a decrease in the net Na^+ flux. The consistent effects on Cl^- movements suggested that a major effect of the two peptides was on anion transport. In the porcine jejunum, large increases in J_{net}^{Cl} and J_{net}^{Res} were noted.[91] McCulloch et al.[50] showed that NPY and galanin reduced the responses of the guinea pig distal colon to neural stimulation. NPY, but not galanin, reduced the secretory responses to VIP and bethanechol, whereas neither somatostatin 201-995 nor human CGRP had any effect on either basal Isc or neurally evoked responses. That NPY reduced the effects of both VIP and cholinergic agonist is interesting, since VIP is generally believed to mediate its effects on adenylate cyclase whereas cholinergic stimulation occurs through a Ca^{2+} pathway.

More recently, Cox et al.[93] have explored the mechanisms underlying the reduction in Isc produced by NPY, PYY, and human pancreatic polypeptide (PYY) on rat jejunum *in vitro*. They also noted dose-dependent decreases in Isc produced by both peptides, with PYY being more potent than NPY. As with the rabbit, neither TTX nor α_2-adrenoceptor blockade altered responses. Pretreatment with a Cl^- channel blocker (diphenylamine 2-carboxylate) and an inhibitor of $Na^+/K^+/2Cl^-$ cotransport (piretanide) reduced responses, again arguing for an effect on anion secretion. Flux studies showed that NPY had no significant effect on Na^+ fluxes, but significantly increased forward flux of Cl (J_{ms}^{Cl}) while decreasing Cl secretion into the lumen (J_{sm}^{Cl}). Interestingly, they had to resort to a more sensitive protocol of more frequent and rapid sampling to measure ion fluxes (particularly Na^+), since large basal fluxes made analysis by conventional procedures difficult. They found that pretreatment with cyclo-oxygenase inhibitors (piroxicam, indomethacin) significantly lowered basal Isc and reduced responses to NPY. However, piroxicam-treated tissues responded to added forskolin with increases in Isc, and NPY reduced the stimulation produced. Thus, the peptide was not a specific inhibitor of prostaglandin-mediated secretion, but appeared to inhibit secretions stimulated by altered cAMP levels.

Biochemical studies have shown the existence of specific binding sites for PYY on rat jejunal epithelium.[94-96] Plasma-membrane-enriched fractions were obtained from epithelial cells from jejunum as well as duodenum, ileum, colon, cecum, and stomach. The binding of [^{125}I]PYY was studied, and the characteristics of binding were explored. The K_d was estimated to be 434 ± 56 pM, and concentration of binding sites was estimated to be 336 ± 41 fmol/mg protein. An important element was the specificity of binding. Binding of the ligand was unaffected by a variety of peptides that interact with jejunal epithelia, including VIP, insulin, glucagon, CCK-octapeptide, PHI, secretin, etc. Pancreatic polypeptide (PP), which shows some similarity to PYY, was effective only at high concentrations. However, NPY (both natural and synthetic) competed well for binding. NPY appeared to be some five times less potent than PYY, again in keeping with the observations from physiological experiments.[94] Interestingly, Gilbert et al.[95] have reported the existence of specific pancreatic

TABLE 2
Summary — NPY/PPY On Intestinal Ion Transport

Neuropeptides under consideration
  NPY/PYY
Location
  NPY — nerve fibers
  PYY — endocrine cells
Effect
  Antisecretory
Mechanism
    Direct (rabbit, guinea pig, rat)
    Indirect (porcine)
    Decrease in Isc
    Increases in net Cl flux (mainly the forward flux)
    Negative coupling to cAMP production
    Effects on Ca^{2+} pathway
Physiological/pathophysiological significance
    Needs definition.

polypeptide binding sites on the vascular surface of the canine small intestine. NPY and PYY inhibited binding, but only at high concentrations — the converse of the effects reported by Laburthe et al.[94] No appreciable binding of PYY could be detected in stomach, cecum, colon, and liver; gut binding was found uniformly in small intestine, with the highest density of binding sites being found in the jejunum and duodenum.

Later studies focused on the effects of NPY and PYY on VIP-stimulated cAMP production.[96] Both peptides reduced cAMP production stimulated not only by VIP, but also by PGE_1, PGE_2, and forskolin. Thus, it was unlikely that there was any inhibition at a receptor level, a point further emphasized by the lack of effect on VIP binding. Both peptides inhibited the efficacy of VIP without altering its potency. Binding to α_2-adrenoceptors was not involved, again in keeping with the functional studies noted earlier. The rank order of potency of peptides to reduce intestinal cAMP levels was similar to that seen with the binding to the PYY-preferring receptor in jejunal epithelia. More detailed structural studies were undertaken using fragments of PYY and NPY. Again, a correlation was found between the inhibition of cAMP production following VIP stimulation and the inhibition of PYY binding. The C-terminal portion of PYY was found to play a much more important role than the N-terminal portion. In a study on vascular smooth muscle cells, Reynolds and Yokola[97] found that NPY inhibited forskolin-stimulated adenylate cyclase activity, but had no effects on phosphoinositide hydrolysis (CPI) or elevation of cytosolic Ca^{2+}. These data suggested that the molecular mechanism underlying the effects of these peptides is their negative coupling to cAMP production system; however, the observations by McCulloch et al.[50] suggest a possible interaction with a Ca^{2+} pathway as well.

Motulsky and Michel[98] have shown recently that NPY alters both cAMP and Ca^{2+} levels in human erythroleukemic cells (HEL). Dose-dependent decreases in forskolin stimulated cAMP accumulation, and increases in intracellular Ca^{2+} were noted. EC_{50}s for both processes were similar, between 2 to 3 nM. Pertussis toxin inhibited both effects, and chelation of extracellular Ca^{2+} did not alter the effects on intracellular Ca^{2+}. These data suggested the involvement of one or more G proteins and perhaps the mobilization of intracellular Ca^{2+}. It is difficult to translate these effects to the situation observed with the enterocyte, since increases in intracellular Ca^{2+} would be expected to increase rather than decrease Isc. Nevertheless, it is interesting that NPY appears to be coupled to two second messenger systems. (See Table 2 for summary.)

The physiological or pathophysiological roles of NPY or PYY are matters of speculation. Given the effects described, one would expect them to have proabsorption or antisecretory effects and function as norepinephrine does. Complex changes in neurotransmitter levels have been reported in experimental streptozotocin-induced diabetes in rats. Schmidt et al.[99] noted that subpopulations of postganglionic sympathetic fibers may be affected differentially. The NPY- and somatostatin-containing nonadrenergic subpopulations of axons are spared relative to these noradrenergic fibers that do not colocalize with the peptides. Belai et al.[100] found different patterns of responses in neurotransmitter levels. Tissue levels of noradrenaline and VIP showed an initial increase followed by a decrease in density of nerve fibers; substance P showed a delayed increase with little evidence of degeneration. NPY-containing nerves were resistant to changes.

Studies have shown mucosal transport defects in streptozotocin-treated rats.[101] Frank secretion was noted which could be rectified by the α_2-adrenoceptor agonist clonidine. There was a denervation supersensitivity as well. How the lack of change in NPY-containing nerve fibers modifies this problem is not clear. Complex changes occur, and it is not easy to translate these observations into the observed clinical picture. Both intractable diarrhea and chronic constipation are seen in diabetes, and the experimental reports provide some explanation for the latter observations.

V. CODA

The last decade has seen a virtual explosion of knowledge concerning neuropeptides. Their role in modulating gastrointestinal function has been explored in a variety of systems. This brief and cursory review has focused largely on the effects of neuropeptides on intestinal ion transport, and readers have often been referred to more elegant and comprehensive reviews for details. The effects of neuropeptides are complex, and a framework is needed at least for heuristic purposes. It was the author's intent to provide one by listing a series of questions which have been used to review current information regarding NPY and PYY.

The more interesting questions are perhaps the most difficult to answer. Given the fact of coexistence, the precise role played by cotransmission in the physiology and pathophysiology of gut function remains to be determined. Interactions between neuropeptides and immunocompetent cells is a fertile field considering the central role played by such cells in inflammatory bowel diseases.[102] It is becoming increasingly evident that the enteric nervous system plays a key role in modulating the responses of the gut epithelium to immune processes.[103] Interactions between the diverse peptides at the final target cell, the enterocyte, need better definition; establishing the molecular mechanism of such interactions would afford better insight into the functioning of the enterocyte itself.

Perhaps the biggest barrier to progress may be a psychological one between the cellular and molecular physiologists with their spectacularly successful reductionist approaches and the systems physiologists who never let us forget that the cells exist in a breathing, living, growing organism. The information obtained on all fronts needs to be better integrated if we are to fully comprehend the role that neuropeptides play in gastrointestinal function. We have, as yet, barely scratched the surface and find ourselves in the position of Voltaire's philosophers: "qui font des systèmes sur la secrete construction de l'univers, sont comme nos voyageurs qui vont a Constantinople, et qui parlent du Serail: Ils n'en ont vu que les dehors, et ils pretendent savoir ce que fait le Sultan avec ses Favorites."[104]

ACKNOWLEDGMENTS

I thank Todd Prior and Betty Guy for their patience in deciphering my scrawl and translating it into WordPerfect. Drs. H. J. Cooke, D. Powell, K. Hubel, D. R. Brown, J. F. Kachur, and S. O'Grady are thanked for sending me reprints and offprints of their work. I trust that I have done justice to their work. I would also like to acknowledge the Medical Research Council (MRC) of Canada for their support.

443

REFERENCES

1. **Gaginella, T. S.**, Neuroregulation of intestinal ion transport, *Trends Pharmacol. Sci.*, September, 397, 1984.
2. **Powell, D. W.**, Ion and water transport in the intestine, in *Physiology of Membrane Disorders*. Andreoli, T. E., Hoffman, J. F., Fanestil, D. D., and Schultz, S. G., Eds., Plenum Press, New York, 1986, 559.
3. **Hubel, K. A.**, Neural control of intestinal ion transport, in *Gastrointestinal and Hepatic Secretions: Mechanism and Control*, Davison, J. S. and Schaffer, E. A., Eds., University of Calgary Press, Alberta, Canada, Calgary, 1988, 175.
4. **Cooke, H. J.**, Neurobiology of the intestinal mucosa, *Gastroenterology*, 90, 1057, 1986.
5. **Schultzberg, M.**, Innervation of the gut, in *Neurochemistry: Modern Methods and Applications*, Panula, P., Paivarinta, H., and Soinila, S., Eds., Alan R. Liss, New York, 1986, 477.
6. **Bridges, R. J. and Rummel, W.**, Mechanistic basis of alterations in mucosal water and electrolyte transport, *Clin. Gastroenterol.*, 15, 491, 1986.
7. **Binder, H. J. and Sandle, G. I.**, Electrolyte absorption and secretion in the mammalian colon, in *Physiology of the Gastrointestinal Tract*, 2nd ed., Johnson, L. R., Ed., Raven Press, New York, 1987, 1389.
8. **Cooke, H. J.**, Neural and humoral regulation of small intestinal electrolyte transport, in *Physiology of the Gastrointestinal Tract*, 2nd ed., Johnson, L. R., Ed., Raven Press, New York, 1987, 1307.
9. **Furness, J. B. and Costa, M.**, *The Enteric Nervous System*, Churchill Livingstone, Edinburgh, 1987.
10. **Ewe, K.**, Intestinal transport in constipation and diarrhea, *Pharmacology*, Suppl. 11, 73, 1988.
11. **Brown D. R., Chandan, R., Quito, F. L., and Seybold, V. S.**, Receptor regulation of ion transport in the intestinal epithelium, *Life Sci.*, 43, 2193, 1988.
12. **Gaginella, T. S. and Kachur, J. F.**, Kinins as mediators of intestinal secretion, *Am. J. Physiol.*, 256, G1, 1989.
13. **Cooke, H. J.**, Role of the "little brain" in the gut in water and electrolyte homeostatis, *FASEB J.*, 3, 127, 1989.
14. **Lundren, O., Svanvik, J., and Jivegard, L.**, Enteric nervous system. I. Physiology and pathophysiology of the intestinal tract, *Dig. Dis. Sci.*, 34, 264, 1989.
15. **Diamond, J. R.**, The epithelial junction: bridge, gate, and fence, *Physiologist*, 20, 10, 1977.
16. **Code, D. F.**, The semantics of the process of absorption, *Perspect. Biol. Med.*, 3, 560, 1960.
17. **O'Grady, S. M., Palfrey, H. C., and Field, M.**, Characteristics and functions of Na-K-Cl cotransport in epithelial tissues, *Am. J. Physiol.*, 253, C177, 1987.
18. **Daniel, E. E., Costa, M., Furness, J. B., and Keast, J. R.**, Peptide neurons in the canine small intestine, *J. Comp. Neurol.*, 237, 227, 1985.
19. **Ekblad, E., Ekman, R., Hakanson, R., and Sundler, F.**, Projections of peptide-containing neurons in rat colon, *Neuroscience*, 27, 655, 1988.
20. **McFadden, D., Zinner, M., and Jaffe, B. M.**, Substance P-induced intestinal secretion of water and electrolytes, *Gut*, 27, 267, 1986.
21. **Yeo, C. J., Couse, N. F., and Zinner, M. J.**, Serotonin and Substance P stimulate intestinal secretion in the isolated perfused ileum, *Surgery*, 105, 86, 1989.
22. **Hubel, K. A.**, The effects of electrical field stimulation and tetrodotoxin on ion transport by the isolated rabbit ileum, *J. Clin. Invest.*, 62, 1039, 1978.
23. **Hubel, K. A.**, Electrical stimulus-secretion coupling in rabbit ileal mucosa, *J. Pharmacol. Exp. Ther.*, 231, 577, 1984.
24. **Cooke, H. J., Nemeth, P. R., and Wood., J. D.**, Histamine action on guinea-pig ileal mucosa, *Am. J. Physiol.*, 246, G372, 1984.
25. **Kuwahara, A., Cooke, H. J., Carey, H. V., Mekhjian, H., Ellison, E. C., and McGregor, B.**, Effects of enteric neural stimulation on chloride transport in human left colon in vitro, *Dig. Dis. Sci.*, 34, 206, 1989.
26. **Zimmerman, T. W. and Binder, H. J.**, Effect of tetrodotoxin on cholinergic agonist-mediated colonic electrolyte transport, *Am. J. Physiol.*, 244, G386, 1983.
27. **McCulloch, C. R. and Cooke, H. J.**, Human alpha-calcitonin gene-related peptide influences colonic secretion by acting on myenteric neurons, *Regul. Peptides*, 24, 87, 1989.
28. **Carey, H. V., Cooke, H. J., and Zafirova, M.**, Mucosal responses evoked by stimulation of ganglion cell somas in the submucosal plexus of the guinea-pig ileum, *J. Physiol. (London)*, 364, 69, 1985.
29. **Andres, H., Bock, R., Bridges, R. J., Rummel, W., and Schreiner, J.**, Submucosal plexus and electrolyte transport across the rat colonic mucosa, *J. Physiol. (London)*. 364, 301, 1985.
30. **Diener, M., Bridges, R. J., Knobloch, S. F., and Rummel, W.**, Neuronally mediated and direct effects of prostaglandins on ion transport in rat colon descendens, *Naunyn-Schmiedeberg's Arch. Pharmacol.*, 337, 74, 1988.
31. **Rangachari, P. K. and McWade, D.**, Epithelial and mucosal preparations of canine proximal colon in Ussing chambers: comparison of reponses. *Life Sci.*, 38, 1641, 1986.

32. **Rangachari, P. K. and McWade, D.,** Histamine stimulation of canine colonic epithelium: potentiation by hydroxylamines, *Eur. J. Pharmacol.,* 135, 331, 1987.

33. **Keenan, C. M. and Rangachari, P. K.,** Eicosanoid interactions in the canine proximal colon, *Am. J. Physiol.,* 256, G673, 1989.

34. **Carey, H. V. and Cooke, H. J.,** Influence of enteric nerves on jejunal mucosal function of the piebald-lethal mouse, *Gastroenterology,* 86, 1040, 1984.

35. **Bridges, R. J., Rack, M., Rummel, W., and Schreiner, J.,** Mucosal plexus and electrolyte transport across the rat colonic mucosa, *J. Physiol.,* 376, 531, 1986.

36. **Sheldon, R. J., Malarchik, M. E., Fox, D. A., Burks, T. F., and Porreca, F.,** Pharmacological characterization of neural mechanisms regulating mucosal ion transport in mouse jejunum, *J. Pharmacol. Exp. Ther.,* 249, 572, 1989.

37. **Werman, R. A.,** Review — criteria for identification of a central nervous system transmitter, *Comp. Biochem. Physiol.,* 18, 745, 1966.

38. **Stead, R. H., Bienenstock, J., and Stanisz, A.,** Neuropeptide regulation of mucosal immunity, *Immunol. Rev.,* 100, 333, 1987.

39. **Mazurek, N., Pecht, I., Teichberg, V. I., and Blumberg, S.,** The role of the N-terminal tetra peptide in the histamine-releasing action of substance P, *Neuropharmacology,* 20, 1025, 1981.

40. **Payan, D. G., Brewster, D. R., and Goetzl, E. J.,** Specific stimulation of human T-lymphocytes by substance P, *J. Immunol.,* 131, 1613, 1983.

41. **Lotz, M., Vaughan, J. H., and Carson, D. A.,** Effect of neuropeptides on production of inflammatory cytokinins by human monocytes, *Science,* 241, 1218, 1988.

42. **Hubel, K. A., Renquist, K. S., and Varley, G.,** Noradrenergic influence on epithelial responses of rabbit ileum to secretagogues, *Am. J. Physiol.,* 256, G919, 1989.

43. **Campbell, G.,** Cotransmission, *Annu. Rev. Pharmacol. Toxicol.,* 27, 51, 1987.

44. **Hokfelt, T., Lundberg, J. M., Schultzberg, M., Johansson, O., Ljungdahl, A., and Rehfeld, J.,** Coexistence of peptides and putative transmitters in neurons, *Adv. Biochem. Psychopharmacol.,* 22, 1, 1980.

45. **Lundberg, J. M., Anggard, A., Fahrenkrug, J., Hokfelt, T., and Mutt, V.,** Vasoactive intestinal polypeptide in cholinergic neurons of exocrine glands: functional significance of co-existing transmitters for vasodilation and secretion, *Proc. Natl. Acad. Sci. (U.S.A.),* 77, 1651, 1980.

46. **Lundberg, J. M.,** Evidence for coexistence of vasoactive intestinal polypeptide (VIP) and acetylcholine in neurons of cat exocrine glands. Morphological, biochemical and functional studies, *Acta Physiol. Scand. Suppl.,* 496, 1, 1981.

47. **Lundberg J. M., Anggard, A., Fahrenkrug, J., Lundgren, G., and Holmstedt, B.,** Corelease of VIP and acetylcholine in relation to blood flow and salivary secretion in cat submandibular salivary gland, *Acta Physiol. Scand.* 115, 525, 1982.

48. **Gronstad, K., Dahlstrom, A., Florence, L., Zinner, M. J., Ahlman, J., and Jaffe, B. M.,** Regulatory mechanisms in endoluminal release of serotonin and substance P from feline jejunum, *Dig. Dis. Sci.,* 32, 393, 1987.

49. **Cooke, H. J., Zafirova, M., Carey, H. V., Walsh, J. H., and Grider, J.,** Vasoactive intestinal polypeptide actions on the guinea-pig intestinal mucosa during neural stimulation, *Gastroenterology,* 92, 361, 1987.

50. **McCulloch, C. R., Kuwahara, A., Condon, C. D., and Cooke, H. J.,** Neuropeptide modification of chloride secretion in guinea-pig distal colon, *Regul. Peptides,* 19, 35, 1987.

51. **Kaczmarek, L. K. and Levitan, I. B.,** What is Neuromodulation?, in *Neuromodulation,* Kaczmarek, L. K. and Levitan, I. B., Eds., Oxford University Press, New York, 1987, 3.

52. **Kow, L. M. and Pfaff, D. W.,** Neuromodulatory actions of peptides, *Annu. Rev. Pharmacol. Toxicol.,* 28, 163, 1988.

53. **Murphy, R. B. and Schuster, D, I.,** Modulation of [^{3}H]-dopamine binding by cholecystokinin octapeptide (CCK 8), *Peptides,* 3, 539, 1982.

54. **Hommer, D.W., and Skirboll, L. R.,** Cholecystokinin-like peptides potentiate apomorphine-induced inhibition of dopamine neurons, *Eur. J. Pharmacol.,* 91, 151, 1983.

55. **Crawley, J. N., Hommer, D. W., and Skirboll, L. R.,** Behavioural and neurophysiological evidence for a facilitatory interaction between co-existing transmitters: cholecystokinin and dopamine, *Neurochemistry,* 6, 755, 1984.

56. **Studler, J. M., Reibaud, M., Herve, D., Blane, G., Glowinski, J., and Tassin J. P.,** Opposite effects of sulfated cholecystokinin on DA-sensitive adenylate cyclase in two areas of the rat nucleus accumbens, *Eur. J. Pharmacol.,* 126, 125, 1986.

57. **Voigt, M., Wang, R. Y., and Westfall, T. C.,** Cholecytokinin octapeptides alter the release of endogenous dopamine from the rat nucleus accumbens *in vitro, J. Pharmacol. Exp. Ther.,* 237, 147, 1986.

58. **Rambaud, J. C. and Modigliani, A.,** Hormones as potential intestinal secretagogues in health and disease, in *Intestinal Secretion,* Turnberg, L., Ed., Smith Kline French, Mundells Welwyn Garden City, 1983, 89.

59. **Vinik, A. and Moattari, A. R.**, Use of somatostatin analog in management of carcinoid syndrome, *Dig. Dis. Sci.*, 34, 14S, 1989.

60. **Bishop, A. E., Polak, J. M., Bryant, M. G., Bloom, S. R., and Hamilton, S.**, Abnormalities of vasoactive intestinal peptide-containing nerves in Crohn's disease, *Gastroenterology*, 79, 853, 1980.

61. **Koch, T. R., Carney, J. A., and Go, V. L. W.**, Distribution and quantitation of gut neuropeptides in normal intestine and inflammatory bowel diseases, *Dig. Dis. Sci.*, 32, 369, 1987.

62. **Mantyth, C. R., Gates, T. S., Zimmerman, R. P., Welton, M. L., Passaro, E. P., Jr., Vigna, S. R., Maggio, J. E., Kruger, L., and Mantyth, P. W.**, Receptor binding sites for substance P, but not substance K or neuromedin K are expressed in high concentrations by arterioles, venules, and lymph nodules in surgical specimens obtained from patients with ulcerative colitis and Crohn's disease, *Proc. Natl. Acad. Sci. U.S.A.*, 85, 3235, 1988.

63. **Mantyth, P. W., Mantyth, C. L., Gates, T., Vigna, S. R., and Maggio, J. E.**, Receptor binding sites for substance P and substance K in the canine gastrointestinal tract and their possible role in inflammatory bowel disease, *Neuroscience*, 25, 817, 1988.

64. **Lucey, M. R. and Yamada, T.**, Biochemistry and physiology of gastrointestinal somatostatin, *Dig. Dis. Sci.*, 34, 5S, 1989.

65. **Krejs, G. J., Browne, R., and Raskin, P.**, Effects of intravenous somatostatin on jejunal absorption of glucose, amino acids, water and electrolytes, *Gastroenterology*, 78, 26, 1980.

66. **Dharmsathaphorn, K., Sherwin, R. S., Cataland, S., Jaffe, B., and Dobbins, J.**, Somatostatin inhibits diarrhoea in the carcinoid syndrome, *Ann. Intern. Med.*, 92, 68, 1980.

67. **Barbezat, G. O. and Reasbede, P. G.** Somatostatin inhibition of glucagon-stimulated jejunal secretion in the dog, *Gastroenterology*, 81, 471, 1980.

68. **Davis, G. R., Camp, R. C., Raskin, P., and Krejs, G. J.**, Effect of somatostatin infusion on jejunal water and electrolyte transport in a patient with secretory diarrhoea due to malignant carcinoid syndrome, *Gastroenterology*, 78, 346, 1980.

69. **Krejs, G. J.**, Effects of somatostatin infusion on VIP-induced transport changes in the human jejunum, *Peptides*, 5, 271, 1984.

70. **Kvols, L. K., Moertel, C. G., O'Connell, M. J., Schutt, A., J., Rubin, J., And Hahn, R.**, Treatment of the malignant carcinoid syndrome, evaluation of a long acting somatostatin analogue, *N. Engl. J. Med.*, 315, 663, 1986.

71. **Cook, D. J., Kelton, J. G., Stanisz, A., and Collins, S. M.**, Somatostatin treatment for cryptosporidial diarrhea in a patient with the acquired immunodeficiency syndrome (AIDS), *Ann. Int. Med.*, 108, 708, 1988.

72. **Fedorak, R. M. and Allen, S.**, Effect of somatostatin analog (SMS201-995) on *in vivo* intestinal fluid transport in rats, *Dig. Dis. Sci.*, 34, 567, 1989.

73. **Berschneider, H. M., Martens, H., and Powell, D. W.**, Effect of BW942C, an enkephalin-like pentapeptide on sodium and chloride transport in rabbit ileum, *Gastroenterology*, 94, 127, 1988.

74. **Dobbins, J., Racusen, L., and Binder, H. J.**, Effect of D-alanine methionine enkephalin amide on ion transport in rabbit ileum, *J. Clin. Invest.*, 66, 19, 1980.

75. **Schiller, L. R., Santa Ana, C. A., Morawski, S. G., and Fordtran, J. S.**, Mechanism of the antidiarrheal effect of loperamide, *Gastroenterology*, 86, 1475, 1984.

76. **McKay, J. S., Linaker, B. D., Higgs, N. B., and Turnberg, L. A.**, Studies of the anti-secretory activity of morphine in the rabbit ileum *in vitro*, *Gastroenterology*, 82, 243, 1982.

77. **Kachur, J. F., Miller, R. J., and Field, M.**, Control of guinea-pig intestinal electrolyte secretion by a delta-opiate receptor, *Proc. Natl. Acad. Sci. U.S.A.*, 77, 2753, 1980.

78. **Beubler, E. and Lembeck, F.**, Inhibition of stimulated fluid secretion in the rat small and large intestine by opiate agonists, *Arch. Pharmacol.*, 306, 113, 1979.

79. **Quito, F. L. and Brown, D. R.**, [D-Ala^2,Met^5]-enkephalinamide: CNS-mediated inhibition of prostaglandin-stimulated intestinal fluid and ion transport in the rat, *Peptides*, 8, 1029, 1987.

80. **Brown, D. R. and Gillespie, M.A.**, Actions of centrally administered neuropeptides on rat intestinal transport: enhancement of ileal absorption by angiotensin 11, *Eur. J. Pharmacol.*, 148, 411, 1988.

81. **Solomon, T. E.**, Pancreatic polypeptide, peptide YY and neuropeptide Y family of regulatory peptides, *Gastroenterology*, 88(3), 838, 1985.

82. **Tatemoto, K. and Mutt, V.**, Isolation of two novel candidate hormones using a chemical method for finding naturally occurring polypeptides, *Nature, (London)*, 285, 417, 1980.

83. **Tatemoto, K., Carlquist, M., and Mutt, V.**, Neuropeptide Y — a novel brain peptide with structural similarities to peptide YY and pancreatic polypeptide, *Nature (London)*, 296, 659, 1982.

84. **Allen, J. M., Hughes, J., and Bloom, S. R.**, Presence, distribution and pharmacological effects of neuropeptide Y in mammalian gastrointestinal tract, *Dig. Dis. Sci.*, 32, 506, 1987.

85. **Saria, A. and Beubler, E.**, Neuropeptide Y (NPY) and peptide YY (PYY) inhibit prostaglandin E_2-induced intestinal fluid and electrolyte secretion in the rat jejunum *in vivo*, *Eur. J. Pharmacol.*, 119, 47, 1985.

86. **MacFadyen, R. J., Allen, J. M., and Bloom, S. R.,** NPY stimulates net absorption across rat intestinal mucosa in vivo, *Neuropeptides,* 7, 219, 1986.

87. **Buell, M. G. and Harding, R. K.,** Effects of peptide YY on intestinal blood flow distribution and motility in the dog, *Regul. Peptides,* 24, 195, 1989.

88. **Keast, J. R., Furness, J. B., and Costa, M.,** Distribution of certain peptide-containing nerve fibres and endocrine cells in the gastrointestinal mucosa in five mammalian species, *J. Comp. Neurol.,* 236, 403, 1985.

89. **Furness, J. B., Costa, M., Emson, P. C., Hakanson, R., Moghimzadah, E., Sundler, F., Taylor, I. L., and Chance, R. E.,** Distribution, pathways and reactions to drug treatment of nerves with neuropeptide Y and pancreatic polypeptide-like immunoreactivity in guinea-pig digestive tract, *Cell Tissue Res.,* 234, 71, 1983.

90. **Freil, D., Miller, R. J., and Walker, M. W.,** Neuropeptide Y: a powerful modulator of epithelial ion transport, *Br. J. Pharmacol.,* 88, 425, 1986.

91. **Brown, D. R., Boster, S. L., and Overend, M. F.,** Neuropeptide Y alters anion transport in porcine distal jejunum: involvement of enteric opioids, *FASEB J.,* 2, 1276(a), 1987.

92. **Hubel, K. A. and Renquist, K. S.,** Effect of neuropeptide Y on ion transport by the rabbit ileum, *J. Pharmacol. Exp. Ther.,* 238, 167, 1986.

93. **Cox, A. M., Cuthbert, A. W., Hakanson, R., and Wahlestedt, C.,** The effect of neuropeptide Y and peptide YY on electrogenic ion transport in rat intestinal epithelium, *J. Physiol.,* 398, 65, 1988.

94. **Laburthe, M., Chenut, B., Rouyer-Fessard, C., Tatemoto, K., Couvineau, A., Servin, A., and Amiranoff, B.,** Interaction of peptide YY with rat intestinal epithelial plasma membranes: binding of the radio-iodinated peptide, *Endocrinology,* 118, 1910, 1986.

95. **Gilbert, W. R., Frank, B. H., Gavin, J. R., III, and Gingerich, R. L.,** Characterisation of specific pancreatic polypeptide receptors of basolateral membranes of the canine small intestine, *Proc. Natl. Acad. Sci. U.S.A.,* 85, 4745, 1988.

96. **Servin A. L., Rouyer-Fessard, C., Balasabramaniam, A., Saint-Pierre, S., and Laburthe, M.,** Peptide YY and neuropeptide Y inhibit vasoactive intestinal peptide stimulated adenosine 3'5'-monophosphate production in rat small intestine: structural requirements of peptides for interacting with peptide YY preferring receptors, *Endocrinology,* 124, 692, 1989.

97. **Reynolds, E. E. and Yokota, S.,** Neuropeptide Y receptor-effector coupling mechanisms in cultured vascular smooth muscle cells, *Biochem. Biophys. Res. Commun.* 151, 919, 1988.

98. **Motulsky, H. J. and Michel, M. C.,** Neuropeptide Y mobilizes Ca^{2+} and inhibits adenylate cyclase in human erythroleukemia cells, *Am. J. Physiol.,* 255, E880, 1988.

99. **Schmidt, R. E., Plurad, D. A., and Roth, K. A.,** Effects of chronic experimental streptozotocin-induced diabetes on the noradrenergic and peptidergic innervation of the rat alimentary tract, *Brain Res.,* 458, 353, 1988.

100. **Belai, A., Lincon, J., Milner, P., and Burnstock, G.,** Progressive changes in adrenergic, serotonergic and peptidergic nerves in proximal colon of streptozotocin-diabetic rats, *Gastroenterology,* 95, 1234, 1988.

101. **Chang, E. B., Fedorak, R. N., and Field, M.,** Experimental diabetic diarrhoea in rats: intestinal mucosal denervation hypersensitivity and treatment with clonidine, *Gastroenterology,* 91, 564, 1986.

102. **Gaginella, T. S. and Kachur, J. F.,** Bradykinin as a mediator of mucosal secretion in IBD., in *Inflammatory Bowel Disease: Current Status and Future Approaches,* McDermott, R. P., Ed., Elsevier, New York, 1988, 383.

103. **Stead, R. H., Perdue, M. H., Blennerhassett, M. G., Kakuta, N., Sestini, P., and Bienenstock, J.,** The innervation of mast cells, in *The Neuroendocrine Immune Network,* Freier, S., Ed., CRC Press, Boca Raton, FL, 1990, 19.

104. **Galison, P. L.,** *How Experiments End,* University of Chicago Press, Chicago, 1987, 178.

Chapter 14

THE ROLE OF NEUROPEPTIDES IN THE NORMAL AND PATHOPHYSIOLOGICAL CONTROL OF BLOOD FLOW

John S. Smeda

TABLE OF CONTENTS

I. Introduction 448

II. Calcitonin Gene-Related Peptide 448
  A. Background 448
    1. Location of CGRP within the Cardiovascular System 449
    2. Pharmacological Action of CGRP on the Vasculature 449
    3. Neuronal Release and Action of CGRP on the
        Vasculature 450
    4. Role of CGRP in the Control of Heart Function 451
    5. The Role of CGRP in the Modulation of Sympathetic
        Nerve Function 452
    6. The Hormonal Action of CGRP on the
        Cardiovasculature 452
    7. The Normal and Pathophysiological Functions of CGRP
        in the Cardiovascular System 452

III. Vasoactive Intestinal Peptide 454
  A. Background 454
    1. Location of VIP within the Cardiovascular System 454
    2. The Pharmacological Action of VIP in the
        Cardiovascular System 455
    3. The Neural Release of VIP and Its Action on the
        Vasculature 455
    4. The Normal and Pathophysiological Function of VIP in
        the Cardiovascular System 456

IV. Neuropeptide Y 458
  A. Background 458
    1. Location of NPY in the Cardiovascular System 458
    2. The Neural Release of NPY from Sympathetic Nerves 458
    3. The Pharmacological Action of NPY on the Vasculature 459
    4. The Effect of NPY on the Contractile Responses of
        Other Vasoconstrictors 459
    5. The Involvement of NPY in the Modulation of
        Noradrenalin Release from the Sympathetic Nerves 460
    6. The Physiological Importance of the Presence of NPY
        within Sympathetic Nerves 460
    7. The Effects of NPY on Heart Function 461
    8. The Normal and Pathophysiological Function of NPY in
        the Vasculature 461

V. Opioid Peptides...462
    A. Location of Opioid Peptides in the Cardiovasculature..................462
    B. The Pharmacology of Opioid Peptides on the Vasculature...............463
    C. Opioid Peptides as Neurotransmitters in the Cardiovascular
        System 464
    D. The Role of Opioid Peptides as Presynaptic Modulators of
        Noradrenalin Release...465
    E. The Normal and Pathophysiological Function of Opioid
        Peptides in the Cardiovascular System 466

VI. Summary 467

Acknowledgment...469

References...469

I. INTRODUCTION

Recently, a great deal of research emphasis has taken place in an attempt to elucidate the role that neuropeptides play in the control of blood flow. In this regard, studies involving immunoreactive labeling and/or quantitative analysis of peptides in a variety of species have indicated that virtually every peptide found to exist within the gut also exists within some segment of the vascular bed. Likewise, when applied topically, nearly all of these peptides exert some type of effect which consists of either a modification of neurotransmission of the perivascular nerves or a direct alteration of vascular tone or heart function. A discussion of the cardiovascular effects of every peptide found within the gut is beyond the scope of this review and in many cases may be more of pharmacological interest than physiological relevance. However, for a small group of peptides, evidence suggests that these peptides not only exist within the innervation of the cardiovascular system, but also can be released from nerves during stimulation to act as neuroeffectors of heart and/or vascular function.

The bulk of this review will concentrate on these peptides. The discussion pertaining to each peptide will center on (1) the neural origin and location of each individual neuro-peptide within the cardiovasculature, (2) the factors governing the control of its release (3) the pre- and postsynaptic action of the peptide and its interaction with other cardiovascular neurotransmitters, and (4) the role the peptide plays in normal cardiovascular function and the putative role of the peptide in circulatory disease. With these discussions, particular emphasis has been placed on analyzing evidence supporting the contention that the peptides discussed are in fact neurotransmitters and peripheral neuroeffectors within at least some segments of the cardiovasculature.

II. CALCITONIN GENE-RELATED PEPTIDE

A. BACKGROUND

The discovery of calcitonin gene-related peptide (CGRP) stems from RNA transcript studies performed by Amara et al.[1] It was noted that the primary RNA transcript of the

DNA segment relating to the calcitonin gene could be processed in a manner that, when translated, produced either calcitonin or CGRP plus a number of terminal peptides. The dominant form of processing was found to differ in different tissues.[1,2] Calcitonin-specific mRNAs and calcitonin predominated in thyroid tissue, whereas CGRP-specific mRNAs and CGRP were primarily produced in the central and peripheral neural tissue.

1. Location of CGRP within the Cardiovascular System

Immunocytochemical methods have been used to demonstrate the existence of CGRP-containing nerves in a wide variety of cardiovascular tissues. Some of these studies are outlined in Table 1.

Double immunostaining techniques, primarily performed on the cerebrovasculature, indicate that within this vasculature some periarterial nerve fibers that contain CGRP also coreact with substance P (SP)[3-5] and neurokinin A antibodies,[6] suggesting that CGRP, SP, and neurokinin A can coexist within the same neurons. Using double-staining immunogold procedures, CGRP and SP have also been colocalized in large granular vesicles present in the trigeminal and dorsal root ganglia as well as in the mesenteric perivascular nerves of guinea pigs.[7]

At least a proportion of the CGRP innervation of the cerebral vasculature emanates from the trigeminal ganglion. Stimulation of this ganglion in humans and cats elevates the plasma levels of CGRP and SP. In the rat, approximately 30% of the trigeminal ganglion cells immunoreact to CGRP.[8] In this species, transection of the sixth division of the trigeminal nerve eliminates both CGRP innervation on the wall of the ipsilateral cerebral arteries and CGRP immunoreactivity in the trigeminal nucleus and cervical spinal cord.[8] Studies involving guinea pigs indicate that certain nuclei within this ganglion contain either only CGRP or CGRP in combination with either neurokinin A or SP.[6] Other studies involving cat middle cerebral arteries indicate that lesions of the trigeminal nerve reduce CGRP levels to measurable quantities, while such treatment halves the quantity of SP present within the artery.[3] This suggests that at least some cell bodies in the trigeminal nucleus of the cat contain only CGRP or substance P.[3,9] CGRP-immunoreactive (CGRP-IR) fibers are also present elsewhere in the central and peripheral nervous systems. Gibson et al.[10] studied the distribution of CGRP immunoreactivity throughout the spinal cord of man and eight animal species. CGRP fibers and terminals were present at all levels of the spinal cord and were most abundant in the dorsal horn, being heavily concentrated in Lissauer's tract, laminae I to III, lamina V, lamina X, and the region around the central canal. In the thoracic segment, CGRP-reactive fibers were densely aggregated around cells of the sympathetic column. Bilateral transection of the greater splanchnic nerve or vagotomy just below the diaphragm decreased the number of CGRP-IR fibers surrounding the hepatic artery and other splanchnic blood vessels.[11] Numerous CGRP-IR cells are also present in the nodose ganglion.[12]

The CGRP innervation present around blood vessels is thought to consist of sensory fibers that convey nociceptive information of thermal, chemical, or mechanical origin to the brain stem.[3,4,11,10] However, as will subsequently be discussed, such fibers can be stimulated antidromically to promote release of CGRP or SP.[13,14]

2. Pharmacological Action of CGRP on the Vasculature

The pharmacological action of topically applied or infused CGRP is to produce a potent vasodilatory effect in all the arterial systems tested to date.[6,9,13,15-17] The ability of CGRP to maximally relax cat middle cerebral arteries precontracted with $10^{-4}\,M$ uridine triphosphate is equal to that of acetylcholine and is far greater than that of vasoative intestinal peptide, bradykinin, or atrial natriuretic factor. In rabbit skin, CGRP is about 1000-fold more potent a vasodilator than adenosine, acetylcholine, or SP and is 10 to 1000 times more potent than isoprenaline. Unlike SP or neurokinin A,[18] vasodilation produced by CGRP is not dependent on the presence of an intact endothelium.[9,16,19]

TABLE 1

Neuropeptide Immunoreactive Innervation Present in the Perivasculature of Various Vascular Beds and within Heart Muscle Tissue

Vascular bed studied	The type of peptide innervation observed in studies involving various species within the corresponding vasculature or heart tissue		
	CGRP	**NPY**	**VIP**
Aorta		Rat,[212] Guinea pig[118,125]	
Brachiocephalic/ subclavian			Cat[49]
Bronchial			Guinea pig[118]
Cerebral	Cat[3,9] Rat[5,8] Mouse[4] Guinea pig[4,6] Human[15]	Human[15,109] Cat[109] Guinea pig[109] Rat[57,109] Mouse[109] Gerbil[214]	Cat[51,52,66] Cow[52] Monkey[52] Human[213] Rabbit[53] Hamster[52] Rat[52,57,59,215,216] Dog[50,52] Pig[52] Gerbil[52] Mouse[52]
Coronary	Rat[23]	Guinea pig[118,121] Rat[121]	
Femoral		Rat[119]	
Hepatic	Rat[11]	Guinea pig[118] Rat[121]	Cat[49] Dog[90]
Iliac/femoral			Cat[49] Guinea pig[98]
Intercostal			Cat[49]
Mesenteric	Rat[13,23]	Cat[121] Human[119,121] Rat[114,212,121] Guinea pig[118,125]	Human[116] Rat[79,114] Cat[49] Guinea pig[98]
Muscle	Rabbit[217]	Rat[121] Pig[121]	
Penile			Dog[104]
Pulmonary		Guinea pig[118,121] Rat[121]	Cat[49,218] Dog[218] Human[220]
Renal	Rat[23,219]	Rat[121,212] Guinea pig[118]	Rat[96,97] Dog[97] Guinea pig[98]
Skin		Human[220]	
Spinal	Human[116]	Human[116]	
Splanchnic	Rat[11]	Guinea pig[118,121] Rat[121]	
Testicular		Guinea pig[118]	Cat[49]
Uterine		Guinea pig[118,221]	
Heart tissue — atrial heart muscle	Guinea pig[22] Rat[23]	Guinea pig[118,125] Rat[212]	

3. Neuronal Release and Action of CGRP on the Vasculature

CGRP can be released from perivascular nerves and can act as a vasoactive neurotransmitter. The best evidence of this comes from studies performed by Kawasaki et al.[13] and Saito et al.[20] In these studies it was demonstrated that the mesenteric vasculature of the rat and the cerebrovasculature of the cat contained a very dense CGRP-IR perivascular nerve plexus.[13,20] Perfusion studies performed by Kawasaki et al.[13] using the isolated mesenteric vasculature demonstrated the presence of vascular relaxation (recorded as a drop in infusion pressure) in response to electrical stimulation of the perivascular nerves. Such responses were observed under conditions where the sympathetic nerves were inactivated via guanethidine and/or 6-hydroxydopamine and the vasculature precontracted with the α-receptor agonist methoxamine. The responses occurred at low frequencies of electrical stimulation (1 to 8 Hz) using short pulses (1 ms duration) and were abolished by tetrodotoxin or capsaicin pretreatment. This indicated that the responses were neural in origin and probably resulted from the release of SP, neurokinin A, or CGRP. In view of the fact that in this vasculature only the infusion of CGRP and not neurokinin A or SP promoted relaxation responses, it was suggested that the capsaicin-sensitive, neural response was mediated via CGRP. In other studies, Saito et al.[20] observed the occurrence of relaxation in response to transmural nerve stimulation in isolated cat middle cerebral arteries precontracted with uridine triphosphate. The relaxation responses were totally inhibited by tetrodotoxin and partially inhibited by capsaicin treatment, but were not affected by guanethidine, atropine, or endothelial removal. Since capsaicin is capable of depleting SP, neurokinins, and CGRP from nerves, but only CGRP is able to promote nonendothelial-dependent relaxation,[18] it was suggested that the capsaicin-sensitive relaxations observed were mediated via CGRP.

Under *in vivo* conditions, the antidromic stimulation of CGRP-containing afferent periarterial nerves could promote a release of CGRP, enabling CGRP to act as a vasodilatory transmitter.[13,21] In this regard, there is evidence to support the view that small-diameter sensory fibers participate in antidromic vasodilation via axonal and/or orthodromic-antidromic reflex action.[14] In the former situation, a stimulus is thought to generate an afferent impulse which travels upward to the point of ramification of the terminal branches of the neuron. The impulses then descend into another branch and travel toward the periphery, leading to the release of dilator from a sensory nerve ending. The stimulus also could initiate an orthodromic impulse, causing a release of transmitter in the spinal cord. A summation of impulses and further central nervous system (CNS) processing could lead to the sensation of pain, whereupon other central impulses traveling down the spine could stimulate CGRP/ SP nerves, causing an antidromic release of transmitter. Considering that electrical stimulation of the trigeminal ganglion released peripheral stores of SP and of CGRP in humans or cats,[14,21] such a system also could be under central control and could be used to effect vasodilatory changes even in the absence of any sensory reflex action.

4. Role of CGRP in the Control of Heart Function

CGRP also may play an important role in controlling heart function. Various studies have indicated the presence of CGRP-IR fibers in guinea pig and rat heart atria,[22-24] where particularly dense innervation exists around the atrial sinus node.[25] Binding studies using rat [^{125}]iodo-CGRP indicate that high levels of specific CGRP binding sites exist in membrane preparations made from the atria of the rat, while lower levels of binding exist in similar ventricular preparations.[26] Electrical transmural stimulation of isolated atrial preparations obtained from guinea pigs[22,25-27] increased both the beat frequency and force of contraction of the atrial muscle. The above responses were not affected by nicotinic, H_1, H_2, or serotonin receptor antagonists, but the positive chronotropic responses (increase in beat frequency) were found to be partially inhibited by capsaicin pretreatment or by the β-receptor antagonist atenolol and totally inhibited when the atrial tissues were incubated with both these agents.[22,25] Other studies measuring the positive inotropic responses (increase in heart contractility) in atropine-treated tissues indicated that they were totally blocked by reserpine plus capsaicin treatment or by atenolol, prazosin, plus capsaicin treatment.[27] Considering that the responses were mimicked by the topical application of CGRP, but not SP, neurokinin A or B, or vasoactive intestinal peptide,[25,27] it was suggested that part of the positive inotropic/chronotropic effects observed during neural stimulation were mediated via the neural release of

CGRP, whereas the balance of the response occurred through the sympathetic release of noradrenaline (NA) acting on atrial β-receptors to produce the positive chronotropic effect[25] and β- plus α_1-receptors to produce the positive inotropic effect.[27]

The positive inotropic effects of CGRP have also been demonstrated *in vitro,* in isolated human atrial tissue obtained from heart surgery patients.[28] *In vivo,* the systemic injection of CGRP into humans produces an increase in the heart rate and a reduction in the duration of the ventricular preinjection period, electromechanical systole, and left ventricular ejection.[28] These observations are consistent with the ability of CGRP to produce positive chronotropic and inotropic effects on the heart. However, such injections are also associated with a decrease in systemic blood pressure which, due to sympathetic reflex action, complicates the interpretation of the observed changes in cardiac function.

5. The Role of CGRP in the Modulation of Sympathetic Nerve Function

In some tissues, CGRP may be capable of modulating sympathetic nerve function. Studies involving the vas deferens indicate that low subvasoactive concentrations of CGRP are capable of inhibiting nerve-mediated noradrenergic contractile responses.[29] Although it was suggested that presynaptic CGRP receptors inhibiting the release of NA could be responsible for these effects,[29] subsequent studies have indicated that concentrations of CGRP inhibiting nerve-mediated contractile responses in the vas deferens did not alter the release of [³H]NA from the noradrenergic nerves.[30] Electrophysiological studies indicate that the inhibitory effects of CGRP are mediated postsynaptically. In these studies it was demonstrated that CGRP did not affect the amplitude of excitatory junction potentials (EJPS), but increased the threshold membrane potential (to more positive values) required to initiate an action potential and slightly reduced the contractile responses produced by topically applied NA and adenosine triphosphate (ATP).[31] It was concluded that CGRP produced its inhibitory effects by suppressing smooth muscle cell (SMC) excitability and excitation-contraction coupling. At present, the electrophysiological effects of CGRP on vacular smooth muscle (VSM) have not been studied. However, it is possible that the mechanisms of CGRP action on blood vessels are similar to those observed in the vas deferens.

6. The Hormonal Action of CGRP on the Cardiovasculature

The levels of CGRP present in the plasma of rats and humans are quite high, ranging between 9.7 and 71 pmol/l in humans (mean 25 ± 1.2)[32] and between 9 and 32 pmol/l in rats.[33] Studies involving young rats have suggested that the circulating levels of CGRP are largely derived from perivascular nerves.[33] Such levels of CGRP, as mentioned previously, surpass the threshold level required to promote relaxation of human cerebral arteries precontracted with $PGF_{2\alpha}$.[15] In humans and experimental animals, the systemic injection of CGRP produces hypotension and other alterations associated with hypotension (tachycardia, elevations in plasma NA and adrenaline) as well as increases in plasma levels of cyclic adenosine monophosphate (cAMP) and cyclic guanosine monophosphate (cGMP).[28] In man, some effects, such as flushing (secondary to vasodilation), diastolic hypotension, tachycardia, and elevations in plasma catecholamines, have been observed at plasma CGRP levels of 56 pmol/l, well within the physiological levels present in humans.[34,35] These results suggest that CGRP not only may act as a vascular neurotransmitter, but also may play a role as a neurohormone affecting vascular beds far distal to the sites of secretion.

7. The Normal and Pathophysiological Functions of CGRP in the Cardiovascular System

To date, malfunctions in CGRP physiology have not been implicated in any specific disease process; however, the potential for CGRP involvement in cardiovascular disease is large. Collation of data collected during surgical procedures indicates that the larger arteries at the base of the brain evoke pain stimuli when they are mechanically distended or otherwise irritated during surgical procedures.[36,37] These arteries are innervated in part, by the ophthalmic division of the trigeminal nerve,[8,37] which has been suggested to be involved in inducing vascular headache and migraine.[37,38] In rats, this nerve has been shown to be a major supplier of CGRP-IR fibers to the proximal segments of the major cerebral arteries.[8] Hence, it has been suggested that CGRP, along with SP, may be one of the transmitters involved in the production of vascular headaches.[8] It is possible that (1) a decrease in the threshold of perception of overdistension by CGRP- and SP-containing sensory nerves; (2) an overactive secretion of CGRP, promoting localized dilation of the cerebral arteries; and/or (3) an elevation in systemic plasma levels of CGRP (which might selectively dilate the large cerebral blood vessels) could potentially contribute to the occurrence of chronic migraine headaches.

CGRP has also been demonstrated to participate in local irritation/inflamatory responses in the skin and eye.[34,39] When injected intradermally into human volunteers, the peptide produces reddening or a flare reaction lasting 12 h.[34] Facial flushing and localized increases in skin temperature have also been noted under conditions of trigeminal neuroglia,[40] after the injection of alcohol into the trigeminal ganglion,[41] and under conditions where the ganglion has been thermocoagulated (for the treatment of tic douloureux).[21] The flushing response is mediated and controlled via a number of mechanisms. The long-lasting reddening observed during the S.C. injection of CGRP is associated with vasodilation and an increase in blood flow to the area as measured using Doppler techniques.[42] In addition, SP coreleased with CGRP (but not CGRP alone) is capable of inducing edema by directly promoting an increase in vascular permeability and inducing mast cell degranulation (the products of which can further increase the extravascular movement of vascular plasma H_2O).[43] In this regard, the presence of CGRP synergistically potentiates the extravascular movement of plasma produced by SP and mast cell products, such as histamine and prostaglandins (PGE_2) as well as leukotriene (B_4) and plasma products such as bradykinin.[43] Finally, mast cell degranulation promoted by the release of SP promotes feedback inhibition of CGRP-induced vasodilation and erythema. This latter response is thought to be mediated via proteases released from the mast cells.[44] These mechanisms could contribute to the protracted erythema seen in response to local infection or injury. Pathologically, the mechanism also could be altered in that the effects of CGRP might be amplified and could contribute to the presence of long-lasting erythematous skin blemishes.

CGRP also could play a role in cardiovascular disease. At present, such aspects have not been fully explored, and it is premature to suggest specific mechanisms of involvement. However, it is worth noting that the neonatal treatment of Kyoto Wistar spontaneously hypertensive rats (SHR) with capsaicin, which should destroy central and peripheral CGRP/ SP-containing nerves, prevents the subsequent development of high blood pressure (at 12 weeks of age) in this animal model of essential hypertension.[45] This suggests that CGRP/ SP nerves play some as yet unclarified role in blood pressure regulation in this animal model.

CGRP-containing nerves also could play a role in controlling the cerebral circulation. The small blood vessels of the brain exhibit myogenic contraction in response to elevation in blood pressure.[46,47] This is thought to be the basis through which the cerebral vasculature autoregulates blood flow; increases in blood pressure which would promote hyperperfusion of the brain tissue are counteracted by increased cerebrovascular constriction, which in turn limits blood flow. CGRP-containing perivascular nerves are localized in the cerebrovasculature in a manner whereby they can effect this process. One can envision the possibility that these nerves could be activated via axonal reflexes or central mechanisms and could serve via vasodilation to modulate and counteract myogenic constriction. If this proves to be the case, such nerves could play important physiological and pathological roles. Physiologically, such nerves could help prevent brain ischemia. Pathologically, overactivation of such nerves could (1) override the myogenic autoregulatory constrictor responses and promote the overperfusion of some brain areas and/or (2) promote increases in vascular permeability within the vessels. Both of these alterations could predispose the brain to hemorrhagic stroke development.

III. VASOACTIVE INTESTINAL PEPTIDE

A. BACKGROUND

Vasoative intestinal peptide (VIP) was isolated from hog gut by Said and Mutt.[48] Subsequent studies, some of which are outlined in Table 1, have indicated the presence of VIP-immunoreactive (VIP-IR) periarterial nerves in a wide variety of vascular beds sampled from humans and experimental animals, as well as within the nonvascular matrix of virtually every internal organ studied to date. VIP-IR nerve fibers have also been observed surrounding some veins (subclavian, axillary, subcapsular, portal mesenteric); however, such nerve fibers are less prevalent in the venous than the arterial system and appear to be absent in many systemic veins.[49]

1. Location of VIP within the Cardiovascular System

The source of the neural projection innervating the blood vessels has not been fully elucidated. However, extensive studies in this area have been performed on the cerebral vasculature of various animals. The bulk of the perivascular VIP-IR cerebrovascular innervation appears to be supplied by autonomic ganglia. Studies involving the extirpation of ganglia in the dog indicate that virtually all the VIP-IR perivascular innervation of cerebral vasculature is supplied by the parasympathetic pterygopalatine (anterior and posterior cerebral circulation), the otic ganglia (major arteries of posterior cerebral circulation), and the sympathetic superior cervical ganglia (mostly the posterior and slightly the anterior circulation).[50] Studies involving other animals are, in part, consistent with these observations.

VIP-IR cell bodies have been located in the superior cervical,[51-53] pterygopalatine (sphenopalatine),[51,52] and otic ganglia[51] in a variety of animals. Other studies involving axonal tracing techniques indicate that periarterial fibers present around the middle cerebral arteries of the cat and rat originate from the pterygopalatine and otic ganglia,[54,55] whereas those surrounding the basilar artery originate from the superior cervical, pterygopalatine, and other ganglia.[54,56] In rats, lesions of these ganglia reduce the VIP-IR nerve fibers surrounding the cerebral arteries,[57,58] However, there is controversy related to this subject area. It is worth noting that some researchers have failed to observe changes in the cerebrovascular VIP-IR innervation upon the bilateral excision of the superior cervical ganglion (rat,[59] cat[60]) and the unilateral excision of the pterygopalatine ganglion or superior cervical ganglion (cat[52]). In addition, in the cat, VIP-IR cell bodies within microganglia have been observed in lingual, chorda tympani nerves within the nerves of the cavernous plexus and tympanic ramus of the glossopharyngeal nerve.[51] These cell bodies do have the potential to innervate the intra- and extracranial arteries of the head. Therefore, the possibility does exist that there may be species differences with respect to the source of VIP innervation of the cerebrovasculature, and it is possible (and likely) that cell bodies other than those present in the superior cervical, pterygopalatine, or otic gangli may be involved in innervating the arteries of the head.

Within the parasympathetic nerves, VIP coexists with the neurotransmitter acetylcholine (ACh).[60-65] VIP and ACh are individually packaged, respectively, in large dense core and small agranular vesicles within cholinergic nerves.[64,66,67] In the sympathetic ganglion, VIP appears to exist in separate cell bodies that differ from those containing neuropeptide Y (NPY) and catecholamines.[63,65] However, some cell bodies within the parasympathetic ganglia (sphenopalatine, otic, ciliary) that are immunoreactive to VIP and/or cholineacetyltransferase also coreact with antiserum to NPY. A very small proportion of these cell bodies contain the adrenergic enzyme tyrosine hydroxylase, but none appear to contain catecholamines.[63] This suggests that some VIP/ACh-containing cholinergic nerves also may be capable of releasing NPY, whereas nerves within the sympathetic ganglion release either NPY plus catecholamines or VIP.

Studies involving the cat sympathetic ganglion indicate that cells containing VIP immunoreactivity also coreact to antisera against peptide histidine isoleucine (PHI).[62] PHI exhibits relative homology to VIP and differs between species (e.g., the human counterpart is termed PHM since the COOH-terminal amino acid is methionine, not isoleucine).[62,68] The coding for both VIP and PHI(M) is present in mRNA transcribed from the VIP precursor gene.[68] Some studies indicate that VIP and PHI are secreted in a 1:1 ratio, suggesting that the processing of PHI occurs after translation.[62] Other studies involving the hypothalamus indicate that some neurons containing PHI-IR are unreactive to VIP antisera, indicating a selective control of either VIP or PHI production.[69]

2. The Pharmacological Action of VIP in the Cardiovascular System

When topically applied to isolated arterial segments, VIP and PHI produce relaxation responses in a variety of arterial systems.[62,70-77] In veins, relaxation responses to topically applied VIP are inconsistently observed within the same tissue type and are absent in some venous tissues.[70] When isolated arteries are precontracted with $PGF_{2\alpha}$, EC_{50} values for VIP relaxation range in the order of 6×10^{-10} to $1 \times 10^{-8} M$.[70,74,75] PHI mimics the relaxation responses produced by VIP[62,74] with respect to the magnitude of relaxation produced;[74] however, PHI exhibits a decreased sensitivity when compared to VIP.[62,74] When compared in the same arterial preparations, EC_{50} values for PHI relaxation in $PGF_{2\alpha}$ precontracted tissue are approximately one log unit higher when compared to VIP.[74] It has been suggested that these differences in potency could be artifactual,[62] since often VIP from one animal source is compared to PHI from another, both of which at times are tested on arterial tissue obtained from a third species. In this regard, porcine VIP (identical to that present in humans) and PHM (the human counterpart of PHI) exhibit equal relaxation sensitivities against precontracted human submandibular arteries.[77]

Studies attempting to ascertain the mechanisms of VIP relaxation in rabbit mesenteric[76] and cat cerebral arteries[78] indicate that VIP relaxation is associated with the stimulation of intracellular cAMP activity[76,78,79] and an elevation in cellular cAMP.[76] Relaxation occurs without a change in the vascular SMC membrane potential (at concentrations $<10^{-7} M$)[76] and is not dependent on the presence of an intact arterial endothelium in virtually all arteries,[66,78,80] except in the rat aorta, where de-endothelialization has been shown to abolish VIP relaxation.[81]

3. The Neural Release of VIP and Its Action on the Vasculature

Various studies have indicated that VIP is capable of acting as a vascular neurotransmitter. A wide variety of studies performed primarily using isolated cat cerebral arteries have indicated the presence of relaxation in response to field stimulation that was tetrodotoxin sensitive,[66,82] resistant to atropine[66,82] or guanethidine,[82] and was not affected by the removal of endothelium.[66] Based on current knowledge, this would suggest that VIP, CGRP, or some other as yet undiscovered peptide was mediating the relaxation responses. Recently, Bevan and his colleagues have provided critical proof with respect to demonstrating that VIP is in fact a vascular neurotransmitter. They did this by showing that the neurogenically mediated relaxation responses observed in isolated cat middle cerebral[82] and lingual[83] arteries could be reversibly inhibited by incubating the preparation with VIP antisera. In other experiments involving intact cats, Gaadsby and MacDonald[84] also demonstrated that stimulation of the locus coeruleus or pterygopalatine ganglion induced intracerebral vasodilation as measured by a drop in common carotid vascular resistance. These studies show that resistance changes were found to be reversibly inhibited by VIP antisera, which, like the work of Bevan and his colleagues,[82,83] suggested that the neurogenic vasodilation of the intra/extracranial arteries was mediated by VIP.

Studies performed on the cat submandibular gland indicate that the stimulation of the parasympathetic innervation entering the gland releases VIP,[62,64] ACh[62,85] and PHI[62,86] into the venous effluent of the organ and increases the blood flow through the organ. The vasodilatory responses observed during parasympathetic stimulation of the organ can be reduced by VIP antisera.[64] Using alterations in blood flow through the organ as indicators of ACh and VIP/PHI release during parasympathetic stimulation, Lundberg and his colleagues and Darke and Smaje showed that blood flow was significantly reduced by atropine and potentiated by eserine (an acetylcholinesterase inhibitor), whereas higher frequencies of stimulation produced increases in blood flow that were not significantly inhibited by atropine.[64,87] It was suggested that the release of ACh or VIP/PHI might be controlled by varying nerve stimulation frequencies, low frequencies releasing ACh and high rates releasing VIP (perhaps in conjunction with PHI).

The release of VIP/ACh is also feedback regulated by presynaptic receptors (for a review see Reference 61). In nerves containing both transmitters, the release of ACh and VIP is inhibited by ACh stimulation of presynaptic muscarinic receptors.[61] For this reason, in the submandibular salivary gland, atropine potentiates the release of VIP during parasympathetic nerve stimulation[64] and leads to the depletion of VIP from the organ.[61,88] Likewise, the presence of VIP acting through presynaptic VIP receptors is thought to inhibit ACh release from parasympathetic nerves.[61] In the cat intestine, the release of VIP during parasympathetic nerve stimulation is decrease by the costimulation of sympathetic nerves entering the gut.[89] The effect of the sympathetic nerve stimulation can be blocked by phentolamine[88] and is thought to be mediated via presynaptic α-receptors on VIPergic nerves.[89] In view of the fact that VIP-containing cell bodies are present within sympathetic ganglia and are in close association with catecholamine nerves that innervate blood vessels,[50,51,53] the potential presence of similar mechanisms in the perivascular nerves could have important implications regarding the control of blood flow.

Studies involving perivascular sympathetic nerves of the hepatic vasculature indicate that although VIP decreases the sympathetically mediated contractile responses, it does not affect the release of [^{3}H]NA from the sympathetic nerves.[90] This suggests that, at least within hepatic arteries, VIP does not exert a presynaptic effect that modifies the release of NA from sympathetic nerves.

4. The Normal and Pathophysiological Function of VIP in the Cardiovascular System

The circulating levels of VIP and PHM in the plasma of normal humans have been measured to be on average, respectively, about 5 and 25 pmol/l. In humans, acute short-term elevations (>20 times normal) in plasma VIP[91] or PHV[92] (a more stable analogue of PHM/PHI) produce only modest changes in cardiovascular parameters. In these studies, the circulating levels of VIP are below the threshold levels required to initiate relaxation responses in isolated arteries, which suggest that in the latter function VIP does not act as a circulating hormone. Other studies performed by Ganong and his colleagues (for a review see Reference 93) suggest that physiological levels of circulating VIP could play a role in the release of renin. Porter et al.[94] observes that, in intact dogs, increasing the plasma levels of VIP (via VIP infusion) from basal levels (30 to 33 pmol/l) to 75 to 130 pmol/l; during a 15-min period promoted a 60 to 100% increase in plasma renin activity. During these experiments, the renal infusion pressure was maintained constant at 110 mmHg via a renal artery clamp, but increases in renal blood flow and creatinine clearance did take place during VIP infusion. However, other studies by Porter et al.[95] indicated that VIP is capable of releasing renin from isolated afferent arteriole-glomerulus preparations suffused with physiological saline.

These latter experiments suggest that VIP-mediated renin release is not a secondary phenomenon resulting from systemic hemodynamic changes produced by VIP, but results from a direct action of VIP on the juxtaglomerular renin release sites. Since humans with VIP-secreting tumors can attain plasma levels of VIP and PHM as high as 600 and 7000 pmol/l, respectively, it is entirely plausible that, under certain pathological conditions, VIP could act as a circulatory hormone affecting renin release. The possibility that VIP-containing renal nerves could be directly involved in stimulating renin release remains controversial. In this regard, although perivascular nerves containing VIP have been located in the renal vasculatures of the rat,[96,97] dog,[97] and guinea pig,[98] stimulation of the renal nerves entering the dog kidney has failed to produce an increase in VIP levels within the venous effluent leaving the kidney.[93,99]

Quantitative and qualitative changes in VIP-containing nerves have been observed in diabetes. A reduction in the density of VIP-IR periarterial innervation has been noted in the arterial and erectile tissue of the penis in the diabetic rats and diabetic humans suffering from impotence.[100] Degeneration of autonomic nerves containing VIP has been observed in the myenteric plexus of diabetic rats.[101] Still other studies have demonstrated that, unlike the situation present in normal rat ileum, field stimulation of ileum obtained from diabetic rats fails to produce a neurogenically mediated release of VIP (but not ACh) from the tissue.[102]

These quantitative and possible functional alterations in VIP innervation could have important physiological implications. Within the cat penis, stimulation of the pelvic and hypogastric nerves stimulates penile erection and VIP output into the penile blood.[103] Other studies involving the dog penis[104] indicate that the intracavernous injection of VIP induces penile erection. In this latter study[104] it was shown that sustained erection produced by the stimulation of the cavernous nerve of the penis could be inhibited by the administration of VIP antisera. It was concluded that VIP is an important neurotransmitter involved in maintaining erection by causing venous outflow restrictions in the penis. In view of the fact that the incidence of impotence in diabetic men is approximately 49%,[105] the decreased density and potentially compromised VIP nerve function within the penile tissue of diabetic men and experimental animals[100] could contribute to their sexual dysfunction.

Decrease in the VIP-IR innervation density has also been noted in the cerebrovasculature of diabetic rats.[106] The consequences of such alterations are unknown. However, considering that VIP has been shown to be a vasodilatory neurotransmitter in cerebral arteries[82] exhibiting myogenic constriction in response to pressure,[107] physiologically VIP could act to oppose and modulate the pressure-dependent myogenic constriction responses in this vasculature. A dysfunction in VIP innervation, such as that observed in diabetics, could eliminate this modulation and promote overconstriction. Since diabetics have an abnormally high incidence of cerebrovascular atherosclerosis,[108] such overconstriction could predispose diabetics to stroke via infarction. In this regard, diabetics do exhibit an above average incidence of stroke.[109] If similar changes were present in the noncerebral peripheral vasculature, such changes also could contribute to the poor peripheral perfusion and the necrotic limb problems associated with diabetes.

Decreases in VIP-IR nerve densities within dog middle cerebral arteries have also been noted after the experimental induction of subarachnoid hemorrhage in dogs.[110,111] Within this model of hemorrhagic stroke, the disappearance of these nerve fibers coincides temporally with the occurrence of cerebral vasospasm, suggesting the possibility that these vasodilatory nerves could be involved in pathological processes associated with cerebrovascular hemorrhage.

Alterations in the circulatory levels of VIP and changes in VIP pharmacology have been noted in heart disease. Humans suffering from heart failure exhibit marked elevations in the circulatory levels of VIP.[112] The ability of VIP to induce a positive inotropic response in isolated perfused heart preparations is reduced in renal hypertensive rats having left ventricular hypertrophy over normotensive control animals having normal hearts.[113] When compared to Kyoto Wistar normotensive control rats, stroke-prone Kyoto Wistar SHR exhibit elevations in VIP-IR periarterial innervation density in superior mesenteric arteries and in portal and mesenteric veins and a decrease in innervation density in the Circle of Willis and superior cerebellar arteries of the brain.[114] At present, the functional consequences of these types of alterations are unknown; however, the observation that alterations in VIP innervation and pharmacology exist in various disease processes suggest that it could be possible that this peptide may be involved in the pathophysiology of various cardiovascular diseases.

IV. NEUROPEPTIDE Y

A. BACKGROUND

Neuropeptide Y (NPY) is a 36-amino-acid peptide initially isolated from pig brains. Cloning studies involving the insertion of cDNA encoding NPY (synthesized from mRNA obtained from human pheochromocytoma) into bacteria revealed that NPY was synthesized from a 97-amino-acid precursor protein which, when cleaved at two sites, produced a 28-amino-acid signal protein, NPY, and a 30-amino-acid COOH-terminal peptide[115] (the last has been termed CPON).[116]

1. Location of NPY in the Cardiovascular System

Neuropeptide Y-immunoreactive (NPY-IR) cell bodies have been located within virtually every sympathetic ganglion studied,[117-120] and NPY-IR periarterial nerve fibers have been observed in both the arteries and veins of a wide variety of vascular beds, some of which are listed in Table 1. The sympathetic origin of NPY-IR perivascular nerves is further demonstrated (1) by their disappearance from the adventitia of blood vessel when sympathetic ganglia are lesioned;[57,109,117,118,121,122] (2) by the release of NPY into the blood stream by various organs/vascular beds when the sympathetic nerves are stimulated;[62,123,124] (3) by the disappearance of NPY immunoreactivity from perivascular nerves when such vascular beds are treated with catecholamine-depleting drugs, such as guanethidine, 6-hydroxydopamine, and in some cases reserpine;[114,118,121,122,125] and (4) by the colocalization of NPY in neurons containing enzymes that synthesize catecholamines.[109,121,122,126] In addition, NPY has also been demonstrated to exist within the cranial parasympathetic nerves of the otic, sphenopalatine, and ciliary ganglia of the rat.[63] Here NPY is present in the absence of catecholamines within cholinergic cell bodies, some of which contain VIP.[63]

Not all sympathetic nerves contain NPY. Sympathetic projections to exocrine elements of certain salivary glands,[62] rat brown fat cells,[120] and cell bodies within the rat stellate ganglion[120] lack NPY. Likewise, in the venous system, the density of innervation of nerves containing NA is greater than those containing NPY.[62] Lundberg and Höklfelt[62] have suggested that the latter heterogeneity may permit the sympathetic nervous system to exert a differential control over arterial and venous blood flow in various regions of the systemic vasculature.

Studies by Fried and his colleagues[126,127] involving cell fractionation/differential centrifugation performed on sympathetic neurons of the cat spleen, celiac ganglion, and rat vas deferens have indicated that NPY occurs only within large dense core vesicles, whereas NA is distributed with and without NPY in large and small dense core vesicles, respectively. Using double immunogold labeling techniques involving gold particles of different sizes, NPY and CPON have been colocated within the same large dense core vesicles of sympathetic nerves of human atria.[116]

2. The Neural Release of NPY from Sympathetic Nerves

The release of NPY from sympathetic nerves is dependent on the frequency at which they are stimulated. Studies involving the perfusion of intact pig spleens have indicated that when the splenic nerve is electrically stimulated by short high-frequency bursts (20 Hz), the release of NPY in relation to NA is much higher than under circumstances where the same number of pulses are administered at a low frequency of stimulation.[128] In these and comparable perfusion studies involving the cat spleen, it was shown that the release of NA into the blood supply of the spleen could be substantially inhibited by reserpine treatment and the release of both NA and NPY could be abolished by guanethidine, suggesting that the release of NPY and NA was mediated by sympathetic nerves.[128-130]

3. The Pharmacological Action of NPY on the Vasculature

The pharmacological effect of topically applied NPY in many arteries and veins is to produce a contractile response,[15,57,131-135] which in some blood vessels is subject to rapid tachyphylaxis.[57] This response, however, is not always observed consistently. Some arteries (rat cerebral, human omental, rabbit femoral) fail to exhibit a contractile response to NPY,[57,118,133] and studies involving the rat cerebrovasculature have indicated that different segments of the same vasculature exhibit either strong or absent degrees of contraction in response to topically applied NPY.[57] When contraction is demonstrated, the EC_{50} values for NPY contraction range from 7×10^{-9} to $5 \times 10^{-6} M$.[15,57,130-133]

In rats, the contractile response of cerebral arteries to NPY has been shown to occur without a change in the VSMC membrane potential,[57] whereas studies involving rat tail arteries indicate the presence of VSMC depolarization in response to NPY136. In cat cerebral arteries, NPY contraction has been shown to be associated with an inhibition of adenylate cyclase activity and a decrease in cAMP accumulation in the VSMCs. In addition, the removal of extracellular Ca^{+2} or the presence of Ca^{+2} channel blockers has been shown to inhibit NPY contractile responses in arteries.[109,131]

4. The Effect of NPY on the Contractile Responses of Other Vasoconstrictors

In many (but not all) arterial and venous segments studied to date, NPY has been shown to potentiate the contractile effects of NA. In these instances, NPY typically decreases EC_{50} values for NA contraction without altering the maximal level of contraction produced. The above effects have been demonstrated in rabbit pulmonary[137] and gastroepiploic arteries[121,137,138] and in rabbit and rat femoral arteries.[121,137-142] The potentiation of contraction by NPY is not a universal phenomenon. Some arteries (rabbit central ear artery,[138] aorta[137]) and most veins (rabbit gastroepiploic, femoral[137,138]) do not exhibit potentiation of NA contraction in the presence of NPY. In arteries that do demonstrate potentiation, such effects can be observed at NPY concentrations as small as $2.5 \times 10^{-10} M$ and virtually always occur at levels below those required to elicit a direct NPY vasoconstriction.[62,137,140,141] The potentiating effects of NPY are not restricted to NA-induced contraction, and NPY has also been demonstrated to potentiate histamine contraction of rabbit basilar, gastroepiploic, and femoral arteries.[138] The ability of NPY to potentiate contractile responses varies with different agonists and tissues. For example, NPY can potentiate histamine but not NA contraction in rabbit femoral[138,140] and gastroepiploic veins;[138] however, in the case of $PGF_{2\alpha}$ contraction, NPY potentiates contraction in the rabbit gastroepiploic vein, but not in the femoral vein.[138] In rabbit femoral and gastroepiploic arteries, NPY potentiates NA contraction, but has no effect on $PGF_{2\alpha}$ contractile responses.[138] These observations suggest that the mechanisms by which NPY potentiates contractile responses are tissue and agonist (or perhaps second messenger) specific and are not a generalized effect. Other studies involving mouse vas deferens have indicated that NPY also can potentiate the contractile responses of ATP (and the stable ATP analogue α-β-methylene ATP),[143] a transmitter thought to be coreleased with NA and NPY during sympathetic nerve stimulation.[144]

In rabbit femoral arteries, potentiation of NA contraction by NPY is not affected by calcium antagonists such as nifedipine or by the brief removal of Ca^{2+} from the physiological saline suffusing the arteries.[137,145] Since this treatment abolishes the direct vasoconstrictive effects of NPY,[109,131] it would suggest that the cellular mechanisms by which NPY mediates vasoconstriction differ from those by which potentiation occurs. Potentiation also is not affected by the inhibition of presynaptic NA uptake (via cocaine)[137] and does not result from alterations in the number or affinity of α_1- and α_2-receptors present on the VSMCs.[151] However, NPY potentiation of NA contraction has been shown to be inhibited via the removal of extracellular Na^+ and/or the inhibition of cellular Na^+/K^+ ATPase via ouabain.[137,145] This latter finding suggests that the mechanisms by which NPY potentiates NA contraction may involve the alteration of Na^+ currents across the VSMC membranes in some vascular tissues.

5. The Involvement of NPY in the Modulation of Noradrenaline Release from the Sympathetic Nerves

NPY also acts presynaptically on sympathetic nerves to inhibit the release of NA during field stimulation. This has been demonstrated by (1) a decrease in [³H]NA overflow during transmural nerve stimulation in vas deferens (-80%), in femoral, basilar, and mesenteric arteries, and in portal veins (-42 to 86%) and heart tissues (-40%);[134,141,146] and (2) via the measurement of NA output from the venous effluent of the isolated perfused mesenteric vascular bed of rats.[147] The inhibitory effects of NPY on NA release are not mediated via either α_1- or α_2-receptors, suggesting that a presynaptic NPY receptor exists which controls the release of NA. In this regard, studies by Wahlestedt et al.[140] indicated that the C-terminal portion of NPY was sufficient for NPY to produce presynaptic inhibitory effects, whereas the full complement of NPY was required to produce direct vasoconstriction (or potentiation of contraction). This would suggest that the pre- and postsynaptic NPY receptors could be of different subtypes.

6. The Physiological Importance of the Presence of NPY within Sympathetic Nerves

Various researchers have speculated as to the physiological importance of NPY within sympathetic nerves.[143,145] It has been noted that, in general, the concentrations of NPY required to elicit direct vasoconstriction are greater than those required to inhibit the release of NA presynaptically, which in turn are greater than the levels required to potentiate the effects of NA-induced contraction postjunctionally.[141,145] Considering that low frequencies of stimulation preferentially release NA + ATP (small dense core vesicles) to NPY (large dense core vesicles), the low levels of NPY produced under such conditions may be sufficient to facilitate/potentiate the contractile effects of NA and NPY, but have minimal effects with respect to the presynaptic inhibition of NA release. Higher frequencies of stimulation would stimulate the release of NPY, raising the ratio of NPY to ATP + NA released. Under such conditions, it has been suggested that NPY could act presynaptically to help conserve NA release,[143] possibly deactivating an excessively active varicosity. In view of the likelihood that not all sympathetic nerve varicosities are active during the passage of an action potential through the plexus,[148] the inactivation of certain overactive varicosities could promote a much more even distribution of transmitter throughout the tissue and as well as a cyclic activity of secretory units, reducing the "transmitter cost" of the contractile process and preventing the depletion of ATP and NA.[143] Since direct vasoconstriction, potentiation of contraction, and even presynaptic inhibition of NA release are not universal phenomena observed in all arteries (see Reference 136), the possibility exists that the ability of an artery to respond in this manner could be organized to meet the individual role of the artery. For example, arteries requiring a rapid on-off "motor" control of blood flow could be adapted in a manner whereby the presynaptic inhibitory/postsynaptic potentiating effects of NPY could be well developed, whereas arteries serving a role where constant tone is required might lack a strong presynaptic NPY-mediated inhibitory effect on NA release. In the latter instance, NPY could serve to potentiate NA + ATP contraction and might accumulate in sufficient concentration within the synapse to produce a direct vasoconstrictor effect on the artery, hence maintaining constant tone.

7. The Effects of NPY on Heart Function

The direct effects of NPY on heart tissue are controversial and inconclusive and may vary with respect to the species studied and the type of experiment performed. Studies involving isolated rabbit hearts receiving perfusate through the coronary vasculature (via the retrograde perfusion of physiological saline through the aorta) at a constant pressure indicate that NPY (bolus injections >20 pmol) decreases myocardial perfusion and initially increases, then subsequently decreases, ventricular contraction in a transient manner, but produces little change in heart rate.[149] Some studies involving isolated guinea pig hearts (vascular perfused at a constant flow) also indicate that NPY increases coronary vascular resistance and decreases ventricular contraction (at concentrations of 2.6×10^{-9} and $1.3 \times 10^{-8}\ M$) with little change in heart rate (at concentrations of 6.6×10^{-10} and $5.2 \times 10^{-8}\ M$).[132] However, other studies have indicated that NPY ($5 \times 10^{-7}\ M$) produces a modest increase in ventricular contraction and heart rate.[134] Likewise, some studies involving isolated (suffused, spontaneously beating) right atria of guinea pigs have indicated that topically applied NPY ($5 \times 10^{-7}\ M$) produced a transient increase in atrial contraction and beat rate,[150] whereas other studies have indicated that the basal tone and force displacement with respect to time, in response to field stimulation, are not altered in papillary muscles isolated from rat, cat, and guinea pig hearts and that the contractility and beat rate of the spontaneously beating atria of guinea pigs were not changed with respect to topically applied NPY (3×10^{-9}, 1×10^{-6} M).[151] There is, however, general agreement in some areas. Unlike the situation in arterial tissue, in heart tissue NPY is unable to potentiate the positive inotropic effects produced by NA;[134,150] like the situation present in arteries, NPY is capable of inhibiting [³H]NA overflow from heart tissue and in this manner is capable of inhibiting the positive inotropic chronotropic effects produced by the stimulation of cardiac sympathetic nerves.[134,150] These observations suggest that in cardiac tissue the primary function of nerve-released NPY is to act as a presynaptic modulator of NA release as opposed to a potentiator of NA action.

8. The Normal and Pathophysiological Function of NPY in the Vasculature

As is the situation with other neuropeptides, the possible pathophysiological roles played by NPY in cardiovascular disease are speculative in nature. Plasma concentrations of NPY are dramatically elevated in certain diseases. In humans with pheochromocytomas and/or ganglion neuroblastomas, plasma levels of NPY can be elevated to 460 pmol/l from normal values of 55 pmol/l.[117] Such concentrations of NPY exceed the threshold levels that have been observed to potentiate contractile responses induced by NA in rat femoral arteries[62] and approach the levels required to constrict coronary vasculature of guinea pigs directly.[132] In this regard, patients with pheochromocytomas exhibit a high incidence of hypertension which represents 0.1% of all hypertensive humans.[109] Since in these patients plasma adrenaline and noradrenaline levels are also elevated,[117] the elevated levels of NPY could enhance the vasoconstrictive effects of the circulating catecholamines and the endogenous action of the sympathetic nerves, thus facilitating hypertension development.

A side effect encountered in patients with pheochromocytomas is the occurrence of angina in the absence of coronary artery disease.[109] This cannot be explained by the constrictive effects of high circulating levels of adrenaline and noradrenaline, since these two agents relax the coronary vasculature via β-receptor stimulation.[152] In view of the high sensitivity of the coronary vasculature to NPY constriction, it is possible that in this disease the high levels of circulating NPY could contribute to the occurrence of angina.

NPY also could play a role in other forms of hypertension. Studies involving adrenalectomized rats treated with corticosteroids indicate the presence of a sixfold elevation in plasma renin activity when compared with similarly treated animals having intact adrenals.[153] In these experiments, it was shown that the infusion of NPY (0.1 µg/min, for 30 min) normalized the plasma renin activity in the rats to levels comparable to those present in animals having intact adrenals. The effects of NPY on renin activity could not be explained on the basis of hemodynamic changes, since in these experiments the infusion of NPY produced only a modest change in blood pressure (+5 mmHg mean arterial pressure) which occurred in both groups of animals; however, a large suppression in renin activity was observed only in the adrenalectomized animals. The mechanism by which NPY effected a decrease in renin activity is unknown, but one possibility could be that NPY is directly involved in the suppression of renin release. NPY-containing nerves have been demonstrated to exist in close proximity to the renal juxtaglomerular apparatus and are in a position where they could effect renin release.[154] In certain forms of high-renin hypertension, such as that produced by the coarctation of the aorta (proximal to the renal arteries) in rats, NPY-containing nerves virtually disappear from the distal regions of the renal afferent arterioles containing the juxtaglomerular apparatus.[154] If such nerves were involved in the inhibition of NPY release, their disappearance could play a role in producing the elevated plasma renin levels observed in this form of hypertension and, hence, contribute to the development of high blood pressure in the animals.

NPY could also possibly play a role in low-renin forms of hypertension. It has long been known that high salt intake represents a risk factor that potentially can predispose an individual to hypertension development.[155] Because plasma renin levels are inversely related to salt intake,[156,157] the renin angiotensin system probably is not involved in these forms of hypertension. In partially nephrectomized animals fed a high salt intake, hypertension is initially associated with a transient increase in plasma volume followed by an increase in peripheral vascular resistance.[158] It has been shown that, in partially nephrectomized rats, plasma levels of NPY, but neither adrenaline nor NA, are elevated in proportion to the salt content of the diet.[159] If NPY released into the plasma were derived from the sympathetic nerves, an intriguing but speculative possibility is that elevations in dietary NaCl might promote a selective increase in the release of NPY in relation to NA, which in turn might potentiate NA-induced contraction and, thus, contribute to the increase in vascular resistance.

NPY is also increased within the plasma of animals subjected to hemorrhage.[160] Here it has been suggested that elevated release of NPY in conjunction with NA and adrenaline could be beneficial in opposing the hypotensive effects of hemorrhage. In this regard, the infusion of NPY into adrenalectomized rats at levels below those capable of producing direct change in blood pressure inhibits hypotension produced in response to endotoxic shock.[161] This latter observation suggests that infusion of NPY could be of benefit in the treatment of hypotension associated with shock.

V. OPIOID PEPTIDES

A. LOCATION OF OPIOID PEPTIDES IN THE CARDIOVASCULATURE

There is evidence available indicating that at least some (if not all) sympathetic nerves contain Leu-enkephalin (Leu-ENK) and/or Met-enkephalin (Met-ENK). Enkephalin-immunoreactive (ENK-IR) cell bodies have been observed within a variety of sympathetic ganglia of guinea pigs and rats[162] and in cat, rat, and guinea pig adrenal glands.[163] ENK-IR nerve fibers have also been observed within dog and bovine vas deferens,[164,165] and the blood vessels associated with the vas deferens and α-neoendorphin-immunoreactive nerve fibers have been found surrounding the coronary vasculature and myocytes of the heart.[165,166] Additional evidence suggesting the presence of ENK within sympathetic nerves is the observation in the vas deferens of various species that the tissue content of ENK varies directly in proportion to the density of sympathetic innervation present[165] and that the release of both ENK and NA from bovine vas deferens decreases in a proportionally similar manner when the tissue is treated with guanethidine.[164]

The vas deferens of a wide variety of species contain greater quantities of Met-ENK than Leu-ENK in a ratio between 1.2 and 2.9:1.[165] Likewise, immunofluorescence studies indicate that the sympathetic ganglia of rats and guinea pigs react to a greater degree and intensity for antibodies against Met-ENK as opposed to Leu-ENK.[162] These studies suggest that Met-ENK, as opposed to Leu-ENK, is the principal enkephalin present within the peripheral nervous system.

Using immunogold electron microscope techniques, Met-ENK has been localized in large (as opposed to small) dense core vesicles in the sympathetic nerves of ox vas deferens.[164] Consistent with this observation, other studies involving cell fractionation, in combination with differential centrifugation, have indicated that cell fractions thought to contain large dense core vesicles also contain high opioid-like immunoreactivity[167] or a large Met- and Leu-ENK content,[168] whereas such activity is absent in fractions thought to contain small dense core vesicles.[167]

B. THE PHARMACOLOGY OF OPIOID PEPTIDES ON THE VASCULATURE

The action of opioids on isolated blood vessels appears dependent on the species and category of blood vessels studied. Studies indicate that endogenous and synthetic opioid compounds, such as morphine, 1-pentazocine, cyclazocine, and levallorphan, are capable of contracting the rat aorta[169,170] and enhancing the spontaneous mechanical activity in the rat portal vein.[170] On the other hand, other studies indicate that natural and synthetic opioids relax rat mesenteric arterioles[170] and arteries,[171] but have little effect on venules.[170] In the case of cat middle cerebral[172] and pial arteries,[173] morphine and/or enkephalins produce vasodilation, whereas morphine elicits contraction in the rat basilar artery.[174] The contractile responses produced in the rat aorta by the synthetic opioid 1-pentazocine can be inhibited by decreasing the Ca^{2+} content of the physiological saline or introducing calcium antagonists (verapamil, SKF-525A) into the media.[169] This suggests that opioid contraction is being produced within this tissue via the use of an external (as opposed to internal) Ca^{2+} source through the activation of potential sensitive Ca^{2+} channels. In cat cerebral arteries it was shown that the application of morphine was associated with a decrease in VSMC membrane resistance, probably produced by an increase in VSMC potassium conductance which concomitantly hyperpolarized the VSMC membrane.[172] Such hyperpolarization in turn was associated with relaxation.[172] On the other hand, contraction of rat basilar arteries by morphine was demonstrated to be associated with an increase in membrane resistance which was produced by a decrease in potassium conductance; this in turn produced VSMC membrane depolarization that coincided with contraction.[174] In summary, it appears that, unlike the action of some agonists which demonstrate pharmacomechanical form of contractile coupling, both contraction and relaxation responses produced by opioids are highly electromechanically coupled.

Various physiological and pharmacological experiments using intact animals and isolated tissues indicate that opioids are capable of activating at least four different types of opiate receptors (for a review see Reference 175). Generally, enkephalins are more potent agonists of the δ-receptor, whereas morphine and its related compounds are more selective for the μ-receptor.[176] In this regard, the synthetic opioids RX783006[177] (termed DAGO) and morphiceptin[176] are highly selective for the μ-receptor, whereas D-Ala^2,D-Leu^5-enkephalin (DADL) and D-Pen^2,D-Pen^5-enkephaline are selective agonists of the δ-receptor.[178] There are no highly selective agonists for the κ receptor. However, benzomorphans, such as ketocyclazocine and bremazocrine, are thought to be more selective at this as opposed to other opioid receptor sites,[179,180] whereas the agonist SKF10,047 is thought to be selective for the σ-opioid receptor.[180] Radioactive ligand binding studies have demonstrated the presence of δ-, μ-, κ-, and σ-receptors within the human brain,[181] and pharmacological studies involving the ileum and vas deferens of various species have demonstrated the presence of δ-, μ-, and κ-receptors on SMCs.[77,175,177,182] Selective antagonists for the individual receptors have not been developed yet. However, naloxone in low concentrations can act selectively at the μ-receptor, whereas the antagonist MR2266 can bind selectively to μ- and κ-receptors while diprenorphine has high binding affinities to μ-, δ-, and κ-receptor sites.[182]

The types of opiate receptors present on the VSMC membrane have not been fully characterized. There is, however, indirect evidence to suggest that at least two or more types of opiate receptors do exist postsynaptically. The observation that both morphine (μ-selective) and enkephalins (δ-selective) relax blood vessels would suggest the presence of μ- and δ-opiate receptors. In this regard, naloxone is a much more potent antagonist of the μ- than the δ-receptor.[177,175] The observation that the application of endogenous or synthetic opioids to isolated arteries results in variations in the ability of naloxone to inhibit the myogenic responses (from partial to total inhibition)[169-172,174] is consistent with the possibility that postsynaptic μ- and δ-receptors could be present. In other studies it was demonstrated that the infusion of Leu-ENK, but not morphine, produces vasoconstriction of the isolated perfused pulmonary vasculature, which cannot be inhibited by naloxone or diprenorphine.[183] Considering that diprenorphine inhibits responses attributable to μ-, δ-, and κ-receptors,[182] the inability of this antagonist to inhibit pulmonary vascular vasoconstriction in response to Leu-ENK would suggest that if these responses were mediated via an opiate receptor the receptor is not of a μ-, δ-, or κ-type.

C. OPIOID PEPTIDES AS NEUROTRANSMITTERS IN THE CARDIOVASCULAR SYSTEM

Studies involving this tissue have demonstrated the release of α-neoendorphin + dynorphin$_{1-8}$ into the suffusing or perfusing solution in response to the stimulation of the acceleran fibers of the heart. In this preparation, NA was also found to be coreleased with the opioid peptides.[166] The treatment of the tissue with the sympathetic neurotoxic agent 6-hydroxydopamine depleted dynorphin$_{1-8}$ α-neoendorphin, and Leu-ENK from the guinea pig heart,[166] thus further confirming that the opioid peptides were located within the sympathetic nerves. In these studies, phentolamine augmented and cocaine inhibited the release of α-neoendorphin during the stimulation of the acceleran fibers.[166] This observation was interpreted as indicating that the release of α-neoendorphin from the sympathetic nerves was under the inhibitory control of presynaptic α- (presumably α$_2$-)receptors for the following reasons. In these experiments, the blockade of the presynaptic α$_2$-receptors with phentolamine might be expected to augment the release of opioid from the sympathetic nerves; on the other hand, the treatment of the tissue with cocaine should inhibit the presynaptic uptake of NA, increasing synaptic NA levels and, thus, augmenting the presynaptic stimulation of α$_2$-receptors. Hence, based on this model, both phentolamine and cocaine might be expected to respectively augment and decrease the release of the opioid from the sympathetic nerves. Studies involving spontaneously beating guinea pig atria indicated that dynorphin$_{1-8}$ exerted a negative inotropic effect on the heart which could be inhibited by κ-receptor-selective antagonists and could not be mimicked via the application of μ- or δ-receptor-selective agonists.[166] These studies strongly imply that opioid compounds coreleased with NA could act as neurotransmitters within the guinea pig heart.

In blood vessels, the most convincing evidence that neurally released opioids exert a postsynaptic effect comes from the work by Kannan and Seip.[171] In these studies it was demonstrated that certain segments of the rat mesenteric artery exhibited relaxation in response to field stimulation. The relaxation responses were inhibited by tetrodotoxin, guanethidine, or naloxone. In addition, topically applied dynorphin$_{1-13}$ also elicited relaxation responses in the tissue that could be inhibited by naloxone. These results suggest that the arterial relaxation observed in response to field stimulation appeared to be mediated via an opioid released from the sympathetic nerves. Additional characterization of the response indicated that relaxation in response to field stimulation could be inhibited by SP or indomethacin at concentrations that did not alter VSMC tone. Since, at the doses used, neither SP or indomethacin altered the relaxation response to topically applied dynorphin$_{1-13}$ substance P and indomethacin were thought to be capable of presynaptically modulating the neural release of the unknown opiate compound.

D. THE ROLE OF OPIOID PEPTIDES AS PRESYNAPTIC MODULATORS OF NORADRENALIN RELEASE

Arterial studies involving the measurement of [^{3}H]NA overflow during neural stimulation, pharmacological studies analyzing neural-mediated contractile responses, and electrophysiological studies have indicated that presynaptic opioid receptors exist on sympathetic nerves.[184-190] When stimulated, opioid receptors inhibit the release of NA and via this mechanism decrease sympathetically mediated vasoconstrictive responses. Various studies have been attempted to characterize the presynaptic opiate receptors.[184-190] Such studies indicate the presence of δ + κ (rabbit iliocolic,[189] tail,[188] jejunal[190] arteries), δ (rabbit mesenteric arteries[186,187]), κ (rabbit pulmonary arteries[184]), and ε (rat tail[185]) presynaptic receptors. These studies have failed to find evidence suggesting the presence of a presynaptic μ-receptor that is capable of modulating NA release. Recent studies have indicated that the presynaptic opioid receptors are most efficient at inhibiting NA release from sympathetic nerves when short trains of pulses are used.[188] In this regard, it has been suggested that the release of greater quantities of NA during repetitive pulses results in the feedback inhibition of NA release via presynaptic α$_2$-receptors. Such inhibition, in turn, is thought to overwhelm the potential inhibition of NA release that might be produced by opioid receptors.[188] Consistent with this view, α$_2$-blockade with yohimbine enhances the ability of opioids to exert a presynaptic inhibition of NA release during repetitive stimulation,[187] and in some tissues κ-receptor-mediated presynaptic inhibition of NA release is only observed when presynaptic α$_2$-receptors are blocked.[190]

At present, it is not known whether opioids coreleased with NA actually limit the release of NA during sympathetic nerve stimulation. In the view of this author, the importance of opioids in this process probably varies from artery to artery and between species. In some preparations (rat tail arteries, rabbit pulmonary arteries, and mesenteric arteries), the addition of 1 μM naloxone has virtually no effect on the degree of nerve-stimulated [^{3}H]NA release, vasoconstriction, or the amplitude of EJPS produced during nerve stimulation.[184,185,187] However, as discussed previously, naloxone at low doses acts more selectively as a μ-receptor antagonist,[175] and such receptors have not been shown to present presynaptically on sympathetic nerves.[184-190] In some studies (i.e., rabbit pulmonary artery), where the effects of 1 vs. 10 μM naloxone have been compared, 10 as opposed to 1 μM levels of naloxone (which might antagonize the δ-receptor to a greater degree) have been particularly effective in enhancing the release of [^{3}H]NA during sympathetic nerve stimulation.[184] This suggests that endogenous opioids coreleased with NA may be involved in modulating NA release. On the other hand, other studies have failed to demonstrate such an effect with the use of 10 μM naloxone or the δ-receptor selective antagonist ICI-154129.[189] In this regard, opioids may not be universally distributed within all sympathetic nerves,[171] and the sympathetic nerves of some blood vessels may have presynaptic opioid receptors, but may not release opioids along with NA. Alternatively, some sympathetic nerves may release opioids only at certain levels of stimulation where the autoinhibition of NA release is dominated by presynaptic α$_2$-receptors, thus masking the presynaptic effects of any opioid coreleased with NA.

E. THE NORMAL AND PATHOPHYSIOLOGICAL FUNCTION OF OPIOID PEPTIDES IN THE CARDIOVASCULAR SYSTEM

The systemic effects of opioids are complex and encompass the activation of the central and peripheral nervous systems as well as opiate receptors on cardiac and vascular muscle cells. In this regard, excellent reviews on the cardiovascular effects of opioids are available (see Reference 178).

Opioids can be released into the circulation, where they can act as hormones. In this regard, Met-ENK is coreleased with catecholamines from the adrenal medulla,[191] while β-endorphin is coreleased into the circulation with adrenocorticotropin from the pituitary gland.[192] It is also possible that opioids such as dynorphin are released from nerves within tissues such as the heart.[166,178] Experiments elucidating the role that these circulating opioid peptides play with regard to normal and pathophysiological cardiovascular function are limited to two general types, which involve the study of either the systemic effects of injected opioids or the effects of the opioid receptor antagonists on normal and pathological physiological function.

In dogs treated with reserpine, splanchnic nerve stimulation promotes a release of Met-ENK and produces a systemic hypotensive effect that (1) is dependent on the frequency of nerve stimulation and (2) can be blocked by naloxone.[191] The systemic injection of Met-ENK into the hind limb vasculature of rabbits also produces a systemic hypotensive response which can be blocked by naloxone. The response also can be blocked by ganglionic blockade and within the region of the hindquarters may result from the presynaptic inhibition of NA release. In addition, however, Met-ENK infused into this area initiates a reflex which increases parasympathetic drive to the heart, producing bradycardia, and depresses sympathetic activity in other systemic vasculatures, thus producing systemic hypotension in the animal.[191] Comparable effects are observed in rats, where the systemic injection of morphine produces hypotension and bradycardia via reflex actions that enhance vagal nerve activity and depress sympathetic motor tone.[193] The peripheral systemic injection of β-endorphins into rats at levels 1/200 to 1/300th of those capable of producing analgesia in rats also depresses blood pressure.[194] In these studies it was demonstrated that the depressor effects of β-endorphin were greatly reduced if the rats were pretreated with *p*-chlorophenylalanine (an agent which depletes serotonin from serotonergic nerves), suggesting that serotonin-containing nerves might be involved in the depressor action of β-endorphin.

In view of the systemic hypotensive effects of opioid peptides, it has been suggested that opioids may be involved in initiating hypotension associated with shock. Humans suffering from shock (secondary to heart failure, sepsis, hemorrhage, and hypovolemia) as well as experimental animals subjected to hemorrhagic or endotoxic shock exhibit very dramatic increases in plasma β-endorphin levels.[195-197] In the case of experimental animal studies, the increase in β-endorphin levels has been shown to coincide chronologically with the development of hypotension.[196] In other studies it has been demonstrated that the mortality of mice subjected to anaphylactic shock increases dramatically if morphine is administered systemically to the animal,[198] whereas the s.c. or intracerebroventricular administration of naloxone decreases the mortality of the animals in a dose-dependent manner.[199] Comparable studies involving surgical stress in rabbits have also indicated that mortality is decreased in animals pretreated with β-endorphin antisera.[200] Still other studies have shown that naloxone reverses hypotension in endotoxic and hemorrhagic shock.[201,202] In this regard, it appears that δ- and κ-receptor but not μ-receptor antagonists are capable of improving cardiac function during shock (for a review see Reference 203).

At present, the specific mechanism through which opiates act to induce hypotensive shock are unknown; however, in certain forms of shock, central and not peripheral opiate receptors are probably involved. This is demonstrated by the observation that animal mortality during endotoxic shock decreases when the opioid receptor antagonist naltrexone is injected intracerebrally into animals, but not under conditions where the drug is administered subcutaneously.[198] Other studies have indicated that, during severe hypotensive hemorrhage in rabbits, renal sympathetic nerve activity actually decreases and that the decrease in activity can be blocked by naloxone.[204] This study suggests that hypotension during shock may be produced via a central nervous system, opiate receptor-mediated decrease in peripheral sympathetic activity.

The involvement of opioids in other disease processes is less clear. In the case of hypertension, studies involving humans have demonstrated suppressed plasma β-endorphin levels in patients with primary hypertension. In this study, clonidine treatment normalized both the plasma β-endorphin levels and the blood pressure of the patients.[205] The direction of change in plasma β-endorphin levels within this study is consistent with what might be expected if β-endorphins were involved in the maintenance of high blood pressure in this form of hypertension. On the other hand, studies involving two-kidney, one-clamp forms of hypertension in rats indicated that naloxone decreased the blood pressure of the animals.[206] Still other studies failed to observe any significant change in blood pressure with naloxone treatment in this model or in rats with genetic hypertension.[207] Studies comparing SHR and Kyoto Wistar normotensive rats (WKY) have failed to demonstrate differences in plasma β-endorphin levels at rest or under stressed conditions or any differences in the ability of β-endorphins to inhibit sympathetic nerve-mediated vasoconstriction in the tail artery.[208] Nevertheless, there is a great deal of evidence that opiate mechanisms pertaining to the central nervous system are altered in SHR when compared to WKY (for a review see Reference 191).

There also is evidence available suggesting that opioids may contribute to the damage produced during head and spinal cord injuries. Studies involving cats have demonstrated that experimental trauma to the cervical spine is associated with hypotension, a reduction in spinal cord blood flow, an increase in plasma levels of β-endorphin, and an accumulation of dynorphin A at the sight of injury.[209] In these studies, naloxone-treated animals were found to have less prominent spinal cord abnormalities and significantly improved neurological recovery when compared to nontreated controls.[209] In other studies involving head-injured cats, the site of the injury as well as other ischemic areas were also found to accumulate dynorphin A. The administration of the κ-opioid receptor-selective antagonist Win 44,441-3 improved the mean arterial pressure, the EEG profile, and the regional cerebral blood flow, reduced the severity of hemorrhage, and improved survival after brain injury.[210] The mechanisms through which opioids potentiate neurological injury are not well understood. However, the accumulation of dynorphin within the neural tissue is an important development, since the intrathecal infusion of dynorphin into the lumbar region of the spine alone can produce hind limb paralysis in rats.[211] To date, an adequate explanation as to the mechanism by which opioids act to enhance neural destruction during injury has not been provided. However, hypotheses have centered around the possibility that opioids might be responsible for producing regional ischemia in the area of the injured region, thus worsening the situation.[210]

VI. SUMMARY

CGRP within at least some nerves is colocalized with SP and neurokinin A. Peripherally, CGRP nerves originate from the axodendritic processes of cell bodies present within the central nervous system, where they are thought to convey sensory information into the CNS. Stimulation of CGRP cell bodies leads to the antidromic release of CGRP into the vasculature, where CGRP mediates vasodilatory effects via nonendothelial-dependent mechanisms. Nerve-released CGRP also produces an increase in atrial heart muscle contractility and beat frequency. CGRP also can alter SMC membrane properties in a manner whereby SMC excitability in response to NA + ATP is reduced.

CGRP also may act as a neurohormone. Upper normal ranges of CGRP levels present in human serum exceed the threshold levels required to relax some cerebral arteries *in vivo*. Malfunctions in CGRP physiology have not been implicated in specific disease processes; however, CGRP has been demonstrated to induce reddening of the skin as well as flare reactions that are associated with edema. In addition, CGRP potentiates the action of SP and other mast cell products, enhancing the extravascular movement of plasma. In this regard, the hypersecretion of CGRP/SP from nerves could be involved in the production of long-lasting erythema.

VIP is predominantly present within parasympathetic autonomic nerves, where it is coreleased with ACh and PHI. VIP-IR nerves are widespread throughout the arterial vasculature and are less prominent within the venous system. Both VIP and PHI produce a nonendothelial-mediated arterial relaxation, and experimental evidence exists indicating that VIP can act as a vascular neurotransmitter.

Evidence suggests that VIP and PHI are released preferentially from the parasympathetic nerves under conditions of high frequencies of nerve stimulation, whereas ACh is released selectively at low stimulation frequencies. Both ACh and VIP release from parasympathetic nerves can be inhibited via the stimulation of presynaptic muscarinic and VIP receptors. In the case of sympathetic nerves, VIP does not act presynaptically to alter NA release, but via postsynaptic effects can inhibit NA contraction of arteries.

Under normal circumstances, the circulatory levels of VIP are low, below the levels that are needed to produce cardiovascular effects. However, in individuals with VIP-secreting tumors, VIP levels can increase dramatically to levels that surpass those required to elicit vasodilation in arteries.

Decreases in VIP innervation and a functional decrease in the release of VIP in response to nerve stimulation have been noted in diabetic humans and in rats. In view of the important role that VIP plays in penile erection, it could be possible that alterations of VIP innervation observed during diabetes could contribute to impotence which is often associated with the disease. In the cerebrovasculature the disappearance of VIP innervation during diabetes might eliminate neurovasodilatory control over the circulation, thus permitting enhanced myogenic constriction. Such an alteration could predispose diabetics to cerebral ischemia and enhance stroke development. A disappearance of VIP innervation has also been observed after cerebral hemorrhage, where it possibly contributes to the occurrence of vasospasm.

NPY is present within virtually all sympathetic nerves, where it is coreleased with NA. In addition, some parasympathetic nerves have also been found to contain NPY. NPY is released preferentially at high frequencies of stimulation and is capable of (1) increasing the contractile sensitivity (i.e., decreasing EC_{50} values) of NA and other agonists, (2) acting presynaptically to inhibit the release of NA from sympathetic nerves, and (3) eliciting a direct vasoconstrictor response. All three types of actions exhibited by NPY do not always exist within the same vascular segment, and it is possible that arteries are endowed selectively with variations of the above characteristics in order to suit their function. In addition, it has been suggested that NPY could serve as a modulator of sympathetic activity, acting to conserve noradrenergic transmitters and to promote a more even distribution of transmitter release throughout the sympathetic plexus.

Under certain pathological conditions (i.e., pheochromocytomas), plasma levels of NPY are increased dramatically. Under these conditions, the elevated levels of NPY could potentiate the contractile effects of adrenaline and NA, which are also elevated and facilitate the initiation of hypertension. In addition, the elevated NPY levels could possibly contract the coronary vasculature and contribute to the symptoms of angina often observed in the disease. The disappearance of NPY nerves from the juxtaglomerular areas of the kidney in high-renin forms of renal hypertension suggests that NPY could possibly play a role in this latter form of hypertension. Likewise, the observation that plasma NPY levels increase with salt loading also suggests a possible role of NPY in high-salt-induced (low-renin) forms of hypertension.

Opioid peptides have been observed within cells of the sympathetic ganglion and adrenal medulla and in nerves surrounding the blood vessels of the vas deferens, coronary arteries, and heart tissue.

The possibility that opioids act as neurotransmitters in the cardiovascular system has not been studied extensively. However, within the heart, experimental evidence is present which demonstrates that α-neoendorphin (1) is present within nerves that innervate the heart, (2) is released by nerve stimulation, (3) is probably sympathetic in origin and is subject to presynaptic inhibitory control, and (4) probably exerts a κ-receptor-mediated negative inotropic effect on the heart. Other studies have also demonstrated that an unknown sympathetic neurally released opioid is capable of relaxing mesenteric arteries. In addition to acting as a neurotransmitter, opioids coreleased with NA also could be involved in the presynaptic inhibition of NA release.

The role that opioids play in cardiovascular disease is unknown; however, studies have shown that the survival rate after shock is improved by the systemic or central injection of opioid antagonists, suggesting that opioids may be involved in this condition. Other studies have demonstrated that the systemic release of opioids and the accumulation of dynorphin within sites of cerebral or spinal neural injury potentiate the degree of damage produced in the area. In these latter instances, opioids complicate the injury by directly or indirectly producing ischemia within the injured site.

ACKNOWLEDGMENT

Dr. Smeda is supported by the Heart and Stroke Foundation of Ontario.

REFERENCES

1. **Amara, S. G., Jonas, V., Rosenfeld, M. G., Ong, E. S., and Evans, R. M.,** Alternative RNA processing in calcitonin gene expression generates mRNAs encoding different polypeptide products, *Nature (London)*, 298, 240, 1982.

2. **Rosenfeld, M. G., Mermod, J.-J., Amara, S. G., Swanson, L. W., Sawchenko, P. E., Rivier, J., Vale, W. W., and Evans, R. M.,** Production of a novel neuropeptide encoded by the calcitonin gene via tissue and specific RNA processing, *Nature (London)*, 304, 129, 1983.

3. **Uddman, R., Edvinsson, L., Ekman, R., Kingman, T., and McCulloch, J.,** Innervation of the feline cerebral vasculature by nerve fibers containing calcitonin gene-related peptide: trigeminal origin and co-existence with substance P, *Neurosci. Lett.*, 62, 131, 1985.

4. **Edvinsson, L., Ekman, R., Janssen, I., McCulloch, J., and Uddman, R.,** Calcitonin gene-related peptide and cerebral blood vessels: distribution and vasomotor effects, *J. Cereb. Blood Flow Metab.*, 7, 720, 1987.

5. **Hanko, J. E., Harebo, J., Kahrstrom, J., Owman, C., and Sundler, F.,** Existence and co-existence of calcitonin gene related peptide (CGRP) and substance P in cerebrovascular nerves and trigeminal ganglion cells, *Acta Physiol. Scand.*, 552, 29, 1986.

6. **Edvinsson, L., Brodin, E., Jansen, I., and Uddman, R.,** Neurokinin A in cerebral vessels characterization localization and effects *in vitro, Regul. Peptides*, 20, 181, 1988.

7. **Gulbenkian, S., Merighi, A., Wharton, J., Varndell, I. M., and Polak, J. M.,** Ultrastructural evidence of the co-existence of calcitonin gene related peptide and substance P in secretory vesicles of peripheral nerves in the guinea pig, *J. Neurocytol.*, 15, 535, 1986.

8. **Tsai, S.-H., Tew, J. M., McLean, J. M., and Shipley, M. T.,** Cerebral arterial innervation by nerve fibers containing calcitonin gene-related peptide (CGRP). I. Distribution and origin of CGRP perivascular innervation in the rat, *J. Comp. Neurol.*, 271, 435, 1988.

9. **Hanko, J., Hardebo, J. E., Kahrstrom, J. K., Owman, C., and Sundler, F.**, Calcitonin gene-related peptide is present in mammalian cerebrovascular nerve fibers and dilates pial and peripheral arteries, *Neurosci. Lett.*, 57, 91, 1985.

10. **Gibson, S. J., Polak, J. M., Bloom, S. R., Sabate, I. M., Mulderry, P. M., Gatei, M. A., McGregor, C. P., Morrison, J. F. B., Kelly, J. S., Evans, R. M., and Rosenfeld, M. G.**, Calcitonin gene related peptide immunoreactivity in the spinal cord of man and eight other species, *J. Neurosci.*, 4, 3101, 1984.

11. **Sasaki, Y., Hayashi, N., Kasahara, A., Matsuda, H., Fusamoto, H., Sato, N., Hillyard, W., Girgis, S., MacIntyre, I., Emson, P. C., Shiosaka, S., Tohyama, M., Shiotani, Y., and Kamada, T.**, Calcitonin gene-related peptide in the hepatic and splanchnic vascular systems of the rat, *Hepatology*, 6, 676, 1986.

12. **Rodrigo, J., Polak, J. M., Fernandez, L., Ghatei, M. A., Mulderry, P., and Bloom, S. R.**, Calcitonin gene related immunoreactive sensory motor nerves of the rat, cat and monkey esophagus, *Gastroenterology*, 88, 441, 1985.

13. **Kawasaki, H., Takasaki, K., Saito, A., and Goto, K.**, Calcitonin gene-related peptide acts as a novel vasodilator neurotransmitter in the mesenteric resistance vessels of rat, *Nature (London)*, 335, 164, 1988.

14. **Lembeck, F. and Gamse, R.**, Substance P in the peripheral sensory processes, *Ciba Found. Symp.*, 91, 35, 1982.

15. **Edvinsson, L., Ekman, R., Jansen, I., Ottosson, A., and Uddman, R.**, Peptide containing nerve fibers in human cerebral arteries: immunocytochemistry, radioimmunoassay and *in vitro* pharmacology, *Ann. Neurol.*, 21, 431, 1987.

16. **Verrecchia, C., Hamel, E., Edvinsson, L., Mackenzie, E. T., and Seylaz, J.**, Role of the endothelium in pial artery responses to several vasoactive peptides, *Acta Physiol. Scand.*, 127 (Suppl. 552), 33, 1986.

17. **Ezra, D., Laurindo, R. M., Goldstein, D. S., Goldstein, R. E., and Feuerstein, G.**, Calcitonin gene-related peptide: a potent modulator of coronary blood flow, *Eur. J. Pharmacol.*, 137, 101, 1987.

18. **D'Orléans-Juste, P., Dion, S., Drapeau, G., and Regoli, D.**, Different receptors are involved in the endothelium-mediated relaxation and the smooth muscle contraction of rabbit pulmonary artery in response to substance P and related neurokinins, *Eur. J. Pharmacol.*, 125, 37, 1985.

19. **Franco-Cerecedo, A., Rudehill, A., and Lundberg, J. M.**, Calcitonin gene-related peptide but not substance P mimics capsaicin-induced coronary vasodilation in the pig, *Eur. J. Pharmacol.*, 142, 235, 1987.

20. **Saito, A., Masaki, T., Uchiyama, Y., Lee, T. J. F., and Goto, K.**, Calcitonin gene related peptide and vasodilator nerves in large cerebral arteries of cats, *J. Pharmacol. Exp. Ther.*, 248, 455, 1989.

21. **Goadsby, P. J., Edvinsson, L., and Ekman, R.**, Release of vasoactive peptides in the extracerebral circulation of humans and the cat during activation of the trigeminovascular system, *Ann. Neurol.*, 23, 193, 1988.

22. **Miyauchi, T., Ishikawa, T., Sugishita, Y., Saito, A., and Goto, K.**, Effects of capsaicin on nonadrenergic noncholinergic nerves in the guinea pig atria: role of calcitonin gene-related peptide as cardiac neurotransmitter, *J. Cardiovasc. Pharmacol.*, 10, 675, 1987.

23. **Mulderry, P. K., Ghatei, M. A., Podrigo, J., Allen, J. M., Rosenfeld, M. G., Polak, J. M., and Bloom, S. R.**, Calcitonin gene related peptide in the cardiovascular tissue of the rat, *Neuroscience*, 14, 947, 1985.

24. **Miyauchi, T., Ishikawa, T., Sugishita, Y., Saito, A., and Goto, K.**, Effect of piperine on calcitonin gene related peptide (CGRP) containing nerves in the isolated rat atria, *Neurosci. Lett.*, 91, 222, 1988.

25. **Saito, A., Kimura, S., and Goto, K.**, Calcitonin gene related peptide as a potential neurotransmitter in guinea pig right atrium, *Am. J. Physiol.*, 250, H693, 1986.

26. **Sigrist, S., Franco-Cereceda, A., Muff, R., Henke, H., Lundberg, J. M., and Fischer, J. A.**, Specific receptor and cardiovascular effects of calcitonin gene related peptide, *Endocrinology*, 119, 381, 1986.

27. **Saito, A., Ishikawa, T., Kimura, S., and Goto, K.**, Role of calcitonin gene-related peptide as a cardiotonic neurotransmitter in guinea pig left atria, *J. Pharmacol. Exp. Ther.*, 243, 731, 1987.

28. **Franco-Cerecedo, A., Gennari, C., Nami, R., Agnusdei, D., Pernow, J., Lundberg, J. M., and Fischer, J. A.**, Cardiovascular effects of calcitonin gene related peptides I and II in man, *Circ. Res.*, 60, 393, 1987.

29. **Ohhashi, T. and Jacobowitz, D. M.**, Effects of calcitonin gene-related peptide on the neuroeffector mechanism of sympathetic nerve terminates in rat vas deferens, *Peptides*, 6, 987, 1985.

30. **Al-Kazwini, S. J., Graig, R. K., and Marshall, I.**, Postjunctional inhibition of contractor responses in the mouse vas deferens by rat and human calcitonin gene related peptides, *Br. J. Pharmacol.*, 88, 173, 1986.

31. **Goto, K., Kimura, S., and Saito, A.**, The inhibitory effect of calcitonin gene related peptide on excitation and contraction of smooth muscle of the rat vas deferens, *J. Pharmacol. Exp. Ther.*, 217, 516, 1987.

32. **Girgis, S. I., MacDonald, D. W. R., Stevenson, J. C., Brevis, P. J. R., Lynch, C., Wimalawansa, S. J., Selt, C. H., Morris, H. R., and MacIntyre, I.**, Calcitonin gene-related peptide a potent vasodilator and major product of the calcitonin gene, *Lancet*, 2, 14, 1985.

33. **Zaidi, M., Bevis, P. J. R., Gigis, S. I., Lynch, C., Stevenson, J. C., and MacIntyre, I.**, Circulating CGRP comes from the perivascular nerves, *Eur. J. Pharmacol.*, 117, 283, 1985.

34. **Zaidi, M., Breimer, L. H., and MacIntyre, I.**, Biology of peptides from the calcitonin genes, *Q.J. Exp. Physiol.*, 72, 371, 1987.

35. **Brown, M. J., and Morice, A. H.**, Clinical pharmacology of vasodilator peptides, *J. Cardiovasc. Pharmacol.*, 10 (Suppl. 12), S82, 1987.

36. **Ray, B. S. and Wolfe, H. G.**, Experimental studies on headache, *Arch. Surg.*, 41, 813, 1940.

37. **Mayberg, M., Langer, R. S., Zervas, N. T., and Moscowitz, M. A.**, Perivascular meningeal projections from the cat trigeminal ganglia: possible pathway for vascular headaches in man, *Science*, 213, 228, 1981.

38. **Moskowitz, M. A.**, The neurobiology of vascular head pain, *Ann. Neurol.*, 16, 157, 1984.

39. **Kroatila, K., Uusitalo, H., and Palkama, A.**, Effect of neurogenic mutation and calcitonin gene-related peptide (CGRP) on ocular blood flow in the rabbit, *Curr. Eye Res.*, 7, 695, 1988.

40. **Jefferson, G.**, Observations on trigeminal neuroglia, *Br. Med. J.*, 2, 879, 1931.

41. **Rowbotham, G. F.**, Observations on the effects of trigeminal denervation, *Brain*, 62, 364, 1939.

42. **Brain, S. D., Tippins, J. R., Morris, H. R., MacIntyre, I., and Williams, T. J.**, Potent vasodilator activity of calcitonin gene-related peptide in human skin, *J. Invest. Dermatol.*, 87, 533, 1986.

43. **Brain, S. D. and Williams, T. J.**, Inflammatory oedema induced by synergism between calcitonin gene related peptide (CGRP) and mediators of increased vascular permeability, *Br. J. Pharmacol.*, 86, 855, 1985.

44. **Brain, S. D. and Williams, T. J.**, Substance P regulates the vasodilator activity of calcitonin gene related peptide, *Nature (London)*, 335, 73, 1986.

45. **Scott, T. M. and Pang, S. C.**, Changes in jejunal arteries in spontaneously hypertensive and normotensive rats following neonatal treatment with capsaicin, *Acta Stereol.*, 2, 123, 1983.

46. **Johnson, P. C.**, Autoregulation of blood flow, *Circ. Res.*, 59, 483, 1986.

47. **Strandgaard, S. and Paulson, O. B.**, Cerebral autoregulation, *Stroke*, 15, 413, 1984.

48. **Said, S. I. and Mutt, V.**, Polypeptide with broad biological activity isolated from the small intestine, *Science*, 169, 1217, 1970.

49. **Uddman, R., Alumets, J., Edvinsson, L., Hakanson, R., and Sundler, F.**, VIP nerve fibers around peripheral blood vessels, *Acta Physiol. Scand.*, 112, 65, 1981.

50. **Uemura, Y., Sugimoto, T., Kikuchi, H., and Mizuno, N.**, Possible origins of cerebrovascular nerve fibers showing vasoactive intestinal polypeptide-like immuno-reactivity: an immunohistochemical study in the dog, *Brain Res.*, 448, 98, 1988.

51. **Gibbins, I. L., Brayden, J. E., and Bevan, J. A.**, Perivascular nerves with immunoreactivity to vasoactive intestinal polypeptide in cephalic arteries of the cat: distribution, possible origins and functional implications, *Neuroscience*, 13, 1327, 1984.

52. **Edvinsson, L., Fahrenkrug, J., Hanko, J., Owman, C., Sundler, F., and Uddman, R.**, VIP (vasoactive intestinal polypeptide)-containing nerves of intracranial arteries in mammals, *Cell Tissue Res.*, 208, 135, 1980.

53. **Uemura, Y., Sugimoto, T., Kikuchi, H., and Mizuno, N.**, Changes of vasoactive intestinal polypeptide like immunoreactivity in cerebrovascular nerve fibers after cervical sympathectomy in the dog, *Neurosci. Lett.*, 82, 6, 1987.

54. **Walters, B. B., Gillespine, S. A., and Moskovitz, M. A.**, Cerebrovascular projections from the sphenopalatine and otic ganglia to the middle cerebral artery of the cat, *Stroke*, 17, 488, 1986.

55. **Suzuki, N., Hardebo, J. E., and Owman, C.**, Origins and pathways of cerebrovascular vasoactive intestinal polypeptide-positive nerves in rat, *J. Cereb. Blood Flow Metab.*, 8, 697, 1988.

56. **Keller, J. T., Beduk, A., and Saunders, M. C.**, Origin of fibers innervating the basilar artery of the cat, *Neurosci. Lett.*, 58, 263, 1985.

57. **Brayden, J. E. Conway, M. A.**, Neuropeptide Y and vasoactive intestinal polypeptide in cerebral arteries of the rat: relationships between innervation pattern and mechanical response, *Regul. Peptides*, 22, 253, 1988.

58. **Hara, H., Hamill, G. S., and Jacobowitz, D. M.**, Origin of cholinergic nerves to the rat major cerebral arteries: coexistence with vasoactive intestinal peptide, *Brain Res. Bull.*, 14, 179, 1985.

59. **Matsuyama, T., Shiosaka, S., Matsumoto, M., Yonda, S., Kumura, K., Abe, H., Hayakawa, T., Inoue, H., and Tohjama, M.**, Overall distribution of vasoactive intestinal peptide-containing nerves on the wall of cerebral arteries: an immunohistochemical study using whole mounts, *Neuroscience*, 10, 89, 1983.

60. **Larsson, L.-I., Edvinsson, L., Fahrenkrug, J., Håkanson, R., Owman, C. H., Schaffalvzky de Muckadell, O., and Sunder, F.**, Immunohistochemical localization of a vasodilatory polypeptide (VIP) in cerebrovascular nerves, *Brain Res.*, 113, 400, 1976.

61. **Bartfai, T., Iverfelt, K., Brodin, E., and Ogren, E. B.**, Functional consequences of co-existence of classical and peptide neurotransmitters, *Prog. Brain Res.*, 68, 321, 1986.

62. **Lundberg, J. M. and Hökfelt, T.,** Multiple co-existence of peptides and classical transmitters in peripheral autonomic and sensory neurons — functional and pharmacological implications, *Prog. Brain Res.,* 68, 241, 1986.

63. **Leblanc, G. G., Trimmer, B. A., and Landis, S. C.,** Neuropeptide Y like immunoreactivity in rat cranial parasympathetic neurons: co-existence with vasoactive intestinal peptide and choline acetyltransferase, *Proc. Natl. Acad. Sci. U.S.A.,* 84, 3511, 1987.

64. **Lundberg, J. M.,** Evidence for coexistence of vasoactive intestinal polypeptide (VIP) and acetylcholine in neurons of cat exocrine glands. Morphological, biochemical and functional studies, *Acta Physiol. Scand.,* 112 (Suppl. 496), 1, 1981.

65. **Lundberg, J. M. and Hökfelt, T.,** Co-existence of peptides and classical neurotransmitters, *Trends Neurosci.,* 6, 325, 1983.

66. **Lee, T. J.-F., Saito, A., and Berezin, I.,** Vasoactive intestinal polypeptide: the potential transmitter for cerebral vasodilation, *Science,* 224, 898, 1984.

67. **Masuko, S. and Chiba, T.,** Projection pathways co-existence of peptides and synaptic organization of nerve fibers in the inferior mesenteric ganglion of the guinea pig, *Cell Tissue Res.,* 253, 507, 1988.

68. **Itoh, N., Obata, K. I., Yanaihara, N., and Okamoto, H.,** Human preprovasoactive intestinal polypeptide contains a novel PHI-27 like peptide PHM-27, *Nature (London),* 304, 547, 1983.

69. **Hökfelt, T., Fahrenkrug, J., Tatemoto, K., Mutt, V., and Werner, S.,** PHI, a VIP like peptide, is present in the rat median eminence, *Acta Physiol. Scand.,* 116, 469, 1982.

70. **Tornebrandt, K., Nobin, A., and Owman, C.,** Contractile and dilatory action of neuropeptides on isolated human mesenteric blood vessels, *Peptides,* 8, 251, 1987.

71. **Sata, T., Misra, H. P., Kubota, E., and Said, S. I.,** Vasoactive intestinal polypeptide relaxes pulmonary artery by endothelium-independent mechanisms, *Peptides,* 7, 225, 1986.

72. **McCulloch, J. and Edvinsson, L.,** Cerebral circulatory and metabolic effects of vasoactive intestinal polypeptide, *Am. J. Physiol.,* 238, H449, 1980.

73. **Hermes, E. P., Kontos, H. A., and Said, S. I.,** Mechanisms of vasoactive intestinal polypeptide on cerebral arterioles, *Am. J. Physiol.,* 239, H765, 1980.

74. **Suzuki, Y., McMaster, D., Lederis, K., and Rorstad, O. P.,** Characterization of the relaxant effects of vasoactive intestinal polypeptide (VIP) and PHI on isolated brain arteries, *Brain Res.,* 332, 9, 1984.

75. **Jansen, I., Edvinsson, L., Jensen, K., Olesen, J., and Uddman, R.,** Neuropeptides in human cerebral and temporal arteries: occurrence and vasomotor responses, in *Advances in Headache Research,* Rose, C. F., Ed., John Libbey, London, 1987.

76. **Itoh, T., Sasaguri, T., Makita, Y., Kanamura, Y., and Kuriyama, H.,** Mechanisms of vasodilation induced by vasoactive polypeptide in rabbit mesenteric artery, *Am. J. Physiol.,* 249, H231, 1985.

77. **Larsson, O., Dunier-Engstrom, M., Lundberg, J. M., Fredholm, B. B., and Anggard, A.,** Effects of VIP, PHM and substance P on blood vessels and secretory elements of the human submandibular gland, *Regul. Peptides,* 13, 329, 1986.

78. **Edvinsson, L., Fredholm, B. B., Hamel, E., Jansen, I., and Verrecchia, C.,** Perivascular peptides relax cerebral arteries concomitant with stimulation of cyclic adenosine monophosphate accumulation or release of an endothelial derived relaxing factor in the cat, *Neurosci. Lett.,* 58, 213, 1985.

79. **Ganz, P., Sandrock, A. W., Landis, S. C., Leopold, J., Gimbrone, M. A., and Alexander, R. W.,** Vasoactive intestinal peptide vasodilation and cyclic AMP generation, *Am. J. Physiol.,* 250, H755, 1986.

80. **D'Orléans-Juste, P., Dion, S., Mizrahi, J., and Regoli, D.,** Effects of peptides and nonpeptides on isolated arterial smooth muscles, role of endothelium, *Eur. J. Pharmacol.,* 114, 9, 1985.

81. **Davis, J. M. and Williams, K. I.,** Endothelial dependent relaxant effects of vasoactive intestinal polypeptide and arachidonic acid in rat aortic strips, *Prostaglandins,* 27, 195, 1984.

82. **Brayden, J. E. and Bevan, J. A.,** Evidence that vasoactive intestinal polypeptide mediates neurogenic vasodilation of feline cerebral arteries, *Stroke,* 17, 1189, 1986.

83. **Bevan, J. A., Moskowitz, M., Said, S. A., and Buga, G.,** Evidence that vasoactive intestinal polypeptide is a dilator transmitter to some cerebral and extracerebral cranial arteries, *Peptides,* 5, 385, 1984.

84. **Gaadsby, P. J. and MacDonald, G. J.,** Extracranial vasodilation mediated by vasoactive intestinal polypeptide (VIP), *Brain Res.,* 329, 285, 1985.

85. **Lundberg, J. M., Änggård, A., Fahrenkrug, J., Lundgren, C., and Holmstedt, B.,** Co-release of VIP and acetylcholine in relation to blood flow and salivary secretion in cat submandibular salivary gland, *Acta Physiol. Scand.,* 115, 525, 1982.

86. **Lundberg, L. M., Fahrenkrug, J., Larsson, O., and Anggard, A.,** Co-release of vasoactive intestinal polypeptide and peptide histidine isoleucine in relation to atropine resistant vasodilation in cat submandibular salivary gland, *Neurosci. Lett.,* 52, 37, 1984.

87. **Darke, A. C. and Smaje, L. H.,** Dependence of functional vasodilation in the cat submaxillary gland upon stimulation frequency, *J. Physiol. (London),* 226, 191, 1972.

88. **Hedlund, B., Abens, J., and Bartfai, T.,** Vasoactive intestinal polypeptide and muscarinic receptors. Supersensitivity induced by long term atropine treatment, *Science,* 220, 519, 1983.

89. Sjöqvist, A. and Fahrenkug, J., Sympathetic nerve activation decreases the release of vasoactive intestinal polypeptide from the feline intestine, *Acta Physiol. Scand.*, 127, 419, 1986.

90. Varga, G., Kiss, J. Z., Papp, M., and Vizi, E. S., Vasoactive intestinal peptide may participate in vasodilation of the dog hepatic artery, *Am. J. Physiol.*, 251, G280, 1986.

91. Palmer, J. B. P., Cuss, F. M., Warren, J. B., Blank, M., Bloom, S. R., and Barnes, P. J., Effect of infused vasoactive intestinal peptide on airway function in normal subjects, *Thorax*, 41, 663, 1986.

92. Chilvers, E. R., Dixon, C. M. S., Yiangou, Y., Bloom, S. R., and Ind, P. W., Effect of peptide histidine valine on cardiovascular and respiratory function in normal subject, *Thorax*, 43, 750, 1988.

93. Ganong, W. F., Special lecture neuropeptides in cardiovascular control, *J. Hypertens.*, 2 (Suppl. 3), 15, 1984.

94. Porter, J. P., Reid, I. A., Said, S. A., and Ganong, W. F., Stimulation of renin release by vasoactive intestinal peptide, *Am. J. Physiol.*, 234, F306, 1982.

95. Porter, J. P., Said, S. I., and Ganong, W. F., Vasoactive intestinal peptide stimulates renin secretion *in vitro*: evidence for a direct action of the peptide on the renal juxtaglomerular cells, *Neuroendocrinology*, 36, 404, 1983.

96. Knight, D. S., Beal, J. A., Yuan, Z. P., and Fournet, T. S., Vasoactive intestinal peptide-immunoreactive nerves in the rat kidney, *Anat. Rec.*, 219, 193, 1987.

97. Barajas, L., Sokoloski, K. N., and Lechago, J., Vasoactive intestinal polypeptide-immunoreactive nerves in the kidney, *Neurosci. Lett.*, 28, 325, 1983.

98. Dhall, V., Cowen, T., Haven, A. J., and Burnstock, G., Perivascular noradrenergic peptide containing nerves show different patterns of change during development and ageing in guinea pig, *J. Auton. Nerv. Syst.*, 16, 109, 1986.

99. Ganong, W. F., Porter, J. P., Bahnson, T. D., and Said, S. I., Peptides and neurotransmitters that affect renin secretion, *J. Hypertens.*, 2 (Suppl. 1), 75, 1984.

100. Crowe, R., Lincoln, J., Blacklay, P. F., Pryor, J. P., Lumley, J. S. P., and Burnstock, G., Vasoactive intestinal peptide-like immunoreactive nerves in diabetic penis, *Diabetes*, 32, 1075, 1983.

101. Loesch, A., Belai, A., Lincoln, J., and Burnstock, G., Enteric nerves in diabetic rats; electron microscopic evidence for neuropathy of vasoactive intestinal polypeptide-containing fibers, *Acta Neuropathol.*, 70, 161, 1986.

102. Belai, A., Lincoln, J., and Burnstock, G., Lack of release of vasoactive intestinal polypeptide and calcitonin gene related peptide during electrical stimulation of enteric nerves in streptozotocin-diabetic rats, *Gastroenterology*, 93, 1034, 1987.

103. Andersson, P. O., Björnberg, J., Bloom, S. R., and Mellander, S., Vasoactive intestinal polypeptide in relation to penile erection in the cat evoked by pelvic and by hypogastric nerve stimulation, *J. Urol.*, 138, 419, 1987.

104. Juenemann, K.-P., Lue, T. F., Luo, J.-A., Jadallah, S. A., Nunes, L. L., and Tanagho, E. A., The role of vasoactive intestinal polypeptide as a neurotransmitter in canine penile erection: a combined *in vivo* and immunohistochemical study, *J. Urol.*, 138, 871, 1987.

105. Kolodny, R. C., Kahn, C. B., Goldstein, H. H., and Barnett, D. M., Sexual dysfunction in diabetic men, *Diabetes*, 23, 306, 1974.

106. Lagrrado, M. L. J., Growe, R., and Burnstock, G., Reduction of nerves containing vasoactive intestinal polypeptide and serotonin, but not neuropeptide Y and catecholamine in cerebral blood vessels of the 8 week streptozotocin-induced diabetic rat, *Blood Vessels*, 24, 169, 1987.

107. Harder, D. R., Pressure-dependent membrane depolarization in cat middle cerebral artery, *Circ. Res.*, 55, 197, 1984.

108. Grunnet, M. L., Cerebrovascular disease: diabetes and cerebral atherosclerosis, *Neurology*, 13, 486, 1963.

109. Edvinsson, L., Copeland, J. R., Emson, P. C., McCulloch, J., and Uddman, R., Nerve fibres containing neuropeptide Y in the cerebrovascular bed, immunocytochemistry, radioimmunoassay and vasomotor effects, *J. Cereb. Blood Flow Metab.*, 7, 45, 1987.

110. Tani, E., Yamagata, S., and Ito, Y., Intercellular granules and vesicles in prolonged cerebral vasospasm, *J. Neurosurg.*, 48, 179, 1978.

111. Gioia, A. E., White, R. P., Bakhtian, B., and Robertson, J. T., Evaluation of the efficacy of intrathecal nimodipine in canine models of cerebral vasospasm, *J. Neurosurg.*, 62, 721, 1985.

112. Clark, A. J. L., Adrian, T. E., McMichael, H. B., and Bloom, S. R., Vasoactive intestinal peptide in shock and heart failure, *Lancet*, 1, 539, 1983.

113. Fouad, F. M., Shimamatsu, K., Said, S. I., and Tarazi, R. C., Inatropic responsiveness in hypertensive left ventricular hypertrophy: impaired inotropic response to glucon and vasoactive intestinal peptide in renal hypertensive rats, *J. Cardiovasc. Pharmacol.*, 8, 398, 1986.

114. Lee, R. M. K. W., Nagahama, M., McKenzie, R., and Daniel, E. E., Peptide containing nerves around blood vessels of stroke prone spontaneously hypertensive rats, *Hypertension*, 11 (Suppl. 1), 117, 1988.

115. Minth, C. D., Bloom, S. R., Polak, J. M., and Dixon, J. E., Cloning, characterization, and DNA sequence of a human cDNA encoding neuropeptide tyrosine, *Proc. Natl. Acad. Sci. U.S.A.*, 81, 4577, 1984.

116. **Wharton, J. and Gulbenkian, S.,** Peptides in the mammalian cardiovascular system, *Experientia,* 43, 821, 1987.
117. **Adrian, T. E., Terenghi, G., Brown, M. J., Allen, J. M., Bacarese-Hamilton, A. J., Polak, J. M., and Bloom, S. R.,** Neuropeptide Y in phaeochromocytomas and ganglioneuroblastomas, *Lancet,* 2, 540, 1983.
118. **Uddman, R., Ekblad, E., Edvinsson, L., Hakansson, R., and Sundler, F.,** Neuropeptide Y like immunoreactivity in perivascular nerves of the guinea pig, *Regul. Peptides,* 10, 243, 1985.
119. **Edvinsson, L., Hakanson, R., Steen, S., Sundler, F., Uddman, R., and Wahlested, C.,** Innervation of human omental arteries and veins and vasomotor responses to noradrenaline, neuropeptide Y, substance P and vasoactive intestinal peptide, *Regul. Peptides,* 12, 67, 1985.
120. **Cannon, B., Nedergaard, J., Lundberg, J. M., Hökfelt, T., Terenius, L., and Goldstein, M.,** Neuropeptide tyrosine (NPY) is co-stored with noradrenaline in vascular but not in parenchymal sympathetic nerves of brown adipose tissue, *Exp. Cell Res.,* 164, 546, 1986.
121. **Ekblad, E., Edvinsson, L., Wahlestedt, C., Uddman, R., Hakanson, R., and Sundler, F.,** Neuropeptide Y coexists and cooperates with noradrenaline in perivascular nerve fibers, *Regul. Peptides,* 8, 225, 1984.
122. **Furness, J. B., Costa, M., Emson, P. C., Hakanson, R., Moghimzadeh, E., Sundler, F., Taylor, I. L., and Chance, R. E.,** Distribution, pathways and reactions to drug treatment of nerves with neuropeptide Y- and pancreatic polypeptide-like immunoreactivity in the guinea pig digestive tract, *Cell Tissue Res.,* 234, 71, 1983.
123. **Rudehill, A., Sollevi, A., Franco-Cereceda, A., and Lundberg, J. M.,** Neuropeptide Y (NPY) and the pig heart: release and coronary vasoconstrictor effects, *Peptides,* 7, 821, 1986.
124. **Lundberg, J. M., Pernow, J., Fried, G., and Anggard, A.,** Neuropeptide Y and noradrenaline mechanisms in relation to reserpine induced impairment of sympathetic neurotransmission in the cat spleen, *Acta Physiol. Scand.,* 131, 1, 1987.
125. **Morris, J. L., Murphy, R., Furness, J. B., and Costa, M.,** Partial depletion of neuropeptide Y from noradrenergic perivascular and cardiac axons by 6 hydroxydopamine and reserpine, *Regul. Peptides,* 13, 147, 1986.
126. **Fried, G., Terenius, L., Hökfelt, T., and Goldstein, M.,** Evidence for the differential localization of noradrenaline and neuropeptide Y in neuronal storage vesicles isolated from rat vas deferens, *J. Neurosci.,* 5, 450, 1985.
127. **Fried, G., Lundberg, J. M., and Theodorsson-Norheim, E.,** Subcellular storage and axonal transport of neuropeptide Y (NPY) in relation to catecholamines in the cat, *Acta Physiol. Scand.,* 125, 145, 1985.
128. **Lundberg, J. M., Rudehill, A., Sollevi, A., Theodorsson-Norheim, E., and Hamberger, B.,** Frequency- and reserpine-dependent chemical coding of sympathetic transmission: differential release of noradrenaline and neuropeptide Y from pig spleen, *Neurosci. Lett.,* 63, 96, 1986.
129. **Lundberg, J. M., Anggard, A., Theodorssan-Norheim, E., and Pernow, J.,** Guanethidine-sensitive release of neuropeptide Y-like immunoreactivity in cat spleen by sympathetic nerve stimulation, *Neurosci. Lett.,* 52, 175, 1984.
130. **Lundberg, J. M., Pernow, J., Fried, G., and Anggard, A.,** Neuropeptide Y and noradrenaline mechanisms in relation to reserpine induced impairment of sympathetic neurotransmission in the cat spleen, *Acta Physiol. Scand.,* 131, 1, 1987.
131. **Fredholm, B. B., Jansen, I., and Edvinsson, L.,** Neuropeptide Y is a potent inhibitor of cyclic AMP accumulation in feline cerebral blood vessels, *Acta Physiol. Scand.,* 124, 467, 1985.
132. **Rioux, F., Bachelard, H., Martel, J.-C., and St. Pierre, S.,** The vasoconstrictor effect of neuropeptide Y and related peptides in the guinea pig isolated heart, *Peptides,* 7, 27, 1986.
133. **Wahlested, C., Yanaihara, N., and Hakanson, R.,** Evidence for different pre- and post-receptors for neuropeptide Y and related peptides, *Regul. Peptides,* 13, 307, 1986.
134. **Franco-Cereceda, A., Lundberg, J. M., and Dahlof, C.,** Neuropeptide Y and sympathetic control of heart contractility and coronary vascular tone, *Acta Physiol. Scand.,* 124, 361, 1985.
135. **Rudehill, A., Sollevi, A., Franco-Cereceda, A., and Lundberg, J. M.,.** Neuropeptide Y (NPY) and the pig heart: release and coronary vasoconstrictor effects, *Peptides,* 7, 821, 1986.
136. **Neild, T. O.,** Actions of neuropeptide Y on innervated and denervated rat tail arteries, *J. Physiol.,* 386, 19, 1987.
137. **Wahlestedt, C., Edvinsson, L., Ekblad, E., and Hakanson, R.,** Neuropeptide Y potentiates noradrenaline-evoked vasoconstriction: mode of action, *J. Pharmacol. Exp. Ther.,* 234, 735, 1985.
138. **Edvinsson, L., Ekblad, E., Hakanson, R., and Wahlestedt, C.,** Neuropeptide Y potentiates the effects of various constrictor agents on rabbit blood vessels, *Br. J. Pharmacol.,* 83, 519, 1984.
139. **Wahlestedt, C. and Hakanson, R.,** Effects of neuropeptide Y (NPY) at the sympathetic neuroeffector junction. Can pre and postjunctional receptors be distinguished, *Med. Biol.,* 64, 85, 1986.
140. **Wahlestedt, C., Yanaihara, N., and Hakanson, R.,** Evidence for different pre- and post-junctional receptors for neuropeptide Y and related peptides, *Regul. Peptides,* 13, 307, 1986.

141. **Pernow, J., Saria, A., and Lundberg, J. M.,** Mechanisms underlying pre- and postjunctional effects of neuropeptide Y in sympathetic vascular control, *Acta Physiol. Scand.,* 126, 239, 1986.

142. **Lundberg, J. M., Pernow, J., Tatemoto, K., and Dahlof, C.,** Pre- and postjunctional effects of NPY on sympathetic control of rat femoral artery, *Acta Physiol. Scand.,* 123, 511, 1985.

143. **Stjarne, L., Lundberg, J. M., and Astrand, P.,** Neuropeptide Y — a cotransmitter with noradrenaline and adenosine 5'-triphosphate in the sympathetic nerves of the mouse vas deferens? A biochemical, physiological and electropharmacological study, *Neuroscience,* 18, 151, 1986.

144. **Burnstock, G. and Sneddon, P.,** Evidence for ATP and noradrenaline as cotransmitters in sympathetic nerves, *Clin, Sci.,* 68 (Suppl. 10), 895, 1985.

145. **Hakanson, R., Wahlestedt, C., Ekblad, E., Edvinsson, L., and Sundler, F.,** Neuropeptide Y: coexistence with noradrenaline. Functional implications, *Prog. Brain Res.,* 68, 279, 1986.

146. **Donoso, V., Silva, M., St. Pierre, S., and Huidobro-Toron, J. P.,** Neuropeptide Y (NPY), an endogenous presynaptic modulator of adrenergic neurotransmission in the rat vas deferens: structural and functional studies, *Peptides,* 9, 545, 1988.

147. **Westfall, T. C., Carpenter, S., Chen, X., Beinfeld, M. C., Naes, L., and Meldrum, M. J.,** Prejunctional and postjunctional effects of neuropeptide Y at the noradrenergic neuroeffector junction of the perfused mesenteric arterial bed of the rat, *J. Cardiovasc. Pharmacol.,* 10, 716, 1987.

148. **Stjarne, L.,** New paradigm: sympathetic neurotransmission by lateral interaction between secretory units, *NIPS,* 1, 103, 1986.

149. **Allen, J. M., Bircham, P. M. M., Edwards, A. V., Tatemoto, K., and Bloom S. R.,** Neuropeptide Y (NPY) reduces myocardial perfusion and inhibits the force of contraction of the isolated perfused heart, *Regul. Peptides,* 6, 247, 1983.

150. **Lundberg, J. M., Hua, X.-Y., and Franco-Cereceda, A.,** Effects of neuropeptide Y (NPY) on mechanical activity and neurotransmission in the heart, vas deferens and urinary bladder of the guinea pig, *Acta Physiol. Scand.,* 121, 325, 1984.

151. **Allen, J. M., Gjorstrup, P., Bjorkman, J. A., Ek, L., Abrahamsson, T., and Bloom, S. R.,** Studies on cardiac distribution and function of neuropeptide Y, *Acta Physiol. Scand.,* 126, 405, 1986.

152. **Rudehill, a., Sollem, A., Franco-Cereceda, A., and Lundberg, J. M.,** Neuropeptide Y (NPY) and the pig heart: release and coronary vasoconstrictive effects, *Peptides,* 7, 821, 1986.

153. **Pfister, A., Waeber, B., Nussberger, J., and Brunner, H. R.,** Neuropeptide Y normalizes renin secretion in adrenalectomized rats without changing blood pressure, *Life Sci.,* 39, 2161, 1986.

154. **Ballesta, J., Lawson, J. A., Pals, D. T., Ludens, J. H., Lee, Y. C., Bloom, S. R., and Polak, J. M.,** Significant depletion of NPY in the innervation of the rat mesenteric, renal arteries and kidneys in experimentally (aorta, coarctation) induced hypertension, *Histochemistry,* 87, 273, 1987.

155. **Freis, E. D.,** Salt, volume and the prevention of hypertension, *Circulation,* 53, 589, 1976.

156. **Blair-West, J. R., Brook, A. H., and Simpson, P. A.,** On the question of body fluid volume or sodium status influencing renin release, *J. Physiol.,* 267, 321, 1977.

157. **Seymour, A. A., Davis, J. O., Freeman, R. H., Deforest, J. M., Rowe, B. P., Stephens, G. A., and Williams, G. M.,** Hypertension produced by sodium depletion and unilateral nephrectomy: a new experimental model, *Hypertension,* 2, 125, 1980.

158. **Coleman, T. G., Cowley, A. W., and Guyton, A. C.,** Experimental hypertension and long term control of arterial pressure, in *Cardiovascular Physiology: MTP International Review of Science,* Vol. 1, Guyton, A. C. and Jones, C. E., Eds., University Park Press, Baltimore, MD, 1974, 259.

159. **Waeber, B., Corder, R., Aubert, J. F., Nussberger, J., Gaillard, R., and Brunner, H. R.,** Influence of sodium intake on circulating levels of neuropeptide Y, *Life Sci.,* 41, 1391, 1987.

160. **Morris, M., Kapoor, V., and Chalmers, J.,** Plasma neuropeptide Y concentration is increased after hemorrhage in conscious rats: relative contributions of sympathetic nerves and the adrenal medulla, *J. Cardiovasc. Pharmacol.,* 9, 541, 1987.

161. **Evequoz, D., Waeber, B., Aubert, J.-F., Flükiger, J.-P., Nussberger, J., and Brunner, H. R.,** Neuropeptide Y prevents the blood pressure fall induced by endotoxin in conscious rats with adrenal medullectomy, *Circ. Res.,* 62, 25, 1988.

162. **Schultzberg, M., Hölkfelt, T., Terenius, L., Elfvin, L.-G., Lundberg, J. M., Brandt, J., Elde, R. P., and Goldstein, M.,** Enkephalin immunoreactive nerve fibers and cell bodies in sympathetic ganglia of guinea-pig and rat, *Neuroscience,* 4, 249, 1979.

163. **Shultzberg, M., Lundberg, J. M., Hökfelt, T., Terenius, L., Brandt, J., Elde, R. P., and Goldstein, M.,** Enkephalin-like immunoreactivity in gland cells and nerve terminals of the adrenal medulla, *Neuroscience,* 3, 1169, 1978.

164. **De Potter, W. P., Coen, E. P., and De Potter, R. W.,** Evidence for the co-existence and co-release of [met] enkephalin and noradrenaline from sympathetic nerves of the bovine vas deferens, *Neuroscience,* 20, 855, 1987.

165. **Douglas, B. H., Duff, R. B., Thureson-Klein, A. K., and Klein, R. L.,** Enkephalin contents reflect large dense cored vesicle population in vasa deferentia, *Regul. Peptides,* 14, 193, 1986.

166. **Archelos, J., Xiang, J. Z., Reinecke, M., and Lang, R. E.,** Regulation of release and function of neuropeptides in the heart, *J. Cardiovasc. Pharmacol.,* 10 (Suppl. 12), S45, 1987.

167. **Wilson, S. P., Klein, R. L., Chang, K. J., Gasparis, M. S., Viveros, O. H., and Yang, W.-H.,** Are opioid peptides cotransmitters in noradrenergic vesicles of sympathetic nerves, *Nature (London),* 288, 707, 1980.

168. **Klein, R. L., Wilson, S. P., Dzielak, D. J., Yang, W.-H., and Viveros, O. H.,** Opioid peptides and noradrenaline co-exist in large dense core vesicles from sympathetic nerves, *Neuroscience,* 7, 2255, 1982.

169. **Lee, C.-H. and Berkowitz, B. A.,** Stereoselective and calcium dependent contractile effects of narcotic antagonist analgesics in the vascular smooth muscle of the rat, *J. Pharmacol. Exp. Ther.,* 198, 347, 1976.

170. **Altura, B. T., Gebewald, A., and Altura, B. M.,** Comparative actions of narcotics on large and microscopic blood vessels, *Fed. Proc.,* 37, 471, 1978.

171. **Kannan, M. S. and Seip, A. E.,** Neurogenic dilatation and constriction of rat superior mesenteric artery *in vitro*: mechanisms and mediators, *Can. J. Physiol. Pharmacol.,* 64, 729, 1986.

172. **Harder, D. A. and Madden, J. A.,** Cellular mechanisms of opiate receptor stimulation in cat middle cerebral artery, *Eur. J. Pharmacol.,* 102, 411, 1984.

173. **Hanko, J., Harebo, J., and Owman, C.,** Vasomotor effects of some neuropeptides on small pial arteries of the cat, in *Cerebral Microcirculation and Metabolism,* Cervos-Navarro, J. and Fritschka, E., Eds., Raven Press, New York, 1981, 157.

174. **Waters, A. and Harder, D. R.,** Electromechanical coupling in rat basilar artery in response to morphine, *Neurosurgery,* 13, 676, 1983.

175. **Lord, J. A., Waterfield, A. A., Hughes, J., and Kosterlitz, H. W.,** Endogenous opioid peptides: multiple agonists and receptors, *Nature (London),* 267, 495, 1976.

176. **Chang, K.-J., Killian, A., Hazum, E., and Cuattrecasus, P.,** Morphiceptin (NH_4 Tyr-Pro-Phe-Pro-$CONH_2$): a potent and specific agonist for morphine (μ) receptors, *Science,* 212, 75, 1981.

177. **Handa, B., Lane, A. C., Lord, J. A. H., Morgan, B. A., Rance, M. J., and Smith, C. F. C.,** Analogues of β-LPH_{61-64} possessing selective agonist activity at μ-opiate receptors, *Eur. J. Pharmacol.,* 70, 531, 1981.

178. **Feuerstein, G. and Siren, A. L.,** The opioid system in cardiac and vascular regulation of normal and hypertensive states, *Circulation,* 75, (Suppl. 1), I-125, 1987.

179. **Romer, D., Buscher, H., Hill, R. C., Maurer, R., Petcher, T. J., Welle, H. B. A., Bakel, H. C. C. K., and Akkerman, A. M.,** Bremazocine: a potent longacting opiate κ-agonist, *Life Sci.,* 27, 971, 1980.

180. **Martin, W. R., Eades, C. G., Thompson, J. A., Huppler, R. E., and Gilbert, R. E.,** The effects of morphine and nalorphine like drugs in non dependent and morphine dependent chronic spinal dog, *J. Pharmacol. Exp. Ther.,* 197, 517, 1976.

181. **Pfeiffer, A., Pasi, A., Mehraein, P., and Herz, A.,** Opiate receptor binding sites in human brain, *Brain Res.,* 248, 87, 1982.

182. **Magnan, J., Paterson, S. J., Tavani, A., and Kosterlitz, H. W.,** The binding spectrum of narcotic analgesic drugs with different agonist and antagonist properties, *Naunyn-Schmiedeberg's Arch. Pharmacol.,* 319, 199, 1982.

183. **Crooks, P. A., Bowdy, B. D., Reinsel, C. N., Iwamoto, E. T., and Gillespie, M. N.,** Structure-activity evidence against opiate receptor involvement in Leu^5-enkephalin induced pulmonary vasoconstriction, *Biochem. Pharmacol.,* 33, 4095, 1984.

184. **Steelhorst, A. and Starke, K.,** Prejunctional opioid receptors in the pulmonary artery of the rabbit, *Arch. Int. Pharmacodyn.,* 281, 298, 1986.

185. **Illes, P., Bettermann, R., Brod, I., and Bucher, B.,** β-Endorphin-sensitive opioid receptors in the rat tail artery, *Naunyn-Schmiedeberg's Arch. Pharmacol.,* 335, 420, 1987.

186. **Illes, P., Ramme, D., and Busse, R.,** Photoelectric measurement of neurogenic vasoconstriction in jejunal branches of the rabbit mesenteric artery reveals the presence of presynaptic opioid δ receptors, *Naunyn-Schmiedeberg's Arch. Pharmacol.,* 335, 701, 1987.

187. **Illes, P., Ramme, D., and Starke, K.,** Presynaptic opioid δ receptors in the rabbit mesenteric artery, *J. Physiol.,* 379, 217, 1986.

188. **Budai, D. and Duckles, S. P.,** Influence of stimulation train length on the opioid-induced inhibition of norepinephrine release in the rabbit ear artery, *J. Pharmacol. Exp. Ther.,* 247, 839, 1988.

189. **Kugelgen, I. V., Illes, P., Wolf, D., and Starke, K.,** Presynaptic inhibitory opioid δ and κ-receptors in a branch of the rabbit ileocolic artery, *Eur. J. Pharmacol.,* 118, 97, 1985.

190. **Ramme, D., Illes, P., Spath, L., and Starke, K.,** Blockade of alpha 2-adrenoreceptors permits the operation of otherwise silent opioid kappa-receptors at the sympathetic axons of rabbit jejunal arteries, *Naunyn-Schmiedeberg's Arch. Pharmacol.,* 334, 48, 1986.

191. **Hanabauer, I., Govoni, S., Majane, E. A., Yang, H. Y. T., and Costa, E.,** *In vivo* regulation of the release of met-enkephalin-like peptides from dog adrenal medulla, *Adv. Biochem. Psychopharmacol.,* 33, 209, 1982.

192. **Giullemin, R., Vargo, T., Rossier, J., Minick, S., Ling, N., Rivier, C., Vale, W., and Bloom, F.,** Beta-endorphin and adrenocorticotropin are secreted concomitantly by the pituitary gland, *Science,* 197, 1367, 1977.

193. **Fennessy, M. R. and Rattray, J. F.,** Cardiovascular effects of intravenous morphine in anaesthetized rat, *Eur. J. Pharmacol.,* 14, 1, 1971.

194. **Lemaire, I., Tseng, R., and Lemaire, S.,** Systemic administration of β-endorphin: potent hypotensive effect involving a seratonergic pathway, *Proc. Natl. Acad. Sci. U.S.A.,* 75, 6240, 1978.

195. **Pasi, A., Moccetti, C., Legler, M., Meuller, J., Foletta, D., Gramsch, C., and Hartmann, H.,** Elevation of blood levels of beta-endorphin-like immunoreactivity in patients with shock, *Res. Commun. Chem. Pathol. Pharmacol.,* 42, 509, 1983.

196. **Hamilton, A. J., Carr, D. B., La Roverre, J. M., and Black, P. M.,** Endotoxic shock elicits greater endorphin secretion than hemorrhage, *Circ. Shock,* 19, 47, 1986.

197. **Chernow, B., Lake, C. R., Teich, S., Moughey, E. H., Meyerhoff, J., Casey, L. C., and Fletcher, J. R.,** Hemorrhagic hypotension increases plasma beta endorphin concentrations in non human primate, *Crit. Care Med.,* 14, 505, 1986.

198. **Amir, S.,** Naloxone improves and morphine exacerbates, experimental shock induced by the release of endogenous histamine by compound 48/80, *Brain Res.,* 297, 187, 1984.

199. **Amir, S.,** Opiate antagonists improve survival in anaphylactic shock, *Eur. J. Pharmacol.,* 80, 161, 1982.

200. **Giuffre, K. A., Vdelsman, R., Listwak, S., and Chrousos, G. P.,** The effects of immune neutralization of corticotropin releasing hormone adrenocorticotropin and β endorphin in surgically stressed rat, *Endocrinology,* 122, 306, 1988.

201. **Holaday, J. W. and Faden, A. I.,** Naloxone reversal of endotoxin hypotension suggests a role of endorphins in shock, *Nature (London),* 275, 450, 1978.

202. **Faden, A. I. and Holaday, J. W.,** Opiate antagonists: a role in the treatment of hypovolemic shock, *Science,* 205, 317, 1979.

203. **Faden, A. I.,** Endogenous opioids, physiologic and pathophysiological actions, *J. Am. Osteopathol. Assoc.,* 84, 129/48, 1984.

204. **Morita, H., Nishida, Y., Motochigawa, H., Uemura, N., Hosomi, H., and Vatner, S. F.,** Opiate receptor-mediated decrease in renal nerve activity during hypotensive hemorrhage in concious rabbits, *Circ. Res.,* 63, 165, 1988.

205. **Kraft, K., Theobald, R., Kolloch, R., and Stumpo, K. O.,** Normalization of blood pressure and plasma concentrations of endorphin and leucine-enkephalin in patients with primary hypertension after treatment with clonidine, *J. Cardiovasc. Pharmacol.,* 10 (Suppl. 12), S147, 1987.

206. **Chen, M., Lee, J., Malvin, R. L., and Huang, B. S.,** Naloxone attenuates development of hypertension in two-kidney one-clip Goldblatt rats, *Am. J. Physiol.,* 255, E839, 1988.

207. **Zamir, N., Simantov, R., and Segal, M.,** Pain sensitivity and opioid activity in genetically and experimentally hypertensive rats, *Brain Res.,* 184, 299, 1980.

208. **Bucher, B., Betterman, R., and Illes, P.,** Plasma concentration and vascular effect of endorphin in spontaneously hypertensive and Wistar Kyoto rats, *Naunyn-Schmiedeberg's Arch. Pharmacol.,* 335, 428, 1987.

209. **Faden, A. I., Jacobs, T. P., Mougey, E., and Holaday, J. W.,** Endorphins in experimental spinal injury: therapeutic effect of naloxone, *Ann. Neurol.,* 10, 326, 1981.

210. **McIntosh, T. K., Hayes, R. L., DeWitt, D. S., Agura, V., and Faden, A. I.,** Endogenous opioids may mediate secondary damage after experimental brain injury, *Am. J. Physiol.,* 253, E565, 1987.

211. **Faden, A. I. and Jacobs, T. P.,** Dynorphin induces partially reversible paraplegia in the rat, *Eur. J. Pharmacol.,* 91, 321, 1983.

212. **Ballesta, J., Lawson, J. A., Pals, D. T., Ludens, J. H., Lee, V. C., Bloom, S. R., and Polak, J. M.,** Significant depletion of NPY in the innervation of the rat mesenteric, renal arteries and kidneys in experimentally (aorta coarctation) induced hypertension, *Histochemistry,* 87, 273, 1987.

213. **Edvinsson, L., Ekman, R., Jansen, I., Ottosson, A., and Uddman, R.,** Peptide-containing nerve fibers in human cerebral arteries. Immunohistochemistry, radioimmunoassay and *in vitro* pharmacology, *Ann. Neurol.,* 21, 431, 1987.

214. **Alfaci, C., Cowen, T., Crockard, N. A., and Burnstock, G.,** Perivascular nerve types supplying cerebral blood vessels of the gerbils, *Acta Physiol. Scand.,* 552, 9, 1986.

215. **Hara, H. and Weir, B.,** Different distributions of substance P and vasoactive intestinal polypeptide in the cerebral arterial innervation in rat and guinea pig, *Anat. Anz.,* 163, 19, 1987.

216. **Hara, H. and Weir, B.,** Pathway of nerves with vasoactive intestinal polypeptide-like immunoreactivity to the major cerebral arteries of the rat, *Cell Tissue Res.,* 251, 275, 1988.

217. **Ohlen, A., Lindbom, L., Staines, W., Hökfelt, T., Cuello, A. C., Fischer, J. A., and Hedovist, P.,** Substance P and calcitonin gene related peptide: immunohistochemical localization and microvascular effects in rabbit skeletal muscle, *Naunyn-Schmiedeberg's Arch. Pharmacol.,* 336, 87, 1987.

218. **Dey, R. D., Shannon, W. A., and Said, S. I.,** Localization of VIP immunoreactive nerves in the airways and pulmonary blood vessels of dogs, cats and human subjects, *Cell Tissue Res.,* 220, 231, 1981.

219. **Kurtz, A., Muff, R., Born, W., Lundberg, J. M., Millberg, B.-I., Gnadinger, M. P., Uehlinger, D. E., Wiedmann, P., Hokfelt, T., and Fischer, J. A.,** Calcitonin gene peptide is a stimulator of renin secretion, *J. Clin. Invest.,* 82, 538, 1988.

220. **Johansson, O.,** A detailed account of NPY-immunoreactive nerves and cells of the human skin. Comparison with VIP-, substance P- and PHI containing structures, *Acta Physiol. Scand.,* 128, 147, 1986.

221. **Morris, J. L., Gibbins, I. L., Furness, J. B., Costa, M., and Murphy, R.,** Co-localization of neuropeptide Y, vasoactive intestinal peptide and dynorphin in non-noradrenergic axons of the guinea pig uterine artery, *Neurosci. Lett.,* 62, 31, 1985.

Chapter 15

NEUROPEPTIDES AND CELL PROLIFERATION

David J. Hill

TABLE OF CONTENTS

I.   Introduction 480

II.   Direct Mitogenic Actions 480

III.   Mechanisms of Cellular Actions 481

IV.   Growth-Promoting Effects of Neuropeptides *In Vivo* 482

V.   Neoplasia 484

VI.   Conclusions 486

References 486

I. INTRODUCTION

Cell proliferation is one aspect of an integrated growth process which also involves tissue differentiation, maturation, and cell migration. Although most apparent during the fundamental tissue organization of embryogenesis and fetal development, these processes are retained to a high degree in adult life during wound repair, ovarian function, hemopoiesis, and the pathogenic proliferation of neoplasia. While the greater concept of tissue growth was originally considered to be under the endocrine control of hormones such as growth hormone, thyroxine, and glucocorticoids, during the last decade it has been recognized that the primary stimuli to cell proliferative events are peptide growth factors. Molecules such as epidermal growth factor (EGF), insulin-like growth factors (IGFs), and platelet-derived growth factor (PDGF) are widely expressed within body tissues and synergize to activate a series of intracellular events culminating in DNA synthesis.[1] These interactions are largely paracrine or autocrine in nature and can reemerge during proliferative disease. The above characteristics are shared by several of the gastrointestinal neuropeptide hormones, suggesting that a separation by nomenclature of neuropeptide and growth factor molecules may be nonsensical on functional grounds. This chapter provides, in overview, evidence to support a role for locally secreted neuropeptides in both normal and neoplastic gastrointestinal and neuronal development. Those peptides to be considered include gastrin releasing peptide (GRP), its amphibian analogue bombesin, and the structurally related neuromedin B; the secretin/glucagon family, including vasoactive intestinal polypeptide (VIP); the gastrin/cholecystokinin (CCK) group of peptides; the tachykinins, including substances P and K; and posterior pituitary peptides such as vasopressin. These hormones are discussed in terms of relationship and structure within Chapter 2 of this volume.

II. DIRECT MITOGENIC ACTIONS

The ability to discriminate between the direct and indirect growth-promoting actions of neuropeptides necessitates the use of defined cell cultures. However, even these manipulated conditions do not allow truly direct actions to be separated from those mediated by the release of other cellular paracrine or autocrine growth factors. Most studies have utilized immortalized fibroblast lines, and extrapolation from these to trophic effects *in vivo* on a variety of cell types must be treated with some caution.

The majority of citations demonstrating direct growth-promoting activities for neuropeptides involve either the GRP/bombesin group of molecules or vasopressin. The first reasoned observation was that of Rozengurt et al.,[2] who demonstrated that vasopressin was mitogenic for Swiss 3T3 mouse fibroblasts, although similar effects had also been observed previously on rat chondrocytes.[3] Later studies found that bombesin also stimulated DNA synthesis in 3T3 cells with a half-maximal concentration on the order of 1 nM,[4] as did the bombesin analogue GRP and the related neuromedin B.[5] Both GRP and bombesin are also mitogenic for normal bronchial epithelial cells, although higher concentrations of approximately 100 nM are necessary, two orders of magnitude greater than that needed for fibroblast cultures.[6]

Direct mitogenic actions of substance P have been shown on smooth muscle cells[7] and dermal fibroblasts,[8] together with a potentiation of prostaglandin E$_2$ release by rheumatoid synoviocytes.[9] Substance P was also demonstrated to promote cell replication by embryonic rat aortal smooth muscle cells,[10] but had no actions on adult rat vascular smooth muscle cells. Payan et al.[11] found that substance P enhanced the proliferation of human blood T-lymphocytes. Collectively, these reports suggest that the release of tachykinins from the peripheral nervous system in skin and other muscle-skeletal tissues may mediate local inflammatory response and contribute to wound healing. This is supported by the recent observation that substance P induced the proliferation of a mouse keratinocyte line.[12] Surprisingly, substance K, which is coexpressed in the same mRNA as substance P as a β-preprotachykinin and released in inflammatory lesions of the skin, was without effect. A role of neuropeptides in wound healing is also supported by the ability of bradykinin, a pain-inducing substance released at sites of tissue change, to promote mouse fibroblast proliferation.[13]

Certain neuropeptides appear to inhibit cellular DNA synthesis or may be bifunctional, depending on cell type and culture conditions. VIP was a potent inhibitor of serum-induced DNA synthesis in cultured smooth muscle cells, and these effects were accompanied by parallel changes in intracellular cAMP.[14] In contrast, VIP initiated DNA synthesis by Swiss 3T3 mouse fibroblasts and synergized with insulin.[15]

III. MECHANISMS OF CELLULAR ACTIONS

Considerable effort has been expended in the determination of the intracellular messenger systems employed by bombesin/GRP during the induction of mitogenesis, particularly in fibroblastic cells. Also of considerable interest has been the relationship between the neuropeptides and the physiologically predominant group of anabolic hormones, the peptide growth factors. Peptide growth factors are considered to act predominantly as paracrine or autocrine agents during normal tissue development,[1] during wound healing,[16] and in certain neoplasias; it is against this background that any physiological actions of neuropeptides upon growth must be superimposed.

A single class of specific receptors for GRP/bombesin has been identified on Swiss 3T3 fibroblasts with a K_d of approximately 1 nM.[17] This was not competed for by VIP or vasopressin or by peptide growth factors such as PDGF, EGF, or insulin. The structure of the bombesin/GRP receptor was analyzed by the affinity cross-linking of cell membrane components to [^{125}I]GRP and separation on sodium dodecyl sulfate polyacrylamide gel electrophoresis (SDS-PAGE). This yielded a single radiolabeled band of 75 to 85 kDa.[17] Separate and specific receptor populations also exist for vasopressin and bradykinin.[13]

During the cycle of cellular replication, growth-arrested or cycling cells must be made competent to reenter the cell cycle at G_1, and agents which facilitate the passage of cells across the G_0/G_1 boundary are referred to as competence factors. For murine fibroblastic cells, PDGF and fibroblast growth factor (FGF) function as such competence factors and are required transiently for initial events in the cell cycle.[18] However, once in G_1 phase these cells require other agents to progress through the sequential uptake of nutrients, protein synthesis, and thymidine kinase induction that culminates in DNA synthesis at S phase. Such "progression" factors include EGF, insulin, and the IGFs. Many of the neuropeptides appear to fall into the category of "competence factors" for fibroblastic cells.

Bombesin, like PDGF and FGF, will act alone as a mitogen for density-arrested 3T3 fibroblasts. However, a synergistic interaction was seen with 1 μg/ml insulin,[4] a concentration at which insulin could be expected to act as a surrogate IGF by cross-reaction with IGF type 1 receptors. Additive effects were noted between bombesin and PDGF or FGF. Vasopressin demonstrated synergistic interaction with EGF, insulin, and PDGF during the induction of 3T3 cell replication, suggesting that it may act as both a competence and a progression factor.[19] The similarity of action between neuropeptides and peptide growth factors was emphasized by the observation that the stimulation of connective tissue cell proliferation by substance K was accompanied by a rapid increase in mRNA encoding the proto-oncogene c-*myc*.[20] Similarly, bombesin and related peptides transiently activated both c-*myc* and c-*fos* proto-oncogenes in fibroblastic cells.[21,22]

Following the binding of mitogenic peptides to cell membrane receptors, a number of intracellular messenger systems transmit appropriate signals to the cellular genome. These are rapid events which trigger the cellular responses necessary to enter and traverse G_1 of the cell cycle and include: (1) changes in intracellular calcium flux and pH, (2) the activation of protein kinase C, (3) an increase in intracellular levels of cyclic AMP, and (4) activation of the phosphatidyl inositol pathways. Bombesin and related peptides stimulate a net input of Na^+ by 3T3 fibroblasts via a Na^+/H^+ channel, causing cytoplasmic alkalization.[23] This is accompanied by a rapid mobilization of Ca^{2+} from intracellular stores, leading to a transient increase in intracellular Ca^{2+} concentrations and an efflux of ions via a Ca^{2+}-ATPase.[24] Inositol triphosphate (IP_3) is able to act as a second messenger for mitogens that stimulate Ca^{2+} efflux and phosphoinositol turnover as a result of phospholipase C activation. Bombesin/GRP have been shown to cause increased phospho-inositol metabolism in 3T3 cells, leading to the formation of IP_3.[25] Both bradykinin and vasopressin also cause rapid increase in intracellular Ca^{2+}.[13]

The hydrolysis of phosphatidyl 4,5-*bis*-phosphate (PIP_2) by phospholipase C within the cell membrane yields diacylglycerol in addition to IP_3. Diacylglycerol, in turn, can activate protein kinase C, which is the initiator of a series of protein phosphorylation events which precede growth factor-induced mitogenesis in many cell types. Studies with fibroblast lines showed that the bombesin family of peptides caused an increase in protein kinase C and the phosphorylation of its substrate.[26] The mitogenic response of cells to bombesin is accompanied by a rapid decrease in the affinity of the EGF receptor population for EGF,[26] a phenomenon also noted for PDGF and FGF.[18] Protein kinase C has been implicated in the regulation by bombesin of the EGF receptor, since depletion of this enzyme from cells by prolonged exposure to phorbol esters, which bind avidly to diacylglycerol and activate protein kinase C, blocks EGF receptor transmodulation.[26] The ability of bombesin to activate transient c-*myc* and c-*fos* proto-oncogene expression in 3T3 cells is mediated by the protein kinase C/Ca^{2+} flux pathway, since these effects are abolished by blockade of protein kinase C or exposure to calcium ionophores which increase intracellular Ca^{2+}.[27]

The activation of adenylate cyclase leading to increased intracellular levels of cAMP is a common pathway preceding cellular mitogenesis. VIP, acting synergistically with insulin, caused a marked accumulation of cAMP in Swiss 3T3 fibroblasts prior to DNA synthesis.[15] This was not accompanied by any change in cytoplasmic Ca^{2+} or in the amounts of protein kinase C. Similarly, VIP and secretin increased cAMP accumulation by cultured rat Schwann cells, while substance P and somatostatin were without effect.[28] It is of interest that the growth-inhibitory actions of VIP on smooth muscle cells were also associated with increased formation of cAMP,[14] demonstrating how the same intracellular pathway can be used by a single ligand to induce dramatically opposite biological effects. The ability of substance P and neurokinin A to promote smooth muscle cell replication was accompanied by increased inositol phosphate metabolism and was associated with Ca^{2+} flux, since calcium and calmodulin inhibitors blocked DNA synthesis.[29]

Therefore, it appears superficially that the neuropeptides utilize two predominant types of intracellular signals during the induction of mitogenesis: either an activation of protein kinase C associated with altered phospho-inositol metabolism as favored by bombesin/GRP, vasopressin, and bradykinin, or the elevation of cAMP as demonstrated for VIP. However, clear distinctions do not exist, and these pathways show considerable interaction. For instance, activation of protein kinase C in 3T3 cells by vasopressin or bombesin, acting through receptors not directly linked to adenylate cyclase, potentiated the accumulation of cAMP seen in response to agents such as forskolin.[30]

IV. GROWTH-PROMOTING EFFECTS OF NEUROPEPTIDES *IN VIVO*

There are a considerable number of reported trophic actions of neuropeptides *in vivo* or in isolated organ systems. It is not possible to conclude from such information whether neuropeptide action is direct or is mediated by the release of secondary hormones, neuro-peptides, or growth factors. There are several documented examples of indirect effects.

While neuropeptides are largely absent from the adult lung and respiratory tubes, a considerable expression of bombesin is seen in the human fetal and neonatal lung.[31,32] This suggests a role for bombesin/GRP during lung development and is supported by:

1.   A reduced bombesin immunoreactivity in the lungs of premature infants suffering respiratory distress syndrome,[32] but an elevation of bombesin in chronic bronchopul-monary dysplasia[33]
2.   The demonstrated mitogenic actions of GRP on isolated bronchial epithelial cells[34]
3.   The elevation of mRNA for GRP between 16 and 30 weeks gestation during the canalicular phase of development characterized by considerable branching of the res-piratory bronchioles and associated development of epithelium and capillary invasion[35]
4.   The demonstration of comparably high levels of GRP expression during small cell lung carcinoma[36]

Ontogenic studies of GRP expression showed a decline in mRNA abundance from 30 weeks gestation to near adult levels at 34 weeks. However, GRP peptide remained elevated in lung tissue until several months after birth; cells demonstrated immunocytochemical staining for GRP peptide despite demonstrating little mRNA following cDNA/mRNA hybridization *in situ*.[35] This suggests that GRP may be stored for long periods within neuroendocrine cells of the lung following synthesis or that secreted peptides may be sequestered by target tissues. The location of mRNA for GRP in the early fetus at points of airway bifurcation suggests a primary role in the budding of new primitive alveoli along the bronchioles. Other vasoactive actions of GRP in the developing lung cannot be ignored. A bronchoconstrictor effect of bombesin in guinea pig was noted,[37] although no vasoactive effects were found in the isolated perfused rat lung.[38]

While it is well established that peripheral administration of GRP/bombesin causes the release of gastrin, CCK, pancreatic polypeptide, insulin, glucagon, and GIP in several mammalian species[39] and increases exocrine secretion from the stomach and pancreas, trophic actions of GRP on the pancreas and intestine have also been observed.[40] Treatment of 1-week-old rats for a period of 6 d with bombesin caused an increase in the weight of stomach, intestine, and pancreas compared to controls. In stomach the heights of the fundic and antral mucosae were elevated, suggesting accelerated gastric mucosal development, as previously observed in the adult rat.[41] An increase in the number of gastrin-secreting cells in the antral mucosa and parietal cells of the fundic mucosa was also characteristic of bombesin-treated animals. Histological examination of the pancreas showed that the increase in pancreatic weight was due to both hypertrophy and hyperplasia of the acinar cell popu-lation. A maturation of the trophic response of the pancreas to bombesin occurs in the neonatal period, since s.c. injection of bombesin was less effective in 3-d-old rats than in 21-d-old animals.[42] However, it cannot be concluded whether the actions of bombesin are direct, since CCK and gastrin, which are mobilized in response to GRP, have also been shown to influence the development of the pancreas and gastrointestinal mucosae. CCK and its analogue, caerulein, stimulated increased weight and DNA content in the pancreas of the adult rat and hamster.[43-45] Similarly, secretin, with or without caerulein, promoted acinar pancreatic growth.[45] While CCK may mediate the trophic actions of bombesin/GRP in the adult rodent, the similar actions of bombesin seen in the newborn are likely to be direct since CCK alone had no actions at this stage of development.[46] Increased pancreatic DNA and protein synthesis were observed in the rat following the chronic administration of pentagastrin,[47] effects blocked by somatostatin.[48] Gastrin has little effect on the development of the gastrointestinal tract in rats prior to weaning, since gastrin receptors are absent, at least in stomach, at this time.[49] However, in the adult rat, gastrin stimulated the growth of the gastric, duodenal, and colonic mucosae.[50] Specific receptors for gastrin have been identified on the human gastric cell line MKN45, for which gastrin is mitogenic both *in vitro* and when xenotransplanted into nude mice.[51] Administration of gastrin to the rat reverses the atrophy of the gastric mucosa seen following antrectomy.[52] A direct trophic role for ectopic GRP present in milk is possible, since the ingestion of bombesin by suckling rats stimulated the proliferation of the intestinal epithelium.[53]

The observation that the vasopressin-deficient Brattleboro rat has an impaired development of the brain which is corrected by vasopressin replacement suggests a function for this neuropeptide in neural maturation.[54] However, it is not clear whether this and other brain neuropeptides are trophic during early neuronal cell growth. Some evidence exists to suggest that, like nerve growth factor, which is essential for the survival of sensory and sympathetic neurons, these neuropeptides are necessary for the maintenance and continued viability of some neuronal tissues. Dissociated neurons from the rat spinal cord enter a period of synoptic activity 1 to 3 weeks into culture, which if blocked chemically results in neuronal death. Exposure to VIP during, but not following, this specific developmental event greatly enhanced cell survival.[55] The expression of several neuropeptides in brain is modulated by nerve growth factor — for example, the presence of mRNA or peptide for substance P and calcitonin-gene related peptide in sensory and sympathetic ganglia or neurons in rat or chick.[56-58] Conversely, substance P regulates nerve growth factor expression in damaged rat saphenous nerves, suggesting a local role for neuropeptides in peripheral nerve regeneration.[59] The degree of importance of neuropeptide action in neuronal maintenance vs. neurotransmitter action remains to be determined.

V. NEOPLASIA

Considerable interest has been generated by the possibility that growth factors and neuropeptides may contribute to the progression or metastasis of neoplastic disease due to their aberrant expression and subsequent action as autocrine or paracrine agents.[16] The findings that a similarly high level of expression of these hormones is often a normal feature of embryonic and fetal development, as already described, raises the possibility that the growth of some cancers involves the inappropriate expression of embryonic developmental peptides. Many tumors of neural and neuroendocrine origin release neuropeptides.[60] However, for only a few tumor types has this been shown to contribute to the development of the cancer.

The best documented case of neuropeptide involvement in neoplasia deals with the abundant expression of bombesin/GRP in small cell lung carcinoma. Comprising approximately 20% of lung cancers, small cell carcinomas show many neuroendocrine characteristics including the presence of neuron-specific enolase, dense core granules, and the secretion of a variety of hormones including vasopressin, parathormone, calcitonin, glucagon, somatostatin, neurotensin, and GRP-like activity.[61,62] These cells are thought to originate from the endocrine Kultschitzky cells.[63] The GRP-like immunoreactivity present within, and secreted by, small cell carcinomas is almost identical to gastrointestinal GRP,[64] and cDNAs corresponding to human GRP have been cloned from a pulmonary endocrine tumor.[36] A functional role for carcinoma-derived GRP is suggested by the presence of bombesin/GRP receptors on about 20% of the cell lines examined.[65] However, this is not a reliable indicator of autocrine/paracrine function, since the release of copious amounts of endogenous GRP may substantially down-regulate the presence of available GRP receptors. More convincing evidence was provided by Cuttitta et al.,[66] who showed that monoclonal antibodies against bombesin/GRP inhibited the clonal growth of small cell lung carcinoma lines *in vitro* and when transplanted into immune-deficient male mice. While some carcinoma lines proliferate in response to exogenous GRP and demonstrate rapid mobilization of intracellular calcium, this is not always the case.[66-69] Attempts to develop clinical tools with which to arrest lung cancer growth have included the use of bombesin antagonists, such as analogues of substance P, which have been shown to inhibit the biological actions of bombesin *in vitro*, or spantide, which inhibits the cellular binding of bombesin.[69] While an inhibition of bombesin/GRP action on small cell carcinoma lines was reported, this may not have been mediated via the GRP receptor, since the effects were not reversible in the presence of increased amounts of bombesin.

Since smooth cell lung carcinomas release a number of neuropeptides and hormones other than GRP, it is likely that some, such as vasopressin, also may contribute to autocrine proliferative loops. In addition, secretin and VIP have been shown to stimulate the release of GRP from several lung carcinoma lines, suggesting the existence of mutually reinforcing actions between secreted peptides.[65,70] Neuropeptides also may contribute to lung tumor metastasis, since bombesin, substance P, and vasopressin were each shown to have chemoattractive properties for smooth cell lung carcinoma cells and human monocytes.[71] The half-maximal effective concentrations were in the femtomole range, two orders of magnitude below those concentrations of peptide necessary to illicit mitogenic responses.

Molecular homogeneity may not exist within the GRP-like molecules released by certain tumors. Some medullary carcinomas of the thyroid contain both large and small forms of GRP that reacted poorly with antiserum against bombesin.[72] Similarly, a high-molecular-weight nonbombesin GRP was located in a small cell lung cancer.[73] Since the mRNA for GRP encodes a peptide precursor with a 95-amino-acid carboxy extension, considerable scope for posttranslational modification exists. Also, alternative splicing of the GRP gene can give rise to at least three mRNA transcripts[36] which vary in the carboxy-terminal extension coding region. Thus, the precursor peptides generated may be modified differentially prior to secretion. At least two of the predicted three GRP forms have now been shown to be expressed in human fetal lung and small cell lung cancer.[74] The biological relevance of GRP peptide heterogeneity cannot be assessed until the actions of these peptides have been compared directly.

Tumors other than those of lung, usually neuroendocrine in nature, also demonstrate possible neuropeptide-dependent growth characteristics. The trophic effects of gastrin, secretin, CCK, and bombesin on normal gut mucosa and pancreas led to an examination of their actions on the growth of gastrointestinal tumors. Endogenous hypergastrinemia induced by antral exclusion or small bowel resection in rats increased DNA synthesis within chemically induced tumors of the colon, demonstrating a likely trophic action of gastrin *in vivo*.[75] Gastrin stimulated the growth of human stomach and colon carcinomas transplanted into nude mice,[76] although growth was inhibited by secretin. Similarly, gastrin was mitogenic *in vitro* on mouse colon carcinoma cells demonstrated to possess gastrin receptors.[77] Studies with rodents showed that gastrin increased the incidence of carcinogen-induced gastric cancer, although conflicting reports also exist.[78,79] The growth of hamster pancreatic duct adenocarcinoma cells was enhanced by secretin and the CCK analogue cerulein.[80] Conversely, CCK octapeptide inhibited the growth of a human cholangiocarcinoma, while somatostatin and its analogues have been used experimentally and clinically to suppress the growth of a number of tumors, including pancreatic adenocarcinomas, insulinomas, and chondrosarcomas.[81-83]

The difficulty of extrapolating from normal to neoplastic physiology was well demonstrated by Alexander et al.,[84] who found that bombesin treatment increased DNA and protein synthesis in the nude mouse pancreas, but inhibited the growth of human ductal pancreatic adenocarcinoma transplanted into similar immune-deficient animals. This suggests that the mitogenic actions of bombesin in the intact host pancreas may have been indirect, mediated by the local release of other neuropeptides or growth factors. This is supported by the ability of bombesin to stimulate the growth of preneoplastic pancreatic acinar cell lesions in carcinogen-treated rats.[85] Other tumors demonstrated to release copious amounts of neuropeptides include human adrenal pheochromocytoma cells which express VIP. Interestingly, the maintenance of normal adult human chromaffin cells, which do not release VIP, for extended periods *in vitro* resulted in similar levels of VIP release to those seen in tumors.[86] The release of VIP from cultured chromaffin cells was potentiated by nerve growth factor. Although not neuroendocrine in origin, primary rat hepatocellular carcinomas induced by nitrosamines showed positive immunostaining for neuron-specific enolase and bombesin.[87] This demonstrates the ability of functionally differentiated cells such as hepatocytes to reprogram their genotypic expression comprehensively in a neural direction during neoplastic disruption. The presence of VIP is a common finding in both ganglion and adenomotous cells of pituitary adenomas.[88]

VI. CONCLUSIONS

The balance of evidence supports a role for certain neuropeptides during normal gastrointestinal growth and development mediated by local paracrine actions in the vicinity of their neuroendocrine sites of synthesis. Peptides such as GRP, VIP, CCK, and gastrin interact with specific cell membrane receptor populations and activate well-recognized series of intracellular messenger systems, including proto-oncogene expression, which precede DNA synthesis. Thus, the mechanisms by which neuropeptides induce mitogenic responses exactly mirror those of the peptide growth factors. The distinction between these groups lies not in mechanisms of action, but in sites of anatomical expression. Peptide growth factors such as EGF, IGFs, and PDGF are expressed during early development,[89] during tissue damage,[90] and in ovarian function[91] by both non-neural epithelium and mesenchymal cell types. Certain neuropeptides appear to represent an equivalent system expressed by neuroectodermal cell populations, although substantial interaction occurs between the two hormone classes. The role of neuropeptides in neoplastic growth is less clear, since neuropeptide expression may simply represent a symptom of uncontrolled proliferation, not a cause. However, in the case of small cell lung carcinoma, there is strong evidence for an autocrine mitogenic role for GRP.

REFERENCES

1. **Hill, D. J.,** Growth factors and their cellular actions, *J. Reprod. Fertil.,* 85, 723, 1989.
2. **Rozengurt, E., Legg, A., and Pettican, P.,** Vasopressin stimulation of mouse 3T3 cell growth, *Proc. Natl. Acad. Sci. U.S.A.,* 76, 1284, 1979.
3. **Miller, R. P., Husain, R., Svensson, M., and Lohin, S.,** Enhancement of [³H-methyl] thymidine incorporation and replication of rat chondrocytes grown in tissue culture by plasma, tissue extracts and vasopressin, *Endocrinology,* 100, 1365, 1977.
4. **Rozengurt, E. and Sinnett-Smith, J.,** Bombesin stimulation of DNA synthesis and cell division in cultures of swiss 3T3 cells, *Proc. Natl. Acad. Sci. U.S.A.,* 80, 2936, 1983.
5. **Zachary, I. and Rozengurt, E.,** High affinity receptors for peptides of the bombesin family in Swiss 3T3 cells, *Proc. Natl. Acad. Sci. U.S.A.,* 76, 7616, 1985.
6. **Willey, J. C., Lechner, J. F., and Harris, C. C.,** Bombesin and the C-terminal tetradecapeptide of gastrin-releasing peptide are growth factors for normal human bronchial epithelial cells, *Exp. Cell Res.,* 153, 245, 1984.
7. **Nilsson, J., Von Euler, A. M., and Dalsgaard, C.-J.,** Stimulation of connective tissue cell growth by substance P and substance K, *Nature (London),* 315, 61, 1985.
8. **Payan, D. G.,** Receptor-mediated mitogenic effects of substance P on cultured smooth muscle cells, *Biochem. Biophys. Res. Commun.,* 130, 104, 1985.

9. **Lotz, M., Carson, D. A., and Vaughan, J. H.,** Substance P activation of rheumatoid synoviocytes: neural pathway in pathogenesis of arthritis, *Science,* 235, 893, 1987.

10. **Mitsuhashi, M. and Payan, D. G.,** The mitogenic effects of vasoactive neuropeptides on cultured smooth muscle cell lines, *Life Sci.,* 40, 853, 1987.

11. **Payan, D. G., Brewster, D. R., and Goetzl, E. J.,** Specific stimulation of human T lymphocytes by substance P, *J. Immunol.,* 131, 1613, 1983.

12. **Tanaka, T., Danno, K., Ikai, K., and Imamura, S.,** Effects of substance P and substance K on the growth of cultured keratinocytes, *J. Invest. Dermatol.,* 90, 399, 1988.

13. **Woll, P. J. and Rozengurt, E.,** Two classes of antagonist interact with receptors for the mitogenic neuropeptides bombesin, bradykinin, and vasopressin, *Growth Factors,* 1, 75, 1988.

14. **Hultgardh-Nilsson, A., Nilsson, J., Johzon, B., and Dalsgaard, C. J.,** Growth-inhibitory properties of vasoactive intestinal peptide, *Regul. Peptides,* 22, 267, 1988.

15. **Zurier, R. B., Kozma, M., Sinnett-Smith, J., and Rozengurt, E.,** Vasoactive intestinal peptide synergistically stimulates DNA synthesis in mouse 3T3 cells: role for cAMP, Ca^{2+} and protein kinase C, *Exp. Cell Res.,* 176, 155, 1988.

16. **Sporn, M. B. and Roberts, A. B.,** Peptide growth factors and inflammation, tissue repair, and cancer, *J. Clin. Invest.,* 78, 329, 1986.

17. **Zachary, I. and Rozengurt, E.,** Identification of a receptor for peptides of the bombesin family in Swiss 3T3 cells by affinity cross-linking, *J. Biol. Chem.,* 262, 3947, 1987.

18. **Van Wyk, J. J., Underwood, L. E., D'Ercole, A. J., Clemmons, D. R., Pledger, W. M., Wharton, W. R., and Loef, E. B.,** Role of somatomedin in cellular proliferation, in *Biology of Normal Human Growth,* Ritzen, M., Aperia, A., Hall, K., Larsson, A., Zetterburg, A., and Zellerstrom, R., Eds., Raven Press, New York, 1981, 223.

19. **Rozengurt, E.,** Early signals in the mitogenic response, *Science,* 234, 161, 1986.

20. **Nilsson, J., Sejerson, T., Nilsson, A. H., and Dalsgaard, C. J.,** DNA synthesis induced by the neuropeptide substance K correlates to the level of *myc*-gene transcripts, *Biochem. Biophys. Res. Commun.,* 137, 167, 1986.

21. **Letterio, J. J., Coughlin, S. R., and Williams, L. T.,** Pertussis toxin — sensitive pathway in the stimulation of *c-myc* expression and DNA synthesis by bombesin, *Science,* 234, 1117, 1986.

22. **Palumbo, A. P., Rossino, P., and Comoglio, P. M.,** Bombesin stimulation of *c-fos* and *c-myc* gene expression in cultures of Swiss 3T3 cells, *Exp. Cell Res.,* 167, 276, 1986.

23. **Mendoza, S. A., Schneider, J. A., Lopez-Rivas, A., Sinnett-Smith, J. W., and Rozengurt, E.,** Early events elicited by bombesin and structurally related peptides in quiescent Swiss 3T3 cells. II. Changes in Na^+ and Ca^{2+} fluxes, Na^+/K^+ pump activity and intra-cellular pH, *J. Cell Biol.,* 102, 2223, 1986.

24. **Takuwa, N., Takuwa, Y., Bollag, W. E., and Rasmussen, H.,** The effects of bombesin on polyphosphoinositide and calcium metabolism in Swiss 3T3 cells, *J. Biol. Chem.,* 262, 182, 1987.

25. **Muir, J. G. and Murray, A. W.,** Bombesin and phorbol esters stimulate phosphatidylcholine hydrolysis by phospholipase C: evidence for a role for protein kinase C, *J. Cell. Physiol.,* 130, 382, 1987.

26. **Zachary, I., Sinnett-Smith, J. W., and Rozengurt, E.,** Early events elicited by bombesin and structurally related peptides in quiescent Swiss 3T3 cells. I. Activation of protein kinase C and inhibition of epidermal growth factor binding, *J. Cell Biol.,* 102, 2211, 1986.

27. **Rozengurt, E. and Sinnett-Smith, J. W.,** Bombesin induction of *c-fos* and *c-myc* proto-oncogenes in Swiss 3T3 cells: significance for the mitogenic response, *J. Cell. Physiol.,* 131, 218, 1987.

28. **Yasuda, T., Sobue, G., Mitsuma, T., and Takahashi, A.,** Peptidergic and adrenergic regulation of the intracellular 3',5'-cyclic adenosine monophosphate content in cultured rat Schwann cells, *J. Neurol. Sci.,* 88, 315, 1988.

29. **Hultgardh-Nilsson, A., Nilsson, J., Jonzon, B., and Dalsgaard, C. J.,** Coupling between inositol phosphate formation and DNA synthesis in smooth muscle cells stimulated with neurokinin A, *J. Cell. Physiol.,* 137, 141, 1988.

30. **Rozengurt, E., Murray, M., Zachary, I., and Collins, M.,** Protein kinase C activation enhances cAMP accumulation in Swiss 3T3 cells: inhibition by pertussis toxin, *Proc. Natl. Acad. Sci. U.S.A.,* 84, 2282, 1987.

31. **Price, J., Penman, E., Bourne, G. L., and Rees, L. H.,** Characterization of bombesin-like immunoreactivity in human fetal lung, *Regul. Peptides,* 7, 315, 1983.

32. **Ghatei, M. A., Sheppard, M. N., Henzen-Logman, S., Blank, M. A., Polak, J. M., and Bloom, S. R.,** Bombesin and vasoactive intestinal polypeptide in the developing lung: marked changes in acute respiratory distress syndrome, *J. Clin. Endocrinol. Metab.,* 57, 1226, 1983.

33. **Johnson, D. E., Lock, J. E., Elde, R. P., and Thompson, T. R.,** Pulmonary neuroendocrine cells in hyaline membrane disease and bronchopulmonary dysplasia, *Pediatr. Res.,* 16, 446, 1982.

34. **Willey, J. C., Lechner, J. F., and Harris, C. C.,** Bombesin and C-terminal tetradecapeptide of gastrin-releasing peptide are growth factors for normal human bronchial epithelial cells, *Exp. Cell Res.,* 153, 245, 1984.

35. **Spindel, E. R., Sunday, M. E., Hofler, H., Wolfe, H. J., Habener, J. F., and Chin, W. W.,** Transient elevation of messenger RNA encoding gastrin-releasing peptide, a putative pulmonary growth factor in human fetal lung, *J. Clin. Invest.,* 80, 1172, 1987.
36. **Sausville, E. A., Lebacq-Verhayden, A. M., Spindel, E. R., Cuttitta, F., Gazdar, A. F., and Battey, J. F.,** Expression of gastrin-releasing peptide gene in human small cell lung cancer, *J. Biol. Chem.,* 261, 2451, 1986.
37. **Impicciatore, M. and Bertaccini, G.,** The broncho constrictor action of the tetradecapeptide bombesin in the guinea pig, *J. Pharm. Pharmacol.,* 25, 872, 1973.
38. **Gillespie, M. N., Reinsel, C. N., and Bowdy, B. D.,** Pulmonary vasoactivity of lung endocrine cell-related peptides, *Peptides,* 5, 21, 1984.
39. **McDonald, T. J.,** The gastrin-releasing polypeptide, *Adv. Metab. Disord.,* 11, 199, 1988.
40. **Lehy, T., Puccio, F., Chariot, J., and Labeille, D.,** Stimulating effect of bombesin on the growth of the gastrointestinal tract and pancreas in suckling rats, *Gastroenterology,* 90, 1942, 1986.
41. **Lehy, T., Accary, J. P., Labeille, D., and Dubrasquet, M.,** Chronic administration of bombesin stimulates antral gastrin cell proliferation in the rat, *Gastroenterology,* 84, 914, 1983.
42. **Pollack, P. F.,** Age and enzyme specificity of the response of developing rat pancreas to the trophic effects of bombesin, *Pancreas,* 4, 101, 1989.
43. **Barrowman, J. A. and Mayston, P. D.,** The trophic influence of cholecystokinin on the rat pancreas, *J. Physiol.,* 238, 73P, 1973.
44. **Solomon, T. E., Peterson, H., Elashoff, J., and Grossman, M. I.,** Interaction of caerulein and secretin on pancreatic size and composition in rat, *Am. J. Physiol.,* 235, E714, 1978.
45. **Dembinski, A. B. and Johnson, L. R.,** Stimulation of pancreatic growth by secretin, caerulein, and pentagastrin, *Endocrinology,* 106, 323, 1980.
46. **Zahavi, I., Kelly, J., and Gall, D. G.,** Role of gastrin and cholecystokinin in the ontogenic development of the gastro-intestinal tract, *Biol. Neonate,* 45, 95, 1984.
47. **Nandi, A. P. and Goltermann, N.,** Chronic administration of pentagastrin. Effects on pancreatic protein and nucleic acid contents and protein synthesis in rats, *Digestion,* 19, 144, 1979.
48. **Morisset, J., Genki, P., Lord, A., and Solomon, T. E.,** Effects of chronic administration of somatostatin on rat exocrine pancreas, *Regul. Peptides,* 4, 49, 1982.
49. **Takeuchi, K., Peitsch, W., and Johnson, L. R.,** Mucosal gastrin receptor. V. Development in newborn rats, *Am. J. Physiol.,* 240, G160, 1981.
50. **Johnson, L. R.,** New aspects of the trophic action of gastrointestinal hormones, *Gastroenterology,* 72, 788, 1981.
51. **Watson, S. A., Durrant, L. G., and Morris, D. L.,** The trophic effects of gastrin on the human gastric cell line, MKN45 *in vitro* and *in vivo, Br. J. Cancer,* 57 (Abstr.), 230, 1988.
52. **Dembinski, A. B. and Johnson, L. B.,** Growth of pancreas and gastrointestinal mucosa in antrectomized and gastrin-treated rats, *Endocrinology,* 105, 769, 1979.
53. **Puccio, F. and Lehy, T.,** Bombesin ingestion stimulates epithelial digestive cell proliferation in suckling rats, *Am. J. Physiol.,* 256, G238, 1989.
54. **Boer, G. J.,** Vasopressin and brain development: studies using the Brattleboro rat, *Peptides,* 6, 49, 1985.
55. **Brenneman, D. E. and Foster, G. A.,** Structural specificity of peptides influencing neuronal survival during development, *Peptides,* 8, 687, 1987.
56. **Hayashi, M., Edgar, D., and Thoenen, H.,** Nerve growth factor changes the relative levels of neuropeptides in developing sensory and sympathetic ganglia of the chick embryo, *Dev. Biol.,* 108, 49, 1985.
57. **MacLean, D. B., Lewis, S. F., and Wheeler, F. B.,** Substance P content in cultured neonatal rat vagal sensory neurons: the effect of nerve growth factor, *Brain Res.,* 457, 53, 1988.
58. **Lindsay, R. M. and Harmar, A. J.,** Nerve growth factor regulates expression of neuropeptide genes in adult sensory neurons, *Nature (London),* 337, 362, 1989.
59. **White, D. M., Ehrhard, P., Hardung, M., Meyer, D. K., Zimmerman, M., and Otten, V.,** Substance P modulates the release of locally synthesized nerve growth factor from rat saphenous nerve neuroma, *Naunyn-Schmiedeberg's Arch. Pharmacol.,* 336, 587, 1987.
60. **Polak, J. M. and Bloom, S. R.,** Pathophysiology of the diffuse endocrine system, in *Systemic Pathology,* Williams, E. D., Ed., Churchill Livingstone, Edinburgh, 1985, 35.
61. **Luster, W., Gropp, C., Kern, H. F., and Hauemann, K.,** Lung tumor cell lines synthesizing peptide hormones established from tumors of four histological types: characterization of the cell lines and analysis of their peptide hormone production, *Br. J. Cancer,* 51, 865, 1985.
62. **Erisman, M. D., Linnoila, R. L., Hernandez, O., Di Augustine, R. P., and Lazarus, L. H.,** Human lung small-cell carcinoma contains bombesin, *Proc. Natl. Acad. Sci. U.S.A.,* 79, 2379, 1982.
63. **Bonikos, D. S. and Bensch, K. G.,** Endocrine cells of bronchial and bronchiolar epithelium, *Am. J. Med.,* 63, 765, 1977.

64. **Yamaguchi, K., Abe, K., Kameya, T., Adachi, I., Taguchi, S., Otsubu, K., and Yaraihara, N.,** Production and molecular size heterogeneity of immunoreactive gastrin-releasing peptide in fetal and adult lungs and primary lung tumours, *Cancer Res.*, 43, 3932, 1983.

65. **Moody, T. W., Russell, E. K., O'Donohue, T. L., Linden, C. D., and Gazdar, A. F.,** Bombesin-like peptides in small cell lung cancer-biochemical characterization and secretion from a cell line, *Life Sci.*, 32, 487, 1983.

66. **Cuttitta, F., Carney, D. N., Mulshine, J., Moody, T. W., Fedorko, J., Fischler, A., and Minna, J. D.,** Bombesin-like peptides can function as autocrine growth factors in human small cell lung cancer, *Nature (London)*, 316, 823, 1985.

67. **Weber, S., Zuckerman, J. E., Bostwick, D. G., Bensch, K. G., Sikic, B. I., and Raffin, T. A.,** Gastrin releasing peptide is a selective mitogen for small cell lung carcinoma *in vitro*, *J. Clin. Invest.*, 75, 306, 1985.

68. **Layton, J. E., Scanlon, D. B., Soveny, C., and Morstyn, G.,** Effects of bombesin antagonists on the growth of small cell lung cancer cells *in vitro*, *Cancer Res.*, 48, 4783, 1988.

69. **Heikkila, R., Trepel, J. B., Cuttitta, F., Neckers, L. M., and Sausville, E. A.,** Bombesin-related peptides induce calcium mobilization in a subset of human small cell lung cancer cell lines, *J. Biol. Chem.*, 262, 16456, 1987.

70. **Sorenson, G. D., Bloom, S. R., Ghatei, M. A., Delprete, S. A., Cate, C. C., and Pettergill, O. S.,** Bombesin production by human small cell carcinoma of the lung, *Regul. Peptides*, 4, 59, 1982.

71. **Ruff, M., Schiffman, E., Terranova, V., and Pert, C. B.,** Neuropeptides are chemoattractants for human tumour cells and monocytes: a possible mechanism for metastasis, *Clin. Immunol. Immunopathol.*, 37, 387, 1985.

72. **Howlett, T. A., Price, J., Hale, A. C., Doniach, I., Rees, L. H., Wass, J. A., and Besser, G. M.,** Pituitary ACTH-dependent Cushing's syndrome due to ectopic production of a bombesin-like peptide by a medullary carcinoma of the thyroid, *Clin. Endocrinol.*, 22, 91, 1985.

73. **MacAuley, V., Joshi, G. P., Everard, M., Smith, E. E., and Millar, J. L.,** A high molecular weight non-bombesin/gastrin releasing peptide growth factor in small cell lung cancer, *Br. J. Cancer*, 56, 791, 1987.

74. **Cuttitta, F., Fedorka, J., Gu, J., Lebaq-Verneyden, A.-M., Linnoila, R. L., and Battey, J. F.,** Gastrin-releasing peptide gene-associated peptides are expressed in normal fetal lung and small cell lung cancer: a novel peptide family found in man, *J. Clin. Endocrinol. Metab.*, 67, 576, 1988.

75. **McGregor, D. B., Jones, R. D., Karlen, D. A., and Romsdahl, M. M.,** Trophic effects of gastrin on colorectal neoplasms in the rat, *Ann. Surg.*, 195, 219, 1982.

76. **Tanake, J., Yamaguchi, T., Takahashi, T., Ogata, N., Bando, K., and Koyama, K.,** Regulatory effects of gastrin and secretin on carcinomas of the stomach and colon, *Tohoku J. Exp. Med.*, 148, 459, 1986.

77. **Winsett, O. E., Townsend, C. M., Singh, P., Glass, E. J., and Thompson, J. C.,** Gastrin stimulates growth of colon cancer, *Surgery*, 99, 302, 1986.

78. **Tahara, E. and Haizuka, S.,** Effect of gastro-entero-pancreatic endocrine hormones on the histogenesis of gastric cancer in rats induced by N-methyl-N^1-nitro-N-nitroso guanidine; with special reference to development of scirrhous gastric cancer, *Gann*, 66, 421, 1975.

79. **Deveney, C. W., Freeman, H., and Way, L. W.,** Experimental gastric carcinogenesis in the rat. Effect of hypergastrinemia and acid secretion, *Am. J. Surg.*, 139, 49, 1980.

80. **Townsend, D. M., Franklin, R. B., Watson, L. C., Glass, E. J., and Thompson, J. C.,** Stimulation of pancreatic cancer by caerulein and secretin, *Surg. Forum*, 32, 228, 1981.

81. **Hudd, C., Euhus, D. M., La Regina, M. C., Herbold, D. R., Palmer, D. C., and Johnson, F. E.,** Effect of cholecystokinin on human cholangio carcinoma into xenografted nude mice, *Cancer Res.*, 45, 1372, 1985.

82. **Upp, J. R., Olson, D., Poston, G. J., Alexander, R. W., Townsend, C. M., and Thompson, J. C.,** Somatostatin analog 201-995 inhibits growth of two human pancreatic adenocarcinomas *in vivo*, *Am. J. Surg.*, 155, 29, 1988.

83. **Reubi, J. C.,** A somatostatin analogue inhibits chondrosarcoma and insulinoma tumour growth, *Acta Endocrinol. (Copenhagen)*, 109, 108, 1985.

84. **Alexander, R. W., Upp, J. R., Poston, G. J., Townsend, C. M., Singh, P., and Thompson, T. C.,** Bombesin inhibits growth of human pancreatic adenocarcinoma in nude mice, *Pancreas*, 3, 297, 1988.

85. **Lhoste, E. F. and Longnecker, D. S.,** Effect of bombesin and caerulein on early stages of carcinogenesis induced by azaserine in the rat pancreas, *Cancer Res.*, 47, 3273, 1987.

86. **Tishcler, A. S., Lee, Y. S., Perlman, R. L., Costopoulas, D., and Bloom, S. R.,** Production of "ectopic" vasoactive intestinal peptide-like immunoreactivity in normal human chromaffin cell cultures, *Life Sci.*, 37, 1881, 1985.

87. **Seglan, P. O., Skomedal, H., Saeter, G., Schwarze, P. E., and Nesland, J. M.,** Neuroendocrine dysdifferentiation and bombesin production in carcinogen-induced hepatocellular rat tumours, *Carcinogenesis,* 10, 21, 1989.

88. **Li, J. Y., Racadot, O., Kujas, M., Kouadri, M., Peillon, R., and Racadot, J.,** Immunocytochemistry of four mixed pituitary adenomas and intra-sellar gangliocytomas associated with different clinical syndromes: acromegaly, amenorrhea-galactorrhea, Cushing's disease and isolated tumoral syndrome, *Acta Neuropathol.,* 77, 320, 1989.

89. **Hill, D. J. and Milner, R. D. G.,** Mechanisms of fetal growth, in *Clinical Paediatric Endocrinology,* Brook, C. G., Ed., 2nd ed., Blackwell Scientific, Oxford, 1988, 3.

90. **Lynch, S. E., Nixon, J. C., Coluin, R. B., and Antoniades, H. N.,** Role of platelet-derived growth factor in wound healing: synergistic effects with other growth factors, *Proc. Natl. Acad. Sci. U.S.A.,* 84, 7696, 1987.

91. **Adashi, E. Y.,** Putative intraovarian regulators, *Semin. Reprod. Endocrinol.,* 7, 1, 1989.

Chapter 16

PERSPECTIVES FOR THE STUDY OF GUT NEUROPEPTIDES

E. E. Daniel

TABLE OF CONTENTS

I. Introduction 492

II. Purpose 492

III. Technical Problems 492

IV. Conceptual Problems 494

References 496

I. INTRODUCTION

As the contents of this volume testify, there has been an exponential growth over the past two decades in our knowledge of the existence, identification, synthesis, degradation, and functions of neuropeptides in the gastrointestinal tract. Examination of the reference lists of these several chapters will reveal that relatively few articles were published before 1970, but that the rate of publication has continued to accelerate ever since. Most early publications related to the hormonal rather than to the paracrine or neural modes of action of peptides (see Chapter 1). It is perhaps not surprising, then, as detailed in Chapter 1, that these hormonal actions are most securely established — for example, gastrin to affect H^+ secretion and parietal cell mass, CCK to affect gallbladder emptying and pancreatic secretion, secretion to affect other aspects of pancreatic secretion, glucagon to affect carbohydrate metabolism, etc. No doubt success in establishing these functions has been facilitated by the availability of more or less agreed upon criteria for establishing a hormonal action (see Reference 1). However, a number of other gut peptides with putative hormonal actions (e.g., motilin to regulate fasting gut motor activity, neurotensin to regulate gastric and intestinal transit in response to fatty meals) have proved more difficult to move from the putative to the established category of physiological function. Moreover, there are few neurotransmitter or paracrine actions of gut peptides which have been established to be physiological.

II. PURPOSE

The purpose of this chapter is to examine whether there are technical reasons (such as lack of highly selective antagonists), conceptual reasons (such as an inappropriate definition of neurotransmitter), or biological reasons (such as the presence of complex and redundant control systems) for the lack of definitely established physiological roles for gut peptides despite the explosion of scientific interest. We also have to consider that the present state of insight may be merely a normal stage in development of understanding the control functions of neuropeptides. This author's hypothesis is that elements of all these explanations are correct, but that the most important factor contributing to the modest rate of progress in identifying normal and described physiological roles of gut peptides relates to a combination of conceptual failures and lack of biological insights.

III. TECHNICAL PROBLEMS

Let us consider each of these in turn, starting with the technical problems limiting our understanding. There are indeed several of these. One relates to the fact that there is lack of clear information about the relevant local concentrations of neuro- or paracrine peptides. When nerve varicosities or paracrine-acting cells release peptides onto receptors in their vicinity, these are at concentrations relevant to neurotransmission or paracrine control. These concentrations have not been measured yet are likely to be much higher than the levels used in organ bath studies, millimolar to molar rather than nanomolar to micromolar. Bevan et al.[2] have calculated, based on structural and mediator release studies of the rather loose nerve-muscle relationship in the adventitia of blood vessels, that norepinephrine concentrations can reach 10 mM in the nerve-muscle junctions. The structural arrangements of the enteric nerve plexuses (see Reference 3) involve much smaller postjunctional volumes and presumably much higher local mediator concentrations, and in the musculature nerve varicosities are often close to muscle cells or even closer (≤ 30 nm) to interstitial cells of Cajal (ICC).[4,5] Clearly we have to consider the possibility that "physiological" peptide concentrations (from nerve releases) may bear little relationship to those normally tested for "identity of action'' or for susceptibility of mediator and candidate substances to antagonists. It could be argued that pre-equilibration of the postjunctional receptors with antagonists would lead to occupation of most receptors and that these receptors would remain occupied during the brief (millisecond to second) period of high mediator concentrations, given that antagonists are characterized by a slow dissociation rate constant from receptors. However, it can be counterargued that there is evidence that some receptors have low affinity for classical antagonists and that in some cases the number of occupied receptors required for response is very low. None of these arguments, however, are based on data from studies of neuro-peptides in gut. There is equal ignorance about actual local levels of substances released in paracrine fashion and their receptors.

Moreover, as discussed later, the local milieu near a nerve varicosity or paracrine cell is likely to contain multiple transmitters, some released from the same nerve ending as the candidate transmitter. Not only does this complicate the technical problem (what active substances are present, relevant, and need to be measured?), but it also raises conceptual problems to be discussed later.

Another technical problem is our lack of ability to determine the locus of relevant receptors for neurohormones and paracrine substances. Neither biochemical nor visual lo-calization techniques have evolved to allow reliable localization of mediators and receptors across synapses and junctions. Only recently (see Chapter 7) have techniques to separate and characterize intestinal muscle and nerve membranes evolved, and these do not provide any information about the geography of mediator release in relation to receptor actions. Light microscopic immunocytochemistry and autoradioautography lack the resolution to provide this information. Ultrastructural autoradioautography, which in theory has the nec-essary resolution, in practice is very difficult to apply. Problems of resolution still exist due to the random spread of radiation and false negative results from method insensitivity, and technical failure should be common and more difficult to detect than false positive results.[6] Ultrastructural immunocytochemistry should be able to overcome these problems, but is severely limited so far by the lack of high-affinity, high-selectivity antibodies to neuropeptide and other peptide receptors.

Inability to localize receptors in relation to release sites is clearly a serious limitation, especially in light of the understanding that there are several loci of receptors to most peptides; they may be presynaptic or postsynaptic in the plexuses and prejunctional or postjunctional in the muscles. Moreover, peptide receptors that affect gastrointestinal function are also located outside the gastrointestinal tract — in prevertebral ganglia, in other neuronal struc-tures, in other endocrine organs, and even in the central nervous system. Thus results of attempts to mimic neurotransmitter or paracrine actions by i.v. infusion will be difficult to interpret whether the results seem positive (was the receptor in the organ that responded?), negative (was sufficient active peptide able to reach the receptors?), or confused (was there more than one set of receptors activated?). Intra-arterial injections may guarantee a local site of action and diminish, but not eliminate, the problem because of the possible multiple sites of any given receptors within the intestine. *In vitro* studies may help, provided the relevant receptors are present in the resultant tissue and are functional. As discussed in Chapter 5, attempts to simplify the system, e.g., using extensively dissected tissues, single muscle cells, or single membranes or molecules, also change it.

The existence of multiple related peptides and receptor subtypes for nearly all neuro-peptides is a further technical complication, since in most cases each of the subtypes can recognize several related peptides. For example, among the mammalian tachykinins, sub-stance P, neurokinin A, and neurokinin B are each recognized by each of their receptors (NK-1, NK-2, and NK-3, respectively). Both Met- or Leu-enkephalin are recognized by μ-, δ-, and κ-opioid receptors, although with different affinities. In the case of tachykinin receptors, affinities at each receptor appear to be varying by less than a log unit for each naturally occurring tachykinin. Multiple related peptides and multiple receptor subtypes complicate identification of neurotransmitter and paracrine actions since both the correct peptide and receptor antagonist must be used in comparing putative to real transmitters (e.g., see Reference 7).

Although various receptors for a family of neuropeptides may or may not be connected to the same second messengers (consider opioid receptors in the myenteric plexus, Chapters 7 and 9) and although many different receptors share the same second messenger system focusing on second messengers rather than on receptors will not ensure that the correct response has been identified for comparison to endogenous mediators.

The most commonly cited technical difficulty in identifying physiological functions of neuropeptides is the lack of potent, highly selective receptor antagonists for each class or subclass of neuropeptide receptors. For example, such antagonists are not yet available for VIP and related peptides, for substance P and related peptides, and for GRP and bombesin-related peptides, etc. In fact, only for opioids and CCK do such agents exist (see Chapters 6, 7, 11, and 12), and no opioid antagonist is really highly selective with respect to receptor subtypes (see Chapters 6 and 7). So far there are few neuropeptide functions in the gastrointestinal tract established unequivocally for opioid or CCK-related peptides using these antagonists. For example, naloxone infused in large doses has no effect on fed motor activity in dogs and only lengthens cycles of the fasted pattern.[8] Provision of appropriate antagonists would accelerate, but would not guarantee, that success in overcoming difficulties in defining paracrine or neuropeptide physiological functions. Antibodies to peptides have been substituted in an attempt to overcome this deficiency. They will combine with a mediator up to the amounts of antibody available in the synaptic cleft or circulation. However, antibodies cannot be used to establish that there is competitive antagonism at a receptor and to define receptor locus or characteristics (pA_2, recognition properties, etc.). Long incubation times with these antibodies are often required, and careful controls for time-dependent changes are needed. False negative as well as false positive results can occur. Availability of real antagonists and their proper use will allow conclusions that a particular neuro- or paracrine peptide does or does not play some role in the sequence of events leading to a physiological response. Such an antagonist may allow false negative results if multiple mediators are involved. It will not permit the determination of the cellular site of action of that neuropeptide or prove that it is the direct mediator of a particular cellular action. A good example of this is the problem of identifying the nonadrenergic, noncholinergic inhibitory (NANCI) mediator to circular muscle including sphincters of the gastrointestinal tract. This problem also illustrates the conceptual and biological problems that must be overcome.

IV. CONCEPTUAL PROBLEMS

VIP has become the leading candidate for the NANCI role with strong credentials up to a point. VIP is present in nerves which project anally to circular muscle (see Reference 3). In several systems, it causes relaxation of circular and sphincteric muscles by a direct action (see References 8 to 11); in several systems, its actions and that of the mediator are both selectively reduced by a high-affinity VIP antibody.[10,11] However, in the system best studied, opossum esophagus circular muscle, the cellular actions of VIP and mediator are different.[12-14] VIP causes modest hyperpolarization in circular muscle, accompanied by a reduction in membrane conductance apparently due to reduced chloride conductance. On the other hand, the mediator causes large hyperpolarizations (inhibitory junction potentials [IJPS]), accompanied by decreased membrane resistance apparently due to increased K^+ conductance.[12,13] Moreover, VIP increases cAMP while the mediator increases cGMP levels.[15] Thus, the cellular actions of VIP and the mediator on circular smooth muscle differ despite the strong evidence that VIP is involved. The use of a highly potent, selective antagonist to block VIP and the mediator would not change these facts. Clearly there is a conceptual and/or a biological problem there; in fact it is both.

In considering what this problem might be, it is worth calling attention to the fact that VIP binding has been demonstrated repeatedly to enterocytes, but not to smooth muscle in the gastrointestinal tract. Radioautography suggests that such receptors are absent or sparse.[16] This may be a clue to the problem. Perhaps the site of mediator and VIP action is not directly on smooth muscle. This author has proposed (reviewed in Reference 8) that VIP actually initiates its inhibitory actions when released from nerves onto a special network of cells, interstitial cells of Cajal (ICC). Cells in this network are mutually connected by gap junctions and are similarly connected to smooth muscles.[4,5] They are preferentially and closely innervated (compared to muscle) by nerves with VIP.[17-19] Recent evidence[19] suggests that these cells respond to VIP by hyperpolarization and conductance increase. Much more evidence is required to establish or invalidate this explanation, but the possibility that it is correct illustrates the need for critical evaluation of concepts and for correct biological information. In this case, the necessary conceptual insight is that the use of gross responses such as contraction or relaxation cannot identify cellular actions at a molecular level; in fact, cellular actions of VIP in smooth muscle are inconsistent with those of the mediator on smooth muscle, and studies in which various antagonists or antibodies block a response cannot resolve this problem. The necessary biological insight is that VIP nerves often innervate ICC rather than smooth muscle and that ICC are coupled to smooth muscle by gap junctions. It should be obvious that without the realization of a conceptual deficiency in the VIP-NANCI mediator hypothesis there would be no incentive to look for alternative possibilities. Also, without the technical ability to localize VIP (currently available) and its receptors at an ultrastructural level (currently difficult), the ability to propose biologically relevant alternate explanations would be severely limited.

An alternative or additional solution to the problem of identifying the NANCI mediator brings us back to one of the technical problems mentioned earlier. There are multiple mediators in all enteric nerves (see References 3 and 15), and multiple mediators may be released when nerves are activated reflexly. Conceivably the cellular actions observed during NANCI nerve activity (large IJPS and increased cGMP) require the combined actions of VIP and another mediator. PHI[20] and CGRP[21] have been proposed as additional mediators. Any additional mediators, too, may act on smooth muscle, in nerves, or on ICC. Before this possibility can be evaluated critically, the nature of such mediators, the extent of their release during NANCI reflex or direct stimulation, the locus of their receptors, and their cellular actions alone and in combination with VIP have to be known. Applications of antagonists to putative additional mediators will be helpful but insufficient to clarify how the system works.

This example can probably be multiplied many fold. The exact roles of substance P and neurokinin A, the diverse functions of enkephalin and dynorphin nerves, the roles of CCK nerves, the function of motilin, and other problems await better technical tools and, more importantly, appropriate conceptual insight based on adequate biological information.

We can anticipate that molecular biological approaches will provide new tools to help solve these problems in the next decade. Especially important will be the information leading to the structure of neuropeptide receptors (see Chapters 7, 8, 11, and 12) and to the nature of neuropeptide processing in various loci (see Chapter 3). These sets of information should provide insight into what neuropeptides are present at important sites of release, allow design of selective antagonists, and provide antibodies to localize their receptors at an ultrastructural level. Selective control over peptide processing from precursor forms would be a valuable tool, but does not seem likely to be available soon. The availability of measurements of local concentrations of neuropeptides also seems unlikely to become available soon. Whatever new tools become available will have to be applied critically, in light of their limitations (asking the right questions), with conceptual clarity, and based on much deeper insight (structural and functional) into the cell biology of nerves, muscles, and endocrine cells of the gastrointestinal tract. During the next 10 years the exponential growth of publications about gut peptides will certainly continue; hopefully we will also see solutions to ongoing problems based on new tools and more critical experimentation. The next decade will also reveal whether we are just at a normal early stage in our physiological insights into peptide control function and whether the rate of growth in understanding will accelerate. The problems noted here, although complex, are really similar in principle to those involved in analysis of the function of neuropeptides in brain. The tools and information that become available in either system will be immediately applicable in the other.

REFERENCES

1. **Grossman, M. L., Robertson, C. R., and Ivy, A. C.,** Proof of a hormonal mechanism for gastric secretion. The humoral transmission of the distention stimulus, *Am. J. Physiol.,* 153, 1, 1948.
2. **Bevan, J. A., Bevan, R. D., and Duckles, S. P.,** Adrenergic regulation of vascular smooth muscle, in: *Handbook of Physiology. The Cardiovascular System Vascular Smooth Muscle,* Section 2, Vol. 2, American Physiology Society, Bethesda, MD, 1980, 515.
3. **Furness, J. B. and Costa, M.,** *The Myenteric Nervous Systems,* Churchill Livingstone, Edinburgh, 1987.
4. **Daniel, E. E. and Posey-Daniel, V.,** Neuromuscular structures in opossum esophagus: role of interstitial cells of Cajal, *Am. J. Physiol.,* 246, G305, 1984.
5. **Berezin, I., Huizinga, J. D., and Daniel, E. E.,** Interstitial cells of Cajal in canine colon: a special communication network at the inner border of the circular muscle, *J. Comp. Neurol.,* 273, 42, 1988.
6. **Leslie, T. M. and Altar, C. A.,** Receptor localization. Ligand autoradiography, in *Receptor Biochemistry and Methodology,* Vol. 13, Ventner, J. C. and Harrison, L. C., Eds., Alan R. Liss, New York, 1988, 217.
7. **Daniel, E. E., Posey-Daniel, V., Jager, L. P., Berezin, I., and Jury, J.,** Structural effects of exposure of smooth muscle in the sucrose gap apparatus, *Am. J. Physiol.,* 252, C77, 1987.
8. **Telford, G. L., Conden, R. E., and Szurszewski, J. H.,** Opioid receptors and the initiation of migrating myoelectric complexes in dogs, *Am. J. Physiol.,* 256, G72, 1989.
9. **Daniel, E. E.,** Nonadrenergic, noncholinergic (NANC) neuronal excitatory interactions with smooth muscle, in: *Calcium and Contractility,* Grover, A. K. and Daniel, E. E., Eds., Humana Press, Clifton, NJ, 1985, 427.
10. **Goyal, R. K., Rattan, S., and Said, S. I.,** VIP as a possible neurotransmitter of non-cholinergic, non-adrenergic inhibitory neurons, *Nature (London),* 288, 378, 1988.
11. **Biancani, P., Walsh, J. H., and Behar, J.,** Vasoactive intestinal polypeptide. A neurotransmitter for lower esophageal sphincter relaxation, *J. Clin. Invest.,* 73, 963, 1984.
12. **Daniel, E. E., Helmy-Elkholy, A., Jager, L. P., and Kannan, M. S.,** Neither a purine nor VIP is the mediator of inhibitory nerves of opossum oesophageal smooth muscle, *J. Physiol. (London),* 336, 243, 1983.
13. **Daniel, E. E., Posey-Daniel, V., Jager, L. P., Berezin, I., and Jury, J.,** Structural effects of exposure of smooth muscle in the sucrose gap apparatus, *Am. J. Physiol.,* 252, C77, 1987.
14. **Daniel, E. E., Jager, L. P., and Jury, J.,** Vasoactive intestine polypeptide and non-adrenergic, non-cholinergic inhibition in lower esophageal sphincter of opossum, *Br. J. Pharmacol.,* 96, 746, 1989.
15. **Torphy, T. J., Fine, C. F., Burman, M., Barnette, M. S., and Ormsbee, H. S., III,** Lower esophageal sphincter relaxation is associated with increased cyclic nucleotide content, *Am. J. Physiol.,* 251, G786, 1986.
16. **Zimmerman, R. P., Gates, T. S., Mantyh, C. R., Vigna, S. R., Boehmer, C. G., and Mantyh, P. W.,** Vasoactive intestinal peptide (VIP) receptors in the canine gastrointestinal tract, *Peptides,* 9, 1241, 1989.
17. **Berezin, I., Sheppard, S., Daniel, E. E., and Yanahari, N.,** Ultrastructural immunocytochemical distribution of VIP-like immunoreactivity in dog ileum, *Regul. Peptides,* 11, 287, 1985.
18. **Berezin, I., Allescher, H.-D., and Daniel, E. E.,** Ultrastructural localization of VIP-immunoreactivity in canine distal oesophagus, *J. Neurocytol.,* 16, 749, 1987.

19. **Berezin, I., Huizinga, H.-D., Farraway, L., and Daniel, E. E.,** Innervation of interstitial cells of Cajal by VIP-containing nerves in canine colon, *Can. J. Physiol. Pharmacol.,* 68, 922, 1990.
20. **Daniel, E. E., Berezin, I., Allescher, H.-D., Manaka, H., and Posey-Daniel, V. P.,** Morphology of the canine pyloric sphincter in relation to function, *Can. J. Physiol. Pharmacol.,* 67, 1580, 1989.
21. **Biancani, P., Beinfeld, M. C., Hillemeier, C., and Behar, J.,** Role of peptide histidine isoleucine in relaxation of cat lower esophageal sphincter, *Gastroenterology,* 97, 1083, 1989.

INDEX

A

ACE, see Angiotensin-converting enzyme
Acetorphan, 275
Acetylcholine (ACh), 134, 151
  bombesin effect on, 418
  CCK stimulation of, 256
  colocalization with other chemicals, 9, 161, 437, 454, 458
  corelease with VIP, 456
  effect of on motility, 344
  Gal interaction with, 416
  nerve action and, 416
  neurokinin receptors and, 194
  opioid inhibition of, 189, 411—413
  peristalsis and, 413
  smooth muscle actions and, 418
  Som inhibition by, 408, 409
  SP mimickry by, 339
  storage of, 312, 313
  VIP interaction with, 405
ACh, see Acetylcholine
ACTH, see Adrenocorticotropic hormones
Adenosine triphosphate (ATP), 63, 183, 452, 460—461
Adenylate cyclase, 30
  effect of on blood flow, 459
  in second messenger systems, 250—253, 258, 261
ADH, see Antidiuretic hormone
Adrenal medulla docasapeptide (BAM), 152, 153, 342
Adrenocorticotropic hormones (ACTH), 13, 21, 137, 434, 467
Affinity chromatography, 237
Affinity gels, 237—239
Alytesin, 54, 417
Amastatin, 275, 296
Amidation, 89, 98
Amine precursor uptake and decarboxylation (APUD), 8—9
Aminopeptidases, 275, 289, -see also- specific types
  in the circulation, 280
  in the GI tract, 283—286
  inactivation and, 294—296
ANF, 279, 281
Angina, 461
Angiotensin, 20, 279—281, 289
Angiotensin-converting enzyme (ACE), 275, 276, 289, 290
  in the circulation, 278, 280
  in the GI tract, 284, 285
  motility and, 363
  peptide inactivation and, 293—299
Antidiarrheal agents, 354
Antidiuretic hormone (ADH), 9, 10
Antiotensinase, 290
APUD, see Amine precursor uptake and decarboxylation

Arphamenine, 275
Asp venom, 61
ATP, see Adenosine triphosphate
Atrial dipeptidyl carboxypeptidase, 290
Atriopeptin III, 279, 281
Atropine, 370, 420
Autonomic nervous system, 8, 67

B

BAM, see Adrenal medulla docasapeptide
Bestatin, 275, 295, 296, 298
Big endothelin, 60
Big gastrin, 34, 35
Biliary system
  CCK effect on, 328
  gastrin effect on, 318
  motilin effect on, 362
  NT effect on, 370
  opioid effect on, 354—356
Binding assays, 233—235
Blood flow, 448—469, -see also- Circulation
  CGRP effect on, 448—455
  NPY effect on, 458—462
  opioid effect on, 462—469
  VIP effect on, 454—458
Blood pressure, 49, -see also- Hypertension; Hypotension
Bombesin, 111, 494
  cell proliferation and, 480, 481, 483—485
  discovery of, 13
  effect of on motility, 336, 353
  GRP dissimilarity to, 54
  microanatomy of, 147, 149
  motor action and, 185
  neurotransmitter role of, 314
  physiological/pathophysiological roles of, 417—419
  structure of, 23, 54—59
Bombesin antagonists, 485
Bradykinin, 280, 282, 289, 290
  cell proliferation and, 481
  half-life of, 281
  inactivation of, 298

C

CAAT boxes, 93—94, 98, 107, 111
CACCC boxes, 94
Caerulein, 6, 13
Calcitonin, 66, 438
Calcitonin gene-related peptide (CGRP), 14, 89, 495
  colocalization
    with CCK, 143, 161, 163
    with ChaT, 143
    with Gal, 163
    with NKA, 449

with NMU, 143, 163
with NPY, 143, 150, 151, 161, 163
with Som, 143, 161, 163, 164
with SP, 137, 156, 164, 449
with VIP, 143, 157, 163
effect of on blood flow, 448—455
ion transport and, 433
microanatomy of, 138—141, 144, 159
neurotransmitter role of, 312, 313
physiological/pathophysiological roles of, 419—
420
structure of, 66—67
Calcium
effect of on blood flow, 459, 463
effect of on motility, 368
effect of on motor action, 182—183, 187
effect of on nerves, 417
effect of on peristalsis, 419
endothelin and, 60
in gene expression and regulation, 108—109
ion transport and, 433, 441
in second messenger systems, 250, 251, 262
CCK and, 254—256
NT and, 259—261
opioids and, 257—258
Som and, 253
VIP and, 251, 252
Xenopus oocytes and, 243, 244
cAMP, see Cyclic adenosine monophosphate
Capsaicin, 137, 451
Captopril, 275, 293—298
Carboxypeptidase, 275, 296
Carcinoids, 59, 339, 438
Carcinoid syndrome, 362, 438
Cardiovascular system
CGRP effect on, 449, 452—454
NPY effect on, 458—462
opioid effect on, 462—469
VIP effect on, 454—458
Casaicin, 143
CAT, see Chloramphenicol acetyltransferase
CCK, see Cholecystokinin
CCK-PZ, see Cholecystokinin-pancreozymin
cDNA, 21, 89, 458, 483
in determination of CCK structure, 39
in determination of CGRP structure, 66
in determination of endothelin structure, 60—62
in determination of Gal structure, 63
in determination of glucagon structure, 27
in determination of GRP structure, 55, 58
in determination of motilin structure, 65
in determination of neurotensin structure, 49
in determination of NT microanatomy, 152
in determination of PP structure, 43, 45
in determination of tachykinin structure, 50, 51
in determination of VIP structure, 31
in gene expression and regulation, 108, 111, 112,
119
Xenopus oocytes and, 243—245
Cecropins, 69
Cell proliferation, 480—486
Central nervous system, 46, 59, 63

Cerulein, 36, 37
cGMP, see Cyclic guanosine monophosphate
CGRP, see Calcitonin gene-related peptide
ChAT, 143, 150, 151, 156, 157, see Choline
acetyltransferase
Chloramphenicol acetyltransferase (CAT), 90, 100
Cholangiocarcinoma, 485
Cholecystokinin (CCK), 89, 289, 492, 494, 495
action mechanisms of, 3
cell proliferation and, 480, 483, 485
in the circulation, 278, 282
colocalization
with CGRP, 143, 145, 161, 163
with ChAT, 145
with Dyn, 145, 163
with Enk, 145, 153, 163
with Gal, 145, 163
with GRP, 145, 149, 163
with NMU, 143, 163
with NPY, 145, 150, 151, 161, 163
with opioid-like peptides, 342
with Som, 145, 161, 163
with SP, 145, 156, 163, 331
with VIP, 145, 157, 163
discovery of, 13
effect of on motility, 319—328, 353
effect of on motor action, 184
effect of on secretin, 29
gastrin and, 37—38, 315, 317
GRP and, 320, 483
half-life of, 279, 281
inactivation of, 290—292
interaction with bombesin, 418
interaction with VIP, 404
ion transport and, 433, 434, 436, 438
microanatomy of, 143—145
molecular form of, 14, 319
neurotransmitter role of, 312—314
release of, 320
Som suppression of, 408
structure of, 20, 22, 23, 35—40
in nonmammalian forms, 40—41
Cholecystokinin-4 (CCK-4), 39, 221
Cholecystokinin-5 (CCK-5), 39, 221
Cholecystokinin-7 (CCK-7), 39, 222, 292
Cholecystokinin-8 (CCK-8), 289
cell proliferation and, 485
discovery of, 6, 8
effect of on motility, 319, 323—325, 328, 339
inactivation of, 292
ion transport and, 434, 440
microanatomy of, 143, 144
release of, 320
structure of, 39—41
Cholecystokinin-12 (CCK-12), 221
Cholecystokinin-17 (CCK-17), 41
Cholecystokinin-18 (CCK-18), 39
Cholecystokinin-21 (CCK-21), 221
Cholecystokinin-22 (CCK-22), 37, 39
Cholecystokinin-25 (CCK-25), 39
Cholecystokinin-33 (CCK-33), 39, 40, 143, 221,
222, 319, 434

Cholecystokinin-34 (CCK-34), 38
Cholecystokinin-36 (CCK-36), 38, 41
Cholecystokinin-39 (CCK-39), 6, 37—40, 221
Cholecystokinin-58 (CCK-58), 39, 40, 143, 221, 328
Cholecystokinin (CCK) antagonists, 320—321
Cholecystokinin (CCK) receptors, 320—321
  ligand binding studies of, 221—222
  motility and, 320
  second messenger systems and, 250, 254—256, 261
Cholecystokinin-8 (CCK-8) receptors, 221, 222
Cholecystokinin-pancreozymin (CCK-PZ), 5—6, 12, 36, 38
Cholera toxins, 254, 261
Chondrosarcomas, 485
Chromatography, 239, 276, -see also- specific types
Chromogranin A, 69
Circulation, -see also- Blood flow
  peptide disappearance in, 278—283
  peptide-inactivating mechanisms in, 277
Colon
  CCK effect on, 144, 327
  gastrin effect on, 318
  motilin effect on, 362
  NT effect on, 370
  opioid effect on, 355, 414
  SP effect on, 339
Colon carcinoma, 485
Constipation, 340, 356—357
Corticotrophin releasing hormone (CRH), 10
CPE-Ala-Ala-Phe-pAB, 275
CPON, see C-terminal 30-amino-acid flanking peptide
CRF, 137, 353
CRH, see Corticotrophin releasing hormone
Crohn's disease, 438
Cryptosporidial infection, 438
C-terminal 30-amino-acid flanking peptide (CPON), 109, 110, 458
Cyclazocine, 463
Cyclic adenosine monophosphate (cAMP), 94, 405—406, 410, 494
  effect of on blood flow, 455, 459
  effect of on motor action, 182, 188
  in gene expression and regulation
    of Gal, 115, 116
    of GRP, 111
    of NMN, 103—105
    of NPY, 108—109
    of NT, 103—105
    of PHI/PHM, 100
    of VIP, 100
  ion transport and, 433, 440, 441
  in second messenger systems, 261
    with NT, 259
    with opioids, 257—258
    with Som, 253, 254
    with VIP, 251, 252
Cyclic guanosine monophosphate (cGMP), 494
  effect of on motility, 325
  effect of on motor action, 182, 188

effect of on nerves, 405—406
ligand binding studies and, 222
in second messenger systems, 255, 258, 259
Cystic fibrosis, 407
Cytoplasmic stability, 96

D

DAP, see Dipeptidyl aminopeptidase
Depolarization
  by bombesin, 417
  by CGRP, 420
  by Gal, 115, 116
  by GRP, 417
  by PHI/PHM, 100
  by Som, 409, 410
  by VIP, 100
Diabetes, 6, 362, 457
Diabetic gastroparesis, 358, 361, 362
Diarrhea, 407, 438, 439
Dipeptidyl aminopeptidase (DAP), 275, 276, 284—286
  in the circulation, 278, 280
  in the GI tract, 286
  peptide inactivation and, 297—298
Diprotin A, 275
Distal gastrointestinal tract, 347—352, 354—356
Diverticulosis, 355
DNA, 14, 56, 90, 91, 93, 95, 449, -see also- cDNA
Dopamine, 408, 412
Dynorphin, 152, 495
  colocalization
    with CCK, 143, 163
    with Enk, 153, 163
    with Gal, 163
    with GRP, 147, 149, 163
    with NMU, 143, 163
    with NPY, 150, 163
    with SP, 163, 331
    with VIP, 157, 163
  effect of on blood flow, 465
  effect of on motility, 343, 344, 352, 355
  ion transport and, 433, 434, 436
  microanatomy of, 152

E

Edema, 453
Eledoisin, 49, 50, 194, 199, 329, 330
EM, see Erythromycin
Enalapril, 298
Endocrine system, 67
Endo-oligopeptidase A, 276, 290
Endopeptidase 24.11, 275, 276, 284—286, 289, 290
  motility and, 363
  peptide inactivation and, 292, 293—295, 297—299
Endopeptidase 24.15, 275, 276, 284—286, 289, 290
Endopeptidase 24.16, 276, 284—286, 290
Endorphin, 408, 434, -see also- specific types

β-Endorphin
    effect of on blood flow, 467, 468
    effect of on motility, 342—344, 348, 350, 356
Endothelin, 23, 59—62
Enkephalin, 89, 495, -see also- specific types
    colocalization
        with CCK, 143, 153, 163
        with Dyn, 153, 163
        with GRP, 149, 163
        with NPY, 150, 153, 163
        with SP, 153, 163
        with VIP, 153, 157, 163
    effect of on blood flow, 462—463
    ion transport and, 433, 439
    microanatomy of, 135, 161
    organ specific actions of, 413—414
Enteric nervous system, 7—8, 59, 63
Enteric neurons, 132, 137
    as multimessenger systems, 161—164
    projection patterns and plasticity in, 159—161
Enteroglucagon, 434, 438
Enterosporption, 430
Erythromycin (EM), 357—358
Esophagus
    CCK effect on, 322—323
    CGRP effect on, 419
    gastrin effect on, 316—317
    motilin effect on, 361
    NT effect on, 366
    SP effect on, 336
Estrogen, 116
ET-1, 61, 62
ET-2, 61—62
ET-3, 61
Eukaryotic gene, 91—98
Exogenous opioids, 348
Exon probes, 90, 91, 92, 96
Exorphines, 342
Extrinsic nervous control, 132

F

Firefly luciferase, 90
FMRF-amide, 68
Foreign genes, 89
Fragments, 198—200

G

GABA, see Gamma aminobutyric acid
Galanin, 89, 97
    colocalization with other chemicals, 143, 145,
        157, 161, 163
    discovery of, 14
    gene expression and regulation in, 114—117
    ion transport and, 433, 434, 440
    microanatomy of, 138, 140, 145—147
    neurotransmitter role of, 312
    physiological/pathophysiological roles of, 415—

417
    precursor proteins in, 116—117
    structure of, 23, 63—64
Gallbladder, 362
Gallstones, 370
GAL-message-associated peptide (GMAP), 116
Gamma aminobutyric acid (GABA), 134, 161
Gastinoma syndrome, 33
Gastric emptying
    CCK effect on, 324—325
    motilin effect on, 361, 362
    NT effect on, 368
    opioid effect on, 348
    secretin effect on, 420—421
    SP effect on, 337, 338
Gastric inhibitory polypeptide (GIP), 6, 483
    discovery of, 6—7, 13
    homology with VIP, 402, 403
    ion transport and, 438
    structure of, 25, 28, 29—31
Gastrimonas, 438
Gastrin, 144, 149, 289, 408, 492
    CCK and, 37—38, 315, 317
    cell proliferation and, 483—484
    in the circulation, 278, 282
    discovery of, 2, 4—5, 12
    effect of on motility, 315—319
    effect of on motor action, 184—185
    half-life of, 279, 281
    inactivation of, 290—292
    mechanisms of function of, 10
    molecular forms of, 315
    neurotransmitter role of, 314
    release of, 316
    structure of, 20, 21, 23, 33—36, 37—38
        in nonmammalian forms, 40—41
Gastrin antagonists, 315—316
Gastrin-4 (G-4), 35
Gastrin-5 (G-5), 35
Gastrin-6 (G-6), 35
Gastrin-14 (G-14), 35
Gastrin-16 (G-16), 34
Gastrin-17 (G-17)
    effect of on motility, 315—317
    inactivation of, 290
    structure of, 33—36
Gastrin-33 (G-33), 35
Gastrin-34 (G-34), 34—36, 315—317
Gastrin I, 290
Gastrin receptors, 250, 315—316
Gastrin-releasing peptide (GRP), 89, 97, 98, 494
    bombesin dissimilarity to, 54
    CCK and, 320, 483
    cell proliferation and, 480, 481, 483—485
    colocalization
        with CCK, 143, 149, 163
        with Dyn, 163
        with Enk, 149, 163
        with opioid-like peptides, 342
        with SP, 147, 163, 331
        with VIP, 147—149, 157, 163

discovery of, 13
effect of on motor action, 185
gene expression and regulation in, 110—113
inactivation of, 298—299
ion transport and, 433
mechanisms of function of, 10
microanatomy of, 133, 138—140, 147—149, 160, 161
neurotransmitter role of, 312, 314
physiological/pathophysiological roles of, 417—419
precursor protein of, 112—113
structure of, 54—59, 66
Gastrin-releasing peptide (GRP) antagonists, 417
Gastroenteropancreatic tumors, 438
Gastrointestinal tract
CCK actions on, 322—328
gastrin actions on, 316—318
neurokinin receptors in, 203—204
neuropeptide-inactivating mechanisms in, 277—278
opioid actions on, 347—352, 353—356
peptidases in, 283—290
SP actions on, 336—340
GC boxes, 93—94
Gel retardation assays, 90
Gene constructs, 89
Gene expression, 89—121
in Gal, 114—117
in GRP, 110—113
in motilin, 117—119
in NMB, 113—114
in NMN, 103—107
in NPY, 107—110
in NT, 103—107
in PHI/PHM, 98—103
time- and tissue-dependent, 91—93
in VIP, 98—103
Gene translation, 96—98
GGAP, see GRP gene-associated peptide
GHRF, see Growth hormone releasing factor
GIP, see Gastric inhibitory peptide
Glicentin, 26, 43, 70, 164
GLP, see Glucagon-like peptide
Glucagon, 492
discovery of, 6, 13
GRP interaction with, 149, 483
homology with VIP, 402, 403
ion transport and, 434, 438
mechanisms of function of, 10
nonmammilian, 31—33
Som suppression of, 408
structure of, 24—29, 70
Glucagon-like peptide (GLP), 14, 25—29, 70
Glucocorticoids, 101, 103—105, 108
GMAP, see GAL-message-associated peptide
G-protein, see Guanine nucleotide-binding protein
GRF, see Growth-hormone releasing factor
Growth factors, 101, 103—105, 108, 480
Growth-hormone releasing factor (GRF), 25, 31—33, 402, 403

GRP, see Gastrin-releasing peptide
GRP gene-associated peptide (GGAP), 57, 58
Guanine nucleotide-binding protein (G-protein), 250, 254
Guanthidine, 464—465

H

Helodermin, 24, 25, 33, 220, 402
Helospectrin, 24, 25, 32—33, 220
Hepatocellular carcinomas, 486
Hexamethonium, 404
High performance liquid chromatography (HPLC), 21, 33, 53, 69, 109, 276
Hirschsprung's disease, 204, 356, 411
Histamine, 134, 161
Histidine, 25, 27, 29
HPLC, see High performance liquid chromatography
5-HT, see 5-Hydroxytryptamine
Hydra head activator, 68—69
5-Hydroxytryptamine (5-HT), 156, 194, 361, 437
Hyperpolarization, 412
Hypertension, 462
Hypotension, 452, 453, 466
Hypothalmic-releasing factors, 10, -see also-specific types
Hyptertrophic pyloric stenosis, 407

I

IAS, see Internal anal sphincter
IBD, see Inflammatory bowel disease
ICS, see Ileocecal sphincter
IGF, see Insulin-like growth factor
Ileocecal sphincter (ICS), 354
Inflammatory bowel disease (IBD), 340, 407, 438
Insulin, 20, 21
GIP and, 7
glucagon and, 6, 25
GRP and, 149, 483
mechanisms of function of, 10
as model for peptide synthesis, 22
neurokinins and, 203
NPY and, 109
Insulin-like growth factor (IGF), 3, 11, 20
Insulinomas, 438, 485
Integrated gastrointestinal tract, 353—354
Interleukin, 436
Internal anal sphincter (IAS), 404
Intrinsic nervous control, 132—135
Intron probes, 90, 92
In vitro studies
on effect of NPY, 151
of NT microanatomy, 152
of peptide motor actions, 182—189
In vivo studies
of effect of NYP, 151
of peptide motor actions, 182—189

of peptides in the circulation, 280
Ion transport, 411, 430—442
  mechanism of, 434—436
  NPY and, 430, 433, 434, 436, 437, 439—442
  PYY and, 430, 439—442
Isotoxins, 61

K

Kassinin, 50, 194, 199, 200, 329, 330
Kelatorphan, 275
Kinetensin, 48, 49

L

β-Lactamase gene, 90
LANT-6, 48, 105, 365
Leucine, 414
Leucosulfakının (LSK), 40—41
Leu-enkephalin, 152, 289, 493
  in the circulation, 280
  effect of on blood flow, 462, 464
  effect of on motility, 342—344, 355, 356
  half-life of, 279
  inactivation of, 294—295
Levallorphan, 463
Levocabastine, 215
Ligand binding studies, 210—223
  of CCK, 221—222
  membrane purification in, 210—211
  of neurokının, 218—219
  of NT, 214—215
  of opioids, 215—218
  of Som, 222—223
  of VIP, 211, 213, 219—221
β-Lipoprotein, 13
Liquid hybridization methods, 90
Lithium, 105
Lıtorin, 417
Little gastrin, 33
Lower esophageal sphıncter (LES)
  bombesin effect on, 418
  CCK effect on, 322—323
  CGRP effect on, 419
  enkephalin effect on, 413
  gastrin effect on, 316—317
  motilin effect on, 361
  NPY effect on, 413
  NT effect on, 366
  opioid effect on, 413
  SP effect on, 336, 4 3
  VIP effect on, 406, 413
LSK, see Leucosulfakının
LY110947, 294

M

Magainins, 69

M cDNA, 117, 118, 119
Membrane purification, 210—211
Messenger sytems, 481
Met-enkephalin, 152, 153, 290, 493
  in the circulation, 280, 282
  colocalization with other chemicals, 156, 164, 413
  effect of on blood flow, 462, 466
  effect of on motility, 342—344, 347, 352, 353, 355, 356
  half-life of, 279, 281
  inactivation of, 294—296
Methionine, 414
Methylation, 95
Migrating myoelectric complex (MMC)
  CCK effect on, 326
  gastrin effect on, 318
  motilin effect on, 358—362
  NT effect on, 370
  opioid effect on, 354, 414
  Som effect on, 409
MMC, see Migrating myoelectric complex
M mRNA, 117
Molecular biology, 39—40, -see also- Receptor isolation studies
Morphiceptin, 355
Morphine
  effect of on blood flow, 463
  effect of on motility, 347, 355, 356
  ion transport and, 439
  ligand binding studies of, 215
  organ specific actions of, 414—415
Motilides, 357—358
Motilin, 89, 495
  discovery of, 7
  effect of on motility, 352, 357—363
  effect of on motor action, 185
  gene expression and regulation in, 117—119
  interaction with opioids, 413
  ion transport and, 434
  microanatomy of, 137
  neurotransmitter role of, 314
  precursor protein in, 118—119
  Som inhibition of, 409
  structure of, 23, 65
Motilin receptors, 358
Motility, 311—371
  CCK effect on, 319—328, 353
  gastrin effect on, 315—319
  motilin effect on, 352, 357—363
  neurokinin effect on, see under Neurokinins
  neurotensin effect on, see under Neurotensin
  neurotransmitter-paracrine substance-hormone and, 311—315
  opioid effect on, see under Opioid peptides
  SP effect on, see under Substance P
  VIP effect on, 352, 353, 362, 404
Motor actions, 182—189
mRNA, 22, 70, 89, 90
  in CGRP, 66, 137
  in endothelin, 60
  in eukaryotic gene, 91, 95—97

in Gal, 63, 97, 114—116, 145
in GRP, 56—58, 97, 98, 110—112, 483
in motilin, 65, 118
in NMB, 59, 113—114
in NMN, 103—105
in NPY, 107—109, 458
in NT, 103—105
in PHI/PHM, 98, 100—102
in tachykinin, 51, 53
in VIP, 31, 96, 98, 100—102
Xenopus oocytes and, 243—245
τ-MSH, 137
Multimessenger systems, 161—164
Mutated genes, 89
Myenteric neurons, 133, 150—153, 156—157
CGRP effect on, 420
GRP/bombesin effect on, 417
ion transport and, 434
projection patterns in, 159—161

N

NA, see Noradrenaline
Naloxone
effect of on blood flow, 464, 465, 467
effect of on motility, 348, 353, 355, 356
effect of on peristalsis, 413
Naltrexone, 466—467
NANC transmitter, see Nonadrenergic noncholinergic transmitter
Neoplasia, 484—486
Nerve growth factor (NGF), 101, 105, 108
Neural activity, 409—410, 412, 416—418
Neuroblastomas, 461
Neuroendocrine tumors, 70
Neurokinin-α, 50, 51
Neurokinin-β, 50
Neurokinin agonists, 198—200
Neurokinin A (NKA), 155, 156, 329—330, 493
characterization of, 13
colocalization with other chemicals, 161, 330, 449
effect of on blood flow, 451
effect of on motility, 337, 338
inactivation of, 298
in intestinal tissues, 196
ion transport and, 433
pathophysiological role of, 340
precursor forms of, 330
structure of, 51, 53, 54
tissue localization of, 333
Xenopus oocytes and, 244
Neurokinin A (NKA) agonists, 203
Neurokinin A (NKA) receptors, 194, 199, 200
in isolated vessels, 197
ligand binding studies of, 218, 219
Neurokinin antagonists, 202—203
Neurokinin B (NKB), 329—330, 493
characterization of, 13
effect of on blood flow, 451
effect of on motility, 338

precursor forms of, 330
structure of, 51, 54
tissue localization of, 333
Xenopus oocytes and, 244
Neurokinin B (NKB) receptors, 194, 199, 200
in intestinal tissues, 196
in isolated vessels, 196, 197
ligand binding studies of, 218
Neurokinin receptors, 194—204
in GI tract, 203—204
in intestinal tissue, 195—196
in isolated vessels, 196—197
ligand binding studies of, 218—219
Neurokinins, -see also- specific types
effect of on blood flow, 451
effect of on motor action, 185—186
ion transport and, 438
motility and, 328—340
actions along GI tract, 336—340
pathophysiological role of, 339—340
Neuromedin B (NMB), 89, 147, 320, 417
cell proliferation and, 480
characterization of, 13
gene expression and regulation in, 113—114
precursor protein of, 113—114
structure of, 55, 58, 59
Neuromedin C (NMC), 110, 147, 320
Neuromedin L (NML), 50—51
Neuromedin N (NMN), 89
characterization of, 13
effect of on motility, 365
gene expression and regulation in, 103—107
structure of, 47—49
Neuromedin U (NMU)
colocalization with other chemicals, 150
microanatomy of, 137, 139—143, 149
structure of, 68
Neuron-specific enolase (NSE), 9
Neuropeptide-inactivating mechanisms, 277—278, 290—299, see also Neuropeptide metabolism; Peptidases
in CCK, 290—292
in gastrin, 290—292
in neurotensin, 292—294
in opioids, 294—296
in tachykinins, 297—298
Neuropeptide K (NPK), 52, 53, 330
Neuropeptide metabolism, 278—299
in the circulation, 278—283
in the GI tract, 283—299
peptidases and, 283—290
Neuropeptide receptors, see Ligand binding studies; Receptor isolation studies; Second messenger systems; specific peptide receptors
Neuropeptide Y (NPY), 89, 95
colocalization
with ACh, 437, 458
with CCK, 143, 150, 151, 161, 163
with CGRP, 143, 150, 151, 161, 163
with ChAT, 150, 151
with CPON, 458

with Dyn, 150, 163
with Enk, 150, 153, 163
with Gal, 145, 163
with NMU, 143, 163
with opioid-like peptides, 342
with Som, 150, 151, 161, 163
with VIP, 150, 151, 157, 161, 163, 437, 454—455, 458
discovery of, 14
effect of on blood flow, 458—462
gene expression and regulation in, 107—110
ion transport and, 430, 433, 434, 436, 437, 439—442
microanatomy of, 150, 159, 161, 164
structure of, 41—42, 44—46
Neurotensin (NT), 89, 152, 289, 290, -see also- specific types
in the circulation, 278, 282
discovery of, 7, 8
effect of on motor action, 186—187
gene expression and regulation in, 103—107
half-life of, 279, 281
inactivation of, 292—294
ion transport and, 433, 434
molecular forms of, 363—365
motility and, 353, 363—371
actions in GI tract, 366—370
preprohormone in, 105—106
release of, 365—366
structure of, 23, 46—49
Neurotensin (NT) receptors
ligand binding studies of, 214—215
motility and, 365
in receptor isolation studies, 234, 238, 240, 241
second messenger systems and, 258—259
Neurotensin-related peptide (NRP), 48, 49
Neurotransmitter-paracrine substance-hormone, 311—315
Neurotransmitters, 2, 3, -see also- specific types
NGF, see Nerve growth factor
NKA, see Neurokinin A
NK-1 antagonists, 202, 203
NK-2 antagonists, 202
NK-3 antagonists, 202
NKB, see Neurokinin B
NK-1 receptors, 186, 194, 493
characterization of, 198
in GI tract, 203
in isolated vessels, 196
ligand binding studies of, 218, 219
motility and, 333, 336
second messenger systems and, 160
Xenopus oocytes and, 244
NK-2 receptors, 186, 194, 197, 493
characterization of, 198—200
in GI tract, 203
in isolated vessels, 196
ligand binding studies of, 218, 219
motility and, 333, 334, 338
second messenger systems and, 160
Xenopus oocytes and, 244, 245

NK-3 receptors, 194, 493
characterization of, 198, 200
in GI tract, 203
in isolated vessels, 197
ligand binding studies of, 218, 219
motility and, 333, 338, 339
second messenger systems and, 160
Xenopus oocytes and, 244
NK-4 receptors, 218
NMB, see Neuromedin B
NMC, see Neuromedin C
NML, see Neuromedin L
NMN, see Neuromedin N
NMU, see Neuromedin U
Nonadrenergic noncholinergic (NANC) transmitter, 182, 194, 336, 337, 494, 495
motility and, 339, 347, 352, 404
neural inhibition and, 406, 407
Noradrenaline (NA)
colocalization with other chemicals, 9, 161
effect of on blood flow, 452, 458, 459—461, 462—465
Norepinephrine, 418, 437
Normorphine, 352
Northern blot analysis, 90, 117
NPK, see Neuropeptide K
NPY, see Neuropeptide Y
NRP, see Neurotensin-related peptide
NSE, see Neuron-specific enolase
NT, see Neurotensin
Nuclear-cytoplasmic transport, 96

O

Oligonucleotides, 90
Opiod-like peptides, 342
Opioid agonists, 344
Opioid antagonists, 344
Opioid peptides, 411—415, 494
colocalization with other chemicals, 412—413
effect of on blood flow, 462—469
effect of on motor action, 188—189
inactivation of, 294—296
microanatomy of, 152—154
motility and, 340—357
mechanisms of action by, 344—345
physiological and pathophysiological role in, 345—347, 356
neurotransmitter role of, 312, 314
sources of, 340—342
structure of, 23
Opioid receptors, 411, -see also- specific types
effect of on motility, 342—344
ligand binding studies of, 215—218
in receptor isolation studies, 232
second messenger systems and, 256—258, 261
δ-Opioid receptors, 411, 493
effect of on blood flow, 463—466
effect of on motility, 342, 344, 347, 352, 353
in GI tract, 215

in receptor isolation studies, 232, 233, 237, 240—242

second messenger systems and, 257

ε-Opioid receptors, 342, 344, 347, 465

κ-Opioid receptors, 411, 493

  effect of on blood flow, 463—466

  effect of on motility, 342—345, 347, 352, 353, 357

  in GI tract, 215

  in receptor isolation studies, 232

  second messenger systems and, 257

μ-Opioid receptors, 411, 493

  effect of on blood flow, 463, 464, 466

  effect of on motility, 342, 344, 345, 347, 352, 353

  in GI tract, 215

  in receptor isolation studies, 232, 238, 240—242

  second messenger systems and, 257

Oxoprolinal, 275

Oxyntomodulin, 13, 26, 402, 403

Oxytocin, 108

P

pAm, see Pyroglutamyl-peptide hydrolase

Pancraastatin, 69

Pancreatic adenocarcinomas, 485

Pancreatic cholera syndrome, 30

Pancreatic polypeptide (PP)

  discovery of, 7, 13

  interaction with GRP, 149, 483

  structure of, 41—49

Pancreozymin (PZ), 20, 36—39, see also Chole-cystokinin-pancreozymin

Parasympathetic nervous system, 8

Penile erection, 457

1-Pentazocine, 463

Peptidase inhibitors, 235—236

Peptidase metabolites, 275—277

Peptidases

  characterization of, 276—277

  in the GI tract, 283—290

  identification of, 275—276

Peptide histidine isoleucine (PHI), 89, 96, 97, 156, 495

  colocalization with other chemicals, 157, 161, 164, 342, 454

  discovery of, 14

  gene expression and regulation in, 98—103

  homology with VIP, 402, 403

  ion transport and, 433, 434, 440

  precursor proteins in, 102—103

  structure of, 25, 31

Peptide histidine isoleucine (PHI) receptors, 220, 250

Peptide histidine methionine (PHM), 89, 96

  colocalization with other chemicals, 454

  effect of on blood flow, 456, 457

  gene expression and regulation in, 98—103

  homology with VIP, 402, 403

  microanatomy of, 156

precursor proteins in, 102—103

structure of, 28, 31, 32

Peptide histidine methionine (PHM) receptors, 220, 250

Peptide histidine valine-42 (PHV-42), 32, 102

Peripheral nervous system, 46, 59

Peristalsis

  CCK effect on, 325

  CGRP effect on, 419

  opioid effect on, 353, 355—356, 411, 413

  Som effect on, 409

  SP effect on, 336

Perisulfakinin (PSK), 41

Pertussis toxin, 253, 254, 257, 258, 261

Phenylalanine, 5

Pheochromocytomas, 340, 461

PHI, see Peptide histidine isoleucine

PHM, see Peptide histidine methionine

Phorbolester (TPA), 100, 115

Phorphoramidon, 298, 299

Phosphoinositides (PI), 250, 253—255, 258, 262

Phospholipase C, 250, 251, 253, 254, 258, 261

Phosphoramidon, 275, 292—295, 297—298

PHV-42, see Peptide histidine valine-42

Phyllolitorin, 58, 59

Phyllomedusin, 330

Physalaemin, 13, 49, 51, 194, 199, 329, 330

PI, see Phosphoinositides

Pituitary adenomas, 486

Polyadenylation, 95

Polypeptide YY (PYY)

  colocalization with other chemicals, 43, 164

  discovery of, 7, 14

  ion transport and, 430, 439—442

  structure of, 41—46

POMC, see Pro-opiomelanocorticotropin

Postproline-cleaving enzyme (PPCE), 275, 284, 286, 289

Posttranscriptional processing, 95

Posttranscriptional regulation, 101—102

Posttranslational processing, 27, 65, 70, 96—98

Potassium, 63, 411, 412

  effect of on motility, 368

  effect of on motor action, 183, 187, 188

  effect of on myenteric nerves, 420

  effect of on nerves, 417

  ion transport and, 438

  in second messenger systems, 257, 259, 261

PP, see Pancreatic polypeptide

PPCE, see Postproline-cleaving enzyme

PPT, see Preprotachykinin

Preproenkephalin A, 341

Prepromotilin, 357

Prepro-opiomelanocortin, 341

Preprotachykinin A precursor (PPT-A), 50—54

Preprotachykinin (PPT), 155

Proenkephalin, 152, 153, 342

Projections, 159

  of CCK neurons, 145

  of CGRP neurons, 143, 144

  of Enk neurons, 153

of Gal neurons, 145
of GRP neurons, 148
of NMU neurons, 150
of VIP neurons, 153, 220
Pro-opiomelanocorticotropin (POMC), 13, 152
Pro-opiomelanocortin, 342
Proproteins, 97—98
Protease inhibitors, 236
Protein kinase A, 261
Protein kinase C
  eukaryotic gene and, 94
  in gene expression and regulation, 100, 108—109,
    115, 116
  in second messenger systems, 255, 261
Proteolytic processing, 70
Proto-oncogene activation, 481
PSK, see Perisulfakinin
Puromycin, 295, 298
Pylorus
  CCK effect on, 323—324
  gastrin effect on, 318
  opioid effect on, 348, 414
  SP effect on, 338
Pyroglutamyl, 5, 22
Pyroglutamyl-ketone ester oxoprolinal, 275
Pyroglutamyl-peptide hydrolase (pAm), 275
PYY, see Polypeptide YY
PZ, see Pancreozymin

R

Radioimmunoassay (RIA), 91
  in detection of peptide molecular forms, 13
  in detection of peptide structure, 21, 32, 35, 48,
    53, 69
  in gene expression and regulation studies, 110,
    111
  in peptidase characterization, 276
Ranatensin, 54, 58, 59, 110, 417
Receptor isolation studies, 232—245
  purification in, see Receptor purification
  solubilization in, 233—237
  Xenopus oocytes and, 243—245
Receptor purification, 236
  choice of receptors and, 232—233
  steps in, 237—241
Receptor solubilization, 233—237
Renal failure, 362
Renin, 456, 457, 462
Reporter genes, 89, 90
RIA, see Radioimmunoassay
RNA, 57, 58, 90, 147, 448—449, -see also- mRNA
  in eukaryotic gene, 91, 95, 96
  in gene expression and regulation, 100—102, 109,
    111
  tachykinin structure and, 53
  Xenopus oocytes and, 243, 244

S

Sarafotoxins, 61

Scorpion venom, 436
Scyliorhinin, 51, 330
Second messenger systems, 250—263
  CCK and, 250, 254—256, 261
  gastrin and, 250
  interactions between, 261—262
  NT and, 258—259
  opioids and, 256—258, 261
  PHI/PHM and, 250
  Som and, 253—254
  SP and, 259—261
  VIP and, 250—253
Secretin, 220, 402, 403
  cell proliferation and, 485
  discovery of, 2—4, 12
  ion transport and, 434, 438, 440
  mechanisms of function of, 10
  nonmammilian, 31—33
  physiological/pathophysiological roles of, 420—
    421
  second messenger systems and, 250
  Som suppression and, 408
  structure of, 20, 24—29
Serotonin, 134
  colocalization with other chemicals, 161
  morphine effects on, 414
  motility and, 339
  as opioid mediator, 412
  SP and, 155, 163—164
Signal peptide, 97
Slot blots, 90
Small bowel obstruction, 407
Small cell lung cancer (SCLC)
  GRP and, 58, 59, 111, 483, 484
  motilin and, 362
Small intestine
  CCK effect on, 325—327
  Gal effect on, 416
  gastrin effect on, 318
  motilin effect on, 361—362
  NT effect on, 368—370
  opioid effect on, 348—352, 414
  Som effect on, 409
  SP effect on, 338—339
Smooth cell lung carcinoma, 485
Smooth muscle
  bombesin effect on, 418—419
  CCK effect on, 256
  CGRP and, 67
  Gal effect on, 63, 416
  NMU effect on, 149
  NPY effect on, 150
  NT effect on, 152
  opioid effect on, 153
  Som effect on, 409
  SP effect on, 480
  tachykinin effect on, 49
  VIP effect on, 402, 405—406
Sodium, 432—433, 440, 459
Somatocrinin, 25, 32
Somatosatinomas, 438
Somatostatin, 6, 20, 89, 149, 155, 203, 419

509

cell proliferation and, 485
colocalization
    with CCK, 143, 161, 163
    with CGRP, 143, 161, 163, 164
    with dopamine, 408
    with Gal, 163
    with NMU, 143, 163
    with NPY, 150, 163
    with NYP, 151, 161
    with SP, 163, 164, 331
discovery of, 7, 12—13
half-life of, 281
ion transport and, 433, 434, 440
microanatomy of, 139, 154—155, 159
neurotransmitter role of, 312
physiological/pathophysiological roles of, 408—
    411
structure of, 23
Somatostatin agonists, 70
Somatostatin receptors, 408
    ligand binding studies of, 222—223
    in receptor isolation studies, 234, 238, 241
    second messenger systems and, 253—254
SP-E receptors, 194, 218
Sphincter ani internus
    gastrin effect on, 318
    motilin effect on, 362
    opioid effect on, 355
    SP effect on, 339
Sphincter of Oddi
    CCK effect on, 328
    gastrin effect on, 318
    motilin effect on, 362
    NT effect on, 370
    opioid effect on, 356
    VIP effect on, 404, 406
SP-P receptors, 194, 218, 219, 333, 338
Stabilizing of soluble receptors, 235
Steroids, 94, 115—116
Stomach
    CCK effect on, 323
    gastrin effect on, 317
    motilin effect on, 361—362
    NT effect on, 366—368
    opioid effect on, 347—348
    SP effect on, 336
Stomach carcinoma, 485
Structure-activity studies, 198—200
Subarachnoid hemorrhage, 457
Substance K, 50, 51, 96, 244, 338, 481
Substance P (SP), 89, 96, 493, 494
    CCK stimulation of, 256, 325
    cell proliferation and, 480—481, 484, 485
    in the circulation, 278—280, 282
    colocalization
        with CCK, 143, 156, 163, 331
        with CGRP, 137, 156, 164, 449
        with ChAT, 156
        with Dyn, 163, 331
        with Enk, 153, 163
        with GRP, 147, 163, 331
        with 5—HT, 156

        with Met-Enk, 156, 164, 413
        with NKA, 161, 330
        with NMU, 143, 163
        with opioid-like peptides, 342
        with opioids, 413
        with PHI, 164
        with Som, 163, 164, 331
        with VIP, 157, 163, 331
    discovery of, 7, 8, 12
    effect of on blood flow, 451, 453
    effect of on motor action, 184—186
    in GI tract, 203
    half-life of, 279, 281
    inactivation of, 297—298
    interaction with bombesin, 417, 418
    interaction with CGRP, 453
    interaction with Gal, 416
    ion transport and, 433, 434, 436—438
    mechanisms of function of, 10
    microanatomy of, 138, 140, 155—156, 159, 163—
        164
    motility and, 328—340
        actions along GI tract in, 336—339
        pathophysiological role of, 339—340
    neurotransmitter role of, 312—314
    precursor forms of, 330
    release of by CCK, 325
    Som inhibition by, 408
    structure of, 46, 49—51, 53, 54
    substrates for, 289, 290
Substance P (SP) antagonists, 202—203, 336, 338,
    339
Substance P (SP) receptors, 199, 202—203
    in intestinal tissues, 195, 196
    in isolated vessels, 196, 197
    ligand binding studies of, 218, 219
    motility and, 333—334
    second messenger systems and, 259—261
Sulfakinins, 38
Sulfates, 5, -see also- specific types
Sympathetic nervous system, 8, 452, 458—459

T

Tachykinins, 13, 194, 200, 493, 494, -see also-
        specific types
    effect of on motor action, 185—186
    inactivation of, 297—298
    structure of, 23, 49—54, 70
TATA boxes, 91, 93—94
    Gal and, 114
    GRP and, 111, 112
    NMN and, 105
    NT and, 105
    PHI/PHM and, 98
    VIP and, 98
Tetrodotoxin (TTX), 151, 155, 156, 186, 189
    effect of on blood flow, 455, 464
    effect of on motility
        with CCK, 323, 324
        with motilin, 361

with NT, 368, 370
with opioids, 344, 345, 347, 352, 354, 355
with SP, 336, 338, 339
effect of on nerves, 405
effect of on peristalsis, 413, 419
effect of on stomach, 317
excitory action and, 420
intestinal reflexes and, 404
ion transport and, 436, 440
neural inhibition and, 407
Thiorphan, 275, 293—298
Threonine, 25
Thyroid carcinomas, 59
Thyrotrophin releasing hormone (TRH), 278, 282,
    289, 290
half-life of, 279, 281
mechanisms of function of, 10
peptide inactivation and, 299
Tic douloureux, 453
TPA, see Phorbolester
Transcriptional activity, 90
Transfection efficiency, 90
TRH, see Thyrotrophin releasing hormone
Trigeminal neuroglia, 453
TTX, see Tetrodotoxin
Tumor necrosis factor, 436
Tyrosine, 5, 6, 22
Tyrosyl, 21

U

Ulcerative colitis, 340, 355, 438
Uperolein, 51
Ussing chamber technique, 434, 435

V

Valosin, 69
Vascular headaches, 453
Vas deferens, 463
Vasoactive intestinal polypeptide (VIP), 89, 96,
    494—495
cell proliferation and, 481, 484—486
in the circulation, 282
colocalization
    with ACh, 454
    with CCK, 143, 157, 163
    with CGRP, 143, 157, 163
    with ChAT, 157
    with Dyn, 157, 163
    with Enk, 153, 157, 163
    with Gal, 145, 157, 161, 163
    with GRP, 147—149, 157, 163
    with NMU, 143, 163
    with NPY, 150, 151, 157, 161, 163, 437, 454—
        455, 458
    with opioid-like peptides, 342

with opioids, 412—413
    with PHI, 157, 161, 454, 455
    with PHM, 455
    with SP, 157, 163, 331
corelease of ACh and, 456
difference between ET and, 62
discovery of, 8
effect of on blood flow, 454—458
effect of on motility, 352, 353, 362, 404
effect of on motor action, 182, 184, 187—189
effect of on nerves, 405
Gal potentiation by, 147
gene expression and regulation in, 98—103
half-life of, 281
intestinal reflexes and, 404
intracellular locus of, 402
ion transport and, 433, 434, 436, 438, 441
mechanisms of function of, 10
microanatomy of, 133, 135, 138—141, 156—160,
    164
in neural inhibition, 406—407
neurotransmitter role of, 314
physiological/pathophysiological roles of, 402—
    408
precursor proteins in, 102—103
similarities to PHI, 31
Som effect on release of, 155, 408
structure of, 25, 28—31, 33
Vasoactive intestinal polypeptide (VIP) antagonists,
    403—404
Vasoactive intestinal polypeptide (VIP) receptors
ligand binding studies of, 211, 213, 219—221
physiological/pathophysiological roles of, 402—
    403
second messenger systems and, 250—253
Vasopressin, 108, 481, 485
Verner-Morrison syndrome, 30
Vertridine, 436
VIP, see Vasoactive intestinal polypeptide
VIPomas, 30, 438

W

WDHA, 30

X

Xenopsin, 47
Xenopus oocytes, 243—245, 334
Xenpsin, 105

Z

Zollinger-Ellison syndrome, 33
Z-pro-prolinal, 275

Printed in the United States
by Baker & Taylor Publisher Services